Understanding and improving crop root function

It is widely recognised that agriculture is a significant contributor to global warming and climate change. Agriculture needs to reduce its environmental impact and adapt to current climate change whilst still feeding a growing population, i.e. become more 'climate-smart'. Burleigh Dodds Science Publishing is playing its part in achieving this by bringing together key research on making the production of the world's most important crops and livestock products more sustainable.

Based on extensive research, our publications specifically target the challenge of climate-smart agriculture. In this way we are using 'smart publishing' to help achieve climate-smart agriculture.

Burleigh Dodds Science Publishing is an independent and innovative publisher delivering high quality customer-focused agricultural science content in both print and online formats for the academic and research communities. Our aim is to build a foundation of knowledge on which researchers can build to meet the challenge of climate-smart agriculture.

For more information about Burleigh Dodds Science Publishing simply call us on +44 (0) 1223 839365, email info@bdspublishing.com or alternatively please visit our website at www.bdspublishing.com.

Related titles:

Achieving sustainable crop nutrition
Print (ISBN 978-1-78676-312-9); Online (ISBN 978-1-78676-314-3, 978-1-78676-315-0)

Advances in Conservation Agriculture Volume 1: Systems and Science
Print (ISBN 978-1-78676-264-1); Online (ISBN 978-1-78676-266-5, 978-1-78676-267-2)

Advances in Conservation Agriculture Volume 2: Practice and Benefits
Print (ISBN 978-1-78676-268-9); Online (ISBN 978-1-78676-270-2, 978-1-78676-271-9)

Managing soil health for sustainable agriculture Volume 1: Fundamentals
Print (ISBN 978-1-78676-188-0); Online (ISBN 978-1-78676-191-0, 978-1-78676-190-3)

Managing soil health for sustainable agriculture Volume 2: Monitoring and management
Print (ISBN 978-1-78676-192-7); Online (ISBN 978-1-78676-195-8, 978-1-78676-194-1)

Improving organic crop cultivation
Print (ISBN 978-1-78676-184-2); Online (ISBN 978-1-78676-186-6, 978-1-78676-187-3)

Chapters are available individually from our online bookshop:
https://shop.bdspublishing.com

climate•SMART•publishing
IN AGRICULTURAL SCIENCE

BURLEIGH DODDS SERIES IN AGRICULTURAL SCIENCE
NUMBER 90

Understanding and improving crop root function

Edited by Emeritus Professor Peter J. Gregory,
University of Reading, UK

burleigh dodds
SCIENCE PUBLISHING

Published by Burleigh Dodds Science Publishing Limited
82 High Street, Sawston, Cambridge CB22 3HJ, UK
www.bdspublishing.com

Burleigh Dodds Science Publishing, 1518 Walnut Street, Suite 900, Philadelphia, PA 19102-3406, USA

First published 2021 by Burleigh Dodds Science Publishing Limited

Library of Congress Control Number: 2020942310

British Library Cataloguing in Publication Data
A catalogue record for this book is available from the British Library

ISBN 978-1-78676-360-0 (Print)
ISBN 978-1-78676-363-1 (PDF)
ISBN 978-1-78676-362-4 (ePub)
ISSN 2059-6936 (print)
ISSN 2059-6944 (online)

DOI 10.19103/AS.2020.0075

Typeset by Deanta Global Publishing Services, Dublin, Ireland

Contents

Series list

Title	Series number
Achieving sustainable cultivation of maize - Vol 1 From improved varieties to local applications Edited by: Dr Dave Watson, CGIAR Maize Research Program Manager, CIMMYT, Mexico	001
Achieving sustainable cultivation of maize - Vol 2 Cultivation techniques, pest and disease control Edited by: Dr Dave Watson, CGIAR Maize Research Program Manager, CIMMYT, Mexico	002
Achieving sustainable cultivation of rice - Vol 1 Breeding for higher yield and quality Edited by: Prof. Takuji Sasaki, Tokyo University of Agriculture, Japan	003
Achieving sustainable cultivation of rice - Vol 2 Cultivation, pest and disease management Edited by: Prof. Takuji Sasaki, Tokyo University of Agriculture, Japan	004
Achieving sustainable cultivation of wheat - Vol 1 Breeding, quality traits, pests and diseases Edited by: Prof. Peter Langridge, The University of Adelaide, Australia	005
Achieving sustainable cultivation of wheat - Vol 2 Cultivation techniques Edited by: Prof. Peter Langridge, The University of Adelaide, Australia	006
Achieving sustainable cultivation of tomatoes Edited by: Dr Autar Mattoo, USDA-ARS, USA & Prof. Avtar Handa, Purdue University, USA	007
Achieving sustainable production of milk - Vol 1 Milk composition, genetics and breeding Edited by: Dr Nico van Belzen, International Dairy Federation (IDF), Belgium	008
Achieving sustainable production of milk - Vol 2 Safety, quality and sustainability Edited by: Dr Nico van Belzen, International Dairy Federation (IDF), Belgium	009
Achieving sustainable production of milk - Vol 3 Dairy herd management and welfare Edited by: Prof. John Webster, University of Bristol, UK	010
Ensuring safety and quality in the production of beef - Vol 1 Safety Edited by: Prof. Gary Acuff, Texas A&M University, USA & Prof. James Dickson, Iowa State University, USA	011
Ensuring safety and quality in the production of beef - Vol 2 Quality Edited by: Prof. Michael Dikeman, Kansas State University, USA	012
Achieving sustainable production of poultry meat - Vol 1 Safety, quality and sustainability Edited by: Prof. Steven C. Ricke, University of Arkansas, USA	013
Achieving sustainable production of poultry meat - Vol 2 Breeding and nutrition Edited by: Prof. Todd Applegate, University of Georgia, USA	014

Introduction

Recent decades have seen a dramatic increase in research on plant roots. A deeper understanding of the complex ways roots interact with soils is making it possible to 'design' roots to optimise nutrient and water uptake in low-input environments as well as deliver other benefits such as improved soil health and reduced nutrient leaching. This collection summarises the wealth of current research in this important area and how it can contribute to more ecologically sustainable and regenerative crop production systems.

Part 1 of this collection reviews recent research on understanding root system architecture and growth together with key interactions in the rhizosphere. Chapters cover topics such as modelling techniques, root architecture, root growth regulators, root anatomy and nutrient acquisition. Chapters also cover plant root interactions with rhizobacteria and the role of arbuscular mycorrhizal fungi in the nutrient cycle. Part 2 assesses how roots respond to biotic stresses such as parasites, root-feeding insects and nematodes. Part 3 summarises what we know about root uptake of water and key nutrients such as nitrogen and phosphorus. The final part of the book explores how this understanding can be used to improve root function. Chapters cover topics such as the genetics of plant root traits, the use of plant growth-promoting rhizobacteria (PGPR) and arbuscular mycorrhizal fungi (AM). Part 4 also discusses improving deep rooting in crop plants, the use of rootstocks to improve root function and crop nutrient-use efficiency and understanding and improving phosphorus acquisition by crop root systems.

Part 1 Analysing root system architecture, growth and interactions with the rhizosphere

Chapter 1 reviews advances in root architectural modeling. Root system architectural (RSA) models have become important tools in root research and plant phenotyping for studying root traits, processes, and interactions with the environment. The models have been used to simulate how various root traits and processes influence water and nutrient uptake. At a more technical level, they have assisted the development of phenotyping technologies, particularly through the production of algorithms for segmenting root systems. Model complexity has also increased to cope with root functioning and contributions of root systems to plant nutrition. This chapter describes first the application of the models to questions in plant biology, breeding, and agronomy, and second the development of the models. It concludes with a short outlook suggesting that models need benchmarking and validation and that new developments are likely to include better descriptions of root plasticity responses and focus on biological interactions among (soil) organisms, including mycorrhizal fungi.

The next chapter discusses the development of crop root architecture and optimization of nutrition acquisition, specifically focusing on rice. Root system architecture (RSA), the spatial configuration of a root system in the soil, is critical for water and nutrient acquisition. Rice generates a root system consisting of seminal, lateral, and crown/adventitious roots, whose growth and development are regulated by plant hormones and can be fine-tuned to meet adverse environmental conditions. Chapter 2 summarises original and more recent findings on genes and molecules identified so far to be involved in rice root development. It also discusses the cellular organization and function of rice roots, as well as the responses of RSA to the availability of water and nutrients.

Chapter 3 examines advances in understanding of plant root growth regulators. Global food production needs to increase by more than 60% by 2050 while concurrently reducing fertiliser applications to crops to reduce greenhouse emissions, production costs and environmental pollution. The chapter explores the role of hormones in nitrogen and phosphate uptake. In particular it focuses on auxin, cytokinin and strigolactones in various nutrient conditions and how they interact with each other to control root architectural responses. Past success with stay-green crop varieties, modified in cytokinin regulation of nitrogen remobilisation, demonstrates the real-world significance of a detailed understanding of hormone networks. The chapter also highlights current limitations in our understanding and exciting directions for the future.

Moving on from Chapter 3, Chapter 4 explores advances in understanding plant root anatomy and nutrient acquisition. The chapter summarises root anatomical traits and trait states, and nutrient acquisition mechanisms, along with some environmental issues affecting nutrient acquisition. Then, the whole range of adaptations of root anatomical traits, and its impact on nutrient acquisition are discussed. Combinations of anatomical traits lead to suggestions for root ideotypes potentially capable of supporting agricultural productivity under different edaphic constraints. Spatiotemporal aerenchyma formation in the various root types of maize under nitrate, phosphate or sulfate deprivation is discussed in a case study.

Chapter 5 assesses advances in understanding plant root hairs in relation to nutrient acquisition and crop root function. Root hairs are found on most terrestrial flowering plant species. They form from epidermal cells at a predetermined distance behind the growing root tip in three main patterns. Their presence, pattern, length, density and function are genetically controlled and numerous genes are expressed solely in root hairs. Their growth and proliferation are attenuated by the environment and root hairs growing in soil are generally shorter and less dense than those in laboratory studies. The chapter begins by discussing root hair physiology and development, followed by a review of phylogenetic and genetic variation in root hairs. It also examines root hair interactions with the environment, then goes on to review trade-offs and interactions with other traits. A discussion on the prospects for utilising root hairs is also included.

Chapter 6 reviews our understanding of plant root interactions with rhizobacteria to improve biological nitrogen fixation in crops. Plant roots have evolved with the presence of rhizobacteria that can colonise the surface or interior of the plant. Some of these rhizobacteria are actively recruited by the plant and carry out particular functions, in particular in nutrient acquisition. Nitrogen-fixing bacteria form associations with many plant species, either as external associations or as symbiotic endophytes. The chapter highlights our current understanding of the molecular determinants of legume nodulation as well as challenges for improvements of biological nitrogen fixation in legumes and non-legumes. There is a need for connecting our knowledge of the molecular regulation of nodulation with field-based studies that take into account the interaction of nodulation with multiple biotic and abiotic constraints. In addition, current approaches for engineering new symbioses are discussed.

The final chapter of Part 1 considers advances in understanding arbuscular mycorrhizal fungal effects on soil nutrient cycling. Arbuscular mycorrhizal (AM) fungi assist with plant acquisition of nutrients, with most studies focussing on the nutrient forms taken up and translocated to the host. Recent studies have focused on how extraradical fungal hyphae can affect nutrient transformations, leaching and movement. However, it is still unclear the extent that nutrient cycling is influenced in agricultural contexts. Chapter 7 summarises how AM fungi can influence specific soil nutrient processes, then focuses on AM fungi in crop production systems and examples of (in)compatibilities in these systems. These research areas are then contextualized with new approaches to understand AM fungal roles in nutrient acquisition and utilisation, which the authors propose will lead to applications in improving nutrient use efficiency of agroecosystems.

Part 2 Root response to biotic threats

The subject of Chapter 8 is advances in understanding plant root response to weedy root parasites. The chapter begins by reviewing host-parasitic weed interactions, focusing specifically on seed dispersal and germination and the possibility of host infection as a consequence of germination. The chapter then moves on to discuss host plant pre-penetration and post-penetration defence mechanisms. It concludes by emphasising the importance of developing management strategies for parasitic weed management.

Chapter 9 assesses the advances in understanding plant root responses to root-feeding insects. The chapter presents an overview of the interactions between plant roots and root-feeding insect herbivores, focussing on changes in growth and physiology and crucially how roots are defended against insect attack. Several previous reviews have covered the ecology and management of insect root herbivores, together with their interactions with the abiotic and biotic soil environment. Therefore, this chapter focuses particularly on advances in our understanding of how plant mutualistic fungi may affect root-herbivores. This is an emerging area of research, with many attendant knowledge gaps, but this is an important component of how plants resist attack by belowground insect herbivores.

The final chapter of Part 2 reviews advances in understanding plant root response to nematode attack. Plant parasitic nematodes are damaging pests on all crops grown across the world. They exploit plants using a range of strategies, ranging from simple browsing ectoparasitism to highly complex biotrophic endoparasites. Chapter 10 begins by highlighting the different modes of parasitism, which is then followed by a review of plant-nematode interactions. The chapter also examines activation and suppression of plant defence responses and discusses how natural resistance can be used to control plant nematodes.

Part 3 Root uptake of nutrients and water

Part 3 begins with a chapter on advances in the understanding of nitrogen uptake by plant roots. Efficient use of nitrogen (N) by plants and particularly crops, is of global importance. In agriculture, high crop yields and protein content depend upon extensive N-inputs, but fertilizer N is costly to the farmer, and overuse can be damaging to the environment. A critical component of optimised usage is efficient capture by crop root systems. Chapter 11 focuses on principal mechanisms of uptake and factors influencing efficiency. Genetic variation in root architecture and in an array of transporters known to be involved in nitrogen capture is detailed. The impacts of abiotic stress factors such as soil structure are described. Finally, prospects and opportunities for crop improvement are discussed.

Chapter 12 looks at the advances in understanding plant root uptake of phosphorus. At a global scale, phosphorus (P) deficiency comprises a large area of cropland, while P has also been used in excess of crop requirements in many other regions. Improved crop P-acquisition efficiency would allow lower target critical soil P values and provide savings in P-fertiliser use. At the same time, it would reduce P lost through erosion, leaching and/or soil sorption. The chapter summarises the progress in research on root traits associated with P acquisition, including root morphology, architecture, biochemistry, colonisation by arbuscular mycorrhizal fungi, and fine root endophytes, and the trade-offs among all these traits. Farming-management practices to improve P acquisition under current intensive agricultural systems are also discussed. The chapter summarises breeding progress in improving P-acquisition efficiency on soils that are both P deficient and P replete through legacy applications, and suggests future directions to improve P acquisition in five key areas.

Part 3 of the book concludes with an examination of advances in understanding plant root water uptake. Water deficit is one of the primary limitations to crop production. Chapter 13 reviews the role of root and rhizosphere hydraulic processes that affect the ability of a plant to extract water from the soil. Prominent features of rhizosphere hydraulic properties are: root shrinkage, alteration of pore geometry in the rhizosphere, effect of mucilage on water retention, hydraulic conductivity and water repellency, root hairs, and mycorrhiza connecting the root surface to the soil matrix. Although our understanding of the mechanisms related to these factors has advanced significantly in the last ten years, the relative importance of these rhizosphere processes for the ability of crops to extract water from the soil and better tolerate drought is still largely unclear. The authors propose that the next research step is to investigate the implications of these rhizosphere processes on crop growth and water use economy and to use this knowledge to grow more resilient crops that match to their environment.

Part 4 Improving root function

The final part of the book opens with a discussion on understanding and exploiting the genetics of plant root traits. Chapter 14 illustrates how genomics and other -omics approaches coupled with new-generation sequencing (NGS) platforms have been deployed to dissect the genetic make-up of RSA traits and better understand their functions, particularly under environmentally constrained conditions that commonly occur in most farmed soils. The major emphasis is devoted to studies during the past two decades in crops with occasional reference to the vast literature from RSA studies conducted in Arabidopsis and other model plants. The chapter also provides examples on how, in some cases, this knowledge is already benefiting farmers and how it can help in reducing the environmental impact of agriculture worldwide.

Chapter 15 focuses on the use of plant-growth-promoting rhizobacteria (PGPR) to improve root function/crop nutrient use efficiency. Bacteria that colonize plant roots and promote plant growth and development, plant growth-promoting rhizobacteria (PGPR), can contribute to more sustainable intensification of agriculture while minimizing detrimental impacts associated with excessive fertilizer applications. The chapter reviews recent research on the use of PGPR as biofertilizers to enhance root function and improve nutrient uptake. PGPR alter root architecture, root metabolism, nutrient use efficiency and enhance plant tolerance to abiotic stresses such as salinity and drought by a variety of mechanisms that are not yet well understood. Beneficial effects observed in the laboratory are not always seen consistently in the field due to varying environmental and complex biotic interactions, limiting the widespread application of PGPR in agriculture. The chapter highlights new research approaches that will facilitate our understanding of this complex community at the molecular level and from a holistic perspective. Applied research to facilitate registration and commercialization of biofertilizers is also considered.

The next chapter examines the use of arbuscular mycorrhizal fungi to improve root function and nutrient use efficiency. Arbuscular mycorrhizal fungi (AMF) form endosymbiosis with over 70% of land plants, including most crops including cereals. These symbioses facilitate resource exchange between partners and can significantly increase plant nutrient uptake and growth, among other benefits. AMF ubiquity in agricultural soils, in addition to the many roles they are known to play in soil health, demands we consider them when discussing crop function. Chapter 16 discusses how AMF are capable of increasing crop acquisition of macro- and micronutrients. It examines further impacts that AMF have on root system architecture, and how this relates to nutrient acquisition. The chapter also highlights reasons why potential benefits of the symbiosis are often not realised and how this influences current perspectives on the utility of AMF. It also discusses aspects of modern agronomy practice which are deleterious to mycorrhizal functioning. Strategies are suggested by which mycorrhizas might be exploited in future highlighting future research priorities.

Chapter 17 focuses on using systems agronomy to exploit deep roots in crops. The chapter reviews strategies to capture benefits from deeper rooting, taking the example of the semi-arid southern Australian wheat belt. The chapter also focuses on the theme of better capturing deep subsoil water with deeper and more effective root systems, and looks at ways of increasing rooting depth, the role of agronomic techniques as well as genetic improvement methods. A discussion on the legacy effects of deeper roots and utilisation of deep soil water is also included, which is then followed by a review of the benefits of scaling up to the whole farm level, specifically large farms in Australia.

The subject of Chapter 18 is the use of rootstocks to improve root function and resource use efficiency. Grafting, a technique to attach genetically different shoots and roots (scions and rootstocks) allows 'designer root systems' to enhance agricultural productivity and sustainability. Rootstocks improve plant nutrient relations by increasing nutrient capture and/or nutrient use efficiency, by multiple mechanisms including altered root morphology. Moreover, rootstocks can enhance water uptake and/or diminish water loss according to changes in root form and function, and root-to-shoot phytohormonal signalling. The chapter begins by focusing on nutrient and water relations to rootstocks, then moves on to discuss how when triggered, changes in phytohormone status can affect rootstock hormone status and in turn, affect shoot processes in plants. A case study on using tomato rootstocks as a viable strategy to overcome abiotic stresses in Ghana is also included.

The final chapter of the book focuses on delivering improved phosphorus acquisition by root systems in pasture and arable crops. Improving the low efficiency of phosphorus (P) use in agriculture is an imperative because P is one of the key nutrients underpinning sustainable intensification of food production and the rock-phosphate reserves, from which P fertilisers are made, are finite. Chapter 19 describes key soil, root and microbial processes that influence P acquisition with a focus on factors that can be managed to ensure optimal use of fertiliser, and development of root systems for improved P acquisition. A case study describing grasslands in southern Australia where the P-balance efficiency of production is very low, mainly because soils are P deficient and moderately to highly P-sorbing, is also included. Progress towards improved P efficiency using pasture legumes with high P-acquisition efficiency is outlined, followed by a discussion on the development of a 'whole-of-system' understanding for effective P acquisition by roots.

Part 1

Analysing root system architecture, growth and interactions with the rhizosphere

Chapter 1

Advances in root architectural modeling

Johannes A. Postma, Forschungszentrum Jülich, Germany; and Christopher K. Black, The Pennsylvania State University, USA

1 Introduction

Mathematical models of plant roots are formulations that quantitatively describe the morphology, size, growth, or physiology of plant roots. These models have a long history of development, and different root models have been presented in the literature. Among the different models, root system architectural (RSA) models are distinguished from other model types by their specific concern for which roots of what types are where in space, and they give arguably the most explicit and detailed representations of the root system (Table 1). The first RSA models were developed at the end of the 1980s (Fitter, 1987; Diggle, 1988; Pagès et al., 1989). At that time the simulated root systems were relatively small and the simulations were mainly concerned with representing the geometric aspects of root architecture. Since those early beginnings, these models have been developed further and the range of applications has expanded. The models have added physiological aspects of root growth and functioning, and thereby in effect have become what we now call functional structural plant models (Vos et al., 2010), albeit with the focus belowground, not aboveground. The functionality and application of RSA models were reviewed by Dunbabin et al. (2013). Since then new model codes have been released and many model studies have been published. The RSA models have been coupled to soil and shoot models, and have acquired more detail and functionality. Significant effort went into improving the code, and the model development typically expanded

http://dx.doi.org/10.19103/AS.2020.0075.02

Table 1 List of cited models

Model software	Citation
ArchiSimple	Pagès et al. (2014)
CRootBox	Schnepf et al. (2018b)
DigR	Barczi et al. (2018)
GroImp	Henke et al. (2014)
OpenSimRoot	Postma et al. (2017)
RSWMS	Javaux et al. (2008)
RootBox	Leitner et al. (2010)
SimRoot	Postma and Lynch (2011b)
SPACSYS	Wu et al. (2007)

from single researchers to development teams. In this chapter, we aim to describe the advances of these models during the last decade (2010–2020) and how the simulation exercises contributed to our scientific understanding of root and plant functioning.

2 Advances in the application of root system architectural (RSA) models

During the last decade, RSA models have been applied to study an increasing number of plant species, root traits, and plant physiological processes. The models have been coupled to increasingly more complex simulations of the (soil) environment (see Section 3) and many papers focus on the interaction between environment and phenotype. This has led to applications in the area of breeding, agronomy, and ecology, although these fields are strictly speaking at a higher scale than single plant models. Currently, our understanding of the function of phenotypic root traits is still fairly limited, but the RSA model studies that we highlight here have demonstrated the complexity of, and provided insights into, phenotype by environment interactions.

2.1 Root traits studied

RSA models contrast with root length density models by explicitly simulating the placement of each root, whereas root length density models typically assume homogeneous distribution of roots within a given soil layer (Postma et al., 2008). Simulating competition for nitrate among different species (maize, bean and squash) (Postma and Lynch, 2012) showed that a homogeneous root distribution gave much greater nitrate uptake than a root distribution determined by root architecture. We can conclude from this that root architecture restricts the foraging capacity of the root. This immediately raises the question as to what the optimal architecture is for the acquisition of various nutrients. Several studies were published in which the sensitivity of nutrient uptake for root architectural parameters was tested. These parameters are usually directly translated to root architectural traits. In many cases, the sensitivity depended on the simulation environment and nutrient simulated, giving some insight into the complex interactions between the environment and root architecture. Additionally, the studies suggested some trade-offs exist for optimizing root architecture toward the acquisition of different nutrients.

A relatively simple way to understand trait is root growth angle of branch- or adventitious roots (Fig. 1). Although various angles might be measured during phenotyping, including insertion angles, set-point angles, and the rate of the gravitropic response (Freschet et al., 2020), the overall angle causes the root system to be either shallower or steeper. Dathe et al. (2016) simulated steeper and shallower root growth angles in maize and showed that the optimum for nitrate uptake depended on the rate of nitrate leaching, with fanned phenotypes (larger variation in root growth angle) achieving relatively good performance in many environments. Already in 2001, Rubio et al. (2001) showed that shallow root growth angles may be advantageous for P uptake by basal roots in bean, and this work was recently followed up by a much larger set of simulations including mechanistic simulation of P uptake and simulating phenotypes that varied both in root growth angle and the number of basal roots (Rangarajan et al., 2018). The study gives support for the existence of

Figure 1 Rendering of a simulated bean and maize root system with indication of various root traits. Simulations were run with OpenSimRoot and rendering was done with Paraview. Figure was reproduced with permission from Postma et al. (2017).

phenotypic variation in both traits, as the optimum is strongly dependent on the environment simulated. Nevertheless, the overall picture is that plants are faced with a trade-off between shallow and deep placement of roots through a change in root growth branching angle and that the trade-off determines the relative uptake of shallow (P) and deep (nitrate) resources. Experimental papers confirming the relation between P and nitrate uptake and root growth angle include greater P uptake by bean phenotypes with shallow basal root growth angles (Miguel et al., 2015) and greater N uptake by deep-rooting maize (Trachsel et al., 2013).

Phenotypic variation in the number of root axes was studied in several papers. We already mentioned the simulation study of Rangarajan et al. (2018) (Fig. 2), in which two or three whorls of basal roots were both the most frequently observed phenotype and the optimal number in many simulation scenarios. Lateral root branching density, which is the inverse of the interbranching

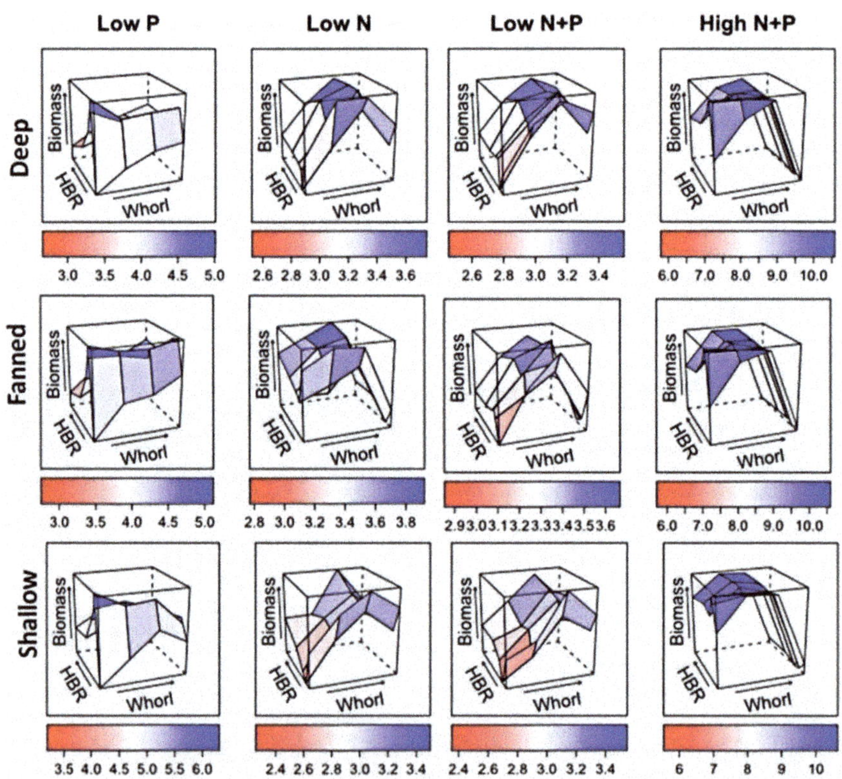

Figure 2 Simulated shoot biomass of 40-day-old bean plants as a function of basal root whorl number (BRWN), number of hypocotyl born roots (HBR), basal root angles type (deep, fanned, shallow), and nutrient availability (low P, low N, low N+P, high N+P). Figure reproduced with permission from Rangarajan et al. (2018).

distance, has been studied experimentally in many species (Pagès, 2019; Freschet et al., 2020). Postma et al. (2014a) simulated observed phenotypic variation for this trait and predicted a trade-off between phenotypes with many short and few long lateral roots. The many short phenotypes would be advantageous for P uptake, whereas the few long phenotypes would benefit nitrate uptake. Experimental confirmation of the results came from two papers studying the same set of genotypes under P deficiency (Jia et al., 2018) and N deficiency (Zhan and Lynch, 2015). With a semi RSA model, Heppell et al. (2015) concluded that higher branching in the topsoil and lower branching in the subsoil could increase P uptake by 42%. In another simulation study, dense root systems were beneficial to quickly capture in-season rainfall but not for exploiting water stored before the growing season (Tron et al., 2015). Recently, Muller et al. (2019) took the concept of branching a step further and distinguished randomness in branching from structured variation. RSA models will be well-suited to simulate these various kinds of variation in branching and might estimate the utility of the different phenotypes in various soil environments. The lateral root branching density not only interacts with the environment but also with other traits. For example, Postma and Lynch (2011a) simulated a positive interaction between the number of branches and the amount of root cortical aerenchyma (discussed further below).

RSA models have also been used to investigate anatomical traits. So far the function of these has been difficult to study, so RSA models have been used to pose and simulate hypotheses on how anatomical traits influence nutrient uptake and plant physiology. Root cortical aerenchyma, the formation of air spaces in the root in this case through programmed cell death, is an interesting example. Aerenchyma formation is negatively correlated to both root respiration and root nutrient content (mol cm^{-3}) and was hypothesized to reduce the metabolic cost of roots, allowing new root growth and thereby greater nutrient uptake (Fan et al., 2003; Postma and Lynch, 2011a,b). The simulation studies suggest large benefits from a metabolically more efficient root system, especially in maize where aerenchyma formation covers a relatively large percentage of the cortex as opposed to bean where aerenchyma formation is much less prevalent and is in part replaced by the loss of the whole cortex (cortical senescence) due to secondary growth. Variation in secondary growth itself, however, may be a form of metabolic efficiency that can be achieved when stressed plants with smaller shoots have reduced water requirements (Strock et al., 2017). Loss of the whole cortex also occurs in barley and several other Poaceae due to programmed cell death rather than secondary growth (Schneider et al., 2018). Cortical senescence is not only associated with reduced respiration and nutrient content but also associated with reduced water uptake and nutrient uptake (Schneider et al., 2017b). Simulating these processes showed that cortical senescence is beneficial when it occurs in the major root axes, but not in the lateral roots (Schneider et al., 2017a). Indeed, cortical senescence in lateral roots was not observed.

One of the most studied anatomical traits in plant nutrition is the formation of root hairs. Different approaches to simulate the function of root hairs have been published. Zygalakis et al. (2011) presented a dual-porosity model for root hairs in which they take account of the interaction between root hair geometry and soil particle geometry. The model simulated greater uptake from P in micropores and showed that the uptake was less sensitive to soil moisture content than previously simulated. Leitner et al. (2010) took a different approach and used homogenization, a mathematical procedure, to derive an analytical solution for the uptake of nutrients by root hairs. Such analytical solutions can compute much faster and thereby be useful when simulating 100 000+ root segments.

Thus far we have treated the architectural traits as static traits, but in reality, many of these traits are influenced by the environment. Simulation models currently include some of this root growth plasticity, even though much about the mechanisms and regulation of root growth plasticity is still unknown. The models often include empirically derived relations between environmental conditions, plant status, and root growth responses (Dunbabin et al., 2011; Tournier et al., 2015; Postma et al., 2017). Root responses are typically defined in terms of tropisms (change in the growth direction), branching (change in the rate of primordia formation), or elongation rate. These empirical relations may be sufficient to study root growth plasticity as a functional trait. Henke et al. (2014) concluded that root growth plasticity in a heterogeneous soil would enhance N uptake. Similarly, Chen et al. (2013) produced a simulation study with root proliferation in response to P banding, which suggested large benefits from the proliferation response.

2.2 Root processes studied

Functional structural plant models (FSPM) not only simulate structure, but also the relevant processes associated with that structure. In this respect, RSA models advanced strongly as they were coupled to models of plant metabolism, nutrient and water uptake. Besides the interactions between structure and environment, the mechanics of the structure and soil were explored.

To simulate plant metabolism, RSA models were coupled to shoot models, albeit not often coupled to geometric shoot models (see Section 3). This coupling allowed simulation of nutrient and carbon balances and source-sink relations, which play an important role in many simulations. The models typically restricted the total carbon consumption to the carbon available from photosynthesis and carbon storage. Carbon consumption is restricted by reducing the growth rate of the root system. Several conclusions resulted from these simulations: (1) plants might cycle dynamically through periods of source and sink limitations, (2) large uncertainty exists with regard to what causes plant growth to be source or sink limited, and (3) oscillatory behavior in source-sink ratios arises when the functional equilibrium responds to current plant status through allocation, which is inherently slow and not corrected for future uptake efficiencies (Postma et al., 2014b). Source-sink models treat a trait like branching density as an increase in the sink term. Assuming no change in the source term, roots will have to stay shorter (Postma et al., 2014a). Although much uncertainty exists around source and sink limitations, especially of nutrient-deficient plants, it is currently one of the easiest ways to understand trade-offs among RSA traits and foraging limitations. Similar to carbon-based source-sink relations, nutrient-use efficiency concepts (defined as the amount of nutrients needed to construct the root system) have been implemented. Greater nutrient-use efficiency (NUE, here the amount of nutrients invested for the amount of nutrients acquired) for N and P, for example, was achieved by simulating root cortical senescence and aerenchyma formation (see the section above). Interestingly, the model was more sensitive to the NUE effects than to the carbon-saving effects, which may suggest that the cost of the root system might be computed in terms of nutrient 'investment costs' (Schneider et al., 2017a; Postma and Lynch, 2011a,b).

There has been much interest in nutrient and water uptake by root systems. Current models for nutrient uptake typically rely on a reaction-diffusion-advection equation for nutrient transport in the soil and a Michaelis-Menten-based nutrient uptake term. It is often said that the models are relatively insensitive to the Michaelis-Menten kinetic parameters, as in many scenarios the soil becomes the limiting factor. York et al. (2016) showed that under competitive conditions the maximum uptake rate, Imax, can give a (small) competitive advantage, and that sensitivity to Imax depends on other RSA traits and soil conditions (Fig. 3). This is an important realization, especially when upscaling from single plants to competitive stands, and to understand interactions among plant species.

The RSA modeling community has been especially active in the area of simulating plant water uptake and transport. A short history of the simulation of water flow from the soil through the root system to the plant starts with representing the flow using electrical analogies (e.g. van den Honert, 1948). Based on these analogies, Alm et al. (1992) developed a numerical model of water transport for an unbranched root. This was extended to branched root systems by Doussan et al. (1998). Javaux et al. (2008) coupled the Doussan model to the well-known SWMS (forerunner of Hydrus) soil water model and thereby enabled further studies of compensatory water uptake and hydraulic lift. Lobet et al. (2014) expanded this work to include water transport through a geometrically explicit shoot. Doussan's equations were originally based on root water potentials which are 'corrected' for gravity, such that gravity does not

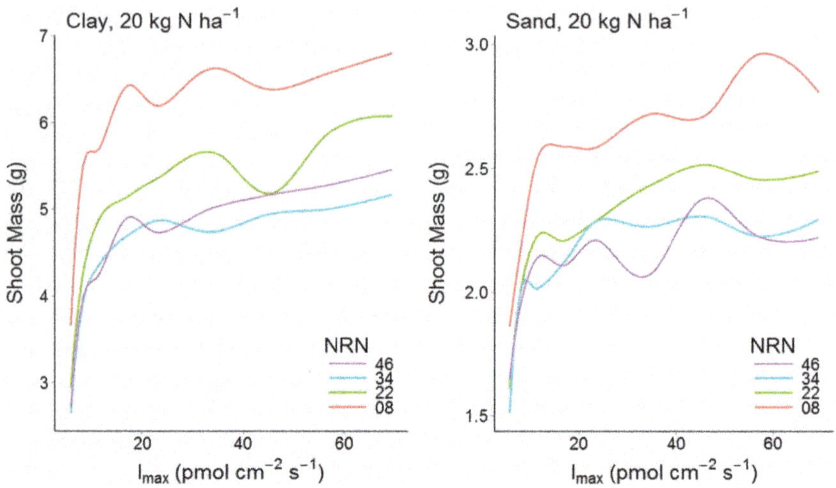

Figure 3 Simulated shoot biomass of maize as a function of Imax (x-axis), nodal root number (NRN), and two soil types (sand and clay). Image printed with permission from L. York. For full study see York et al. (2016).

appear in his equations. Schnepf et al. (2018b) present a version that introduces gravity into the equation. Meunier et al. (2017b) and Meunier et al. (2017a) present a hybrid analytical-numerical solution for a branched root system.

Thus far, model studies suggested that there is no optimal root architecture for water uptake, as it all depends on the environment (Tron et al., 2015). This is by itself not surprising, as water and roots need to coincide in time and space. This means that deep rooting is only beneficial in environments with deep water available, and with deep water recharge during rainy seasons. However, the model studies gave more precise insights into these interactions between environment and root phenotype. A long-standing issue in plant hydrology has been whether root to shoot signaling occurs via hormonal transport from root to shoot. Huber et al. (2014) concluded that shoot responses to soil drying might be explained just as well or better by a hydraulic signal (a change in xylem water potentials as the soil is drying) than by hormonal transport.

2.3 Trait synergies and integrated phenotypes

One of the great challenges in understanding the utility of a trait is that interactions among traits can result in both synergistic and antagonistic effects on root functioning. For example, Miguel et al. (2015) demonstrated experimentally that bean phenotypes with either long root hairs or shallow basal root angles had greater growth on low P soils. However, phenotypes that had both long root hairs and shallow root angles had far greater growth than phenotypes with only one of these traits. Such interactions were easily explored with RSA models. Examples are found for tiller number interacting synergistically with root cortical senescence (Schneider et al., 2017a) and lateral root branching density positively interacting with aerenchyma (Postma and Lynch, 2011a). Integrated phenotypes that combine variations in multiple traits and may represent real-world phenotypes can be explored further. Such integrated phenotypes were simulated by Rangarajan et al. (2018) and York et al. (2016) (Figs. 2 and 3). Phenotypic variation is not a continuum but tends to group in what might be termed root system types. Bodner et al. (2013) worked out such types based on statistics and these types were simulated to study water uptake in different scenarios (Leitner et al., 2014b; Tron et al., 2015).

2.4 Simulating the soil environment

Plants in the natural environment grow their roots in a complex, dynamic, heterogeneous soil matrix. At the beginning of the decade, most RSA models largely ignored this environmental complexity to focus on simulating growth under uniform and nonrestrictive conditions. These simplifying assumptions were made partly to simplify initial model development, but also because model representations of many soil processes (even uncoupled from RSA models) were still under development (Vereecken et al., 2016). Further, most RSA models were initially parameterized using data from experiments conducted under glasshouse or laboratory conditions at the mesocosm scale or smaller, meaning many of the parameters available for calibration were, in fact, observed under uniform and nonrestrictive conditions. Therefore, credit for progress in modeling the soil environment must go as much to the researchers developing improved methods for field measurements that can be used for model validation as to those writing down the model code.

Whether root growth responses to soil hardness should be thought of as primarily driven by average responses or by local heterogeneity (i.e. roots grow through cracks and biopores) is an unresolved research question. It is therefore not surprising that modeling of the interactions between soil hardness and root growth has taken two differing directions: (i) investigating how root architecture responds to hard soil at the macroscopic scale, and (ii) trying to represent soil heterogeneity at the scale of the root tip. de Moraes et al. (2018) used RootBox to simulate the effect of soil strength and water status on soybean root-extension rates. They calculated soil strength using a soil-specific exponential function of bulk density and water content, and additionally assumed that root growth was affected directly by soil matric potential (independently of the effect of water status on soil strength) according to the classic Feddes et al. (2001) root water uptake function. One intriguing feature of the de Moraes model is that by adjusting its empirical parameters it can account for the macroscopic changes in root architecture attributable to soil biopores without requiring highly detailed simulations of individual pores. Similarly, SPACSYS (Wu et al., 2007) has included a term for the effect of soil impedance on root growth since the initial description of the model. This term can only be varied at the root class level and requires user calibration, but when sufficiently detailed soil profile data are available it has been used to reproduce the observed root distribution of field-grown wheat (Bingham and Wu, 2011). By contrast, representation of biopores in R-SWMS takes an explicitly detailed approach. The magnitude and direction of each root tip's growth vector is scaled by the frictional forces from the immediately surrounding soil, represented as a tensor (array of force vectors) computed from the conductance (inverse of penetration resistance) of the eight nearest soil grid nodes (Landl et al., 2017). This approach allows simulation of, in three-dimensional space directed, root responses to individual pores down to the scale of the simulated soil mesh. Using this approach, Landl et al. (2019) showed that root penetration of strong soil layers via large-diameter biopores is likely a substantial benefit to net plant water uptake, even when accounting for a loss of water uptake caused by incomplete contact between the root and the pore wall.

2.5 Simulating rhizosphere processes and soil biota

Rhizosphere models are typically at a scale below RSA models, simulating single root tips or segments and the rhizosphere around. These models have been reviewed by Darrah et al. (2006). RSA models, however, have been used to upscale the rhizosphere models to whole-plant level. Although some difficulty exists with respect to overlap between rhizospheres of different roots, this problem is typically mitigated by setting the outer boundary to the mean mid-distance between roots and by the assumption that overlap is relatively small (Postma and Lynch, 2011a, 2012). Alternatively, the rhizosphere models are upscaled to two- or three-dimensional models including several root segments. This approach was taken by de Parseval et al. (2017) who specifically simulated facilitation as it may occur between roots differing in the amount of organic acid exudation. Nevertheless, reduced dimensionality of the models allows the rhizosphere to be simulated with greater detail around the root. The utility of organic exudates was thus simulated by Schnepf et al. (2012) who coupled a citrate exudation model to a P uptake model. The more computational intensive three-dimensional approach was used by Fang et al. (2019) who coupled the RSA model RootBox to the eSTOMP chemistry model. The coupled model thereby simulated the chemical interaction between different solutes and the uptake of multiple nutrients. The detail around the root, including the importance of root diameter and root hair geometry, however, is typically lost in the three-dimensional approaches which treat single root segments as sinks with a given surface area.

Rhizosphere processes not only include chemistry, but also microbial activity. Thus far, little has been achieved in modeling these systems, in part because they are not well understood. In studies simulating the mycorrhizal system, Schnepf et al. (2016) and Schnepf et al. (2011) simulated the hyphal network of mycorrhizal fungi around a root to estimate the uptake of P by the hyphal network. Additionally, the SPACSYS model has been used to produce field-scale representations of microbially mediated processes including N_2O emissions (Wu et al., 2015) and N fixation (Liu et al., 2013), but to our knowledge these were only validated with a 1D root profile and have not been extended to the full three-dimensional RSA model.

2.6 Application of RSA models in breeding

RSA models have application in breeding. Thus far, the application has been mostly limited to ideotype development by simulating trait-environment interactions and integrated phenotypes (see above). This is possible for an increasing number of crop species including lupin, maize, bean, squash, rice, wheat, barley, and the model species Brachypodium (Liu et al., 2012; Postma and Lynch, 2012; see for example Dunbabin et al., 2013). Beyond the species level, parameterization at the genotype level has allowed the understanding of functional aspects associated with specific genotypes (Postma and Lynch, 2011a; York et al., 2015; see, for example, Fang et al., 2019). York et al. (2015) used the RSA model SimRoot to show that changes in root phenotype that were introduced during a 100-year timeline for breeding were associated with greater N capture. So far, however, the direct application of models in breeding has been limited, possibly because of a perception that RSA models have not yet been sufficiently validated (Koevoets et al., 2016; Ndour et al., 2017). However, an encouraging counterpoint to this perception comes from Zhao et al. (2017), who noted that observations of pea seedling root phenotypes had little predictive power when correlated directly to mature root traits, but that using these same seedling traits to parameterize RootBox simulations resulted in predicted root lengths that corresponded well with observations of the mature plants. This suggests immediate applications of models to aid phenotyping, and we expect a greater role for these models in breeding in the future, as outlined by Lynch (2007), Tracy et al. (2020) and in detail by Ndour et al. (2017) for breeding for drought tolerance.

2.7 Application of RSA models in agronomy and ecology

Although agronomy and ecology are strictly speaking at a higher scale than these plant models, they have been used to understand plant-plant interactions and in the ecological realm, species diversity. By simulating stands of multiple plants, RSA models are useful to understand how plant traits affect root competition (Fig. 4). Multiple plant stands are featured in CRootBox (Schnepf et al., 2018b) and OpenSimRoot (Postma et al., 2017). Using the later, Rangarajan et al. (2018) showed that the optimal number of basal roots in bean was lower when the planting density was higher. Such interactions between plant density and root architecture are currently poorly understood (Hecht et al., 2019), but of importance when breeding for high production at high plant densities and also for the translation of single plant phenotypes to an agricultural context (Hecht et al., 2020).

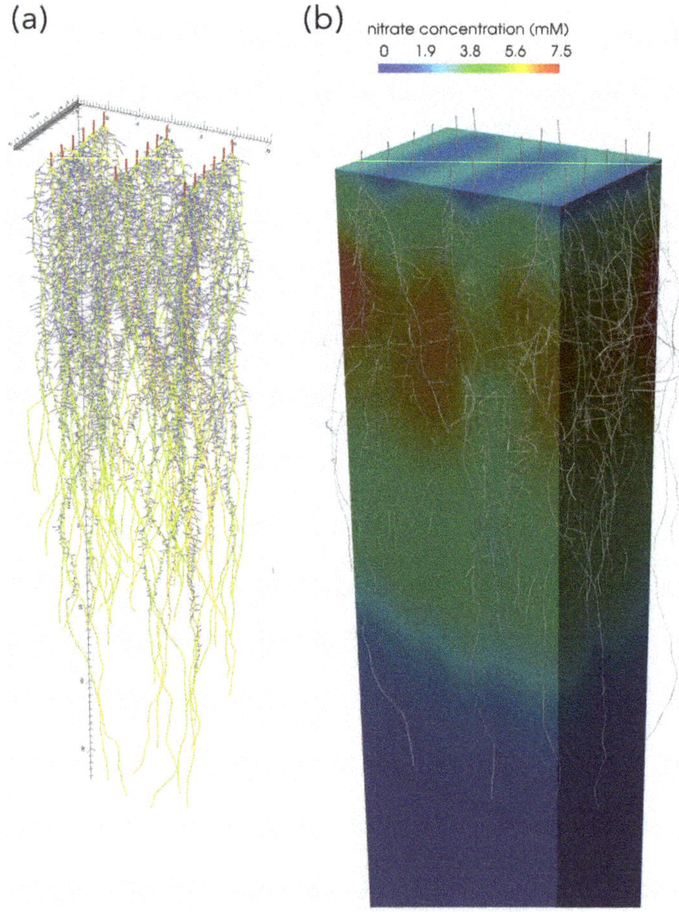

Figure 4 Simulation of barley growing at a high planting density. Roots were simulated with OpenSimRoot and rendered with Paraview. (a) Roots with pseudo colors indicating different root types in order of diameter laterals (blue), seminals (yellow), nodals (green), and mesocotyl (red). (b) Simulation of nitrate depletion by a high planting density of barley. For pseudo colors see legend.

Future sustainable agriculture should be biodiverse. Intercropping studies are of renewed interest as robotics might solve long-standing issues in the mechanization of such cropping systems. Postma and Lynch (2012) simulated the competition between maize bean and squash root systems for nitrate and phosphorus and concluded that the architectures and acquisition strategies of these plants are complementary. The maize-bean-squash intercropping system is an ancient one, and the simulation study merely provides some mechanistic insights into the functioning of the intercrop as observed experimentally (Zhang et al., 2014). However, Evers et al. (2019) proposed that the functional structural plant models can be used to explore many more intercropping schemes and combinations, in order to reduce the number of systems tested experimentally. Such studies not only have utility in agriculture, but will also contribute to our ecological understanding of biodiversity. Pagès et al. achieved much progress in this area by developing a relatively simple RSA model such that they were able to parameterize it for a wide range of species (Pagès et al., 2014; for example, see Pagès and Picon-Cochard, 2014). This work in the future might be linked to databases of functional trait variation in ecology in order to explain the variation in functional root traits.

2.8 Application of RSA models for refining experimental measurements

Once an RSA model has been validated to produce a realistic three-dimensional root system, it immediately becomes useful not just for asking questions about root architecture, but also as a tool for producing known root architectures that can be used to calibrate measurement techniques. For example Lobet et al. (2017) used ArchiSimple to create a ground-truth library of images of root systems with known dimensions and then used these images for training of an image analysis tool. Similarly, Leitner et al. (2014a) used RootBox models to improve segmentation of neutron radiography images of living root systems by adding weight to detected paths that lie within the growth trajectory expected by RootBox and downweighting as probable errors the paths that do not. Schulz et al. (2013) tested the presented segmentation algorithm for roots in MRI images and used artificial generated images with varying resolution and noise levels to ground-truth the algorithm and find acceptable values. The images were generated with SimRoot. Rao et al. (2019) used modeled electrical conductivity of root-soil systems in order to interpret measurements.

Similarly, simulated root systems provide a ready method for testing field sampling methods. While this is not new (c.f. Pagès and Bengough, 1997), it continues to be a fruitful way of interpreting the reliability and bias of core, trench, and minirhizotron samples (Miguel et al., 2015; Wu et al., 2018; Morandage et al., 2019). Pagès et al. (2012) took a different approach and compared the output of an RSA model against a collection of possible root-length density models fitted to the same data used to parameterize the RSA model, and showed that the results were largely congruent. This was reassuring and, given the differing goals and parameterizations of RSA and RLD models, not a foregone conclusion.

3 Advances in model development

Model development in the last decade has been dominated by a desire for implementing an ever-increasing number of processes, more precise presentation of morphology, and inclusion of the environment. This required greater flexibility in the code, but also lead to more complex and extensive models. Communicating what the model do has become ever more difficult. Nearly all codes were developed modularly and adopted forms of object-oriented programming in order to accommodate team development and included more flexible implementation of new algorithms. This has led to model names such as CRootBox and OpenSimRoot which, in practice, are rather names for code bases with which different models can be constructed, depending on modules and parameters used. Modularity, in this case, leads to the encapsulation of different parts of the model, not necessarily transparency about what the model does. Few models have trackable public histories or backward compatibility and it is even a question as to what extent published results can be repeated, assuming licenses do not restrict access to the code. Because of this, there is a need for more transparency, simplification and better, more formal ways for parameterization of the models.

3.1 Continued model coupling

Multiscale-multiphysics is a current buzz-word that carries over into systems biology and visions have been expounded to simulate from gene regulatory networks up to the crop (Marshall-Colon et al., 2017). Root architectural models are supposed to play an important scale and discipline-bridging role in this vision (Zhu et al., 2016). At the lower end of the scale, it has been explored how multi-cell models of roots and root tips, and the cellular networks that they simulate, can be upscaled to the whole-plant level (Baldazzi et al., 2012; Band et al., 2012). At even greater scales, Warren et al. (2015) explored how root models can be coupled to terrestrial biosphere models.

Modeling across scales and coupling processes that function at different scales is, however, a difficult task. This challenge was well summarized by Koevoets et al. (2016) in a discussion about using RSA models to advance crop breeding:

> Current models, however, are often not easy to integrate [into breeding programs]. When developing a model, the general challenge is to make it comprehensive,

> widely applicable and simple. [. . .] As soon as models tend to be more widely applicable or incorporate more conditions, they tend to become more complex and the number of parameters increases. This decreases the ease of interpretation and especially the ease of integration into a larger model (including soil and plant performance models).

One of the challenges currently being explored is the coupling of roots to soil models. In this coupling, the geometries of both models do not necessarily match and matching them would, because of the deforming nature of root growth, require a computationally intensive solution (Postma et al., 2008). Adaptive meshing was tried in the RSWMS model (Schröder et al., 2009), but abandoned as it did not seem to provide more speed or accuracy than a static mesh. A challenge is posed not only by the geometry but also by the required resolution of the mesh around the roots. To circumvent these issues, Mai et al. (2018) coupled a high-resolution rhizosphere model around the roots to a much coarser three-dimensional soil-domain model. The work demonstrated that in doing so, the sink term at the higher scale might be improved, as well as the outer boundary condition at the lower scale, but that conflicting assumptions at both scales remain.

A more straightforward development of model coupling is the coupling of RSA models to FSPM of the shoot. An early approach was presented by Lobet et al. (2014) who used a coupled model to simulate water transport from roots to leaves, and more recently Fang et al. (2019) have taken a similar approach using different component models and adding nutrient ion flux as well as water flow. The advances in FSPM-shoot models were recently reviewed by Evers et al. (2018). Coupling of shoot and root seems a logical step forward and will enable a much better understanding of shoot-root interactions than thus far possible.

3.2 Modeling platforms to aid in defining new models

To facilitate the development of coupled models, different software platforms have been developed which are able to couple various submodels. From an IT perspective, this is best achieved through what is called loose coupling (Fig. 5). Loose coupling is achieved when different components of software (in this case different models) are coupled over a standardized programming interface, without the components having any specific information about the implementation of the other component. This means that the data exchange between models is abstracted, such that one model can have implementation changes without the need to reprogram other models. This enables independent development, and crucially independent testing, of the different coupled components. It also facilitates the extension of the software by introducing new models and coupling them over the standardized programming interface. Software platforms for simulation typically implement some form of this

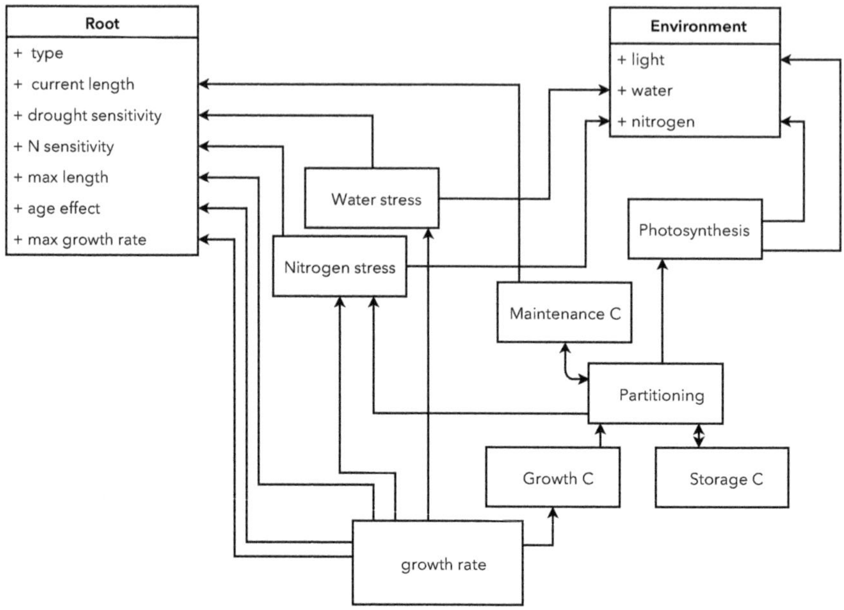
Root
+ type
+ current length
+ drought sensitivity
+ N sensitivity
+ max length
+ age effect
+ max growth rate
Environment
+ light
+ water
+ nitrogen
Water stress
Photosynthesis
Nitrogen stress
Maintenance C
Partitioning
Growth C
Storage C
growth rate

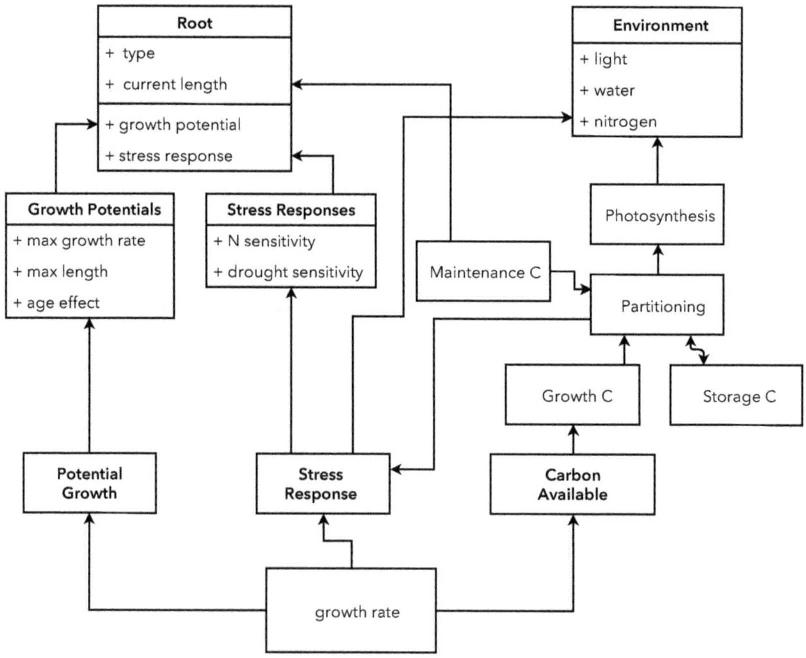
Root
+ type
+ current length
+ growth potential
+ stress response
Environment
+ light
+ water
+ nitrogen
Growth Potentials
+ max growth rate
+ max length
+ age effect
Stress Responses
+ N sensitivity
+ drought sensitivity
Photosynthesis
Maintenance C
Partitioning
Growth C
Storage C
Potential Growth
Stress Response
Carbon Available
growth rate

modularity concept and various platforms have been developed. Various FSPM and RSA models have been constructed using such software platforms for simulation (SPFS).

CrossTalk (Draye and Pagès, 2006) is such an SPFS and was used to construct a coupled model of root and shoot architecture (Lobet et al., 2014). The coupling interface in CrossTalk defines the simulation theme as it requires a model for the root, the shoot, and the environmental components. OpenAlea (Pradal et al., 2008) is a general-purpose platform for constructing FSPMs and has a large code base containing many models. It has been used in various publications in diverse ways, but development activity on its Git repository seems to have stalled in recent years. AMAPstudio (Griffon and de Coligny, 2014) with AMAPsim (Barczi et al., 2008) describes a scene, but the implementation is more abstract and extensible. The software includes a root architectural component, called DigR (Barczi et al., 2018). The approach taken in AMAPsim may be similar to approaches found in the gaming industry where the scene is made up of different geometric models, all expressed in data structures that are defined in the programming interface. This approach allows interactive editing of the scene, and to incorporate geometric models extracted from three-dimensional measurements, including laser-derived point clouds. The designers of AMAPsim see this as a clear advantage over OpenAlea's more developer and process-focused approach (Griffon and de Coligny, 2014).

L-systems have long been used to construct FSPMs (Prusinkiewicz and Lindenmayer, 1990). Although useful as a formalism to construct plant geometries, L-systems do not define the functional aspects of these models. As an alternative, 'Growth Grammar' was developed to have a complete formalism to describe FSPMs (Kurth and Lanwert, 2011). To our knowledge, GroImp (Kniemeyer, 2008) is the only SPFS that implements 'Growth Grammar'. It is mostly used to simulate shoot architectures, however an RSA model was implemented by Henke et al. (2014).

◀ **Figure 5** Conceptual diagrams illustrating 'loose' versus 'tight' coupling between model components. Shown are two possible implementations of a hypothetical model calculating the elongation of a single root based on assimilate supply, genetic potential, and carbon partitioning factors. Each box represents a group of related calculations or values (which would likely be implemented as a class), and arrows indicate information lookup (e.g. read x -> y as 'x uses y'). Top panel: Tightly coupled design. Each calculation requires detailed knowledge of the internal details of other modules. Bottom panel: Loosely coupled design. Modules share information through standardized interfaces 'Potential Growth', 'Stress Response', and 'Carbon Available', thus isolating the growth rate calculation from the details of, for example, where and how parameters are stored or whether the potential growth rate of the current root is limited more by developmental factors or by root type parameters. Note that both implementations ultimately perform the same set of calculations; and thoughtful code design can improve maintainability and interoperability even within a set of mathematical framework.

The majority of the SPFSs discussed above are either limited to integrating models written in a single language or rely on generic cross-language bindings. By contrast Yggdrasil (Lang, 2019) provides interprocess communication through message-passing, thus allowing models written in different languages to synchronize information at each timestep even when running in parallel with each other. This tool is part of the larger 'Crops in Silico' project (Marshall-Colon et al., 2017), an ambitious SFPS that aims to support coupling plant models across all scales from molecular to ecosystem. As part of this ambition, Yggdrasil's developers claim that it will be extended to support automatic runtime coupling of ODE models, effectively allowing models that are written separately to be solved as a single model. Although this automatic coupling facility remains to be seen, if successful it should provide a major step forward in model coupling and co-simulation.

Many of the models that are coupled to construct an FSPM can and, in our opinion, should be mathematically defined as conditional algebraic equations, or as often will be the case, ordinary and partial differential equations (ODE, PDE). Too often they are only defined as programmed algorithms with discontinuities and time-step dependencies hidden. The models can be regarded as a system of coupled differential equations. In numerical mathematics, it is well known that the stability of such systems is strongly problem-dependent. Finding the right solution can be challenging, and loose coupling further complicates this as it introduces the numerical challenges associated with co-simulation. Most SPFS ignore this and often rely on a single step, first order, explicit coupling method. OpenSimRoot was specifically developed as an SPFS for coupling ODEs using predictor-corrected methods. This, in theory, should provide more robustness and reduce numerical errors, but is no guarantee that the right solution will be found. Thus far, users have little way to know how accurate the solution is other than checking mass balances and sensitivity for time and space discretization. These issues are not specific to RSA models, and present a challenge to nearly all dynamic models in biology. Numerical accuracy has been much more of a concern in areas of physics and soil science, possibly in part because plant biology is plagued by other formidable sources of uncertainty.

In addition to large multi-model platforms, recent advances in web technology have allowed the development of a variety of new interfaces for individual models, in some cases even allowing users to run the entire model inside their browser with no further computing resources required. For example, CRootBox itself is a compiled C++ application, but its developers also maintain shinyRootBox (Schnepf et al., 2018a), a live interactive modeling environment, that allows users to explore predefined parameter sets and experiment with new parameter values by adjusting sliders in a web browser while the simulation updates in real time (Fig. 6).

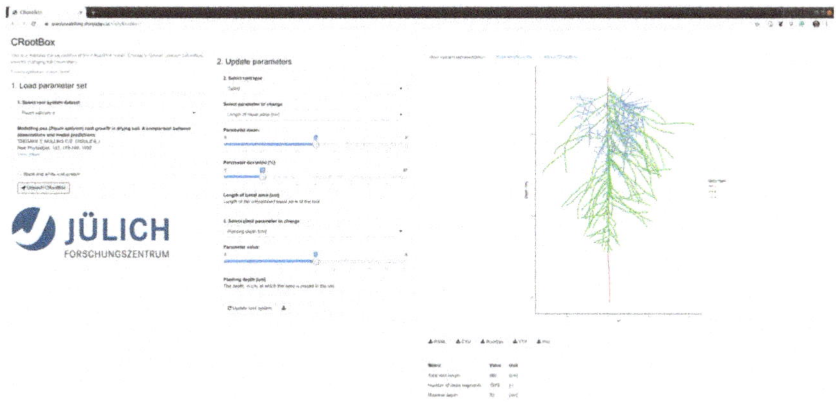

Figure 6 Screenshot of Shiny web app for generating root system architectures with CRootBox (https://plantmodelling.shinyapps.io/shinyRootBox/).

3.3 Simplification of models

The ever-increasing complexity of RSA models has led to important work toward their simplification. Specifically, Archisimple (Pagès et al., 2014) is an RSA model implemented in the scripting language R and has a reduced parameter set. This was achieved by assuming correlations among certain parameters. Archisimple takes advantage of experimentally observed correlations for branching patterns (Pagès, 2014), lateral root traits (Wu et al., 2016), and axial root traits (Wu et al., 2014). Sensitivity analysis is more easily achieved with a model that has a reduced parameter set (Pagès, 2011) and the relation between root-length density profiles and RSA models could be determined (Pagès et al., 2012), as well as studies of the stochasticity of root systems (Pagès et al., 2013).

Although RSA models constructed with, for example, Crootbox and OpenSimRoot are often complex, their modularity allows simple models to be built. This raises a relevant question concerning what complexity can be left out. In general, this is answered through sensitivity analysis (e.g., Schnepf et al., 2018a; Morandage et al., 2019). However, complete sensitivity analyses of complex parameter sets are prohibitively large, so too often the analysis is performed on a subset of parameters already suspected to be sensitive. This leads to the conclusion that the processes included significantly influence results and therefore should not be ignored. For example, Thomas and Pollen-Bankhead (2010) concluded that root failure calculations cannot ignore soil type when parameterizing root friction angle. This demonstrates that there is still significant uncertainty in how to define RSA models and that different concepts varying in complexity are likely to be developed.

One way to reduce the RSA model is to see if the modeled root distributions can be represented by simpler continuous models containing fewer parameters.

Different approaches are found in Zhang et al. (2014) and Bonneu et al. (2012). But since these models are not FSPM, they fall outside the scope of this review.

3.4 Better handling of input and output

In addition to uncertainty about which model parameters to include, RSA model operators suffer from uncertainty in deciding which values to assign to the parameters. This task is easier when parameters correspond to directly measurable root traits (e.g. root diameter in mm rather than a scaling factor in arbitrary units), but is still challenging. The de facto standard for most models has been to begin from an existing parameterization for the species of interest, then to measure plants or perform a meta-analysis (e.g. Pagès et al., 2012). Further, models differ in which parameters they need and in the units, names, and interpretations of the parameters they share, which means parameters developed for one model are of limited value for simulating the same plant using a different model.

Several researchers have proposed systems for converting image data (scanned root systems, rhizoboxes, minirhizotrons, and MRI scans) into model parameters. One pipeline relies on the partially automated extraction of images into a machine-readable format, for example, RSML (Lobet et al., 2015), followed by the computation of relevant parameters from the RSML input using the R package archiDART (Delory et al., 2016). Several RSA models can also now read and write RSML directly (Postma et al., 2017; Schnepf et al., 2018b), in some cases allowing simulations to begin from a digitized real architecture and continue growing as if this architecture had been generated by the model. An alternate pipeline for image data is to estimate the parameters through a reverse-modeling approach, as Garré et al. (2012) did for a minirhizotron dataset that was too sparse to fully constrain the model, and as somewhat more elaborately done by Ziegler et al. (2019) using an approximate Bayesian computation that increased the reliability of the procedure on noisy datasets.

In any case, the rise of more detailed inputs leads to the need to rethink what constitutes a valid output. Comparisons between observed and predicted traits require that the model be parameterized for the same scale and using the same experimental treatments as the observations, whether that is whole-field biomass, root length density (e.g. Bingham and Wu, 2011), root length of individual plants (Postma and Lynch, 2011b), or detailed time series of root placement and water flow (Koch et al., 2019).

This handling of input and output is one area where modeling platforms (discussed above) have potential to help. An example from the field of ecological modeling, and of which we are not aware of a root-modeling equivalent, is PEcAn ('Predictive Ecosystem Analyzer'; LeBauer et al., 2013), a meta-analytic data-assimilation pipeline built for ecosystem forecasting models that is intended to make it easy to assimilate all available data about an experimental site, species, or climate dataset, update it when new observations become available, and feed it to many models at once. If a similar platform existed for parameterizing and comparing RSA models, it would greatly reduce the effort needed to run existing models with new data and to evaluate the same scenario using multiple RSA models. Such multi-model comparisons could greatly aid benchmarking efforts (Schnepf et al., 2019) by showing divergence or convergence between predictions from models expected to be equivalent, and could also be used for scaling results up or down (to the extent that is valid for the available inputs) by running detailed and generalized models alongside each other.

4 Current challenges and future directions

Future RSA model development will likely be focused on the inclusion of more mechanisms and concepts. These will be multiscale (from cell to crop), but also at the plant scale will include, among others, root-shoot coupling, rhizosphere processes, plasticity responses, and root sensing and signaling. A curiously ignored subject in this area is plant protection. This is probably because model development has typically ignored interactions among organisms possibly because such interactions are still not understood very well. Despite this, there is a need for new concepts and their implementation. Although model complexity will continue increasing by introduction of new processes, the time and mental capacity of model developers will not, and the quality of the code bases must be tested and maintained. Fortunately the software industry has also developed and many tools are now available to ease the quality-assurance process. For example, every code change in OpenSimRoot is tested using continuous integration (CI) functionality built into its GitLab repository. The number of tests, however, is limited and needs extension. Benchmarking is one way of testing that the code basis can achieve sufficient accuracy, but this has so far not been widely attempted. The first benchmarks for RSA models have been described (Schnepf et al., 2019), but benchmarking requires that models are to some extent made comparable. Currently, there is large variation between models both in the algorithms they use and in the assumptions they make. Documentation and unification will need to be main themes in the years to come. Besides benchmarking, validation of the predictions is necessary. Although a good agreement has been achieved between the utility of traits simulated and the experimental genotypic contrast studies, more detailed validation is lacking.

Lack of validation raises questions about the usability of the models for nonspecialists, but the number of researchers that would like to test a hypothesis in silico is increasing. Nevertheless, the current software has steep learning curves. Graphical User Interface (GUI) development, better documentation, and adopting modern development practices may help.

5 Conclusions

In the last decade, RSA models moved from close-sourced, single-user or lab developments to open-source community-driven development. This was necessary as the models have incorporated much more complex processes. RSA models have proven to be of relevance to many functional questions in plant nutrition and plant phenotyping. They have become useful tools for testing the functional benefits of RSA traits for nutrient and water acquisition. However, many challenges remain. Conceptually, the models do not represent the rhizosphere very well and the biological interactions with soil microbes are poorly represented if at all. Furthermore, validation of the models is still scattered, and benchmarking of model components to arrive at an agreed level of quality is needed.

6 Acknowledgements

We thank Larry York for providing Fig. 3. This research was institutionally funded by the Helmholtz Association (POF III Program - Research Field Key Technologies - Key Technologies for the Bioeconomy). C. K. B. acknowledges funding by the Advanced Research Projects Agency-Energy (ARPA-E), US Department of Energy, under Award Number DE-AR0000821.

7 Where to look for further information

In Section 4 we outlined future directions which will be balanced by a tension between on the one side an emphasis on integrating more processes and on the other side an attempt to reach a wider public by making the codes easier to understand and more accessible to other researchers. Benchmarking is necessary to guarantee quality development of complex codes, as well as build confidence with code users. Currently, the forefront in RSA modeling development lies in Europe with leading institutions: Forschungszentrum Juelich (Germany), INRA (France) and UClouvain (Germany).

8 References

Alm, D. M., Cavelier, J. and Nobel, P. S. (1992). A finite-element model of radial and axial conductivities for individual roots: development and validation for two desert succulents. *Annals of Botany* 69(1): 87–92.

Baldazzi, V., Bertin, N., de Jong, H. and Génard, M. (2012). Towards multiscale plant models: integrating cellular networks. *Trends in Plant Science* 17(12): 728–736.

Band, L. R., Fozard, J. A., Godin, C., Jensen, O. E., Pridmore, T., Bennett, M. J. and King, J. R. (2012). Multiscale systems analysis of root growth and development: modeling beyond the network and cellular scales. *The Plant Cell* 24(10): 3892–3906.

Barczi, J. F., Rey, H., Caraglio, Y., de Reffye, P., Barthélémy, D., Dong, Q. X. and Fourcaud, T. (2008). AmapSim: a structural whole-plant simulator based on botanical knowledge and designed to host external functional models. *Annals of Botany* 101(8): 1125–1138.

Barczi, J. F., Rey, H., Griffon, S. and Jourdan, C. (2018). DigR: a generic model and its open source simulation software to mimic three-dimensional root-system architecture diversity. *Annals of Botany* 121(5): 1089–1104.

Bingham, I. J. and Wu, L. (2011). Simulation of wheat growth using the 3D root architecture model SPACSYS: validation and sensitivity analysis. *European Journal of Agronomy* 34(3): 181–189.

Bodner, G., Leitner, D., Nakhforoosh, A., Sobotik, M., Moder, K. and Kaul, H. P. (2013). A statistical approach to root system classification. *Frontiers in Plant Science* 4: 292.

Bonneu, A., Dumont, Y., Rey, H., Jourdan, C. and Fourcaud, T. (2012). A minimal continuous model for simulating growth and development of plant root systems. *Plant and Soil* 354(1–2): 211–227.

Chen, Y. L., Dunbabin, V. M., Postma, J. A., Diggle, A. J., Siddique, K. H. M. and Rengel, Z. (2013). Modelling root plasticity and response of narrow-leafed lupin to heterogeneous phosphorus supply. *Plant and Soil* 372(1–2): 319–337.

Darrah, P. R., Jones, D. L., Kirk, G. J. D. and Roose, T. (2006). Modelling the rhizosphere: a review of methods for "upscaling" to the whole-plant scale. *European Journal of Soil Science* 57(1): 13–25.

Dathe, A., Postma, J. A., Postma-Blaauw, M. B. and Lynch, J. P. (2016). Impact of axial root growth angles on nitrogen acquisition in maize depends on environmental conditions. *Annals of Botany* 118(3): 401–414.

de Moraes, M. T., Bengough, A. G., Debiasi, H., Franchini, J. C., Levien, R., Schnepf, A. and Leitner, D. (2018). Mechanistic framework to link root growth models with weather and soil physical properties, including example applications to soybean growth in Brazil. *Plant and Soil* 428(1–2): 67–92.

de Parseval, H., Barot, S., Gignoux, J., Lata, J.-C. and Raynaud, X. (2017). Modelling facilitation or competition within a root system: importance of the overlap of root depletion and accumulation zones. *Plant and Soil* 419(1–2): 97–111.

Delory, B. M., Baudson, C., Brostaux, Y., Lobet, G., du Jardin, P., Pagès, L. and Delaplace, P. (2016). archiDART: an R package for the automated computation of plant root architectural traits. *Plant and Soil* 398(1–2): 351–365.

Diggle, A. J. (1988). ROOTMAP–a model in three-dimensional coordinates of the growth and structure of fibrous root systems. *Plant and Soil* 105(2): 169–178.

Doussan, C., Pagès, L. and Vercambre, G. (1998). Modelling of the hydraulic architecture of root systems: an integrated approach to water absorption - Model description. *Annals of Botany* 81(2): 213–223.

Draye, X. and Pagès, L. (2006). CrossTalk: a simulation platform for the linking of existing soil, plant and atmosphere models. In: International Symposium on Plant Growth Modeling and Applications. IEEE Computer Society, Los Alamitos, CA, pp 93–100.

Dunbabin, V. M., Airey, M., Diggle, A. J., Renton, M., Rengel, Z., Armstrong, R., Chen, Y. and Siddique, K. H. M. (2011). Simulating the interaction between plant roots, soil water and nutrient flows, and barriers and objects in soil using ROOTMAP. In: 19th International Congress on Modelling and Simulation, Perth, Australia, 12–16. http://mssanz.org.au/modsim2011.

Dunbabin, V. M., Postma, J. A., Schnepf, A., Pagès, L., Javaux, M., Wu, Lianhai, Leitner, D., Chen, Y. L., Rengel, Z. and Diggle, A. J. (2013) Modelling root-soil interactions using three-dimensional models of root growth, architecture and function. *Plant and Soil* 372(1-2): 93-124.

Evers, J. B., Letort, V., Renton, M. and Kang, M. (2018). Computational botany: advancing plant science through functionalStructural plant modelling. *Annals of Botany* 121(5): 767-772.

Evers, J. B., van der Werf, W., Stomph, T. J., Bastiaans, L. and Anten, N. P. R. (2019). Understanding and optimizing species mixtures using functionalStructural plant modelling. *Journal of Experimental Botany* 70(9): 2381-2388.

Fan, M., Zhu, J., Richards, C., Brown, K. M. and Lynch, J. P. (2003). Physiological roles for aerenchyma in phosphorus-stressed roots. *Functional Plant Biology* 30(5): 493-506.

Fang, Y., Yabusaki, S. B., Ahkami, A. H., Chen, X. and Scheibe, T. D. (2019). An efficient three-dimensional rhizosphere modeling capability to study the effect of root system architecture on soil water and reactive transport. *Plant and Soil* 441(1-2): 33-48. doi: 10.1007/s11104-019-04068-z.

Feddes, R. A., Hoff, H., Bruen, M., Dawson, T., de Rosnay, P., Dirmeyer, P., Jackson, R. B., Kabat, P., Kleidon, A., Lilly, A. and Pitman, A. J. (2001). Modeling root water uptake in hydrological and climate models. *Bulletin of the American Meteorological Society* 82(12): 2797-2809.

Fitter, A. H. (1987). An architectural approach to the comparative ecology of plant root systems. *New Phytologist* 106: 61-77.

Freschet, G., Pagès, L., Iversen, C., Comas, L., Rewald, B., Roumet, C., Klimešová, J., Zadworny, M., Poorter, H., Postma, J., Adams, T. S., Bagniewska-Zadworna, A., Blancaflor, E. B., Brunner, I., Cornelissen, J. H. C., Garnier, E., Gessler, A., Hobbie, S. E., Lambers, H., Meier, I. C., Mommer, L., Picon-Cochard, C., Rose, L., Ryser, P., Scherer-Lorenzen, M., Soudzilovskaia, N. A., Stokes, A., Sun, T., Valverde-Barrantes, O. J., Weemstra, M., Weigelt, A., Wurzburger, N., York, L. M., Batterman, S. A., Bengough, A. G., Gomes de Moraes, M., Janeček, Š., Salmon, V., Tharayil, N. and McCormack, M. L. (2020). A starting guide to root ecology: strengthening ecological concepts and standardizing root classification, sampling, processing and trait measurements. *New Phytologist*, under review.

Garré, S., Pagès, L., Laloy, E., Javaux, M., Vanderborght, J. and Vereecken, H. (2012) Parameterizing a dynamic architectural model of the root system of spring barley from minirhizotron data. *Vadose Zone Journal* 11: 0.

Griffon, S. and de Coligny, F. (2014). AMAPstudio: an editing and simulation software suite for plants architecture modelling. *Ecological Modelling* 290: 3-10.

Hecht, V. L., Nagel, K. A., Temperton, V., Van Dusschoten, D. V., Rascher, U., Léon, J. and Postma, J. A. (2020). Translation from lab to field: from single-plant-based selection for larger root system to agronomic traits and performance under competitive field conditions in spring barley. *Plant and Soil*.

Hecht, V. L., Temperton, V. M., Nagel, K. A., Rascher, U., Pude, R. and Postma, J. A. (2019). Plant density modifies root system architecture in spring barley (*Hordeum vulgare* L.) through a change in nodal root number. *Plant and Soil* 439(1-2): 179-200.

Henke, M., Sarlikioti, V., Kurth, W., Buck-Sorlin, G. H. and Pagès, L. (2014). Exploring root developmental plasticity to nitrogen with a three-dimensional architectural model. *Plant and Soil* 385(1-2): 49-62.

Heppell, J., Talboys, P., Payvandi, S., Zygalakis, K. C., Fliege, J., Withers, P. J. A., Jones, D. L. and Roose, T. (2015). How changing root system architecture can help tackle a reduction in soil phosphate (P) levels for better plant P acquisition. *Plant, Cell and Environment* 38(1): 118–128.

Huber, K., Vanderborght, J., Javaux, M., Schröder, N., Dodd, I. C. and Vereecken, H. (2014). Modelling the impact of heterogeneous rootzone water distribution on the regulation of transpiration by hormone transport and/or hydraulic pressures. *Plant and Soil* 384(1–2): 93–112.

Javaux, M., Schroeder, T., Vanderborght, J. and Vereecken, H. (2008). Use of a three-dimensional detailed modeling approach for predicting root water uptake. *Vadose Zone Journal* 7(3): 1079–1088.

Jia, X., Liu, P. and Lynch, J. P. (2018). Greater lateral root branching density in maize improves phosphorus acquisition from low phosphorus soil. *Journal of Experimental Botany* 69(20): 4961–4970.

Kniemeyer, O. (2008). Design and implementation of a graph grammar based language for functional-structural plant modelling. (PhD Thesis). Brandenburg University of Technology. Cottbus-Senftenberg, Germany.

Koch, A., Meunier, F., Vanderborght, J., Garré, S., Pohlmeier, A. and Javaux, M. (2019). FunctionalStructural root-system model validation using a soil MRI experiment. *Journal of Experimental Botany* 70(10): 2797–2809.

Koevoets, I. T., Venema, J. H., Elzenga, J. T. M. and Testerink, C. (2016). Roots withstanding their environment: exploiting root system architecture responses to abiotic stress to improve crop tolerance. *Frontiers in Plant Science* 7: 1335. doi: 10.3389/fpls.2016.01335.

Kurth, W. and Lanwert, D. (2011). Grammar-based models and fractals. In: Jopp, F., Reuter, H. and Breckling, B. (Eds), *Modelling Complex Ecological Dynamics*. Springer, Berlin Heidelberg, pp 147–161.

Landl, M., Huber, K., Schnepf, A., Vanderborght, J., Javaux, M., Bengough, A. G. and Vereecken, H. (2017). A new model for root growth in soil with macropores. *Plant and Soil* 415: 99–116.

Landl, M., Schnepf, A., Uteau, D., Peth, S., Athmann, M., Kautz, T., Perkons, U., Vereecken, H. and Vanderborght, J. (2019). Modeling the impact of biopores on root growth and root water uptake. *Vadose Zone Journal* 18: 180196.

Lang, M. (2019). Yggdrasil: a Python package for integrating computational models across languages and scales. *in silico Plants* 1(1). doi: 10.1093/insilicoplants/diz001.

LeBauer, D. S., Wang, D., Richter, K. T., Davidson, C. C. and Dietze, M. C. (2013). Facilitating feedbacks between field measurements and ecosystem models. *Ecological Monographs* 83(2): 133–154.

Leitner, D., Felderer, B., Vontobel, P. and Schnepf, A. (2014a). Recovering root system traits using image analysis exemplified by two-dimensional neutron radiography images of lupine. *Plant Physiology* 164(1): 24–35.

Leitner, D., Klepsch, S., Bodner, G. and Schnepf, A. (2010). A dynamic root system growth model based on L-systems. *Plant and Soil* 332(1–2): 177–192.

Leitner, D., Meunier, F., Bodner, G., Javaux, M. and Schnepf, A. (2014b). Impact of contrasted maize root traits at flowering on water stress tolerance simulation study. *Field Crops Research* 165: 125–137.

Liu, K. N., Yang, Q. L., Ge, Z. Y. and Liu, X. G. (2012). Simulation of *Jatropha Curcas* L. Root in response to salt stress based on 3D visualization. *Advanced in Materials Research* 518-523: 5330-5334.

Liu, Y., Wu, L., Watson, C. A., Baddeley, J. A., Pan, X. and Zhang, L. (2013). Modeling biological dinitrogen fixation of field pea with a process-based simulation model. *Agronomy Journal* 105(3): 670-678.

Lobet, G., Koevoets, I. T., Noll, M., Meyer, P. E., Tocquin, P., Pagès, L. and Périlleux, C. (2017). Using a structural root system model to evaluate and improve the accuracy of root image analysis pipelines. *Frontiers in Plant Science* 8: 447. doi: 10.3389/fpls.2017.00447.

Lobet, G., Pagès, L. and Draye, X. (2014). A modeling approach to determine the importance of dynamic regulation of plant hydraulic conductivities on the water uptake dynamics in the soil-plant-atmosphere system. *Ecological Modelling* 290: 65-75.

Lobet, G., Pound, M. P., Diener, J., Pradal, C., Draye, X., Godin, C., Javaux, M., Leitner, D., Meunier, F., Nacry, P., Pridmore, T. P. and Schnepf, A. (2015). Root system markup language: toward a unified root architecture description language. *Plant Physiology* 167(3): 617-627.

Lynch, J. P. (2007). Roots of the second green revolution. *Australian Journal of Botany* 55(5): 493-512.

Mai, T. H., Schnepf, A., Vereecken, H. and Vanderborght, J. (2018). Continuum multiscale model of root water and nutrient uptake from soil with explicit consideration of the 3D root architecture and the rhizosphere gradients. *Plant and Soil*. doi: 10.1007/s11104-018-3890-4.

Marshall-Colon, A., Long, S. P., Allen, D. K., Allen, G., Beard, D. A., Benes, B., von Caemmerer, S., Christensen, A. J., Cox, D. J., Hart, J. C., Hirst, P. M., Kannan, K., Katz, D. S., Lynch, J. P., Millar, A. J., Panneerselvam, B., Price, N. D., Prusinkiewicz, P., Raila, D., Shekar, R. G., Shrivastava, S., Shukla, D., Srinivasan, V., Stitt, M., Turk, M. J., Voit, E. O., Wang, Y., Yin, X. and Zhu, X. G. (2017). Crops in silico: generating virtual crops using an integrative and multi-scale modeling platform. *Frontiers in Plant Science* 8: 1-7.

Meunier, F., Couvreur, V., Draye, X., Zarebanadkouki, M., Vanderborght, J. and Javaux, M. (2017a). Water movement through plant roots exact solutions of the water flow equation in roots with linear or exponential piecewise hydraulic properties. *Hydrology and Earth System Sciences* 21(12): 6519-6540.

Meunier, F., Draye, X., Vanderborght, J., Javaux, M. and Couvreur, V. (2017b). A hybrid analytical-numerical method for solving water flow equations in root hydraulic architectures. *Applied Mathematical Modelling* 52: 648-663.

Miguel, M. A., Postma, J. A. and Lynch, J. P. (2015). Phene synergism between root hair length and basal root growth angle for phosphorus acquisition. *Plant Physiology* 167(4): 1430-1439.

Morandage, S., Schnepf, A., Leitner, D., Javaux, M., Vereecken, H. and Vanderborght, J. (2019). Parameter sensitivity analysis of a root system architecture model based on virtual field sampling. *Plant and Soil* 438(1-2): 101-126.

Muller, B., Guédon, Y., Passot, S., Lobet, G., Nacry, P., Pagès, L., Wissuwa, M. and Draye, X. (2019). Lateral roots: random diversity in adversity. *Trends in Plant Science* 24(9): 810-825.

Ndour, A., Vadez, V., Pradal, C. and Lucas, M. (2017). Virtual plants need water too: functional-structural root system models in the context of drought tolerance breeding. *Frontiers in Plant Science* 8: 1577. doi: 10.3389/fpls.2017.01577.

Pagès, L. (2011). Links between root developmental traits and foraging performance. *Plant, Cell and Environment* 34(10): 1749-1760.

Pagès, L. (2014). Branching patterns of root systems: quantitative analysis of the diversity among dicotyledonous species. *Annals of Botany* 114(3): 591-598.

Pagès, L. (2019). Analysis and modeling of the variations of root branching density within individual plants and among species. *Frontiers in Plant Science* 10: 1020. doi: 10.3389/fpls.2019.01020.

Pagès, L., Bécel, C., Boukcim, H., Moreau, D., Nguyen, C. and Voisin, A.-S. (2014). Calibration and evaluation of ArchiSimple, a simple model of root system architecture. *Ecological Modelling* 290: 76-84.

Pagès, L. and Bengough, A. G. (1997). Modelling minirhizotron observations to test experimental procedures. *Plant and Soil* 189: 81-89.

Pagès, L., Bruchou, C. and Garré, S. (2012). Links between root length density profiles and models of the root system architecture. *Vadose Zone Journal* 11(4). doi: 10.2136/vzj2011.0152.

Pagès, L., Jordan, M. O. and Picard, D. (1989). A simulation model of the three-dimensional architecture of the maize root system. *Plant and Soil* 119(1): 147-154.

Pagès, L. and Picon-Cochard, C. (2014). Modelling the root system architecture of Poaceae. Can we simulate integrated traits from morphological parameters of growth and branching? *New Phytologist* 204(1): 149-158.

Pagès, L., Xie, J. and Serra, V. (2013). Potential and actual root growth variations in root systems: modeling them with a two-step stochastic approach. *Plant and Soil* 373(1-2): 723-735.

Postma, J. A., Dathe, A. and Lynch, J. P. (2014a). The optimal lateral root branching density for maize depends on nitrogen and phosphorus availability. *Plant Physiology* 166(2): 590-602.

Postma, J. A., Jaramillo, R. E. and Lynch, J. P. (2008). Towards modeling the function of root traits for enhancing water acquisition by crops. In: Ahuja, L., Reddy, V., Saseendran, S. and Yu, Q. (Eds), *Response of Crops to Limited Water: Understanding and Modeling Water Stress Effects on Plant Growth Processes*. ASA-CSSA-SSSA, Madison, WI, pp 251-275.

Postma, J. A., Kuppe, C., Owen, M. R., Mellor, N., Griffiths, M., Bennett, M. J., Lynch, J. P. and Watt, M. (2017). OpenSimRoot: widening the scope and application of root architectural models. *New Phytologist* 215(3): 1274-1286.

Postma, J. A. and Lynch, J. P. (2012). Complementarity in root architecture for nutrient uptake in ancient maize/bean and maize/bean/squash polycultures. *Annals of Botany* 110(2): 521-534.

Postma, J. A. and Lynch, J. P. (2011a). Root cortical aerenchyma enhances the growth of maize on soils with suboptimal availability of nitrogen, phosphorus, and potassium. *Plant Physiology* 156(3): 1190-1201.

Postma, J. A. and Lynch, J. P. (2011b). Theoretical evidence for the functional benefit of root cortical aerenchyma in soils with low phosphorus availability. *Annals of Botany* 107(5): 829-841.

Postma, J. A., Schurr, U. and Fiorani, F. (2014b). Dynamic root growth and architecture responses to limiting nutrient availability: linking physiological models and experimentation. *Biotechnology Advances* 32(1): 53-65.

Pradal, C., Dufour-Kowalski, S., Boudon, F., Fournier, C. and Godin, C. (2008). OpenAlea: a visual programming and component-based software platform for plant modelling. *Functional Plant Biology* 35(10): 751-760.

Prusinkiewicz, P. and Lindenmayer, A. (1990). *The Algorithmic Beauty of Plants, The Virtual Laboratory*. Springer-Verlag, New York.

Rangarajan, H., Postma, J. A. and Lynch, J. P. (2018). Co-optimization of axial root phenotypes for nitrogen and phosphorus acquisition in common bean. *Annals of Botany* 122(3): 485–499.

Rao, S., Meunier, F., Ehosioke, S., Lesparre, N., Kemna, A., Nguyen, F., Garré, S. and Javaux, M. (2019). Impact of maize roots on SoilRoot electrical conductivity: a simulation study. *Vadose Zone Journal* 18(1): 0.

Rubio, G., Walk, T., Ge, Z., Yan, X., Liao, H. and Lynch, J. P. (2001). Root gravitropism and below-ground competition among neighbouring plants: a modelling approach. *Annals of Botany* 88(5): 929–940.

Schneider, H. M., Postma, J. A., Wojciechowski, T., Kuppe, C. and Lynch, J. P. (2017a). Root cortical senescence improves growth under suboptimal availability of N, P, and K. *Plant Physiology* 174(4): 2333–2347.

Schneider, H. M., Wojciechowski, T., Postma, J. A., Brown, K. M., Lücke, A., Zeisler, V., Schreiber, L. and Lynch, J. P. (2017b). Root cortical senescence decreases root respiration, nutrient content and radial water and nutrient transport in barley. *Plant, Cell and Environment* 40(8): 1392–1408.

Schneider, H. M., Wojciechowski, T., Postma, J. A., Brown, K. M. and Lynch, J. P. (2018). Ethylene modulates root cortical senescence in barley. *Annals of Botany* 122(1): 95–105.

Schnepf, A., Black, C. K., Couvreur, V., Delory, B. M., Doussan, C., Koch, A., Koch, T., Javaux, M., Landl, M., Leitner, D., Lobet, G., Mai, T. H., Meunier, F., Petrich, L., Postma, J. A., Priesack, E., Schmidt, V., Vanderborght, J., Vereecken, H. and Weber, M. (2019). Call for participation: collaborative benchmarking of functional-structural root architecture models. The case of root water uptake. *Frontiers in Plant Science* 11. doi: 10.3389/fpls.2020.00316.

Schnepf, A., Huber, K., Landl, M., Meunier, F., Petrich, L. and Schmidt, V. (2018a). Statistical characterization of the root system architecture model CRootBox. *Vadose Zone Journal* 17(1): 0.

Schnepf, A., Leitner, D. and Klepsch, S. (2012). Modeling phosphorus uptake by a growing and exuding root system. *Vadose Zone Journal* 11(3). doi: 10.2136/vzj2012.0001.

Schnepf, A., Leitner, D., Klepsch, S., Pellerin, S. and Mollier, A. (2011). Modelling phosphorus dynamics in the soil-plant system. In: Bünemann, E., Oberson, A. and Frossard, E. (Eds), *Phosphorus in Action*. *Soil Biology*, vol. 26. Springer, Berlin Heidelberg, pp 113–133.

Schnepf, A., Leitner, D., Landl, M., Lobet, G., Mai, T. H., Morandage, S., Sheng, C., Zörner, M., Vanderborght, J. and Vereecken, H. (2018b). CRootBox: a structuralFunctional modelling framework for root systems. *Annals of Botany* 121(5): 1033–1053.

Schnepf, A., Leitner, D., Schweiger, P. F., Scholl, P. and Jansa, J. (2016). L-system model for the growth of arbuscular mycorrhizal fungi, both within and outside of their host roots. *Journal of the Royal Society Interface* 13: 20160129.

Schröder, T., Tang, L., Javaux, M., Vanderborght, J., Körfgen, B. and Vereecken, H. (2009). A grid refinement approach for a three-dimensional soil-root water transfer model. *Water Resources Research* 45(10): W10412.

Schulz, H., Postma, J. A., van Dusschoten, D., Scharr, H. and Behnke, S. (2013). Plant root system analysis from MRI images. In: Csurka, G., Kraus, M., Laramee, R. S., Richard, P. and Braz, J. (Eds), *Computer Vision, Imaging and Computer Graphics: Theory and Application*. Springer, Berlin Heidelberg, pp 411–425.

Strock, C. F., de la Riva, L. M. and Lynch, J. (2017). Reduction in root secondary growth as a strategy for phosphorus acquisition. *Plant Physiology* 01583.2017.

Thomas, R. E. and Pollen-Bankhead, N. (2010). Modeling root-reinforcement with a fiber-bundle model and Monte Carlo simulation. *Ecological Engineering* 36(1): 47–61.

Tournier, P.-H., Hecht, F. and Comte, M. (2015). Finite element model of soil water and nutrient transport with root uptake: explicit geometry and unstructured adaptive meshing. *Transport in Porous Media* 106(2): 487–504.

Trachsel, S., Kaeppler, S. M., Brown, K. M. and Lynch, J. P. (2013). Maize root growth angles become steeper under low N conditions. *Field Crops Research* 140: 18–31.

Tracy, S. R., Nagel, K. A., Postma, J. A., Schneider, H., Wasson, A. and Watt, M. (2020). Crop improvement from phenotyping roots: highlights reveal expanding opportunities. *Trends in Plant Science* 25(1): 105–118.

Tron, S., Bodner, G., Laio, F., Ridolfi, L. and Leitner, D. (2015). Can diversity in root architecture explain plant water use efficiency? A modeling study. *Ecological Modelling* 312: 200–210.

van den Honert, T. H. (1948). Water transport in plants as a catenary process. *Discussions of the Faraday Society* 3: 146–153. doi: 10.1039/DF9480300146.

Vereecken, H., Schnepf, A., Hopmans, J. W., Javaux, M., Or, D., Roose, T., Vanderborght, J., Young, M. H., Amelung, W., Aitkenhead, M., Allison, S. D., Assouline, S., Baveye, P., Berli, M., Brüggemann, N., Finke, P., Flury, M., Gaiser, T., Govers, G., Ghezzehei, T., Hallett, P., Hendricks Franssen, H. J., Heppell, J., Horn, R., Huisman, J. A., Jacques, D., Jonard, F., Kollet, S., Lafolie, F., Lamorski, K., Leitner, D., McBratney, A., Minasny, B., Montzka, C., Nowak, W., Pachepsky, Y., Padarian, J., Romano, N., Roth, K., Rothfuss, Y., Rowe, E. C., Schwen, A., Šimůnek, J., Tiktak, A., Van Dam, J., van der Zee, S. E. a. T. M., Vogel, H. J., Vrugt, J. A., Wöhling, T. and Young, I. M. (2016). Modeling soil processes: review, key challenges, and new perspectives. *Vadose Zone Journal*. doi: 10.2136/vzj2015.09.0131.

Vos, J., Evers, J. B., Buck-Sorlin, G. H., Andrieu, B., Chelle, M. and de Visser, P. H. B. (2010). Functional-structural plant modelling: a new versatile tool in crop science. *Journal of Experimental Botany* 61(8): 2101–2115.

Warren, J. M., Hanson, P. J., Iversen, C. M., Kumar, J., Walker, A. P. and Wullschleger, S. D. (2015). Root structural and functional dynamics in terrestrial biosphere models - evaluation and recommendations. *New Phytologist* 205(1): 59–78.

Wu, J., Pagès, L., Wu, Q., Yang, B. and Guo, Y. (2014). Three-dimensional architecture of axile roots of field-grown maize. *Plant and Soil* 387(1–2): 363–377.

Wu, L., McGechan, M. B., McRoberts, N., Baddeley, J. A. and Watson, C. A. (2007). SPACSYS: integration of a 3D root architecture component to carbon, nitrogen and water cycling model description. *Ecological Modelling* 200(3–4): 343–359.

Wu, L., Rees, R. M., Tarsitano, D., Zhang, X., Jones, S. K. and Whitmore, A. P. (2015). Simulation of nitrous oxide emissions at field scale using the SPACSYS model. *Science of the Total Environment* 530–531: 76–86.

Wu, Q., Pagès, L. and Wu, J. (2016). Relationships between root diameter, root length and root branching along lateral roots in adult, field-grown maize. *Annals of Botany* 117(3): 379–390.

Wu, Q., Wu, J., Zheng, B. and Guo, Y. (2018). Optimizing soil-coring strategies to quantify root-length-density distribution in field-grown maize: virtual coring trials using 3-D root architecture models. *Annals of Botany* 121(5): 809–819.

York, L. M., Galindo-Castañeda, T., Schussler, J. R. and Lynch, J. P. (2015). Evolution of US maize (*Zea mays* L.) root architectural and anatomical phenes over the past 100 years corresponds to increased tolerance of nitrogen stress. *Journal of Experimental Botany* 66(8): 2347–2358.

York, L. M., Silberbush, M. and Lynch, J. P. (2016). Spatiotemporal variation of nitrate uptake kinetics within the maize (*Zea mays* L.) root system is associated with greater nitrate uptake and interactions with architectural phenes. *Journal of Experimental Botany* 67(12): 3763–3775.

Zhan, A. and Lynch, J. P. (2015). Reduced frequency of lateral root branching improves N capture from low-N soils in maize. *Journal of Experimental Botany* 66(7): 2055–2065.

Zhang, H., Zhu, Y. and Chen, H. (2014). Root growth model: a novel approach to numerical function optimization and simulation of plant root system. *Soft Computing* 18(3): 521–537.

Zhao, J., Bodner, G., Rewald, B., Leitner, D., Nagel, K. A. and Nakhforoosh, A. (2017). Root architecture simulation improves the inference from seedling root phenotyping towards mature root systems. *Journal of Experimental Botany* 68(5): 965–982.

Zhu, X. G., Lynch, J. P., LeBauer, D. S., Millar, A. J., Stitt, M. and Long, S. P. (2016). Plants in silico: why, why now and what? – an integrative platform for plant systems biology research. *Plant, Cell and Environment* 39(5): 1049–1057.

Ziegler, C., Dyson, R. J. and Johnston, I. G. (2019). Model selection and parameter estimation for root architecture models using likelihood-free inference. *Journal of the Royal Society Interface* 16(156): 20190293.

Zygalakis, K. C., Kirk, G. J. D., Jones, D. L., Wissuwa, M. and Roose, T. (2011). A dual porosity model of nutrient uptake by root hairs. *New Phytologist* 192(3): 676–688.

Chapter 2

The development of crop root architecture and optimization of nutrition acquisition: the case of rice

Wei Xuan, Nanjing Agricultural University, China; Yuanming Xie, Nanjing Agricultural University, China and Ghent University and VIB-UGent Center for Plant Systems Biology, Belgium; and Tom Beeckman, Ghent University and VIB-UGent Center for Plant Systems Biology, Belgium

1 Introduction

Rice (*Oryza sativa. L*) is a staple food feeding over half of the world's population. Increasing its yield is a major challenge to achieve sustainable agricultural development that can keep pace with an ever-increasing world population in the background of limited land resources. As is the case for most plants, rice roots perform essential functions during development, including anchoring and water and nutrient uptake from the soil.

Rice is a monocotyledonous plant that develops a root system primarily comprising seminal roots (including a radical/primary root and two or more embryonic crown roots), postembryonic crown roots (also known as adventitious roots), and lateral roots that are formed along the seminal and crown roots in a radial pattern. The ability of rice to efficiently take up and use soil resources, even in unfavorable conditions such as water or nutrient deficit, has been considered to be a consequence of its ideally adapted root system architecture

http://dx.doi.org/10.19103/AS.2020.0075.03

(RSA). Parameters that greatly contribute to the performance of RSA include: root growth rate, the rate at which lateral roots and crown roots are formed, and the angles at which lateral roots and crown roots grow in response to gravity.

Over the past decades, molecular and genetic approaches have been applied to identify the main genetic determinants governing these root developmental parameters, and have revealed a complex gene regulatory network. This network involves an endogenous genetic program controlled by plant hormones and other signaling pathways, as well as external environmental factors. In this chapter, we focus in on the cellular and tissue organization of rice root architecture, and review the present genetic regulators of root development in rice that were identified using forward and reverse genetic approaches, such as QTL analysis, mutagenesis and transcriptomics. Where possible, the underlying mechanisms such as the involvement of hormones and genes in rice RSA adaptation to nutrient and drought stresses are further discussed.

2 Rice root architecture and cellular organization of the rice root

2.1 Morphology of the rice root system

The development of the rice root system initiates as early as the radical/primary root develops during embryogenesis before it breaks out from the seed coat, followed by the emergence of embryonic crown roots (see Fig. 1a). The meristem of the radical/primary root is characterized by a highly active cell division, facilitating a fast elongation of the roots to explore for water and nutrients in the soil. Unlike seminal roots, lateral roots and crown roots develop post-embryonically from a set of pericycle cells in the seminal roots and at the stem base of tillers, respectively (see Fig. 1b).

Interestingly, the appearance of rice seminal roots ceases shortly after seed germination, whereas the crown roots are continuously produced along both the main stem and the tillers during the entire rice life cycle and predominantly contribute to the building of the RSA. Thus, crown roots become more important than the seminal roots in water and nutrient uptake at the later development stages, and their developmental pattern determines the mature RSA.

2.2 Tissue organization of rice roots

Along the longitudinal axis, rice roots contain the normal developmental zones typical of angiosperm roots, with from tip to base: a meristematic zone, an elongation zone and a differentiation zone. The meristematic zone is made up of undifferentiated cells with high division activity, which is subsequently activated to increase cell length when they enter the elongation zone. Therefore,

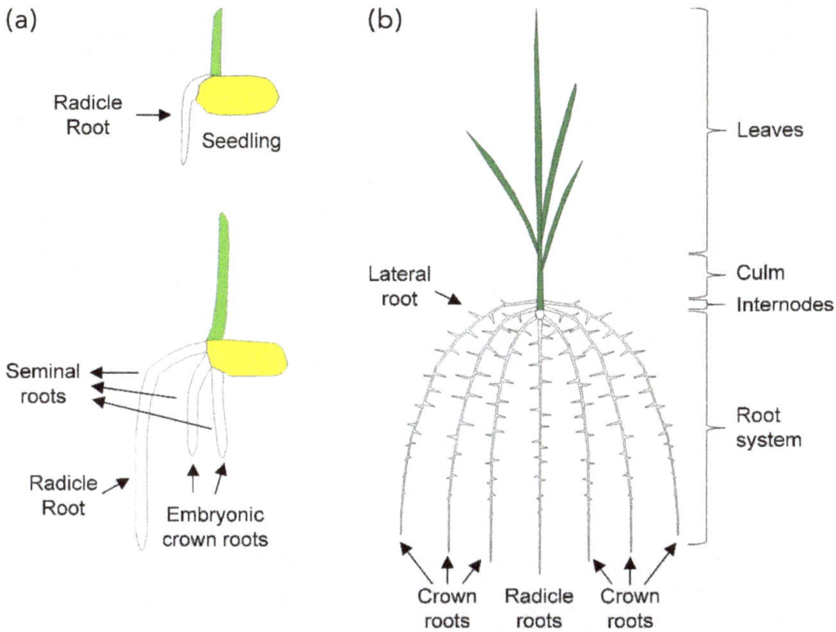

Figure 1 Simplified scheme of a rice root system at different seedlings stages. (a) Root system of young rice seedlings. Upper panels show a radical root emerged from the seed coat after germination; lower panel represents the embryonic root system of rice seedlings. (b) Fibrous root system of rice plant at the tillering stage.

the meristematic activity and subsequent cell elongation determine the length of seminal roots. In the differentiation zone, the root cells become mature and differentiate into specialized cell types with specific functions.

The radial structure of a mature rice root comprises the following tissues from outside to inside: epidermis, exodermis, sclerenchyma, cortex, endodermis, pericycle, and stele (see Fig. 2a). These tissues have been suggested to be originally generated from three tiers of stem cell initials surrounding the quiescent center (QC) (Rebouillat et al. 2009). The most root-ward tier of the initials produces the lateral root cap and central root cap (including columella), the middle tier forms epidermis and endodermis initial cells, and the most shoot-ward tier develops the stele. The cortex, sclerenchyma, and exodermis cell layers are produced after several successive rounds of periclinal cell division of endodermal cells.

In the mature root, the stele displays a vascular cylinder with a poly-arch structure as the central metaxylem vessel is surrounded by multiple alternating xylem and phloem poles (see Fig. 2a) (Mellor et al. 2019). In addition to a layer of endodermal cells, rice roots contain several cell layers of cortex cells surrounding the stele, which are involved in the formation of lysigenous

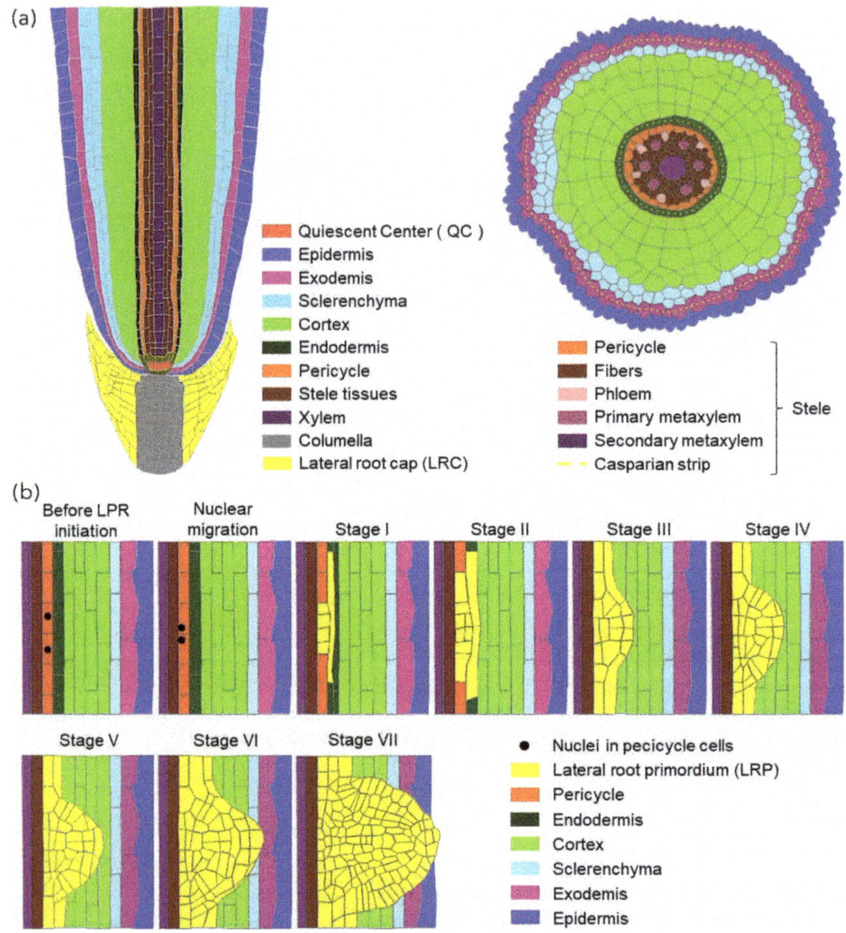

Figure 2 Schematic representation of the rice root anatomy and lateral root primordium (LRP) stages. (a) Longitudinal and transverse sections of a radicle/primary root. (b) Longitudinal sections of the different stages of lateral root primodium (LRP) formation.

aerenchyma. Lysigenous aerenchyma is a tissue containing enlarged air-filled cavities resulting from the lysis of root cortical cells that underwent programmed cell death. The cavities enable gas diffusion between shoots and roots to alleviate hypoxia stress (Jackson et al. 1985, Shiono et al. 2011, Yamauchi et al. 2013). Aerenchyma is either constitutively formed under aerobic conditions, or induced by limited oxygen supply under waterlogging conditions.

Surrounding the stele, ring-like cell wall impregnations in the endodermal cell layer, known as Casparian strips, are formed which act as a protective barrier for selective material exchange between roots and their environment (Geldner 2013). Noticeably, unlike *Arabidopsis,* rice roots also form extra Casparian strips in the exodermis (see Fig. 2a) (Cai et al. 2011, Enstone et al. 2002), which might contribute to micronutrient uptake, salt tolerance, and defense against harmful microorganisms (Reddy et al. 2017, Wang et al. 2019b). In rice, formation of Casparian strips in the endodermis is under control of OsCASP1, a homologue to *Arabidopsis* AtCASP1 (Wang et al. 2019b). *OsCASP1* is specifically expressed in the endodermis of mature root regions, and OsCASP1-mediated Casparian strip formation is required to control the uptake of mineral elements (i.e. calcium and silicon) and to contribute to root growth. However, OsCASP1 does not regulate the formation of Casparian strip at the exodermis (Wang et al. 2019b), implying the existence of a different regulatory mechanism of Casparian strip formation between endodermis and exodermis.

Together, epidermis, exodermis, and sclerenchyma form the outer cell layers of rice roots, exterior to the cortical cell layer. The exodermis becomes the functional epidermis at the moment the epidermis is sloughed off, while the sclerenchyma consists of a single layer of cells that are smaller but more suberized as compared to the exodermal cells. These outer tissue layers do not contribute to water uptake, but are important for radial oxygen homeostasis by maintaining the structure of the aerenchyma (Ranathunge et al. 2004, Henry et al. 2012, Colmer et al. 1998).

The root tip is protected by the root cap. This cap is frequently replaced following cell death and detachment of the outermost layers that become easily damaged while roots are penetrating the soil (Barlow 2002). The root cap comprises a central root cap or columella and a lateral root cap (see Fig. 2a). It arises from a set of root cap stem cells (~ 13 cells) located below the QC (Wang et al. 2014a). As the central root cap stem cells divide anticlinally to develop a columella, the outer stem cells generate the lateral root cap by several rounds of periclinal and anticlinal cell divisions. Noticeably, rice root cap cells produce a thick cell wall at the boundary between the root cap and the root meristem, which was referred to as a cap junction (Wang et al. 2014a, Clowes and Juniper 1964). The formation of the cap junction produces a closed root meristem in rice resulting in possible detachment of the rest of the root tip while confronting stresses.

2.3 Lateral root formation

The developmental process of lateral root formation in rice has been described previously (Orman-Ligeza et al. 2013). In comparison to *Arabidopsis*, in which lateral roots initiate from a restricted number of 5–10 xylem pole pericycle cells, rice lateral root initials not only develop from a set of phloem pole pericycle cells, but also from the overlying endodermal cells. It remains, however, unclear how many pericycle and endodermal cells are involved at the onset of lateral root formation in rice.

At the morphological level, lateral root formation in rice shares a similar developmental pattern with that found in *Arabidopsis*, and includes two major steps: (1) priming and initiation (stage I), and (2) primordium development (following stage I until it emerges as a lateral root) (see Fig. 2b). At the first step, a pair of pericycle cells in line with each other along the longitudinal axis of the root are activated to divide anticlinally for 1–2 times to generate a stage I primordium. Before division, the nuclei of the paired pericycle cells are observed to migrate toward the common cell wall (Kawata and Shibayama 1965), a process also observed during lateral root initiation in Arabidopsis (De Rybel et al. 2010). Subsequently, the central pericycle cells at the stage I primordium swell, and divide periclinally, anticlinally, and tangentially to form a dome-shaped primordium. Meanwhile, the overlying endodermal cells of stage II primordia also initiate several rounds of anticlinal divisions, but remain as a single layer overlying the continuously expanding lateral root primordium. During the development of the lateral root primordium, the surrounding cortex and epidermal cells adjust their cell volume/shape or loss cell wall connection to accommodate primordium growth, eventually allowing the emergence of lateral roots. This event has also been reported in *Arabidopsis* to be mediated by cell-to-cell communication between the pericycle and the endodermal cell layer (Vermeer et al. 2014). However, rice roots contain more cortical cell layers than Arabidopsis roots, so it is unclear whether an identical signaling pathway is activated during lateral root emergence in rice.

2.4 Crown/adventitious root formation

Crown/adventitious roots originate from the shoot stem base, and their development is characterized by seven stages (Itoh et al. 2005). The development of crown roots shares similar morphological changes with lateral root development. It also begins when a group of initial cells becomes activated to undergo formative cell divisions to develop a primordium, which then breaks through the epidermis of stem nodes to emerge as a crown root.

It is also important to note that crown roots develop not only from pericycle cells, but also from various cell types adjacent to the vascular tissue, including phloem and xylem parenchyma cells, or even interfascicular cambium cells close to phloem cells (Bellini et al. 2014, Inukai et al. 2005). However, it is unclear whether the formation of lateral and crown roots shares a common regulatory network.

3 Hormone regulation of rice root development

The development of an increasing toolbox for genetic studies in rice has led to the identification of many of the core components of the biosynthesis and signaling pathways for all key plant hormones. With those components at hand, the research focus progressively shifted toward understanding how the different hormones work in a coordinated fashion to orchestrate complex biological processes. This shift was further encouraged by the results of classical physiological and genetic experiments that clearly showed the importance of the interaction between multiple signals in a number of critical developmental processes. Among the most striking pieces of evidence for the existence of crosstalk between hormones were the repeated reports on well-characterized auxin mutants displaying reduced ethylene responses in roots (Stepanova and Alonso 2019, Stepanova et al. 2007). Despite these well-documented observations, the molecular mechanism behind these intriguing relations remained largely unknown. In the following paragraphs, we give an overview for each major plant hormone for which experimental data are available to support its role in rice root development.

3.1 Auxin

For decades, the plant hormone auxin has been demonstrated to be the most important regulator of root development in plants. In rice, several genes associated with auxin biosynthesis have been identified to play roles in regulating the elongation of seminal roots, and the formation of lateral and crown roots. Transgenic rice seedlings, overexpressing the auxin biosynthesis gene *OsYUCCA1*, increased auxin content and auxin signaling *in vivo*, leading to a shorter seminal root but inducing more crown roots (Yamamoto et al. 2007). On the other hand, a defective root phenotype, a reduced auxin content and auxin-dependent signaling were observed in mutants of *OsTryptophan Deficient Dwarf1* (*TTD1*) and *OsFISH BONE* (*OsFIB*), encoding proteins homologous to, respectively, Arabidopsis anthranilate synthase β-subunit and *TAA1/TAR1/TAR2* genes; both of these genes are involved in auxin biosynthesis (Yoshikawa et al. 2014, Sazuka et al. 2009).

Auxin is perceived by its receptor F-box protein TIR1 and AFBs which results in the activation of the canonical AUX/IAA-ARF-signaling cascade. Overexpression of microRNA *miR393a/b*, which targets OsTIR1 and OsAFB2, the rice homologues to the Arabidopsis auxin receptor TIR1 and AFB2, repressed auxin-signaling levels and root sensitivity to exogenous auxin (Bian et al. 2012). Meanwhile, its overexpression lines produced longer seminal roots, and more lateral roots but less crown roots as compared to wild type, implying an opposite effect of auxin receptors on seminal root and crown root development. Consistently, blocking auxin signaling by overexpression of *OsIAA1*, a member of *Aux/IAA* family in rice that interacts with OsTIR1 and OsAFB2 also increased seminal root elongation and lateral root formation, but reduced crown root formation and root sensitivity to auxin (Song et al. 2009). However, the gain-of-function mutants of *OsIAA11* and *OsIAA13*, in which

OsIAA proteins are stabilized, were found to produce very few lateral root primordia and lateral roots, but still have normal crown root growth, suggesting the two IAA proteins are essential for lateral root initiation (Kitomi et al. 2012, Zhu et al. 2012). A more severe defective root phenotype was observed in the *osiaa23* gain-of-function mutant, which produces neither lateral roots nor crown roots and is agravitropic (Jun et al. 2011). Remarkably, *OsIAA23* is specifically expressed in the quiescent center (QC) cells, moreover QC and root cap are both impaired in the *osiaa23* mutant, suggesting an essential role of OsIAA23 for QC identity maintenance in roots. In addition, mutants of the auxin response factor genes, *OsARFs,* also exhibited dramatic root phenotypes. Knocking out *OsARF12* alone, or together with *OsARF25*, both reduced auxin signaling and transport in the root, in turn inhibiting seminal root growth (Qi et al. 2012). Meanwhile, the loss-of-function mutant of *OsARF16* did not exhibit altered root growth at the young seedling stage under normal conditions, but root growth was hyposensitive to auxin treatment (Shen et al. 2013). These results suggest a central role of auxin signaling components on regulating rice root development. However, the different corresponding mutants produced diverse root phenotypes, which might be ascribed to differential regulation pathways downstream of OsAUX/IAAs and OsARFs.

Auxin transport is also involved in rice root development as application of two potent auxin efflux inhibitors, 1-*N*-naphthylphthalamic acid (NPA), and 2-[4-(diethylamino)-2-hydroxybenzoyl] benzoic acid (BUM) both resulted in a substantial reduction of seminal root elongation and lateral root formation (Chen et al. 2018b). In rice, twelve auxin efflux carrier genes *OsPINs*, homologous to Arabidopsis *PINs*, have been identified and found to be expressed in different root tissues (Wang et al. 2009, Miyashita et al. 2010). Among them, knock down of *OsPIN1a* delayed the emergence and reduced the number of crown roots (Xu et al. 2005). *OsPIN2* has been shown to share a similar expression pattern with its Arabidopsis homologue *AtPIN2* in roots and mediates shoot-ward auxin transport. A missense mutant of *OsPIN2/OsLARGE ROOT ANGLE1* resulted in the loss of root response to gravity, but did not affect the primary root and crown root formation (Wang et al. 2018). Transport activity of OsPINs is regulated by the ADP-ribosylation factor (OsGNOM1) and OsPINONID (OsPID), whereas *Osgnom1* mutant and *OsPID* overexpression both reduced polar auxin transport in rice roots, thus inhibiting root sensitivity to gravity and the formation of lateral roots and crown roots (Morita and Kyozuka 2007, Liu et al. 2009). Nevertheless, it is unclear if the apical or basal polarity of OsPIN protein at the plasma membrane is regulated by OsGNOM1 and OsPID in rice. Additionally, auxin influx carriers also play a role in regulating rice root development. *OsAUX1*, the rice homologue to the Arabidopsis auxin influx carrier *AUX1*, is strongly expressed in the root cap and epidermis, and facilitates auxin uptake in the root to regulate seminal root elongation, lateral root initiation, and root gravitropism (Yu et al. 2015a, Giri et al. 2018, Zhao et al. 2015a).

3.2 Cytokinin

Cytokinins (CK), another class of plant hormones, have also been suggested to be essential for root development, but negatively regulate root development by counteracting auxin signaling (Su et al. 2011, Moubayidin et al. 2009). The role of CK on rice root development and the molecular interaction between CK and auxin have been assessed.

The CK-activated signaling cascade typically comprises three functional modules: a sensory histidine kinase (HK), a histidine phosphotransfer protein (HP), and a response regulator (RR). In rice, a putative histidine kinase HK1 was found to regulate the circumnutation radius of the root tip by activating the canonical CK signaling pathway (Lehner et al. 2018), while another CK receptor gene *OsCYTOKININ TOLERANT 1* (*OsCKT1*), which encodes *histidine kinase 6* (*OsHK6*), regulates root sensitivity to exogenous CKs, but does not affect root growth of young seedlings (Ding et al. 2017). Two rice authentic HPs, *OsAHP1* and *OsAHP2*, are also involved in regulating CK signaling and root development. Knocking down both of them increased lateral root formation and also suppressed zeatin-inhibited root growth (Sun et al. 2014b). In addition, type-A RRs seem to trigger diverse root responses in rice. Overexpression of *OsRR6*, a type-A RR, was also found to repress root development (Hirose et al. 2007). In contrast, overexpressing *OsRR3* and *OsRR5* both increased seminal root length and lateral root number (Cheng et al. 2010), while overexpressing *OsRR1* and *OsRR2* promoted crown root formation (Kitomi et al. 2011, Zhao et al. 2015b). Expression of *OsRR1* is also induced by auxin through the activation of an OsAP2/ERF transcription factor, which is downstream of the OsAUX/IAA-OsARF signaling module (Kitomi et al. 2011). *OsRR2* has also been identified as a target of *OsERF3*, an auxin-induced regulator of crown root formation in rice (Zhao et al. 2015b). These results imply a possible interaction between CK and auxin during rice root development.

Moreover, rice root development also involves the metabolism pathway of CK. In plants, CK biosynthesis is mainly mediated by isopentenyltransferase (IPT) and cytokinin nucleoside 5′-monophosphate phosphoribohydrolase, while its degradation is largely catalyzed by CYTOKININ OXIDASE/DEHYDROGENASE (CKX) (Hwang et al. 2012). Constitutively overexpressing five OsIPT genes (*OsIPT1-OsIPT4*, *OsIPT7*) has been found to repress root development (Sakamoto et al. 2006), whereas one of the rice OsCKXs, OsCKX4, positively regulates rice crown root development (Gao et al. 2014). *OsCKX4* is highly expressed in seminal and lateral roots, and its expression is directly trans-activated by OsARF25 and type-A OsRR2 and OsRR5 (Gao et al. 2014). Enhanced expression of *OsCKX4* in roots promotes crown root formation, whereas knock down of its expression leads to decreased crown root formation. However, it remains elusive how these CK metabolism-associated genes regulate rice root development.

3.3 Ethylene

In plants, the gaseous hormone ethylene has also been shown to regulate root development, mainly by interfering with the auxin-signaling pathway and polar auxin transport (Swarup et al. 2007, Ruzicka et al. 2007, Lewis et al. 2011, Stepanova et al. 2007). In rice, exogenously applying ethylene or its precursor, 1-aminocyclopropane-1-carboxylic acid (ACC), and overexpression of *OsACC SYNTHAE 1* (*OsACS1*), both restricted seminal root elongation (Ma et al. 2013, Ma et al. 2014, Qin et al. 2019), suggesting a negative role in this process. The underlying mechanism of ethylene-mediated root elongation has been investigated via genetic approaches.

From a mutagenesis screen, seven rice mutants with reduced seminal root sensitivity to ethylene were isolated and named as *mao hu zi* (*mhz*, Chinese word for whiskers) (Ma et al. 2013). Among them, root growth of the *mhz7* mutant was shown to be hyposensitive to ethylene. *MHZ7* encodes a membrane protein homologous to *Arabidopsis* ETHYLENE INSENSITIVE 2 (EIN2), a central component of the ethylene signaling pathway (Ma et al. 2013). *mhz6* corresponds to a mutation in *OsETHYLENE INSENSITIVE3-LIKE 1* (*OsEIL1*), a homologue of *ETHYLENE INSENSITIVE3* (*EIN3*), a master transcriptional regulator of ethylene signaling in Arabidopsis (Yang et al. 2015). In contrast, the *mhz5* mutant phenotype was found to result from loss of function of a carotenoid isomerase that is required for carotenoid biosynthesis (Yin et al. 2015). The *mhz4* mutant also encodes a membrane protein orthologous to *Arabidopsis ABA4*, which acts on the conversion of zeaxanthin to neoxanthin in the ABA biosynthesis pathway (Ma et al. 2014). Interestingly, both *mhz4* and *mhz5* mutants had higher ethylene accumulation but reduced ABA content and exogenously supplying ABA could rescue the seminal root elongation of these mutants. ABA content and the expression of ABA biosynthesis pathway genes were also reduced in the gain-of-function mutant of *OsETHYLENE RESPONSIVE FACOTR 2* (*OsERF2*), in which seminal root elongation and the ethylene signal were inhibited (Xiao et al. 2016). These findings suggest the requirement of ABA in ethylene-induced inhibition of rice root growth. Another ethylene-insensitive mutant *mhz2* is defective in gravity response and displays a soil-surface rooting phenotype (Chen et al. 2018a, Hanzawa et al. 2013). Genetic evidence showed that *mhz2* encodes the E3 ubiquitin ligase soil-surface rooting 1 (*OsSOR1*), and regulates ethylene-mediated rice root growth by modulating the stability of AUX/IAA proteins *in vivo* (Chen et al. 2018a).

Ethylene-mediated root development also involves other signaling components. A DOF transcription factor, OsDOF15, was found to directly bind to the promoter of *OsACS1* to repress the biosynthesis of ethylene, thereby regulating root meristem activity and seminal root elongation (Qin et al. 2019). Ethylene might also be involved in CK-mediated root development. Treatments of lovastatin (an inhibitor of CK biosynthesis) and kinetin (KT; a synthetic CK) resulted in high ethylene accumulation and reduction of root elongation, which could be reversed by applying aminoethoxyvinylglycine (AVG), an inhibitor of ethylene biosynthesis (Zou et al. 2018). An OsERF transcription factor, OsCROWN ROOTLESS5 (OsCRL5), was found to act downstream of OsARF and activate the expression of two type-ARR genes, *OsRR1* and *OsRR2*, in turn regulating crown root initiation in rice (Kitomi et al. 2011). Altogether, these studies strongly suggest crosstalk between ethylene and other hormones during its regulation of seminal root elongation.

In addition, ethylene has also been implicated in regulating the formation of lysigenous aerenchyma, which results from the dying and lysis of root cortical cells and is essential for rice roots to survive under oxygen-deficient conditions in waterlogged soil (Yamauchi et al. 2016, Yukiyoshi and Karahara 2014, Yamauchi et al. 2015). During aerenchyma formation, ethylene biosynthesis is induced. The formation of lysigenous aerenchyma can be promoted by the ethylene precursor ACC, but repressed in the presence of an ethylene perception inhibitor 1-methylcyclopropene (1-MCP).

3.4 Abscisic acid

As discussed previously, ABA is also involved in rice seminal root development by interacting with ethylene. It has also been reported that ABA treatment caused global changes of rice root architecture, including root tip swelling, increased root hairs, and abnormal lateral root branching. These ABA-triggered root responses are largely dependent on an increase of cytoplasmic Ca^{2+} upon ABA treatment (Chen et al. 2006). The ABA signaling cascade comprises its receptor PYR1/PYL/RCAR, the subclass A type 2C protein phosphatases (PP2Cs), and SNF1-RELATED PROTEIN KINASE 2 (SnRK2s), which phosphorylate and activate bZIP transcription factors to activate ABA responses (Ma et al. 2009, Park et al. 2009, Umezawa et al. 2009, Cutler et al. 2010). In rice, some of the core ABA signaling pathway components are also suggested to mediate root development. For instance, overexpression of *OsABI-LIKE2* (*OsABIL2*), a member of the rice clade A PP2C family, was found to repress the growth of seminal roots, crown roots, and root hairs, but to induce lateral root elongation (Li et al. 2015a). In contrast, *OsStress/ABA-activated protein kinase 10* (*OsSAPK10*) positively regulates root hair elongation as its overexpression results in increased root hair elongation (Wang et al. 2017). The effects of

OsABIL2 and OsSAPK10 on root hair development both require polar auxin transport and auxin biosynthesis. Another ABA signaling component, OsABI5-Like1, was identified to be required for maintaining rice root sensitivity to ABA, and it mediates the root sensitivity to auxin through regulating the expression of AUX/IAA genes (Yang et al. 2011). As with ethylene, ABA seems to regulate rice root development through interaction with auxin signaling.

3.5 Strigolactones

Strigolactones (SLs) are recently identified as novel plant hormones that regulate shoot branching in plants, and recent studies also reveal a potential role of SLs on rice root development. In rice *dwarf* mutants *d10* and *d14*, which are defective in SL biosynthesis and signaling respectively, reduced meristem activity and shorter crown roots were observed. Exogenous application of GR24, a synthetic SL analog, restored crown root elongation in the *d10* mutant but not in the *d14* mutant (Arite et al. 2012). However, both seminal root elongation and lateral root formation were not affected in both mutants. These results suggest that SLs positively regulate crown root elongation but have no effect on seminal root elongation. Further studies revealed that SLs regulate crown root formation by restraining polar auxin transport as a stronger auxin responsive DR5 signal and auxin content were detected in the crown root tip of *d10* and *d3* mutants (Sun et al. 2015). SL-regulated auxin transport is also suggested to be involved in nitrogen- and phosphate-deficiency-dependent seminal root and lateral root development (Sun et al. 2014a).

3.6 Other plant hormones

Other plant hormones, such as brassinosteroids (BRs), gibberellic acid (GA), salicylic acid (SA), and jasmonic acid (JA) have also been implicated in rice root development. BR treatment resulted in a reduction of rice root elongation (Tong et al. 2014), and the rice dwarf mutant *Osbrassinosteroid-dependent 1* (*Osbrd1*), which harbors a 0.2-kb deletion in the coding sequence of the OsBR-6-oxidase (OsBR6ox) encoding gene, was defective in BR biosynthesis and exhibited less and shorter crown roots, suggesting the involvement of BR in promoting root development (Mori et al. 2002).

Interestingly, BR-inhibited root growth could be associated with altered GA metabolism and signaling. The GA-deficient mutants (*d18* and *d35*), decreased-GA-signaling mutants (*gid1* and *gid2*), and an enhanced-GA-signaling mutant (*slr1*) all exhibited root growth that was less sensitive to BL than the wild-type (WT) (Tong et al. 2014). Meanwhile, a rice GA-deficiency mutant *Osshoebox* (*Osshb*) had a small meristem and a short seminal root, which could be rescued by exogenous application of GA. *OsSHB* encodes an AP2/ERF transcription factor that could directly trans-activate the expression of the GA biosynthesis gene *KS1*, which is required for maintaining endogenous GA content and the elongation and proliferation of root meristematic cells (Li et al. 2015b). These findings thus imply a potential role of GA on rice root development.

SA and JA are also involved in rice root elongation. Exogenously applying SA was found to reduce root meristem activity and root elongation, which is suggested to act through the regulation of redox homeostasis in the root tip (Xu et al. 2017). JA was shown to affect rice lateral root formation by activating heme oxygenase1 (HO1)-mediated carbon monoxide (CO) production (Chen et al. 2012, Hsu et al. 2013). Additionally, a natural metabolite, β-Cyclocitral, which is derived from β-carotene, has also been shown to have a hormone-linked function on rice root stem cell activity (Dickinson et al. 2019), though the molecular mechanism remains elusive.

4 Control of rice root development by other pathways

Beside plant hormones, rice root development has also been shown to be regulated by genes involved in a range of signaling, metabolic, and physiological pathways. However, the effect of these genes on root development and the activation of their specific downstream components act again through the interaction with plant hormones.

4.1 Pathways involved in seminal root elongation

The elongation of roots is coordinated by cell division in the root meristem and subsequent elongation of these cells when they reach the elongation zone. The length of roots is an important trait and determines the capacity of roots to take up water and nutrients. Because of its significance for plant growth and stress tolerance, a number of studies aimed at identifying the molecular mechanism underlying rice root elongation have been conducted, mainly through forward genetic approaches.

An early study succeeded in isolating several recessive mutants with short seminal roots (Inukai et al. 2001). Interestingly, the short root phenotype of these mutants resulted either from restricted cell elongation in the mature root region or from reduced apical meristem size, suggesting seminal root elongation in rice might involve different genes acting on different developmental processes.

A large number of mutants with short seminal roots have been isolated in rice. Several of these mutants are affected by genes involved in sucrose metabolism and sugar-based protein modification, including *OsGNA1* (Jiang et al. 2005), *OsCYT-INV1* (Jia et al. 2008), *OsDGL1* (Qin et al. 2013), and *OsMOGS* (Wang et al. 2014b). Besides the short root phenotype, *Osgna1* and *Oscyt-inv1* mutants also exhibited abnormal cell shape, which might be a result of disordered microtubule arrays. Interestingly, *OsCYT-INV1* encodes an alkaline/neutral invertase that converts sucrose into glucose and fructose, and the defective root phenotype of the *Oscyt-inv1* mutant could be rescued by exogenous sucrose supply, indicating the essential role of sugar metabolism for maintaining seminal root growth.

Amino acid metabolism has also been linked to seminal root elongation as the loss-of-function mutant of *OsASL1*, which encodes an argininosuccinate lyase acting on the final step of arginine biosynthesis, produced a short seminal root (Xia et al. 2014). The retarded root elongation of *Osasl1* mutants could be restored by supplying arginine, suggesting that a high level of endogenous arginine is required for normal root growth in rice. Noticeably, *asl1* mutants exhibited unaltered radical tissue organization of the root meristem but merely reduced the activity of the root apical meristem. These results indicated that arginine metabolism is required for sustaining meristem activity. Meanwhile, disruption of a broad-spectrum amino acid transporter lysine/histidine transporter (LHT1) also led to a shorter seminal root as compared to WT (Wang et al. 2019a), highlighting a critical role of amino acid in regulating seminal root elongation in rice.

Genes associated with cell wall synthesis, β-oxidation pathway, and other physiological pathways have also been implicated in the regulation of seminal root elongation. For instance, *OsGLU3*, which encodes a putative membrane-bound endo-1,4-β-glucanase and affects cell wall cellulose content, and *OsEXPENSION8* (*OsEXPA8*), which regulates cell wall stiffness and cell swelling, are identified to mediate seminal root elongation by moderating cell wall biosynthesis (Zhang et al. 2012, Wang et al. 2014c). *OsABNORMAL INFLORESCENCE MERISTEM 1* (*OsAIM1*), which encodes a 3-hydroxyacyl-CoA dehydrogenase, a key β-oxidation pathway enzyme, is required for seminal root elongation as its loss-of-function mutant produced a short seminal root (Xu et al. 2017). In *Arabidopsis*, the beta-oxidation pathway is involved in auxin biosynthesis and contributes to auxin-dependent root development (Strader et al. 2011, Xuan et al. 2015), whereas it affects rice seminal root elongation through the regulation of SA-dependent ROS homeostasis in the root meristem. Meanwhile, a glutamic acid receptor-like gene, *OsGLR3.1*, regulates seminal root elongation by controlling programmed cell death in the root cap and apical root meristem (Li et al. 2006). GatB, which encodes a subunit of tRNA-dependent amidotransferase (AdT) involved in Gln-$tRNA^{Gln}$ synthesis, also regulates root meristem division and cell elongation by affecting mitochondrial ATP production as well as mitochondrial structure (Qin et al. 2016). Furthermore, rice DNA TOPOISOMERASE 1 (OsTOP1), a regulator of DNA topology during RNA transcription and DNA replication, is also essential for maintaining seminal root elongation and gravitropism by preventing the over-accumulation of inherent R-loops at auxin-associated gene loci (Shafiq et al. 2017). This study highlights the existence of epigenetic regulatory control on rice seminal root elongation.

4.2 Pathways regulating lateral root formation

In rice, as in all angiosperms, lateral root formation is dominated by the plant hormone auxin. Stabilized forms of AUX/IAA proteins and disruption of auxin transport both completely abolish lateral root initiation (see Section 3).

A set of lateral root defective rice mutants, *lateral rootless 1* (*lrt1*), *lrt2*, *arm1*, *arm2*, and *lateral root formation 1* (*alf1*) has been isolated from mutagenesis screens, without identification of the causal genes except for *lrt2* (Chhun et al. 2003a,b, Debi et al. 2003, Wang et al. 2006, Faiyue et al. 2010). These mutants also exhibit less sensitivity to auxin treatments, implying their role in lateral root formation might also be associated with auxin signaling.

The function of a series of lateral root-regulating genes, including *OsNARROW LEAF 2* (*OsNAL2*), *OsNAL3*, *OsIPK2*, and *OsCYCLOPHILIN 2* (*OsCYP2*)/*OsLRT2* has been suggested to be mediated by multiple auxin pathways. For instance, *OsNAL2* and *OsNAL3* are paralogs that encode an identical OsWUSCHEL-RELATED HOMEOBOX 3A (OsWOX3A) transcriptional activator, and co-regulate the formation and elongation of lateral roots by regulating the expression of auxin transporter genes *OsPIN1* and *OsPIN2* (Cho et al. 2013). OsIPK2 is an inositol polyphosphate kinase and negatively regulates lateral root formation by stabilizing OsAUX/IAA proteins and preventing their proteasomal degradation, which in turn blocks auxin signaling. *OsCYP2* encodes a cyclophilin-type peptidylprolyl cis/trans isomerase and its knock-out version is responsible for the lateral root defective phenotype in the *lrt2* mutant (Zheng et al. 2013). OsCYP2 catalyzes the peptidyl-prolyl cis/trans isomerization of OsIAA11, and thus facilitates the binding of OsIAA11 to the auxin receptor OsTIR1 resulting in its proteasomal degradation and activation of auxin signaling required for lateral root initiation (Jing et al. 2015). Furthermore, a C2H2 type zinc finger protein OsZFP was also found to directly interact with OsCYP2 to regulate auxin signaling and lateral root formation (Cui et al. 2017).

However, in rice, several lateral root-regulating genes are not associated with auxin but rather with other plant hormones (e.g. CK, ABA, SL, and JA), or even act independently of plant hormones. For example, from a forward genetic screen, a lateral root-defective mutant was isolated and identified to encode *ORIGIN RECOGNITION COMPLEX SUBUNIT 3* (*OsORC3*), a regulator of DNA replication. *OsORC3* promotes the formative cell divisions of lateral root primordia and their emergence as lateral roots (Chen et al. 2013). In another study, *OsSHORT LATERAL ROOT LENGTH 1* (*OsSLL1*) was identified to specifically regulate lateral root formation (Shelley et al. 2013). *OsSLL1* encodes a member of the stearoyl-acyl carrier protein fatty acid desaturase family, indicating a possible link between fatty acid metabolism and lateral root elongation.

4.3 Pathways regulating crown root formation

As with lateral roots, the formation of crown roots is also under the control of plant hormones; their regulatory roles were discussed in Section 3. Crown root-regulating genes also act downstream or interact with auxin pathways. In rice, *Osadventitious rootless1/Oscrown rootless1 (Osarl1/Oscrl1)* was the first isolated rice mutant lacking crown roots and is defective in crown root initiation (Liu et al. 2005, Inukai et al. 2005). Its mutant phenotype results from the loss-of-function of a gene encoding an OsASYMMETRICLEAVES2 (AS2)/LATERAL ORGAN BOUNDARIES (LOB) domain transcription factor, which is regulated by auxin and directly targeted by OsARF1 (Inukai et al. 2005). Another two identified crown rootless mutants, *Oscrl4* and *Osclr6*, are also suggested to regulate crown root formation by acting on auxin signaling pathways. *Oscrl4* encodes *OsGNOM*, a regulator of the auxin efflux carrier OsPIN1 (Liu et al. 2009), while *Oscrl6* was mapped to a CHD (chromodomain helicase DNA-binding) family gene, which moderates the expression of OsAUX/IAA genes (Wang et al. 2016). OsPINs-mediated auxin transport was also shown to be negatively regulated by the activity of a Ca^{2+}-independent Ser/Thr kinase OsRPK1, thereby negatively regulating crown root formation and elongation (Zou et al. 2014). In contrast, *OsNAL1*, which encodes a putative trypsin-like serine/cysteine protease, promotes crown root formation through the activation of auxin transport (Cho et al. 2014). Furthermore, OsCAND1, a homologue to Arabidopsis CULLIN-ASSOCIATED AND NEDDYLATION-DISSOCIATED 1 (CAND1), also mediates crown root emergence through the regulation of auxin-dependent cell division activity in the primordium (Wang et al. 2011).

Ethylene and CK are also essential for crown root development in rice as suggested by studies on *Oscrown rootless5* (*Oscrl5*) and *root enhancer1* (*ren1-D*) mutants (Gao et al. 2014, Kitomi et al. 2011). *Oscrl5* failed to produce crown roots and carries a missense mutation on an AP2/ERF transcription factor, whereas *ren1-D* exhibited high frequency of crown roots by activating the expression of *OsCKX4*. Interestingly, a WUSCHEL-related homeobox gene *OsWOX11* plays an important role in ethylene- and CK-regulated crown root development (Zhao et al. 2009, 2015b; Zhou et al. 2017). The loss-of-function or down-regulation of OsWOX11 caused a significant reduction in the number and length of crown roots, while its overexpression enhanced their growth. Further studies revealed that OsWOX11 interacts with OsERF3 to directly repress the expression of *OsRR2*, a type-A cytokinin response regulator, in the crown root primordium by binding to its promoter (Zhao et al. 2015b, Zhao et al. 2009).

More recently, microRNAs are also suggested to mediate crown root formation. A *lower crown root number1* (*lcrn1*) mutant, which produces few crown roots at all development stages, was identified to carry a point mutation in a transcription factor gene *OsSQUAMOSA PROMOTER BINDING PROTEIN-LIKE3* (*OsSPL3*) (Shao et al. 2019). Further studies showed this mutation in *OsSPL3* disrupted the rice specific microRNA156 (OsmiR156)-activated cleavage of the *OsSPL3* transcripts. In rice, *OsSPL3* negatively regulates crown root formation by trans-activating the expression of *OsMADS50*, which suppresses auxin signaling and transport. Furthermore, the expression level of *OsmiR156* is down-regulated in a crown root defective mutant *Oscrd1*, which encodes a gene required for the proper accumulation of miRNAs *in vivo* (Zhu et al. 2019). However, the regulation of *OsCRD1* on pre-miRNAs is not specific to *OsmiR156*, implying the employment of other microRNAs in regulating crown root formation.

5 Adaptation of RSA to environmental conditions

Under natural conditions, plants have to optimize their root system architecture to efficiently perceive and uptake heterogeneously distributed nutrients and water, and also to cope with or escape from environmental stresses. In *Arabidopsis*, limited supply of nutrients, such as nitrogen and phosphate, can activate root foraging responses, visualized by the changes in root length, growth direction and branching (Gruber et al. 2013, Giehl and von Wiren 2014). Water availability can also affect root growth direction and lateral root positioning towards water sources (Orosa-Puente et al. 2018, Bao et al. 2014). In contrast, other environmental factors, such as salt and metals, cause osmotic and ionic stress, and constrain root growth at very early developmental stages (Tu et al. 2014). Alterations in RSA by nutrient and water availabilities have also been observed in rice, and appear to be regulated by complex signaling pathways. In the following, we discuss the plasticity of rice RSA responses to water and nutrient stress and the genes involved in these processes.

5.1 Nitrogen nutrition

Nitrogen (N) is an essential and limiting macronutrient for plants and N availability has a large effect on rice root development and growth. Low N conditions result in elongation of rice seminal roots, a behavior known as foraging response (Xuan et al. 2017, Chen et al. 2018b). Rice mostly takes up ammonium-derived nitrogen in a paddy field but root responses differ depending on whether the nitrogen source is ammonium- or nitrate-derived. While ammonium treatment inhibited seminal root elongation in a dose-dependent manner, nitrate only slightly affected seminal root elongation but promoted lateral root formation (Hirano et al. 2008, Sun et al. 2014a).

In Arabidopsis, the ammonium-triggered root response is mainly mediated by ammonium transporters (AMTs) and its regulator CIPK23 (Straub et al. 2017). In rice, an ammonium transporter mutant *Osamt1.1*, decreased ammonium uptake by the roots, but could not alleviate ammonium-inhibited root growth (Li et al. 2016). In contrast, a serine/threonine/tyrosine protein kinase gene OsACTPK1, which negatively regulates the activity of OsAMT1;2 rather than OsAMT1;1, reduced ammonium uptake and seminal root elongation (Beier et al. 2018), suggesting the involvement of other ammonium transporters in regulating root development (see Fig. 3a). Plant hormones have also been implicated in the mediation of the ammonium-inhibited seminal root elongation (see Fig. 3a). For instance, under high ammonium supply, auxin biosynthesis and the expression of auxin metabolism-associated genes were largely repressed at the root tip of ammonium-sensitive cultivars, while cultivars with higher auxin biosynthesis showed tolerance to high ammonium levels at

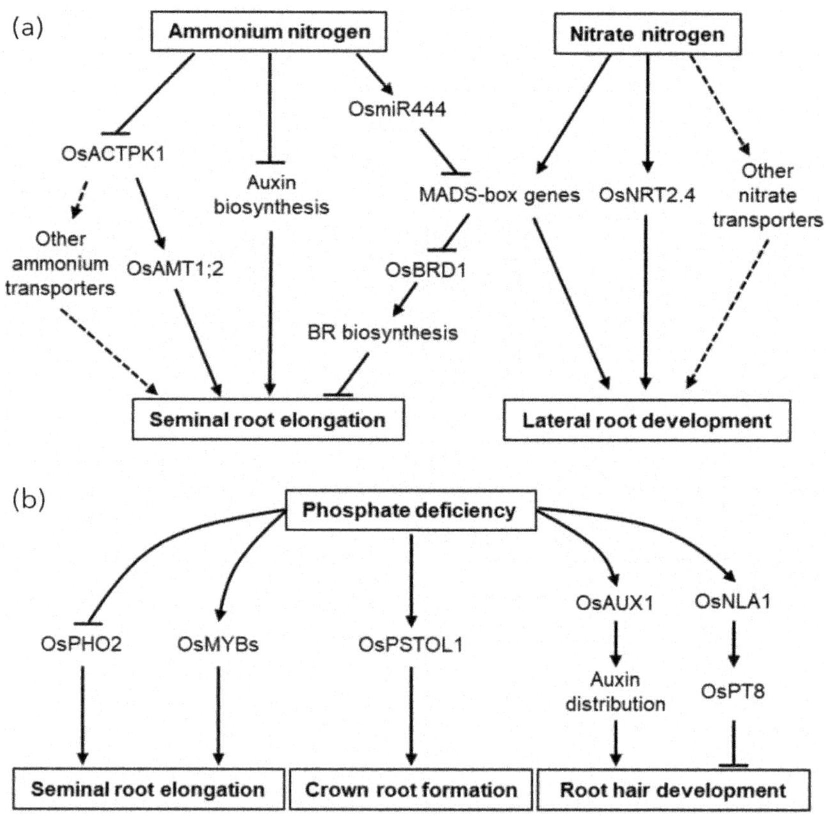

Figure 3 Schematic representation of the molecular regulation of nitrogen and phosphate effects on rice root development. (a) The signaling molecules involved in ammonium- and nitrate-regulated root development. (b) Genes involved in phosphate deficiency-mediated root development. Arrows represent positive regulation; lines ending in a flat head indicate negative regulation; dotted lines implicate the potential regulation pathways.

the level of seminal root elongation (Di et al. 2018). High ammonium supply also triggers the biosynthesis of BR through the activation of OsmiR444-OsMADS-box-OsBRD1 (OsBR-deficient dwarf1, encoding a key BR biosynthetic gene) signaling cascade. Interruption of BR by a BR biosynthesis inhibitor BRZ alleviated the ammonium-inhibited root elongation, indicating the involvement of BR in root response to high ammonium supply (Jiao et al. 2019).

Interestingly, rice root tips were induced to coil when high ammonium was supplied, indicating ammonium might also affect root gravitropism. This typical root response has been attributed to increased ammonium uptake and N-associated metabolism under the control of a transcription factor OsINDETERMINATE DOMAIN 10 (OsIDD10), which activates the transcription of the ammonium transporter gene *OsAMT1;2* and the glutamate dehydrogenase encoding gene by binding to the cis-element motif in their promoter regions (Xuan et al. 2013).

The mechanism underlying nitrate-regulated lateral root formation has also been investigated in rice. Seven putative quantitative trait loci (QTL) have been suggested to contribute 9–15% of the nitrate-dependent lateral root phenotype respectively (Huang et al. 2004), though the responsive genes associated with these QTL have not been identified yet. Interestingly, two rice MADS-box transcription factors, OsMADS25 and OsMADS57, together with their specific repressor OsmiR444a, have been identified to act as a signaling module, to mediate nitrate-dependent lateral root development through regulating nitrate uptake and transport (Yu et al. 2015b, Huang et al. 2019, Yan et al. 2014) (see Fig. 3a). In rice, nitrate uptake and transport are facilitated by a group of high and low-affinity nitrate transporters (Xuan et al. 2017). Among the identified nitrate transporters, a dual affinity nitrate transporter OsNRT2.4 positively regulated lateral root formation and elongation under both low and high nitrate conditions (Wei et al. 2018). The role of other nitrate transporters on root development remains unclear (see Fig. 3a). In addition, nitric oxide and SLs signals also mediate nitrate-dependent root development by moderating auxin transport and signaling (Sun et al. 2014a, 2018).

5.2 Phosphate nutrition

Phosphorus is immobile in soil and plant roots have to alter their RSA to forage for available P sources. In rice, low P conditions slightly affect seminal root elongation and lateral root formation, but stimulate root hair elongation and crown root initiation and elongation, through activating various signaling pathways (see Fig. 3b) (Hu et al. 2011, Dai et al. 2012, Zhou et al. 2008). *OsPHOSPHORUS-STARVATION TOLERANCE 1* (+) was identified from QTL analyses to confer rice tolerance to P deficiency by promoting seminal root elongation and crown root formation, suggesting an optimized RSA is critical for rice to survive under P deficiency conditions (Gamuyao et al. 2012). A detailed analysis of P-associated root foraging responses has been performed in diverse rice varieties (Vejchasarn et al. 2016). In this study, low P conditions were shown to generally trigger the elongation of seminal roots and root hairs, but reduced the number of lateral roots and crown roots, suggesting a high-sensitivity of RSA to P deficiency.

Among these root parameters, the activation of root hair elongation by P-deficiency has been observed in many plant species (Brown et al. 2013). In rice, the P-deficiency-induced root hair foraging response is mediated by the auxin efflux carrier OsAUX1, which facilitates shoot-ward auxin movement from the root apex via the lateral root cap to elongation and differentiation zones (Giri et al. 2018). P-deficiency-induced root hair elongation was repressed in the overexpression line of a phosphate transporter gene *OsPht1;8* (*OsPT8*), which reduces auxin sensitivity in response to low P (Jia et al. 2017). OsNITROGEN LIMITATION ADAPTATION 1 (OsNLA1), a rice homologue to Arabidopsis NLA1, also regulates P-deficiency-mediated root hair responses by moderating *OsPTs* transcripts and protein degradation in roots (Yang et al. 2017).

In Arabidopsis, primary root elongation exhibited strong sensitivity to low P-supply, but in rice seminal root elongation was only slightly affected under P-deficiency. A significant increase of seminal root elongation under low P-conditions was observed in the overexpression line of *OsPHR2* and a R2R3-type transcription factor gene *OsMYB2P-1*, as well as in the mutants of *OsPHO2* and *OsMYB1*, two P-deficiency responsive genes (Gu et al. 2017, Zhou et al. 2008, Hu et al. 2011, Dai et al. 2012). These lines accumulated high P in roots under P-deficiency, implying a possible link between P-homeostasis and seminal root elongation. P-deficiency-regulated seminal root elongation might also involve the plant hormone SLs and the signaling molecule NO, which act in the same signaling pathway to regulate root meristem activity under P-deficiency (Sun et al. 2016).

5.3 Water availability

Water availability is a serious limiting factor constraining rice development and yield. Under water-limiting conditions, plant roots have to adjust their RSAs to avoid drought stress and to promote water use efficiency. Therefore, several root parameters, including depth, width, and branching of the root system, have been proposed to be critical for plant tolerance to drought stress (Henry et al. 2012, Manschadi et al. 2006).

A number of QTL analyses have been performed on different rice populations to probe the genes that are related to these root parameters under drought stress (Champoux et al. 1995, Price et al. 2002, Kamoshita et al. 2002, Yue et al. 2006, Uga et al. 2008). A QTL *qFSR4* controlling root volume (depth and width) under drought stress was mapped to a 38-kb interval on chromosome 4, and *OsNAL1*, a crown root regulator, has been predicted to be responsible for this QTL (Ding et al. 2011). Six major rice QTLs controlling rice deep rooting, *DRO1* (*DEEPER ROOTING 1*), *DRO2*, *DRO3*, *DRO4*, *DRO5*, and *qSOR1* (a QTL for SOIL SURFACE ROOTING 1) were also identified from different mapping populations with a basket-based root phenotyping method (Uga et al. 2011, Uga et al. 2012, 2013b, 2015, Kitomi et al. 2015). Among them, *DRO1* has been mapped to a single 1-bp deletion in the coding sequence of an unknown gene, and appears to function downstream of auxin signaling thereby affecting the root growth angle (Uga et al. 2013a). Introducing a functional *DRO1* allele could produce a significant larger root system with deep roots, thus promoting tolerance to drought stress. However, DRO1 cannot fully explain the difference in RSA between the parental lines IR64 (shallow rooting) and Kinandang Patong (deep rooting), implying the involvement of other QTLs in regulating root depth. A major QTL *qRDR-2*, a region possessing seven SNPs, has also been suggested to be associated with the deep rooting trait (Lou et al. 2015).

Several other genes have been implicated in the regulation of root development under drought stress. A putative basic helix-loop-helix (bHLH) transcription factor gene, *OsbHLH120*, is transcriptionally regulated by PEG and ABA and mediates drought stress-regulated root thickness and elongation in rice (Li et al. 2015c). From a genome-wide association study (GWAS) with 529 rice accessions, *OsNAL1* and a novel OsJAZ1 gene, which mediate rice spikelet development, were identified to positively regulate root depth and crown root formation under drought stress (Li et al. 2017). In this study, several root regulators, such as *DRO1*, *OsPID*, *OsWOX11*, *OsRR2*, were also suggested to be associated with root traits in response to drought stress. In addition, root-specific overexpression of *OsERF71* or three NAC domain family genes, *OsNAC5*, *OsNAC9* and *OsNAC10*, all confer drought tolerance by promoting increased seminal root diameters (Redillas et al. 2012, Jeong et al. 2010, Lee et al. 2016). Further study revealed that these genes regulate root morphology through the activation of lignin biosynthesis pathways and cell wall modifications.

6 Future trends in research

Understanding of the genetic and molecular mechanism of plant root system architecture has the potential to provide the genetic basis for the development of new rice root ideotypes that may be better adapted to variable growing conditions thereby ensuring sustainable rice yields. So far, multiple molecular and genetic signaling components have been identified to determine or to be involved in the shaping of rice by affecting different root parameters.

However, our knowledge of rice root development is still fragmentary. The plethora of root-regulating genes and molecules described earlier were mostly investigated in independent studies, with the molecular links among these regulators largely undetermined. Forward genetic approaches, such as QTL analysis and mutagenesis, are powerful for identifying genes with major effects on root growth and development, but not sufficient for in-depth analysis of gene functions. Further studies will be needed that can make use of multiple molecular and genetic resources (e.g. transcriptome, genome sequence, meta-QTL analysis), combined with advanced techniques for a more comprehensive analysis of gene interactions during rice root development and growth. Roots develop from a small group of cells (e.g. stem cells and pericycle cells), and the genes involving the priming and activation of these cells still remain to be discovered. In a previous study, laser capture microdissection has been successfully used to obtain 40 different cell types from rice root and shoot tissues for gene expression analysis (Jiao et al. 2009). Further genetic analysis on a specific cell type has the potential to discover specific signaling pathways associated with corresponding root development processes.

All roots are branching organs that are regulated by plant hormones. At the anatomical level, they share a common cell division pattern and tissue organization during successive developmental stages. However, genetic evidence suggests only a partially shared regulatory mechanism between different types of root. For instance, the *Osiaa11* mutant is only defective in the initiation of lateral roots rather than crown roots, whereas several crown root defective mutants, including *Osarl1*, *Osarl15,* and *ren1-D*, do not show defects in lateral root formation. Differences in gene regulation might contribute to the restricted expression of these genes in tissues associated either with lateral or crown roots, or still to be uncovered molecular mechanisms might explain the differences between lateral and crown root formation.

As described, environmental factors are also able to shape rice RSA through the activation of multiple endogenous signaling components. However, previous studies were mostly carried out to investigate a single environmental factor. Emerging evidence has suggested that environmental factors can interact to affect plant development. For instance, nitrate can trigger the degradation of a phosphate signaling repressor SPX4 to activate nitrogen and phosphate responses (Hu et al. 2019), while drought stress can limit nitrogen and phosphate utilization (He and Dijkstra 2014). Additionally, light is essential as a signal for crown root formation in rice (Lin and Sauter 2018), and has been shown to regulate nitrogen- and phosphate-mediated root responses in *Arabidopsis* (Zheng et al. 2019, Chen et al. 2016). All these observations indicate a strong interaction among environmental factors during their regulation of plant development and root development in particular. Therefore, the big challenge for further research in rice root development will be to get insight into the many interacting of environmental conditions that contribute to the shaping of rice root system architecture.

7 Where to look for further information

The following articles provide a good overview of the subject:

- Meng, F., Xiang, D., Zhu, J., Li, Y. and Mao, C. (2017) 'Molecular Mechanisms of Root Development in Rice', *Rice*, 12.
- Rebouillat, J., Dievart, A., Verdeil, J. L., Escoute, J., Giese, G., Breitler, J. C., Gantet, P., Espeout, S., Guiderdoni, E. and Perin, C. (2009) 'Molecular Genetics of Rice Root Development', *Rice*, 2(1), 15–34.
- Orman-Ligeza, B., Parizot, B., Gantet, P. P., Beeckman, T., Bennett, M. J. and Draye, X. (2013) 'Post-embryonic root organogenesis in cereals: branching out from model plants', *Trends In Plant Science,* 18(8), 464–467.

Key research in this area can be found at the following organizations:

- The International Joint Laboratory LMI RICE, https://sites.google.com/site/lmiricevn/the-team.
- Université de Montpellier, IRD, UMR DIADE, Montpellier, France, http://diade.ird.fr/en/teams/ceres-team.html.
- Rice Genome Annotation Project, http://rice.plantbiology.msu.edu/.
- International Rice Institute, https://www.irri.org/.

8 Acknowledgement

We are grateful to the Research Foundation – Flanders (FWO, Belgium), Bilateral Research Cooperation with MOST (China) (2016YFE0109900), National Natural Science Foundation (31672223 and 31822047), the Fundamental Research Funds for the Central Universities (KJYQ201903 and KYT201802), and Innovative Research Team Development Plan of the Ministry of Education of China (No. IRT_17R56).

9 References

Arite, T., Kameoka, H. and Kyozuka, J. (2012). Strigolactone positively controls crown root elongation in rice, *Journal of Plant Growth Regulation* 31(2), 165–172.

Bao, Y., Aggarwal, P., Robbins, N. E., Sturrock, C. J., Thompson, M. C., Tan, H. Q., Tham, C., Duan, L. N., Rodriguez, P. L., Vernoux, T., Mooney, S. J., Bennett, M. J. and Dinneny, J. R. (2014). Plant roots use a patterning mechanism to position lateral root branches toward available water, *Proceedings of the National Academy of Sciences of the United States of America* 111(25), 9319–9324.

Barlow, P. W. (2002). The root cap: cell dynamics, cell differentiation and cap function, *Journal of Plant Growth Regulation* 21(4), 261-286.

Beier, M. P., Obara, M., Taniai, A., Sawa, Y., Ishizawa, J., Yoshida, H., Tomita, N., Yamanaka, T., Ishizuka, Y., Kudo, S., Yoshinari, A., Takeuchi, S., Kojima, S., Yamaya, T. and Hayakawa, T. (2018). Lack of ACTPK1, an STY kinase, enhances ammonium uptake and use, and promotes growth of rice seedlings under sufficient external ammonium, *Plant Journal: for Cell and Molecular Biology* 93(6), 992-1006.

Bellini, C., Pacurar, D. I. and Perrone, I. (2014). Adventitious roots and lateral roots: similarities and differences, *Annual Review of Plant Biology* 65(65), 639-666.

Bian, H. W., Xie, Y. K., Guo, F., Han, N., Ma, S. Y., Zeng, Z. H., Wang, J. H., Yang, Y. N. and Zhu, M. Y. (2012). Distinctive expression patterns and roles of the miRNA393/TIR1 homolog module in regulating flag leaf inclination and primary and crown root growth in rice (Oryza sativa), *New Phytologist* 196(1), 149-161.

Brown, L. K., George, T. S., Dupuy, L. X. and White, P. J. (2013). A conceptual model of root hair ideotypes for future agricultural environments: what combination of traits should be targeted to cope with limited P availability?, *Annals of Botany* 112(2), 317-330.

Cai, X., Chen, T., Zhou, Q., Xu, L., Qu, L., Hua, X. and Lin, J. (2011). Development of Casparian strip in rice cultivars, *Plant Signaling and Behavior* 6(1), 59-65.

Champoux, M. C., Wang, G., Sarkarung, S., Mackill, D. J., Otoole, J. C., Huang, N. and Mccouch, S. R. (1995). Locating genes associated with root morphology and drought avoidance in rice via linkage to molecular markers, *Theoretical and Applied Genetics* 90(7-8), 969-981.

Chen, C. W., Yang, Y. W., Lur, H. S., Tsai, Y. G. and Chang, M. C. (2006). A novel function of abscisic acid in the regulation of rice (Oryza sativa L.) root growth and development, *Plant and Cell Physiology* 47(1), 1-13.

Chen, H., Ma, B. A., Zhou, Y., He, S. J., Tang, S. Y., Lu, X., Xie, Q., Chen, S. Y. and Zhang, J. S. (2018a). E3 ubiquitin ligase SOR1 regulates ethylene response in rice root by modulating stability of Aux/IAA protein, *Proceedings of the National Academy of Sciences of the United States of America* 115(17), 4513-4518.

Chen, X., Shi, J., Hao, X., Liu, H., Shi, J., Wu, Y., Wu, Z., Chen, M., Wu, P. and Mao, C. (2013). OsORC3 is required for lateral root development in rice, *Plant Journal: for Cell and Molecular Biology* 74(2), 339-350.

Chen, X. B., Yao, Q. F., Gao, X. H., Jiang, C. F., Harberd, N. P. and Fu, X. D. (2016). Shoot-to-root mobile transcription factor HY5 coordinates plant carbon and nitrogen acquisition, *Current Biology* 26(5), 640-646.

Chen, Y. H., Chao, Y. Y., Hsu, Y. Y., Hong, C. Y. and Kao, C. H. (2012). Heme oxygenase is involved in nitric oxide- and auxin-induced lateral root formation in rice, *Plant Cell Reports* 31(6), 1085-1091.

Chen, Y. Q., Xie, Y. M., Song, C. H., Zheng, L. L., Rong, X., Jia, L. T., Luo, L., Zhang, C., Qu, X. X. and Xuan, W. (2018b). A comparison of lateral root patterning among dicot and monocot plants, *Plant Science: an International Journal of Experimental Plant Biology* 274, 201-211.

Cheng, X., Jiang, H., Zhang, J., Qian, Y., Zhu, S. and Cheng, B. (2010). Overexpression of type-A rice response regulators, OsRR3 and OsRR5, results in lower sensitivity to cytokinins, *Genetics and Molecular Research* 9(1), 348-359.

Chhun, T., Taketa, S., Tsurumi, S. and Ichii, M. (2003a). The effects of auxin on lateral root initiation and root gravitropism in a lateral rootless mutant Lrt1 of rice (Oryza sativa L.), *Plant Growth Regulation* 39(2), 161-170.

Chhun, T., Taketa, S., Tsurumi, S. and Ichii, M. (2003b). Interaction between two auxin-resistant mutants and their effects on lateral root formation in rice (Oryza sativa L.), *Journal of Experimental Botany* 54(393), 2701–2708.

Cho, S. H., Yoo, S. C., Zhang, H., Lim, J. H. and Paek, N. C. (2014). Rice NARROW LEAF1 regulates leaf and adventitious root development, *Plant Molecular Biology Reporter* 32(1), 270–281.

Cho, S. H., Yoo, S. C., Zhang, H., Pandeya, D., Koh, H. J., Hwang, J. Y., Kim, G. T. and Paek, N. C. (2013). The rice narrow leaf2 and narrow leaf3 loci encode WUSCHEL-related homeobox 3A (OsWOX3A) and function in leaf, spikelet, tiller and lateral root development, *New Phytologist* 198(4), 1071–1084.

Clowes, F. A. L. and Juniper, B. E. (1964). The fine structure of the quiescent centre and neighbouring tissues, *Journal of Experimental Botany* 15(3), 622–623.

Colmer, T. D., Gibberd, M. R., Wiengweera, A. and Tinh, T. K. (1998). The barrier to radial oxygen loss from roots of rice (Oryza sativa L.) is induced by growth in stagnant solution, *Journal of Experimental Botany* 49(325), 1431–1436.

Cui, P., Liu, H. B., Ruan, S. L., Ali, B., Gill, R. A., Ma, H. S., Zheng, Z. F. and Zhou, W. J. (2017). A zinc finger protein, interacted with cyclophilin, affects root development via IAA pathway in rice, *Journal of Integrative Plant Biology* 59(7), 496–505.

Cutler, S. R., Rodriguez, P. L., Finkelstein, R. R. and Abrams, S. R. (2010). Abscisic acid: emergence of a core signaling network, *Annual Review of Plant Biology* 61, 651–679.

Dai, X. Y., Wang, Y. Y., Yang, A. and Zhang, W. H. (2012). OsMYB2P-1, an R2R3 MYB transcription factor, is involved in the regulation of phosphate-starvation responses and root architecture in rice, *Plant Physiology* 159(1), 169–183.

De Rybel, B., Vassileva, V., Parizot, B., Demeulenaere, M., Grunewald, W., Audenaert, D., Van Campenhout, J., Overvoorde, P., Jansen, L., Vanneste, S., Moller, B., Wilson, M., Holman, T., Van Isterdael, G., Brunoud, G., Vuylsteke, M., Vernoux, T., De Veylder, L., Inze, D., Weijers, D., Bennett, M. J. and Beeckman, T. (2010). A novel aux/IAA28 signaling cascade activates GATA23-dependent specification of lateral root founder cell identity, *Current Biology* 20(19), 1697–1706.

Debi, B. R., Mushika, J., Taketa, S., Miyao, A., Hirochika, H. and Ichii, M. (2003). Isolation and characterization of a short lateral root mutant in rice (Oryza sativa L.), *Plant Science* 165(4), 895–903.

Di, D. W., Sun, L., Zhang, X. N., Li, G. J., Kronzucker, H. J. and Shi, W. M. (2018). Involvement of auxin in the regulation of ammonium tolerance in rice (Oryza sativa L.), *Plant and Soil* 432(1–2), 373–387.

Dickinson, A. J., Lehner, K., Mi, J. N., Jia, K. P., Mijar, M., Dinneny, J., Al-Babili, S. and Benfey, P. N. (2019). β-Cyclocitral is a conserved root growth regulator, *Proceedings of the National Academy of Sciences of the United States of America* 116(21), 10563–10567.

Ding, W. N., Tong, H. S., Zheng, W. J., Ye, J., Pan, Z. C., Zhang, B. T. and Zhu, S. H. (2017). Isolation, characterization and transcriptome analysis of a cytokinin receptor mutant Osckt1 in rice, *Frontiers in Plant Science* 8, 88.

Ding, X. P., Li, X. K. and Xiong, L. H. (2011). Evaluation of near-isogenic lines for drought resistance QTL and fine mapping of a locus affecting flag leaf width, spikelet number, and root volume in rice, *Theoretical and Applied Genetics* 123(5), 815–826.

Enstone, D. E., Peterson, C. A. and Ma, F. S. (2002). Root endodermis and exodermis: structure, function, and responses to the environment, *Journal of Plant Growth Regulation* 21(4), 335–351.

Faiyue, B., Vijayalakshmi, C., Nawaz, S., Nagato, Y., Taketa, S., Ichii, M., Al-Azzawi, M. J. and Flowers, T. J. (2010). Studies on sodium bypass flow in lateral rootless mutants lrt1 and lrt2, and crown rootless mutant crl1 of rice (Oryza sativa L.), *Plant, Cell and Environment* 33(5), 687-701.

Gamuyao, R., Chin, J. H., Pariasca-Tanaka, J., Pesaresi, P., Catausan, S., Dalid, C., Slamet-Loedin, I., Tecson-Mendoza, E. M., Wissuwa, M. and Heuer, S. (2012). The protein kinase Pstol1 from traditional rice confers tolerance of phosphorus deficiency, *Nature* 488(7412), 535-539.

Gao, S. P., Fang, J., Xu, F., Wang, W., Sun, X. H., Chu, J. F., Cai, B. D., Feng, Y. Q. and Chu, C. C. (2014). CYTOKININ OXIDASE/DEHYDROGENASE4 integrates cytokinin and auxin signaling to control rice crown root formation, *Plant Physiology* 165(3), 1035-1046.

Geldner, N. (2013). Casparian strips, *Current Biology* 23(23), R1025-R1026.

Giehl, R. F. H. and von Wiren, N. (2014). Root nutrient foraging, *Plant Physiology* 166(2), 509-517.

Giri, J., Bhosale, R., Huang, G. Q., Pandey, B. K., Parker, H., Zappala, S., Yang, J., Dievart, A., Bureau, C., Ljung, K., Price, A., Rose, T., Larrieu, A., Mairhofer, S., Sturrock, C. J., White, P., Dupuy, L., Hawkesford, M., Perin, C., Liang, W. Q., Peret, B., Hodgman, C. T., Lynch, J., Wissuwa, M., Zhang, D. B., Pridmore, T., Mooney, S. J., Guiderdoni, E., Swarup, R. and Bennett, M. J. (2018). Rice auxin influx carrier OsAUX1 facilitates root hair elongation in response to low external phosphate, *Nature Communications* 9(1), 1408.

Gruber, B. D., Giehl, R. F. H., Friedel, S. and von Wiren, N. (2013). Plasticity of the Arabidopsis Root system under nutrient deficiencies, *Plant Physiology* 163(1), 161-179.

Gu, M., Zhang, J., Li, H. H., Meng, D. Q., Li, R., Dai, X. L., Wang, S. C., Liu, W., Qu, H. Y. and Xu, G. H. (2017). Maintenance of phosphate homeostasis and root development are coordinately regulated by MYB1, an R2R3-type MYB transcription factor in rice, *Journal of Experimental Botany* 68(13), 3603-3615.

Hanzawa, E., Sasaki, K., Nagai, S., Obara, M., Fukuta, Y., Uga, Y., Miyao, A., Hirochika, H., Higashitani, A., Maekawa, M. and Sato, T. (2013). Isolation of a novel mutant gene for soil-surface rooting in rice (Oryza sativa L.), *Rice* 6(1), 30.

He, M. and Dijkstra, F. A. (2014). Drought effect on plant nitrogen and phosphorus: a meta-analysis, *New Phytologist* 204(4), 924-931.

Henry, A., Cal, A. J., Batoto, T. C., Torres, R. O. and Serraj, R. (2012). Root attributes affecting water uptake of rice (Oryza sativa) under drought, *Journal of Experimental Botany* 63(13), 4751-4763.

Hirano, T., Satoh, Y., Ohki, A., Takada, R., Arai, T. and Michiyama, H. (2008). Inhibition of ammonium assimilation restores elongation of seminal rice roots repressed by high levels of exogenous ammonium, *Physiologia Plantarum* 134(1), 183-190.

Hirose, N., Makita, N., Kojima, M., Kamada-Nobusada, T. and Sakakibara, H. (2007). Overexpression of a type-A response regulator alters rice morphology and cytokinin metabolism, *Plant and Cell Physiology* 48(3), 523-539.

Hsu, Y. Y., Chao, Y. Y. and Kao, C. H. (2013). Methyl jasmonate-induced lateral root formation in rice: the role of heme oxygenase and calcium, *Journal of Plant Physiology* 170(1), 63-69.

Hu, B., Jiang, Z., Wang, W., Qiu, Y., Zhang, Z., Liu, Y., Li, A., Gao, X., Liu, L., Qian, Y., Huang, X., Yu, F., Kang, S., Wang, Y., Xie, J., Cao, S., Zhang, L., Wang, Y., Xie, Q., Kopriva, S. and

Chu, C. (2019). Nitrate-NRT1.1B-SPX4 cascade integrates nitrogen and phosphorus signalling networks in plants, *Nature Plants* 5(4), 401-413.

Hu, B., Zhu, C. G., Li, F., Tang, J. Y., Wang, Y. Q., Lin, A. H., Liu, L. C., Che, R. H. and Chu, C. C. (2011). LEAF TIP NECROSIS1 plays a pivotal role in the regulation of multiple phosphate starvation responses in rice, *Plant Physiology* 156(3), 1101-1115.

Huang, G., Yi, K. K., Wu, Y. R., Zhu, L., Mao, C. Z. and Wu, P. (2004). QTLs for nitrate induced elongation and initiation of lateral roots in rice (Oryza sativa L.), *Plant and Soil* 263(1), 229-237.

Huang, S. J., Liang, Z. H., Chen, S., Sun, H. W., Fan, X. R., Wang, C. L., Xu, G. H. and Zhang, Y. L. (2019). A transcription factor, OsMADS57, regulates long-distance nitrate transport and root elongation, *Plant Physiology* 180(2), 882-895.

Hwang, I., Sheen, J. and Muller, B. (2012). Cytokinin signaling networks, *Annual Review of Plant Biology* 63(63), 353-380.

Inukai, Y., Miwa, M., Nagato, Y., Kitano, H. and Yamauchi, A. (2001). RRLI, RRL2 and CRL2 loci regulating root elongation in rice, *Breeding Science* 51(4), 231-239.

Inukai, Y., Sakamoto, T., Ueguchi-Tanaka, M., Shibata, Y., Gomi, K., Umemura, I., Hasegawa, Y., Ashikari, M., Kitano, H. and Matsuoka, M. (2005). Crown rootless1, which is essential for crown root formation in rice, is a target of an AUXIN RESPONSE FACTOR in auxin signaling, *Plant Cell* 17(5), 1387-1396.

Itoh, J.-I., Nonomura, K.-I., Ikeda, K., Yamaki, S., Inukai, Y., Yamagishi, H., Kitano, H. and Nagato, Y. (2005). Rice plant development: from zygote to spikelet, *Plant and Cell Physiology* 46(1), 23-47.

Jackson, M. B., Fenning, T. M. and Jenkins, W. (1985). Aerenchyma (gas-space) formation in adventitious roots of rice (Oryza-sativa-L) is not controlled by ethylene or small partial pressures of oxygen, *Journal of Experimental Botany* 36(171), 1566-1572.

Jeong, J. S., Kim, Y. S., Baek, K. H., Jung, H., Ha, S. H., Do Choi, Y., Kim, M., Reuzeau, C. and Kim, J. K. (2010). Root-specific expression of OsNAC10 improves drought tolerance and grain yield in rice under field drought conditions, *Plant Physiology* 153(1), 185-197.

Jia, H. F., Zhang, S. T., Wang, L. Z., Yang, Y. X., Zhang, H. Y., Cui, H., Shao, H. F. and Xu, G. H. (2017). OsPht1;8, a phosphate transporter, is involved in auxin and phosphate starvation response in rice, *Journal of Experimental Botany* 68(18), 5057-5068.

Jia, L. Q., Zhang, B. T., Mao, C. Z., Li, J. H., Wu, Y. R., Wu, P. and Wu, Z. C. (2008). OsCYT-INV1 for alkaline/neutral invertase is involved in root cell development and reproductivity in rice (Oryza sativa L.), *Planta* 228(1), 51-59.

Jiang, H. W., Wang, S. M., Dang, L., Wang, S. F., Chen, H. M., Wu, Y. R., Jiang, X. H. and Wu, P. (2005). A novel short-root gene encodes a glucosamine-6-phosphate acetyltransferase required for maintaining normal root cell shape in rice, *Plant Physiology* 138(1), 232-242.

Jiao, X., Wang, H., Yan, J., Kong, X., Liu, Y., Chu, J., Chen, X., Fang, R. and Yan, Y. (2019). Promotion of BR biosynthesis by miR444 is required for ammonium-triggered inhibition of root growth, *Plant Physiology* 182(3), 1454-1466.

Jiao, Y., Tausta, S. L., Gandotra, N., Sun, N., Liu, T., Clay, N. K., Ceserani, T., Chen, M., Ma, L., Holford, M., Zhang, H. Y., Zhao, H., Deng, X. W. and Nelson, T. (2009). A transcriptome atlas of rice cell types uncovers cellular, functional and developmental hierarchies, *Nature Genetics* 41(2), 258-263.

Jing, H. W., Yang, X. L., Zhang, J., Liu, X. H., Zheng, H. K., Dong, G. J., Nian, J. Q., Feng, J., Xia, B., Qian, Q., Li, J. Y. and Zuo, J. R. (2015). Peptidyl-prolyl isomerization targets rice Aux/IAAs for proteasomal degradation during auxin signalling, *Nature Communications* 6, 7395.

Jun, N., Gaohang, W., Zhenxing, Z., Huanhuan, Z., Yunrong, W. and Ping, W. (2011). OsIAA23-mediated auxin signaling defines postembryonic maintenance of QC in rice, *Plant Journal: for Cell and Molecular Biology* 68(3), 433–442.

Kamoshita, A., Wade, L. J., Ali, M. L., Pathan, M. S., Zhang, J., Sarkarung, S. and Nguyen, H. T. (2002). Mapping QTLs for root morphology of a rice population adapted to rainfed lowland conditions, *Theoretical and Applied Genetics* 104(5), 880–893.

Kawata, S. and Shibayama, H. (1965). On the lateral root primordia formation in the crown roots of rice plants, *Japanese Journal of Crop Science* 33(4), 423–431.

Kitomi, Y., Inahashi, H., Takehisa, H., Sato, Y. and Inukai, Y. (2012). OsIAA13-mediated auxin signaling is involved in lateral root initiation in rice, *Plant Science: an International Journal of Experimental Plant Biology* 190, 116–122.

Kitomi, Y., Ito, H., Hobo, T., Aya, K., Kitano, H. and Inukai, Y. (2011). The auxin responsive AP2/ERF transcription factor CROWN ROOTLESS5 is involved in crown root initiation in rice through the induction of OsRR1, a type-A response regulator of cytokinin signaling, *Plant Journal: for Cell and Molecular Biology* 67(3), 472–484.

Kitomi, Y., Kanno, N., Kawai, S., Mizubayashi, T., Fukuoka, S. and Uga, Y. (2015). QTLs underlying natural variation of root growth angle among rice cultivars with the same functional allele of DEEPER ROOTING 1, *Rice* 8, 16.

Lee, D. K., Jung, H., Jang, G., Jeong, J. S., Kim, Y. S., Ha, S. H., Do Choi, Y. and Kim, J. K. (2016). Overexpression of the OsERF71 transcription factor alters rice root structure and drought resistance, *Plant Physiology* 172(1), 575–588.

Lehner, K. R., Taylor, I., McCaskey, E. N., Jain, R., Ronald, P. C., Goldman, D. I. and Benfey, P. N. (2018). A histidine kinase gene is required for large radius root tip circumnutation and surface exploration in rice, *bioRxiv*. doi: https://doi.org/10.1101/437012.

Lewis, D. R., Negi, S., Sukumar, P. and Muday, G. K. (2011). Ethylene inhibits lateral root development, increases IAA transport and expression of PIN3 and PIN7 auxin efflux carriers, *Development* 138(16), 3485–3495.

Li, C., Tang, Z., Wei, J., Qu, H., Xie, Y. and Xu, G. (2016). The OsAMT1.1 gene functions in ammonium uptake and ammonium-potassium homeostasis over low and high ammonium concentration ranges, *Journal of Genetics and Genomics = Yi Chuan Xue Bao* 43(11), 639–649.

Li, C. X., Shen, H. Y., Wang, T. and Wang, X. L. (2015a). ABA regulates subcellular redistribution of OsABI-LIKE2, a negative regulator in ABA signaling, to control root architecture and drought resistance in Oryza sativa, *Plant and Cell Physiology* 56(12), 2396–2408.

Li, J., Zhu, S. H., Song, X. W., Shen, Y., Chen, H. M., Yu, J., Yi, K. K., Liu, Y. F., Karplus, V. J., Wu, P. and Deng, X. W. (2006). A rice glutamate receptor-like gene is critical for the division and survival of individual cells in the root apical meristem, *Plant Cell* 18(2), 340–349.

Li, J. T., Zhao, Y., Chu, H. W., Wang, L. K., Fu, Y. R., Liu, P., Upadhyaya, N., Chen, C. L., Mou, T. M., Feng, Y. Q., Kumar, P. and Xu, J. (2015b). SHOEBOX modulates root meristem size in rice through dose-dependent effects of gibberellins on cell elongation and proliferation, *PLOS Genetics* 11(8), e1005464.

Li, J. Z., Han, Y. C., Liu, L., Chen, Y. P., Du, Y. X., Zhang, J., Sun, H. Z. and Zhao, Q. Z. (2015c). qRT9, a quantitative trait locus controlling root thickness and root length in upland rice, *Journal of Experimental Botany* 66(9), 2723–2732.

Li, X. K., Guo, Z. L., Lv, Y., Cen, X., Ding, X. P., Wu, H., Li, X. H., Huang, J. P. and Xiong, L. Z. (2017). Genetic control of the root system in rice under normal and drought stress conditions by genome-wide association study, *PLoS Genetics* 13(7), e1006889.

Lin, C. and Sauter, M. (2018). Control of adventitious root architecture in rice by darkness, light, and gravity, *Plant Physiology* 176(2), 1352–1364.

Liu, H. J., Wang, S. F., Yu, X. B., Yu, J., He, X. W., Zhang, S. L., Shou, H. X. and Wu, P. (2005). ARL1, a LOB-domain protein required for adventitious root formation in rice, *Plant Journal: for Cell and Molecular Biology* 43(1), 47-56.

Liu, S. P., Wang, J. R., Wang, L., Wang, X. F., Xue, Y. H., Wu, P. and Shou, H. X. (2009). Adventitious root formation in rice requires OsGNOM1 and is mediated by the OsPINs family, *Cell Research* 19(9), 1110–1119.

Lou, Q. J., Chen, L., Mei, H. W., Wei, H. B., Feng, F. J., Wang, P., Xia, H., Li, T. M. and Luo, L. J. (2015). Quantitative trait locus mapping of deep rooting by linkage and association analysis in rice, *Journal of Experimental Botany* 66(15), 4749–4757.

Ma, B., He, S. J., Duan, K. X., Yin, C. C., Chen, H., Yang, C., Xiong, Q., Song, Q. X., Lu, X., Chen, H. W., Zhang, W. K., Lu, T. G., Chen, S. Y. and Zhang, J. S. (2013). Identification of rice ethylene-response mutants and characterization of MHZ7/OsEIN2 in distinct ethylene response and yield trait regulation, *Molecular Plant* 6(6), 1830–1848.

Ma, B., Yin, C. C., He, S. J., Lu, X., Zhang, W. K., Lu, T. G., Chen, S. Y. and Zhang, J. S. (2014). Ethylene-induced inhibition of root growth requires abscisic acid function in rice (Oryza sativa L.) seedlings, *PLoS Genetics* 10(10), e1004701.

Ma, Y., Szostkiewicz, I., Korte, A., Moes, D., Yang, Y., Christmann, A. and Grill, E. (2009). Regulators of PP2C phosphatase activity function as abscisic acid sensors, *Science* 324(5930), 1064–1068.

Manschadi, A. M., Christopher, J., Devoil, P. and Hammer, G. L. (2006). The role of root architectural traits in adaptation of wheat to water-limited environments, *Functional Plant Biology* 33(9), 823–837.

Mellor, N., Vaughan-Hirsch, J., Kumpers, B. M. C., Help-Rinta-Rahko, H., Miyashima, S., Mahonen, A. P., Campilho, A., Kings, J. R. and Bishopp, A. (2019). A core mechanism for specifying root vascular patterning can replicate the anatomical variation seen in diverse plant species, *Development* 146(6). doi:10.1242/dev.172411.

Miyashita, Y., Takasugi, T. and Ito, Y. (2010). Identification and expression analysis of PIN genes in rice, *Plant Science* 178(5), 424–428.

Mori, M., Nomura, T., Ooka, H., Ishizaka, M., Yokota, T., Sugimoto, K., Okabe, K., Kajiwara, H., Satoh, K., Yamamoto, K., Hirochika, H. and Kikuchi, S. (2002). Isolation and characterization of a rice dwarf mutant with a defect in brassinosteroid biosynthesis, *Plant Physiology* 130(3), 1152–1161.

Morita, Y. and Kyozuka, J. (2007). Characterization of OsPID, the rice ortholog of PINOID, and its possible involvement in the control of polar auxin transport, *Plant and Cell Physiology* 48(3), 540–549.

Moubayidin, L., Di Mambro, R. and Sabatini, S. (2009). Cytokinin-auxin crosstalk, *Trends in Plant Science* 14(10), 557–562.

Orman-Ligeza, B., Parizot, B., Gantet, P. P., Beeckman, T., Bennett, M. J. and Draye, X. (2013). Post-embryonic root organogenesis in cereals: branching out from model plants, *Trends in Plant Science* 18(8), 459–467.

Orosa-Puente, B., Leftley, N., von Wangenheim, D., Banda, J., Srivastava, A. K., Hill, K., Truskina, J., Bhosale, R., Morris, E., Srivastava, M., Kumpers, B., Goh, T., Fukaki, H., Vermeer, J. E. M., Vernoux, T., Dinneny, J. R., French, A. P., Bishopp, A., Sadanandom,

A. and Bennett, M. J. (2018). Root branching toward water involves posttranslational modification of transcription factor ARF7, *Science* 362(6421), 1407–1410.

Park, S. Y., Fung, P., Nishimura, N., Jensen, D. R., Fujii, H., Zhao, Y., Lumba, S., Santiago, J., Rodrigues, A., Chow, T. F. F., Alfred, S. E., Bonetta, D., Finkelstein, R., Provart, N. J., Desveaux, D., Rodriguez, P. L., McCourt, P., Zhu, J. K., Schroeder, J. I., Volkman, B. F. and Cutler, S. R. (2009). Abscisic acid inhibits type 2C protein phosphatases via the PYR/PYL family of START proteins, *Science* 324(5930), 1068–1071.

Price, A. H., Steele, K. A., Moore, B. J. and Jones, R. G. W. (2002). Upland rice grown in soil-filled chambers and exposed to contrasting water-deficit regimes II. Mapping quantitative trait loci for root morphology and distribution, *Field Crops Research* 76(1), 25–43.

Qi, Y. H., Wang, S. K., Shen, C. J., Zhang, S. N., Chen, Y., Xu, Y. X., Liu, Y., Wu, Y. R. and Jiang, D. A. (2012). OsARF12, a transcription activator on auxin response gene, regulates root elongation and affects iron accumulation in rice (Oryza sativa), *New Phytologist* 193(1), 109–120.

Qin, C., Cheng, L. M., Zhang, H. H., He, M. L., Shen, J. Q., Zhang, Y. H. and Wu, P. (2016). OsGatB, the subunit of tRNA-dependent amidotransferase, is required for primary root development in rice, *Frontiers in Plant Science* 7, 599.

Qin, C., Li, Y. Y., Gan, J., Wang, W. X., Zhang, H. H., Liu, Y. and Wu, P. (2013). OsDGL1, a homolog of an oligosaccharyltransferase complex subunit, is involved in N-glycosylation and root development in rice, *Plant and Cell Physiology* 54(1), 129–137.

Qin, H., Wang, J., Chen, X. B., Wang, F. F., Peng, P., Zhou, Y., Miao, Y. C., Zhang, Y. O., Gao, Y. D., Qi, Y. D., Zhou, J. H. and Huang, R. F. (2019). Rice OsDOF15 contributes to ethylene-inhibited primary root elongation under salt stress, *New Phytologist* 223(2), 798–813.

Ranathunge, K., Kotula, L., Steudle, E. and Lafitte, R. (2004). Water permeability and reflection coefficient of the outer part of young rice roots are differently affected by closure of water channels (aquaporins) or blockage of apoplastic pores, *Journal of Experimental Botany* 55(396), 433–447.

Rebouillat, J., Dievart, A., Verdeil, J. L., Escoute, J., Giese, G., Breitler, J. C., Gantet, P., Espeout, S., Guiderdoni, E. and Perin, C. (2009). Molecular genetics of rice root development, *Rice* 2(1), 15–34.

Reddy, I. N. B. L., Kim, B. K., Yoon, I. S., Kim, K. H. and Kwon, T. R. (2017). Salt tolerance in rice: focus on mechanisms and approaches, *Rice Science* 24(3), 123–144.

Redillas, M. C. F. R., Jeong, J. S., Kim, Y. S., Jung, H., Bang, S. W., Choi, Y. D., Ha, S. H., Reuzeau, C. and Kim, J. K. (2012). The overexpression of OsNAC9 alters the root architecture of rice plants enhancing drought resistance and grain yield under field conditions, *Plant Biotechnology Journal* 10(7), 792–805.

Ruzicka, K., Ljung, K., Vanneste, S., Podhorska, R., Beeckman, T., Friml, J. and Benkova, E. (2007). Ethylene regulates root growth through effects on auxin biosynthesis and transport-dependent auxin distribution, *Plant Cell* 19(7), 2197–2212.

Sakamoto, T., Sakakibara, H., Kojima, M., Yamamoto, Y., Nagasaki, H., Inukai, Y., Sato, Y. and Matsuoka, M. (2006). Ectopic expression of KNOTTED1-like homeobox protein induces expression of cytokinin biosynthesis genes in rice, *Plant Physiology* 142(1), 54–62.

Sazuka, T., Kamiya, N., Nishimura, T., Ohmae, K., Sato, Y., Imamura, K., Nagato, Y., Koshiba, T., Nagamura, Y., Ashikari, M., Kitano, H. and Matsuoka, M. (2009). A rice tryptophan

deficient dwarf mutant, tdd1, contains a reduced level of indole acetic acid and develops abnormal flowers and organless embryos, *Plant Journal: for Cell and Molecular Biology* 60(2), 227-241.
Shafiq, S., Chen, C., Yang, J., Cheng, L., Ma, F., Widemann, E. and Sun, Q. (2017). DNA topoisomerase 1 prevents R-loop accumulation to modulate auxin-regulated root development in rice, *Molecular Plant* 10(6), 821-833.
Shao, Y. L., Zhou, H. Z., Wu, Y. R., Zhang, H., Lin, J., Jiang, X. Y., He, Q. J., Zhu, J. S., Li, Y., Yu, H. and Mao, C. Z. (2019). OsSPL3, an SBP-domain protein, regulates crown root development in rice, *Plant Cell* 31(6), 1257-1275.
Shelley, I. J., Nishiuchi, S., Shibata, K. and Inukai, Y. (2013). SLL1, which encodes a member of the stearoyl-acyl carrier protein fatty acid desaturase family, is involved in cell elongation in lateral roots via regulation of fatty acid content in rice, *Plant Science: an International Journal of Experimental Plant Biology* 207, 12-17.
Shen, C. J., Wang, S. K., Zhang, S. N., Xu, Y. X., Qian, Q., Qi, Y. H. and Jiang, de A. (2013). OsARF16, a transcription factor, is required for auxin and phosphate starvation response in rice (Oryza sativa L.), *Plant, Cell and Environment* 36(3), 607-620.
Shiono, K., Ogawa, S., Yamazaki, S., Isoda, H., Fujimura, T., Nakazono, M. and Colmer, T. D. (2011). Contrasting dynamics of radial O-2-loss barrier induction and aerenchyma formation in rice roots of two lengths, *Annals of Botany* 107(1), 89-99.
Song, Y. L., You, J. and Xiong, L. Z. (2009). Characterization of OsIAA1 gene, a member of rice Aux/IAA family involved in auxin and brassinosteroid hormone responses and plant morphogenesis, *Plant Molecular Biology* 70(3), 297-309.
Stepanova, A. N. and Alonso, J. M. (2019). From ethylene-auxin interactions to auxin biosynthesis and signal integration, *Plant Cell* 31(7), 1393-1394.
Stepanova, A. N., Yun, J., Likhacheva, A. V. and Alonso, J. M. (2007). Multilevel interactions between ethylene and auxin in Arabidopsis roots, *Plant Cell* 19(7), 2169-2185.
Strader, L. C., Wheeler, D. L., Christensen, S. E., Berens, J. C., Cohen, J. D., Rampey, R. A. and Bartel, B. (2011). Multiple facets of Arabidopsis seedling development require indole-3-butyric acid-derived auxin, *Plant Cell* 23(3), 984-999.
Straub, T., Ludewig, U. and Neuhauser, B. (2017). The kinase CIPK23 inhibits ammonium transport in Arabidopsis thaliana, *Plant Cell* 29(2), 409-422.
Su, Y. H., Liu, Y. B. and Zhang, X. S. (2011). Auxin-cytokinin interaction regulates meristem development, *Molecular Plant* 4(4), 616-625.
Sun, H. W., Bi, Y., Tao, J. Y., Huang, S. J., Hou, M. M., Xue, R., Liang, Z. H., Gu, P. Y., Yoneyama, K., Xie, X. N., Shen, Q. R., Xu, G. H. and Zhang, Y. L. (2016). Strigolactones are required for nitric oxide to induce root elongation in response to nitrogen and phosphate deficiencies in rice, *Plant, Cell and Environment* 39(7), 1473-1484.
Sun, H. W., Feng, F., Liu, J. and Zhao, Q. Z. (2018). Nitric oxide affects rice root growth by regulating auxin transport under nitrate supply, *Frontiers in Plant Science* 9, 659.
Sun, H. W., Tao, J. Y., Hou, M. M., Huang, S. J., Chen, S., Liang, Z. H., Xie, T. N., Wei, Y. Q., Xie, X. N., Yoneyama, K., Xu, G. H. and Zhang, Y. L. (2015). A strigolactone signal is required for adventitious root formation in rice, *Annals of Botany* 115(7), 1155-1162.
Sun, H. W., Tao, J. Y., Liu, S. J., Huang, S. J., Chen, S., Xie, X. N., Yoneyama, K., Zhang, Y. L. and Xu, G. H. (2014a). Strigolactones are involved in phosphate- and nitrate-deficiency-induced root development and auxin transport in rice, *Journal of Experimental Botany* 65(22), 6735-6746.
Sun, L. J., Zhang, Q., Wu, J. X., Zhang, L. Q., Jiao, X. W., Zhang, S. W., Zhang, Z. G., Sun, D. Y., Lu, T. G. and Sun, Y. (2014b). Two rice authentic histidine phosphotransfer

proteins, OsAHP1 and OsAHP2, mediate cytokinin signaling and stress responses in rice, *Plant Physiology* 165(1), 335–345.
Swarup, R., Perry, P., Hagenbeek, D., Van Der Straeten, D., Beemster, G. T. S., Sandberg, G., Bhalerao, R., Ljung, K. and Bennett, M. J. (2007). Ethylene upregulates auxin biosynthesis in Arabidopsis seedlings to enhance inhibition of root cell elongation, *Plant Cell* 19(7), 2186–2196.
Tong, H. N., Xiao, Y. H., Liu, D. P., Gao, S. P., Liu, L. C., Yin, Y. H., Jin, Y., Qian, Q. and Chu, C. C. (2014). Brassinosteroid regulates cell elongation by modulating gibberellin metabolism in rice, *Plant Cell* 26(11), 4376–4393.
Tu, Y., Jiang, A. M., Gan, L., Hossain, M., Zhang, J. M., Peng, B., Xiong, Y. G., Song, Z. J., Cai, D. T., Xu, W. F., Zhang, J. H. and He, Y. C. (2014). Genome duplication improves rice root resistance to salt stress, *Rice* 7(1), 15.
Uga, Y., Hanzawa, E., Nagai, S., Sasaki, K., Yano, M. and Sato, T. (2012). Identification of qSOR1, a major rice QTL involved in soil-surface rooting in paddy fields, *Theoretical and Applied Genetics* 124(1), 75–86.
Uga, Y., Kitomi, Y., Yamamoto, E., Kanno, N., Kawai, S., Mizubayashi, T. and Fukuoka, S. (2015). A QTL for root growth angle on rice chromosome 7 is involved in the genetic pathway of DEEPER ROOTING 1, *Rice* 8, 8.
Uga, Y., Okuno, K. and Yano, M. (2008). QTLs underlying natural variation in stele and xylem structures of rice root, *Breeding Science* 58(1), 7–14.
Uga, Y., Okuno, K. and Yano, M. (2011). Dro1, a major QTL involved in deep rooting of rice under upland field conditions, *Journal of Experimental Botany* 62(8), 2485–2494.
Uga, Y., Sugimoto, K., Ogawa, S., Rane, J., Ishitani, M., Hara, N., Kitomi, Y., Inukai, Y., Ono, K., Kanno, N., Inoue, H., Takehisa, H., Motoyama, R., Nagamura, Y., Wu, J. Z., Matsumoto, T., Takai, T., Okuno, K. and Yano, M. (2013a). Control of root system architecture by DEEPER ROOTING 1 increases rice yield under drought conditions, *Nature Genetics* 45(9), 1097–1102.
Uga, Y., Yamamoto, E., Kanno, N., Kawai, S., Mizubayashi, T. and Fukuoka, S. (2013b). A major QTL controlling deep rooting on rice chromosome 4, *Scientific Reports* 3(1), 3040.
Umezawa, T., Sugiyama, N., Mizoguchi, M., Hayashi, S., Myouga, F., Yamaguchi-Shinozaki, K., Ishihama, Y., Hirayama, T. and Shinozaki, K. (2009). Type 2C protein phosphatases directly regulate abscisic acid-activated protein kinases in Arabidopsis, *Proceedings of the National Academy of Sciences of the United States of America* 106(41), 17588–17593.
Vejchasarn, P., Lynch, J. P. and Brown, K. M. (2016). Genetic variability in phosphorus responses of rice root phenotypes, *Rice* 9(1), 29.
Vermeer, J. E., von Wangenheim, D., Barberon, M., Lee, Y., Stelzer, E. H., Maizel, A. and Geldner, N. (2014). A spatial accommodation by neighboring cells is required for organ initiation in Arabidopsis, *Science* 343(6167), 178–183.
Wang, H., Taketa, S., Miyao, A., Hirochika, H. and Ichii, M. (2006). Isolation of a novel lateral-rootless mutant in rice (Oryza sativa L.) with reduced sensitivity to auxin, *Plant Science* 170(1), 70–77.
Wang, J. R., Hu, H., Wang, G. H., Li, J., Chen, J. Y. and Wu, P. (2009). Expression of PIN genes in rice (Oryza sativa L.): tissue specificity and regulation by hormones, *Molecular Plant* 2(4), 823–831.
Wang, L. K., Chu, H. W., Li, Z. Y., Wang, J., Li, J. T., Qiao, Y., Fu, Y. R., Mou, T. M., Chen, C. L. and Xu, J. (2014a). Origin and development of the root cap in rice, *Plant Physiology* 166(2), 603–613.

Wang, L. L., Guo, M. X., Li, Y., Ruan, W. Y., Mo, X. R., Wu, Z. C., Sturrock, C. J., Yu, H., Lu, C. G., Peng, J. R. and Mao, C. Z. (2018). LARGE ROOT ANGLE1, encoding OsPIN2, is involved in root system architecture in rice, *Journal of Experimental Botany* 69(3), 385-397.

Wang, S. K., Xu, Y. X., Li, Z. L., Zhang, S. N., Lim, J. M., Lee, K. O., Li, C. Y., Qian, Q., Jiang, A. and Qi, Y. H. (2014b). OsMOGS is required for N-glycan formation and auxin-mediated root development in rice (Oryza sativa L.), *Plant Journal: for Cell and Molecular Biology* 78(4), 632-645.

Wang, T., Li, C. X., Wu, Z. H., Jia, Y. C., Wang, H., Sun, S. Y., Mao, C. Z. and Wang, X. L. (2017). Abscisic acid regulates auxin homeostasis in rice root tips to promote root hair elongation, *Frontiers in Plant Science* 8, 1121.

Wang, X. F., He, F. F., Ma, X. X., Mao, C. Z., Hodgman, C., Lu, C. G. and Wu, P. (2011). OsCAND1 is required for crown root emergence in rice, *Molecular Plant* 4(2), 289-299.

Wang, X. H., Yang, G. Z., Shi, M. X., Hao, D. L., Wei, Q. X., Wang, Z. G., Fu, S., Su, Y. H. and Xia, J. X. (2019a). Disruption of an amino acid transporter LHT1 leads to growth inhibition and low yields in rice, *BMC Plant Biology* 19(1), 268.

Wang, Y., Ma, N. N., Qiu, S. C., Zou, H. Y., Zang, G. C., Kang, Z. H., Wang, G. X. and Huang, J. L. (2014c). Regulation of the alpha-expansin gene OsEXPA8 expression affects root system architecture in transgenic rice plants, *Molecular Breeding* 34(1), 47-57.

Wang, Y. H., Wang, D., Gan, T., Liu, L. L., Long, W. H., Wang, Y. L., Niu, M., Li, X. H., Zheng, M., Jiang, L. and Wan, J. M. (2016). CRL6, a member of the CHD protein family, is required for crown root development in rice, *Plant Physiology and Biochemistry* 105, 185-194.

Wang, Z. G., Yamaji, N. K., Huang, S., Zhang, X., Shi, M. X., Fu, S., Yang, G. Z., Feng, J. F. and Xia, J. X. (2019b). OsCASP1 is required for Casparian strip formation at endodermal cells of rice roots for selective uptake of mineral elements, *Plant Cell* 31(11), 2636-2648.

Wei, J., Zheng, Y., Feng, H., Qu, H., Fan, X., Yamaji, N., Ma, J. F. and Xu, G. (2018). OsNRT2.4 encodes a dual-affinity nitrate transporter and functions in nitrate-regulated root growth and nitrate distribution in rice, *Journal of Experimental Botany* 69(5), 1095-1107.

Xia, J. X., Yamaji, N., Che, J., Shen, R. F. and Ma, J. F. (2014). Normal root elongation requires arginine produced by argininosuccinate lyase in rice, *Plant Journal: for Cell and Molecular Biology* 78(2), 215-226.

Xiao, G. Q., Qin, H., Zhou, J. H., Quan, R. D., Lu, X. Y., Huang, R. F. and Zhang, H. W. (2016). OsERF2 controls rice root growth and hormone responses through tuning expression of key genes involved in hormone signaling and sucrose metabolism, *Plant Molecular Biology* 90(3), 293-302.

Xu, L., Zhao, H. Y., Ruan, W. Y., Deng, M. J., Wang, F., Peng, J. R., Luo, J., Chen, Z. X. and Yi, K. K. (2017). ABNORMAL INFLORESCENCE MERISTEM1 functions in salicylic acid biosynthesis to maintain proper reactive oxygen species levels for root meristem activity in rice, *Plant Cell* 29(3), 560-574.

Xu, M., Zhu, L., Shou, H. X. and Wu, P. (2005). A PIN1 family gene, OsPIN1, involved in auxin-dependent adventitious root emergence and tillering in rice, *Plant and Cell Physiology* 46(10), 1674-1681.

Xuan, W., Audenaert, D., Parizot, B., Moller, B. K., Njo, M. F., De Rybel, B., De Rop, G., Van Isterdael, G., Mahonen, A. P., Vanneste, S. and Beeckman, T. (2015). Root cap-derived auxin pre-patterns the longitudinal axis of the Arabidopsis Root, *Current Biology* 25(10), 1381–1388.

Xuan, W., Beeckman, T. and Xu, G. H. (2017). Plant nitrogen nutrition: sensing and signaling, *Current Opinion in Plant Biology* 39, 57–65.

Xuan, Y. H., Priatama, R. A., Huang, J., Je, B. I., Liu, J. M., Park, S. J., Piao, H. L., Son, D. Y., Lee, J. J., Park, S. H., Jung, K. H., Kim, T. H. and Han, C. D. (2013). Indeterminate domain 10 regulates ammonium-mediated gene expression in rice roots, *New Phytologist* 197(3), 791–804.

Yamamoto, Y., Kamiya, N., Morinaka, Y., Matsuoka, M. and Sazuka, T. (2007). Auxin biosynthesis by the YUCCA genes in rice, *Plant Physiology* 143(3), 1362–1371.

Yamauchi, T., Shimamura, S., Nakazono, M. and Mochizuki, T. (2013). Aerenchyma formation in crop species: a review, *Field Crops Research* 152, 8–16.

Yamauchi, T., Shiono, K., Nagano, M., Fukazawa, A., Ando, M., Takamure, I., Mori, H., Nishizawa, N. K., Kawai-Yamada, M., Tsutsumi, N., Kato, K. and Nakazono, M. (2015). Ethylene biosynthesis is promoted by very-long-chain fatty acids during lysigenous aerenchyma formation in rice roots, *Plant Physiology* 169(1), 180–193.

Yamauchi, T., Tanaka, A., Mori, H., Takamure, I., Kato, K. and Nakazono, M. (2016). Ethylene-dependent aerenchyma formation in adventitious roots is regulated differently in rice and maize, *Plant, Cell and Environment* 39(10), 2145–2157.

Yan, Y. S., Wang, H. C., Hamera, S., Chen, X. Y. and Fang, R. X. (2014). miR444a has multiple functions in the rice nitrate-signaling pathway, *Plant Journal: for Cell and Molecular Biology* 78(1), 44–55.

Yang, C., Ma, B., He, S. J., Xiong, Q., Duan, K. X., Yin, C. C., Chen, H., Lu, X., Chen, S. Y. and Zhang, J. S. (2015). MAOHUZI6/ETHYLENE INSENSITIVE3-LIKE1 and ethylene INSENSITIVE3-LIKE2 regulate ethylene response of roots and coleoptiles and negatively affect salt tolerance in rice, *Plant Physiology* 169(1), 148–165.

Yang, J., Wang, L., Mao, C. Z. and Lin, H. H. (2017). Characterization of the rice NLA family reveals a key role for OsNLA1 in phosphate homeostasis, *Rice* 10(1), 52.

Yang, X., Yang, Y. N., Xue, L. J., Zou, M. J., Liu, J. Y., Chen, F. and Xue, H. W. (2011). Rice ABI5-Like1 regulates abscisic acid and auxin responses by affecting the expression of ABRE-containing genes, *Plant Physiology* 156(3), 1397–1409.

Yin, C. C., Ma, B. A., Collinge, D. P., Pogson, B. J., He, S. J., Xiong, Q., Duan, K. X., Chen, H., Yang, C., Lu, X., Wang, Y. Q., Zhang, W. K., Chu, C. C., Sun, X. H., Fang, S., Chu, J. F., Lu, T. G., Chen, S. Y. and Zhang, J. S. (2015). Ethylene responses in rice roots and coleoptiles are differentially regulated by a carotenoid isomerase-mediated abscisic acid pathway, *Plant Cell* 27(4), 1061–1081.

Yoshikawa, T., Ito, M., Sumikura, T., Nakayama, A., Nishimura, T., Kitano, H., Yamaguchi, I., Koshiba, T., Hibara, K. I., Nagato, Y. and Itoh, J. I. (2014). The rice FISH BONE gene encodes a tryptophan aminotransferase, which affects pleiotropic auxin-related processes, *Plant Journal: for Cell and Molecular Biology* 78(6), 927–936.

Yu, C. L., Sun, C. D., Shen, C. J., Wang, S. K., Liu, F., Liu, Y., Chen, Y. L., Li, C. Y., Qian, Q., Aryal, B., Geisler, M., Jiang, de A. and Qi, Y. H. (2015a). The auxin transporter, OsAUX1, is involved in primary root and root hair elongation and in Cd stress responses in rice (Oryzasativa L.), *Plant Journal: for Cell and Molecular Biology* 83(5), 818–830.

Yu, C. Y., Liu, Y. H., Zhang, A. D., Su, S., Yan, A., Huang, L., Ali, I., Liu, Y., Forde, B. G. and Gan, Y. B. (2015b). MADS-box transcription factor OsMADS25 regulates root development through affection of nitrate accumulation in rice, *PLoS ONE* 10(8), e0135196.

Yue, B., Xue, W. Y., Xiong, L. Z., Yu, X. Q., Luo, L. J., Cui, K. H., Jin, D. M., Xing, Y. Z. and Zhang, Q. F. (2006). Genetic basis of drought resistance at reproductive stage in rice: separation of drought tolerance from drought avoidance, *Genetics* 172(2), 1213–1228.

Yukiyoshi, K. and Karahara, I. (2014). Role of ethylene signalling in the formation of constitutive aerenchyma in primary roots of rice, *AoB Plants* 6, plu043.

Zhang, J. W., Xu, L., Wu, Y. R., Chen, X. A., Liu, Y., Zhu, S. H., Ding, W. N., Wu, P. and Yi, K. K. (2012). OsGLU3, a putative membrane-bound endo-1,4-beta-glucanase, is required for root cell elongation and division in rice (Oryza sativa L.), *Molecular Plant* 5(1), 176–186.

Zhao, H. M., Ma, T. F., Wang, X., Deng, Y. T., Ma, H. L., Zhang, R. S. and Zhao, J. (2015a). OsAUX1 controls lateral root initiation in rice (Oryza sativa L.), *Plant, Cell and Environment* 38(11), 2208–2222.

Zhao, Y., Cheng, S. F., Song, Y. L., Huang, Y. L., Zhou, S. L., Liu, X. Y. and Zhou, D. X. (2015b). The interaction between rice ERF3 and WOX11 promotes crown root development by regulating gene expression involved in cytokinin signaling, *Plant Cell* 27(9), 2469–2483.

Zhao, Y., Hu, Y. F., Dai, M. Q., Huang, L. M. and Zhou, D. X. (2009). The WUSCHEL-related homeobox gene WOX11 is required to activate shoot-borne crown root development in rice, *Plant Cell* 21(3), 736–748.

Zheng, H. K., Li, S. J., Ren, B., Zhang, J., Ichii, M., Taketa, S., Tao, Y. Z., Zuo, J. R. and Wang, H. (2013). LATERAL ROOTLESS2, a cyclophilin protein, regulates lateral root initiation and auxin signaling pathway in rice, *Molecular Plant* 6(5), 1719–1721.

Zheng, Z., Wang, Z., Wang, X. and Liu, D. (2019). Blue light triggered-chemical reactions underlie phosphate deficiency-induced inhibition of root elongation of Arabidopsis seedlings grown in Petri dishes, *Molecular Plant* 12(11), 1515–1523.

Zhou, J., Jiao, F. C., Wu, Z. C., Li, Y. Y., Wang, X. M., He, X. W., Zhong, W. Q. and Wu, P. (2008). OsPHR2 is involved in phosphate-starvation signaling and excessive phosphate accumulation in shoots of plants, *Plant Physiology* 146(4), 1673–1686.

Zhou, S., Jiang, W., Long, F., Cheng, S., Yang, W., Zhao, Y. and Zhou, D. X. (2017). Rice homeodomain protein WOX11 recruits a histone acetyltransferase complex to establish programs of cell proliferation of crown root meristem, *Plant Cell* 29(5), 1088–1104.

Zhu, J., Li, Y., Lin, J., Wu, Y., Guo, H., Shao, Y., Wang, F., Wang, X., Mo, X., Zheng, S., Yu, H. and Mao, C. (2019). CRD1, an Xpo1 domain protein, regulates miRNA accumulation and crown root development in rice, *Plant Journal: for Cell and Molecular Biology* 100(2), 328–342.

Zhu, Z. X., Liu, Y., Liu, S. J., Mao, C. Z., Wu, Y. R. and Wu, P. (2012). A gain-of-function mutation in OsIAA11 affects lateral root development in rice, *Molecular Plant* 5(1), 154–161.

Zou, X., Shao, J. W., Wang, Q., Chen, P. S., Zhu, Y. C. and Yin, C. X. (2018). Supraoptimal cytokinin content inhibits rice seminal root growth by reducing root meristem size and cell length via increased ethylene content, *International Journal of Molecular Sciences* 19(12), 4051.

Zou, Y., Liu, X. Y., Wang, Q., Chen, Y., Liu, C., Qiu, Y. and Zhang, W. (2014). OsRPK1, a novel leucine-rich repeat receptor-like kinase, negatively regulates polar auxin transport and root development in rice, *Biochimica et Biophysica Acta* 1840(6), 1676–1685.

Chapter 3

Advances in understanding plant root growth regulators

Findimila Dio Ishaya and Amanda Rasmussen, University of Nottingham, UK

1 Introduction

Global food production needs to increase by more than 60% by 2050 (FAO, 2016) in the face of climate change alongside population increases and changes in diet. Concurrently we need to reduce fertiliser applications to crops to reduce greenhouse emissions, production costs and environmental pollution. The agriculture sector generates one-fifth of global greenhouse gas emissions (FAO, 2016) and reducing nitrogen applications will reduce emissions created during fertiliser synthesis and from soil nitrous oxide emissions post-application. Together with leaching, these nitrogen losses mean only about 50% of applied nitrogen is captured by crops (Brackin et al., 2015). Phosphate is also an essential but limited mined resource used in fertilisers (Crombez et al., 2019). However the low mobility of phosphate means that only 10–25% of phosphate applied to crops is taken up (Crombez et al., 2019, Hasan et al., 2016, Zhang et al., 2014). These losses in nitrogen and phosphate are environmentally undesirable, wasteful of resources and financially costly. This means we need to improve crop-nutrient use efficiency to ensure sustainability of our food supply.

Nitrogen and phosphate uptake are both controlled and regulated by hormone pathways ((Crombez et al., 2019, Xuan et al., 2017) for recent reviews).

http://dx.doi.org/10.19103/AS.2020.0075.04

Understanding these pathways opens opportunities for breeding crops with improved growth under deficient conditions, reducing the need for fertilizers. After briefly introducing nitrogen and phosphate uptake and growth responses, this chapter will focus on auxin, cytokinin and strigolactones as key interacting hormones involved with these important nutrients.

1.1 Nitrogen availability and uptake

Nitrogen is present in soil as inorganic nitrate and ammonium, and organic (carbon-containing) amino acids, peptides and proteins (Brackin et al., 2015). As reviewed elsewhere (Xuan et al., 2017) nitrate uptake is driven by four families of nitrate transporters: Nitrate Transporter 1 family (NRT/NPF), Nitrate transporter 2 Family (NRT2/NAR2), CHLORIDE CHANNEL FAMILY (CLC) and the SLOW ACTION ANION CHANNEL-ASSOCIATED HOMOLOGUES (SLAC/SLAH) but only NRT1 and NRT2 families are notably involved in root nitrate uptake (Gu et al., 2018, Lupini et al., 2016). NRT1.1(NPF6.3/CHL1) which is mainly expressed in the epidermis of Arabidopsis young root tips, is nitrate-inducible and is a dual affinity nitrate transporter which means it shows affinity for high or low soil nitrate levels based on its phosphorylation status (Kant, 2018, Shahzad and Amtmann, 2017, Xuan et al., 2017, Zhao et al., 2018a). NRT1.1 is also a transporter for auxin and is central to many of the hormone and root architecture regulation processes discussed below. The phosphorylated type mediates auxin transport and nitrate uptake during low soil nitrogen conditions (Han and Hwang, 2018, Kant, 2018, Shahzad and Amtmann, 2017, Zhao et al., 2018a). In maize, AtNPF6.3 has four putative homologs (ZmNPF6.4 (ZmNRT1.1A), ZmNPF6.6 (ZmNRT1.1B), ZmNPF6.7 (ZmNRT1.1C) and ZmNPF6.5 (ZmNRT1.1D) (Dechorgnat et al., 2018, Plett et al., 2018) and each of these was shown to have different tissue specificity and transcriptional responses to nitrogen availability (Plett et al., 2018).

In contrast to NRT1.1, NRT1.2 (NPF4.6) is localised in both young root tips and mature regions of Arabidopsis root and is constitutively expressed (Plett et al., 2018). Orthologues of NRT1.2 have been identified in every sequenced cereal species suggesting evolutionarily conserved roles for this group (Plett et al., 2018). Additionally, in maize, ZmNRT1.2 did not show a difference in transcript levels in plants grown in low and high soil nitrate until the late reproductive phase where expression levels differ between treatments (Garnett et al., 2013, Plett et al., 2018).

Another group of transporters present in most angiosperms are the NRT2 family which transport nitrates in the company of NAR2.1 (NRT3.1) (with the exception of NRT2.7 (Gu et al., 2018, Kant, 2018, O'Brien et al., 2016)). For example, OsNRT2 family genes (OsNRT2.1, OsNRT2.2, OsNRT2.3, OsNRT2.5) partner with OsNRT3.1 to control nitrate acquisition by rice at various nitrate concentrations (Gu et al., 2018). NRT2.1 and NRT2.2 have been suggested to be a product of gene duplication as they share 90.4% sequence identity and are localised in tandem on chromosome 1 (Plett et al., 2018). AtNRT2.1 was shown to be the predominant high affinity transporter under diverse environments while AtNRT2.2 contributes comparatively little to nitrate uptake (Lupini et al., 2016, Plett et al., 2018). Demonstrating the importance of these transporter groups, when AtNRT2.1 is knocked out, AtNRT2.2 increased contributing more to the high-affinity transport system (Lupini et al., 2016) and consistent with the idea of conserved roles, this has also been observed in cereals (Plett et al., 2018).

Plants also have ammonium transporters in the root epidermis to take up ammonium from the soil (Xuan et al., 2017). Ammonium uptake and transport is mediated by the ammonium transporter/methylammonium permease/rhesus (AMT/MEP/Rh) transporter family with AMT1 family being the major entry pathway of root uptake of ammonium ions (Gu et al., 2013, Hawkesford et al., 2012, Miller and Chapman, 2011). Just like for nitrate, the ammonium transport system consists of a low- and high-affinity transport system (Dechorgnat et al., 2018, Garnett et al., 2015). In low soil ammonium levels plants require members of the AMT gene family (Dechorgnat et al., 2018). Recent identification of AMF1 proteins in soybean and yeast depicts a promising candidate for low-affinity ammonium transport (Dechorgnat et al., 2018). As with nitrate transporters, AMT expression is increased by nitrogen starvation and suppressed by ammonium fertilisation (Xuan et al., 2017).

Plants also have specialised amino acid transporters such as the amino acid transporter family (ATF: including Amino acid permeases and lysine histadine transporters (Nacry et al., 2013, Rentsch et al., 2007), Lysine Histidine Transporter 1 (LHT1) in Arabidopsis (Ganeteg et al., 2017), amino acid polyamine-choline (APC), mitochondrial carrier family (MCF), plastid outer envelope protein (OEP16) and divalent anion Na+ symporter (DASS) (Rentsch et al., 2007)). Transporters have also been identified for Ureids (drug/metabolite transporter (DMT) and DUR (Nacry et al., 2013)) and peptides (peptide transporter (PTR), oligopeptide transporter (OPT) and ATP binding cassette (ABC); reviewed in (Rentsch et al., 2007)). However little is known about how these transporter families interact with signalling molecules.

Soil nitrate fluctuates greatly in time and space, so plants have adapted to quickly modulate their nitrate uptake ability in response to internal N status and external N availability (Bouguyon et al., 2016, Sun et al., 2017, Xuan et al., 2017). In support of this, many of the signalling pathways discussed below occur within a few hours of nitrogen sensing. Some of the developmental changes that occur in response to nitrogen availability include changes in primary or seminal root growth and lateral root initiation (Xuan et al., 2017). These modulations allow effective nitrogen foraging of the soil environment.

1.2 Phosphate availability and uptake

Phosphorus is the second most important macronutrient for plants and is involved in the formation of nucleic acid, membrane build-up and stability, energy metabolism, DNA synthesis, photosynthesis, phospholipids and phosphoproteins, conversion of sugars to starch, and nutrients movement in plants (Hasan et al., 2016, Sakuraba et al., 2018, Zhang et al., 2019). Phosphate is taken up by roots via phosphate transporters (PHT; reviewed in (Crombez et al., 2019)). In Arabidopsis the PHT1 family has 9 high affinity members responsible for 70% of root phosphate uptake (Niu et al., 2013). Interestingly about 20% of phosphate uptake happens via PHT1.1 and PHT1.4 expressed exclusively in the root cap cells (Crombez et al., 2019). Phosphate can be mobilised in the plant via phloem and shoot-to-root phosphorus transport is regulated via PHO2 (Niu et al., 2013).

Phosphate starvation responses in plants consist of modification of root system architecture through regulation of primary and seminal root growth, lateral root initiation and root hair formation and growth (Fang Zhu et al., 2018, Niu et al., 2013, Zhang et al., 2014). Since phosphate is hard to extract from soils, low phosphate nutrition also changes root exudates to help free phosphate from the soil matrix and to increase connections with mycorrhizal fungi (Fang Zhu et al., 2018, Niu et al., 2013, Zhang et al., 2014).

The specific patterns of response are species (and line) specific (Chevalier et al., 2003, Niu et al., 2013, Péret et al., 2014) and nutrient-form-specific (Gruber et al., 2013, Kellermeier et al., 2014, Linkohr et al., 2002). This means there are different strategies for nutrient deficiency responses that could be captured for improving crop nutrient use efficiency. However, before this is possible, we need to understand the regulatory networks controlling plant responses to nutrient signals.

2 Auxin

Auxin is involved in many plant developmental and response processes (for recent review, see Blakeslee et al., 2019). Given the importance of auxin in regulating many aspects of primary, lateral and adventitious root initiation, growth and development (Bouguyon et al., 2016, Han and Hwang, 2018, Leyser, 2018, Luo et al., 2018, Steffens and Rasmussen, 2016, Zhao et al., 2018a) it is unsurprising that any set of conditions requiring an adjustment in root architecture will involve this central hormone.

2.1 Auxin biosynthesis, transport and signalling

Auxin distribution among plant tissues is controlled by biosynthesis, inactivation and transport pathways. Biosynthesis is closely regulated to ensure favourable cellular concentration in specific tissues and at different developmental stages (Han and Hwang, 2018). Multiple pathways have been postulated to contribute to the *de novo* auxin biosynthesis, but the predominant pathway is the highly conserved two-step TAA/YUC (multigene families) route (Blakeslee et al., 2019). Here, tryptophan aminotransferases catalyses YUCCA (YUC), Flavin-dependent monooxygenases to produce auxin in the form of Indole Acetic Acid (IAA) from tryptophan (Blakeslee et al., 2019).

Most auxin is synthesised in the shoot (although some is also produced in roots) and transported by auxin transporters including AUXIN RESISTANT 1/ LIKE AUX1 (Aux/LAX) family (influx carriers) and ABC (ATP binding cassette) and PIN family (efflux carriers) (Han and Hwang, 2018, Krouk, 2016, Motte and Beeckman, 2019, Sun et al., 2018). These carriers define the position at which auxin signalling is triggered and they regulate auxin transport to achieve auxin maxima and tissue-specific auxin responses (Han and Hwang, 2018, Sun et al., 2018).

Auxin signalling involves 3 protein families, the F-BOX TRANSPORT INHIBITIOR RESPONSE1/AUXIN SIGNALLING F-BOX PROTEIN (TIR1/AFB) auxin co-receptors, the Auxin/INDOLE-3-ACETIC ACID (Aux/IAA), transcriptional repressors and the AUXIN RESPONSE FACTOR (ARF) transcriptional factors (Lavy and Estelle, 2016). Auxin enters the nucleus and binds to TIR1 which then rapidly degrades Aux/IAA which releases ARFs from suppression by Aux/IAA and modulates auxin-response gene expression (Han and Hwang, 2018, Lavy and Estelle, 2016).

2.2 Nitrogen and auxin

The biosynthesis, transport and accumulation of auxin is different in different N regimes in a range of species including monocots like maize and rice, legumes like soybean, and dicots like Arabidopsis (Krouk, 2016, Wang et al., 2019). In some studies, a decrease in nitrate supply enhanced translocation of auxin from shoot to root (de Jong et al., 2014, Krouk, 2016). But several studies have also shown a reduction in auxin transport in rice grown in low N soils (Sun et al., 2015, Xu et al., 2015). Thus the mechanism that controls auxin signalling pathways and changes in carbon and nitrogen metabolism remains unclear (Han and Hwang, 2018, Wang et al., 2019).

As described in the introduction, NRT1.1 (CHL1/NPF6.3) is a nitrate transporter important in sensing external nitrate availability (Bouguyon et al., 2016, Krouk, 2016). AtNRT1.1 functions as an auxin efflux carrier in addition to its role in nitrate transport and is highly expressed in lateral root primordia, root cap and epidermis where auxin transport controls lateral root development (Kiba et al., 2011, Xuan et al., 2017). The auxin influx action of NRT1.1 is inhibited by nitrate in a concentration-dependent manner

(Bouguyon et al., 2016, O'Brien et al., 2016). In low soil nitrate conditions, NRT1.1 levels increased auxin export from lateral root primordia to inhibit auxin accumulation and prevent further primordia development (Bouguyon et al., 2015). In high nitrogen soils, NRT1.1 was repressed by nitrate leading to accumulation of auxin in the root tip to promote lateral root growth (Kiba et al., 2011).

NRT1.1-induced auxin transport in low nitrate supply also increases AFB3 (auxin signalling) transcript level and activates NAC4 (NAC DOMAIN CONTAINING PROTEIN) (Armijo and Gutiérrez, 2017, O'Brien et al., 2016). NAC4 is a transcription factor involved in the regulation of lateral root initiation through the inhibition of microRNA393 (O'Brien et al., 2016). This miRNA393 was also activated by decreased nitrate and post transcriptionally repressed AFB3 gene expression (O'Brien et al., 2016). This modulation integrates external inorganic nitrogen sources and the internal nitrogen status of the plant to control auxin signalling and downstream developmental responses (Armijo and Gutiérrez, 2017, O'Brien et al., 2016).

NRT1.1 is not the only nitrate transporter in roots. NRT2.1 is the predominant inducible component of the high affinity transport system (Kant, 2018, Kiba et al., 2011, O'Brien et al., 2016) often expressed during N starvation (Kant, 2018, Kiba et al., 2011). In contrast to NRT1.1 the nitrate affinity of AtNRT2 is always high and there is no affinity for auxin transport (Gu et al., 2018, Kant, 2018).

NRT2.1 is involved in lateral root development, but the underlying mechanism remains to be unravelled (O'Brien et al., 2016). However, Yu and colleagues (Yu et al., 2016) reported that in rice the NRT2 family regulates nitrate-dependent lateral foot formation by interacting with NAR2 (NRT3.1). They reported that NRT3.1 regulated polar auxin transport from shoot to root by fine-tuning the transcription of PIN family of auxin efflux transporters. Modulation of PIN proteins in response to low nitrogen has been shown in a range of species (O'Brien et al., 2016, Yu et al., 2019). In turn, this affects how much auxin reaches the roots and is thus available to regulate root architecture.

Nitrogen fertilisation was central to the green revolution, and is now a focus for improving efficiency and minimising greenhouse emissions. For this reason it is particularly important to note that the types of nitrogen fertiliser applied result in distinct signalling and growth responses. Nitrate and ammonium applied together enhanced auxin biosynthesis in maize, by increased levels of phosphoenolpyruvate and tryptophan compared to a single nitrate or ammonium fertiliser source (Wang et al., 2019). This increased auxin content was also found in leaves of rice grown in drought when fertilised with both nitrate and ammonium compared to individual nitrogen sources (Cao et al., 2018). The authors inferred that the increased auxin content could be attributed to increased carbohydrate allocation to shoots than to roots in mixed nitrogen treatments, making the shoots an active growth site, and requiring higher levels of auxin.

It has been shown that localised patches of ammonium supply to nitrogen stressed plants robustly triggered lateral root branching while repressing primary and lateral root elongation (Liu and von Wirén, 2017). In Arabidopsis, this inhibition of primary root growth occurred via suppression of cell proliferation in the root apical meristem and reduced cell size, while the structure and activity of the stem cell niche remained stable (Liu and von Wirén, 2017). Ammonium transporters (AMTs) and GDP-Mannose pyrophosphorylase might also be involved in root architecture responses to local soil ammonium levels (Ueda et al., 2017). However, detailed studies of the processes controlling ammonium-induced changes in root architecture are lacking.

2.3 Phosphate and auxin

Auxin participates in the low phosphate response by adjusting root system architecture and to a lesser extent, promotion of arbuscular mycorrhizal fungi associations. However there is evidence that changes in root-available phosphate also affects auxin sensitivity.

Low-phosphate soils increased auxin sensitivity of pericycle cells resulting in lateral root initiation (Nath and Tuteja, 2016, Zhang et al., 2014) while auxin levels were unaffected (Nath and Tuteja, 2016). However, phosphate also influences auxin biosynthesis, transport and signalling pathways to regulate root responses (Hasan et al., 2016, Zhang et al., 2019). In these studies phosphate starvation activated the expression of TIR1, the auxin receptor, which triggered ARF7 and ARF19 to induce downstream responses (Ham et al., 2018, Niu et al., 2013, Zhang et al., 2014). Low external phosphate activated root hair-expressed auxin-inducible transcription factors ARF19, RSL2 and RSL4, and lack of these genes affected the low phosphorus root hair response in Arabidopsis (Rahul et al., 2018, Zhang et al., 2019).

The interaction between phosphate and auxin signalling has also been found in rice (Shen et al., 2013, Wang et al., 2014, Zhang et al., 2014). The knockout of OsARF16 led to the loss of auxin sensitivity to auxin and phosphate stress response in rice primary roots, lateral roots and root hairs (Shen et al., 2013). Also involved in the auxin-phosphorus interaction, OsARF12 knockout mutants had increased phosphorus concentration than WT triggered by increased phosphate transporter levels (Wang et al., 2014). OsARF12 expression was also inhibited by low phosphate levels (Wang et al., 2014). This demonstrates that OsARF16 and OsARF12 are associated with auxin signalling and phosphate stress-responsive genes (see also review (Zhang et al., 2014).

Root responses to low phosphate conditions are also controlled via auxin transport (Zhang et al., 2019). In rice, the auxin influx transporter OsAUX1 promoted root hair elongation in response to low soil phosphorus (Zhang et al., 2019); in maize, root growth was improved by promoting auxin accumulation in root tips through overexpressing ZmPIN1a(Zhang et al., 2019). Cluster roots (discussed elsewhere) produced in Proteaceae and lupins in response to low phosphate are also controlled by auxin transport (Cheng et al., 2011, Niu et al., 2013).

Although there are clear interactions between phosphate levels and auxin signalling it should be noted that although the Arabidopsis auxin signalling mutants (*axr1-3*, *axr2-1* and *axr4-1*) show phosphate starvation symptoms such as lateral root and root hair elongation, *axr1-3* responded the same as wild type during phosphate stress (Nath and Tuteja, 2016). This means phosphate interacts with auxin downstream or independently of *AXR1,2* and *4*.

3 Cytokinin

Cytokinins are a class of mobile adenine derivatives involved in many plant growth and development processes (Liu et al., 2019, Kiba et al., 2011) including as secondary messengers involved in nutrient signals and responses to biotic and abiotic stress (Kiba et al., 2011, Landrein et al., 2018).

3.1 Cytokinin biosynthesis, transport and signalling

Cytokinins are classified into four main groups: N6-(Δ2-Isopenyl) adenine (iP), trans-zeatin(tZ), cis-Zeatin (cZ), and dihydrogezeatin (DHZ) (Liu et al., 2019). It was earlier believed that cytokinin was only synthesised in the roots, but more recently it has been accepted that IP-type cytokinins are produced throughout the plant, while Z-type are predominantly produced in roots (Gu et al., 2018, Kang et al., 2017, Kieber and Schaller, 2018).

The process of cytokinin biosynthesis begins with an ADP/ATP precursor to which a prenyl group derived from dimethylallyl diphosphate is added. The process is catalysed by isopentenyl transferase (IPT) encoded by a multigene family; the resulting IP ribosides can then be transformed to various IP-type conjugates or transferred to the roots where they can be converted to Z-types of cytokinin. This process involves hydroxylation of the isoprenoid side chain via cytochrome P450 enzymes, CYP735A1 and CYP735A2 (Liu et al., 2019, Kiba et al., 2019, Kieber and Schaller, 2018, Pavlů et al., 2018). The free base, ip-type and tZ-type, active forms of cytokinin are synthesised from their ribotide conjugates in a single enzymatic step catalysed by LONELY GUY (LOG) family cytokinin nucleoside 5′-monophosphate phosphoribohydrolases (Liu et al., 2019, Kang et al., 2017, Kieber and Schaller, 2018). Active cytokinin can also be regulated through conjugation to glucose or through irreversible fractionalisation by cytokinin oxidases (CKXs) (Kieber and Schaller, 2018).

Cytokinins can be transported via the xylem and phloem, while three cytokinin transporter families (PURINE PERMEASES (PUPs), EQUILIBRATIVE NUCLEOSIDE TRANSPORTER (ENTs) and the G SUBFAMILY ATP-BINDING CASSETTE (ABCG)) assist both long- and short-distance movements (Gu et al., 2018, Kieber and Schaller, 2018, Veselov et al., 2018) PUPs are also known as proton-coupled multiphasic transporters that convey kinetin, zeatin (and some zeatin ribosides) and adenine precursors intracellularly and specifically to vascular tissues. ENTs regulate transport of hydrophilic nucleoside substrates including riboside forms of TZ and IP. The ATP-binding cassette transporter, AtABCG14 is expressed in the pericycle and stele of roots and conveys TZ-type cytokinin to the xylem sap (Gu et al., 2018, Kang et al., 2017, Veselov et al., 2018). Interruption of ABCG14 led to a 90% decrease in xylem cytokinin levels which reduced shoot growth (Kieber and Schaller, 2018).

Cytokinin perception is a two-step process beginning with membrane-localised histidine-kinases receptors which are transduced via a His-Asp phosphorelay to induce a family of transcription factors in the nucleus (Kiba et al., 2019, Kieber and Schaller, 2018, Sakakibara et al., 2006). Histidine kinases direct the specificity of cytokinin regulation for different physiological mechanisms (Danilova et al., 2017, Kieber and Schaller, 2018). In Arabidopsis, AHK2, AHK3, and AHK4/CRE1 are families of receptors that initiate cytokinin response (Liu et al., 2019, Danilova et al., 2017). These receptors phosphorylate the downstream phosphortransmitters which subsequently move to the nucleus and phosphorylate type-B modulators (Danilova et al., 2017, Kieber and Schaller, 2018). The phosphorylated type-B response regulator proteins activate the transcription of downstream target proteins of plant response to cytokinin with type-A negative modulators (Danilova et al., 2017). In addition, type-A response regulators are involved in a negative feedback loop (Kieber and Schaller, 2018). Cytokinin induces type-A response regulators by both directly stabilising type-A response regulators and via type-B response regulators (Kieber and Schaller, 2018).

Cytokinin signalling cascades are a two-way component that is activated during changes in temperature, nutrients level, chemoatractants and osmotic conditions (Osugi and Sakakibara, 2015, Pavlů et al., 2018).

3.2 Nitrate and cytokinin

The functions of cytokinin in long-distance signals is to communicate root nitrogen status and control N assimilation affecting shoot metabolism and overall plant development (O'Brien et al., 2016, Pavlů et al., 2018, Veselov et al., 2018). During conditions where shoot nitrogen is adequate , cytokinin acts as a satiety signal of nitrogen to inhibit nitrate uptake in roots by repressing AtNRT2.1, AtNRT2.2, AtNRT1.1 and AtNRT1.5 (xylem loaders)

(Gu et al., 2018, Nath and Tuteja, 2016, Pavlů et al., 2018) and upregulating AtNRT2.7, AtNRT1.4 and AtNRT1.7 (nitrate partitioners and transporters) (Gu et al., 2018). The suppression of 'xylem loaders' requires the cytokinin receptor (AHK4/CRE1/WOL CCRE1) and AHK3, implying His-Asp phosphorelay-mediated cytokinin signalling (Gu et al., 2018, Kiba et al., 2011). The increase in cytokinin biosynthesis gene expression results in increased production of tZ-type cytokinins in the root which are then transported to the shoot via ABCG14 (Poitout et al., 2018).

Heterogeneous soil nitrate supply and nitrate inducible *IPT3* and *IPT5* genes can influence entire plant development by co-ordinating biosynthesis of different active forms of cytokinin including trans–zeatin (Roy, 2018). For example the ratio of trans-zeatin and trans-zeatin ribosides in the xylem sap increased in response to nitrate availability (Kieber and Schaller, 2018). This implies that plants control the ratio of these cytokinin forms to fine-tune shoot growth responses to variable environments (Kieber and Schaller, 2018).

Evidence from other species suggests this pathway may be conserved across families. Cytokinin biosynthesis increased in xylem sap, roots and shoots of maize in response to nitrate availability (Kiba et al., 2011). Similarly *OsIPT4, OsIPT5, OsIPT7* and *OsIPT8* expression were upregulated in rice when exogenous mixed nitrate and ammonium were applied accompanied by increased cytokinin levels (Gu et al., 2018). There is also a close correlation between nitrogen level and cytokinin content of barley, tobacco and *Urtica dioica* (Kiba et al., 2011) while nitrate promoted root-to-shoot transport of cytokinin in tobacco (Wang et al., 2019). Interestingly although ammonium can increase cytokinin levels (Gu et al., 2018), only nitrate is able to induce genes related to A-type Arabidopsis response regulator (ARR) proteins (O'Brien et al., 2016) and cytokinin response factor (CRFs) (Pavlů et al., 2018). However, the full molecular mechanism and gene network involved in the interaction between nitrate and cytokinin is not fully understood (O'Brien et al., 2016).

3.3 Phosphate and cytokinin

Not only is cytokinin involved in nitrate signalling, but it is also decreased under low phosphate conditions (Zhang et al., 2014) via downregulation of IPT3 (Pavlů et al., 2018). Cytokinin signalling components such as the cytokinin receptor AHK4 (Pavlů et al., 2018) and CYTOKININ RESPONSE 1(CRE1) (Niu et al., 2013) are also down regulated during phosphate-deficient conditions (Niu et al., 2013, Pavlů et al., 2018). Similarly when cytokinin levels are reduced, the expression of phosphate transporters is increased to improve uptake (Pavlů et al., 2018). When phosphate is resupplied to roots *IPT3, CRF5* and *CRF6* are upregulated (Pavlů et al., 2018)

Since low cytokinin levels are associated with phosphate stress, responsive genes such as *AtIPS1*, *At4* and *AtIPT1* are also inhibited by higher levels of cytokinin (Niu et al., 2013, Zhang et al., 2014). Similarly, in Arabidopsis and rice, exogenous cytokinin suppressed phosphate uptake and starvation responses probably by mobilising phosphate from shoot tissues (Pavlů et al., 2018, Zhang et al., 2014). This may temporarily minimise phosphate stress signalling leading to relief of deficiency symptoms, including rescuing shoot growth repression (Pavlů et al., 2018, Zhang et al., 2014).

Although low cytokinin is reported in soils with low phosphate the story is more complex. Silva-Navas et al. (2019) demonstrated that cZ, cZR and iPR cytokinins are increased while tZ, tZR and iP are decreased in plants grown in low phosphate conditions. They suggested that the change in the ratio of cZ:tZ cytokinins was responsible for regulating the change in root:shoot ratio observed in plants grown on low phosphate. Further they showed that this response required IPT9, while the change in internal phosphate levels required AHK3 and AHK4 cytokinin-signalling components (Silva-Navas et al., 2019). As seen in the interaction between cytokinin and nitrate responses (Kieber and Schaller, 2018), and in the response of cuttings to root removal (Rasmussen et al., 2015) this provides further evidence that the ratio between different cytokinin types is important for regulation of plant developmental responses to the environment.

4 Strigolactones

Strigolactones are a recently identified group of hormones that regulate shoot branching (Gomez-Roldan et al., 2008, Umehara et al., 2008), parasitic seed germination (Matusova et al., 2005, Xie et al., 2008), mycorrhizal associations (Akiyama, 2014) and root development (Mayzlish-Gati et al., 2012, Rasmussen et al., 2012). Under nutrient-deficient conditions, mycorrhizal fungi improve nutrient acquisition for their plant partners, so it's unsurprising that exudation of plant signals such as strigolactones increase when nitrogen or phosphorus are limiting. Here we summarise the latest understanding of strigolactone biosynthesis and signalling, responses to nitrogen and phosphorus conditions and interactions with other hormones already discussed in previous sections.

4.1 Strigolactone biosynthesis, transport and signalling

Strigolactone synthesis begins with the production of carlactone from carotenoids via DWARF27 isomerase (D27), CAROTENOID CLEAVAGE DIOXYGENASE7 and 8 (CCD7, CCD8) sequentially (Alder et al., 2012). Carlactone is then acted on by MORE AXILLARY BRANCHING 1 (MAX1) to make active strigolatones (Crawford et al., 2010). Strigolactone production can occur anywhere in the plant but the highest expression levels of biosynthesis genes are found in roots (Machin et al., 2020). Root-derived strigolactone is transported in the xylem to the shoot where it inhibits the outgrowth of buds that would become branches (Machin et al., 2020).

Perception occurs when strigolactone binds to DWARF14 (D14; an α/β-hydrolase) allowing interaction with the SCF-type E3 ubiquitin ligase complex which includes the F-box protein D3/MAX2. This complex then targets the SMAX1-LIKE (SMXL) protein family for degradation releasing the brakes on strigolactone responses (for recent review see Lv et al., 2018, Machin et al., 2020).

It's interesting to note that Karrikins, the active component in smoke that triggers seed germination after fire, contain a similar active D-ring in the molecule. The receptor complex for karrikins is similar to that in strigolactones with Karrikin binding to KAI2 (an α/β-hydrolase) which then interacts with SCF^{MAX2} and SMAX1 (Machin et al., 2020).

Strigolactone biosynthesis and signalling respond to the nitrogen and phosphorus status of the soil environment to increase mycorrhizal associations and regulate biomass allocation. The following sections will explore the latest understanding of strigolactones and their interactions with other hormones in response to soil-nutrient conditions. Since in many cases nitrogen and phosphorus are studied together with strigolactones, they will be explored together, and discrepancies between the two will be highlighted. The final section will describe the latest understanding of how strigolactones interact with other hormones in response to nutrient conditions.

4.2 Strigolactone response to nitrogen or phosphorus

Because of their importance in triggering parasitic weed seed germination, the exudation of strigolactones from plant roots has been measured under different soil conditions (Trabelsi et al., 2017). Both nitrogen and phosphorus deficiencies increased strigolactone production and exudation (López-Ráez et al., 2008, Sun et al., 2016, Trabelsi et al., 2017, Yoneyama et al., 2015a, Yoneyama et al., 2012, Yoneyama et al., 2007) while potassium deficiency had no effect on strigolactone production in Faba bean (Trabelsi et al., 2017).

Strigolactone root exudation profiles are of great interest because parasitic weed seeds respond to specific strigolactone molecules (Yoneyama et al., 2015a) while mycorrhizae appear to respond to the group generally (Yoneyama et al., 2015a). These molecular profiles demonstrate that not only are there species-specific differences in strigolactone types (Yoneyama et al., 2012) but also variations within species (e.g. between parasitic weed susceptible and non-susceptible lines (Jamil et al., 2011, Yoneyama et al., 2015b) and in response to different conditions (Jamil et al., 2011, Sun et al., 2014, Yoneyama et al., 2015b, Yoneyama et al., 2012, Yoneyama et al., 2013). For a list of plants and strigolactone molecule responses, see Table 1 but briefly, most strigolactones are increased under low phosphorus conditions (and in many cases increased more in low P than in low N soil (Yoneyama et al., 2015a, Yoneyama et al., 2012, Yoneyama et al., 2013)). The exception to this is 5-Deoxystrigol in sorghum which was only increased under low N (Yoneyama et al., 2013) and Orobanchyl acetate in Chinese milk vetch which was not affected by either low N or low P (Yoneyama et al., 2012). Most molecules also increased in response to low external N (albeit sometimes lesser than to low P (Yoneyama et al., 2015a, Yoneyama et al., 2012, Yoneyama et al., 2013)) except Orobanchol in alfalfa and tomato which was only increased in response to low P (Yoneyama et al., 2012). Complementing this nutrient-driven strigolactone production, replacing nutrients in the soil reduced the exudation of strigolactones (Jamil et al., 2011, Yoneyama et al., 2013) and strigolactone biosynthesis not just in roots in contact with the nutrient source but also in distal roots indicating a systemic strigolactone response (Yoneyama et al., 2015b).

Given the increase in strigolactone production under low soil nitrogen conditions it is no surprise that low nitrogen conditions increase the expression of strigolactone biosynthesis genes (Ito et al., 2017, Sun et al., 2016, Sun et al., 2014, Xu et al., 2015). Supporting the suggested role of strigolactones in response to nitrogen and phosphorus deficiency, the rice biosynthesis mutants d10, d27 and d3 all had reduced root responses compared to WT (Arite et al., 2012, Sun et al., 2014) and all except the d3 signalling mutant had increased responses when resupplied with synthetic strigolactone GR24 (Sun et al., 2014). This difference was due to differences in meristematic zone with d10 and d14 having fewer cells in the meristematic region (meaning a shorter meristematic zone) compared to wildtype roots grown in low phosphate (Arite et al., 2012).

As discussed above, low nitrogen conditions increased the expression of nitrogen transporters including ammonium transporters (OsAMT1.1, OsAMT1.2, and OsAMT1.3) and amino acid transporters (OsAAP1-1, OsAAP1-2, OsAAP10, and OsLHT1) (Luo et al., 2018). This increased expression is reduced or absent in the strigolactone mutants d3 (deficient strigolactone signalling) and d10 (deficient strigolactone biosynthesis) (Luo et al., 2018) suggesting strigolactones are involved not only in growth responses but also in nutrient transport.

4.3 Strigolactone interaction with other hormones in nutrient deficient conditions

To summarise how hormones interact to control plant responses we have collated the presented pathways for nitrogen responses into Fig. 1. The interactions for phosphate are quite similar. It should be noted that not every hormone interaction presented here have been experimentally demonstrated in

Table 1 Production of strigolactone types under low N or low P soil conditions. Suscept means lines that were more susceptible to parasitic weed infection, while non-suscept represents lines less susceptible to parasitic weeds. Data collated from Yoneyama et al., 2015a, 2012, 2013; Jamil et al., 2011.

Strigolactone molecule	Plant species	Response to low N	Response to low P
Sorgomol	Maize	Increased	Increased more than -N
	Chinese milk vetch	Increased	Increased
	Sorghum	Increased	Increased
5-Deoxystrigol	Maize	Increased	Increased more than -N
	Chinese milk vetch	Increased	Increased more than -N
	Sorghum	Increased only to -N	
SL1 (unidentified)	Maize	Increased	Increased more than -N
SL2 (unidentified	Maize	Increased	Increased
Orobanchyl acetate	Milk vetch	No response	No response
	Lettuce	Increased	Increased
	Marigold	Increased	Increased
Orobanchol	Alfalfa	No response	Increased
	Tomato	No response	Increased
	Wheat	Increased	Increased more than -N
	Marigold	Increased	Increased more than -N
	Lettuce	Increased	Increased
	Rice (suscept)		Increased
	Rice (not suscept)	No response	No response
2′-epi-5-Deoxystrigol	Rice (suscept)		Increased
	Rice (not suscept)		Increased
SL1 (unidentified)	Rice (suscept)	Increased	Increased more than -N
	Rice (not suscept)		Increased
SL2 (unidentified)	Rice (suscept)	Increased	Increased more than -N
	Rice (not suscept)		Increased
SL3 (unidentified)	Rice (suscept)	Increased	Increased more than -N
	Rice (not suscept)	Increased	Increased

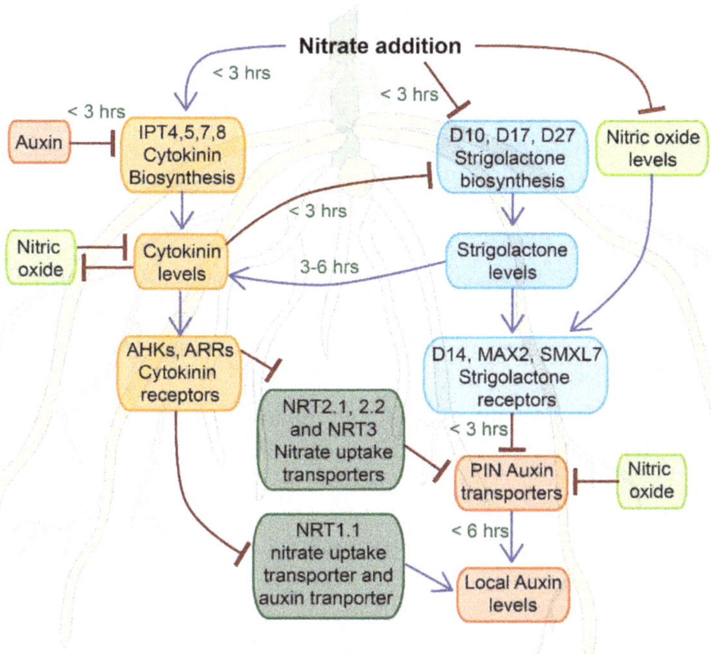

Figure 1 Hormone interactions controlling nitrate responses in plants. Arrows (blue) represent positive interactions while flat-ended symbols (red) represent inhibition. The green times are the how long after nitrate addition a change in response is seen (based on Xu et al., 2015).

the context of every other hormone. However evidence for each interaction has been demonstrated experimentally and this represents our best understanding at time of writing.

One of the quickest interactions with strigolactones is with nitric oxide. Nitric oxide (NO) is a signalling molecule involved in root organogenesis, growth and development (Sun et al., 2018). The two NO production pathways in plants are NO synthase-like (NOS-Like) and nitrate reductase (NR) (Sun et al., 2018). In low nitrogen or low phosphorus conditions, nitric oxide (a gaseous hormone) levels increased (Sun et al., 2018) promoting expression of strigolactone signalling genes (D14, D53) but not strigolactone biosynthesis genes (D10, D17 or D27; (Sun et al., 2016)). Strigolactone genes were shown to be needed for NO induced elongation of seminal roots in response to low N and low P in rice (Sun et al., 2018). This means low nutrient conditions not only increase strigolactone levels but also sensitivity.

As mentioned previously, cytokinin levels are decreased under low nitrogen conditions but within 3 hours of nitrate fertilisation there was an increase in cytokinin biosynthesis gene expression followed by an increase in cytokinin levels (Xu et al., 2015). The fast decrease in strigolactone biosynthesis gene expression after nutrient addition (also within 3 hours) could also be a result of cytokinin inhibition of these genes as exogenous cytokinin treatments inhibited D10, D17 and D27 expression (Xu et al., 2015).

Strigolactones have been shown to regulate auxin transport and therefore local auxin levels (de Jong et al., 2014, Sun et al., 2018, Sun et al., 2015, Sun et al., 2014, Xu et al., 2015). Under low nitrogen or low phosphorus conditions auxin transport in rice and Arabidopsis was decreased (Sun et al., 2018, Sun et al., 2014, Xu et al., 2015). Using a split root setup, nitrogen fertilisation to one side of the roots reduced auxin in the roots in contact with the high N while phosphorus fertilisation had no effect on endogenous root auxin levels (Yoneyama et al., 2015b).

Auxin also interacts with nitric oxide. Under low nitrogen or low phosphorus conditions nitric oxide reduced auxin levels (Sun et al., 2018) via reduced polar auxin transport (Sun et al., 2018, Tognetti et al., 2017). The reduction in auxin in the seminal roots then increased root growth (Manoli et al., 2016, Sami et al., 2018, Sun et al., 2018). Similarly there is evidence for mutual inhibition between cytokinin and nitric oxide levels via post-transcriptional modifications (Feng et al., 2013, Lehotai et al., 2016, Liu et al., 2013). The timing of these interactions in response to changes in soil nutrient levels, and hence where these interactions sit on the proposed hormone pathway has yet to be confirmed.

This area of hormone networks is a growing field and the coming years will likely unravel further interactions between these and other hormones involved in environmental responses including ethylene, gibberellins and abscisic acid.

5 Using plant growth regulators to control root development for improved nutrient uptake

In the previous sections we have outlined some of the key hormone responses to nitrogen and phosphate availability. In this final section we explore how hormone treatments can control root development or improve nutrient uptake.

5.1 Manipulating auxin

Exogenous auxin can enhance lateral root growth and root hair density as well as suppressing primary root growth in Arabidopsis, parallel to what occurs in low phosphate conditions (Ham et al., 2018, Zhang et al., 2014). Under sufficient phosphate conditions, Indole 3-acetic acid (auxin) treatments increased the density and elongation of root hairs nearly threefold, but under low phosphate, when root hair density and length are already high, unsurprisingly less effect was seen (Niu et al., 2013).

Similarly, 1-Napthalenacetic acid (NAA; auxin) treatments increased maize lateral root density under both low and adequate phosphate (Zhang et al., 2019). There was also an increase in shoot biomass and although the phosphate concentration in the plant tissues was unchanged between control and NAA treatments, the total phosphate per plant increased with NAA treatments in low soil phosphate conditions (Zhang et al., 2019). Inhibiting auxin transport with NPA had the opposite effect (Zhang et al., 2019). In Arabidopsis under low phosphorus conditions, using a different auxin transport inhibitor, triiodobenzoic acid (TIBA), increased auxin levels and auxin sensitivity in the root meristem (Hasan et al., 2016, López-Bucio et al., 2002). This in turn repressed primary root elongation (Hasan et al., 2016, López-Bucio et al., 2002).

5.2 Manipulating cytokinin

Cytokinin has been widely used to improve crop responses to environmental stress by reducing shoot senescence. The leaf life cycle begins as a sink for nitrogen during growth and chlorophyll production, before becoming the primary carbon source for export to other organs. The onset of senescence marks a transition from carbon capture to nitrogen remobilisation (Thomas and Ougham, 2014). Functional stay-green crop varieties (sometimes associated with higher yields under stressful conditions) delay this transition which is tightly linked to cytokinin regulation of senescence (Thomas and Ougham, 2014).

Some examples of stay-green plants include constitutive signalling such as *ore12-1* gain of function Arabidopsis mutant with constitutive *AHK3* receptor (Thomas and Ougham, 2014) and ARR2 constitutive expression (type B response regulator) (Thomas and Ougham, 2014, review). Similarly, increased cytokinin levels via a chimeric transgene *AtMYB32xs-p::IPT* which increased IPT expression in wheat delayed senescence and increased yield under drought stress (Joshi et al., 2019). The Wennong stay-green wheat also has increased cytokinin levels (Luo et al., 2018). Exogenous applications of the synthetic cytokinin 6-benzyladenine also delayed leaf senescence (Gu et al., 2018).

Delaying senescence as described above has implications for grain filling and subsequent grain yield. Higher soil N before cereal crop heading can increase cytokinin in plant leaves and grains at the time of grain filling. This increases cell number, cell division activity in the grain, enhanced grain yield and nitrogen use efficiency. But shortage of soil N decreases cytokinin concentration leading to senescence (Gu et al., 2018). Cytokinin applications can overcome growth limiting effects caused by N stress in *Plantago major* (Kiba et al., 2011)and wheat (Cai et al., 2018). However delaying senescence too long (by excessive soil N, or cytokinin manipulation) can cause poor grain filling by inadequate nutrient remobilisation and unused carbohydrates in straw (Gu et al., 2018).

The interaction with water is also important. Alternate wetting and drying improved crop yield by increased endogenous leaf cytokinin concentrations after re-watering and increased xylem tZR levels (Gu et al., 2018). This enhanced photosynthesis and nitrate assimilation during N stress and increased the re-opening of the stomata as a result of positive water potential following the dry period.

As described previously, cytokinins both regulate root architecture and coordinate the expression of nutrient transport and uptake genes (Kieber and Schaller, 2018). Over expression of IPT, controlled by the promoter of a drought-inducible gene in rice, overcame the drought-caused inhibition of nitrate uptake and glutamine synthesis-dependent N assimilation (Xuan et al., 2017). When cytokinin was exogenously applied to plants, it triggered plant growth by increasing nitrogen uptake (Small and Degenhardt, 2018). Cytokinin also modulated plant responses to light, water and abiotic and biotic stress (Small and Degenhardt, 2018). Together these provide evidence that the interaction between cytokinin and nitrogen remobilisation can be important in crop breeding for improved efficiency in low nutrients and other stressful environmental conditions.

Manipulating cytokinin in plants may help to improve nutrient use efficiency and yield. However the complex interactions between different cytokinin types demonstrate the clear need for further genetic and pharmacological investigations.

6 Future trends in research

There is a growing need to improve fertiliser acquisition and utilisation for improved crop yield due to the negative environmental and economic impact from excessive use of synthetic nitrogen and phosphate fertilisers. Previously, the quest to understand how nutrients are captured by roots led to the discoveries of transporters implicated in nutrient uptake as well as internal maintenance of the plant nutrient status. However in order to reduce fertiliser requirements, improvements in uptake of organic (carbon containing) nutrient sources is an important aspect of crop breeding requiring future research. Only comparatively recently have transporters for organic nutrients (such as amino acids and peptides) been identified (Ganeteg et al., 2017, Nacry et al., 2013, Rentsch et al., 2007), and even less is known about how hormone-signalling pathways are involved.

Additionally, most studies on improving nutrient uptake have focussed on factors directly influencing membrane transport with less consideration given to hormone signalling, which integrates tissue nutrient-status, soil-nutrient status and nutrient assimilation and translocation via these transporters. Recently the NRT1 nitrate transporter was also identified as a sensor for soil nitrogen status integrating auxin and nutrient signals (Han and Hwang, 2018, Kant, 2018, Shahzad and Amtmann, 2017, Zhao et al., 2018b). This has important implications for improvements in root responses to soil-nutrient availability including optimised root growth patterns (regulated in part by auxin) and transporter levels. Future research identifying root molecular sensors for organic nitrogen and other macro and micro nutrients will provide new ways to improve nutrient efficient crops. These yet-to-be-identified sensors may also integrate hormone signals as is the case for the NRT1 protein and understanding the role of plant hormones in the functions, localisation and abundance of these transporters will be necessary to harness new improvement strategies. This is because these sensors and integrators of nutrient-hormone pathways will respond to whole plant nutrient status dynamically meaning unless whole plant responses are adjusted; it is likely that the plant will compensate via feedback pathways and redundancy. To this end, integrated studies linking the membrane transport proteins with whole plant hormone signalling of nutrient status (both of shoot and in the root sensing of the soil environment) are required.

At a more fundamental level, aspects of posttranslational regulation of hormone signalling are not fully understood. New research is beginning to demonstrate that gene expression alone is insufficient to make lasting improvements because of posttranslational changes (e.g. that described above for cytokinin and nitric oxide; (Feng et al., 2013, Lehotai et al., 2016, Liu et al., 2013). This means efforts should be put in place to broaden our understanding of how phosphorylation and proteolysis affects hormonal regulation of nutrient uptake especially during low external nutrient conditions.

As is true of past research, many of the above suggestions will be easier to first study in model species. However caution is required in making generalisations from dicot models to monocot crops, or from annual models to perennial crops. Thus, it is crucial to study manipulation of nutrient transporters by genetics (either traditional breeding or more targeted genetic modification) and pharmacological supply of hormones in the crop of interest. Novel techniques such as CRISPR are revolutionising our ability to manipulate crop genes, and in future these will prove valuable when GM crops become more widely accepted. In the meantime hormone treatments (such as auxin manipulators) are widely used in horticulture from improving root development on cuttings and as herbicides. This means better understanding of the quantity, ratios and timings of hormone applications in relation to plant age for improved nutrient uptake, together with targeted release options using nanotechnology options adapted from medical research, could have immediate applications for major food crop species.

Understanding nutrient dynamics within plants across their lifecycle is extremely important. As described in Section 5.2 when plants switch from growth to senescence (which coincides with grain filling in annual cereal crops) nutrient uptake is reduced and remobilisation from leaves begins. Delaying this switch such as via manipulation of cytokinins can improve crop survival in stressful conditions (as harnessed in current stay-green varieties; (Joshi et al., 2019, Liu et al., 2018)). Future research into hormone signalling driving nutrient remobilisation (potentially without delaying other aspects of senescence such as grain filling) could result in crop varieties with higher efficiency of internal nutrient reuse – beneficial to low nutrient conditions or when nutrients are less available such as during times of drought.

There is currently a push in some crops to switch from annual to perennial growth patterns. As mentioned above it is important to study nutrient uptake and dynamics across the life cycle. This is even less understood in perennial crops as the switch to seed/fruit production is followed by further vegetative growth. How this affects nutrient uptake and remobilisation is essential if new more efficient varieties are to be developed. In addition, crops are not limited to cereals. Important fruit and nut crops are perennial trees and have been propagated via cuttings and using shoots grafted onto rootstocks. Much of this work has been to minimise susceptibility to soil pathogens while maximising fruit-shoot ratios. However almost nothing is known about how the hormone signalling within the shoot affects the nutrient uptake of the root stocks. Since there are likely to be differences in hormone levels or ratios between the varieties used for shoot and root stocks, it is very important to understand how nutrient status signals in one affects the processes in the other.

The interaction between soil biota and roots is important for all of the examples provided above. Positive interactions with microbes and mycorrhizae can improve nutrient uptake while also influencing the hormone-signalling pathways, in turn affecting transporters and root architectural changes. Similarly, root exudates, used by mycorrhizae to detect roots (including strigolactones), are also co-opted by parasitic weeds to trigger seed germination which decreases crop productivity. Because of the negative implications for productivity, many places use practices to trigger parasitic seed germination prior to seeding in order to minimise the prevalence during crop growth. However these processes are also likely to have negative impacts on beneficial soil biota. Similarly the long-term breeding of crops in optimal nutrient conditions has reduced the need for beneficial biotic interactions to improve nutrient uptake. To improve nutrient efficiency while reducing signalling for negative interactions, a better and more integrated understanding of root-rhizosphere communities is required. This includes identifying beneficial species or communities for nutrient dynamics to understanding the signalling involved or resulting from these interactions.

7 Conclusion

In this chapter we have explored the role of hormones on nitrogen and phosphate uptake and responses. In particular we focussed on auxin, cytokinin and strigolactones in each nutrient condition and how they interact to control root architectural responses. The details of these hormone networks are still being explored, and aspects of the proposed network we present in Fig. 1 based on the latest studies still need to be confirmed. Similarly, other hormones involved in environmental responses such as gibberellins, ethylene and ABA are also important, but their role in this network is yet to be confirmed.

Perhaps the best example of hormone manipulation of nutrition to improve crops is cytokinin in stay-green varieties. As described above, increasing cytokinin (either genetically or with cytokinin treatments) delays the transition from carbon assimilation (and hence growth) to nitrogen remobilisation. As noted, delaying senescence too long can also have negative implications for grain filling. However the evidence of improvements in certain situations and in response to other environmental stresses highlights the potential advantages of using nutrient-related hormone signalling pathways for crop improvement.

Despite the urgent need for improved crops, we have demonstrated here that there is a lot of scope for progress to happen via improved understanding of hormone regulation of nutrient dynamics. The real-world impact could in part be realised via genetic modification of appropriate pathways, while targeted pharmaceutical applications at appropriate plant developmental stages may provide immediate benefits once understood. The most important way forward is for trans-disciplinary research. This includes integrated molecular and physiological understanding of signals and processes from the soil to the plant and of nutrient status of the shoot back to the root. Similarly it requires adaptation of medical technologies such as nanocarriers used to deliver targeted treatments in specific tissues, for use of targeted delivery of hormones to specific locations at the right moment.

In summary, hormone control of nutrient dynamics is a key area of future crop advancements with new molecular and physiological techniques making integrated studies and manipulation of whole plant nutrient dynamics in crops an exciting reality.

8 Where to look for further information

8.1 Recent nitrogen reviews

- Xuan, W., Beeckman, T. and Xu, G. (2017). Plant nitrogen nutrition: sensing and signaling, *Current Opinion in Plant Biology*. 39; 57–65.

- O'Brien, José A., Vega, A., Bouguyon, E., Krouk, G., Gojon, A., Coruzzi, G. and Gutiérrez, Rodrigo A. (2016). Nitrate transport, sensing, and responses in plants, *Molecular Plant*. 9; 837–856.
- Kiba, T. and Krapp, A. (2016). Plant nitrogen acquisition under low availability: regulation of uptake and root architecture. *Plant and Cell Physiology*. 57; 707–714.

8.2 Recent phosphate reviews

- Crombez, H., Motte, H. and Beeckman, T. (2019). Tackling plant phosphate starvation by the roots, *Developmental Cell*. 48; 599–615.
- Ham, B.-K., Chen, J., Yan, Y. and Lucas, W. J. (2018). Insights into plant phosphate sensing and signaling, *Current Opinion in Biotechnology*. 49; 1–9.

8.3 Recent auxin reviews

- Blakeslee, J. J., Spatola Rossi, T. and Kriechbaumer, V. (2019). Auxin biosynthesis: spatial regulation and adaptation to stress, *Journal of Experimental Botany*. 70; 5041.
- Han, S. and Hwang, I. (2018). Integration of multiple signaling pathways shapes the auxin response, *Journal of Experimental Botany*. 69; 189–200.
- Lavy, M. and Estelle, M. (2016). Mechanisms of auxin signaling, *Development (Cambridge, England)*. 143; 3226.

8.4 Recent cytokinin reviews

- Gu, J., Li, Z., Mao, Y., Struik, P., Zhang, H., Liu, L., Wang, Z. and Yang, J. (2018), Roles of nitrogen and cytokinin signals in root and shoot communications in maximising of plant productivity and their agronomic applications, *Plant Science*. 274; 320–331.
- Kieber, J. J. and Schaller, G. E. (2018), Cytokinin signaling in plant development, *Development*. 145; 1–7, dev149344.

8.5 Recent strigolactone review

- Machin, D. C., Hamon-Josse, M. and Bennett, T. (2020). Fellowship of the rings: a saga of strigolactones and other small signals, *New Phytologist*. 225; 621–636.

9 References

Akiyama, K. (2014). Chemical identification and functional analysis of apocarotenoids involved in the development of arbuscular mycorrhizal symbiosis, *Bioscience, Biotechnology, and Biochemistry* 71(6); 1405–1414.

Alder, A., Jamil, M., Marzorati, M., Bruno, M., Vermathen, M., Bigler, P., Ghisla, S., Bouwmeester, H., Beyer, P. and Al-Babili, S. (2012). The path from β-carotene to carlactone, a strigolactone-like plant hormone, *Science (New York, N.Y.)* 335(6074); 1348–1351.

Arite, T., Kameoka, H. and Kyozuka, J. (2012). Strigolactone positively controls crown root elongation in rice, *Journal of Plant Growth Regulation* 31(2); 165–172.

Armijo, G. and Gutiérrez, R. A. (2017). Emerging players in the nitrate signaling pathway, *Molecular Plant* 10(8); 1019–1022.

Blakeslee, J. J., Spatola Rossi, T. and Kriechbaumer, V. (2019). Auxin biosynthesis: spatial regulation and adaptation to stress, *Journal of Experimental Botany* 70(19); 5041–5049.

Bouguyon, E., Brun, F., Meynard, D., Kubeš, M., Pervent, M., Leran, S., Lacombe, B., Krouk, G., Guiderdoni, E., Zažímalová, E., Hoyerová, K., Nacry, P. and Gojon, A. (2015). Multiple mechanisms of nitrate sensing by Arabidopsis nitrate transceptor NRT1.1, *Nature Plants* 1; 15015.

Bouguyon, E., Perrine-Walker, F., Pervent, M., Rochette, J., Cuesta, C., Benkova, E., Martinière, A., Bach, L., Krouk, G., Gojon, A. and Nacry, P. (2016). Nitrate controls root development through posttranscriptional regulation of the NRT1.1/NPF6.3 transporter/sensor, *Plant Physiology* 172(2); 1237–1248.

Brackin, R., Näsholm, T., Robinson, N., Guillou, S., Vinall, K., Lakshmanan, P., Schmidt, S. and Inselsbacher, E. (2015). Nitrogen fluxes at the root-soil interface show a mismatch of nitrogen fertilizer supply and sugarcane root uptake capacity, *Scientific Reports* 5; 15727.

Cai, T., Meng, X., Liu, X., Liu, T., Wang, H., Jia, Z., Yang, D. and Ren, X. (2018). Exogenous hormonal application regulates the occurrence of wheat tillers by changing endogenous hormones, *Frontiers in Plant Science* 9; 1886.

Cao, X., Zhu, C., Zhong, C., Hussain, S., Zhu, L., Wu, L. and Jin, Q. (2018). Mixed-nitrogen nutrition-mediated enhancement of drought tolerance of rice seedlings associated with photosynthesis, hormone balance and carbohydrate partitioning, *Plant Growth Regulation* 84(3); 451–465.

Cheng, L., Bucciarelli, B., Shen, J., Allan, D. and Vance, C. P. (2011). Update on lupin cluster roots. Update on white lupin cluster root acclimation to phosphorus deficiency, *Plant Physiology* 156(3); 1025–1032.

Chevalier, F., Pata, M., Nacry, P., Doumas, P. and Rossignol, M. (2003). Effects of phosphate availability on the root system architecture: large-scale analysis of the natural variation between Arabidopsis accessions, *Plant, Cell and Environment* 26(11); 1839–1850.

Crawford, S., Shinohara, N., Sieberer, T., Williamson, L., George, G., Hepworth, J., Müller, D., Domagalska, M. A. and Leyser, O. (2010). Strigolactones enhance competition between shoot branches by dampening auxin transport, *Development (Cambridge, England)* 137(17); 2905–2913.

Crombez, H., Motte, H. and Beeckman, T. (2019). Tackling plant phosphate starvation by the roots, *Developmental Cell* 48(5); 599–615.

Danilova, M. N., Doroshenko, A. S., Zabrodin, D. A., Kudryakova, N. V., Oelmüller, R. and Kusnetsov, V. V. (2017). Cytokinin membrane receptors modulate transcript accumulation of plastid encoded genes, *Russian Journal of Plant Physiology* 64(3); 301-309.

De Jong, M., George, G., Ongaro, V., Williamson, L., Willetts, B., Ljung, K., Mcculloch, H. and Leyser, O. (2014). Auxin and strigolactone signaling are required for modulation of Arabidopsis shoot branching by nitrogen supply, *Plant Physiology* 166(1); 384-395.

Dechorgnat, J., Francis, K. L., Dhugga, K. S., Rafalski, J. A., Tyerman, S. D. and Kaiser, B. N. (2018). Root ideotype influences nitrogen transport and assimilation in maize, *Frontiers in Plant Science* 9; 531.

Fang Zhu, X., Sheng Zhao, X., Wu, Q. and Fang Shen, R. (2018). Abscisic acid is involved in root cell wall phosphorus remobilization independent of nitric oxide and ethylene in rice (Oryza sativa), *Annals of Botany* 121(7); 1361-1368.

FAO (2016). *The State of Food and Agriculture: Climate Change Agriculture and Food Security*. http://www.fao.org/3/a-i6030e.pdf.

Feng, J., Wang, C., Chen, Q., Chen, H., Ren, B., Li, X. and Zuo, J. (2013). S-nitrosylation of phosphotransfer proteins represses cytokinin signalling, *Nature Communications* 4; 1529.

Ganeteg, U., Ahmad, I., Jämtgård, S., Aguetoni-Cambui, C., Inselsbacher, E., Svennerstam, H., Schmidt, S. and Näsholm, T. (2017). Amino acid transporter mutants of Arabidopsis provides evidence that a non-mycorrhizal plant acquires organic nitrogen from agricultural soil, *Plant, Cell and Environment* 40(3); 413-423.

Garnett, T., Conn, V., Plett, D., Conn, S., Zanghellini, J., Mackenzie, N., Enju, A., Francis, K., Holtham, L., Roessner, U., Boughton, B., Bacic, A., Shirley, N., Rafalski, A., Dhugga, K., Tester, M. and Kaiser, B. N. (2013). The response of the maize nitrate transport system to nitrogen demand and supply across the lifecycle, *New Phytologist* 198(1); 82-94.

Garnett, T., Plett, D., Conn, V., Conn, S., Rabie, H., Rafalski, J. A., Dhugga, K., Tester, M. A. and Kaiser, B. N. (2015). Variation for N uptake system in maize: genotypic response to N supply, *Frontiers in Plant Science* 6; 936.

Gomez-Roldan, V., Fermas, S., Brewer, P. B., Puech-Pagès, V., Dun, E. A., Pillot, J. P., Letisse, F., Matusova, R., Danoun, S., Portais, J. C., Bouwmeester, H., Bécard, G., Beveridge, C. A., Rameau, C. and Rochange, S. F. (2008). Strigolactone inhibition of shoot branching, *Nature* 455(7210); 189-194.

Gruber, B. D., Giehl, R. F. H., Friedel, S. and Von Wirén, N. (2013). Plasticity of the Arabidopsis root system under nutrient deficiencies, *Plant Physiology* 163(1); 161-179.

Gu, J., Li, Z., Mao, Y., Struik, P. C., Zhang, H., Liu, L., Wang, Z. and Yang, J. (2018). Roles of nitrogen and cytokinin signals in root and shoot communications in maximizing of plant productivity and their agronomic applications, *Plant Science* 274; 320-331.

Gu, R., Duan, F., An, X., Zhang, F., Von Wirén, N. and Yuan, L. (2013). Characterization of AMT-mediated high-affinity ammonium uptake in roots of maize (*Zea mays* L.), *Plant and Cell Physiology* 54(9); 1515-1524.

Ham, B. K., Chen, J., Yan, Y. and Lucas, W. J. (2018). Insights into plant phosphate sensing and signaling, *Current Opinion in Biotechnology* 49; 1-9.

Han, S. and Hwang, I. (2018). Integration of multiple signaling pathways shapes the auxin response, *Journal of Experimental Botany* 69(2); 189-200.

Hasan, M. M., Hasan, M. M., Teixeira Da Silva, J. A. and Li, X. (2016). Regulation of phosphorus uptake and utilization: transitioning from current knowledge to practical strategies, *Cellular and Molecular Biology Letters* 21; 7.

Hawkesford, M., Horst, W., Kichey, T., Lambers, H., Schjoerring, J., Moller, I. S. and White, P. (2012). Functions of micronutrients. In: *Mineral Nutrition of Higher Plants* Marschiner, P. E. (Ed.). Academic Press.

Ito, S., Yamagami, D., Umehara, M., Hanada, A., Yoshida, S., Sasaki, Y., Yajima, S., Kyozuka, J., Ueguchi-Tanaka, M., Matsuoka, M., Shirasu, K., Yamaguchi, S. and Asami, T. (2017). Regulation of strigolactone biosynthesis by gibberellin signaling, *Plant Physiology* 174(2); 1250–1259.

Jamil, M., Charnikhova, T., Cardoso, C., Jamil, T., Ueno, K., Verstappen, F., Asami, T. and Bouwmeester, H. J. (2011). Quantification of the relationship between strigolactones and Striga hermonthica infection in rice under varying levels of nitrogen and phosphorus, *Weed Research* 51(4); 373–385.

Joshi, S., Choukimath, A., Isenegger, D., Panozzo, J., Spangenberg, G. and Kant, S. (2019). Improved wheat growth and yield by delayed leaf senescence using developmentally regulated expression of a cytokinin biosynthesis gene, *Frontiers in Plant Science* 10; 1285.

Kang, J., Lee, Y., Sakakibara, H. and Martinoia, E. (2017). Cytokinin transporters: GO and STOP in signaling, *Trends in Plant Science* 22(6); 455–461.

Kant, S. (2018). Understanding nitrate uptake, signaling and remobilisation for improving plant nitrogen use efficiency, *Seminars in Cell and Developmental Biology* 74; 89–96.

Kellermeier, F., Armengaud, P., Seditas, T. J., Danku, J., Salt, D. E. and Amtmann, A. (2014). Analysis of the root system architecture of Arabidopsis provides a quantitative readout of crosstalk between nutritional signals, *The Plant Cell* 26(4); 1480–1496.

Kiba, T., Kudo, T., Kojima, M. and Sakakibara, H. (2011). Hormonal control of nitrogen acquisition: roles of auxin, abscisic acid, and cytokinin, *Journal of Experimental Botany* 62(4); 1399–1409.

Kiba, T., Takebayashi, Y., Kojima, M. and Sakakibara, H. (2019). Sugar-induced de novo cytokinin biosynthesis contributes to Arabidopsis growth under elevated CO, *Scientific Reports* 9(1); 7765.

Kieber, J. J. and Schaller, G. E. (2018). Cytokinin signaling in plant development, *Development* 145(4); dev149344.

Krouk, G. (2016). Hormones and nitrate: a two-way connection, *Plant Molecular Biology* 91(6); 599–606.

Landrein, B., Formosa-Jordan, P., Malivert, A., Schuster, C., Melnyk, C. W., Yang, W., Turnbull, C., Meyerowitz, E. M., Locke, J. C. W. and Jönsson, H. (2018). Nitrate modulates stem cell dynamics in Arabidopsis shoot meristems through cytokinins, *Proceedings of the National Academy of Sciences of the United States of America* 115(6); 1382–1387.

Lavy, M. and Estelle, M. (2016). Mechanisms of auxin signaling, *Development (Cambridge, England)* 143(18); 3226–3229.

Lehotai, N., Feigl, G., Koós, Á., Molnár, Á., Ördög, A., Pető, A., Erdei, L. and Kolbert, Z. (2016). Nitric oxide-cytokinin interplay influences selenite sensitivity in Arabidopsis, *Plant Cell Reports* 35(10); 2181–2195.

Leyser, O. (2018). Auxin signaling, *Plant Physiology* 176(1); 465–479.

Linkohr, B. I., Williamson, L. C., Fitter, A. H. and Leyser, H. M. O. (2002). Nitrate and phosphate availability and distribution have different effects on root system architecture of Arabidopsis, *Plant Journal* 29(6); 751–760.

Liu, C.-J., Zhao, Y. and Zhang, K. (2019). Cytokinin transporters: multisite players in cytokinin homeostasis and signal distribution, *Frontiers in Plant Science* 10, 693.

Liu, H., Ren, X., Zhu, J., Wu, X. and Liang, C. (2018). Effect of exogenous abscisic acid on morphology, growth and nutrient uptake of rice (Oryza sativa) roots under simulated acid rain stress, *Planta* 248(3); 647–659.

Liu, W.-Z., Kong, D.-D., Gu, X.-X., Gao, H.-B., Wang, J.-Z., Xia, M., Gao, Q., Tian, L.-L., Xu, Z.-H., Bao, F., Hu, Y., Ye, N.-S., Pei, Z.-M. and He, Y.-K. (2013). Cytokinins can act as suppressors of nitric oxide in Arabidopsis, *Proceedings of the National Academy of Sciences* 110(4); 1548–1553.

Liu, Y. and Von Wirén, N. (2017). Ammonium as a signal for physiological and morphological responses in plants, *Journal of Experimental Botany* 68(10); 2581–2592.

López-Bucio, J., Hernández-Abreu, E., Sánchez-Calderón, L., Nieto-Jacobo, M. F., Simpson, J. and Herrera-Estrella, L. (2002). Phosphate availability alters architecture and causes changes in hormone sensitivity in the Arabidopsis root system, *Plant Physiology* 129(1); 244–256.

López-Ráez, J. A., Charnikhova, T., Gómez-Roldán, V., Matusova, R., Kohlen, W., De Vos, R., Verstappen, F., Puech-Pages, V., Bécard, G., Mulder, P. and Bouwmeester, H. (2008). Tomato strigolactones are derived from carotenoids and their biosynthesis is promoted by phosphate starvation, *New Phytologist* 178(4); 863–874.

Luo, Y., Tang, Y., Zhang, X., Li, W., Chang, Y., Pang, D., Xu, X., Li, Y. and Wang, Z. (2018). Interactions between cytokinin and nitrogen contribute to grain mass in wheat cultivars by regulating the flag leaf senescence process, *The Crop Journal* 6(5); 538–551.

Lupini, A., Mercati, F., Araniti, F., Miller, A. J., Sunseri, F. and Abenavoli, M. R. (2016). NAR2.1/NRT2.1 functional interaction with NO_3^- and H^+ fluxes in high-affinity nitrate transport in maize root regions, *Plant Physiology and Biochemistry*, 102; 107–114.

Lv, S., Zhang, Y., Li, C., Liu, Z., Yang, N., Pan, L., Wu, J., Wang, J., Yang, J., Lv, Y., Zhang, Y., Jiang, W., She, X. and Wang, G. (2018). Strigolactone-triggered stomatal closure requires hydrogen peroxide synthesis and nitric oxide production in an abscisic acid-independent manner, *New Phytologist* 217(1); 290–304.

Machin, D. C., Hamon-Josse, M. and Bennett, T. (2020). Fellowship of the rings: a saga of strigolactones and other small signals, *New Phytologist* 225(2); 621–636.

Manoli, A., Trevisan, S., Voigt, B., Yokawa, K., Baluška, F. and Quaggiotti, S. (2016). Nitric oxide-mediated maize root apex responses to nitrate are regulated by auxin and strigolactones, *Frontiers in Plant Science* 6; 1269.

Matusova, R., Rani, K., Verstappen, F. W. A., Franssen, M. C. R., Beale, M. H. and Bouwmeester, H. J. (2005). The strigolactone germination stimulants of the plant-parasitic Striga and Orobanche spp. are derived from the carotenoid pathway, *Plant Physiology* 139(2); 920–934.

Mayzlish-Gati, E., De-Cuyper, C., Goormachtig, S., Beeckman, T., Vuylsteke, M., Brewer, P. B., Beveridge, C. A., Yermiyahu, U., Kaplan, Y., Enzer, Y., Wininger, S., Resnick, N., Cohen, M., Kapulnik, Y. and Koltai, H. (2012). Strigolactones are involved in root response to low phosphate conditions in Arabidopsis, *Plant Physiology* 160(3); 1329–1341.

Miller, A. J. and Chapman, N. (2011). *Transporters Involved in Nitrogen Uptake and Movement* Hawkesford, M. and Barraclough, B. (Eds). Chichester: John Wiley & Sons.

Motte, H. and Beeckman, T. (2019). The evolution of root branching: increasing the level of plasticity, *Journal of Experimental Botany* 70(3); 785–793.

Nacry, P., Bouguyon, E. and Gojon, A. (2013). Nitrogen acquisition by roots: physiological and developmental mechanisms ensuring plant adaptation to a fluctuating resource, *Plant and Soil* 370(1–2); 1–29.

Nath, M. and Tuteja, N. (2016). NPKS uptake, sensing, and signaling and miRNAs in plant nutrient stress, *Protoplasma* 253(3); 767–786.

Niu, Y. F., Chai, R. S., Jin, G. L., Wang, H., Tang, C. X. and Zhang, Y. S. (2013). Responses of root architecture development to low phosphorus availability: a review, *Annals of Botany* 112(2); 391–408.

O'Brien, J. A., Vega, A., Bouguyon, E., Krouk, G., Gojon, A., Coruzzi, G. and Gutiérrez, R. A. (2016). Nitrate transport, sensing, and responses in plants, *Molecular Plant* 9(6); 837–856.

Osugi, A. and Sakakibara, H. (2015). Q&A: how do plants respond to cytokinins and what is their importance?, *BMC Biology* 13; 102.

Pavlů, J., Novák, J., Koukalová, V., Luklová, M., Brzobohatý, B. and Černý, M. (2018). Cytokinin at the crossroads of abiotic stress signalling pathways, *International Journal of Molecular Sciences* 19(8); 2450.

Péret, B., Desnos, T., Jost, R., Kanno, S., Berkowitz, O. and Nussaume, L. (2014). Root architecture responses: in search of phosphate, *Plant Physiology* 166(4); 1713–1723.

Plett, D. C., Holtham, L. R., Okamoto, M. and Garnett, T. P. (2018). Nitrate uptake and its regulation in relation to improving nitrogen use efficiency in cereals, *Seminars in Cell and Developmental Biology* 74; 97–104.

Poitout, A., Crabos, A., Petřík, I., Novák, O., Krouk, G., Lacombe, B. and Ruffel, S. (2018). Responses to systemic nitrogen signaling in Arabidopsis roots involve-zeatin in shoots, *The Plant Cell* 30(6); 1243–1257.

Rahul, B., Jitender, G., Bipin, K. P., Ricardo, F. H. G., Anja, H., Richard, T., Jekaterina, T., Nicola, L., Meredith, H., Kamal, S., Afaf, R., Ute, V., Jose, A., Anna, S., Jeonga, Y., Karin, L., Kathleen, M. B., Jonathan, P. L., Liam, D., Teva, V., Anthony, B., Darren, W., Nicolaus, W. v., Malcolm, J. B. and Ranjan, S. (2018), A mechanistic framework for auxin dependent Arabidopsis root hair elongation to low external phosphate, *Nature Communications* 9; 1–9.

Rasmussen, A., Hosseini, S. A., Hajirezaei, M. R., Druege, U. and Geelen, D. (2015). Adventitious rooting declines with the vegetative to reproductive switch and involves a changed auxin homeostasis, *Journal of Experimental Botany* 66(5); 1437–1452.

Rasmussen, A., Mason, M. G., De Cuyper, C., Brewer, P. B., Herold, S., Agusti, J., Geelen, D., Greb, T., Goormachtig, S., Beeckman, T. and Beveridge, C. A. (2012). Strigolactones suppress adventitious rooting in Arabidopsis and pea, *Plant Physiology* 158(4); 1976–1987.

Rentsch, D., Schmidt, S. and Tegeder, M. (2007). Transporters for uptake and allocation of organic nitrogen compounds in plants, *FEBS Letters* 581(12); 2281–2289.

Roy, S. (2018). Nitrate ahoy! Shoot cytokinin signals integrate growth responses with nitrogen availability, *The Plant Cell* 30(6); 1169–1170.

Sakakibara, H., Takei, K. and Hirose, N. (2006). Interactions between nitrogen and cytokinin in the regulation of metabolism and development, *Trends in Plant Science* 11(9); 440–448.

Sakuraba, Y., Kanno, S., Mabuchi, A., Monda, K., Iba, K. and Yanagisawa, S. (2018). A phytochrome-B-mediated regulatory mechanism of phosphorus acquisition, *Nature Plants* 4(12); 1089–1101.

Sami, F., Faizan, M., Faraz, A., Siddiqui, H., Yusuf, M. and Hayat, S. (2018). Nitric oxide-mediated integrative alterations in plant metabolism to confer abiotic stress tolerance, NO crosstalk with phytohormones and NO-mediated post translational modifications in modulating diverse plant stress, *Nitric Oxide* 73; 22–38.

Shahzad, Z. and Amtmann, A. (2017). Food for thought: how nutrients regulate root system architecture, *Current Opinion in Plant Biology* 39; 80–87.

Shen, C., Wang, S., Zhang, S., Xu, Y., Qian, Q., Qi, Y. and Jiang, de A. (2013). OsARF16, a transcription factor, is required for auxin and phosphate starvation response in rice (*Oryza sativa* L.), *Plant, Cell and Environment* 36(3); 607–620.

Silva-Navas, J., Conesa, C. M., Saez, A., Navarro-Neila, S., Garcia-Mina, J. M., Zamarreño, A. M., Baigorri, R., Swarup, R. and Pozo, J. C. (2019). Role of cis-zeatin in root responses to phosphate starvation, *New Phytologist* 224(1); 242–257.

Small, C. C. and Degenhardt, D. (2018). Plant growth regulators for enhancing revegetation success in reclamation: a review, *Ecological Engineering* 118; 43–51.

Steffens, B. and Rasmussen, A. (2016). The physiology of adventitious roots, *Plant Physiology* 170(2); 603–617.

Sun, H., Bi, Y., Tao, J., Huang, S., Hou, M., Xue, R., Liang, Z., Gu, P., Yoneyama, K., Xie, X., Shen, Q., Xu, G. and Zhang, Y. (2016). Strigolactones are required for nitric oxide to induce root elongation in response to nitrogen and phosphate deficiencies in rice, *Plant, Cell and Environment* 39(7); 1473–1484.

Sun, H., Feng, F., Liu, J. and Zhao, Q. (2017). The interaction between auxin and nitric oxide regulates root growth in response to iron deficiency in rice, *Frontiers in Plant Science* 8; 2169.

Sun, H., Feng, F., Liu, J. and Zhao, Q. (2018). Nitric oxide affects rice root growth by regulating auxin transport under nitrate supply, *Frontiers in Plant Science* 9; 659.

Sun, H., Li, J., Song, W., Tao, J., Huang, S., Chen, S., Hou, M., Xu, G. and Zhang, Y. (2015). Nitric oxide generated by nitrate reductase increases nitrogen uptake capacity by inducing lateral root formation and inorganic nitrogen uptake under partial nitrate nutrition in rice, *Journal of Experimental Botany* 66(9); 2449–2459.

Sun, H., Tao, J., Liu, S., Huang, S., Chen, S., Xie, X., Yoneyama, K., Zhang, Y. and Xu, G. (2014). Strigolactones are involved in phosphate- and nitrate-deficiency-induced root development and auxin transport in rice, *Journal of Experimental Botany* 65(22); 6735–6746.

Thomas, H. and Ougham, H. (2014). The stay-green trait, *Journal of Experimental Botany* 65(14); 3889–3900.

Tognetti, V. B., Bielach, A. and Hrtyan, M. (2017). Redox regulation at the site of primary growth: auxin, cytokinin and ROS crosstalk, *Plant, Cell and Environment* 40(11); 2586–2605.

Trabelsi, I., Yoneyama, K., Abbes, Z., Amri, M., Xie, X., Kisugi, T., Kim, H. I., Kharrat, M. and Yoneyama, K. (2017). Characterization of strigolactones produced by Orobanche foetida and Orobanche crenata resistant faba bean (*Vicia faba* L.) genotypes and effects of phosphorous, nitrogen, and potassium deficiencies on strigolactone production, *South African Journal of Botany* 108; 15–22.

Ueda, Y., Konishi, M. and Yanagisawa, S. (2017). Molecular basis of the nitrogen response in plants, *Soil Science and Plant Nutrition* 63(4); 329–341.

Umehara, M., Hanada, A., Yoshida, S., Akiyama, K., Arite, T., Takeda-Kamiya, N., Magome, H., Kamiya, Y., Shirasu, K., Yoneyama, K., Kyozuka, J. and Yamaguchi, S. (2008). Inhibition of shoot branching by new terpenoid plant hormones, *Nature* 455(7210); 195–200.

Veselov, S. Y., Timergalina, L. N., Akhiyarova, G. R., Kudoyarova, G. R., Korobova, A. V., Ivanov, I., Arkhipova, T. N. and Prinsen, E. (2018). Study of cytokinin transport from shoots to roots of wheat plants is informed by a novel method of differential localization of free cytokinin bases or their ribosylated forms by means of their specific fixation, *Protoplasma* 255(5); 1581-1594.

Wang, P., Wang, Z., Pan, Q., Sun, X., Chen, H., Chen, F., Yuan, L. and Mi, G. (2019). Increased biomass accumulation in maize grown in mixed nitrogen supply is mediated by auxin synthesis, *Journal of Experimental Botany* 70(6); 1859-1873.

Wang, S., Zhang, S., Sun, C., Xu, Y., Chen, Y., Yu, C., Qian, Q., Jiang, D. A. and Qi, Y. (2014). Auxin response factor (Os ARF 12), a novel regulator for phosphate homeostasis in rice (Oryza sativa), *New Phytologist* 201(1); 91-103.

Xie, X., Yoneyama, K., Kusumoto, D., Yamada, Y., Takeuchi, Y., Sugimoto, Y. and Yoneyama, K. (2008). Sorgomol, germination stimulant for root parasitic plants, produced by sorghum bicolor, *Tetrahedron Letters*. 49(13); 2066-2068.

Xu, J., Zha, M., Li, Y., Ding, Y., Chen, L., Ding, C. and Wang, S. (2015). The interaction between nitrogen availability and auxin, cytokinin, and strigolactone in the control of shoot branching in rice (*Oryza sativa* L.), *Plant Cell Reports* 34(9); 1647-1662.

Xuan, W., Beeckman, T. and Xu, G. (2017). Plant nitrogen nutrition: sensing and signaling, *Current Opinion in Plant Biology* 39; 57-65.

Yoneyama, K., Arakawa, R., Ishimoto, K., Kim, H. I., Kisugi, T., Xie, X., Nomura, T., Kanampiu, F., Yokota, T., Ezawa, T. and Yoneyama, K. (2015a). Difference in Striga-susceptibility is reflected in strigolactone secretion profile, but not in compatibility and host preference in arbuscular mycorrhizal symbiosis in two maize cultivars, *New Phytologist* 206(3); 983-989.

Yoneyama, K., Kisugi, T., Xie, X., Arakawa, R., Ezawa, T., Nomura, T. and Yoneyama, K. (2015b). Shoot-derived signals other than auxin are involved in systemic regulation of strigolactone production in roots, *Planta* 241(3); 687-698.

Yoneyama, K., Xie, X., Kim, H. I., Kisugi, T., Nomura, T., Sekimoto, H., Yokota, T. and Yoneyama, K. (2012). How do nitrogen and phosphorus deficiencies affect strigolactone production and exudation?, *Planta* 235(6); 1197-1207.

Yoneyama, K., Xie, X., Kisugi, T., Nomura, T. and Yoneyama, K. (2013). Nitrogen and phosphorus fertilization negatively affects strigolactone production and exudation in sorghum, *Planta* 238(5); 885-894.

Yoneyama, K., Yoneyama, K., Takeuchi, Y. and Sekimoto, H. (2007). Phosphorus deficiency in red clover promotes exudation of orobanchol, the signal for mycorrhizal symbionts and germination stimulant for root parasites, *Planta* 225(4); 1031-1038.

Yu, P., Gutjahr, C., Li, C. and Hochholdinger, F. (2016). Genetic control of lateral root formation in cereals, *Trends in Plant Science* 21(11); 951-961.

Yu, P., Hochholdinger, F. and Li, C. (2019). Plasticity of lateral root branching in maize, *Frontiers in Plant Science* 10; 363.

Zhang, X., Wang, B., Zhao, Y., Zhang, J. and Li, Z. (2019). Auxin and GA signaling play important roles in the maize response to phosphate deficiency, *Plant Science* 283; 177-188.

Zhang, Z., Liao, H. and Lucas, W. J. (2014). Molecular mechanisms underlying phosphate sensing, signaling, and adaptation in plants, *Journal of Integrative Plant Biology* 56(3); 192-220.

Zhao, L., Liu, F., Crawford, N. M. and Wang, Y. (2018a). Molecular regulation of nitrate responses in plants, *International Journal of Molecular Sciences* 19(7); 2039-2057.

Zhao, Y., Liu, Z., Duan, F., An, X., Liu, X., Hao, D., Gu, R., Wang, Z., Chen, F. and Yuan, L. (2018b). Overexpression of the maize ZmAMT1;1a gene enhances root ammonium uptake efficiency under low ammonium nutrition, *Plant Biotechnology Reports* 12(1); 47-56.

Chapter 4

Advances in understanding plant root anatomy and nutrient acquisition

Dimitris L. Bouranis and Styliani N. Chorianopoulou, Agricultural University of Athens, Greece

1 Introduction

The main function of roots is water and nutrient uptake, which is highly dependent on anatomical structures, growth conditions and plant age, while nutrient availability influences root architectural considerations (Gregory 2011; Hawkesford 2011; Sparks and Benfey 2017).

This chapter deals with the adaptations of root anatomical traits and their impact on nutrient acquisition by the adapted root anatomy (Fig. 1). To this end, a brief overview of the root anatomical traits and nutrient acquisition mechanisms is provided and the environmental issues affecting nutrient acquisition are discussed.

2 Root anatomical traits

The structural features of root systems are root system architecture, root morphology and root anatomy. Root system architecture refers to the spatial arrangement of a root system within the growth substrate, root morphology to

http://dx.doi.org/10.19103/AS.2020.0075.05

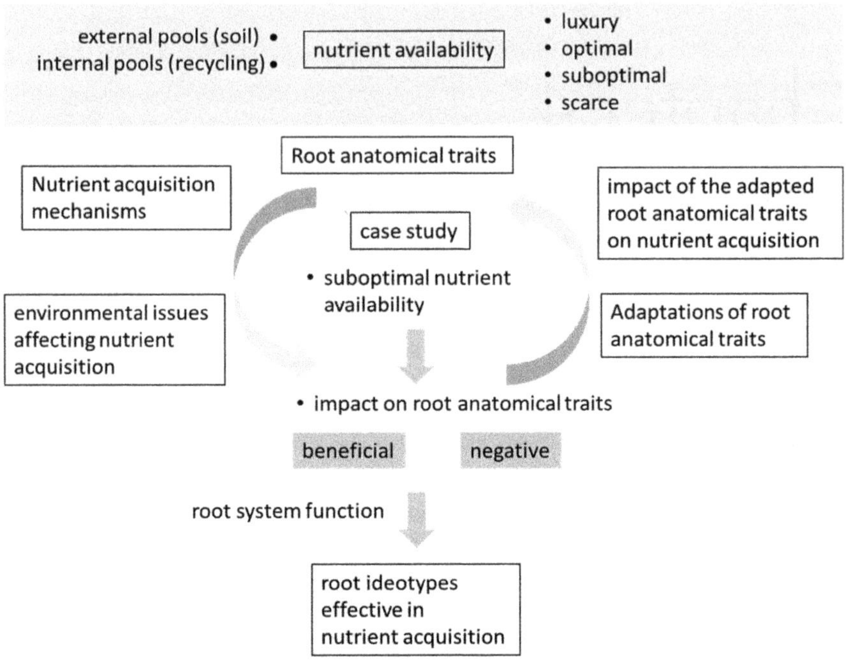

Figure 1 The concept of the chapter.

the surface features of the root (e.g. root hairs) and root anatomy to the internal cellular organization and structure.

A cross section of a root axis consists of the central area of the stele (or central cylinder) and the area of the cortex (or cortical area) that embraces the stele. The stele contains the vascular bundles consisting of the xylem and the phloem. The cortex consists of a cell layer, that of the endodermis, the structure of which forms a barrier to the radial flow of water and nutrients, and several layers of parenchymatous cortical cells, the cortical parenchyma, which is divided into three concentric regions: (1) the inner cortical region close to endodermis; (2) the middle of the cortex or mesodermis; and (3) the epidermis plus two or three layers beneath it (the hypodermis) forming the outer cortical region or exodermis.

In a cross section, the living cortical area (LCA) comprises several living parenchymatous cells arranged in concentric files of parenchyma cells, the number of which is referred to as the cortical cell file number (CCFN). Cortical cell files are formed by several successive asymmetrical periclinal divisions in the root apical meristem. The number of such periclinal divisions varies among species, genotypes, and root types, and generates differences in CCFN whose variation along the longitudinal axis of a root represents the radial patterning in the root apical meristem.

The cortical cell size (CCS) also varies substantially both among and within species. In a given species and tissue, CCS is the result of the coordinated control of cell division and cell growth, both of which are under genetic control. CCS increases due to cytoplasmic growth (the net accumulation of macromolecules and cellular organelles) and cell expansion (the increase in cell volume caused by the enlargement of the vacuole). The LCA is determined by CCLN between the epidermis and the endodermis as well as CCS.

If a lateral root is present in a cross section of the root axis, this comprises a new developing or developed root axis, perpendicular to the parental one, that crosses the area from the endodermis outward and is surrounded by the aforementioned cortical regions (York et al. 2013; Chimungu and Lynch 2014; Chimungu et al. 2014a,b, 2015a,b).

3 The nutrient acquisition mechanisms

The main function of the root is to acquire nutrients along with water. As a nutrient is a solute, water is needed for nutrient capture by roots. On their path from the root periphery to its center, water and solutes can be radially transported through three different pathways: (1) the apoplastic, (2) the symplastic and (3) the trans-cellular pathway. As solutes can be beneficial or harmful to the plant, a plant must have the ability to control its internal environment. This is linked to the selective uptake of necessary nutrients and exclusion of potentially harmful phytotoxic compounds. Selectivity in the root-rhizosphere communication is related to the presence of various anatomical root traits, such as the existence of apoplastic barriers.

The apoplastic pathway – In this pathway a nutrient travels through the cell walls. The longest pathway for this form of transport would be through the cell wall space of epidermal and cortical cells, entering the symplast only by uptake at the plasma membrane of an endodermal cell.

The symplastic pathway – In the symplastic pathway a nutrient enters into a cell through influx transporters and then travels through the interconnected cytosol of root cells, which is referred to as the cell-to-cell pathway. The longest symplastic pathway would be via uptake into the protoplast of the root hair cells of the epidermis, repeated transport through the cytoplasmic bridges of plasmodesmata and export into the xylem.

The trans-cellular pathway – The trans-cellular pathway requires repeated uptake and efflux of nutrients through the concerted action of influx and efflux transporters, a possibility that has been substantiated by a number of recent publications showing strictly polar distribution of numerous trasporters in root cells. This pathway is the combination of the two other pathways (Barberon et al. 2016; Tylová et al. 2017; Kreszies et al. 2019).

4 Environmental issues affecting nutrient acquisition

Nutrient deficiency along with water deficiency (or drought) are global constraints to agricultural productivity. Fertilizer and water application as agricultural inputs are suboptimal in many areas of the world as most smallholder farmers have limited access to fertilizers, while intensive fertilization is a primary energy, economic, and environmental cost of agricultural production. Globally, water for irrigation is decreasing both in quantity due to aquifer depletion and quality due to salinization. Climate change is projected to intensify the importance of crop water and nutrition stress, as a combined result of altered precipitation patterns along with soil degradation.

Water and nutrients are soil resources and root traits are of primary importance in soil resource capture. Nutrients vary in their mobility and location within the soil profile. Nitrogen is acquired from three different sources in the soil: nitrate, ammonium and amino acids. Nitrate is highly mobile and can leach deep in the soil profile. Ammonium and potassium are less mobile nutrients in the soil than nitrate. Phosphorus is an extremely immobile nutrient, due to its formation of insoluble and immobile complexes within the soil, but it can still be acquired by the roots.

This variation in resource mobility is important to consider when defining the ideal root system architecture (ideotype) for nutrient acquisition. Root architectural traits regulate the placement of roots in distinct soil regions and subsequently affect the acquisition of mobile and immobile nutrients. The acquisition of mobile resources (e.g. nitrate, along with water) is enhanced by root phenotypes that increase the exploration of deeper soil layers. Over the growing season, mobile resources may be acquired from deep soil layers, but immobile resources typically have the greatest availability in shallow soil layers.

Root anatomical traits have important roles in soil resource capture. Soil exploration by the root and acquisition of soil resources requires energy. Several root anatomical traits influence the energy cost of soil exploration by changing the proportion of respiring and non-respiring root tissues, thus affecting the carbon and nutrient cost of tissue construction and maintenance. Root respiration during growth, maintenance and nutrient uptake are major components of root metabolic costs. The cost of root construction is assumed to be a once-only cost incurred when the root is formed. In contrast, maintenance costs accumulate over time and can quickly exceed initial construction costs. Therefore, maintenance costs are important determinants of root energy cost. The two costs are tightly coupled. Traits and trait states that reduce the metabolic cost of soil exploration improve the acquisition of limiting soil resources. Such traits include the production of an optimal number of root axes, biomass allocation to metabolically efficient root classes and reduced tissue respiration.

Efficient use of any nutrient by a crop comprises two functional processes: acquisition efficiency and utilization efficiency. Nutrient capture is essentially a root trait and to be fully expressed it requires adequate sinks for temporary storage or final deposition of the acquired nutrients. Efficient acquisition will depend on root architecture, root functions in terms of transporters and exudates, and the presence or absence of symbiotic associations. Active, highly efficient transporter systems are required for uptake into root cells, expressed in the plasma membranes of root cells, root tips, root hairs or in associated organisms (i.e. mycorrhiza).

A well-developed shallow root system will be ideal for intercepting fertilizer applications. Deeper roots assume importance with the depletion of surface nutrients, as water near the surface becomes limiting and restricts uptake of nutrients. In the case of high water supply, deeper roots assume importance for the interception of nutrients that would be potentially leached from the soil profile. Local proliferation of roots in response to nutrient supply is also observed. Another strategy to enhance capture (root architecture and function) and hence acquisition is to enhance mechanisms for increasing bioavailability of nutrients. The energy costs of root growth and maintenance should include the investment of limiting resources, as well as carbon costs. Therefore, understanding traits that enhance plant performance under nutrient-limited and water-limited conditions is important in increasing the productivity, resilience and sustainability of global agriculture (Hawkesford 2011; Lynch 2013; York et al. 2013; Lynch 2015; Schneider and Lynch 2018).

5 Adaptations of root anatomical traits

There is a whole range of adaptation of root anatomical traits in response to environmental issues (Fig. 2).

5.1 Forming apoplastic barriers in root anatomical structures: lignification and suberization deposits

Nature of apoplastic barriers – The apoplastic pathway can be blocked by suberin lamellae (SL) and Casparian strips (CS) in endodermal and exodermal cell walls. Both SL and CS are distinct subcellular structures.

Suberin lamellae – Suberin is a hydrophobic polymer composed of polyaliphatic and polyaromatic domains. The aliphatic monomers are primary alcohols, fatty acids, a-ω dicarboxylic acids and ω-hydroxy acids. The aromatic monomers are ferulic and coumaric acids. Suberin is deposited as a secondary cell wall in the form of lamellae between the plasma membrane and the primary cell wall (Fig. 3).

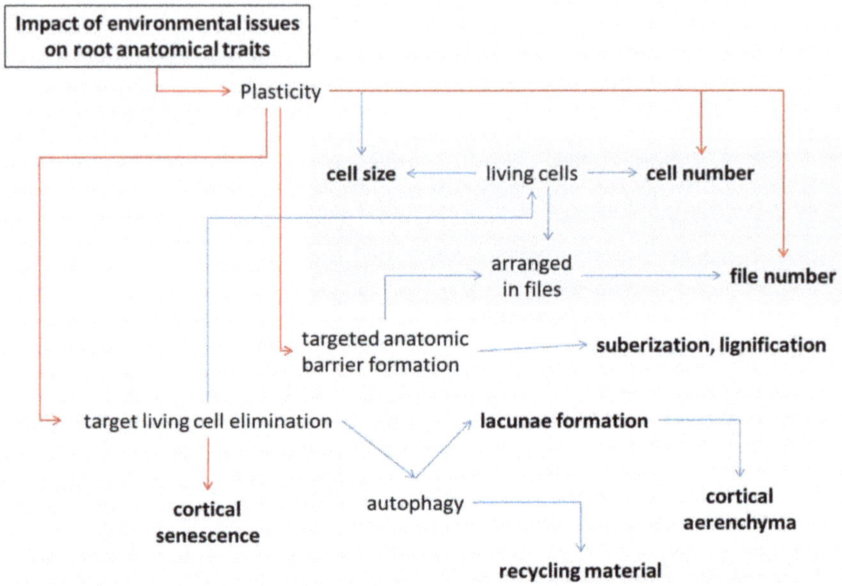

Figure 2 The adaptation range of root anatomical traits in response to environmental issues. The grey frame includes the functional characteristics of the cortex. Blue lines represent constitutive effects, while red lines represent environmental effects that induce adaptations toward different trait states.

Casparian strips (or bands) – CS are mainly composed of lignin and partly of suberin polymers. Lignin consists of syringyl, guaiacyl and p-hydroxy phenol monomers which form a complex aromatic biopolymer. CS can be described as a primary wall intermingled with a mixture of lignin and suberin and the chemical nature of the strip is quite similar to that of a lignified cell wall.

The CS cover the entire endodermal cell, firmly attached to the plasma membrane of the endodermal cell at all points. As a result there are no intercellular spaces between the cells of the endodermis, and therefore no movement can occur through the radial walls. The CS matures simultaneously with the maturation of the protoxylem elements, that is, the first developed xylem, in growing roots. At maturity it forms a continuum from the main root to the lateral root and encloses the vascular tissue of the main and lateral roots. After its inception CS undergoes considerable growth, and enlargement is correlated with the development of the endodermal cells (Figs 4 and 5; Dickison 2000).

5.2 Providing protection to the cortex

The two barrier regions of the cortex are the endodermis and exodermis.

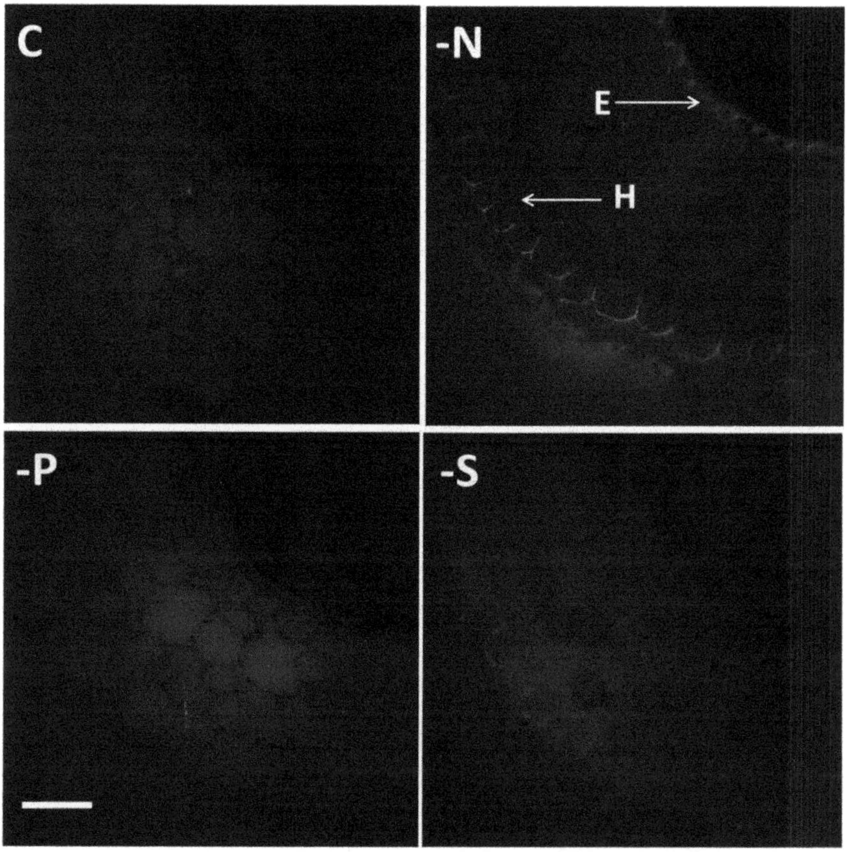

Figure 3 Freehand cross sections of nodal roots (first whorl) of maize plants grown for 12 days in a complete nutrient solution (C), or in a nutrient solution deprived of nitrate (-N), sulphate (-S) or phosphate (-P). Sections were performed in the middle of the root sector carrying lateral roots (LR), stained with neutral red and viewed under a fluorescence microscope (excitation wavelength 450–490 nm) for suberin detection. Suberin fluoresced yellow/green. H: Hypodermis, E: Endodermis. Scale bar = 100 μm (authors photo archive).

5.2.1 Endodermis: the internal barrier

Substances entering or leaving the stele must pass through the protoplast of the endodermal cells. The endodermis thus functions (i) to regulate the inward flow of water and nutrients into the stele and (ii) to prevent outward leakage. The endodermis is the primary barrier to radial movement so that the developmental state of the endodermis is critical.

The primary endodermis refers to the developmental stage in which the CS is represented by a band-like deposition within the anticlinal and periclinal

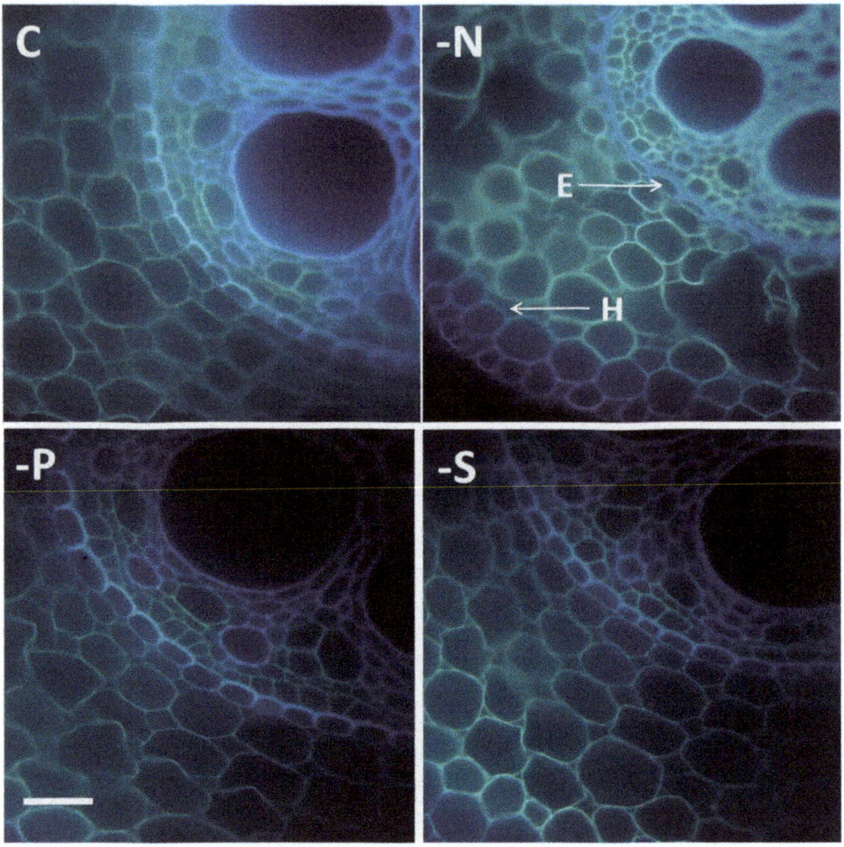

Figure 4 Freehand cross sections of nodal roots (first whorl) of maize plants grown for 12 days in a complete nutrient solution (C), or in a nutrient solution deprived of nitrate (-N), sulphate (-S) or phosphate (-P). Sections were performed in the middle of the root sector carrying lateral roots (LR) and viewed in an ammonium buffer pH10 under a fluorescence microscope (excitation wavelength 365 nm). Lignin's precursors fluoresced green, lignin fluoresced blue. H: Hypodermis, E: Endodermis. Scale bar = 100 μm (authors photo archive).

endodermal walls. In a transverse section, the strip is detectable as dark and partially thickened points within the anticlinal walls. A small number of endodermal cells may remain thin-walled and weakly suberized; these cells are known as passage cells, located opposite to the protoxylem poles.

The secondary endodermis is distinguished by the deposition of a thin SL onto the inner surface of the anticlinal and periclinal walls.

The tertiary endodermis develops when additional layers composed of lignified wall materials are added onto the SL of the secondary endodermis. This is the final developmental stage, and can result in endodermal cells with extremely thick secondary cell walls. The tertiary stage of the endodermis is

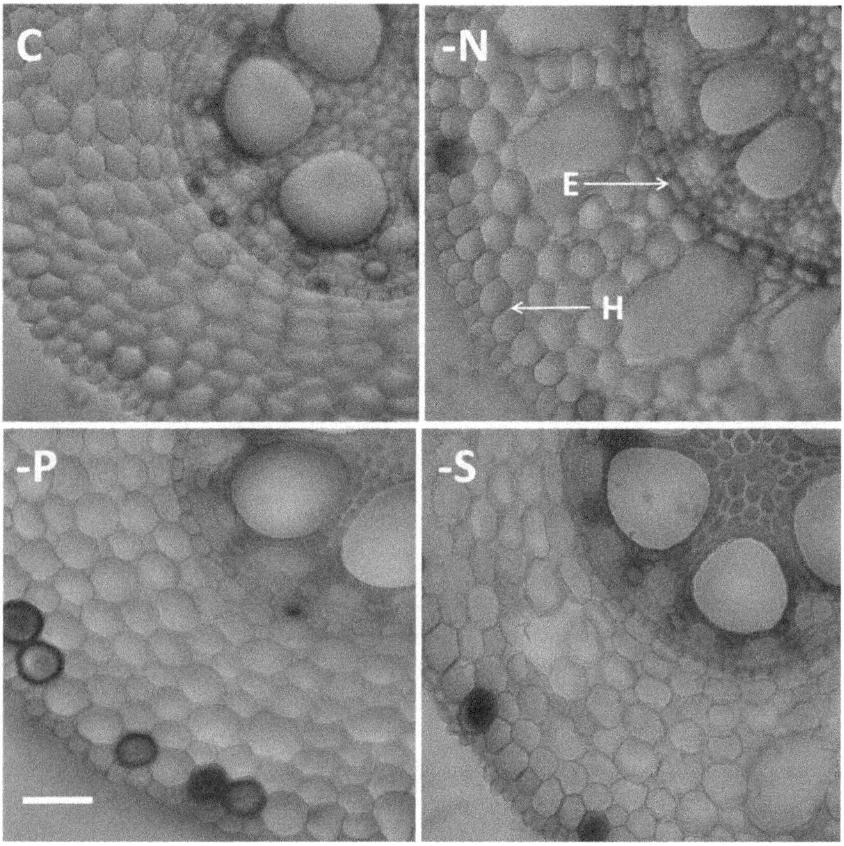

Figure 5 Freehand cross sections of nodal roots (first whorl) of maize plants grown for 12 days in a complete nutrient solution (C), or in a nutrient solution deprived of nitrate (-N), sulphate (-S) or phosphate (-P). Sections were performed in the middle of the root sector carrying lateral roots (LR) and stained with phloroglucinol-HCl (Wiesner test) for lignin detection. Lignins stained red. H: Hypodermis, E: Endodermis. Scale bar = 100 μm (authors photo archive).

best interpreted as it is composed of two distinct developmental events. In the first of those, only cellulose is deposited onto the inner surfaces of the radial and transverse walls, whereas in the final stage phenolic compounds are deposited as wall inclusions. The endodermis of many monocots has thickened inner tangential and radial walls and a thin tangential wall.

Endodermal suberization occurs in a switch-like manner, in which individual endodermal cells rapidly suberize in an apparently random fashion. This leads to a zone of patchy suberization that later develops into a zone of continuous suberization.

The endodermis forms a paracellular apoplastic transport barrier that defines two separate spaces within a root. The inner space is the nutrient-transporting stele with its network of interconnected xylem tubes. The outer space is the cortex, the intercellular (apoplastic) space of which is connected to the soil.

The first state of endodermis differentiation is an absorbing endodermis, an actively absorbing cell layer. The second state of differentiation forms a protective endodermis mainly or exclusively protective in nature. The protective endodermis is characterized by the formation of the hydrophobic suberin polymer over the entire surface. The nature of protective endodermis is exemplified by the fact that in many plant species, environmental stresses (such as water deficiency) can lead to 'cortical dieback' leaving the endodermis as the outermost protective layer of the root (Dickison 2000; Barberon et al. 2016; Seago et al. 2005; Chimungu et al. 2015a).

5.2.2 Exodermis: the external barrier

The exodermis is a common, although non-obligatory, apoplastic barrier of the outer cortex of angiosperm roots with variable structures among plant taxons. The exodermis commonly differs from the endodermis in its pattern of differentiation. Exodermal differentiation is clearly subjected to dynamic impacts from the environment (Tylová et al. 2017).

5.2.3 Comparison of the two barriers in maize roots

In maize roots CS form synchronously, as dot-like structures, within the entire endodermis close to the root apex; these structures later increase in width. The SL are deposited later, but less synchronously. In contrast, the entire maturation of the maize exodermis is 'patchy' in both radial and longitudinal directions. Even older root structures may only possess some exodermal cells with detectable CB (Tylová et al. 2017).

5.3 Adapting the cortical area

Traits and trait states that reduce the energy costs of soil exploration may improve the capture of limiting soil resources. The deployment of root phenotypes with greater metabolic efficiency of soil exploration represents a novel, unexploited paradigm to develop crops with greater resource efficiency and resilience.

Plasticity in living cortical area – Lines with high or low LCA have been distinguished, so that decreasing of LCA could be an adaptation to the adverse soil environment. LCA can be useful as a predictor of root respiration, soil exploration and tolerance.

Plasticity in root cortical cell file number - CCFN is a key determinant of LCA and an important determinant of root metabolic cost (Chimungu et al. 2014a). Lines with high or low CCFN have been studied. Maize, for example, has substantial genetic variation for root architecture and anatomical traits. Maize, like other monocots, has no secondary growth in its roots. CCNF varies substantially among maize genotypes in the range of 7–16 in *Zea* spp including maize landraces and wild *Zea* spp. Genotypes with reduced CCNF had greater rooting depth, greater acquisition of deep soil water, better plant water status, better growth and better yield. It has been proposed that reduced CCFN may be a useful adaptation to drought. Decreased CCNF is associated with reduced root respiration. This respiratory pattern reflects the effect of decreasing the proportion of metabolically active cells in the cortex, thus increasing the proportion of non-respiring tissues such as aerenchyma and xylem vessels. Reduced CCNF is associated with increased rooting depth in the field under water deficiency, but it does not affect rooting depth in well-watered conditions. The utility of reduced CCNF may be useful in the acquisition of nitrogen in leaching environments and might be a feature of other plant species, especially graminaceous species lacking secondary root growth (rice, wheat, barley, oat, sorghum and millet) (Hochholdinger 2009; Lynch 2013; Burton et al. 2013a,b; Chimungu et al. 2014a,b).

Plasticity in root cortical cell size - CCS is an anatomical trait that affects the energy costs of soil exploration by affecting tissue respiration (Chimungu et al. 2014b). It has been proposed that large cortical cells may also substantially reduce root respiration, because they have a higher ratio of vacuolar to cytoplasmic volume and hence reduced respiration per unit of tissue volume than smaller cells. Large CCS would decrease the metabolic costs of root growth and maintenance, both in terms of the carbon cost or root respiration and the nutrient content of living tissue, by increasing the ratio of vacuolar to cytoplasmic volume. Large CCS increases drought tolerance by reducing root metabolic costs, permitting greater root growth and water acquisition. Small cells have a greater density of cell walls per volume, providing rigidity and strength, and thus are more resistant to buckling and deflection than large cells. This is particularly important for root penetration in hard soil, which is common under drought. Large cells are important in reducing the metabolic cost for soil exploration, permitting root growth into deeper soil regions and thereby enhancing water acquisition under drought. Large CCS may also interact with other traits that enhance root penetration in hard soils, such as root diameter and root hairs, and may synergistically enhance root penetration under combined stress of drought and mechanical impedance (Lynch 2013; Chimungu et al. 2014a,b).

5.4 Remobilizing nutrients: autophagy and autophagy-like processes

Autophagy comprises an essential catabolic pathway for growth and development, as it is a major pathway for degradation and recycling of cytoplasmic material, including individual proteins, aggregates and entire organelles. Autophagy involves degradation and intracellular trafficking, with conserved key functions in cellular homeostasis, development and stress responses in plants. It is activated in response to environmental cues or developmental stimuli, but relatively little is known regarding the molecular mechanisms involved in autophagy and the regulation of this complex pathway.

An outline of the autophagy pathway in plants is as follows: Following induction by the signal or a combination of signals (developmental or environmental ones), autophagosome formation is initiated by nucleation of the pre-autophagosomal structure (the phagophore assembly site (PAS)) at the outer surface of the endoplasmic reticulum. The expanding phagophore engulfs cellular targets in a selective or non-selective manner. The completed and maturated autophagosome fuses with the vacuole to release the inner autophagic body for hydrolytic degradation of the sequestered cargo. The autophagy pathway relies on a core machinery of autophagy-related genes (ATG) that act in functional units during the sequential steps of autophagosome biogenesis. Autophagy induction and nucleation involves the ATG1 complex that is negatively controlled by the target of rapamycin (TOR) kinase. Inhibition of TOR under certain conditions such as starvation leads to ATG1 complex activation and phagophore formation that also involves the phosphatidylinositol 3-kinase (LI3K) complex. ATG9 vesicles and related complex components are likely required for membrane delivery to the growing phagophore and autophagosome progression from the ER. Phagophore expansion and autophagosome completion further relies on two ubiquitine-like conjugation pathways that involve the E1 ligase-like ATG7, E2 ligase-like ATG3 and E3 ligase-like ATG12-ATG5-ATG16 complex to mediate ATG8 lipidation and subsequent insertion into the phagophore membrane (Üstün et al. 2017; Avin-Wittenberg et al. 2018).

5.5 Forming lacunae in the cortex

Root cortex can contain air-filled spaces (lacunae) among the living cells, called aerenchyma. The following types (or patterns) of lacunae have been distinguished in angiosperms (Seago et al. 2005):

Lysigeny (L) is the collapse and death of cells to create lacunae coupled with cell separations that usually precede cell collapse. At least three distinct patterns of lysigenous origin of lacunae have been identified: radial (rL), tangential (tL) and packet lysigeny (pL).

Expansigeny (E) is the expansion of intercellular spaces into lacunae by cell division and cell enlargement, without further cell separations and without cell collapse or death. Expansigeny is of two types: honeycomb expansigeny (E) and radial expansigeny (rE). The continued enlargement of the initial tiny intercellular spaces into slightly enlarged intercellular air spaces leaves a condition in which there is no aerenchyma (nA), that is, no lanucae cortex. As roots age there may be a few schizogenous or lysigenous spaces.

Schizogeny (S) is the expansion of intercellular spaces into lacunae along radial sectors between radially adjacent cells or sectors of cells. The separation of cells takes place along their common walls in order to join radially or obliquely aligned intercellular spaces. Combinations of schizogeny and some expansigenous lacunae is mixed aerenchyma (mA) which can be found in the cortex of some Fabaceae species. The separation of cells along radial walls, followed by deaths of some cells, especially cells which produce crystals and then break apart, is termed schizo-lysigeny (SL). An angiosperm phylogeny tree coupled with the distribution of aerenchyma types has been published (Seago et al. 2005). Not all families form lacunae in the root cortex. Among monocots, Poales create S, SL, rL, tL or nA, while among eudicots, Fabales form S, mA or nA (Seago et al. 2005).

5.6 Forming lacunae by targeted cell elimination processes: root cortical aerenchyma

Root cortical aerenchyma (RCA) is a type of programmed cell death (PCD) that replaces living cortical cells with (air-filled) lacunae (Fig. 6). RCA is a PCD that follows a cell elimination program. RCA is usually considered to be an adaptation to hypoxia, but it can also be induced by suboptimal availability of nitrogen, phosphorus, sulfur and water (Drew et al. 2000; Evans 2003; Bouranis et al. 2007, Zhu et al. 2010; Yang et al. 2012; Yamauchi et al. 2013; Jaramillo et al. 2013; Dickman et al. 2017).

PCD is an integral part of plant development and of responses to abiotic stress or pathogens. Although the molecular understanding of PCD regulation and execution is insufficient to define precise biochemical pathways, it is possible to classify PCD scenarios based on morphological criteria. Although the morphology of plant PCD is in some cases well-characterized and molecular mechanisms controlling plant PCD are beginning to emerge, there is still confusion about the classification of PCD in plants. van Doorn et al. (2011) suggested a classification based on morphological criteria. According to this classification, the use of the term 'apoptosis' is not justified in plants, but at least two classes of PCD can be distinguished, 'vacuolar cell death' and 'necrosis'.

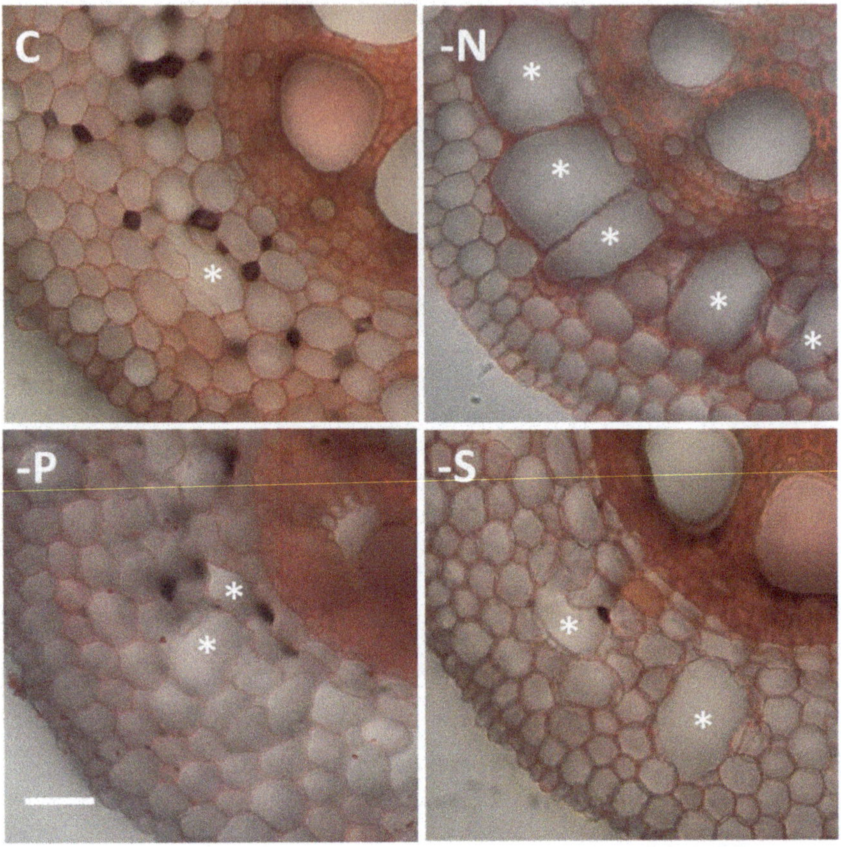

Figure 6 Freehand cross sections of nodal roots (first whorl) of maize plants grown for 12 days in a complete nutrient solution (C), or in a nutrient solution deprived of nitrate (-N), sulphate (-S) or phosphate (-P). Sections were performed in the middle of the root sector carrying lateral roots (LR) and stained with Congo red. Aerenchymatous spaces (*) were apparent in the root cortex. Scale bar = 100 μm (authors photo archive).

During vacuolar cell death, the cell contents are removed by a combination of autophagy-like processes and release of hydrolases from collapsed lytic vacuoles. Vacuolar cell death is common during tissue and organ formation and elimination. Necrosis is characterized by early rupture of the plasma membrane, shrinkage of the protoplast and absence of vacuolar cell death features. Necrosis is typically found under abiotic stress.

Reape and McCabe (2013) argued that classifying all types of PCD into the aforementioned two categories is premature and does not fully take into account of different forms of cell death that may operate in plant cells. Instead, they suggest that the condensed protoplast morphology associated with many instances of PCD may come about due to an active cell death-associated cellular retraction rather than simple rupture of the plasma membrane. They argue that it is important to be able to distinguish between cells that have died having undergone this protoplast retraction and those which have died without protoplast retraction. This differentiation is essential as the control of these two types of death may differ in several respects (Reape and McCabe 2008, 2013).

Plasticity in root cortical aerenchyma - It has been hypothesized that RCA is an adaptation to low nutrient availability by reducing the energy cost of soil exploration. By replacing living cells with lacunae, RCA significantly reduces the respiration and nutrient requirement of root tissue, permitting greater root growth and greater acquisition of soil resources for a given metabolic investment. The trade-offs associated with RCA formation are poorly understood. The formation of RCA reduces root respiration and nutrient content by converting living tissue to air volume (Burton et al. 2013a,b; Lynch 2015).

5.7 Root cortical senescence: another type of PCD

Root cortical senescence (RCS) is a type of programmed cell death in cortical cells of several Poaceae species, including triticale, barley, rye, oats, Kentucky blue grass, perennial ryegrass, blue grama grass and wheat. Dicotyledonous species do not undergo RCS.

RCS begins in the epidermis and progresses to inner cortical cell files. The cortex of axial roots directly adjacent to lateral roots, and lateral roots themselves, do not undergo senescence. RCS progresses basipetally and is anatomically distinct from the formation of root cortical aerenchyma and the loss of the root cortex due to secondary growth in dicots.

Cortical cells of seminal and nodal roots in several Poaceae species can senesce very early in plant growth, leaving the tips of the root axes and lateral roots to continue to grow.

RCS begins in outer cortical cell layers, directly behind the zone of anucleate root hairs and progresses inwards and basipetally resulting in anucleate cortical cells. Eventually, nearly all cortical cell layers in nodal and seminal roots from approximately 2 cm behind the root apex to approximately 3 cm from the root-shoot junction will become anucleate. The development of RCS is reduced or delayed in the axial root around the base of lateral roots, and is most pronounced on the side of the axis where the lateral emerges. The delayed RCS effect caused by lateral roots is temporary as with increasing distance behind the apex the number of anucleate cell layers around the base of lateral roots progressively increases. The development of RCS is influenced by cultivar (Schneider et al. 2017a,b; Schneider and Lynch 2018; Henry and Deacon 1981).

6 Impact of adapted root anatomical traits on nutrient acquisition

The aforementioned adaptations have to effectively support plant growth and maintenance under various environmental circumstances. Hence the question: How effective is the adaptation of root anatomical traits in the case of nutrient acquisition? In response, the following prerequisites must be fulfilled: effective uptake and delivery of nutrients in a secure internal environment (Fig. 7).

6.1 How effective is nutrient uptake by the adapted root surfaces?

Maximizing the absorbing surface area along with increasing the density of transporters per unit surface area must be achieved. Root length should be increased by investing the least possible amount of carbon and energy in order to exploit the surrounding soil.

Lateral roots provide surface area, so that they are extremely important. They represent a large part of a complex root system's absorptive surface and should provide the greatest surface area along with the least investments of biomass. Branching of laterals affects a plant's ability to absorb necessary nutrients, while at the same time the entry of harmful compounds must be

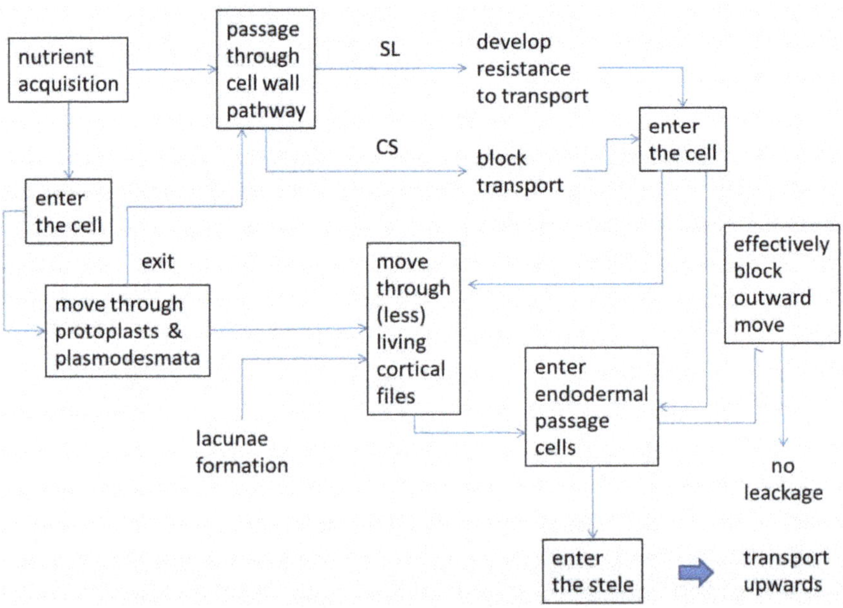

Figure 7 Outline of the impact of the adapted root anatomical traits on nutrient acquisition. SL: deposition of suberin lamellae, CS: formation of Casparian strips.

avoided. This is of critical importance considering plant performance under changeable environmental conditions with consequences on agronomic yields and food quality. Therefore, both the cortex of axial roots directly adjacent to lateral roots, as well as the cortex of the lateral roots themselves, do not perform cell elimination programs and the formation of lacunae is avoided, reduced or delayed, especially in the axial root cortex around the base of lateral roots, on the side of the axis where the lateral root emerges. Root hairs are also needed to increase absorbing surface area, so that root hair-bearing epidermal cells and suberized epidermis are desirable traits. Plasticity as a feasible adaptation of plants to the heterogenic soil environment is even more pronounced in the exodermis. Information about exodermis (and endodermis) differentiation in the lateral roots of herbaceous plants is highly fragmentary, as the majority of studies have focused on the main roots (Tylová et al. 2017).

6.2 How effectively are the nutrients delivered through the thinned root cortical tissues?

Once taken up, the nutrient has to be effectively delivered. This involves two processes: part of the uptake is invested to support root elongation and growth, maintenance and activity, and the remainder is transported to the stele. Delivery of nutrients involves water as the solvent and implies turgid cells. Keeping water within the tissue is crucial in the case of nutrient deficiencies coupled with drought. The delivery process should be performed by minimalizing the delivery cost, reducing the delivery pathways, securing inward movement and avoiding leakage, as well as recycling of nutrients from internal pools.

Reduction of cortex where possible, while keeping the rest of the cortical area active, is crucial for effective delivery. The stele must be kept intact, so that minimalizing cortex is the preferred strategy in monocots. Respiring versus non-respiring tissues must be compromised. In the cortex and at the tissue level, respiring tissues include living, metabolically active cells. Non-respiring tissues include lacunae. At the cell level, cytoplasmic volume versus vacuolar volume must be compromised, too. In terms of nutrient delivery, nutrient flow takes place though the intact cell files that connect exodermis to endodermis.

van der Weele et al. (1996) reported the existence of water in the aerenchymatous spaces of roots. They showed the presence of liquid filling some of the aerenchymatous spaces, while other spaces of the same root contained air. The liquid was water with few detectable ions. Similar liquid was present in small intercellular spaces within the spoke-like radial files of cells between the large spaces, or between remnants of collapsed cell walls at the edges of the large spaces. Therefore, they proposed that regions of roots with high diffusivity are those in which some of the aerenchyma spaces are filled with water. The origin of this liquid has been studied, too.

Overall the following scenario emerges: increased lacunae volume increases the percentage of non-respiring cortical tissue, thus decreasing the respiring cortical area. Reduced cortical area is translated to reduced root respiration, that is, reduced energy cost for maintenance. Increased cell size can be achieved by increasing the cytoplasmic volume, and a parallel decrease of the vacuolar volume. In this case, the energy cost per unit of tissue volume decreases. Increased lacunae volume decreases the number of cell cortical files, so that nutrient delivery occurs effectively through the remaining functional files, constructed of cells larger in size with more cytoplasmic volume. Carbon investment has been reduced along with energy cost.

Lacunae formation through PCD implies prior nutrient recycling. Through autophagy, the internal pool of nutrients is released and mobilized thus entering the delivery process.

It is generally accepted that autophagy is involved in nutrient recycling and that it is induced in nutrient-starved tissues. Both carbon and nitrogen starvation are known to induce autophagy. Links between autophagy and other nutrient deficiencies are less-documented, but it is not known what triggers the induction of autophagy-related genes during limitation of specific nutrients. However, at least for carbon, nitrogen and sulfur starvation autophagy is probably TOR-dependent (see Section 5.4). It is considered that starvation induces bulk autophagy of cytoplasmic components for nutrient remobilization. The diversity of the cytoplasmic components destined for degradation by autophagy suggests that in addition to C and N molecules, many other mineral nutrients could be released by the process. These compounds can then be used for the cell's metabolism to sustain respiration among other functions, or available to the whole organism after remobilization. It is not yet documented whether P, Fe, Zn, S or K can be recycled through autophagy (Üstün et al. 2017; Avin-Wittenberg et al. 2018).

As regards RCS, nutrient deficiencies do not affect the pattern or where RCS begins relative to the root apex, rather the rate of RCS development. The connection between nutrient deficiencies and the rate of RCS has been discussed by Schneider and Lynch (2018). For instance, RCS was accelerated by suboptimal availability of nitrogen and phosphorus. Increased endodermal suberization may occur after the development of RCS in order to protect the root from pathogen invasion and radial water and nutrient loss after the senescence of cortical and epidermal cells. Reallocation of nutrients after RCS has important functional implications for plant growth particularly under edaphic stress. Simulation results demonstrated that remobilization of N, K and P after RCS was always beneficial and increased growth in nutrient-dependent fashion under suboptimal availability of the nutrient (Schneider and Lynch 2018).

6.3 How effectively are the thinned root tissues protected and defended?

This question has been addressed in detail by Barberon et al. (2016). Suberization is an adaptive response to nutrient stresses. Moreover, the formation of suberin has been shown to be enhanced by both abiotic and biotic stresses. Suberin establishes a barrier for uptake from the apoplast into the endodermis. SL eventually coat the entire endodermal surface but do not affect the apoplastic transport across the endodermis that is blocked by CS. The hydrophobicity of suberin often leads to the assumption that the formation of SL represents an effective, qualitative block for the uptake of charged, possibly polar molecules. This assumption has been challenged in the case of water transport, where it was proposed that suberization causes a partial resistance to water flow, which is dependent on the degree of suberization (Schreiber 2010). Quantitative regulation of suberin in response to certain deficiencies fits the idea of suberization introducing a resistance rather than a block.

Nutrient delivery occurs in water, so plant roots sense water deficiency and cope with it. The barriers' structural features (CS, SL and tertiary walls) are established by intrinsic mechanisms and finely tuned in response to the (deficient) environment. The suberized layer of the root system responds to a range of nutrient stresses, and its deposition is regulated by stress hormones, revealing a level of functional and anatomical plasticity in adult roots. Suberization can be enhanced or reversed depending on nutrient availabilities. This stress-hormone-regulated plasticity allows adaptive regulation of root function (Barberon et al. 2016).

Suberization of endodermal cells has often been confounded with CS formation, as there were often assumed to be suberized structures. This led to the assumption that suberization somehow affects both the paracellular barrier properties of the endodermis as well as the direct uptake of nutrient into the endodermal cytoplasm. The function of CS has been distinguished from that of SL. Regulation of suberization specifically affects the apoplastic pathway as well as the coupled trans-cellular pathway, whereas it leaves the symplastic pathway unaltered (Barberon et al. 2016; Tylová et al. 2017; Kreszies et al. 2019).

Based on the above broad picture, it can be stated that plant root anatomy, that is, a structure, and nutrient acquisition, that is, a process, are strongly interlinked, and a whole array of anatomical root trait states are interconnected to provide effective delivery of nutrient to the stele for long-distance transport (Fig. 3). Effective acquisition requires the integration of all trait states involved, coupled with an effective system to acquire and deliver (i) the environmental information and (ii) the information of the internal nutritional status in order to react properly and in a timely manner.

7 Case study: root anatomy of maize under nutrient shortage

7.1 Root anatomy of maize under nutrient shortage

Maize (*Zea mays* L.) is a major cereal crop widely cultivated around the world. Therefore it serves as a model C4 graminaceous plant with an enormous information base. Its root system comprises two parts, the embryonic and the postembryonic ones. The embryonic root system includes one primary root and a variable number of seminal roots. The postembryonic root system is constructed by shoot-borne roots distinguished in two categories, the crown roots and the brace roots. The crown roots are created at consecutive underground nodes, while the brace roots occur at consecutive aboveground nodes of the shoot. Lateral roots emerge from all root types. The postembryonic shoot-borne root system becomes dominant later in development and, along with its lateral roots, is responsible for water and nutrient uptake (Bouranis et al. 2008).

The anatomy of a maize root is typical of a monocotyledonous plant. Within the aforementioned root types the anatomical structures are very similar, but they are initiated from different tissues during embryonic and postembryonic development (Hochholdinger et al. 2004). Mature primary and seminal roots as well as shoot-borne roots exhibit a central cylinder (protostele) with many ribs of xylem and the pericycle forms the outermost layer of the central cylinder. The ground tissue consists of one layer of endodermal tissue with the suberized and often lignified casparian band. The parenchymatous cortex tissue consists of several cell layers. The outermost cell layer is formed by the epidermis (or rhizodermis), which consists of root-hair-forming trichoblasts and non-root-hair-forming atrichoblasts. Maize roots do not show secondary growth of the root (Hochholdinger et al. 2004).

In older roots the epidermis is short-lived, replaced by a lignified and/or suberized exodermis, which develops from the outermost cells of the cortex and forms an additional casparian band. In the above-ground-formed brace roots the epidermis persists and forms a protective cuticula. The root apical meristem has a closed organization with three distinct layers of initials. The longitudinal structure of the maize root includes the following partially overlapping specialized zones of development: the root cap, the root apical meristem, the distal elongation zone, the elongation zone and the maturation zone (Ishikawa and Evans 1995, Bouranis et al. 2008).

Root anatomy of maize under nutrient shortage has been studied extensively (Bouranis et al. 2003, 2006; Postma and Lynch 2011a,b; Yamauchi et al. 2011; Burton et al. 2013a,b; Hu et al. 2014; Saengwilai et al. 2014; Chimungu et al. 2014a,b, 2015a,b; Galindo-Castañeda et al. 2018; Díaz et al. 2018; Yang et al. 2019; Wang et al. 2019). Nitrate (N), phosphate (P) or sulphate (S) deprivation causes aerenchyma formation in maize roots (Siyiannis et al. 2012).

The S-deprivation reduced shoot growth and enhanced root proliferation. Under S-deprivation, the first aerenchymatous spaces were apparent in the middle of the cortex, while in the fully developed aerenchyma only a few files of cells bridged hypodermis to endodermis and stele of roots. Moreover, S deprivation induced a progressive thickening of the cell walls of the endodermis. In the cortex with fully developed aerenchyma, cell wall thickness of the endodermis was increased by 68% (Bouranis et al. 2003).

As regards the distribution of aerenchyma along the root axis, aerenchyma was found in the cortex of the 77% of the root length, particularly in the region of emerging or developing lateral roots. The basal and apical sectors contained no aerenchyma. Moreover, no aerenchyma connection was found in the shoot. Lignified epidermal layers were located at the basal and the proximal sectors. S-deprivation resulted in shorter lateral roots in the upper sectors and in a limited extension of the lignified layers toward the next lateral root-bearing sector. Lateral root proliferation was accompanied by spatially localized induced cell death in the cortex of developing young maize adventitious roots (Bouranis et al. 2006).

The start time, location and extent of aerenchyma formation differed depending upon the specific deprivation. P-deprivation produced aerenchyma to the least extent, N-deprivation to the greatest and S-deprivation in between. Under N-deprivation, the cell elimination program started almost immediately after nutrient withdrawal and produced extensive aerenchyma in the root apex (A) sector. Under P- and S-deprivations, there was a delay in the production of aerenchyma for some days. The intensity of aerenchyma formation in the A sector was less compared to that of the N-deprivation. In the sector carrying the emerging lateral roots (the ELR sector) the intensity of aerenchyma also differed with the same order: control < P- < S- < N-deprivation. As regards the distribution of aerenchyma within the sector that carries lateral roots (the LR sector), it was expanded more to the base following the same order. Aerenchyma in young maize roots usually reaches a maximum of 30–40% of the total cortical area (He et al. 1996; Siyiannis et al. 2012). Without exception, the basal sector had no aerenchyma.

The PCD induced by S- or P- or N-deprivation is a nutritional PCD (nPCD) with both common features and characteristic differences among the cases. According to Siyiannis et al. (2012), nPCD was characterized by higher oxygen consumption and lower ATP concentration at day 2 of nutrient deprivation, indicating a perturbation in the energetics of the root compared to the control. It has been proposed that this perturbation could play a role in the initiation of nPCD. The values of oxygen consumption were higher in all three deprivations compared with control. The nPCD that resulted in lysigenous aerenchyma formation in the cortex of N-, S- or P-deprived maize roots may be induced by perturbations in energy status of the roots, while ROS are likely to serve dual roles as signals and eliminators.

At the same time and apart from aerenchyma in the cortex of crown roots under S- deficiency, aerenchyma was also found in S-deprived mesocotyl near the scutellar node (Maniou et al. 2014). Interestingly, this was also the case for the mesocotyl of control plants, while the deprivation reduced aerenchyma formation by 7.4%. In both cases, the location of aerenchyma along the mesocotyl axis coincided with the presence of mesocotyl roots (MR), which were not uniformly allocated along the mesocotyl axis. Our data (Maniou et al. 2014) suggest that mesocotyl cortical aerenchyma supports the MR, which are lateral relative to the mesocotyl axis. The mesocotyl behaves as a root axis. Under deprivation, the root system contained five seminal roots (SR; instead of four under full nutrition)

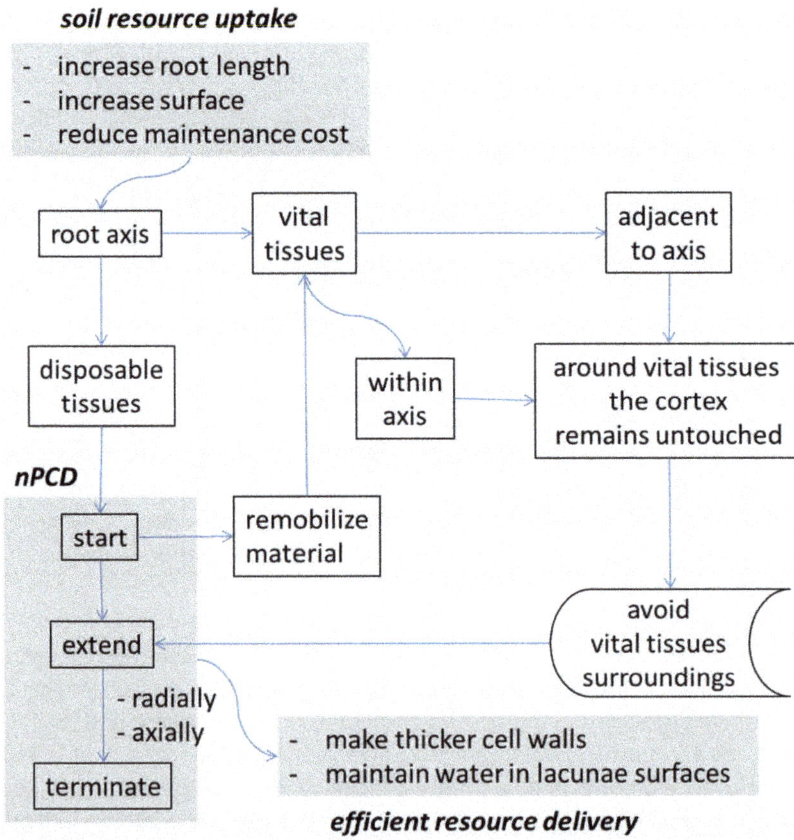

Figure 8 Under the circumstances the vital tissues must be protected and maintained. The extension of the nutritional programmed cell death (nPCD) process in relation to the vital tissues.

and three mesocotyl roots (MR; instead of two at day 10 and four at day 19 under full nutrition), a finding that probably explains the reduced aerenchyma formation in the mesocotyl: less root axes with more invested dry mass. All root axes that were attached to the scutellar node had no aerenchyma at the vicinity of the scutellar node, that is, their base, probably because the scutellar node is a trafficking center. This is another proof that aerenchyma formed by nPCD is not formed for the transfer of oxygen, that is, is not a ventilating trait.

These findings suggest that trophic aerenchyma formation in maize follows a strategy at the whole plant level (Fig. 8). This strategy includes the protection of vital tissues, such as the crown and the scutellar node, by preventing PCD in them. The scutellar node is temporary during plant's development and may be rapidly destroyed, while crown is a vital organ and as such is protected. Therefore, the basal zones of the attached organs are not subject to trophic aerenchyma formation. nPCD is involved in nutrient cycling; it has been shown that this mechanism plays a role in nitrogen remobilization and because it is a non-specific mechanism it might also control remobilization of nutrients (Pottier et al. 2014). The lateral roots are among the vital organs that need to be protected, because around them the cortex remains untouched.

8 Summary

Based on the above, the arsenal of adaptations of root anatomical traits is effective in ensuring nutrient acquisition. Fulfilling all prerequisites, that is, effective uptake of nutrients and effective delivery of them in a secure internal environment, is the cheapest possible way. The ultimate goal is to deliver nutrients first to the organs of vital importance such as the lateral roots and then to the aboveground plant parts. The deployment of anatomical traits and trait states of roots that can exploit effectively the problematic soils and their heterogeneity provides a research area with great significance for developing crops with root systems capable of water and nutrient capture in such soils.

9 Future trends in research

9.1 Root phenotypes for improved nutrient capture

Nutrient-efficient crops are a solution to the two grand challenges of modern agriculture: improving food security while reducing environmental impacts. Nutrient-efficient crops could provide benefits at all input levels. According to Lynch (2019), although our current understanding is sufficient to deploy root phenotypes for improved nutrient capture in crop breeding, this complex topic does not receive the resources it merits in either applied or basic plant biology. Renewed emphasis on these topics is needed to develop the nutrient-efficient crops urgently needed in global agriculture.

9.2 Root anatomical traits, software and modelling

OPENSIMROOT is an open-source, functional plant model and mathematical description of root growth and function. Its functionality has been described and broadens the benefits of root modelling to the plant science community. OPENSIMROOT is an extended version of SIMROOT, established to simulate root system architecture, nutrient acquisition and plant growth. OPENSIMROOT has a plugin, modular infrastructure, coupling single plant and crop stands to soil nutrient and water transport models. It estimates the value of root traits for water and nutrient acquisition in environments and plant species. It allows upscaling from root anatomy to plant community to estimate the following: resource cost of developmental and anatomical traits, trait synergisms and interspecies root competition. It integrates plant phenotypic data with environmental metadata to support experimental designs and to gain a mechanistic understanding at system scales (Burton et al. 2012; Postma et al. 2017).

RootScan is a program for semi-automated image analysis of anatomical traits in root cross sections. The software offers considerable improvements in the amount and quality of data, ease of use, and time needed for data collection. RootScan permits phenotypic scoring of physiologically and agronomically important traits on a large number of genotypes (Burton et al. 2012).

9.3 What is the situation with other nutrient deficiencies?

The available information does not extend to every nutrient deficiency or combination of them. Therefore, the area is still open for research.

10 Where to look for further information

- Plant Roots: The Hidden Half, Fourth Edition, Amram Eshel, Tom Beeckman, CRC Press, 2013.
- Annals of Botany, SPECIAL ISSUE: Matching Roots to Environment (Volume 112, Issue 2, July 2013): https://academic.oup.com/aob/issue/112/2
- Plant Physiology, Plant Roots Focus Collection: http://www.plantphysiol.org/plant_roots.
- Lynch Lab: https://plantscience.psu.edu/research/labs/roots.
- Hochholdinger lab: https://www.hochholdinger-lab.uni-bonn.de/cfg/crop-functional-genomics/people/hochholdinger.
- Sparks lab: https://sites.udel.edu/canr-sparkslab/.
- International Society of Root Research: https://www.rootresearch.org/.
- International Plant Phenotyping Network: https://www.plant-phenotyping.org/IPPN_home.

- Plant Image Analysis: http://plant-image-analysis.org/.
- Plant Illustrations Repository: https://figshare.com/authors/Plant_Illustrations/3773596.

11 References

Avin-Wittenberg, T., Baluska, F., Bozhkov, P. V., Elander, P. H., Fernie, A. R., Galili, G., Hassan, A., Hofius, D., Isono, E., Le Bars, R., Masclaux-Daubresse, C., Minina, E. A., Peled-Zehavi, H., Coll, N. S., Sandalio, L. M., Satiat-Jeunemaitre, B., Sirko, A., Testillano, P. S. and Batoko, H. (2018). Autophagy-related approaches for improving nutrient use efficiency and crop yield protection, *Journal of Experimental Botany* 69(6), 1335–1353.

Barberon, M., Vermeer, J. E., De Bellis, D., Wang, P., Naseer, S., Andersen, T. G., Humbel, B. M., Nawrath, C., Takano, J., Salt, D. E. and Geldner, N. (2016). Adaptation of root function by nutrient-induced plasticity of endodermal differentiation, *Cell* 164(3), 447–459.

Bouranis, D. L., Chorianopoulou, S. N., Siyiannis, V. F., Protonotarios, V. E. and Hawkesford, M. J. (2003). Aerenchyma formation in roots of maize during sulphate starvation, *Planta* 217(3), 382–391.

Bouranis, D. L., Chorianopoulou, S. N., Kollias, C., Maniou, P., Protonotarios, V. E., Siyiannis, V. F. and Hawkesford, M. J. (2006). Dynamics of aerenchyma distribution in the cortex of sulfate-deprived adventitious roots of maize, *Annals of Botany* 97(5), 695–704.

Bouranis, D. L., Chorianopoulou, S. N., Siyiannis, V. F., Protonotarios, V. E. and Hawkesford, M. J. (2007). Lysigenous aerenchyma development in roots – triggers and cross-talks for a cell elimination program, *International Journal of Plant Developmental Biology* 1(1), 127–140.

Bouranis, D. L., Buchner, P., Chorianopoulou, S. N., Hopkins, L., Protonotarios, V. E., Siyiannis, V. F. and Hawkesford, M. J. (2008), Responses to sulfur limitation in maize. In: Khan, N. A., et al. (Eds), *Sulfur Assimilation and Abiotic Stress in Plants'*, Springer-Verlag, Berlin, Heidelberg, 1–19.

Burton, A. L., Williams, M., Lynch, J. P. and Brown, K. M. (2012). RootScan: software for high-throughput analysis of root anatomical traits, *Plant and Soil* 357(1–2), 189–203.

Burton, A. L., Lynch, J. P. and Brown, K. M. (2013a). Spatial distribution and phenotypic variation in root cortical aerenchyma of maize (Zea mays L.), *Plant and Soil* 367(1–2), 263–274.

Burton, A. L., Brown, K. M. and Lynch, J. P. (2013b). Phenotypic diversity of root anatomical and architectural traits in Zea species, *Crop Science* 53(3), 1042–1055.

Chimungu, J. G., Brown, K. M. and Lynch, J. P. (2014a). Reduced root cortical cell file number improves drought tolerance in maize, *Plant Physiology* 166(4), 1943–1955.

Chimungu, J. G., Brown, K. M. and Lynch, J. P. (2014b). Large root cortical cell size improves drought tolerance in maize, *Plant Physiology* 166(4), 2166–2178.

Chimungu, J. G. and Lynch, J. P. (2014). Root traits for improving nitrogen acquisition efficiency. In: Ricroch, A., et al. (Eds), *Plant Biotechnology: Experience and Future Prospects'*, Chapter 15, Springer International Publishing, Switzerland, 3.

Chimungu, J. G., Loades, K. W. and Lynch, J. P. (2015a). Root anatomical phenes predict root penetration ability and biomechanical properties in maize (*Zea mays*), *Journal of Experimental Botany* 66(11), 3151–3162.

Chimungu, J. G., Maliro, M. F. A., Nalivata, P. C., Kanyama-Phiri, G., Brown, K. M. and Lynch, J. P. (2015b). Utility of root cortical aerenchyma under water limited conditions in tropical maize (*Zea mays* L.), *Field Crops Research* 171, 86–98.

Díaz, A. S., Aguiar, G. M., Pereira, M. P., Mauro de Castro, E. M., Magalhães, P. C. and Pereira, F. J. (2018). Aerenchyma development in different root zones of maize genotypes under water limitation and different phosphorus nutrition, *Biologia Plantarum* 62(3), 561–568.

Dickison, W. C. (2000). *Integrative Plant Anatomy*, Academic Press, UK.

Dickman, M., Williams, B., Li, Y., Figueiredo, P. and Wolpert, T. (2017). Reassessing apoptosis in plants, *Nature Plants* 3(10), 773–779.

Drew, M. C., He, C. J. and Morgan, P. W. (2000). Programmed cell death and aerenchyma formation in roots, *Trends in Plant Science* 5(3), 123–127.

Evans, D. E. (2003). Aerenchyma formation, *New Phytologist* 161(1), 35–49.

Galindo-Castañeda, T., Brown, K. M. and Lynch, J. P. (2018). Reduced root cortical burden improves growth and grain yield under low phosphorus availability in maize, *Plant, Cell and Environment* 41(7), 1579–1592.

Gregory, P. J. (2011). Crop root systems and nutrient uptake from soils. In: Hawkesford, M. J. and Barraclough, P., *The Molecular and Physiological Basis of Nutrient Use Efficiency in Crops*, Wiley-Blackwell, Chichester, 21–45.

Hawkesford, M. J. (2011), An overview of nutrient use efficiency and strategies for crop improvement. In: Hawkesford, M. J. and Barraclough, P., *The Molecular and Physiological Basis of Nutrient Use Efficiency in Crops*, Wiley-Blackwell, 5–19.

He, C.-J., Morgan, P. W. and Drew, M. C. (1996). Transduction of an ethylene signal is required for cell death and lysis in the root cortex of maize during aerenchyma formation induced by hypoxia, *Plant Physiology* 112, 463–472.

Henry, C. M. and Deacon, J. W. (1981). Natural (non-pathogenic) death of the cortex of wheat and barley seminal roots, as evidenced by nuclear staining with acridine orange, *Plant and Soil* 60(2), 255–274.

Hochholdinger, F., Woll, K., Sauer, M. and Dembinsky, D. (2004). Genetic dissection of root formation in maize (*Zea mays*) reveals root-type specific developmental programmes, *Annals of Botany* 93(4), 359–368.

Hochholdinger, F. (2009). The maize root system: morphology, anatomy, and genetics. In: Bennetzen, J. L. and Hake, S. C. (Eds), *Handbook of Maize: Its Biology*, Springer, New York, NY.

Hu, B., Henry, A., Brown, K. M. and Lynch, J. P. (2014). Root cortical aerenchyma inhibits radial nutrient transport in maize (*Zea mays*), *Annals of Botany* 113(1), 181–189.

Ishikawa, H. and Evans, M. L. (1995). Specialized zones of development in roots, *Plant Physiology* 109, 725–727.

Jaramillo, R. E., Nord, E. A., Chimungu, J. G., Brown, K. M. and Lynch, J. P. (2013). Root cortical burden influences drought tolerance in maize, *Annals of Botany* 112(2), 429–437.

Kreszies, T., Shellakkutti, N., Osthoff, A., Yu, P., Baldauf, J. A., Zeisler-Diehl, V. V., Ranathunge, K., Hochholdinger, F. and Schreiber, L. (2019). Osmotic stress enhances suberization of apoplastic barriers in barley seminal roots: analysis of chemical, transcriptomic and physiological responses, *New Phytologist* 221(1), 180–194.

Lynch, J. P. (2013). Steep, cheap and deep: an ideotype to optimize water and N acquisition by maize root systems, *Annals of Botany* 112(2), 347–357.

Lynch, J. P. (2015). Root phenes that reduce the metabolic costs of soil exploration: opportunities for 21st century agriculture, *Plant, Cell and Environment* 38(9), 1775–1784.

Lynch, J. P. (2019). Root phenotypes for improved nutrient capture: an underexploited opportunity for global agriculture, *New Phytologist* 223(2), 548–564.

Maniou, F., Chorianopoulou, S. N. and Bouranis, D. L. (2014). New insights into trophic aerenchyma formation strategy in maize (Zea mays L.) organs during sulphate deprivation, *Frontiers in Plant Science* 5, 581.

Postma, J. A. and Lynch, J. P. (2011a). Root cortical aerenchyma enhances the growth of maize on soils with suboptimal availability of nitrogen, phosphorus, and potassium, *Plant Physiology* 156(3), 1190–1201.

Postma, J. A. and Lynch, J. P. (2011b). Theoretical evidence for the functional benefit of root cortical aerenchyma in soils with low phosphorus availability, *Annals of Botany* 107(5), 829–841.

Postma, J. A., Kuppe, C., Owen, M. R., Mellor, N., Griffiths, M., Bennet, M. J., Lynch, J. P. and Watt, M. (2017). OPENSIMROOT: widening the scope and application of root architectural models, *New Phytologist* 215(3), 1274–1286.

Pottier, M., Masclaux-Daubresse, C., Yoshimoto, K. and Thomine, S. (2014). Autophagy as a possible mechanism for micronutrient remobilization from leaves to seeds, *Frontiers in Plant Science* 5, Article 11.

Reape, T. J. and McCabe, P. F. (2008). Apoptotic-like programmed cell death in plants, *New Phytologist* 180(1), 13–26.

Reape, T. J. and McCabe, P. F. (2013). Commentary: the cellular condensation of dying plant cells: programmed retraction or necrotic collapse?, *Plant Science* 207, 135–139.

Saengwilai, P., Nord, E. A., Chimungu, J. G., Brown, K. M. and Lynch, J. P. (2014). Root cortical aerenchyma enhances nitrogen acquisition from low-nitrogen soils in maize, *Plant Physiology* 166(2), 726–735.

Schneider, H. M., Postma, J. A., Wojciechowski, T., Kuppe, C. and Lynch, J. P. (2017a). Root cortical senescence improves growth under suboptimal availability of N, P, and K, *Plant Physiology* 174(4), 2333–2347.

Schneider, H. M., Wojciechowski, T., Postma, J. A., Brown, K. M., Lücke, A., Zeisler, V., Schreiber, L. and Lynch, J. P. (2017b). Root cortical senescence decreases root respiration, nutrient content and radial water and nutrient transport in barley, *Plant, Cell and Environment* 40(8), 1392–1408.

Schneider, H. M. and Lynch, J. P. (2018). Functional implications of root cortical senescence for soil resource capture, *Plant and Soil* 423(1–2), 13–26.

Schreiber, L. (2010). Transport barriers made of cutin, suberin and associated waxes, *Trends in Plant Science* 15(10), 546–553.

Seago, J. L., Marsh, L. C., Stevens, K. J., Soukup, A., Votrubova, O. and Enstone, D. E. (2005). A re-examination of the root cortex in wetland flowering plants with respect to aerenchyma, *Annals of Botany* 96(4), 565–579.

Siyiannis, V. F., Protonotarios, V. E., Zechmann, B., Chorianopoulou, S. N., Müller, M., Hawkesford, M. J. and Bouranis, D. L. (2012). Comparative spatiotemporal analysis of root aerenchyma formation processes in maize due to sulphate, nitrate or phosphate deprivation, *Protoplasma* 249(3), 671–686.

Sparks, E. E. and Benfey, P. N. (2017). The contribution of root systems to plant nutrient acquisition. In: *Plant Macronutrient Use Efficiency*. Chapter 5,. Copyright 2017 Elsevier Inc.

Tylová, E., Pecková, E., Blascheová, Z. and Soukup, A. (2017). Casparian bands and suberin lamellae in exodermis of lateral roots: an important trait of roots system response to abiotic stress factors, *Annals of Botany* 120(1), 71–85.

Üstün, S., Hafrén, A. and Hofius, D. (2017). Autophagy as a mediator of life and death in plants, *Current Opinion in Plant Biology* 40, 122–130.

van der Weele, C. M., Canny, M. J. and McCully, M. E. (1996). Water in aerenchyma spaces in roots. A fast diffusion path for solutes, *Plant and Soil* 184(1), 131–141.

van Doorn, W. G., Beers, E. P., Dangl, J. L., Frankling-Tong, V. E., Gallois, P., Hara-Nishimura, I., Jones, A. M., Kawai-Yamada, M., Lam, E., Mundy, J., Mur, L. A., Petersen, M., Smertenko, A., Taliansky, M., Van Breusegem, F., Wolpert, T., Woltering, E., Zhivotovsky, B. and Bozhkov, P. V. (2011). Morphological classification of plant cell deaths, *Cell Death and Differentiation* 18(8), 1241–1246.

Wang, Z., Mou, C., Chen, H., Chen., J., Wang, P., Yin, M., Fu, L., Li, Z., Su, F. and Hong, L. (2019). Effect of root cortical aerenchyma of hybrid corn (*Zea mays* L.) on maize resilience after rehydration under extreme drought, *Polish Journal of Environmental Studies* 28(3), 1413–1423.

Yamauchi, T., Rajhi, I. and Nakazono, M. (2011). Lysigenous aerenchyma formation in maize root is confined to cortical cells by regulation of genes related to generation and scavenging of reactive oxygen species, *Plant Signaling and Behavior* 6(5), 759–761.

Yamauchi, T., Shimamura, S., Nakazono, M. and Mochizuki, T. (2013). Aerenchyma formation in crop species: a review, *Field Crops Research* 152, 8–16.

Yang, J. T., Schneider, H. M., Brown, K. M. and Lynch, J. P. (2019). Genotypic variation and nitrogen stress effects on root anatomy in maize are node-specific, *Journal of Experimental Botany* 70(19), 5311–5325.

Yang, X., Li, Y., Ren, B., Ding, L., Gao, C., Shen, Q. and Guo, S. (2012). Drought-induced root aerenchyma formation restricts water uptake in rice seedlings supplied with nitrate, *Plant and Cell Physiology* 53(3), 495–504.

York, L. M., Nord, E. A. and Lynch, J. P. (2013). Integration of root phenes for soil resource acquisition, *Frontiers in Plant Science* 4, Article 355.

Zhu, J., Brown, K. M. and Lynch, J. P. (2010). Root cortical aerenchyma improves the drought tolerance of maize (Zea mays L.), *Plant, Cell and Environment* 33(5), 740–749.

Chapter 5

Advances in understanding plant root hairs in relation to nutrient acquisition and crop root function

Timothy S. George and Lawrie K. Brown, The James Hutton Institute, UK; and A. Glyn Bengough, University of Dundee, UK

1 Introduction

Root hairs are virtually ubiquitous in flowering plants, being absent from only a few orders of the angiosperms tested thus far. They are known to be a very ancient trait in the development of the earliest terrestrial plants, but it is clear that some species rely more heavily on root hairs than others for acquisition of resources (Honkanen and Dolan 2016). Root hairs massively increase the interaction between the plant and the soil, with early estimates by Dittmer (1937) that a single rye plant had 14 billion root hairs that provided a potential surface area in contact with the soil of 400 m^2. With this in mind, it is clear that root hairs are critical in understanding the interface between plants and soil, which is a key zone in the cycling of water and elements on a global scale (White et al. 2013). Influx channels and transporters for phosphate and potassium, water, ammonium, nitrate, urea, calcium, peptides and sulphate have been found in root hairs (reviewed in Libault et al. 2010). Root hairs play a particularly critical role in the acquisition of nutrients considered to be immobile, such as phosphorus and potassium, as they increase the volume

http://dx.doi.org/10.19103/AS.2020.0075.06

of the soil explored. They also allow the bridging of air-filled voids, gain access into smaller voids inaccessible to roots, stabilise soils and are critical in structural formation. Collectively, these functions greatly decrease the length of the path of diffusion of immobile nutrients to the root surface for uptake. Root hairs also play a role in modulating the properties and composition of the rhizosphere. Root hairs are involved in the exudation of an array of organic compounds, including organic acids, amino acids, proteins, polysaccharides, phenolics and secondary metabolites including volatiles. There is a variation in the length, density and longevity of root hairs, between and within species, suggesting that root hair length and density are under genetic control. Several quantitative trait loci (QTLs) controlling root hair abundance have been identified in a range of species (Wang et al. 2004; Yan et al. 2004; Zhu et al. 2005). The anatomy of root hairs falls into three different types, for all angiosperms, based on the relative position of root hair cells on the epidermis (Datta et al. 2011; Honkanen and Dolan 2016). Interactions between root hairs and soil lead to trade-offs between this trait and others and an optimum root hair length can be ascertained for different conditions (Brown et al. 2013b). Notwithstanding this, the root hair zone of plants is critical for the utilisation of water and nutrients and for the interaction of plants with microorganisms. They are fundamental in many functions of the rhizosphere that have an impact on issues of global significance such as food security, agricultural sustainability, evolution of greenhouse gas emissions and storage of carbon belowground. In this chapter we will highlight our current understanding on the physiology and development of root hairs, discuss their genetic variation, their function and interactions with soil. We will look specifically at their role in biophysical interactions with soil and discuss the impacts of this on the ability of the plants to acquire nutrients and water. Finally, we will discuss the role of root hairs on crop performance in the field.

2 Root hair physiology and development

Root hairs develop as long extensions from the root epidermal cells. After the formation of an initial bulge at the distal end of the epidermal cell, the root hair structure elongates by tip growth. In Fig. 1i we see the typical development of the root hair zone in the more mature part of the root tip of the plant. Behind the root cap the meristematic zone is the site of new cell differentiation and growth pushing the root cap ahead of it. Behind this is the elongation zone and then the differentiation zone, where cells age and differentiate into root hair cells, those performing different functions. The difference in age between the cells in the meristematic zone and differentiation zone is a matter of days. Once differentiated the root hairs fully elongate, function and then wither and die or cease to function, but remain in place. The time course for the functioning

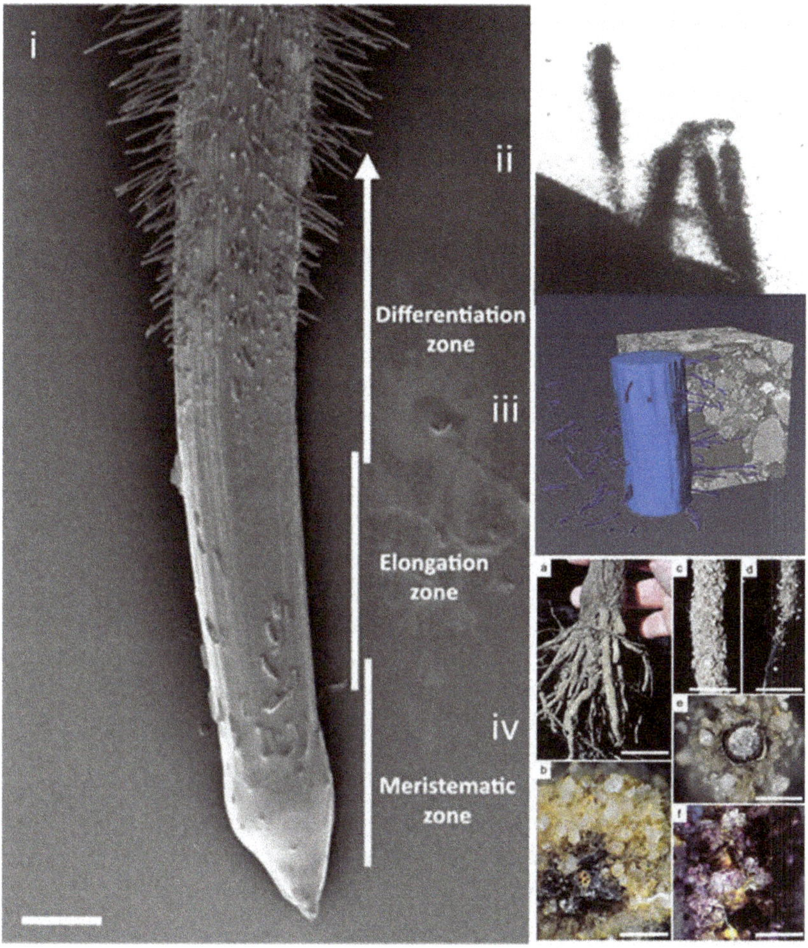

Figure 1 Images of root hair development and interaction with the soil. (i) Classical image of root hair development in the differentiation zone some distance back from the root tip; (ii) root hairs captured using autoradiography; (iii) synchrotron X-ray CT image of root hairs growing in soil; and (iv a–f) a range of images showing the complexity of root hairs interacting with real soil (Figures reproduced with the permission of Datta et al. 2011; Belford and Preston 1961; Köebernick et al. 2017; Smith et al. 2011).

of the root hair zone remains to be satisfactorily quantified, but could feasibly range from a few days to weeks or months.

Textbook illustrations of the formation of the root hair zone (e.g. Fig. 1i) are generally produced in controlled conditions optimal for unimpeded root growth and development. Roots growing in field soils contend with physical, chemical and biological stresses. Also, it is apparent that in some species, and under certain circumstances, root hairs can form on a number of root types and all along the length of the roots and still function for long periods on mature plants (Goron et al. 2015; Nestler et al. 2016).

Some of the earliest images of root hairs grown *in situ* in soil were captured using autoradiographs of roots labelled with ^{14}C isotopes, as can be seen in Fig. 1ii. This early work showed that the rapid wall synthesis occurred over a length of about 120 μm back from the tip of the root hair cells (Belford and Preston 1961). They also demonstrated that the microfibrillar material deposited as the cytoskeleton was cellulose and that over the hemispherical tip of the hair the microfibrils are arranged at random. The lamella carrying these microfibrils reaches almost the tip, while the inner wall lamella has microfibrils running longitudinally. Both these lamella are encrusted with material ejected from the root tip. This early work also suggests that the extreme tip of the hair is the site of cellulose synthesis and rapid growth in the wall area (Belford and Preston 1961).

More recently we have been able to image root hairs non-destructively, *in situ,* in soil using synchrotron X-ray CT scanning as seen in Fig. 1iii (Köebernick et al. 2017; Keyes et al. 2013). Here it is possible to see the intimate interaction between the root hairs and the soil, and derive the root hair length and density *in-situ*. However, currently it is only possible to capture images like this on young roots grown in very small volumes of soil, and the reality of root hair growth in field soil is an order of magnitude more complex than this. Figure 1iv (a–f) shows some excellent examples of the interaction between root hairs and soil on roots from plants grown in the field. These illustrate the complexity of this interaction and suggest that the many studies on root hairs performed on model plant species in hydroponics may not give us all the answers to understand the role of root hairs in acquisition of resources and tolerance to stress in the real-world environment. We will go on to discuss this in more detail in the later sections.

Elongation of root hairs occurs by growth of the tip, with cytoplasmic organelles and cytoskeletal elements showing a polarized distribution: the apical dome of the root hair contains many vesicles, associated with cell-wall synthesis (Peterson and Farquhar 1996; Gilroy and Jones 2000; Grierson et al. 2014). Root hair development begins with swelling of the cell wall on the outer surface of the epidermis, expanding relatively slowly at first, but accelerating to up to 2 μm min^{-1} at its maximum (Honkanen and Dolan 2016). Much research into root hair elongation has been focussed on the relationship between cellulose microfibril deposition and the pattern of cortical microtubules, the role of the cytoskeleton and the distribution of calcium in the root hair controlling cytoplasmic streaming. For example, experiments with actin interfering drugs suggest that the actin cytoskeleton is an important factor in the establishment of cell polarity and in the maintenance of the tip growth at the apex of the growing root hair (Ringli et al. 2002). The essential role of specific proteins (GTP-binding) in regulating intercellular membrane trafficking and maintaining polar growth has also been studied (Song et al. 2006; Zhang et al. 2018b) and has been shown to be essential for the development of root hair elongation (Yoo et al. 2018). Cytoplasmic Ca is known to regulate cytoskeletal rearrangements and vesicular trafficking in root hairs (Zhang et al. 2018b). Early research discovered that a Ca gradient existed in the growing root hairs (Cormack et al. 1963), and it was later shown that this was most pronounced in the rapidly elongating root hairs (Wymer et al. 1997). The establishment of a Ca gradient in root hairs is critical for root hair elongation, and its intensity dictates the final length of the root hairs (Gilroy and Jones 2000; Grierson et al. 2014).

Much work has been performed on the morphological distribution of root hairs in many plant species in the first half of the last century (reviewed by Cormack 1962 and more recently by Grierson et al. 2014) and this seminal work has been built on in recent years by a number of researchers (Dolan et al. 1994; Datta et al. 2011; Honkanen and Dolan 2016). This has led to a consensus on the morphology of root hairs, which generally fall into three types of root hair distribution on the root epidermis (Fig. 2; Datta et al. 2011). In type 1 patterning of root hair development all epidermal cells start off morphologically identical and any cell in the epidermis can differentiate into a root hair cell. The relative numbers of hair cells and non-hair cells varies depending on environmental conditions, leading to the development of roots without hairs under some conditions and almost complete coverage of root hairs under others (Datta et al. 2011). Root hair cells can also develop from specialized epidermal cells called trichoblasts, which differentiate into root hair cells and are morphologically and functionally different from other epidermal cells (atrichoblasts). Before root hair initiation, trichoblasts are smaller and more densely cytoplasmic than the surrounding atrichoblasts that differentiate as hairless epidermal cells (Leavitt 1904). Trichoblasts develop in two distinct patterns (Fig. 2) – trichoblasts can alternate with atrichoblasts along longitudinal epidermal cell files (type 2 development) or trichoblasts and atrichoblasts can develop in separate longitudinal files (type 3 development). The type 2 alternating pattern of cell types along a file results from asymmetry in the last cell division just before epidermal cells leave the meristematic zone (Datta et al. 2011). Asymmetry of this division results in the formation of two daughter cells that differ in size and will divide no further; the relatively small cell forms a trichoblast and the larger cell develops as an atrichoblast. In contrast, in type 3 development trichoblasts and atrichoblasts are differentiated early in development in the rootward regions of the meristem and develop in distinct cell files (Datta et al. 2011). The phylogenetic variation in these morphological aspects of root hair formation will be discussed in the next section.

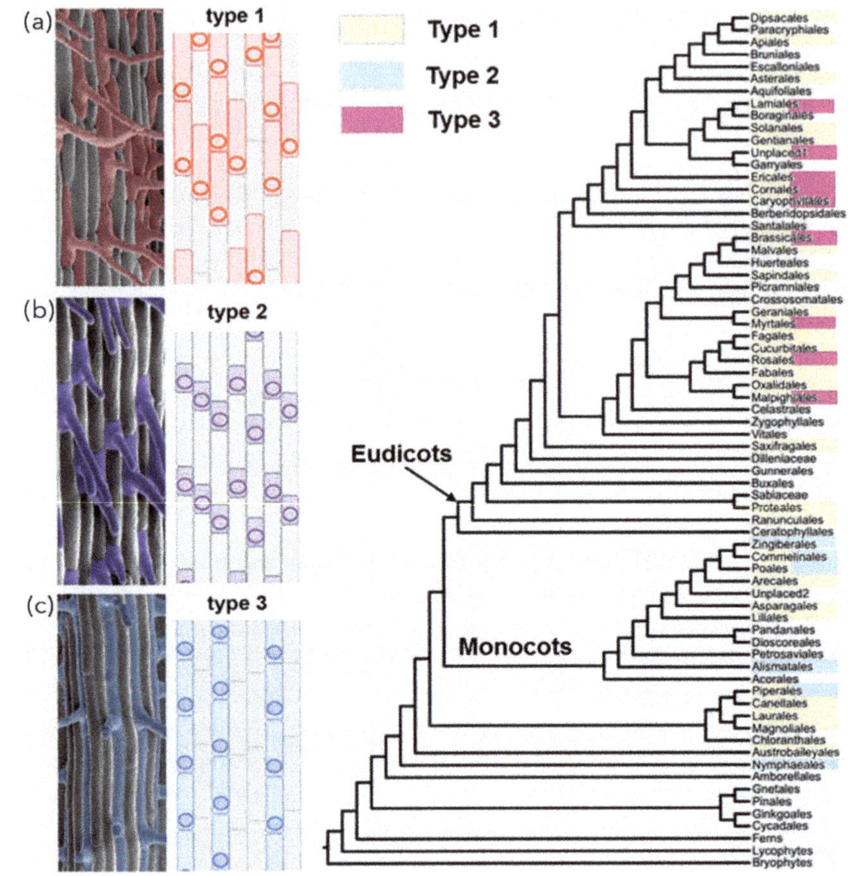

Figure 2 Illustration of root hair cell file distribution types and the phylogeny of this phenomenon in the angiosperms (Figure adapted from and reproduced with permission by Datta et al. 2011 and Huang 2016).

3 Phylogenetic and genetic variation in root hairs

Many of the root hair traits described are genetically controlled, highly heritable and show considerable genotypic and species variation. The overlaying of highly heritable root hair traits like root hair patterning (discussed in Section 2) onto the Angiosperm Phylogeny Group (2009) can provide indications to the possible lineage of root hair development in the angiosperms (Fig. 2).

The development patterns of root hairs are described above and seen in Fig. 2. Type 1 development, where root hairs can develop from any epidermal cells, is found in many plant species across the monocots and eudicots. In contrast, type 2 development, where trichoblasts develop into root hairs alternately along a cell file, is found mainly in old species of land plants like

Lycopsida or Sphenopsida (Datta et al. 2011) in many monocot species and some families of dicots. Type 2 development can be further sub-divided depending on the polarity of the asymmetric cell division where the smaller daughter cell may form either in the shootward position in magnoliids and in some monocots (Cutter and Feldman 1970; Cutter and Hung 1972; Clowes 2000). In contrast, the smaller daughter cells formed in the rootward position are found in closely related monocot families, such as Restionaceae, Juncaceae, Cyperaceae and some Poaceae (Sinnott and Bloch 1939; Reeder and von Maltzahn 1953; Row and Reeder 1957; Clowes 2000). The distribution of these two types of cell division across the phylogeny of angiosperms is indicative that the type 2 pattern evolved at least three times independently, once in the magnoliids and twice among the monocots (APG III 2009). Different again are the type 3 root hairs, which form a striped pattern, where files of trichoblasts and atrichoblasts develop longitudinally (Cormack 1947; Dolan et al. 1994; Galway et al. 1994; Clowes 2000; Pemberton et al. 2001). This type is only present in the eudicots and can be found among many families of the Caryophylales, Brassicales and in a few species in other diverse families including the Boraginaceae (Clowes 2000; Pemberton et al. 2001). A possible variant of type 3 development is found in *Echium plantagineum* L. which develops root hairs in the trichoblasts and follows this up with a second wave of hair development in the atrichoblasts (Tsai et al. 2003). Consequently, all epidermal cells develop root hairs in this case. It appears that type 3 trichoblast development evolved many times during eudicot evolution (APG III 2009).

Considering these findings that Huang (2016) and Huang et al. (2017) suggest, it may be that the type 1 pattern is the most primitive form and evolved into type 2 and type 3 patterns at a later time during land plants' adaptation. However, because both type 2 and type 3 patterns have been found in multiple lineages, where they co-exist with type 1 pattern, it is equally possible that these lineages lost or gained the root hair patterning independently. Our understanding of root hair evolution, development and function in grasses (type 1 and 2) is of importance, given that many of the world's crop plants are monocots (Poales). To date much work has been done on Arabidopsis (type 3) to identify genes associated with root hairs. The underlying genetic networks which are revealed will provide new molecular tools to allow adaptations to the environment in which root hairs are central players (Shane et al. 2005; Abdolzadeh et al. 2010), but if only limited to Arabidopsis, the research will ignore two major types of root hair formation and their associated genetic diversity. Efforts now need to be extended into grasses which would have large implications on future crop breeding as roots hairs are important contributors to yield (Gahoonia and Nielsen 2003) and allow plants to deal with various abiotic stresses like nutrient deficiency and water availability (Datta et al. 2011), as we will see later.

Large variations in root hair length (Fig. 3) reported by Brown et al. (2017) in both monocot and eudicot species were found throughout the angiosperms, ranging from the complete absence of root hairs in *Allium* spp., to lengths of over 1.4 mm in *Hordeum vulgare* and *Silene latifolia*. Kidd et al. (2016) reported root hair length variation in pasture legumes and grasses where there was over a 6-fold variation in root hair length among the species. Grasses had the longest root hairs (>0.65 mm), *Lupinus* spp., *Ornithopus* spp. and *B. pelecinus* also had relatively long root hairs (>0.40 mm), while many *Trifolium* spp. had root hairs shorter than 0.2 mm. Images of such inter species variation in root hair length and density can be seen in Fig. 3, which illustrates how root hair length and density can vary markedly between species in the zone of differentiation.

The highly heritable nature of root hair traits means that they can be screened for, and many studies have measured root hairs and root hair density using traditional methods, as described by Gahoonia et al. (1997). Such methods, which involve microscopy and imaging followed by image analysis are very time-consuming. Roots, with apparently intact root hairs, can be washed readily from intact field samples of sandy soils, provided care is taken to properly saturate the soil samples before using gentle agitation. Both length and density of root hairs present vary greatly even within root samples of a single genotype. Quantifying the density of root hairs (number per unit length) and length of root hairs (measured relative to the root surface) is a daunting task, because of the large variation both within and between samples. Ranking scales, where root hair density of samples is visually compared with standard images on a progressive scale, can be very valuable in speeding up the process and can be used semi-quantitatively. This approach was applied to wheat and barley seedlings allowing root hair density in three dimensions to be predicted

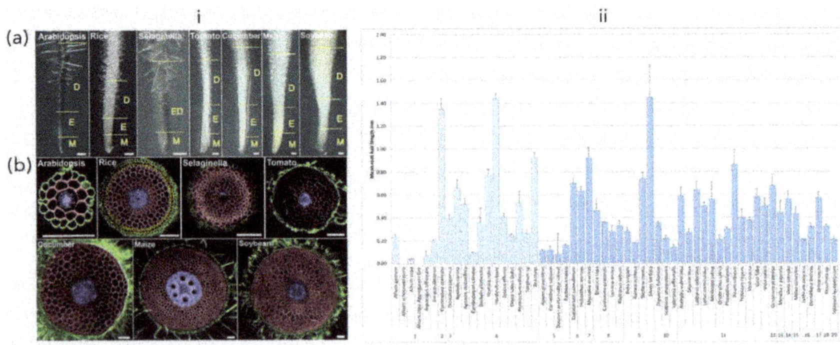

Figure 3 Interspecific variation in root hair density and length in a range of monocot and dicot species (i) Images showing patterns of root hair differentiation zones and cross-sectional density of root hairs for a range of species and (ii) root hair length of a number of both monocot and dicot species (Figures reproduced with permission from Huang and Schiefelbein 2015 and Brown et al. 2017).

successfully from the two-dimensional visual ranking score, using a quadratic regression that accounted for 95% of the variation (i.e. R2 = 95%; Haling et al. 2010).

More recently researchers have employed the 'rhizosheath' mass per unit root length as a proxy for root hair length, which facilitates faster screening of large populations to identify genetic variation (George et al. 2014; Delhaize et al. 2015; James et al. 2016; Adu et al. 2017; Pang et al. 2017). The rhizosheath is the mass of soil that adheres strongly to roots on excavation (George et al. 2014). As the size of the rhizosheath is dependent on root hair traits, the visual assessment and measurement of the rhizosheath has been used as a visual indicator for underlying root hair traits, while imposing minimal disturbance to the root system (Delhaize et al. 2012; Haling et al. 2010)

Rhizosheath size has also been used in phylogenetic studies to investigate the evolution of this belowground phenomenon which is associated with root hair traits. Smith et al. (2011) studied species within one family, while Brown et al. (2017) screened 57 species covering 19 orders throughout the angiosperms. The current data do not provide a detailed understanding of the evolution of the rhizosheath trait in the angiosperms, but it is likely that because the prevalence of the trait differs markedly between closely related orders, this trait evolved independently in different branches of the angiosperm phylogeny in response to contrasting environmental stresses (Davies et al. 2004). Combining phylogenetic screens of various root hair traits can help to build a clearer picture of the heredity of beneficial traits and be of value when looking for the genes associated with root hairs for future breeding. Large-scale studies combining phylogenetic and transcriptome analyses to define and compare root hair genes from seven diverse vascular plant species, including eudicots, monocots and a lycophyte have been performed. A major finding of these was that most root hair genes are similar in structure and expression in all species tested, suggesting that a core programme for root hair development and function is conserved across the vascular plants.

Rhizosheath size has been found to have good correlations with root hair length, in the case of wheat (James et al. 2016), cocksfoot and phalaris (Haling et al. 2010), but to a lesser extent in barley (George et al. 2014) and maize (Adu et al. 2017). Brown et al. (2017) demonstrated good correlations for root hairs in multiple species where root hair length was less than 0.28 mm but the relationship appears to break down as root hairs become longer, confirming other factors affecting rhizosheath formation and size, for example, exudates or microbial associations play an important role. James et al. (2016) also employed rhizosheath as a surrogate for root hair length to show that it is one factor that can improve the phosphorus acquisition efficiency of wheat grown on acid soils. Brown et al. (2017) proposed a consistent methodology for rhizosheath measurement and Yeates (2018) has developed a rhizosheath selection tool using perennial ryegrass as an aid to rapid selective screening of large populations, but it is fair to say that rhizosheath is not a universal proxy for the assessment of root hair length.

Given the importance of root hair length and density in the acquisition of nutrients and water by the root system (see later sections) it is interesting to note that there is a variation in these traits both within and between species. Differences in root hair traits between different species (Dittmer 1949; Gahoonia et al. 1997; Brown et al. 2017) have been reported, suggesting that root hair length and density are under genetic control. It is not only interspecies genetic variation in root hair traits that is of interest to researchers investigating the genetics behind them, but also the intraspecies variation. Large variation has been shown in root hair lengths in individual species and effort has been invested in the mapping of QTLs associated with root hairs (Wissuwa et al. 2009). Several QTLs controlling root hair abundance have been identified in *Glycine max* L., *Phaseolus vulgaris* L. and *Zea mays* L. (Wang et al. 2004; Yan et al. 2004; Zhu et al. 2005; Liu et al. 2017b; Horn et al. 2016). Work on barley has also shown complex genetic control of root hair development, revealing that at least nine loci are responsible (Chmielewska et al. 2014). Transcriptomic research in rice shows more than 400 root hair preferential genes (Moon et al. 2018).

There are complex sets of genes, regulators and biosynthetic pathways involved in the development and growth of root hairs in plants (Schiefelbein 2000; Bibikova and Gilroy 2002; Jones et al. 2006; Grierson et al. 2014; Vijayakumar et al. 2016; Moon et al. 2019). The gene regulatory network of vascular plants is similar to that for rhizoids in more primitive plants such as mosses, suggesting that these networks evolved in the earliest land plants (Jones and Dolan 2012) and this is backed up by fossil evidence of root hairs in primitive plants (Honkanen and Dolan 2016). With more than 800 genes exclusively expressed in root hairs and more than 5000 genes preferentially expressed (Hey et al. 2017), the complexity is manifest. In addition, there are epigenetic controls on root hair formation and development through regulation by chromatin modification (Zhu et al. 2017). Much of the research into the genetic control of root hairs has been performed in the model species *Arabidopsis thaliana*, but many of the genes have orthologs in other species including common cereals (Hey et al. 2017).

Many of the genes upregulated in root hairs are associated with energy metabolism reflecting the large requirement for energy in root hair growth (Hey et al. 2017). However, other gene families that have been implicated in root hair development include those involved in initiation, growth, longevity and death of root hair cells (Wada et al. 2015; Vijayakumar et al. 2016; Tanaka et al. 2017). A range of genes have been shown to be involved in the development of the cytoskeleton and cell-wall morphogenesis through α-tubulin expression (Bao et al. 2001; Ringli et al. 2002), LRR-extensin cell wall proteins (Baumberger et al. 2003) and in promoting increased growth in length of the root hairs through auxin transport to the root hair differentiation zone (Bhosale et al. 2018). In addition, there are genes, such as those that regulate receptor kinases, which are involved in the negative regulation of root hair length by moderating root cell wall-associated signals and limiting the length to which root hairs are able to grow (Hwang et al. 2016).

Phytohormone biosynthetic pathways involved in cytokinin, jasmonate, ethylene and auxin production and interactions between these are also involved in regulating the development of root hairs (Liu et al. 2018b; Masucci and Schiefelbein 1996; Rahman et al. 2002; Shi and Zhu 2002; Zhu et al. 2006; Honkanen and Dolan 2016; Salazar-Henao et al. 2016; Zhang et al. 2016). If these phytohormones are perturbed they have been shown to interfere with the agravitropic and lateral response in roots and root hair elongation (Santelia et al. 2005; Rigas et al. 2013) and interfere with the plants' root hair response to environmental cues such as P deficiency (Zhang et al. 2003). In fact, when plants are subjected to microgravity critical genes regulating phytohormone sensing required for tip growth are downregulated suggesting that gravity is critical for sustained root hair growth (Kwon et al. 2015). Root hair density can also be influenced by a range of genes including those involved in nitrate signalling where increased nitrate stimulates increased density through nitrate signalling genes directly regulating genes specifying the fate of root hair cells (Canales et al. 2017) as was also the case for genes regulated by phosphate availability (Stetter et al. 2017). Similarly, genes involved in root hair development and length have been shown to be regulated by phosphate availability (Kirchner et al. 2018). Interactions between root hairs and environmental cues will be discussed further in a later section.

Then there are those genes that have only been shown to promote root hair growth, such as genes involved in cytokinin response associated with roots (Cheng et al. 2016) or in the production of basic Helix-Loop-Helix (bHLH) transcription factors in a number of species including barley (Gajewska et al. 2018; Honkanen and Dolan 2016; Karas et al. 2009). One such bHLH transcription factor (RSL4) has been shown to be necessary for root hair growth and elongation in arabidopsis. Moreover, it has been shown that this transcript is produced in a pulse at root hair initiation and that its degradation to a baseline level determines the length of the root hair (Datta et al. 2015). Importantly, the intensity of the pulse and therefore the length of time over which elongation occurs are determined by environmental factors such as P availability (Honkanen and Dolan 2016). Similarly, RSL orthologs in wheat have been shown to be involved in root hair length determination (Zhang et al. 2018a) and have dynamically changed during allopolyploid evolution in wheat (Han et al. 2017) in that polyploidisation has increased the copy number of

RSL genes, expression and root hair length from diploid ancestors to hexaploid bread wheat (Honkanen and Dolan 2016; Stetter et al. 2015). Other genes in rice have been shown to promote root hair growth by mobilising auxin from the root tip and redistributing it to the root hair differentiation zone, where it is involved in regulating root hair length (Giri et al. 2018). In barley, β-expansin genes have been shown to be involved in the regulation of root hair length under drought (He et al. 2015).

A useful development for access to information on the genetic control of root hair phenotypes is the development of the database of root hair genomics, iRootHair (Kwasniewski et al. 2013). This database includes information on >150 root hair-related genes that have been identified in a number of dicot and monocot species along with their putative orthologs in higher plants with sequenced genomes.

We have seen that the patterning of root hair development is under genetic control and that physiology of root hairs with respect to length and density are also genetically controlled. However, root hairs have also been shown to be sensitive to environmental conditions such as nutrient deficiencies and water availability (Bates and Lynch 1996; Yan et al. 2004). Examples of this plasticity of root hairs to abiotic stresses have been described in barley by Brown et al. (2012) and Haling et al. (2011) and in maize by Zhu et al. (2010) and will be discussed in the next sections of this chapter.

4 Root hair interactions with the environment

4.1 Rhizosphere development and growth of root hairs in soil

As described earlier, root hairs initially emerge as epidermal cells and cease expanding to reach their final length at the back of the root elongation zone: in *Arabidopsis thaliana*, timelapse particle image velocimetry of confocal images of roots showed that cell expansion ceased at the point that root hairs emerged (Bengough et al. 2010). The length of the elongation zone varies according to species and growth conditions, with larger fast-growing roots generally having longer elongation zones. The length of the bare root apex (before root hairs emerge, causing soil particles to adhere) was up to 14 mm for fast-growing 1 mm diameter primary maize roots, but this decreased linearly with elongation rate, to only 2–3 mm in sandy loam soil compacted to 1.5 gcm^{-3} dry bulk density, with matric potential of −1.2 MPa (Schmidt et al. 2012).

Once root hairs emerge, they elongate at a rate that depends on their local environment and, particularly, the soil matric potential and the soil physical resistance to deformation at the root hair scale. Root hairs grown in soil from −5 to −10 kPa matric potential were shorter in barley (Haling et al. 2013) and in maize (Mackay and Barber 1987), than for roots grown in soil from −30 to −175 kPa. An increase in soil dry-bulk density from 1.2 to 1.7 gcm^{-3} halved the length of root hairs in barley, associated with increasing penetrometer resistance from 0.03 to 4.45 MPa (Haling et al. 2013). This is a smaller decrease than would normally be expected for mechanically impeding the elongation of a primary root (Bengough et al. 2011), and arises perhaps because of the relatively small size of root hairs. Due to the thin (e.g. 10 μm) diameter of root hairs, the resistance experienced by a root hair will be very variable at the scale of the root hair tip, according to the amount of pore-space available between particles. In soil, sand particles have a diameter between 6 and 200-fold greater than the diameter of the root hair, probably constraining root hair extension largely to the pre-existing continuous pore space in the soil. Given that a root hair-sized pore will drain at a matric potential of approximately −30 kPa, it may be that the root hairs proliferate most rapidly within this drained humid pore space, rather than within a completely water-filled pore.

4.2 Root hair function in structured soil

Root hairs have long been recognised as providing a proportionately large extra surface area of contact between the root and the soil. This aspect is logically of greatest value when roots are growing through soil that contains a large proportion of pore space, or when roots are growing within relatively porous regions of an otherwise dense soil. In field soils that are macroscopically structured, containing substantial cracks, channels and biopores, roots often grow preferentially within such macropores (Tardieu 1988; Passioura 1991; Stirzaker et al. 1996). Within such a macropore environment root-soil contact may occur partially or not at all, along regions of the root length, unless the root can somehow secure itself to the pore wall. Endoscope studies of roots within biopores have revealed that different species may adopt contrasting strategies to maintain contact, with axes of barley tracking or spiralling along biopore walls, often with the proliferation of root hairs helping to adhere the root surface to the pore wall (Athmann et al. 2013). In contrast, oil seed rape main axes tended to grow more centrally within macropores, with many fine lateral roots achieving contact with the pore walls (Athmann et al. 2013).

The degree of mechanical adhesion between roots and soil that is provided by root hairs has been tested mechanically using a peel test, and by pullout tests of roots of maize from predrilled holes (Czarnes et al. 1999; Bengough et al. 2016). The degree of anchorage that root hairs afforded depended on soil texture and density, probably because this influenced the way that root hairs became entangled and interlocked with the soil matrix. The anchorage of root hairs in sandy loam soil was greatest at a dry bulk density of 1.2 gcm^{-3}: in dense soil (1.5 gcm^{-3}) root hairs did not penetrate the matrix sufficiently to become well-anchored, whereas in very loose soil (1.0 gcm^{-3}), the soil matrix had insufficient strength to provide much anchorage (Bengough et al. 2016).

The role of root hairs in mechanical anchorage of roots to the surrounding soil and to pore-wall surfaces should aid roots to grow across gradients in soil strength, from loose soil into hard soil, and from pores into the bulk soil matrix. This has been shown experimentally using hairless mutants and hairy wild-type roots of both barley and maize in repacked soil cores (Haling et al. 2013; Bengough et al. 2016). Eighty-eight percent of wild type barley roots penetrated from a loose soil layer (dry bulk density 1.2 gcm^{-3}) into a dense layer (1.7 gcm^{-3}), as compared with only 1% of the hairless barley mutant (Haling et al. 2013). Similarly, wild type primary roots of maize penetrated from artificial macropores into bulk soil at a density of 1.2 gcm^{-3} more effectively than hairless mutants, although this effect was lost at a density of 1.5 gcm^{3} because root hairs did not anchor sufficiently in the dense soil matrix. Overall, root hairs increased the root tip anchorage by up to five-fold as compared with hairless root mutants (Bengough et al. 2006).

During a diurnal cycle, roots may undergo substantial shrinkage in their diameter during periods of transpiration if the root-soil interface becomes a main limiting hydraulic resistance to transpiration. Theoretical models of this process have predicted that water inflow rate may be slowed to only one-third of the rate of that in a fully hydrated root, due to the shrinkage-associated loss of contact (Nye 1994). Decrease in diameter was first observed experimentally using timelapse photography of a single lateral root of cotton, where the diameter shrank by up to 40% in the afternoon when the plant was transpiring rapidly (Huck et al. 1970). Interestingly this much-cited study of an individual lateral root remained the only direct observation of root shrinkage for some 39 years until the use of X-ray microtomography for this purpose in 2009 (Carminati et al. 2009), although other less direct evidence supported the suggestion: for example, shrinkage in diameter by up to 60% was observed in excised segments of wheat roots under air drying (Cole and Alston 1974), although this represents a relatively severe treatment in a relatively artificial situation.

Substantial root shrinkage must decrease contact between roots and the surrounding soil, or macropore wall, unless root hairs maintain contact and anchor the root to the soil. The advent of high-resolution microtomography has enabled the degree of close contact to be quantified in soil-grown roots (Schmidt et al. 2012), and the degree of root shrinkage and gap formation to be imaged dynamically (Carminati et al. 2009, 2013). Shrinkage of a lupin taproot in very sandy soil (0.92 gg^{-1} sand) was readily apparent and associated with gap formation at the root surface (Carminati et al. 2009). Further work by the same research group suggested that the gap forms as a consequence of the decrease in water supply by the rhizosphere driving a decrease in the water potential of the root tissue (Carminati et al. 2013). Much smaller gaps were observed around lateral roots, possibly due to slower water flux demand associated with the lateral roots maintaining the more hydrated root tissue. The role of root hairs in resisting gap formation was not specifically identified in these papers, although it is likely that more viable root hairs may be present on the lateral roots than on the main axis.

The role of root hairs in maintaining close hydraulic contact has been demonstrated clearly in an interesting series of experiments by Carminati et al. (2017), using a root pressure chamber. Hairless roots of mutant barley growing in sandy loam soil required a greater differential pressure to maintain the transpiration rate than wild type barley with root hairs present. This indicates that the root hairs minimised the drop in water potential at the root-soil interface, probably by maintaining better root-soil contact than was present in the hairless roots. Although root hairs lose viability in the older regions of the root, root hairs in field soils may remain physically present and presumably mechanically functional for many months (White and Kirkegaard 2010). In a fascinating study of root form in deep field soil, root hair density on wheat roots present in structural macropores of a dense Australian subsoil decreased approximately exponentially with increasing root-soil contact: where the root had 20% or less contact with the pore wall, there were 6–12 root hairs per mm, whereas if contact was more than 50% there were only 0–2 root hairs per mm (White and Kirkegaard 2010). Given that more than 85% of roots in the subsoil were located within pores and cracks, root hairs may play a substantial role in creating improved contact with the pore walls and improving uptake rates of deeper water reserves.

4.3 Resource uptake and carbon exudation

Root hairs, on many species, are initiated and grow longer and denser in response to the deficiency of many nutrients and water. It has been demonstrated that the length of root hairs is greatly affected by the concentration of many elements and that root hairs are the site of uptake (Gilroy and Jones 2000) of a range of nutrients including Ca, Mg, K, NH_4, NO_3, Fe, Mn, Zn, Cl and PO_4 (Liu et al. 2017a; Tanaka et al. 2014; Grierson et al. 2014; Gilroy and Jones 2000; Ewens and Leigh 1985; Klinsawang et al. 2018; Peterson and Farquhar, 1996). The availability of Ca, for example, has a profound effect on root hair length, with Ca-deficient plants having much shorter root hairs (Peterson and Farquhar 1996). This is likely to be due to the requirement of Ca gradients within the root hair cell which drives root hair elongation, but could also be due to the lack of competition against toxic elements such as Al. It has also been demonstrated that the concentration of nitrate is more effective at regulating root hair length and density than ammonium (Vatter et al. 2015) and that root hairs are the sites of uptake of both NO_3 and NH_4 with a number of specific N transporters expressed in root hairs (Gilroy and Jones 2000).

The most often studied role of root hairs in nutrient acquisition is for immobile nutrients, such as phosphate (P). The driving force for P uptake in plants is provided by the electrochemical gradient across the plasma membrane, which is generated by H^+-ATPase whose gene expression in root hairs has been demonstrated, specifically occurring in developing root hairs and reduced in mature root hair zones (Gilroy and Jones 2000). Many species are shown to have P-regulated root hair production, including Arabidopsis, barley and rice (Gahoonia et al. 1997; Bates and Lynch 2000; Brown et al. 2012; Tanaka et al. 2014; Nestler and Wissuwa 2016; Wada et al. 2015). The expression of root hairs in P-deficient plants has been shown to make up >90% of the root surface area and increase P uptake by 60% (Gilroy and Jones 2000). This increase in root hair density is related not only to many more epidermal cells forming root hairs in root hair-forming cell files, but also an increase in the number of root hair-forming cell files (Grierson et al. 2014). It has also been demonstrated that there is an intraspecific variation in the response in root hairs to P, with genotypes of wheat with sparse short root hairs under sufficient P conditions being the genotypes that react the most to P-deficiency (Stetter et al. 2015) and some genotypes having increased root hair density, while others have increased root hair length with P deficiency (Wang et al. 2016). In addition to the role of P availability in initiating root hairs and influencing their morphology, it has also been shown that a number of other phenomenon are induced in root hairs under P deficiency, including cell cycle control, putative plastid-to-nucleus signalling, pathogen defence, reprogramming of cell wall-related carbohydrate metabolism and chromatin remodelling (Salazar-Henao et al. 2016).

One of the key functions of root hairs is to be the site of transporters for P uptake (Daram et al. 1998; Schünmann et al. 2004) and the release of P-active exudates (Zhang et al. 2018c; Giles et al. 2018) (Fig. 4i,ii). In Fig. 4ii we see the expression of a phosphate transporter promoting sequence cloned from arabidopsis expressed solely in the root hair forming cells, confirming the role of root hairs in P acquisition from the soil (Schünmann et al. 2004). Notwithstanding this, it has been suggested that the expression of P transporters in root hairs alone does not increase P-acquisition, but that it is more effective when in combination with other P-responsive traits (Brown et al. 2013b; Zhang et al. 2018c). The additional role of root hairs as the site of exudation of P active compounds, such as phosphatase, can be seen in Fig. 4iii.

It is clear from studies of root hairless mutants in several plant species that the presence of root hairs is critical for acquisition of P and plants lacking root hairs are at a disadvantage with regard to growth and P uptake relative to their wild-type equivalents (Gahoonia et al. 1997; Bates and Lynch 2000; Brown et al. 2012; Klamer et al. 2019). However, it has also been shown that root hairs contribute most from the specific patches in soil where root hairs

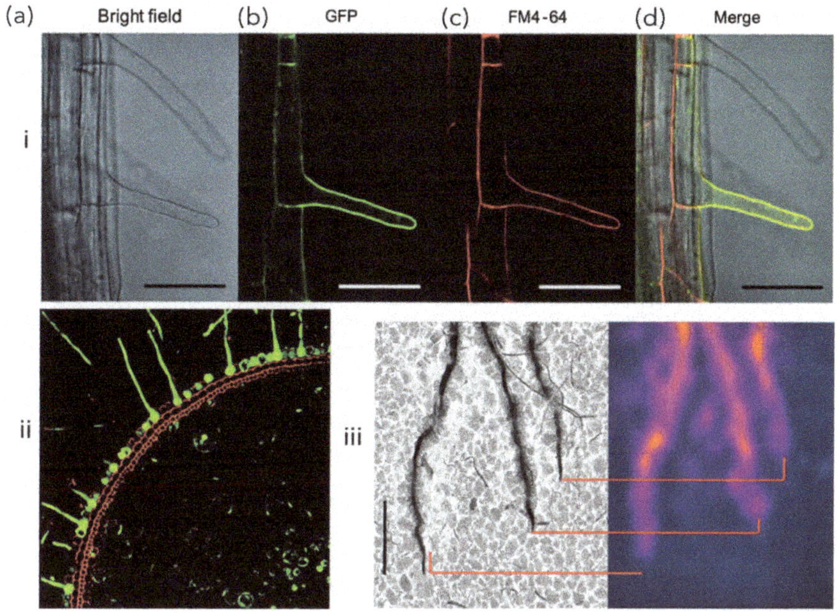

Figure 4 Functions of root hairs. Images showing sites of expression of (i) organic acid exudate transporters, (ii) nutrient transporters, and (iii) release of exudates in root hair zones (images reproduced with permission from Maruyama et al. 2019; Schünmann et al 2004; Giles et al. 2018).

proliferate and orientations are optimised (Keyes et al. 2013, 2017). Importantly, root hairs have been shown to be critical for P uptake in compacted soils with root hairs giving the roots greater ability to penetrate the hard soil and access greater soil resources (Haling et al. 2013). Moreover, it has also been shown that the absence of root hairs is not compensated for by the upregulation of other P-responsive traits, such as longer roots with more laterals or enhanced mycorrhizal symbiosis (Brown et al. 2012, 2013a; Klamer et al. 2019).

Root hair length and density can also be affected by the presence of toxic elements, such as Cd and Al, which tend to reduce the effectiveness of root hairs (Delhaize et al. 2012; Kohanová et al. 2018). Also, responses to excess salt (NaCl) lead to bladder like root hairs in some species and a reduced ability of root hairs to form entry points for rhizobium in some leguminous species (Peterson and Farquhar 1996). The genetic response to different limitations or toxicities is different and, interestingly, the pattern of root hair response can be quite different between stresses (Müller and Schmidt 2004). For example, the root hair density of plants is increased by both P and Fe deficit, but for the former this is caused by the initiation of more root hair-forming cells, while in the latter it is due to greater frequency of root hairs with two tips (Müller and Schmidt 2004).

It has been demonstrated in a number of species, such as wheat and soybean, that root hair length is decreased in extremely acidic soils which have a large proportion of the cation exchange capacity taken by Al ions or in hydroponics with toxic levels of Al added (Brady et al. 1993; Haling et al. 2010; Delhaize et al. 2012). Root hair response can be affected by pH, and the changes in pH in both hydroponics and soils can have impacts on the length and density of root hairs (Ewens and Leigh 1985; Inoue and Hirota 2000). It has also been demonstrated that root hairs are the site of outward rectifying transporters of root exudates involved in tolerance to toxic elements (Tanaka et al. 2014; Maruyama et al. 2019). In Fig. 4i we see the expression of the promoter of an aluminium tolerance gene which works by exuding either citrate or malate to the rhizosphere, dependent on the plant species (Maroyama et al. 2018). In Fig. 4iii it is apparent that there is a strong localisation of phosphatase activity with the roots when grown in soil, but closer examination shows that this activity is only associated with the root hair zone, this could imply that the root hairs are the site of exudation of phosphatases, similar to citrate and malate in the previous image, or it could indicate that the root hair zones are sites of enhanced microbial activity and the phosphatases are of microbial origin (Giles et al. 2018). Taken together, this demonstrates well the role of root hairs in interactions with the soil, both as a source of compounds and a sink of elements. In addition, it is clear that the root hair zone is coincident with a zone of functional exudate accumulation. These sites of exudation may be promoted by an excess of a toxic element, but also by the deficiency of an essential element such as P (Delhaize et al. 2009).

In addition to acquisition of essential elements and tolerance to toxicity, root hairs are also implicated in the acquisition of water (Tanaka et al. 2014), although the hydraulic properties of root hairs in comparison to other root types does not single them out as being specialised for water uptake (Peterson and Farquhar 1996). It has been demonstrated that root hairs are longer in dry soils (Haling et al. 2014) and facilitate the uptake of water by substantially reducing the drop in matric potential at the interface between root and soil in rapidly transpiring plants (Carminati et al. 2017). In contrast, when root hairs are grown in hydroponics with PEG (an osmoticum that is used to simulate drought stress) root hairs have been shown to be reduced in diameter and length (Robin et al. 2015). Further evidence of the role of root hairs in water relations is provided by imaging studies, where it has been demonstrated that the soil physical conditions of the rhizosphere are altered compared to the soil just beyond this zone (Köebernick et al. 2017, 2019) by the presence of root hairs. This is manifest in rhizosphere soils being more porous in the presence of root hairs than when root hairs are absent, suggesting potential for easier extraction of water from soil in the presence of root hairs.

Other environmental cues have been shown to regulate root hair production and function. Elevated carbon monoxide has been shown to increase root hair length and density in tomatoes, due to the interference in phytohormone signalling in the plants (Guo et al. 2009). Also, soil strength has impacts on root hairs, leading to reduced root hair length (Haling et al. 2014) as discussed previously. Interestingly, temperature and light and the interaction between the two have been shown to affect the pattern of distribution of root hairs on tomato roots (Tsai et al. 2004). The diverse range of environmental cues affecting root hair development and growth suggest that they are especially susceptible to the influence of environmental variability and so that environmental change may have large impacts on root hair phenotypes.

4.4 Impact of root hairs on rhizosphere properties

Root hairs modify the spatial variation in critical characteristics of the soil environment in the rhizosphere. These impacts are highlighted in studies comparing genotypes with and without root hairs and of root hairs on soil properties. Root hairs affect the biology, physical conditions, chemistry and biochemistry of the rhizosphere: either changing the local root environment (Köebernick et al. 2017) or extending the effect of the root to a greater volume of soil (Holz et al. 2018; Giles et al. 2018). Some of these phenomena are highlighted in Fig. 5.

Models predict that root growth in soil will decrease the porosity of the soil at the root-soil interface due to compaction, whereas observations made using synchrotron X-ray CT, show that while this is the case with root hairless mutants, those genotypes with root hairs actually increase the porosity of the soil at the root soil interface (Fig. 5i). This has implications for the biophysical environment, such as water relations and oxygenation of the soil. The combination of root hairs and different root exudates (Naveed et al. 2017) has profound impacts on the ability of plants to acquire water and nutrients (Naveed et al. 2019). Oxygenation also changes the rhizosphere environment for microbial growth and function and may account for some of the differences seen in the microbial community structure in the rhizosphere of genotypes with root hairs compared to those without root hairs (Robertson-Albertyn et al. 2017). Root hairs affect the distribution of root exudates within the soil volume (Fig. 5ii). The distance moved by exuded carbon is greatly enhanced by the presence of root hairs (Holz et al. 2018). This is also seen with phosphatase activity in the rhizosphere (Fig. 5iii), where its activity is greater and more widely distributed in the rhizosphere of barley genotypes with root hairs than those without root hairs (Giles et al. 2018) and this has also been demonstrated for enzymes exuded by wheat (Ma et al. 2018). Species with more and longer root hairs have a larger rhizosphere than those with few root hairs. Composition of exudates from root hairless mutants is different to those from genotypes with root hairs, such that root hairs may impact the activity, species composition and

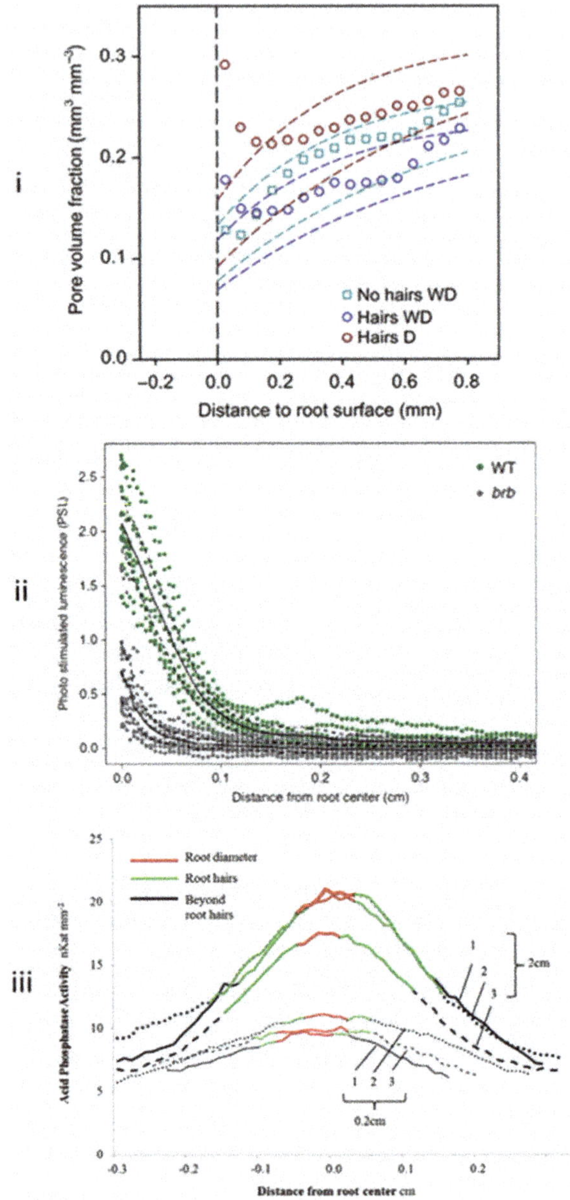

Figure 5 Spatial variation in physical and chemical characteristics with distance away from roots with and without root hairs (i) changes in porosity with distance from the root in the rhizosphere of plants derived from synchrotron X-ray CT images, (ii) changes in the concentration of exuded carbon compounds with distance from roots derived from isotope tracer studies and (iii) changes in phosphatase enzyme activity with distance from the root derived from soil zymography studies (Figures used with permission from Köebernick et al. 2017; Holz et al. 2018 and Giles et al. 2018).

function of the rhizosphere microbes (Robertson-Albertyn et al. 2017; Marasco et al. 2018; Giles et al. 2018; Naveed et al. 2019), partly through exuded carbon acting as a source of energy for the rhizosphere microbes.

5 Trade-offs and interactions with other traits

A cost-benefit analysis of root traits suggests that root hairs have the greatest potential for P acquisition relative to their cost of production, as compared with other common root traits (Brown et al. 2013b). The greatest gains in P-uptake efficiency are likely to be through increased length and longevity of root hairs rather than changes in hair density. However, the greatest gains are likely to be made when complementary traits are combined appropriately and negative trade-offs are avoided (Brown et al. 2013b). If appropriate root hair phenotypes can be combined with architectural, anatomical and biochemical traits, such that more root hair zones are produced in surface soils, where P resources are found, then the benefits for uptake are likely to be enhanced (Brown et al. 2013b). Increasing the number of root hairs can also lead to competition between individual root hairs for resources and therefore reduce their effectiveness (Peterson and Farquhar 1996; Haling et al. 2010). Longer, denser root hairs can also lead to greater competition between plants of the same species and less competition between species (Müller and Bartelheimer 2013) – with potential impacts for intensive farming of single-genotype monocultures.

Root hairs are the site of entry into plants for endophytic microorganisms, such as nitrogen-fixing bacteria, other beneficial endophytic bacteria such as Frankia, yeasts and mycorrhizal and dark septate fungi (Peterson and Farquhar 1996; Mercado-Blanco and Prieto 2012; Zou et al. 2019; Zahran and Sprent 1986). Entry occurs at points of weakness in the cytoskeleton such as the tip or the junction between root hair and epidermis where there may be a lesion point (Mercado-Blanco and Prieto 2012; Zahran and Sprent 1986). The ability to infect the plant through the root hairs is also time and species dependent. For example, rhizobium can only infect root hairs of sorghum within 5 hours of root hair initiation, whereas in clover this can occur at any time including in mature root hairs (Peterson and Farquhar 1996).

The role of root hairs in the infection of roots with mycorrhizal fungi is species specific (Zou et al. 2019). When plants lose root hairs through mutation, they are more likely to form mycorrhizal symbioses (Brown et al. 2013a; Li et al. 2014), suggesting that the presence of arbuscular mycorrhizal fungi (AMF) is compensating for the lack of root hairs. However, it has also been noted that inoculation with mycorrhizae in some species supresses the growth and density of root hairs and this is likely due to perturbation in auxin signalling in the plant (Zou et al. 2019). The association between species with prolific root hairs lacking AMF was first noted over a century ago (Höeveler 1892) and has been confirmed many times since (Zou et al. 2019). Similarly, the production of cluster roots (bottlebrush formations on roots made of a proliferation of lateral roots with root hairs) and dauciform roots (zones of intense root hair proliferation) in species which are adapted to extremely P-deficient soils are generally found on non-mycorrhizal species, suggesting a superior ability to acquire P by the root hairs compared to the AMF hyphae (Lambers et al. 2018). Therefore, one of the key trade-offs with root hairs is in relation to the function of mycorrhizal hyphae.

Many studies demonstrate that there is a trade-off between root hairs and mycorrhizae and that this compensation by mycorrhizae for root hairs can be complete (Li et al. 2014), while others suggest that root hairs are more effective at acquiring nutrients than mycorrhizal hyphae (Brown et al. 2013a); this may be due to different species of AMF primarily infecting the plant roots. While mycorrhizae may not fully replace root hairs for P-acquisition, they are effective at replacing root hairs for water uptake (Zou et al. 2019). In contrast to most of the research performed on annual crops or model plants, in woody perennial species it was shown that root hair density was increased in mycorrhizal plants (Wu et al. 2016) and this was associated with an increased expression of expansin genes (Liu et al. 2018a). It has also been shown that plants with coarse or woody root systems are more likely to be mycorrhizal and that some species with prolific root hairs are highly mycorrhizal (Zou et al. 2019). Importantly, it is also thought that mycorrhizal fungi will be more effective at acquiring resources, being able to explore smaller soil voids and a greater volume of soil than root hairs (Peterson and Farquhar 1996). When considering ectomycorrhizae, root hairs tend to get incorporated into the fungal mantle that covers the root epidermis. However, in cases where roots have particularly prolific root hairs the ectomycorrhizae may be prevented from forming a fungal sheath (Peterson and Farquhar 1996; Zou et al. 2019). So, the trade-off between root hairs and mycorrhizae is likely to be complicated.

There may be complementarity between the production of root hairs and the production of root exudates with the production of a large surface area for uptake combined with a pulse of root exudation, such as phosphatase (Fig. 4iii and 5iii) or organic acids, allowing effective mining of P-rich patches in soils (Giles et al. 2018; Kidd et al. 2016; Brown et al. 2013b; Zhang et al. 2018c). Overall, these lines of evidence suggest that plants either rely on AMF for acquisition of P (and maybe other resources) or they produce prolific root hair zones combined with a pulse of exudation, with the latter strategy being more effective in extreme P-deficiency situations (Lambers et al. 2018).

Trade-offs are also apparent with respect to the acquisition of water: on root hair mutants the absence of root hairs can be compensated for by other traits (Dodd and Diatloff 2016). Barley genotypes without root hairs maintained plant water status by producing a larger root system compared to the wild type with root hairs. Under combined P and water deficit the root hairless plants required >40% bigger root systems to compensate for the lack of root hairs. This was also demonstrated for root hair mutants in maize (Klamer et al. 2019).

6 Prospects for utilising root hairs

Root hairs influence the biology, physics and chemistry of the rhizosphere and the associated ability of the plant to utilise soil resources. This has a large impact on the ability of plants to cope with stresses associated with environmental change and changes in the management of cropping systems. Modelling exercises have demonstrated that the presence of root hairs cause a depletion of phosphorus from soil, particularly in the surface horizons (Fig. 6; Ruiz et al. 2019). This means that over the medium term (5–10 years) root hairs could be used to manipulate P use and availability at a field scale. It also demonstrates that root hairs could use some of the P accumulated in soils over decades of fertilisation. This would, in turn, reduce the need for P fertiliser and the load of P in the environment thereby reducing the pollution of watercourses.

Many of the concepts raised in this chapter give impetus to crop breeders to make root hairs a target for their breeding programmes. Such programmes should exploit the large genotypic variation in root hair traits associated with enhanced nutrient and water acquisition. Importantly, it has been observed that many traits present in traditional landraces and older cultivars may have been bred out of modern-day genotypes, which are adapted to high-input, intensive agricultural systems (Bingham et al. 2012). Reintroduction of genes regulating

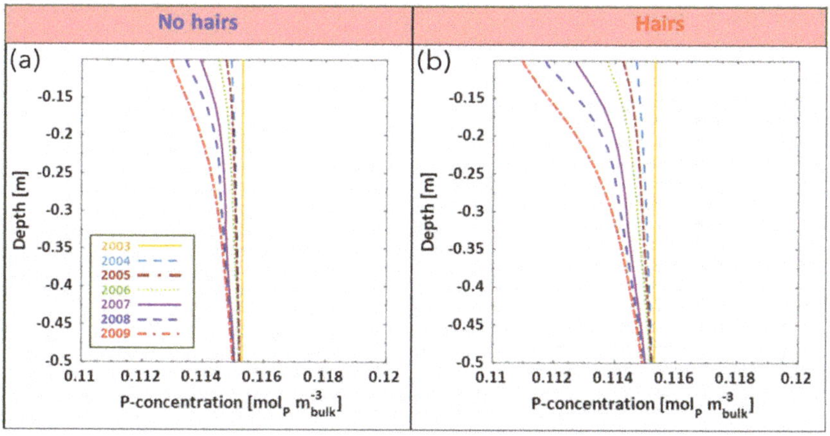

Figure 6 Long-term impact of root hairs on the soil fertility. Soil P concentration after multiple growing seasons (a) illustrates the relative impact that hairless mutants would have on the soil P compared to wild type hairy roots (b). Each curve depicts the end of the year P distribution along the soil depth assuming an initially uniform distribution at the beginning end of 2003 and simulating successive seasons up to the end of 2009 (figure reproduced with the permission of Ruiz et al. 2019).

adaptive traits such as root hairs may become of increasing value in the current context of finite resources and the need to increase the productivity of poorer soils. Under severe P-deficiency the combined benefits of increased number and longevity of root hairs are likely to increase P acquisition. This can be complimented by increased exudation of P-liberating compounds. Breeding programmes which involve the integration of P absorption, solubilisation and mineralisation traits are likely to produce the greatest benefits to P acquisition. Developments in root hairs which improve the efficiencies by which humans utilise finite resources such as phosphate, have potential to make a contribution on a global scale in a future of population growth and environmental change.

7 Acknowledgement

This chapter was written with the financial support of the Rural and Environment Science and Analytical Services Division of the Scottish Government.

8 Where to look for further information

8.1 Further reading

- Emons, A. M. and Ketelaar, T (Eds). 2009. *Root Hairs*. Springer Science & Business Media.
- Gregory, P. J. 2008. *Plant Roots: Growth, Activity and Interactions with the Soil*. John Wiley & Sons.
- Ridge, R. W. and Emons, A. M. (Eds). 2012. *Root Hairs: Cell and Molecular Biology*. Springer Science & Business Media.

8.2 Key journals and conferences

- *Annals of Botany.*
- *Functional Plant Biology.*
- International Society of Root Research Conference (ISRR).
- *New Phytologist.*
- *Plant and Soil.*
- Plant Biology Europe Conference.
- Rhizosphere Conference.
- Society of Experimental Biology Conferences.

8.3 Major international research projects

- iRoot hair information database http://www.iroothair.org/.
- Rhizosphere by Design, UK, https://www.soilsecurity.org/rhizosphere-by-design/.

- Root anatomy for abiotic stress tolerance https://plantscience.psu.edu/research/labs/roots/projects/copy_of_root-architecture.
- ChAnnels and Reactive Oxygen species in Legume root hair: role in Symbiosis with Rhizobium - CAROLS https://anr.fr/Project-ANR-11-BSV7-0010.

9 References

Abdolzadeh, A., Wang, X., Veneklaas, E. J. and Lambers, H. 2010. Effects of phosphorus supply on growth, phosphate concentration and cluster-root formation in three *Lupinus* species. *Annals of Botany* 105(3), 365–74. doi:10.1093/aob/mcp297.

Adu, M. O., Asare, P. A., Yawson, D. O., Ackah, F. K., Amoah, K. K., Nyarko, M. A. and Andoh, D. A. 2017. Quantifying variations in rhizosheath and root system phenotypes of landraces and improved varieties of juvenile maize. *Rhizosphere* 3(3), 29–39. doi:10.1016/j.rhisph.2016.12.004.

Angiosperm Phylogeny Group. 2009. An update of the Angiosperm Phylogeny Group classification for the orders and families of flowering plants: APG III. *Botanical Journal of the Linnean Society* 161(2), 105–21. doi:10.1111/j.1095-8339.2009.00996.x.

Athmann, M., Kautz, T., Pude, R. and Koepke, U. 2013. Root growth in biopores-evaluation with in situ endoscopy. *Plant and Soil* 371(1-2), 179–90. doi:10.1007/s11104-013-1673-5.

Bao, Y., Kost, B. and Chua, N. H. 2001. Reduced expression of α-tubulin genes in *Arabidopsis thaliana* specifically affects root growth and morphology, root hair development and root gravitropism. *The Plant Journal: for Cell and Molecular Biology* 28(2), 145–57. doi:10.1046/j.1365-313x.2001.01142.x.

Bates, T. R. and Lynch, J. P. 1996. Stimulation of root hair elongation in *Arabidopsis thaliana* by low phosphorus availability. *Plant, Cell and Environment* 19(5), 529–38. doi:10.1111/j.1365-3040.1996.tb00386.x.

Bates, T. R. and Lynch, J. P. 2000. Plant growth and phosphorus accumulation of wild type and two root hair mutants of *Arabidopsis thaliana* (Brassicaceae). *American Journal of Botany* 87(7), 958–63. doi:10.2307/2656994.

Baumberger, N., Steine,M., Ryser, U., Keller, B. and Ringli, C. 2003. Synergistic interaction of the two paralogous *Arabidopsis* genes LRX1 and LRX2 in cell wall formation during root hair development. *The Plant Journal: for Cell and Molecular Biology* 35(1), 71–81. doi:10.1046/j.1365-313x.2003.01784.x.

Belford, D. S. and Preston, R. D. 1961. The structure and growth of root hairs. *Journal of Experimental Botany* 12(1), 157–68. doi:10.1093/jxb/12.1.157.

Bengough, A. G., Bransby, M. F., Hans, J., McKenna, S. J., Roberts, T. J. and Valentine, T. A. 2006. Root responses to soil physical conditions; growth dynamics from field to cell. *Journal of Experimental Botany* 57(2), 437–47. doi:10.1093/jxb/erj003.

Bengough, A. G., Hans, J., Bransby, M. F. and Valentine, T. A. 2010. PIV as a method for quantifying root cell growth and particle displacement in confocal images. *Microscopy Research and Technique* 73(1), 27–36. doi:10.1002/jemt.20749.

Bengough, A. G., McKenzie, B. M., Hallett, P. D. and Valentine, T. A. 2011. Root elongation, water stress, and mechanical impedance: a review of limiting stresses and beneficial root tip traits. *Journal of Experimental Botany* 62(1), 59–68. doi:10.1093/jxb/erq350.

Bengough, A. G., Loades, K. and McKenzie, B. M. 2016. Root hairs aid soil penetration by anchoring the root surface to pore walls. *Journal of Experimental Botany* 67(4), 1071-8. doi:10.1093/jxb/erv560.

Bhosale, R., Giri, J., Pandey, B. K., Giehl, R. F. H., Hartmann, A., Traini, R., Truskina, J., Leftley, N., Hanlon, M., Swarup, K., Rashed, A., Voß, U., Alonso, J., Stepanova, A., Yun, J., Ljung, K., Brown, K. M., Lynch, J. P., Dolan, L., Vernoux, T., Bishopp, A., Wells, D., von Wirén, N., Bennett, M. J. and Swarup, R. 2018. A mechanistic framework for auxin dependent Arabidopsis root hair elongation to low external phosphate. *Nature Communications* 9(1), 1409. doi:10.1038/s41467-018-03851-3.

Bibikova, T. and Gilroy, S. 2002. Root hair development. *Journal of Plant Growth Regulation* 21(4), 383–415. doi:10.1007/s00344-003-0007-x.

Bingham, I. J., Karley, A. J., White, P. J., Thomas, W. T. B. and Russell, J. R. 2012. Analysis of improvements in nitrogen use efficiency associated with 75 years of barley breeding. *European Journal of Agronomy* 42, 49–58. doi:10.1016/j.eja.2011.10.003.

Brady, D. J., Edwards, D. G., Asher, C. J. and Blamey, F. P. C. 1993. Calcium amelioration of aluminium toxicity effects on root hair development in soybean [*Glycine max* (L.) Merr.]. *New Phytologist* 123(3), 531–8. doi:10.1111/j.1469-8137.1993.tb03765.x.

Brown, L. K., George, T. S., Thompson, J. A., Wright, G., Lyon, J., Hubbard, S. F. and White, P. J. 2012. What are the implications of variation in root hair length on P-limited yield in barley (*Hordeum vulgare* L.)? *Annals of Botany* 110, 319–28.

Brown, L. K., George, T. S., Barrett, G. E., Mclaren, S., Hubbard, S. F. and White, P. J. 2013a. Interactions between root hair length and arbuscular mycorrhizal (AM) colonisation in phosphorus deficient barley (*Hordeum vulgare* L.). *Pant and Soil* 372, 195–205.

Brown, L. K., George, T. S., Dupuy, L. X. and White, P. J. 2013b. A conceptual model of root hair ideotypes for future agricultural environments: what combination of traits should be targeted to cope with limited P availability? *Annals of Botany* 112(2), 317–30. doi:10.1093/aob/mcs231.

Brown, L. K., George, T. S., Neugebauer, K. and White, P. J. 2017. The rhizosheath–a potential trait for future agricultural sustainability occurs in orders throughout the angiosperms. *Plant and Soil* 418(1–2), 115–28. doi:10.1007/s11104-017-3220-2.

Canales, J., Contreras-López, O., Álvarez, J. M. and Gutiérrez, R. A. 2017. Nitrate induction of root hair density is mediated by TGA 1/TGA 4 and CPC transcription factors in *Arabidopsis thaliana*. *The Plant Journal: for Cell and Molecular Biology* 92(2), 305–16. doi:10.1111/tpj.13656.

Carminati, A., Vetterlein, D., Weller, U., Vogel, H. J. and Oswald, S. E. 2009. When roots lose contact. *Vadose Zone Journal* 8(3), 805–9. doi:10.2136/vzj2008.0147.

Carminati, A., Vetterlein, D., Köebernick, N., Blaser, S., Weller, U. and Vogel, H.-J. 2013. Do roots mind the gap? *Plant and Soil* 367(1–2), 651–61. doi:10.1007/s11104-012-1496-9.

Carminati, A., Passioura, J. B., Zarebanadkouki, M., Ahmed, M. A., Ryan, P. R., Watt, M. and Delhaize, E. 2017. Root hairs enable high transpiration rates in drying soils. *New Phytologist* 216(3), 771–81. doi:10.1111/nph.14715.

Cheng, S., Zhou, D. X. and Zhao, Y. 2016. WUSCHEL-related homeobox gene WOX11 increases rice drought resistance by controlling root hair formation and root system development. *Plant Signaling and Behavior* 11(2), e1130198. doi:10.1080/15592324.2015.1130198.

Chmielewska, B., Janiak, A., Karcz, J., Guzy-Wrobelska, J., Forster, B. P., Nawrot, M., Rusek, A., Smyda, P., Kedziorski, P., Maluszynski, M. and Szarejko, I. 2014. Morphological,

genetic and molecular characteristics of barley root hair mutants. *Journal of Applied Genetics* 55(4), 433–47. doi:10.1007/s13353-014-0225-x.
Clowes, F. A. L. 2000. Pattern in root meristem development in angiosperms. *New Phytologist* 146(1), 83–94. doi:10.1046/j.1469-8137.2000.00614.x.
Cole, P. J. and Alston, A. M. 1974. Effect of transient dehydration on absorption of chloride by wheat roots. *Plant and Soil* 40(1), 243–7. doi:10.1007/BF00011429.
Cormack, R. G. H. 1947. A comparative study of developing epidermal cells in white mustard and tomato roots. *American Journal of Botany* 34(6), 310–4. doi:10.1002/j.1537-2197.1947.tb12994.x.
Cormack, R. G. H. 1962. Development of root hairs in angiosperms. II. *The Botanical Review* 28(3), 446–64. doi:10.1007/BF02868690.
Cormack, R. G. H., Lemay, P. and Maclachlan, G. A. 1963. Calcium in the root-hair wall. *Journal of Experimental Botany* 14(2), 311–5. doi:10.1093/jxb/14.2.311.
Cutter, E. G. and Feldman, L. J. 1970. Trichoblasts in hydrocharis. I. Origin. Differentiation, dimensions and growth. *American Journal of Botany* 57(2), 190–201. doi:10.1002/j.1537-2197.1970.tb09807.x.
Cutter, E. G. and Hung, C. Y. 1972. Symmetric and asymmetric mitosis and cytokinesis in the root tip of *Hydrocharis morsus-ranae* L. *Journal of Cell Science* 11(3), 723–37.
Czarnes, S., Hiller, S., Dexter, A. R., Hallett, P. D. and Bartoli, F. 1999. Root: soil adhesion in the maize rhizosphere: the rheological approach. *Plant and Soil* 211(1), 69–86. doi: 10.1023/A:1004656510344.
Daram, P., Brunner, S., Persson, B. L., Amrhein, N. and Bucher, M. 1998. Functional analysis and cell-specific expression of a phosphate transporter from tomato. *Planta* 206(2), 225–33. doi:10.1007/s004250050394.
Datta, S., Kim, C. M., Pernas, M., Pires, N. D., Proust, H., Tam, T., Vijayakumar, P. and Dolan, L. 2011. Root hairs: development, growth and evolution at the plant-soil interface. *Plant and Soil* 346(1–2), 1–14. doi:10.1007/s11104-011-0845-4.
Datta, S., Prescott, H. and Dolan, L. 2015. Intensity of a pulse of RSL4 transcription factor synthesis determines Arabidopsis root hair cell size. *Nature Plants* 1(10), 15138. doi:10.1038/nplants.2015.138.
Davies, T. J., Barraclough, T. G., Chase, M. W., Soltis, P. S., Soltis, D. E. and Savolainen, V. 2004. Darwin's abominable mystery: insights from a supertree of the angiosperms. *Proceedings of the National Academy of Sciences of the United States of America* 101(7), 1904–9. doi:10.1073/pnas.0308127100.
Delhaize, E., Taylor, P., Hocking, P. J., Simpson, R. J., Ryan, P. R. and Richardson, A. E. 2009. Transgenic barley (*Hordeum vulgare* L.) expressing the wheat aluminium resistance gene (TaALMT1) shows enhanced phosphorus nutrition and grain production when grown on an acid soil. *Plant Biotechnology Journal* 7(5), 391–400. doi:10.1111/j.1467-7652.2009.00403.x.
Delhaize, E., James, R. A. and Ryan, P. R. 2012. Aluminium tolerance of root hairs underlies genotypic differences in rhizosheath size of wheat (*Triticum aestivum*) grown on acid soil. *New Phytologist* 195(3), 609–19. doi:10.1111/j.1469-8137.2012.04183.x.
Delhaize, E., Rathjen, T. M. and Cavanagh, C. R. 2015. The genetics of rhizosheath size in a multiparent mapping population of wheat. *Journal of Experimental Botany* 66(15), 4527–36. doi:10.1093/jxb/erv223.
Dittmer, H. J. 1937. A quantitative study of the roots and root hairs of a winter rye plant (*Secale cereale*). *American Journal of Botany* 24(7), 417–20. doi:10.1002/j.1537-2197.1937.tb09121.x.

Dittmer, H. J. 1949. Root hair variations in plant species. *American Journal of Botany* 36(2), 152–5. doi:10.1002/j.1537-2197.1949.tb05241.x.

Dodd, I. C. and Diatloff, E. 2016. Enhanced root growth of the brb (bald root barley) mutant in drying soil allows similar shoot physiological responses to soil water deficit as wild-type plants. *Functional Plant Biology* 43(2), 199–206. doi:10.1071/FP15303.

Dolan, L., Duckett, C. M., Grierson, C., Linstead, P., Schneider, K., Lawson, E., Dean, C., Poething, S. and Roberts, K. 1994. Clonal relationships and cell patterning in the root epidermis of *Arabidopsis*. *Development* 120, 2465–74.

Ewens, M. and Leigh, R. A. 1985. The effect of nutrient solution composition on the length of root hairs of wheat (*Triticum aestivum* L.). *Journal of Experimental Botany* 36(5), 713–24. doi:10.1093/jxb/36.5.713.

Gahoonia, T. S. and Nielsen, N. E. 2003. Phosphorus (P) uptake and growth of a root hairless barley mutant (bald root barley, brb) and wild type in low-and high-P soils. *Plant, Cell and Environment* 26(10), 1759–66. doi:10.1046/j.1365-3040.2003.01093.x.

Gahoonia, T. S., Care, D. and Nielsen, N. E. 1997. Root hairs and phosphorus acquisition of wheat and barley cultivars. *Plant and Soil* 191(2), 181–8. doi:10.1023/A:1004270201418.

Gajewska, P., Janiak, A., Kwasniewski, M., Kędziorski, P. and Szarejko, I. 2018. Forward genetics approach reveals a mutation in bHLH transcription factor-encoding gene as the best candidate for the root hairless phenotype in barley. *Frontiers in Plant Science* 9, 1229. doi:10.3389/fpls.2018.01229.

Galway, M. E., Masucci, J. D., Lloyd, A. M., Walbot, V., Davis, R. W. and Schiefelbein, J. W. 1994. The TTG gene is required to specify epidermal cell fate and cell patterning in the *Arabidopsis* root. *Developmental Biology* 166(2), 740–54. doi:10.1006/dbio.1994.1352.

George, T. S., Brown, L. K., Ramsay, L., White, P. J., Newton, A. C., Bengough, A. G., Russell, J. and Thomas, W. T. 2014. Understanding the genetic control and physiological traits associated with rhizosheath production by barley (*Hordeum vulgare*). *New Phytologist* 203(1), 195–205. doi:10.1111/nph.12786.

Giles, C. D., Dupuy, L., Boitt, G., Brown, L. K., Condron, L. M., Darch, T., Blackwell, M. S. A., Menezes-Blackburn, D., Shand, C. A., Stutter, M. I., Lumsdon, D. G., Wendler, R., Cooper, P., Wearing, C., Zhang, H., Haygarth, P. M. and George, T. S. 2018. Root development impacts on the distribution of phosphatase activity: improvements in quantification using soil zymography. *Soil Biology and Biochemistry* 116, 158–66. doi:10.1016/j.soilbio.2017.08.011.

Gilroy, S. and Jones, D. L. 2000. Through form to function: root hair development and nutrient uptake. *Trends in Plant Science* 5(2), 56–60. doi:10.1016/s1360-1385(99)01551-4.

Giri, J., Bhosale, R., Huang, G., Pandey, B. K., Parker, H., Zappala, S., Yang, J., Dievart, A., Bureau, C., Ljung, K., Price, A., Rose, T., Larrieu, A., Mairhofer, S., Sturrock, C. J., White, P., Dupuy, L., Hawkesford, M., Perin, C., Liang, W., Peret, B., Hodgman, C. T., Lynch, J., Wissuwa, M., Zhang, D., Pridmore, T., Mooney, S. J., Guiderdoni, E., Swarup, R. and Bennett, M. J. 2018. Rice auxin influx carrier OsAUX1 facilitates root hair elongation in response to low external phosphate. *Nature Communications* 9(1), 1408. doi:10.1038/s41467-018-03850-4.

Goron, T. L., Watts, S., Shearer, C. and Raizada, M. N. 2015. Growth in Turface® clay permits root hair phenotyping along the entire crown root in cereal crops and demonstrates that root hair growth can extend well beyond the root hair zone. *BMC Research Notes* 8(1), 143. doi:10.1186/s13104-015-1108-x.

Grierson, C., Nielsen, E., Ketelaarc, T. and Schiefelbein, J. 2014. Root hairs. *The Arabidopsis Book/American Society of Plant Biologists* 12, e0172. doi:10.1199/tab.0172.

Guo, K., Kong, W. W. and Yang, Z. M. 2009. Carbon monoxide promotes root hair development in tomato. *Plant, Cell and Environment* 32(8), 1033–45. doi:10.1111/j.1365-3040.2009.01986.x.

Haling, R. E., Simpson, R. J., Delhaize, E., Hocking, P. J. and Richardson, A. E. 2010. Effect of lime on root growth, morphology and the rhizosheath of cereal seedlings growing in an acid soil. *Plant and Soil* 327(1–2), 199–212. doi:10.1007/s11104-009-0047-5.

Haling, R. E., Simpson, R. J., Culvenor, R. A., Lambers, H. and Richardson, A. E. 2011. Effect of soil acidity, soil strength and macropores on root growth and morphology of perennial grass species differing in acid-soil resistance. *Plant, Cell and Environment* 34(3), 444–56. doi:10.1111/j.1365-3040.2010.02254.x.

Haling, R. E., Brown, L. K., Bengough, A. G., Young, I. M., Hallett, P. D., White, P. J. and George, T. S. 2013. Root hairs improve root penetration, rootsoil contact, and phosphorus acquisition in soils of different strength. *Journal of Experimental Botany* 64(12), 3711–21. doi:10.1093/jxb/ert200.

Haling, R. E., Brown, L. K., Bengough, A. G., Valentine, T. A., White, P. J., Young, I. M. and George, T. S. 2014. Root hair length and rhizosheath mass depend on soil porosity, strength and water content in barley genotypes. *Planta* 239(3), 643–51. doi:10.1007/s00425-013-2002-1.

Han, H., Wang, H., Han, Y., Hu, Z., Xin, M., Peng, H., Yao, Y., Sun, Q. and Ni, Z. 2017. Altered expression of the TaRSL2 gene contributed to variation in root hair length during allopolyploid wheat evolution. *Planta* 246(5), 1019–28. doi:10.1007/s00425-017-2735-3.

He, X., Zeng, J., Cao, F., Ahmed, I. M., Zhang, G., Vincze, E. and Wu, F. 2015. HvEXPB7, a novel β-expansin gene revealed by the root hair transcriptome of Tibetan wild barley, improves root hair growth under drought stress. *Journal of Experimental Botany* 66(22), 7405–19. doi:10.1093/jxb/erv436.

Hey, S., Baldauf, J., Opitz, N., Lithio, A., Pasha, A., Provart, N., Nettleton, D. and Hochholdinger, F. 2017. Complexity and specificity of the maize (*Zea mays* L.) root hair transcriptome. *Journal of Experimental Botany* 68(9), 2175–85. doi:10.1093/jxb/erx104.

Holz, M., Zarebanadkouki, M., Kuzyakov, Y., Pausch, J. and Carminati, A. 2018. Root hairs increase rhizosphere extension and carbon input to soil. *Annals of Botany* 121(1), 61–9. doi:10.1093/aob/mcx127.

Honkanen, S. and Dolan, L. 2016. Growth regulation in tip-growing cells that develop on the epidermis. *Current Opinion in Plant Biology* 34, 77–83. doi:10.1016/j.pbi.2016.10.006.

Horn, R., Wingen, L. U., Snape, J. W. and Dolan, L. 2016. Mapping of quantitative trait loci for root hair length in wheat identifies loci that co-locate with loci for yield components. *Journal of Experimental Botany* 67(15), 4535–43. doi:10.1093/jxb/erw228.

Höveler, W. J. 1892. Ueber die Verwerthung des Humus bei der Ernährung der chlorophyllführenden Pflanzen. Inaugural-Dissertation von Wilhelm Höveler, Gebrüder Borntraeger.

Huang, L. 2016. Comparative analysis of gene expression in the developing roots and root hairs across vascular plants (Doctoral dissertation).

Huang, L. and Schiefelbein, J. 2015. Conserved gene expression programs in developing roots from diverse plants. *The Plant Cell* 27(8), 2119–32. doi:10.1105/tpc.15.00328.

Huang, L., Shi, X., Wang, W., Ryu, K. H. and Schiefelbein, J. 2017. Diversification of root hair development genes in vascular plants. *Plant Physiology* 174(3), 1697–712. doi:10.1104/pp.17.00374.

Huck, M. G., Klepper, B. and Taylor, H. M. 1970. Diurnal variations in root diameter. *Plant Physiology* 45(4), 529–30. doi:10.1104/pp.45.4.529.

Hwang, Y., Lee, H., Lee, Y. S. and Cho, H. T. 2016. Cell wall-associated ROOT HAIR SPECIFIC 10, a proline-rich receptor-like kinase, is a negative modulator of *Arabidopsis* root hair growth. *Journal of Experimental Botany* 67(6), 2007–22. doi:10.1093/jxb/erw031.

Inoue, Y. and Hirota, K. 2000. Low pH-induced root hair formation in lettuce (*Lactuca sativa* L. cv. Grand Rapids) seedlings: determination of root hair-forming site. *Journal of Plant Research* 113(3), 245–51. doi:10.1007/PL00013864.

James, R. A., Weligama, C., Verbyla, K., Ryan, P. R., Rebetzke, G. J., Rattey, A., Richardson, A. E. and Delhaize, E. 2016. Rhizosheaths on wheat grown in acid soils: phosphorus acquisition efficiency and genetic control. *Journal of Experimental Botany* 67(12), 3709–18. doi:10.1093/jxb/erw035.

Jones, V. A. and Dolan, L. 2012. The evolution of root hairs and rhizoids. *Annals of Botany* 110(2), 205–12. doi:10.1093/aob/mcs136.

Jones, M. A., Raymond, M. J. and Smirnoff, N. 2006. Analysis of the root-hair morphogenesis transcriptome reveals the molecular identity of six genes with roles in root-hair development in *Arabidopsis*. *The Plant Journal: for Cell and Molecular Biology* 45(1), 83–100. doi:10.1111/j.1365-313X.2005.02609.x.

Karas, B., Amyot, L., Johansen, C., Sato, S., Tabata, S., Kawaguchi, M. and Szczyglowski, K. 2009. Conservation of lotus and *Arabidopsis* basic helix-loop-helix proteins reveals new players in root hair development. *Plant Physiology* 151(3), 1175–85. doi:10.1104/pp.109.143867.

Keyes, S. D., Daly, K. R., Gostling, N. J., Jones, D. L., Talboys, P., Pinzer, B. R., Boardman, R., Sinclair, I., Marchant, A. and Roose, T. 2013. High resolution synchrotron imaging of wheat root hairs growing in soil and image-based modelling of phosphate uptake. *New Phytologist* 198(4), 1023–9. doi:10.1111/nph.12294.

Keyes, S. D., Zygalakis, K. C. and Roose, T. 2017. An explicit structural model of root hair and soil interactions parameterised by synchrotron X-ray computed tomography. *Bulletin of Mathematical Biology* 79(12), 2785–813. doi:10.1007/s11538-017-0350-x.

Kidd, D. R., Ryan, M. H., Haling, R. E., Lambers, H., Sandral, G. A., Yang, Z., Culvenor, R. A., Cawthray, G. R., Stefanski, A. and Simpson, R. J. 2016. Rhizosphere carboxylates and morphological root traits in pasture legumes and grasses. *Plant and Soil* 402(1–2), 77–89. doi:10.1007/s11104-015-2770-4.

Kirchner, T. W., Niehaus, M., Rössig, K. L., Lauterbach, T., Herde, M., Küster, H. and Schenk, M. K. 2018. Molecular background of Pi deficiency-induced root hair growth in *Brassica carinata*–a fasciclin-like arabinogalactan protein is involved. *Frontiers in Plant Science* 9, 1372. doi:10.3389/fpls.2018.01372.

Klamer, F., Vogel, F., Li, X., Bremer, H., Neumann, G., Neuhäuser, B., Hochholdinger, F. and Ludewig, U. 2019. Estimating the importance of maize root hairs in low phosphorus conditions and under drought. *Annals of Botany* 124(6), 961-8. doi:10.1093/aob/mcz011.

Klinsawang, S., Sumranwanich, T., Wannaro, A. and Saengwilai, P. 2018. Effects of root hair length on potassium acquisition in rice (*Oryza sativa* L.). *Applied Ecology and Environmental Research* 16(2), 1609-20. doi:10.15666/aeer/1602_16091620.

Köebernick, N., Daly, K. R., Keyes, S. D., George, T. S., Brown, L. K., Raffan, A., Cooper, L. J., Naveed, M., Bengough, A. G., Sinclair, I., Hallett, P. D. and Roose, T. 2017. High-resolution synchrotron imaging shows that root hairs influence rhizosphere soil structure formation. *New Phytologist* 216(1), 124-35. doi:10.1111/nph.14705.

Köebernick, N., Daly, K. R., Keyes, S. D., Bengough, A. G., Brown, L. K., Cooper, L. J., George, T. S., Hallett, P. D., Naveed, M., Raffan, A. and Roose, T. 2019. Imaging microstructure of the barley rhizosphere: particle packing and root hair influences. *New Phytologist* 221(4), 1878-89. doi:10.1111/nph.15516.

Kohanová, J., Martinka, M., Vaculík, M., White, P. J., Hauser, M. T. and Lux, A. 2018. Root hair abundance impacts cadmium accumulation in *Arabidopsis thaliana* shoots. *Annals of Botany* 122(5), 903-14. doi:10.1093/aob/mcx220.

Kwasniewski, M., Nowakowska, U., Szumera, J., Chwialkowska, K. and Szarejko, I. 2013. iRootHair: a comprehensive root hair genomics database. *Plant Physiology* 161(1), 28-35. doi:10.1104/pp.112.206441.

Kwon, T., Sparks, J. A., Nakashima, J., Allen, S. N., Tang, Y. and Blancaflor, E. B. 2015. Transcriptional response of *Arabidopsis* seedlings during spaceflight reveals peroxidase and cell wall remodeling genes associated with root hair development. *American Journal of Botany* 102(1), 21-35. doi:10.3732/ajb.1400458.

Lambers, H., Albornoz, F., Kotula, L., Laliberté, E., Ranathunge, K., Teste, F. P. and Zemunik, G. 2018. How belowground interactions contribute to the coexistence of mycorrhizal and non-mycorrhizal species in severely phosphorus-impoverished hyperdiverse ecosystems. *Plant and Soil* 424(1-2), 11-33. doi:10.1007/s11104-017-3427-2.

Leavitt, R. G. 1904. Trichomes of the root in vascular cryptogams and angiosperms. *Proceedings of the Boston Society of Natural History* 3, 1904.

Li, T., Lin, G., Zhang, X., Chen, Y., Zhang, S. and Chen, B. 2014. Relative importance of an arbuscular mycorrhizal fungus (*Rhizophagus intraradices*) and root hairs in plant drought tolerance. *Mycorrhiza* 24(8), 595-602. doi:10.1007/s00572-014-0578-3.

Libault, M., Brechenmacher, L., Cheng, J., Xu, D. and Stacey, G. 2010. Root hair systems biology. *Trends in Plant Science* 15(11), 641-50. doi:10.1016/j.tplants.2010.08.010.

Liu, M., Liu, X. X., He, X. L., Liu, L. J., Wu, H., Tang, C. X., Zhang, Y. S. and Jin, C. W. 2017a. Ethylene and nitric oxide interact to regulate the magnesium deficiency-induced root hair development in *Arabidopsis*. *New Phytologist* 213(3), 1242-56. doi:10.1111/nph.14259.

Liu, M., Rathjen, T., Weligama, K., Forrest, K., Hayden, M. and Delhaize, E. 2017b. Analysis of aneuploid lines of bread wheat to map chromosomal locations of genes controlling root hair length. *Annals of Botany* 119(8), 1333-41. doi:10.1093/aob/mcx030.

Liu, C. Y., Wang, P., Zhang, D. J., Zou, Y. N., Kuča, K. and Wu, Q. S. 2018a. Mycorrhiza-induced change in root hair growth is associated with IAA accumulation and expression of EXPs in trifoliate orange under two P levels. *Scientia Horticulturae* 234, 227-35. doi:10.1016/j.scienta.2018.02.052.

Liu, M., Zhang, H., Fang, X., Zhang, Y. and Jin, C. 2018b. Auxin acts downstream of ethylene and nitric oxide to regulate magnesium deficiency-induced root hair development in *Arabidopsis thaliana*. *Plant and Cell Physiology* 59(7), 1452–65. doi:10.1093/pcp/pcy078.

Ma, X., Zarebanadkouki, M., Kuzyakov, Y., Blagodatskaya, E., Pausch, J. and Razavi, B. S. 2018. Spatial patterns of enzyme activities in the rhizosphere: effects of root hairs and root radius. *Soil Biology and Biochemistry* 118, 69–78. doi:10.1016/j.soilbio.2017.12.009.

Mackay, A. D. and Barber, S. A. 1987. Effect of cyclic wetting and drying of a soil on root hair-growth of maize roots. *Plant and Soil* 104(2), 291–3. doi:10.1007/BF02372544.

Marasco, R., Mosqueira, M. J., Fusi, M., Ramond, J. B., Merlino, G., Booth, J. M., Maggs-Kölling, G., Cowan, D. A. and Daffonchio, D. 2018. Rhizosheath microbial community assembly of sympatric desert speargrasses is independent of the plant host. *Microbiome* 6(1), 215. doi:10.1186/s40168-018-0597-y.

Maruyama, H., Sasaki, T., Yamamoto, Y. and Wasaki, J. 2019. AtALMT3 is involved in malate efflux induced by phosphorus deficiency in *Arabidopsis thaliana* root hairs. *Plant and Cell Physiology* 60(1), 107–15. doi:10.1093/pcp/pcy190.

Masucci, J. D. and Schiefelbein, J. W. 1996. Hormones act downstream of TTG and GL2 to promote root hair outgrowth during epidermis development in the *Arabidopsis* root. *The Plant Cell* 8(9), 1505–17. doi:10.1105/tpc.8.9.1505.

Mercado-Blanco, J. and Prieto, P. 2012. Bacterial endophytes and root hairs. *Plant and Soil* 361(1–2), 301–6. doi:10.1007/s11104-012-1212-9.

Moon, S., Chandran, A. K. N., An, G., Lee, C. and Jung, K. H. 2018. Genome-wide analysis of root hair-preferential genes in rice. *Rice* 11(1), 48. doi:10.1186/s12284-018-0241-2.

Moon, S., Cho, L. H., Kim, Y. J., Gho, Y. S., Jeong, H. Y., Hong, W. J., Lee, C., Park, H., Jwa, N. S., Dangol, S., Chen, Y., Park, H., Cho, H. S., An, G. and Jung, K. H. 2019. RSL class II transcription factors guide the nuclear localization of RHL1 to regulate root hair development. *Plant Physiology* 179(2), 558–68. doi:10.1104/pp.18.01002.

Müller, B. and Bartelheimer, M. 2013. Interspecific competition in *Arabidopsis thaliana*: root hairs are important for competitive effect, but not for competitive response. *Plant and Soil* 371(1–2), 167–77. doi:10.1007/s11104-013-1675-3.

Müller, M. and Schmidt, W. 2004. Environmentally induced plasticity of root hair development in *Arabidopsis*. *Plant Physiology* 134(1), 409–19. doi:10.1104/pp.103.029066.

Naveed, M., Brown, L. K., Raffan, A. C., George, T. S., Bengough, A. G., Roose, T., Sinclair, I., Köebernick, N., Cooper, L., Hackett, C. A. and Hallett, P. D. 2017. Plant exudates may stabilize or weaken soil depending on species, origin and time. *European Journal of Soil Science* 68(6), 806–16. doi:10.1111/ejss.12487.

Naveed, M., Ahmed, M. A., Benard, P., Brown, L. K., George, T. S., Bengough, A. G., Roose, T., Köebernick, N. and Hallett, P. D. 2019. Surface tension, rheology and hydrophobicity of rhizodeposits and seed mucilage influence soil water retention and hysteresis. *Plant and Soil* 437(1), 65–81. doi:10.1007/s11104-019-03939-9.

Nestler, J. and Wissuwa, M. 2016. Superior root hair formation confers root efficiency in some, but not all, rice genotypes upon P deficiency. *Frontiers in Plant Science* 7, 1935. doi:10.3389/fpls.2016.01935.

Nestler, J., Keyes, S. D. and Wissuwa, M. 2016. Root hair formation in rice (*Oryza sativa* L.) differs between root types and is altered in artificial growth conditions. *Journal of Experimental Botany* 67(12), 3699–708. doi:10.1093/jxb/erw115.

Nye, P. H. 1994. The effect of root shrinkage on soil-water inflow. *Philosophical Transactions of the Royal Society of London. Series B: Biological Sciences* 345(1314), 395–402. doi:10.1098/rstb.1994.0117.

Pang, J., Ryan, M. H., Siddique, K. H. M. and Simpson, R. J. 2017. Unwrapping the rhizosheath. *Plant and Soil* 418(1–2), 129–39. doi:10.1007/s11104-017-3358-y.

Passioura, J. B. 1991. Soil structure and plant-growth. *Australian Journal of Soil Research* 29(6), 717–28. doi:10.1071/SR9910717.

Pemberton, L. M., Tsai, S. L., Lovell, P. H. and Harris, P. J. 2001. Epidermal patterning in seedling roots of eudicotyledons. *Annals of Botany* 87(5), 649–54. doi:10.1006/anbo.2001.1385.

Peterson, R. L. and Farquhar, M. L. 1996. Root hairs: specialized tubular cells extending root surfaces. *The Botanical Review* 62(1), 1–40. doi:10.1007/BF02868919.

Rahman, A., Hosokawa, S., Oono, Y., Amakawa, T., Goto, N. and Tsurumi, S. 2002. Auxin and ethylene response interactions during Arabidopsis root hair development dissected by auxin influx modulators. *Plant Physiology* 130(4), 1908–17. doi:10.1104/pp.010546.

Reeder, J. R. and von Maltzahn, K. 1953. Taxonomic significance of root-hair development in the Gramineae. *Proceedings of the National Academy of Sciences of the United States of America* 39(7), 593–8. doi:10.1073/pnas.39.7.593.

Rigas, S., Ditengou, F. A., Ljung, K., Daras, G., Tietz, O., Palme, K. and Hatzopoulos, P. 2013. Root gravitropism and root hair development constitute coupled developmental responses regulated by auxin homeostasis in the *Arabidopsis* root apex. *New Phytologist* 197(4), 1130–41. doi:10.1111/nph.12092.

Ringli, C., Baumberger, N., Diet, A., Frey, B. and Keller, B. 2002. ACTIN2 is essential for bulge site selection and tip growth during root hair development of *Arabidopsis*. *Plant Physiology* 129(4), 1464–72. doi:10.1104/pp.005777.

Robertson-Albertyn, S., Alegria Terrazas, R., Balbirnie, K., Blank, M., Janiak, A., Szarejko, I., Chmielewska, B., Karcz, J., Morris, J., Hedley, P. E., George, T. S. and Bulgarelli, D. 2017. Root hair mutations displace the barley rhizosphere microbiota. *Frontiers in Plant Science* 8, 1094. doi:10.3389/fpls.2017.01094.

Robin, A., Uddin, M. and Bayazid, K. 2015. Polyethylene glycol (PEG)-treated hydroponic culture reduces length and diameter of root hairs of wheat varieties. *Agronomy* 5(4), 506–18. doi:10.3390/agronomy5040506.

Row, H. C. and Reeder, J. R. 1957. Root-hair development as evidence of relationships among genera of *Gramineae*. *American Journal of Botany* 44(7), 596–601. doi:10.1002/j.1537-2197.1957.tb10582.x.

Ruiz, S., Köebernick, N., Duncan, S., McKay-Fletcher, D., Scotson, C., Boghi, A., Marin, M., Bengough, A. G., George, T. S., Brown, L. K., Hallett, P. D. and Roose, T. 2019. Significance of root hairs at the field scale – modelling root water and phosphorus uptake under different field conditions. *Plant and Soil*. doi:10.1007/s11104-019-04308-2.

Salazar-Henao, J. E., Lin, W. D. and Schmidt, W. 2016. Discriminative gene co-expression network analysis uncovers novel modules involved in the formation of phosphate deficiency-induced root hairs in *Arabidopsis*. *Scientific Reports* 6, 26820. doi:10.1038/srep26820.

Santelia, D., Vincenzetti, V., Azzarello, E., Bovet, L., Fukao, Y., Düchtig, P., Mancuso, S., Martinoia, E. and Geisler, M. 2005. MDR-like ABC transporter AtPGP4 is involved in auxin-mediated lateral root and root hair development. *FEBS Letters* 579(24), 5399–406. doi:10.1016/j.febslet.2005.08.061.

Schiefelbein, J. W. 2000. Constructing a plant cell. The genetic control of root hair development. *Plant Physiology* 124(4), 1525–31. doi:10.1104/pp.124.4.1525.

Schmidt, S., Bengough, A. G., Gregory, P. J., Grinev, D. V. and Otten, W. 2012. Estimating root-soil contact from 3D X-ray microtomographs. *European Journal of Soil Science* 63(6), 776–86. doi:10.1111/j.1365-2389.2012.01487.x.

Schünmann, P. H., Richardson, A. E., Smith, F. W. and Delhaize, E. 2004. Characterization of promoter expression patterns derived from the Pht1 phosphate transporter genes of barley (*Hordeum vulgare* L.). *Journal of Experimental Botany* 55(398), 855–65. doi:10.1093/jxb/erh103.

Shane, M. W., Dixon, K. W. and Lambers, H. 2005. The occurrence of dauciform roots amongst Western Australian reeds, rushes and sedges, and the impact of phosphorus supply on dauciform-root development in *Schoenus unispiculatus* (*Cyperaceae*). *New Phytologist* 165(3), 887–98. doi:10.1111/j.1469-8137.2004.01283.x.

Shi, H. and Zhu, J. K. 2002. SOS4, a pyridoxal kinase gene, is required for root hair development in *Arabidopsis*. *Plant Physiology* 129(2), 585–93. doi:10.1104/pp.001982.

Sinnott, E. W. and Bloch, R. 1939. Cell polarity and the differentiation of root hairs. *Proceedings of the National Academy of Sciences of the United States of America* 25(5), 248–52. doi:10.1073/pnas.25.5.248.

Smith, R. J., Hopper, S. D. and Shane, M. W. 2011. Sand-binding roots in *Haemodoraceae* global survey and morphology in a phylogenetic context. *Plant and Soil* 348(1–2), 453–70. doi:10.1007/s11104-011-0874-z.

Song, X. F., Yang, C. Y., Liu, J. and Yang, W. C. 2006. RPA, a class II ARFGAP protein, activates ARF1 and U5 and plays a role in root hair development in *Arabidopsis*. *Plant Physiology* 141(3), 966–76. doi:10.1104/pp.106.077818.

Stetter, M. G., Schmid, K. and Ludewig, U. 2015. Uncovering genes and ploidy involved in the high diversity in root hair density, length and response to local scarce phosphate in *Arabidopsis thaliana*. *PLoS ONE* 10(3), e0120604. doi:10.1371/journal.pone.0120604.

Stetter, M. G., Benz, M. and Ludewig, U. 2017. Increased root hair density by loss of WRKY6 in *Arabidopsis thaliana*. *PeerJ* 5, e2891. doi:10.7717/peerj.2891.

Stirzaker, R. J., Passioura, J. B. and Wilms, Y. 1996. Soil structure and plant growth: impact of bulk density and biopores. *Plant and Soil* 185(1), 151–62. doi:10.1007/BF02257571.

Tanaka, N., Kato, M., Tomioka, R., Kurata, R., Fukao, Y., Aoyama, T. and Maeshima, M. 2014. Characteristics of a root hair-less line of *Arabidopsis thaliana* under physiological stresses. *Journal of Experimental Botany* 65(6), 1497–512. doi:10.1093/jxb/eru014.

Tanaka, N., Uno, H., Okuda, S., Gunji, S., Ferjani, A., Aoyama, T. and Maeshima, M. 2017. SRPP, a cell wall protein is involved in development and protection of seeds and root hairs in *Arabidopsis thaliana*. *Plant and Cell Physiology* 58(4), 760–9. doi:10.1093/pcp/pcx008.

Tardieu, F. 1988. Analysis of the spatial variability of maize root density I. Effect of wheel compaction on the spatial arrangement of roots. *Plant and Soil* 107(2), 259–66. doi:10.1007/BF02370555.

Tsai, S. L., Harris, P. J. and Lovell, P. H. 2003. The root epidermis of *Echium plantagineum* L.: a novel type of pattern based on the distribution of short and long root hairs. *Planta* 217(2), 238–44. doi:10.1007/s00425-003-0989-4.

Tsai, S. L., Harris, P. J. and Lovell, P. H. 2004. Bands of root hairs are produced in tomato (*Lycopersicon esculentum*) in response to specific combinations of thermoperiods and photoperiods. *New Zealand Journal of Crop and Horticultural Science* 32(1), 121–9. doi:10.1080/01140671.2004.9514286.

Vatter, T., Neuhäuser, B., Stetter, M. and Ludewig, U. 2015. Regulation of length and density of *Arabidopsis* root hairs by ammonium and nitrate. *Journal of Plant Research* 128(5), 839–48. doi:10.1007/s10265-015-0733-8.
Vijayakumar, P., Datta, S. and Dolan, L. 2016. ROOT HAIR DEFECTIVE SIX-LIKE4 (RSL4) promotes root hair elongation by transcriptionally regulating the expression of genes required for cell growth. *New Phytologist* 212(4), 944–53. doi:10.1111/nph.14095.
Wada, Y., Kusano, H., Tsuge, T. and Aoyama, T. 2015. Phosphatidylinositol phosphate 5-kinase genes respond to phosphate deficiency for root hair elongation in *Arabidopsis thaliana*. *The Plant Journal: for Cell and Molecular Biology* 81(3), 426–37. doi:10.1111/tpj.12741.
Wang, L., Liao, H., Yan, X., Zhuang, B. and Dong, Y. 2004. Genetic variability for root hair traits as related to phosphorus status in soybean. *Plant and Soil* 261(1/2), 77–84. doi:10.1023/B:PLSO.0000035552.94249.6a.
Wang, Y. S., Jensen, L. S. and Magid, J. 2016. Differential responses of root and root hair traits of spring wheat genotypes to phosphorus deficiency in solution culture. *Plant, Soil and Environment* 62(12), 540–6. doi:10.17221/485/2016-PSE.
White, R. G. and Kirkegaard, J. A. 2010. The distribution and abundance of wheat roots in a dense, structured subsoil - implications for water uptake. *Plant, Cell and Environment* 33(2), 133–48. doi:10.1111/j.1365-3040.2009.02059.x.
White, P. J., George, T. S., Gregory, P. J., Bengough, A. G., Hallett, P. D. and McKenzie, B. M. 2013. Matching roots to their environment. *Annals of Botany* 112(2), 207–22. doi:10.1093/aob/mct123.
Wissuwa, M., Mazzola, M. and Picard, C. 2009. Novel approaches in plant breeding for rhizosphere-related traits. *Plant and Soil* 321(1–2), 409–30. doi:10.1007/s11104-008-9693-2.
Wu, Q. S., Liu, C. Y., Zhang, D. J., Zou, Y. N., He, X. H. and Wu, Q. H. 2016. Mycorrhiza alters the profile of root hairs in trifoliate orange. *Mycorrhiza* 26(3), 237–47. doi:10.1007/s00572-015-0666-z.
Wymer, C. L., Bibikova, T. N. and Gilroy, S. 1997. Cytoplasmic free calcium distributions during the development of root hairs of *Arabidopsis thaliana*. *The Plant Journal: for Cell and Molecular Biology* 12(2), 427–39. doi:10.1046/j.1365-313x.1997.12020427.x.
Yan, X., Liao, H., Beebe, S. E., Blair, M. W. and Lynch, J. P. 2004. QTL mapping of root hair and acid exudation traits and their relationship to phosphorus uptake in common bean. *Plant and Soil* 265(1–2), 17–29. doi:10.1007/s11104-005-0693-1.
Yeates, R. M. 2018. Development of a Rhizosheath Selection Tool for screening perennial ryegrass (*Lolium perenne* L.) for reduced root competition against white clover (*Trifolium repens* L.) for soil phosphorus. Doctoral dissertation. The University of Waikato, New Zealand.
Yoo, C. M., Naramoto, S., Sparks, J. A., Khan, B. R., Nakashima, J., Fukuda, H. and Blancaflor, E. B. 2018. Deletion analysis of AGD1 reveals domains crucial for plasma membrane recruitment and function in root hair polarity. *Journal of Cell Science* 131(2), jcs203828. doi:10.1242/jcs.203828.
Zahran, H. H. and Sprent, J. I. 1986. Effects of sodium chloride and polyethylene glycol on root-hair infection and nodulation of *Vicia faba* L. plants by *Rhizobium leguminosarum*. *Planta* 167(3), 303–9. doi:10.1007/BF00391332.
Zhang, Y. J., Lynch, J. P. and Brown, K. M. 2003. Ethylene and phosphorus availability have interacting yet distinct effects on root hair development. *Journal of Experimental Botany* 54(391), 2351–61. doi:10.1093/jxb/erg250.

Zhang, S., Huang, L., Yan, A., Liu, Y., Liu, B., Yu, C., Zhang, A., Schiefelbein, J. and Gan, Y. 2016. Multiple phytohormones promote root hair elongation by regulating a similar set of genes in the root epidermis in *Arabidopsis*. *Journal of Experimental Botany* 67(22), 6363–72. doi:10.1093/jxb/erw400.

Zhang, C., Simpson, R. J., Kim, C. M., Warthmann, N., Delhaize, E., Dolan, L., Byrne, M. E., Wu, Y. and Ryan, P. R. 2018a. Do longer root hairs improve phosphorus uptake? Testing the hypothesis with transgenic *Brachypodium distachyon* lines overexpressing endogenous RSL genes. *New Phytologist* 217(4), 1654–66. doi:10.1111/nph.14980.

Zhang, Q., van Wijk, R., Zarza, X., Shahbaz, M., van Hooren, M., Guardia, A., Scuffi, D., García-Mata, C., Van den Ende, W., Hoffmann-Benning, S., Haring, M. A., Laxalt, A. M. and Munnik, T. 2018b. Knock-down of *Arabidopsis* PLC5 reduces primary root growth and secondary root formation while overexpression improves drought tolerance and causes stunted root hair growth. *Plant and Cell Physiology* 59(10), 2004–19. doi:10.1093/pcp/pcy120.

Zhang, D. J., Yang, Y. J., Liu, C. Y., Zhang, F. and Wu, Q. S. 2018c. Root hair growth and development in response to nutrients and phytohormones. In: Giri, B., Prasad, R. and Varma, A. (Eds), *Root Biology*. Soil Biology (vol. 52). Springer, Cham, Switzerland, pp. 65–84. doi:10.1007/978-3-319-75910-4_3.

Zhu, J., Kaeppler, S. M. and Lynch, J. P. 2005. Mapping of QTL controlling root hair length in maize (*Zea mays* L.) under phosphorus deficiency. *Plant and Soil* 270(1), 299–310. doi:10.1007/s11104-004-1697-y.

Zhu, C., Gan, L., Shen, Z. and Xia, K. 2006. Interactions between jasmonates and ethylene in the regulation of root hair development in *Arabidopsis*. *Journal of Experimental Botany* 57(6), 1299–308. doi:10.1093/jxb/erj103.

Zhu, J., Zhang, C. and Lynch, J. P. 2010. The utility of phenotypic plasticity of root hair length for phosphorus acquisition. *Functional Plant Biology* 37(4), 313–22. doi:10.1071/FP09197.

Zhu, Y., Rong, L., Luo, Q., Wang, B., Zhou, N., Yang, Y., Zhang, C., Feng, H., Zheng, L., Shen, W. H., Ma, J. and Dong, A. 2017. The histone chaperone NRP1 interacts with WEREWOLF to activate GLABRA2 in *Arabidopsis* root hair development. *The Plant Cell* 29(2), 260–76. doi:10.1105/tpc.16.00719.

Zou, Y. N., Zhang, D. J., Liu, C. Y. and Wu, Q. S. 2019. Relationships between mycorrhizas and root hairs. *Pakistan Journal of Botany* 51(2), 2. doi:10.30848/PJB2019-2(39).

Chapter 6

Understanding plant-root interactions with rhizobacteria to improve biological nitrogen fixation in crops

Ulrike Mathesius, Australian National University, Australia; Jian Jin, La Trobe University, Australia and Chinese Academy of Sciences, China; Yansheng Li, Chinese Academy of Sciences, China; and Michelle Watt, Forschungszentrum Juelich GmbH, Germany and University of Melbourne, Australia

1 Introduction

Biological nitrogen fixation is the biggest contributor to nitrogen input into crop plants other than synthetic nitrogen fertilisers, estimated to fix 20–22 million tons of nitrogen per year in agricultural production systems alone (Herridge et al., 2008). Excessive nitrogen-fertiliser efflux results in severe environmental problems, necessitating alternative and more sustainable approaches to improve plant nitrogen nutrition (Sutton et al., 2011; Stevens, 2019). This can be achieved partly by changing land-management practices (Cassman et al., 2002), by improving plant nitrogen-use efficiency (Garnett et al., 2009), but also by improving and extending the provision of biologically fixed nitrogen by bacteria (Mus et al., 2016). Only certain bacteria ('diazotrophs') can fix nitrogen into a usable form for the plant (ammonia), and symbioses of plants with nitrogen-fixing bacteria have evolved multiple times between different host plants and bacteria (Delaux et al., 2015). These include ferns (e.g. the water fern Azolla and cyanobacteria colonising leaf cavities; Peters and Meeks,

http://dx.doi.org/10.19103/AS.2020.0075.09

1989), gymnosperms (e.g. Cycads and cyanobacteria contained in coralloid roots, Costa and Lindblad, 2002) and angiosperms (e.g. Gunnera and Nostoc cyanobacteria contained in stem glands; Bergman et al., 1992), actinorhizal species forming root nodules with actinomycetes and legumes forming root nodules with rhizobia (Franche et al., 2009). The legume-rhizobia symbiosis is by far the most agronomically important symbiosis and will be the major focus of this chapter. However, actinorhizal symbioses also play important roles in many ecosystems, as well as providing interesting evolutionary insights into root nodulation (Santi et al., 2013). In addition, many plants form non-symbiotic associations with nitrogen-fixing bacteria that colonise the rhizosphere, root surface or intracellular spaces (Baldani et al., 1997; Reinhold-Hurek and Hurek, 1998).

In addition to the direct benefit of nitrogen fixation by rhizobia and other rhizobacteria (defined loosely here as bacteria associated with root surfaces or as root endophytes), host plants also benefit from colonisation of diverse nitrogen-fixing rhizobacteria through alteration of abiotic stress resistance (Goh et al., 2013), disease tolerance (Zamioudis and Pieterse, 2012) and altered root architecture (Goh et al., 2016).

Efforts to improve nitrogen-fixing root associations include the direct transfer of the nitrogenase complex to plants, the transfer of genes required for nitrogen-fixing symbioses to non-hosts, improvements to nitrogen fixation in existing legume-rhizobia symbioses and improvements in nitrogen-fixing associations in non-legumes (Geurts et al., 2016; Mus et al., 2016).

2 Understanding existing interactions of plants with nitrogen-fixing rhizobacteria

2.1 Requirements for successful legume-rhizobia symbioses

Numerous studies and reviews have elaborated the known molecular components required for nodulation in legumes, reviewed by Oldroyd and Downie (2008); Murray (2011); Oldroyd et al. (2011) and briefly summarised below. Most, but not all, legumes form nitrogen-fixing symbioses with rhizobia, a group of soil bacteria that comprises mainly members of the α-proteobacteria as well as some genera of β-proteobacteria (Peix et al., 2015). The only non-legume known to form nitrogen-fixing nodules with rhizobia is *Parasponia andersonii*, which interacts with the broad host range *Rhizobium* NGR234. Within the different species of legumes and their diverse symbiotic partners, a surprising diversity of infection and nodule types can be found (Masson-Boivin et al., 2009).

The molecular dialogue between legume host and their rhizobial symbionts starts with the exudation of flavonoids from legume roots, with specific flavonoid structures being recognised by a rhizobial transcription factor, NodD.

Therefore, flavonoids are some of the main determinants of bacterial host range (Liu and Murray, 2016). Upon activation by flavonoids, NodD activates many other nodulation genes, some of which synthesise Nod factors, which are necessary for nodulation and also determine host-range specificity. Flavonoids also act as chemo-attractants to the rhizobia as well as interfering with bacterial cell-to-cell communication via quorum-sensing molecules (Pérez-Montaño et al., 2011; Hassan and Mathesius, 2012).

Quorum-sensing signalling is necessary at the early stages of symbiosis to coordinate rhizobial behaviours such as motility and exopolysaccharide synthesis; it also controls infection efficiency and bacteroid colonisation at later stages (Sanchez-Contreras et al., 2007). Plants can detect specific quorum-sensing signals (QSS) from different bacteria, leading to wide ranging molecular and physiological changes in the root (Mathesius et al., 2003). In fact, the perception of QSS by the host plant prior to formation of the symbiosis might 'prime' the plant to prepare for symbiosis early on, as pre-treatment of the host with QSS specifically from its symbiont enhances nodule numbers (Veliz-Vallejos et al., 2014; Palmer et al., 2016).

Most rhizobia require Nod factors, specific lipochitin oligosaccharides (LCOs), to initiate nodulation. However, some rhizobia are able to nodulate certain legume hosts without Nod factors (Giraud et al., 2007). Interestingly, some, although not all, of these strains require functional type III secretion systems, possibly to use effectors to downregulate defence responses rather than using Nod factors (Okazaki et al., 2016). LCO production has also been found in some species of Frankia, the symbionts in actinorhizal symbioses (Persson et al., 2015), although this does not appear to be a universal feature of Frankia bacteria.

Following Nod factor perception by specific LysM-type receptors, a cascade of intracellular signals is initiated in root hairs, including specific signatures of calcium spiking and activation of a number of transcription factors required for symbiosis (Oldroyd et al., 2011). Several of the early nodulation genes required for legume nodulation are also used in *Parasponia* and in the actinorhizal symbioses (e.g. Gherbi et al., 2008; Op den Camp et al., 2011). Nod factor signalling is necessary to regulate host-immune responses and controls the expression of a large number of defence-related genes required for successful infection (Zipfel and Oldroyd, 2017; further explained below). In addition, correct perception of other rhizobial molecules is required, for example, of exopolysaccharides, which are bound by a specific receptor subsequent to Nod factor perception (Kawaharada et al., 2015).

Rhizobia usually start to infect the root hairs located just behind the root tip, although nodule initiation can also start at sites of lateral root emergence (Mathesius et al., 2000). Following formation of an infection pocket, the root hair tip invaginates to initiate an infection thread, which elongates and progresses to the cortex, where rhizobia start to be taken up by plant cells by a process similar to endocytosis. Correct defence signalling and changes to the cell wall and cytoskeleton are essential to coordinate this crucial infection step (Murray, 2011). However, in some legumes, rhizobia do not infect through infection threads but infect intercellularly through crack entry, which may have been an ancestral infection mechanism that was made more specific by evolution of infection threats (Sprent, 2008).

Concurrent with the infection, Nod factor signal transduction leads to the activation of cytokinin signalling in the root cortex, which is necessary and sufficient for initiation of cortical cell divisions (Gamas et al., 2017). Cytokinin also controls the transport and accumulation of auxin, a plant hormone essential for the control of cell division and differentiation to initiate nodule development. This is mediated by the action of flavonoids, likely different to the ones acting as Nod gene inducers (Wasson et al., 2006; Ng et al., 2015). Dividing cortical and pericycle cells progress to form a nodule, which is then invaded by rhizobia by a mechanism resembling endocytosis. Intracellular housing of the symbiont has been proposed as a central unifying feature of nodulating plants, requiring fine control over cell-wall dissolution and pressure on the plant membrane (Parniske, 2018). Differentiation of rhizobia into bacteroids is a prerequisite for nitrogen fixation. Inside a mature nodule, bacteroids are surrounded by a plant membrane and supplied with organic acids by the host, which provides a carbon and an energy source to sustain nitrogen fixation. At the same time, the host controls the transport of other essential elements to the bacteroids, including iron and molybdenum required by nitrogenase (White et al., 2007). Bacteroids have lost many of the biosynthetic functions of free-living rhizobia, and thus depend on the host for these metabolites. Homocitrate is an example for a metabolite strictly required by bacteroids from the host (Clark et al., 2014), and thus efficient transport of such metabolites into the nodules is essential and under host control. The host also controls the differentiation of bacteroids inside the nodule by producing a large number of bacteriocidal peptides (NCR peptides), which compatible rhizobia are resistant to if they contain specific genes including bacA and specific exopolysaccharides (Kondorosi et al., 2013; Maroti et al., 2015). This mechanism is important for ongoing selection of specific symbionts in the nodule, as opposed to other bacteria, although many different species of bacteria manage to co-inhabit the inside of nodules (Peix et al., 2015).

2.1.1 *Autoregulation of nodulation (AON)*

Because of the energetic cost of forming nodules and providing symbionts with sufficient quantities of carbon and micronutrients, legumes have mechanisms to restrict nodulation, both intrinsically after the formation of new nodules as well as in response to sufficient nitrogen supply from existing nodules or from the soil. Systemic autoregulation of nodulation has been reviewed previously (Mortier et al., 2012; Reid et al., 2011b; Ferguson et al., 2019), and is summarised briefly here: Upon perception of rhizobia and cytokinin signalling in the root, a peptide of the CLE (CLAVATA3/ENDOSPERM SURROUNDING REGION) family is synthesised in the root, which is subsequently arabinosylated and transported to the shoot in the xylem, where it is perceived by a receptor-like kinase (Nodulation Autoregulation Receptor Kinase (NARK)). This leads to the translocation of another signal, likely a microRNA (Tsikou et al., 2018), although it involves changes in shoot-to-root auxin (van Noorden et al., 2006) and possibly cytokinin transport (Sasaki et al., 2014; Mens et al., 2018) back to the root to inhibit further nodulation. The same NARK receptor is necessary in the root to perceive nitrate-induced CLE peptides, leading to a local inhibition of nodulation when sufficient nitrate is available (Reid et al., 2011a).

In addition, an autoregulation of infection (AOI) mechanism has been identified (Sorroche et al., 2019), in which bacteria that have already infected the root prevent infection from other bacteria in an ethylene-dependent way. This mechanism is independent of AON.

While many of the genes necessary for nodulation in legumes have been identified and characterised, it is not well known to what extent the variation in expression or function of these genes determines the ultimate efficiency of the symbiosis.

2.2 Crosstalk of the legume nodulation programme with immunity and plant development

Rhizobia are not only found in association with legumes, but are highly abundant in rhizosphere soil and even as relatively abundant surface colonisers and endophytes on Arabidopsis roots (Lundberg et al., 2012). A survey of over 1300 strains of rhizobia isolated from different environments has shown that many of the rhizobia associating with non-legumes do not encode nitrogen-fixation genes, and that gene clusters required for nodulation and nitrogen fixation may have been acquired later during evolution, once symbioses with legumes appeared (Garrido-Oter et al., 2018). The initial interactions of rhizobia with non-legumes were more likely commensal, with effects of those root-associated rhizobia on root elongation as well as on plant-immune responses (see below). Rhizobia have adapted to being general root and rhizosphere colonisers in legumes and non-legumes, with 2% of genes estimated to be functioning mainly in rhizosphere colonisation (Salas et al., 2017; Garrido-Oter et al., 2018). It will be interesting to identify plant traits that specifically attract and retain rhizobia or allow root colonisation on non-legumes and to compare this to mechanism of rhizobial attraction and colonisation in legumes. This might inform future efforts to engineer nitrogen-fixation capability into rhizobia that have evolved to interact with non-legumes.

One of the crucial adaptations that rhizobia have acquired in their interaction with legumes is the ability to modulate the host-immune response, which has likely evolved from plant-pathogen signalling (Gourion et al., 2015; Zipfel and Oldroyd, 2017). When rhizobia are first encountered by the root of a legume host, like other bacteria, they are thought to be initially perceived as potential pathogens, while shortly later, there is evidence that compatible rhizobia downregulate host defence responses in order to successfully invade the root (El Yahyaoui et al., 2004; Kouchi et al., 2004; Maunoury et al., 2010; Moreau et al., 2011). The initial defence response is likely due to the recognition of microbial-associated molecular patterns (MAMPs), common epitopes of bacteria and fungi, including flagellin, chitin fragments or surface (lipo)polysaccharides of bacteria and fungi. However, rhizobia have evolved changes in the MAMP structures such that they evade recognition from some of the MAMPs. For example, the flagellin protein of rhizobia has mutated to a degree that it is not recognised by its specific host with a typical defence response (Lopez-Gomez et al., 2012). Similarly, while the lipopolysachides of rhizobia stimulate a defence response in the non-legume tobacco, they do not cause this response in their legume host, suggesting that legumes have fine-tuned the recognition of their symbionts (Albus et al., 2001). The downregulation of defence responses is dependent on both Nod factor signalling (Mitra and Long, 2004), as well as recognition of rhizobial surface molecules, for example, lipopolysaccharides and exopolysaccharides. For example, mutants with incorrect exopolysaccharide structures cannot infect their legume host efficiently (Jones et al., 2008). Exopolysaccharides also play a role at later stages of the symbiosis by protecting rhizobia from NCR peptides (Arnold et al., 2018). In addition, legumes can use R genes to recognise unspecific rhizobia, leading to a defence response preventing infection (Yang et al., 2010). In some rhizobia, the type III secretion system has been shown to be important in facilitating the transfer of rhizobial proteins, for example, nodulation outer proteins (NOPs), into the host cell, which can affect the host range of the symbiont (Staehelin and Krishnan, 2015). Type III secretion systems are commonly used by plant pathogens to inject effector proteins that prevent immune responses in a host, another argument for the similarity between pathogens and symbionts (Deakin and Broughton, 2009). In fact, the type III secretion system might play more fundamental roles in balancing the outcomes of root-bacterial interactions from pathogenic to symbiotic lifestyle. A pathogen, *Ralstonia solanacearum*, was converted to a symbiont infecting the legume Mimosa by transferring a symbiotic plasmid from a broad host-range symbiont, *Cupriavidus taiwanenis*, as well as selecting for mutants that were able to infect the host, which emerged to be mutations in the type III secretion system, likely restricting the transport of effectors into the host that would activate effector-triggered immunity (Guan et al., 2013). Collectively, these early immune responses are all likely to contribute to the specificity of the symbiosis of host legumes with their symbionts; this is particularly important to prevent unspecific infection by pathogens in cells where the immune responses are downregulated.

Similarly, in non-legumes, Nod factors are perceived by a receptor (LYK3, similar in structure to the LysM-type Nod factor receptor in legumes) and this leads to a repression of the plant innate immune response triggered by the MAMP flagellin-22 (Liang et al., 2013). Similar to Nod factor application, inoculation of rhizobia onto Arabidopsis roots has also shown that rhizobia reduce MAMP-triggered immunity responses (judged from gene expression analysis) in Arabidopsis, in a manner opposite to responses to the MAMP flagellin-22 (Garrido-Oter et al., 2018). It has been speculated that this recognition system in Arabidopsis may be predating other recognition processes for chitin-derived signals, including mycorrhization, fungal pathogen infection and Nod factor-based nodulation (Liang et al., 2013). Since lysine motif (LysM) proteins, which generally recognise glycans with an *N*-acetyl-glucosamine backbone, play a role in both Nod factor signalling and immune responses in plants, a tight crosstalk between these pathways is evident (Gourion et al., 2015; Zipfel and Oldroyd, 2017). The perception of Nod factors by non-legumes and downregulation of MAMP responses could have significant implications for the concurrent interactions with other fungal or bacterial pathogens/commensals.

In addition to the local and probably transient effect of rhizobial infection on the local immune response at the infection site, rhizobial infection also causes other local and long-distance changes to defence responses that can affect the outcomes of concurrent infection with pathogens (Pieterse et al., 2014; Reitz et al., 2000). Rhizobia have been demonstrated to alleviate plant infection and damage from a number of plant pathogens (Das et al., 2017), although this is not specific to rhizobia and also occurs in response to a number of other plant growth-promoting rhizobacteria (Compant et al., 2005, 2010). This antagonism can sometimes be mediated by the production of antibiotics/toxins by rhizobia that directly kill a fungal pathogen (Raaijmakers et al., 2002). In addition, rhizobia can induce the production of phytoalexins in inoculated plants, for example isoflavonoids and pterocarpans (Dakora, 2003). The systemic effects of rhizobia on infections of other microbes in distant tissues are very similar to the systemic regulation of nodule numbers - possibly via a similar mechanism that affects defence responses through root-shoot-root signalling (Zamioudis and Pieterse, 2012). The full extent of the implications of this for other plant-microbe interactions has yet to be revealed.

Specific evidence of crosstalk of nodulation with immunity comes from studies of nodulation mutants infected with plant pathogens (Rey et al., 2013; 2015). For example, the Nod factor perception (*nfp*) mutant of *Medicago truncatula* is more susceptible to the oomycete pathogen *Aphanomyces euteiches* and the fungal pathogen *Colletotrichum trifolii* (Rey et al., 2013). In contrast, overexpression of *NFP* increased resistance to these pathogens. This suggests that the Nod factor receptor can perceive pathogen signals (i.e. likely *N*-acetyl-glucosamine components of fungal cell walls) as well as Nod factors, with important implications for the evolution and ecology of nodulation in legumes. Several, but not all, of the proteins acting downstream of Nod factor perception also affect susceptibility to the oomycete pathogen *Phytophthora palmivora* (Rey et al., 2015). Similarly, the ethylene insensitive *skl* mutant, which is hypernodulated by rhizobia, is hyperinfected by other pathogens and mycorrhizal symbionts (Penmetsa et al., 2008). Thus, engineering of the Nod factor perception pathway into non-legumes may have off-target effects for pathogen responses.

In addition, inoculation of legumes with rhizobia has effects on root architecture, as a likely result of how nodule development may have evolved from lateral root formation. Rhizobia not only synthesise plant hormones but also significantly affect the plant hormone balance in their hosts in order to regulate nodule development (Ferguson and Mathesius, 2014). This includes hormones involved in other aspects of root development like auxin, cytokinin, ethylene, strigolactones, brassinosteroids, gibberellins and abscisic acid. Several of these hormones not only affect root development per se, but also integrate developmental responses with environmental factors (Huault et al., 2014). For example, peptide signalling via the CEP (C-terminally encoded peptides) class of regulatory peptides regulates both lateral root and nodule development and integrates these in response to the nitrogen supply (Taleski et al., 2018).

Rhizobial Nod factors directly influence the formation of lateral roots in legume hosts, an effect mediated by auxin responses in the plant (Herrbach et al., 2017). Interestingly, the effect of Nod factors on lateral root formation even occurs in the non-legume *Brachypodium distachion*, suggesting that it occurs via a nodulation-independent mechanism (Buendia et al., 2019). However, nodules and lateral roots also clearly require specific hormonal concentrations, with lateral root formation favoured under high auxin and low cytokinin ratios, while nodule formation is favoured under high cytokinin and low auxin conditions (Bensmihen, 2015; Kohlen et al., 2018). While some nodules, especially those in actinorhizal symbioses, are modified lateral roots, legume nodules originate mainly from the cortex, and this can be regulated independently of lateral roots, which usually originate from the pericycle. A crucial invention of legumes that has allowed cortical nodule development was the acquisition of cortical cytokinin responses that are absent in non-legumes and non-nodulating legumes (Gauthier-Coles et al., 2019).

Rhizobia also alter the root architecture of legumes, as well as altering architecture responses to available nitrogen, and this depends on the presence of functional nodulation genes (Goh al., 2016, 2019). Similarly, rhizobia affect the architecture of non-legume root systems. For instance, in Arabidopsis, infection of roots with rhizobia altered root length and root hair elongation, which can be partly mediated by auxin-response genes in Arabidopsis (Poitout et al., 2017). A large-scale study with many different rhizobial strains showed that root elongation is a conserved response of Arabidopsis colonised by rhizobia - and this does not involve nodulation or nitrogen-fixation genes (Garrido-Oter et al., 2018). It is possible that root elongation is a result of reduced immunity responses in rhizobia-colonised plants, since there is a trade-off between growth and immunity (Garrido-Oter et al., 2018). This is likely the case for legumes as well. These changes in root architecture may affect root-system functioning of rhizobia-infected roots in the field, although it remains to be shown what the direct advantages of these root-architecture changes are in the field or in complex ecosystems.

Lastly, there is a complex interaction of nodulation with the rhizosphere and endosphere microbiome, with specific microbial community structures being driven by the presence of nodules on legume roots (Zgadzaj et al., 2016). This will entail further complex interactions of the microbiome on the plant, which will have to be elucidated in detail in the future.

In summary, our insights into the evolution of nodulation in legumes have shown that rhizobia have 'learnt' to intimately modulate the host-immune response as well as interfering with developmental pathways. This means that future improvements or engineering of nitrogen-fixing symbioses into non-legumes is likely to not only affect the ability for nitrogen fixation, but also alter the ability of these plants to interact with their environment via changes to plant-pathogen responses, the plant microbiome and root architecture.

2.3 Contribution of rhizobacteria to nitrogen fixation for non-legumes

There are many examples of interactions of nitrogen-fixing bacteria, including rhizobia, with non-legumes that lead to nitrogen fixation and benefits the host. Some of these are more efficient than others, and many are not stable or efficient under stressful environmental conditions. Specific interactions between rhizobia and non-legumes have been documented, in particular, for rice (Pittol et al., 2016). For example, a strain of *Rhizobium leguminosarum* bv. *trifollii* isolated from rice plants rotated with berseem clover over hundreds of years in Egypt, has significant beneficial effects on root and shoot weight, as well as root depth, and has been observed to colonise rice tissue endophytically (Yanni et al., 2001). During the interaction, rhizobia infection affects hundreds of genes, including gene induction in the shoot at a distance from the infecting rhizobia, especially those related to cell cycle, photosynthesis, carbohydrate metabolism and cell-wall modification. This suggests specific recognition of the rhizobia by rice, followed by extensive modification of plant metabolism as a result (Wu et al., 2018). Similar to the positive effect of rhizobia on rice, growth-promoting effects and colonisation of rhizobia have been documented in wheat, including in field experiments (Yanni et al., 2016). It has been noted though that efficient use of rhizobia on non-legumes in the field will require specific selection of rhizobia that are beneficial to both non-legumes and legumes grown in rotation, as some strains are beneficial to one host but can be pathogenic to the other (Yanni and Dazzo, 2010).

Another example of a successful interaction of nitrogen-fixing bacteria with non-legumes in the field is an association with a particular indigenous landrace of maize with nitrogen-fixing bacteria (although it was not shown whether these include rhizobia in this case), which contribute a significant amount of biologically fixed nitrogen to the plant (van Deynze et al., 2018). This cultivar had been grown by indigenous Mexicans under low-nitrogen conditions for a long time, which was likely a contributing factor for evolving an association with nitrogen-fixing bacteria. The bacteria exclusively reside on adventitious aerial root tips that exude large amounts of sugar-rich mucilage. This likely provides a sufficient carbon source for nitrogen fixation as well as providing a relatively oxygen-protected niche (van Deynze et al., 2018).

Several other free-living nitrogen-fixing bacteria, for example, *Azetobacter*, *Azospirillum*, *Azoarcus* or *Herbaspirillum*, form associations with grasses, many of them endophytic, although the bacteria remain apoplastic (Reinhold-Hurek and Hurek, 1998, 2011; Santi et al., 2013). Again, root mucilage has been found to attract some of these diazotrophs and retain them in the root (Compant et al., 2010). Similar to legumes, flavonoids contained in root exudates of grasses were found to be regulating gene expression in the associated diazotrophs (Tadra-Sfeir et al., 2011). In further similarity to legume-rhizobia symbioses, attachment of associative bacteria onto cereal roots is mediated by exopolysaccharides (Jofré et al., 1998). Cell wall digestions facilitates the infection of the root by the endophytes (e.g. Reinhold-Hurek et al., 2006), and infection points are typically the zone just behind the growing root tip or sites of lateral root emergence (Reinhold-Hurek and Hurek, 2011). There is evidence also for moderate suppression of defence responses by endophytes, some of which have lost type III secretion systems, possibly to avoid detection by the plant through injection of effectors (Krause et al., 2006). While some of the plant growth promotion observed in these associations is due to nitrogen fixation, the bacteria also produce plant hormones, which could indirectly improve plant growth by altering root architecture (summarised in Santi et al., 2013).

In summary, efficient host-rhizobacteria interactions will need to fine-tune physical, chemical and biological traits. First, signals need to be exchanged to attract rhizobia and modulate defence responses. Secondly, the host needs to establish a physical niche for the microbe that reduces competition with commensals or pathogens (in the simplest case purely apoplastic) and provides an environment that protects nitrogenase from high oxygen concentrations. Thirdly, the host needs to supply an available and abundant carbon source for the bacteria.

3 Improvements and limitations to nitrogen-fixing associations and engineering new symbioses

3.1 Limitations and improvements of legume nitrogen fixation - agronomical considerations using soybean as an example

Breeding efforts towards more efficient nitrogen-fixation in legumes (i.e. higher yield and seed-nitrogen content) have included approaches at better selection of rhizobia as well as improvements in legume breeding and improvement of crop-management practices (Terpolilli et al., 2012). However, breeding or successful engineering of better nitrogen-fixing host-symbiont associations has not usually taken account of our knowledge of the molecular determinants of nodulation.

One important limiting factor in many legumes grown in the field is the infection with suboptimal rhizobia strains. Rhizobia are sensitive to environmental stress, especially pH, salt and drought. Therefore, selection of stress-resistant rhizobia has been important in improving inoculants. This has been extensively reviewed by Dwivedi et al. (2015) and will not be covered in detail here. The effectiveness of rhizobial inoculation depends on how the inoculated rhizobia are able to compete with native soil rhizobia (Hirsch, 2010). However, the application of the rhizobial inoculants is still problematic due to variations in the soil environment (Andrews et al., 2009). In addition, the root tips move away from the zone of inoculation around the seed so there is a narrow spatiotemporal window when the rhizobia are fresh and competitive to infect the root hairs behind the tips present in the same zone. Even if legume seeds are inoculated with optimal strains that have been selected for the host and the particular environment, they are often replaced over time by less efficiently nitrogen-fixing strains and become parasitic inside nodules - demanding carbon while not fixing nitrogen (Hollowell et al., 2016). One reason for this loss of specificity is that the plant host cannot determine the nitrogen-fixation efficiency of a strain until the bacteria have colonised the nodule. However, host sanctions are activated once nitrogen-fixing nodules have been formed and detected to be inefficient (Kiers et al., 2003). These sanctions can be directed towards individual cells inside a nodule, resulting in programmed cell death of inefficiently nitrogen-fixing cells of nodules infected with a mixture of efficient and inefficient symbionts (Regus et al., 2017). This underlines the importance of continued inoculation of crop legumes with highly efficient rhizobial strains.

There is also substantial genetic variation in nitrogen-fixation traits within and between legumes species that can be exploited to improve nitrogen fixation (Hardarson and Atkins, 2003). Improvements on the plant side have involved QTL mapping of nitrogen-fixation traits (nodule numbers, nodule weight, amount of fixed nitrogen in leaves and seeds, ability to fix nitrogen in the presence of inorganic nitrogen) in a number of legume crops, and their integration into breeding programmes (Yang et al., 2019; Ramongolalaina et al., 2018). However, we still lack a broad understanding of what traits ultimately determine highly efficient symbioses and through which mechanisms they act.

In legume cropping systems, N management is a major determinant of the contribution of biological nitrogen fixation (BNF), which is crucial to maintaining photosynthesis, plant biomass accumulation and seed production (Lawlor and Cornic, 2002; Jin et al., 2010a). BNF can be the dominant contributor to N uptake, besides soil inorganic N, including fertiliser N applied to the field (Tamagno et al., 2017). A range of 0–94% of total N content in aboveground biomass of legumes originates from BNF with an average of 55% (Ciampitti and Salvagiotti, 2018). With 108 studies in soybean documented between 1966 and 2006, BNF contributed 50–60% of N uptake and 0–337 kg N ha^{-1} (Salvagiotti et al., 2008). In annual legume crops, plants always take a period to establish fully functional nodules after germination. This weakness in BNF at the seedling stage may limit plant growth in low-N soils. Thus, if N from seed reserves and labile N (NO_3^-/NH_4^+) from the soil cannot satisfy the plant N demand prior to the establishment of functional nodules in low-N soils, a small amount of N application is beneficial to stimulate root growth of legume crops and facilitate the formation of functional nodules (Sprent and Thomas, 1984; McKenzie et al., 2001).

During the whole growth season, whether symbiotically fixed N plus soil inorganic N satisfy the N demand for the legume production depends on soil fertility (Gelfand and Robertson, 2015), temperature (George et al., 1988), soil management (Oberson et al., 2007), abundance of effective indigenous rhizobia (Hungria et al., 2015) and their interactions (George et al., 1988; Santachiara et al., 2017). For example, Schipanski et al. (2010) stated high soil inorganic N levels and/or excessive soil moisture led to a decrease in N fixation. Soil N concentrations are one of the main determinants of BNF, as nitrogen availability inhibits nodulation at the early stages as well as at later, nitrogen-fixing stages (Streeter, 1988). Conversely, BNF is increased under low-N conditions. For example, Li et al. (2016b) found that, in Mollisols with a high fertilization history, reducing the N-fertiliser input considerably increased N_2 fixation during the seed-filling period and consequently maintained seed yield of soybean. The greater contribution of N_2 fixation to N assimilation during this growth period was likely to meet N demand for maintaining soybean yield when fertiliser N supply was reduced. More allocation of photosynthetic C to roots and enhanced nodulation would greatly contribute to the alteration of N acquisition pattern under such conditions. The increase in nodule density under low-N fertilisation was due to the abundant diversity of nitrogen-fixing rhizobia as large farming areas have been continuously planted with soybean over years on this soil type (Jin et al., 2013; Chen et al., 2007). The infection of rhizobia in soybean is probably enhanced when available N in soil cannot meet the plant N demand (Lindström et al., 2015). Therefore, native N status, rhizobia activity and N demand during growth stages should be considered for N fertilisation strategies to enhance BNF.

In agricultural systems, the improvement of crop yield is the priority of farming practice and breeding. Improvement of legume yield is largely attributed to increases in N uptake and its partitioning efficiency. Anglade et al. (2015) found that total N accumulation in the shoot was significantly correlated with fixed N in forage and grain legumes. Collino et al. (2015) indicated that BNF was stronger as soybean yield potential increased, and van Kessel and Hartley (2000) suggested that high-yielding soybean environments should be accompanied by strong BNF activity.

The capacity for BNF has been enhanced with the genetic improvement in legume crops. In a field experiment with more than 50 soybean cultivars released from 1950 to 2006 in Northeast China, the demand for nitrogen accumulation in the seed continuously increased over the period due to the genetic improvement of yield (Jin et al., 2010a). Nodulation ability has been enhanced considerably over years of cultivar releases. Nodule number and fresh weight have significantly increased over time. Moreover, photosynthetic parameters like leaf N concentration were significantly correlated with nodule weight, indicating that nodulation might drive biomass accumulation in newly released cultivars (Jin et al., 2011).

The mechanism of nodulation improvement during the genetic improvement of soybean cultivars may partly be related to (iso)flavonoids that play an essential role in nodulation host range and efficiency (Maj et al., 2010; Liu and Murry, 2016). A number of previous studies reported that seed yield of soybean was positively correlated to isoflavone content (Morrison et al., 2008; Seguin et al., 2004; Primomo et al., 2005). This is supported by in vitro studies showing that increases in flavonoid concentrations improve nodulation efficiency of rhizobia (Maj et al., 2010). However, there might also be an indirect beneficial effect of these flavonoids as defence compounds against pathogens (Hassan and Mathesius, 2012). Therefore, in the selection for higher-yielding cultivars, greater concentration of isoflavones in the roots may favour increases of nodulation, and a greater increase in BNF. It will be interesting to determine how the plant flavonoid content and composition affects cultivar-rhizobia matching under field conditions and whether it contributes to stronger selection of higher fixing strains. Clarifying these aspects will help to limit the input of N fertilisers for high N-efficient cultivars in farming systems by maximising N-fixation and consequently satisfying the N demand for the improvement and stabilisation of yield. In addition, it is important to match the timing of nitrogen fixation with that of seed filling (Mourtzinis et al., 2018). For example, the ability of rhizobia to form nodules on lateral roots rather than the root crown during the seed-filling stage might be important to provide fixed N at this crucial stage (Hardarson and Atkins, 2003). How this late formation of nodules on lateral roots is regulated is not known.

Finally, the current scenario of climate change presents one of the most significant and challenging issues for future food production (Howden et al., 2007). Elevated CO_2 normally promotes C3-plant growth and increases nutrient demands, especially N (Butterly et al., 2016). However, lack of N supply limits legume crop growth and yield in response to elevated CO_2. The plant response may greatly depend on the ability for efficient N_2 fixation. For example, in different soybean cultivars grown under elevated CO_2, greater increase of BNF was found as compared to fertiliser-derived N and soil-derived N, especially during the initial seed-fill stage and up to maturity (Li et al., 2017). Symbiotic N_2 fixation dominantly contributed to N gain under elevated CO_2. Cultivars with greater BNF during the late reproductive stages may exhibit greater yield response to increased CO_2, highlighting the importance of selecting the high-BNF cultivars as an adaptation to climate change. Furthermore, abiotic stress is a significant limitation to nodulation in crop legumes and improvements in legume responses, and adaptation to abiotic stress should be a priority to indirectly enhance BNF and crop yield. Particular limitations include drought and temperature stress as well as phosphorus nutrition (Dwivedi et al., 2015; Valentine et al., 2018).

3.2 Limitations in combining our knowledge on molecular aspects of nodulation with the uncertainties of soil complexities

Irrespective of the large increase in our understanding of the molecular determinants of legume nodulation, it has been difficult to translate much of this information to improvements of nitrogen fixation in the field. While we now know many of the genes and signals necessary for successful symbiosis, we are lacking an understanding of how the expression or abundance of these genes and signals scales with the efficiency of nitrogen fixation, and how legume genotypes that differ in nitrogen-fixation efficiency are distinguished at a molecular level. In addition, many of the molecular studies have been carried out under (semi)sterile lab or glasshouse conditions, often eliminating the heterogeneity of the soil environment.

One example of complex interactions in cropping systems is the stimulation of flavonoid exudation and nodulation in a legume by intercropping with maize (Li et al., 2016a), highlighting the fact that management systems clearly affect the efficiency of the symbiosis. In addition, studies will have to include multipartite interactions of roots with multiple microbes, both symbionts and pathogens, to understand the complexities of signalling found in soil-grown plants (Lace and Ott, 2018).

It is possible to use space and time concepts to highlight the importance and need for research on soil-grown roots (Watt et al., 2006b; Kuzyakov and Blagodatskaya, 2015). The efficacy of rhizobia and other rhizobacteria that are applied with the seed as a seed treatment depends on five steps, outlined in Mathesius and Watt (2011). The first step depends on events outside the root, in the soil, and these events are poorly studied relative to the detailed knowledge about events in the root acquired using molecular methods and modern multi-omics of agar-grown roots. Here we expand on the dependence of the rhizobia and root and their exchanged signals aligning to initiate a nodule, outside the root. This interaction at the macroscale is shown in Fig. 1

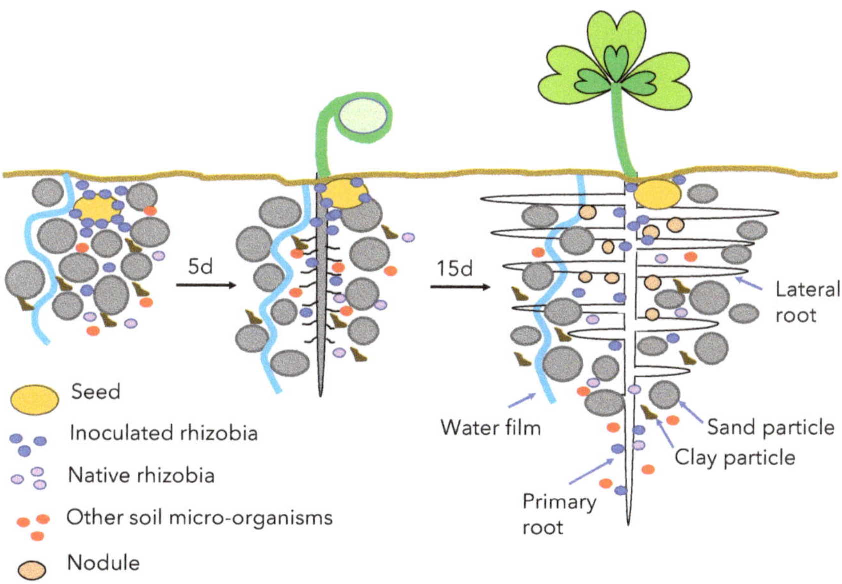

Figure 1 Space-time model for the interaction of seed-inoculated rhizobia with the developing root. Successful colonisation of the root with rhizobia depends on the location of rhizobia on the seed surface and their movement to the susceptible root zone. The symbiosis is also dependent on the diffusion of signals in the soil, which are influenced by soil porosity, composition, pH and water availability.

and depends on co-location of the seed-applied rhizobia or native rhizobia in the region of developing root hairs or junction points between the parent axis and a branch root. The highest probability of this occurrence is at the base of the root system, within the first 15 days of growth. This hot point in time depends on soil moisture and soil type. Soil moisture should not be too dry or too wet, to optimise signal exchange between the root and the rhizobia. If the applied bacteria are not functional in this critical window and migrate at least 3–4 cm below the seed, nodulation events can be expected to be low, or will need to rely on native rhizobia.

Interactions between roots and rhizobia or rhizobacteria depend on distances because the time to travel between distances is proportional to the distance squared (Watt et al., 2006b). It takes a thousand times longer for signals to exchange between a root surface and a bacterium that is 1 mm away as compared to 0.01 mm away. The time is influenced strongly by the soil texture, with sandy soil having 1000 times less surface area than a clay-dominated soil (http://passel.unl.edu/pages/informationmodule.php?idinformationmodule=1130447039&topicorder=3&maxto=10). Time to diffuse is also greatly influenced by moisture status of the soil. Nichol and Silk (2001) showed that the pH changed much faster and five times farther around a root growing in agar than a root growing in sand. Interestingly QS signal exchange between bacteria on leaf surfaces was higher at lower-moisture contents than higher-moisture contents (Dulla and Lindow, 2008). This presents a paradox and an explanation as to why roots have relied on root hairs to extend to form contact with rhizobia, especially in drier soil conditions. The soil pH is influenced by the charges on soil particles and root uptake and exudation close to the root (Hinsinger et al., 2009). The pH greatly influences the functionality of QS molecules and binding of signals to surfaces (Wang and Leadbetter, 2005). In summary, beyond the rhizoplane biofilm, about 20 μm thick (Watt et al., 2006a), the root has little control of contact and efficacy of association with microorganisms, unless it forms new surfaces through root hairs, branch roots or fungal associations (Mogge et al., 2000). These space-time diagrammatic models require data from real or near-real experimental systems.

3.3 Extension and improvement of nitrogen-fixing symbioses for non-legumes

The long-term goal of many of the studies on legume-rhizobia symbiotic signalling has been to transfer the ability for nodulation to non-legumes. Because nodulation requires so many processes to be coordinated, from infection and control of defence responses to nodule organogenesis and efficient nutrient exchange in the nodule, this has not been realised yet. Clearly, the symbiosis confers a number of benefits to legume hosts, especially under low nitrogen and high CO_2 conditions. However, the symbiosis also carries costs

- for example, the provision of large amounts of fixed carbon by the host as well as indirect effects of the presence of nodulation genes on pathogen resistance. Therefore, it may not be surprising that many legumes have lost the ability to form the symbiosis, likely through mutations of crucial nodulation genes. It remains unclear whether the symbiosis has evolved once and then been lost several times or has evolved several times independently (Doyle, 2016), although the currently accepted scenario is a single gain of predisposition for nodulation in a clade of plants encompassing Fabales, Fagales, Cucurbitales and Rosales, followed by multiple independent losses of the symbiosis (van Velzen et al., 2019). Apart from the disadvantages of hosting 'cheaters' and potentially allowing infection by pathogens, the drop in CO_2 concentrations in the earth's atmosphere from ~1000 ppm about 100 million years ago (mya), at the time nodulation likely evolved, to less than 400 ppm about 25 mya is a likely contributor to a loss of nodulation ability, as nitrogen fixation confers a better advantage under high C:N ratios. If nodulation has evolved only once, the question arises as to why it has not happened in other species. This question has not been answered yet.

Various approaches have been initiated to try and transfer the ability to fix nitrogen - either directly or through symbiosis with rhizobia - to non-legume crops. One avenue is to efficiently express the nitrogenase complex in plants to allow direct nitrogen fixation. This has been hampered by the large number of proteins required to be transferred to make nitrogenase functional. Another problem is to protect nitrogenase from oxygen. So far, it has been possible to express a large number of nitrogenase genes in the mitochondrial matrix of yeast and tobacco, a compartment likely to support a reasonably low oxygen environment for nitrogenase (López-Torrejón et al., 2016; Allen et al., 2017), or into the plastid genome (Ivleva et al., 2016). However, effective nitrogen fixation via nitrogenase expressed in a crop plant remains a future challenge.

An alternative approach is to improve the effectiveness of existing nitrogen-fixing associations of rhizobacteria and non-legumes (Santi et al., 2013). Rhizobacteria that colonise non-legumes need to accumulate in sufficient numbers and in root niches suitable for efficient nitrogen fixation. These interactions could be strengthened through (1) selection or directed evolution of the best rhizobacteria (bacterial consortia) for the interaction; (2) attraction of the bacteria by root signals; (3) sustained nurturing of the bacteria by the host, possibly through exudation; (4) improving the transfer of fixed N from the rhizobacteria to the plant; and (5) provision of a physical niche with a low-oxygen environment to prevent inactivation of nitrogenase but with sufficient oxygen to allow oxidative phosphorylation by rhizobia to produce ATP. The physical niche is also important in free-living, surface associated nitrogen-fixing bacteria because fixed nitrogen can be lost in the soil due to the presence of denitrifying bacteria (De Bruijn et al., 1995). While certain strains of rhizobia have been found to colonise non-legumes apoplastically, their infection mechanism and persistence in root and shoot tissue and sometimes also in nodule-like structures initiated by these rhizobia remains unclear (De Bruijn et al., 1995; Chi et al., 2005; Yanni et al., 2001).

Considering the large effect of rhizobial strains on legume nodulation, efforts to select better performing strains of rhizobia or other nitrogen-fixing bacteria seems an additional priority. It has been observed that apart from Nod factor diversity and flavonoid responsiveness, the type III secretion system of rhizobia has a large influence over the rhizobial host range (Deakin and Broughton, 2009; Okazaki et al., 2016) and therefore might be a good target for expanding the host range of rhizobia for non-legumes. In addition, other signals and receptors from rhizobia might have to be optimised, including NodD, exopolysaccharides, NOPs, and others. Alternatively, the non-rhizobial rhizosphere might be able to be engineered with small plasmids conferring specific symbiotic competencies for colonisation of non-legumes (Unay and Perret, 2019). Synthetic approaches at engineering new signalling circuits between hosts and symbiont to specifically turn on bacterial genes have also been demonstrated in Medicago and barley (Geddes et al., 2019). This study engineered plants to synthesise a new signal, the rhizopine scyllo-inosamine, that was perceived by engineered rhizobium biosensors. Rhizopines are usually synthesised only by certain species of rhizobia inside nodules, from where they can be exuded into the soil and used as a nutrient source for con-specific rhizobia that are capable of metabolising them. The use of rhizopines as plant signals to microbes demonstrates that new plant-microbe signalling circuits can be engineered to make symbioses more efficient. It might be possible to engineer novel signals into plants that entice symbionts to release fixed nitrogen more efficiently, to enhance niche colonization or to activate other rhizosphere functions like phosphorus solubilization. Being able to induce synthesis of these signals is essential to activate microbial functions in space and time and to avoid the energetic costs of constitutive signal production (Geddes et al., 2019).

The most ambitious approach is to transfer a suite of nodulation genes to non-legumes. However, despite the identification of many genes necessary for nodulation and nitrogen fixation in legumes, it is first not clear which genes need to be transferred, and secondly, coordinating their expression in space and time will be a challenge. Comparative genomics studies have identified most legume genes in non-legumes. There are only three genes that have likely been lost in non-nodulating species; these are the essential nodulation genes *NFR/NFP* (Nod factor receptor genes), *NIN* (a central transcription factor) and *RPG* (a gene required for infection thread formation) (Griesmann et al., 2018). However, some nodulation-related genes from legumes may show similarity to genes in non-legumes without carrying the same functions in non-legumes

(Griesmann et al., 2018). In some cases, a clear loss of essential nodulation genes has been documented when comparing closely related nodulating and non-nodulating species. For example, genome comparison of the nodulating species *Parasponia* with its non-nodulating relative *Trema* has shown that *Trema* lost *NFR/NFP*, *NIN* and *RPG* (van Velzen et al., 2018). It is likely that many of the other genes required for nodulation will be present and functional in non-legumes because many of the common symbiosis genes are necessary for the very widespread mycorrhizal associations (Oldroyd, 2013), as well as for actinorhizal associations and nodulation in *Parasponia* (Geurts et al., 2016). However, this might not be sufficient to result in a functional symbiosis. For example, endophytic colonisation of rice by the nitrogen-fixing endophyte *Azoarcus* did not activate the common symbiosis pathway, even though the genes are present in rice (Chen et al., 2015). In addition, when several common nodulation genes, including Nod factor receptors, were transferred to non-legumes, this was not sufficient to initiate symbiosis (Untergasser et al., 2012), although this did not include *RPG* or the correct promoter sequence of *NIN*, which was only identified recently (Liu et al., 2019). One explanation is that it is challenging to coordinate a novel nodulation programme in non-legumes with existing developmental and defence pathways that may have evolved with different hormonal sensitivities. Examples of different hormonal functions and sensitivities in legumes compared with non-legumes have been demonstrated for ABA (Liang and Harris, 2005) and cytokinin (Gauthier-Coles et al., 2019).

4 Conclusion

4.1 What makes a good nitrogen-fixing symbiont?

The essential genes needed for bacteria to become nitrogen-fixing symbionts are nodulation (Nod factor synthesis; *nod/nol/noe*) and nitrogen-fixation (*nif*) genes, which typically are located on plasmids that can be spread easily within genera, but not as easily between unrelated genera (Remigi et al., 2016). In addition, the symbiont genome needs to make multiple adjustments for life as a symbiont, including expression of the right surface polysaccharides, secretion systems, metabolic genes, transporters and genes enabling defence against plant NCR peptides, for example (Masson-Boivin and Sachs, 2018). Clearly, being a good provider of fixed nitrogen is a high priority to establish an efficient symbiosis and benefit the host. In addition, the ability to survive challenging abiotic soil conditions is important, as well as competing or cooperating with other rhizosphere organisms, efficiently colonising the root and being able to benefit from root exudates. At the early stages of symbiosis, being able to recognise and respond to host signals is essential. The ability to infect the emerging root and modulate the host-immune response to avoid defence responses is equally important, both at the early infection stage and during later stages when NCR peptides target rhizobia in mature nodules (Fig. 2).

4.2 What makes a good host for nitrogen-fixing bacteria?

Host plants are likely to be successful partners if they are able to specifically recognise their symbionts. This includes recognition of Nod factors, surface epitopes like exo- and lipopolysaccharides, quorum-sensing signals, and effector proteins injected through type III secretion systems. Exudation of correct mixtures of flavonoids to attract and activate rhizobia for symbiosis is essential, and increased flavonoid production can improve nodulation (Maj et al., 2010). Exudation of other signals from the root is an important target for manipulation of activity of rhizosphere microbes (Geddes et al., 2019). In addition, the ability to adjust defence responses during infection is crucial to allow infection only by a compatible strain. The ability to adjust nodule numbers to the nitrogen supply of the soil or from more mature nodules reduces the cost of making nodules that are not needed. At the nitrogen-fixation stage, provision of sufficient but not too much carbon to nodules is essential for ATP production by rhizobia. Just as important is the supply of other nutrients to nodules, for example, iron and molybdenum for nitrogenase. Conversely, efficient export of fixed nitrogen from nodules, its transport to the shoot and allocation to leaves or seeds are important

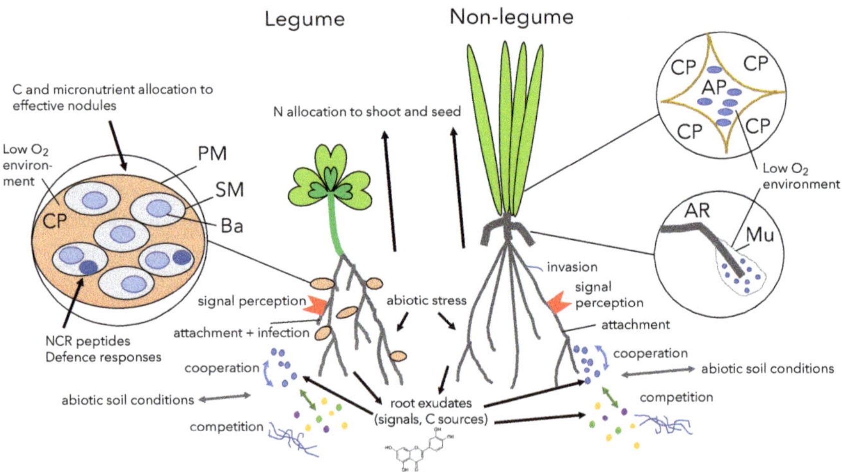

Figure 2 Physical, chemical and biological factors determining the effectiveness of biological nitrogen fixation in symbioses or associations between host plants and nitrogen-fixing bacteria. AP - apoplast; AR - adventitious root; Ba - bacteroids; CP - cytoplasm of plant; Mu - mucilage; SM - symbiosome membrane. Nitrogen-fixing bacteria are shown in blue; other rhizosphere bacteria are shown in yellow, green and purple. Within a nodule, more or less efficient N_2-fixing bacteria are shown in different shades of blue.

for balancing improved photosynthesis with grain yield. For all phenotypes, while many have been studied in isolation and in artificial environments, the ability to nodulate effectively in complex soil environments is important. For example, while high nitrogen fixation is certainly desirable, high N content in leaves also makes highly efficient nitrogen-fixing plants more palatable to herbivores, which can eliminate the benefit from better nitrogen fixation (Simonsen and Stinchcombe, 2014). Similarly, while plant signals and exudates are optimised for their specific symbionts, pathogens can exploit these signals to achieve infection. For example, changes in flavonoid root exudates that control nodulation may also alter plant-pathogen interactions (Hassan and Mathesius, 2012). In addition, susceptibility to nodulation has been shown to also predispose the plant to infection by parasitic nematodes (Wood et al., 2018). Therefore, efficient nitrogen-fixing symbioses can have trade-offs that have to be managed. In soil environments, there are likely many concurrent infections occurring with symbionts and pathogens, and one of the important objectives for the future will be to connect the molecular knowledge we have on nitrogen fixation and nodulation in legumes with the complex environment composed of physical, chemical, biological and environmental variation occurring in field soils.

5 Acknowledgements

We thank the Grains Research and Development Corporation and the Australian Research Council and for funding through grant # IH1401000013 and DP190101817 to UM.

6 Where to look for further information

The chapter cites a number of key review articles in the area. In addition, applied research in this area can be found at the following organisations:

- Food and Agriculture Organization of the United Nations (http://www.fao.org/home/en/).
- International Crops Research Institute for the semi-arid tropics (https://www.icrisat.org/).
- Engineering Nitrogen Symbiosis for Africa (https://www.ensa.ac.uk/).
- Centre for Rhizobium Studies (http://www.murdoch.edu.au/Research-capabilities/Centre-for-Rhizobium-Studies/).
- Nitrogen fixation research at EMBRAPA – Brazilian Agricultural Research Corporation (https://www.embrapa.br/en/tema-fixacao-biologica-de-nitrogenio).
- Global Pulse Confederation and International year of pulses (https://iyp2016.org/).

7 References

Albus, U., Baier, R., Holst, O., Puhler, A. and Niehaus, K. (2001), 'Suppression of an elicitor-induced oxidative burst reaction in *Medicago sativa* cell cultures by *Sinorhizobium meliloti* lipopolysaccharides', *New Phytol.*, 151(3), 597–606.

Allen, R. S., Tilbrook, K., Warden, A. C., Campbell, P. C., Rolland, V., et al. (2017), 'Expression of 16 nitrogenase proteins within the plant mitochondrial matrix', *Front. Plant Sci.*, 8, 1–14.

Andrews, M., Lea, P. J., Raven, J. A. and Azevedo, R. A. (2009), 'Nitrogen use efficiency. 3. Nitrogen fixation: genes and costs', *Ann. Appl. Biol.*, 155(1), 1–13.

Anglade, J., Billen, G. and Garnier, J. (2015), 'Relationships for estimating N_2 fixation in legumes: incidence for N balance of legume-based cropping systems in Europe', *Ecosphere*, 6(3), 1–24.

Arnold, M. F. F., Penterman, J., Shabab, M., Chen, E. J. and Walker, G. C. (2018), 'Important late-stage symbiotic role of the *Sinorhizobium meliloti* exopolysaccharide succinoglycan', *J. Bacteriol.*, 200(13), e00665-17.

Baldani, J. I., Caruso, L., Baldani, V. L. D., Goi, S. R. and Dobereiner, J. (1997), 'Recent advances in BNF with non-legume plants', *Soil Biol. Biochem.*, 29(5–6), 911–22.

Bensmihen, S. (2015), 'Hormonal control of lateral root and nodule development in legumes', *Plants-Basel*, 4(3), 523–47.

Bergman, B., Johansson, C. and Soderback, E. (1992), 'The *Nostoc-Gunnera* symbiosis', *New Phytol.*, 122(3), 379–400.

Buendia, L., Maillet, F., O'Connor, D., van de-Kerkhove, Q., Danoun, S., et al. (2019), 'Lipo-chitooligosaccharides promote lateral root formation and modify auxin homeostasis in *Brachypodium distachyon*', *New Phytol.*, 221(4), 2190–202.

Butterly, C. R., Phillips, L. A., Wiltshire, J. L., Franks, A. E., Armstrong, R. D., et al. (2016), 'Long-term effects of elevated CO_2 on carbon and nitrogen functional capacity of microbial communities in three contrasting soils', *Soil Biol. Biochem.*, 97, 157–67.

Cassman, K. G., Dobermann, A. and Walters, D. T. (2002), 'Agroecosystems, nitrogen-use efficiency, and nitrogen management', *Ambio*, 31(2), 132–40.

Chen, P., Sneller, C. H., Purcell, L. C., Sinclair, T. R., King, C. A., et al. (2007), 'Registration of soybean germplasm lines R01-416F and R01-581F for improved yield and nitrogen fixation under drought stress', *J. Plant Regist.*, 1(2), 166–7.

Chen, X., Miche, L., Sachs, S., Wang, Q., Buschart, A., et al. (2015), 'Rice responds to endophytic colonization which is independent of the common symbiotic signaling pathway', *New Phytol.*, 208(2), 531–43.

Chi, F., Shen, S. H., Cheng, H. P., Jing, Y. X., Yanni, Y. G., et al. (2005), 'Ascending migration of endophytic rhizobia, from roots to leaves, inside rice plants and assessment of benefits to rice growth physiology', *Appl. Environ. Microbiol.*, 71(11), 7271–8.

Ciampitti, I. A. and Salvagiotti, F. (2018), 'New insights into soybean biological nitrogen fixation', *Agron. J.*, 110(4), 1185–96.

Clarke, V. C., Loughlin, P. C., Day, D. A., and Smith, P. M. C. (2014), 'Transport processes of the legume symbiosome membrane.', *Front. Plant Sci.*, 5, 699.

Collino, D. J., Salvagiotti, F., Perticari, A., Piccinetti, C., Ovando, G., et al. (2015), 'Biological nitrogen fixation in soybean in Argentina: relationships with crop, soil, and meteorological factors', *Plant Soil*, 392(1–2), 239–52.

Compant, S., Clement, C. and Sessitsch, A. (2010), 'Plant growth-promoting bacteria in the rhizo- and endosphere of plants: their role, colonization, mechanisms involved and prospects for utilization', *Soil Biol. Biochem.*, 42(5), 669–78.

Compant, S., Duffy, B., Nowak, J., Clement, C. and Barka, E. A. (2005), 'Use of plant growth-promoting bacteria for biocontrol of plant diseases: principles, mechanisms of action, and future prospects', *Appl. Environ. Microbiol.*, 71(9), 4951-9.

Costa, J.-L. and Lindblad, P. (2002), 'Cyanobacteria in symbiosis with cycads', in: Rai, A. N., Bergman, B. and Rasmussen, U. (Eds), *Cyanobacteria in Symbiosis*. Springer, Dordrecht, the Netherlands, pp. 195-205.

Dakora, F. D. (2003), 'Defining new roles for plant and rhizobial molecules in sole and mixed plant cultures involving symbiotic legumes', *New Phytol.*, 158(1), 39-49.

Das, K., Prasanna, R. and Saxena, A. K. (2017), 'Rhizobia: a potential biocontrol agent for soilborne fungal pathogens', *Folia Microbiol.*, 62(5), 425-35.

Deakin, W. J. and Broughton, W. J. (2009), 'Symbiotic use of pathogenic strategies: rhizobial protein secretion systems', *Nat. Rev. Microbiol.*, 7(4), 312-20.

De Bruijn, F. J., Jing, Y. and Dazzo, F. B. (1995), 'Potential and pitfalls of trying to extend symbiotic interactions of nitrogen-fixing organisms to presently nonnodulated plants, such as rice', *Plant Soil*, 174(1-2), 225-40.

Delaux, P. M., Radhakrishnan, G. and Oldroyd, G. (2015), 'Tracing the evolutionary path to nitrogen-fixing crops', *Curr. Opin. Plant Biol.*, 26, 95-9.

den Camp, R. O., Streng, A., De Mita, S., Cao, Q. Q., Polone, E., et al. (2011), 'LysM-type mycorrhizal receptor recruited for rhizobium symbiosis in nonlegume *Parasponia*', *Science*, 331(6019), 909-12.

Doyle, J. J. (2016), 'Chasing unicorns: nodulation origins and the paradox of novelty', *Am. J. Bot.*, 103(11), 1865-8.

Dulla, G. and Lindow, S. E. (2008), 'Quorum size of *Pseudomonas syringae* is small and dictated by water availability on the leaf surface', *Proc. Natl. Acad. Sci. USA*, 105(8), 3082-7.

Dwivedi, S. L., Sahrawat, K. L., Upadhyaya, H. D., Mengoni, A., Galardini, M., et al. (2015), 'Advances in host plant and rhizobium genomics to enhance symbiotic nitrogen fixation in grain legumes', in: Sparks, D. L. (Ed.), *Advances in Agronomy*, Elsevier, pp. 1-116.

El Yahyaoui, F., Kuster, H., Ben Amor, B., Hohnjec, N., Puhler, A., et al. (2004), 'Expression profiling in *Medicago truncatula* identifies more than 750 genes differentially expressed during nodulation, including many potential regulators of the symbiotic program', *Plant Physiol.*, 136(2), 3159-76.

Ferguson, B. J. and Mathesius, U. (2014), 'Phytohormone regulation of legume-rhizobia interactions', *J. Chem. Ecol.*, 40(7), 770-90.

Ferguson, B. J., Mens, C., Hastwell, A. H., Zhang, M. B., Su, H. A., et al. (2019), 'Legume nodulation: the host controls the party', *Plant Cell Environ.*, 42(1), 41-51.

Franche, C., Lindstrom, K. and Elmerich, C. (2009), 'Nitrogen-fixing bacteria associated with leguminous and non-leguminous plants', *Plant Soil*, 321(1-2), 35-59.

Gamas, P., Brault, M., Jardinaud, M. F. and Frugier, F. (2017), 'Cytokinins in symbiotic nodulation: when, where, what for?', *Trends Plant Sci.*, 22(9), 792-802.

Garnett, T., Conn, V. and Kaiser, B. N. (2009), 'Root based approaches to improving nitrogen use efficiency in plants', *Plant Cell Environ.*, 32(9), 1272-83.

Garrido-Oter, R., Nakano, R. T., Dombrowski, N., Ma, K. W., McHardy, A. C., et al. (2018), 'Modular traits of the rhizobiales root microbiota and their evolutionary relationship with symbiotic rhizobia', *Cell Host Microbe*, 24(1), 155-67.

Gauthier-Coles, C., White, R. G. and Mathesius, U. (2019), 'Nodulating legumes are distinguished by a sensitivity to cytokinin in the root cortex leading to pseudonodule development', *Front. Plant Sci.*, 9, 01901.

Geddes, B. A., Paramasivan, P., Joffrin, A., Thompson, A. L., Christensen, K., et al. (2019), 'Engineering transkingdom signalling in plants to control gene expression in rhizosphere bacteria', *Nat. Commun.*, 10, 3430.

Gelfand, I. and Robertson, G. P. (2015), 'A reassessment of the contribution of soybean biological nitrogen fixation to reactive N in the environment', *Biogeochemistry*, 123(1-2), 175-84.

George, T. and Singleton, P. W. (1992), 'Nitrogen assimilation traits and dinitrogen fixation in soybean and common bean', *Agron. J.*, 84(6), 1020-8.

George, T., Singleton, P. W. and Bohlool, B. B. (1988), 'Yield, soil nitrogen uptake, and nitrogen fixation by soybean from four maturity groups grown at three elevations', *Agron. J.*, 80(4), 563-7.

Geurts, R., Xiao, T. T. and Reinhold-Hurek, B. (2016), 'What does it take to evolve a nitrogen-fixing endosymbiosis?', *Trends Plant Sci.*, 21(3), 199-208.

Gherbi, H., Markmann, K., Svistoonoff, S., Estevan, J., Autran, D., et al. (2008), 'SymRK defines a common genetic basis for plant root endosymbioses with arbuscular mycorrhiza fungi, rhizobia, and *Frankiabacteria*', *Proc. Natl. Acad. Sci. USA*, 105(12), 4928-32.

Giraud, E., Moulin, L., Vallenet, D., Barbe, V., Cytryn, E., et al. (2007), 'Legumes symbioses: absence of *Nod* genes in photosynthetic bradyrhizobia', *Science*, 316(5829), 1307-12.

Goh, C. H., Vallejos, D. F. V., Nicotra, A. B. and Mathesius, U. (2013), 'The impact of beneficial plant-associated microbes on plant phenotypic plasticity', *J. Chem. Ecol.*, 39(7), 826-39.

Goh, C. H., Nicotra, A. B. and Mathesius, U. (2016), 'The presence of nodules on legume root systems can alter phenotypic plasticity in response to internal nitrogen independent of nitrogen fixation', *Plant Cell Environ.*, 39(4), 883-96.

Goh, C. H., Nicotra, A. B. and Mathesius, U. (2019), 'Genes controlling legume nodule numbers affect phenotypic plasticity responses to nitrogen in the presence and absence of rhizobia', *Plant Cell Environ.*, 42(5), 1747-57.

Gourion, B., Berrabah, F., Ratet, P. and Stacey, G. (2015), 'Rhizobium-legume symbioses: the crucial role of plant immunity', *Trends Plant Sci.*, 20(3), 186-94.

Griesmann, M., Chang, Y., Liu, X., Song, Y., Haberer, G., et al. (2018), 'Phylogenomics reveals multiple losses of nitrogen-fixing root nodule symbiosis', *Science*, 361(6398), eaat1743.

Guan, S. H., Gris, C., Cruveiller, S., Pouzet, C., Tasse, L., et al. (2013), 'Experimental evolution of nodule intracellular infection in legume symbionts', *ISME J.*, 7(7), 1367-77.

Hardarson, G. and Atkins, C. (2003), 'Optimising biological N_2 fixation by legumes in farming systems', *Plant Soil*, 252(1), 41-54.

Hassan, S. and Mathesius, U. (2012), 'The role of flavonoids in root-rhizosphere signalling: opportunities and challenges for improving plant-microbe interactions', *J. Exp. Bot.*, 63(9), 3429-44.

Herrbach, V., Chirinos, X., Rengel, D., Agbevenou, K., Vincent, R., et al. (2017), 'Nod factors potentiate auxin signaling for transcriptional regulation and lateral root formation in *Medicago truncatula*', *J. Exp. Bot.*, 68(3), 568-82.

Herridge, D. F., Peoples, M. B. and Boddey, R. M. (2008), 'Global inputs of biological nitrogen fixation in agricultural systems', *Plant Soil*, 311(1-2), 1-18.

Hinsinger, P., Bengough, A. G., Vetterlein, D. and Young, I. M. (2009), 'Rhizosphere: biophysics, biogeochemistry and ecological relevance', *Plant Soil*, 321(1-2), 117-52.

Hirsch, A. M. (2010), 'How rhizobia survive in the absence of a legume host, a stressful world indeed', in: Seckbach, J. and Martin, G. (Eds), *Symbioses and Stress*. Springer, Dordrecht, pp. 375-91.

Hollowell, A. C., Regus, J. U., Turissini, D., Gano-Cohen, K. A., Bantay, R., et al. (2016), 'Metapopulation dominance and genomic-island acquisition of *Bradyrhizobium* with superior catabolic capabilities', *Proc. R. Soc. B-Biol. Sci.*, 283(1829), 20160496.

Howden, S. M., Soussana, J. F., Tubiello, F. N., Chhetri, N., Dunlop, M., et al. (2007), 'Adapting agriculture to climate change', *Proc. Natl. Acad. Sci. USA*, 104(50), 19691-6.

Huault, E., Laffont, C., Wen, J. Q., Mysore, K. S., Ratet, P., et al. (2014), 'Local and systemic regulation of plant root system architecture and symbiotic nodulation by a receptor-like kinase', *PLoS Genet.*, 10(12), e1004891.

Hungria, M., Menna, P. and Delamuta, J. R. M. (2015), '*Bradyrhizobium*, the ancestor of all rhizobia: phylogeny of housekeeping and nitrogen-fixation genes', in: Bruijn, F. J. d. (Ed.), *Biological Nitrogen Fixation*. Wiley, New Jersery, pp. 191-202.

Ivleva, N. B., Groat, J., Staub, J. M. and Stephens, M. (2016), 'Expression of active subunit of nitrogenase via integration into plant organelle genome', *PLoS ONE*, 11(8), e0160951.

Jin, J., Liu, X. B., Wang, G. H., Mi, L., Shen, Z. B., et al. (2010a), 'Agronomic and physiological contributions to the yield improvement of soybean cultivars released from 1950 to 2006 in Northeast China', *Field Crops Res.*, 115(1), 116-23.

Jin, J., Wang, G. H., Liu, X. B., Mi, L., Li, Y. H., et al. (2010b), 'Genetic improvement of yield shapes the temporal and spatial root morphology of soybean (*Glycine max*) grown in northeast China', *N. Z. J. Crop Hortic. Sci.*, 38(3), 177-88.

Jin, J., Liu, X. B., Wang, G. H., Liu, J. D., Mi, L. A., et al. (2011), 'Leaf nitrogen status as a main contributor to yield improvement of soybean cultivars', *Agron. J.*, 103(2), 441-8.

Jin, J., Liu, X. B., Wang, G. H., Chen, X. L., Yu, Z. H., et al. (2013), 'Effect of phosphorus application on hierarchical lateral root morphology and phosphorus acquisition in soybean', *J. Plant Nutr.*, 36(10), 1578-89.

Jofré, E., Rivarola, V., Balegno, H. and Mori, G. (1998), 'Differential gene expression in *Azospirillum brasilense* Cd under saline stress', *Can. J. Microbiol.*, 44(10), 929-36.

Jones, K. M., Sharopova, N., Lohar, D. P., Zhang, J. Q., VandenBosch, K. A., et al. (2008), 'Differential response of the plant *Medicago truncatula* to its symbiont *Sinorhizobium meliloti* or an exopolysaccharide-deficient mutant', *Proc. Natl. Acad. Sci. USA*, 105(2), 704-9.

Kawaharada, Y., Kelly, S., Nielsen, M. W., Hjuler, C. T., Gysel, K., et al. (2015), 'Receptor-mediated exopolysaccharide perception controls bacterial infection', *Nature*, 523(7560), 308.

Kiers, E. T., Rousseau, R. A., West, S. A. and Denison, R. F. (2003), 'Host sanctions and the legume-rhizobium mutualism', *Nature*, 425(6953), 78-81.

Kohlen, W., Ng, J. L. P., Deinum, E. E. and Mathesius, U. (2018), 'Auxin transport, metabolism, and signalling during nodule initiation: indeterminate and determinate nodules', *J. Exp. Bot.*, 69(2), 229-44.

Kondorosi, E., Mergaert, P. and Kereszt, A. (2013), 'A paradigm for endosymbiotic life: cell differentiation of *Rhizobium* bacteria provoked by host plant factors', *Annu. Rev. Microbiol.*, 67, 611–28.

Kouchi, H., Shimomura, K., Hata, S., Hirota, A., Wu, G. J., et al. (2004), 'Large-scale analysis of gene expression profiles during early stages of root nodule formation in a model legume, *Lotus japonicus*', *DNA Res.*, 11(4), 263–74.

Krause, A., Ramakumar, A., Bartels, D., Battistoni, F., Bekel, T., et al. (2006), 'Complete genome of the mutualistic, N_2-fixing grass endophyte *Azoarcus* sp. strain BH72', *Nat. Biotechnol.*, 24(11), 1385–91.

Kuzyakov, Y. and Blagodatskaya, E. (2015), 'Microbial hotspots and hot moments in soil: concept & review', *Soil Biol. Biochem.*, 83(1), 184–99.

López-Torrejón, G., Jiménez-Vicente, E., Buesa, J. M., Hernandez, J. A., Verma, H. K., et al. (2016), 'Expression of a functional oxygen-labile nitrogenase component in the mitochondrial matrix of aerobically grown yeast', *Nat. Commun.*, 7, 11426.

Lace, B. and Ott, T. (2018), 'Commonalities and differences in controlling multipartite intracellular infections of legume roots by symbiotic microbes', *Plant Cell Physiol.*, 59(4), 666–77.

Lawlor, D. W. and Cornic, G. (2002), 'Photosynthetic carbon assimilation and associated metabolism in relation to water deficits in higher plants', *Plant Cell Environ.*, 25(2), 275–94.

Li, B., Li, Y. Y., Wu, H. M., Zhang, F. F., Li, C. J., et al. (2016a), 'Root exudates drive interspecific facilitation by enhancing nodulation and N_2 fixation', *Proc. Natl. Acad. Sci. USA*, 113(23), 6496–501.

Li, Y. S., Liu, X. B., Wang, G. H., Yu, Z. H., Mathesius, U., et al. (2016b), 'Shift in origin of plant nitrogen alters carbon and nitrogen assimilation during reproductive stages of soybean grown in a Mollisol', *Crop Pasture Sci.*, 67(8), 872–80.

Li, Y. S., Yu, Z. H., Liu, X. B., Mathesius, U., Wang, G. H., et al. (2017), 'Elevated CO_2 increases nitrogen fixation at the reproductive phase contributing to various yield responses of soybean cultivars', *Front. Plant Sci.*, 8, 1546.

Liang, Y. and Harris, J. M. (2005), 'Response of root branching to abscisic acid is correlated with nodule formation both in legumes and nonlegumes', *Am. J. Bot.*, 92(10), 1675–83.

Liang, Y., Cao, Y. R., Tanaka, K., Thibivilliers, S., Wan, J. R., et al. (2013), 'Nonlegumes respond to rhizobial nod factors by suppressing the innate immune response', *Science*, 341(6152), 1384–7.

Lindström, K., Aserse, A. A. and Mousavi, S. A. (2015), 'Evolution and taxonomy of nitrogen-fixing organisms with emphasis on rhizobia', in: Bruijn, F. J. d. (Ed.), *Biological Nitrogen Fixation*, Wiley and Sons, Jew Jersey, pp. 21–38.

Liu, C. W. and Murray, J. D. (2016), 'The role of flavonoids in nodulation host-range specificity: an update', *Plants-Basel*, 5(3), 33.

Liu, J., Rutten, L., Limpens, E., van der Molen, T., van Velzen, R., et al. (2019), 'A remote cis-regulatory region is required for NIN expression in the pericycle to initiate Medicago nodule primordium formation', *Plant Cell* 31(1), 68–83.

Lopez-Gomez, M., Sandal, N., Stougaard, J. and Boller, T. (2012), 'Interplay of flg22-induced defence responses and nodulation in *Lotus japonicus*', *J. Exp. Bot.*, 63(1), 393–401.

Lundberg, D. S., Lebeis, S. L., Paredes, S. H., Yourstone, S., Gehring, J., et al. (2012), 'Defining the core *Arabidopsis thaliana* root microbiome', *Nature*, 488(7409), 86.

Maj, D., Wielbo, J., Marek-Kozaczuk, M. and Skorupska, A. (2010), 'Response to flavonoids as a factor influencing competitiveness and symbiotic activity of *Rhizobium leguminosarum*', *Microbiol. Res.*, 165(1), 50–60.

Maroti, G., Downie, J. A. and Kondorosi, E. (2015), 'Plant cysteine-rich peptides that inhibit pathogen growth and control rhizobial differentiation in legume nodules', *Curr. Opin. Plant Biol.*, 26, 57–63.

Masson-Boivin, C. and Sachs, J. L. (2018), 'Symbiotic nitrogen fixation by rhizobia - the roots of a success story', *Curr. Opin. Plant Biol.*, 44, 7–15.

Masson-Boivin, C., Giraud, E., Perret, X. and Batut, J. (2009), 'Establishing nitrogen-fixing symbiosis with legumes: how many rhizobium recipes?', *Trends Microbiol.*, 17(10), 458–66.

Mastrodomenico, A. T. and Purcell, L. C. (2012), 'Soybean nitrogen fixation and nitrogen remobilization during reproductive development', *Crop Sci.*, 52(3), 1281–9.

Mathesius, U. and Watt, M. (2011), 'Rhizosphere signals for plant-microbe interactions: implications for field-grown plants', in: Lüttge, U. E., Beyschlag, W., Büdel, B. and Francis, D. (Eds), *Progress in Botany 72. Progress in Botany (Genetics - Physiology - Systematics - Ecology)*, vol. 72. Springer, Berlin, Heidelberg, pp. 125–61.

Mathesius, U., Weinman, J. J., Rolfe, B. G. and Djordjevic, M. A. (2000), 'Rhizobia can induce nodules in white clover by "hijacking" mature cortical cells activated during lateral root development', *Mol. Plant-Microbe Interact.*, 13(2), 170–82.

Mathesius, U., Mulders, S., Gao, M. S., Teplitski, M., Caetano-Anolles, G., et al. (2003), 'Extensive and specific responses of a eukaryote to bacterial quorum-sensing signals', *Proc. Natl. Acad. Sci. USA*, 100(3), 1444–9.

Maunoury, N., Redondo-Nieto, M., Bourcy, M., Van de Velde, W., Alunni, B., et al. (2010), 'Differentiation of symbiotic cells and endosymbionts in *Medicago truncatula* nodulation are coupled to two transcriptome-switches', *PLoS ONE*, 5(3), e9519.

McKenzie, R. H., Middleton, A. B., Solberg, E. D., DeMulder, J., Flore, N., Clayton, G. W., and Bremer, E. (2001), 'Response of pea to rhizobia inoculation and starter nitrogen in Alberta. ', *Can. J. Plant Sci.*, 81, 637–643.

Mens, C., Li, D. X., Haaima, L. E., Gresshoff, P. M. and Ferguson, B. J. (2018), 'Local and systemic effect of cytokinins on soybean nodulation and regulation of their isopentenyl transferase (IPT) biosynthesis genes following rhizobia inoculation', *Front. Plant Sci.*, 9, 1150.

Mitra, R. M. and Long, S. R. (2004), 'Plant and bacterial symbiotic mutants define three transcriptionally distinct stages in the development of the *Medicago truncatula/ Sinorhizobium* meliloti symbiosis', *Plant Physiol.*, 134(2), 595–604.

Mogge, B., Loferer, C., Agerer, R., Hutzler, P. and Hartmann, A. (2000), 'Bacterial community structure and colonization patterns of *Fagus sylvatica* L. ectomycorrhizospheres as determined by fluorescence *in situ* hybridization and confocal laser scanning microscopy', *Mycorrhiza*, 9(5), 271–8.

Moreau, S., Verdenaud, M., Ott, T., Letort, S., de Billy, F., et al. (2011), 'Transcription reprogramming during root nodule development in *Medicago truncatula*', *PLoS ONE*, 6(1), e16463.

Morrison, M. J., Cober, E. K., Saleem, M. F., McLaughlin, N. B., Fregeau-Reid, J., et al. (2008), 'Changes in isoflavone concentration with 58 years of genetic improvement of short-season soybean cultivars in Canada', *Crop Sci.*, 48(6), 2201–8.

Mortier, V., Holsters, M. and Goormachtig, S. (2012), 'Never too many? How legumes control nodule numbers', *Plant Cell Environ.*, 35(2), 245–58.

Mourtzinis, S., Kaur, G., Orlowski, J. M., Shapiro, C. A., Lee, C. D., et al. (2018), 'Soybean response to nitrogen application across the United States: a synthesis-analysis', *Field Crops Res.*, 215, 74–82.

Murray, J. D. (2011), 'Invasion by invitation: rhizobial infection in legumes', *Mol. Plant-Microbe Interact.*, 24(6), 631–9.

Mus, F., Crook, M. B., Garcia, K., Costas, A. G., Geddes, B. A., et al. (2016), 'Symbiotic nitrogen fixation and the challenges to its extension to nonlegumes', *Appl. Environ. Microbiol.*, 82(13), 3698–710.

Ng, J. L. P., Hassan, S., Truong, T. T., Hocart, C. H., Laffont, C., et al. (2015), 'Flavonoids and auxin transport inhibitors rescue symbiotic nodulation in the *Medicago truncatula* cytokinin perception mutant *cre1*', *Plant Cell*, 27(8), 2210–26.

Nichol, S. A. and Silk, W. K. (2001), 'Empirical evidence of a convection-diffusion model for pH patterns in the rhizospheres of root tips', *Plant Cell Environ.*, 24(9), 967–74.

Oberson, A., Nanzer, S., Bosshard, C., Dubois, D., Mader, P., et al. (2007), 'Symbiotic N_2 fixation by soybean in organic and conventional cropping systems estimated by ^{15}N dilution and ^{15}N natural abundance', *Plant Soil*, 290(1–2), 69–83.

Okazaki, S., Tittabutr, P., Teulet, A., Thouin, J., Fardoux, J., et al. (2016), '*Rhizobium*-legume symbiosis in the absence of Nod factors: two possible scenarios with or without the T3SS', *ISME J.*, 10(1), 64–74.

Oldroyd, G. E. D. (2013), 'Speak, friend, and enter: signalling systems that promote beneficial symbiotic associations in plants', *Nat. Rev. Microbiol.*, 11(4), 252–63.

Oldroyd, G. E. D. and Downie, J. M. (2008), 'Coordinating nodule morphogenesis with rhizobial infection in legumes', *Annu. Rev. Plant Biol.*, 59, 519–46.

Oldroyd, G. E. D., Murray, J. D., Poole, P. S. and Downie, J. A. (2011), 'The rules of engagement in the legume-rhizobial symbiosis', *Annu. Rev. Genet.*, 45, 119–44.

Pérez-Montaño, F., Guasch-Vidal, B., González-Barroso, S., López-Baena, F. J., Cubo, T., et al. (2011), 'Nodulation-gene-inducing flavonoids increase overall production of autoinducers and expression of N-acyl homoserine lactone synthesis genes in rhizobia', *Res. Microbiol.*, 162(7), 715–23.

Palmer, A. G., Mukherjee, A., Stacy, D. M., Lazar, S., Ane, J. M., et al. (2016), 'Interkingdom responses to bacterial quorum sensing signals regulate frequency and rate of nodulation in legume-rhizobia symbiosis', *Chembiochem*, 17(22), 2199–205.

Parniske, M. (2018), 'Uptake of bacteria into living plant cells, the unifying and distinct feature of the nitrogen-fixing root nodule symbiosis', *Curr. Opin. Plant Biol.*, 44, 164–74.

Peix, A., Ramirez-Bahena, M. H., Velazquez, E. and Bedmar, E. J. (2015), 'Bacterial associations with legumes', *Crit. Rev. Plant Sci.*, 34(1–3), 17–42.

Penmetsa, R. V., Uribe, P., Anderson, J., Lichtenzveig, J., Gish, J. C., et al. (2008), 'The *Medicago truncatula* ortholog of *Arabidopsis EIN2*, sickle, is a negative regulator of symbiotic and pathogenic microbial associations', *Plant J.*, 55(4), 580–95.

Persson, T., Battenberg, K., Demina, I. V., Vigil-Stenman, T., Heuvel, B. V., et al. (2015), '*Candidatus* Frankia datiscae Dg1, the actinobacterial microsymbiont of *Datisca glomerata*, expresses the canonical nod genes *nodABC* in symbiosis with its host plant', *PLoS ONE*, 10(5), e0127630.

Peters, G. and Meeks, J. (1989), 'The *Azolla-Anabaena* symbiosis: basic biology', *Annu. Rev. Plant Biol.*, 40(1), 193–210.

Pieterse, C. M., Zamioudis, C., Berendsen, R. L., Weller, D. M., Van Wees, S. C., et al. (2014), 'Induced systemic resistance by beneficial microbes', *Annu. Rev. Phytopathol.*, 52, 347–75.

Pittol, M., Durso, L., Valiati, V. H. and Fiuza, L. M. (2016), 'Agronomic and environmental aspects of diazotrophic bacteria in rice fields', *Ann. Microbiol.*, 66(2), 511-27.

Poitout, A., Martinière, A., Kucharczyk, B., Queruel, N., Silva-Andia, J., et al. (2017), 'Local signalling pathways regulate the *Arabidopsis* root developmental response to *Mesorhizobium loti* inoculation', *J. Exp. Bot.*, 68(5), 1199-211.

Primomo, V. S., Poysa, V., Ablett, G. R., Jackson, C.-J., Gijzen, M., et al. (2005), 'Mapping QTL for individual and total isoflavone content in soybean seeds', *Crop Sci.*, 45(6), 2454-64.

Raaijmakers, J. M., Vlami, M. and De Souza, J. T. (2002), 'Antibiotic production by bacterial biocontrol agents', *Anton. Leeuwen.*, 81(1-4), 537.

Ramongolalaina, C., Teraishi, M. and Okumoto, Y. (2018), 'QTLs underlying the genetic interrelationship between efficient compatibility of *Bradyrhizobium* strains with soybean and genistein secretion by soybean roots', *PLoS ONE*, 13(4), e0194671.

Regus, J. U., Quides, K. W., O'Neill, M. R., Suzuki, R., Savory, E. A., et al. (2017), 'Cell autonomous sanctions in legumes target ineffective rhizobia in nodules with mixed infections', *Am. J. Bot.*, 104(9), 1299-312.

Reid, D. E., Ferguson, B. J. and Gresshoff, P. M. (2011a), 'Inoculation-and nitrate-induced CLE peptides of soybean control NARK-dependent nodule formation', *Mol. Plant-Microbe Interact.*, 24(5), 606-18.

Reid, D. E., Ferguson, B. J., Hayashi, S., Lin, Y.-H. and Gresshoff, P. M. (2011b), 'Molecular mechanisms controlling legume autoregulation of nodulation', *Ann. Bot.*, 108(5), 789-95.

Reinhold-Hurek, B. and Hurek, T. (1998), 'Life in grasses: diazotrophic endophytes', *Trends Microbiol.*, 6(4), 139-44.

Reinhold-Hurek, B. and Hurek, T. (2011), 'Living inside plants: bacterial endophytes', *Curr. Opin. Plant Biol.*, 14(4), 435-43.

Reinhold-Hurek, B., Maes, T., Gemmer, S., Van Montagu, M. and Hurek, T. (2006), 'An endoglucanase is involved in infection of rice roots by the not-cellulose-metabolizing endophyte *Azoarcus* sp. strain BH72', *Mol. Plant-Microbe Interact.*, 19(2), 181-8.

Reitz, M., Rudolph, K., Schröder, I., Hoffmann-Hergarten, S., Hallmann, J., et al. (2000), 'Lipopolysaccharides of *Rhizobium etli* strain G12 act in potato roots as an inducing agent of systemic resistance to infection by the cyst nematode *Globodera pallida*', *Appl. Environ. Microbiol.*, 66(8), 3515-18.

Remigi, P., Zhu, J., Young, J. P. W. and Masson-Boivin, C. (2016), 'Symbiosis within symbiosis: evolving nitrogen-fixing legume symbionts', *Trends Microbiol.*, 24(1), 63-75.

Rey, T., Nars, A., Bonhomme, M., Bottin, A., Huguet, S., et al. (2013), 'NFP, a LysM protein controlling Nod factor perception, also intervenes in *Medicago truncatula* resistance to pathogens', *New Phytol.*, 198(3), 875-86.

Rey, T., Chatterjee, A., Buttay, M., Toulotte, J. and Schornack, S. (2015), '*Medicago truncatula* symbiosis mutants affected in the interaction with a biotrophic root pathogen', *New Phytol.*, 206(2), 497-500.

Salas, M. E., Lozano, M. J., López, J. L., Draghi, W. O., Serrania, J., et al. (2017), 'Specificity traits consistent with legume-rhizobia coevolution displayed by *Ensifer meliloti* rhizosphere colonization', *Environ. Microbiol.*, 19(9), 3423-38.

Salvagiotti, F., Cassman, K. G., Specht, J. E., Walters, D. T., Weiss, A., et al. (2008), 'Nitrogen uptake, fixation and response to fertilizer N in soybeans: a review', *Field Crops Res.*, 108(1), 1-13.

Sanchez-Contreras, M., Bauer, W. D., Gao, M., Robinson, J. B. and Allan Downie, J. (2007), 'Quorum-sensing regulation in rhizobia and its role in symbiotic interactions with legumes', *Phil. Trans. R. Soc. B Biol. Sci.*, 362(1483), 1149–63.

Santachiara, G., Borrás, L., Salvagiotti, F., Gerde, J. A. and Rotundo, J. L. (2017), 'Relative importance of biological nitrogen fixation and mineral uptake in high yielding soybean cultivars', *Plant Soil*, 418(1–2), 191–203.

Santi, C., Bogusz, D. and Franche, C. (2013), 'Biological nitrogen fixation in non-legume plants', *Ann. Bot.*, 111(5), 743–67.

Sasaki, T., Suzaki, T., Soyano, T., Kojima, M., Sakakibara, H., et al. (2014), 'Shoot-derived cytokinins systemically regulate root nodulation', *Nat. Commun.*, 5, 4983.

Schipanski, M., Drinkwater, L. and Russelle, M. (2010), 'Understanding the variability in soybean nitrogen fixation across agroecosystems', *Plant Soil*, 329(1–2), 379–97.

Seguin, P., Zheng, W., Smith, D. L. and Deng, W. (2004), 'Isoflavone content of soybean cultivars grown in eastern Canada', *J. Sci. Food Agric.*, 84(11), 1327–32.

Simonsen, A. K. and Stinchcombe, J. R. (2014), 'Herbivory eliminates fitness costs of mutualism exploiters', *New Phytol.*, 202(2), 651–61.

Sorroche, F., Walch, M., Zou, L., Rengel, D., Maillet, F., et al. (2019), 'Endosymbiotic *Sinorhizobium meliloti* modulate *Medicago* root susceptibility to secondary infection via ethylene', *New Phytol.*, 223(3), 1505–15.

Sprent, J. I. (2008), '60Ma of legume nodulation. What's new? What's changing?', *J. Exp. Bot.*, 59(5), 1081–4.

Sprent, J. and Thomas, R. (1984), 'Nitrogen nutrition of seedling grain legumes: some taxonomic, morphological and physiological constraints', *Plant, Cell Environ.*, 7(9), 637–45.

Staehelin, C. and Krishnan, H. B. (2015), 'Nodulation outer proteins: double-edged swords of symbiotic rhizobia', *Biochem. J.*, 470(3), 263–74.

Stevens, C. J. (2019), 'Nitrogen in the environment', *Science*, 363(6427), 578–80.

Streeter, J. (1988), 'Inhibition of legume nodule formation and N_2 fixation by nitrate', *Crit. Rev. Plant Sci.*, 7(1), 1–23.

Sutton, M. A., Oenema, O., Erisman, J. W., Leip, A., van Grinsven, H., et al. (2011), 'Too much of a good thing', *Nature*, 472(7342), 159.

Tadra-Sfeir, M., Souza, E., Faoro, H., Müller-Santos, M., Baura, V., et al. (2011), 'Naringenin regulates expression of genes involved in cell wall synthesis in *Herbaspirillum seropedicae*', *Appl. Environ. Microbiol.*, 77(6), 2180–3.

Taleski, M., Imin, N. and Djordjevic, M. A. (2018), 'CEP peptide hormones: key players in orchestrating nitrogen-demand signalling, root nodulation, and lateral root development', *J. Exp. Bot.*, 69(8), 1829–36.

Tamagno, S., Balboa, G. R., Assefa, Y., Kovács, P., Casteel, S., et al. (2017), 'Nutrient partitioning and stoichiometry in soybean: a synthesis-analysis', *Field Crops Res.*, 200, 18–27.

Terpolilli, J. J., Hood, G. A. and Poole, P. S. (2012), 'What determines the efficiency of N_2-fixing *Rhizobium*-legume symbioses?', *Adv. Microb. Physiol.*, 60, 325–89.

Tsikou, D., Yan, Z., Holt, D. B., Abel, N. B., Reid, D. E., et al. (2018), 'Systemic control of legume susceptibility to rhizobial infection by a mobile microRNA', *Science*, 362(6411), 233–6.

Unay, J. and Perret, X. (2019), 'Synthetic plasmids to challenge symbiotic nitrogen fixation between rhizobia and legumes', in: Reinhardt, D. and Sharma, A. (Eds), *Methods in Rhizosphere Biology Research. Rhizosphere Biology*. Springer, Singapore, pp.3–18.

Untergasser, A., Bijl, G. J., Liu, W., Bisseling, T., Schaart, J. G., et al. (2012), 'One-step *Agrobacterium* mediated transformation of eight genes essential for *Rhizobium* symbiotic signaling using the novel binary vector system pHUGE', *PLoS ONE*, 7(10), e47885.

Valentine, A. J., Benedito, V. A. and Kang, Y. (2018), 'Legume nitrogen fixation and soil abiotic stress: from physiology to genomics and beyond', *Annu. Plant Rev.* 42, 207–48.

van Deynze, A., Zamora, P., Delaux, P.-M., Heitmann, C., Jayaraman, D., et al. (2018), 'Nitrogen fixation in a landrace of maize is supported by a mucilage-associated diazotrophic microbiota', *PLoS Biol.*, 16(8), e2006352.

van Kessel, C. and Hartley, C. (2000), 'Agricultural management of grain legumes: has it led to an increase in nitrogen fixation?', *Field Crops Res.*, 65(2–3), 165–81.

van Noorden, G. E., Ross, J. J., Reid, J. B., Rolfe, B. G. and Mathesius, U. (2006), 'Defective long-distance auxin transport regulation in the *Medicago truncatula super numeric* nodules mutant', *Plant Physiol.*, 140(4), 1494–506.

van Velzen, R., Holmer, R., Bu, F. J., Rutten, L., van Zeijl, A., et al. (2018), 'Comparative genomics of the nonlegume *Parasponia* reveals insights into evolution of nitrogen-fixing rhizobium symbioses', *Proc. Natl. Acad. Sci. USA*, 115(20), E4700–9.

van Velzen, R., Doyle, J. J. and Geurts, R. (2019), 'A resurrected scenario: single gain and massive loss of nitrogen-fixing nodulation', *Trends Plant Sci.*, 24(1), 49–57.

Veliz-Vallejos, D. F., van Noorden, G. E., Yuan, M. Q. and Mathesius, U. (2014), 'A *Sinorhizobium meliloti*-specific N-acyl homoserine lactone quorum-sensing signal increases nodule numbers in *Medicago truncatula* independent of autoregulation', *Front. Plant Sci.*, 5, 551.

Wang, Y. J. and Leadbetter, J. R. (2005), 'Rapid acyl-homoserine lactone quorum signal biodegradation in diverse soils', *Appl. Environ. Microbiol.*, 71(3), 1291–9.

Wasson, A. P., Pellerone, F. I. and Mathesius, U. (2006), 'Silencing the flavonoid pathway in *Medicago truncatula* inhibits root nodule formation and prevents auxin transport regulation by rhizobia', *Plant Cell*, 18(7), 1617–29.

Watt, M., Hugenholtz, P., White, R. and Vinall, K. (2006a), 'Numbers and locations of native bacteria on field-grown wheat roots quantified by fluorescence in situ hybridization (FISH)', *Environ. Microbiol.*, 8(5), 871–84.

Watt, M., Silk, W. and Passioura, J. (2006b), 'Rates of root and organism growth, soil conditions, and temporal and spatial development of the rhizosphere', *Ann. Bot.*, 97(5), 839–55.

White, J., Prell, J., James, E. K. and Poole, P. (2007), 'Nutrient sharing between symbionts', *Plant Physiol.*, 144(2), 604–14.

Wood, C. W., Pilkington, B. L., Vaidya, P., Biel, C. and Stinchcombel, J. R. (2018), 'Genetic conflict with a parasitic nematode disrupts the legume-rhizobia mutualism', *Evol. Lett.*, 2(3), 233–45.

Wu, Q. Q., Peng, X. J., Yang, M. F., Zhang, W. P., Dazzo, F. B., et al. (2018), 'Rhizobia promote the growth of rice shoots by targeting cell signaling, division and expansion', *Plant Mol. Biol.*, 97(6), 507–23.

Yang, S. M., Tang, F., Gao, M. Q., Krishnan, H. B. and Zhu, H. Y. (2010), 'R gene-controlled host specificity in the legume-rhizobia symbiosis', *Proc. Natl. Acad. Sci. USA*, 107(43), 18735–40.

Yang, Q., Yang, Y. Q., Xu, R. N., Lv, H. Y. and Liao, H. (2019), 'Genetic analysis and mapping of QTLs for soybean biological nitrogen fixation traits under varied field conditions', *Front. Plant Sci.*, 10, 75.

Yanni, Y. G. and Dazzo, F. B. (2010), 'Enhancement of rice production using endophytic strains of *Rhizobium leguminosarum* bv. trifolii in extensive field inoculation trials within the Egypt Nile delta', *Plant Soil*, 336(1-2), 129-42.

Yanni, Y. G., Rizk, R. Y., Abd El-Fattah, F. K., Squartini, A., Corich, V., et al. (2001), 'The beneficial plant growth-promoting association of *Rhizobium leguminosarum* bv. trifolii with rice roots', *Aust. J. Plant Physiol.*, 28(9), 845-70.

Yanni, Y. G., Dazzo, F. B., Squartini, A., Zanardo, M., Zidan, M. I., et al. (2016), 'Assessment of the natural endophytic association between *Rhizobium* and wheat and its ability to increase wheat production in the Nile delta', *Plant Soil*, 407(1-2), 367-83.

Zamioudis, C. and Pieterse, C. M. J. (2012), 'Modulation of host immunity by beneficial microbes', *Mol. Plant-Microbe Interact.*, 25(2), 139-50.

Zgadzaj, R., Garrido-Oter, R., Jensen, D. B., Koprivova, A., Schulze-Lefert, P., et al. (2016), 'Root nodule symbiosis in *Lotus japonicus* drives the establishment of distinctive rhizosphere, root, and nodule bacterial communities', *Proc. Natl. Acad. Sci. USA*, 113(49), E7996-8005.

Zipfel, C. and Oldroyd, G. E. D. (2017), 'Plant signalling in symbiosis and immunity', *Nature*, 543(7645), 328-36.

Chapter 7

Advances in understanding arbuscular mycorrhizal fungal effects on soil nutrient cycling

Haiyang Zhang and Jeff R. Powell, Western Sydney University, Australia

1 Introduction

Arbuscular mycorrhizal (AM) fungi (subphylum: Glomeromycotina) can form beneficial partnerships with many crop and horticultural species, with the exceptions mainly being lupins and species from the Brassicaceae. In these partnerships, AM fungi obtain and transfer nutrients from the soil to the plant in exchange for photosynthetically fixed carbon (C). As such, the AM partnership is considered a 'nutritional symbiosis', in which two organisms form an intimate association with one (the 'symbiont') living inside the other (the 'host'). Exchange occurs within the root cortex where fungal hyphae penetrate cortical cells and form arbuscules - branched or coiled structures supporting a high shared surface area between fungal and plant cell membranes (Smith and Read, 2010). Plant and fungal transport proteins are responsible for the exchange of C and nutrients across the apoplast, and much research has been focussed on the structure, expression and functional attributes of these proteins (Behie and Bidochka, 2014, Smith and Smith, 2011). Harder to study, however, is the hyphal mycelium that extends out into the soil and accesses nutrients that are then transported to the plant. Several mechanisms exist by which AM fungi can influence nutrient availability in cropping systems, including direct effects associated with uptake, transfer and immobilisation of mineral nutrients and indirect effects associated with nutrient transformations by other soil micro-organisms. Current work has mainly focussed on cycling and uptake of nitrogen

http://dx.doi.org/10.19103/AS.2020.0075.10

(N) (Veresoglou et al., 2012a) and phosphorus (P) (Smith et al., 2011), although there is potential for other elements such as zinc (Coccina et al., 2019).

Despite increasing evidence for AM fungi to contribute to these mechanisms in controlled experiments, there is little understanding of how important these mechanisms are in cropping systems and whether they can be translated to produce cropping systems with improved nutrient-use efficiency. These partnerships and their associated benefits are sensitive to agronomic factors, for example, the level of fertilisation and tillage employed in a cropping system (Ryan and Graham, 2018). This suggests that some aspects of modern agriculture may be incompatible with realizing benefits from AM fungi, which has led to debate as to whether AM fungi should be a target for agronomic management (Ryan et al., 2019, Rillig et al., 2019, Rillig et al., 2016) and has highlighted limitations of research into AM fungal contributions in agronomic systems (Ryan et al., 2019, Ryan and Graham, 2018). The role of diversity is one aspect in which researchers have had limited success communicating outcomes, with agronomists and crop scientists typically considering AM fungi as a composite exhibiting invariable behaviour despite knowledge among ecologists that AM fungi differ in their ability to increase nutrient uptake (Hodge and Fitter, 2010, Leigh et al., 2009) and higher AM fungal community diversity can lead to increased nutrient uptake and reduced leaching (van der Heijden et al., 1998).

This chapter provides an overview of recent additions to the state-of-the-art in our understanding of how AM fungi modify nutrient availability in soil, specifically concerning the roles that fungal ecology and physiology may play during the processes of nutrient acquisition and transformation. The chapter then proposes research frontiers with the promise to identify opportunities for exploiting AM fungi for improving nutrient-use efficiency. Advancing towards these frontiers requires a greater understanding of the morphological and biochemical variation that exists among diverse AM fungi and the consequences of this variation for spatial and temporal distribution of nutrient availability.

2 Current understanding of AM fungi and nutrient cycling

2.1 Nutrient uptake and transfer

2.1.1 Inorganic N and P

Much of the N in soil is in organic forms, transformed by saprotrophic microbes into inorganic N via mineralisation. Ammonium (NH_4^+) and nitrate (NO_3^-) are the main forms of inorganic N taken up by AM fungi, with a higher rate of NH_4^+ uptake than NO_3^- observed by Tanaka and Yano (2005) studying maize-associated *Glomus aggregatum*; in that study, extraradical hyphae contained 10 more times NH_4^+ than NO_3^-. This ratio may reflect a greater efficiency associated with NH_4^+ taken up for N transport: absorbed NO_3^- must be transformed into NH_4^+ in hyphae before assimilation into arginine, which is then transported throughout the mycelium (Jin et al., 2005). Once transported to the intraradical hyphae and the point of transport to the plant, arginine is broken down via the urease cycle and NH_4^+ transferred to the plant: Tian et al. (2010) observed a three-fold increase in arginine in AM fungal-associated plant roots.

Phosphorus plays a key role in virtually all metabolic processes in plants, and particularly in photosynthesis and respiration, but it is also a major limiting nutrient in many terrestrial ecosystems (Lambers et al., 2008, Schachtman et al., 1998). Orthophosphate (PO_4^{3-}) is the most abundant P form taken up by plants via roots (Schachtman et al., 1998) or AM fungi (Smith et al., 2011). Extraradical hyphae of AM fungi extend beyond the rhizosphere and can scavenge PO_4^{3-} from large volumes of soil, then rapidly translocate this P to intercellular hyphae and highly branched arbuscules inside the root for transport directly into root cortical cells (Smith et al., 2011). Before this long-distance transfer, PO_4^{3-} is converted to polyphosphate and then converted back into inorganic P prior to transfer to the plant (Smith et al., 2011, Harrison, 2005).

2.1.2 Organic N and P

AM fungi possess a relatively weak exo-enzymatic ability as compared to saprotrophic fungi and other types of (ecto-, ericoid and orchid) mycorrhizal fungi (Talbot et al., 2008). However, AM fungi have been shown to accelerate the decomposition of organic matter by other micro-organisms and to access nutrients derived from organic sources. Hodge and Fitter (2010) showed that the extraradical hyphae of *Glomus hoi* were able to access N from complex organic forms derived from ^{13}C- and ^{15}N- labelled ryegrass litter, with three-times more ^{15}N transferred to mycorrhizal plants than found in plants without AM fungi. Leigh et al. (2009) observed that AM fungi producing high abundances of extraradical hyphae (*G. intraradices* vs. *G. hoi* in their study) can stimulate organic N uptake. Cooperation between AM fungi and a soil bacterium, *Rahnella aquatilis*, has been recently shown to increase plant uptake of P from an organic source (phytate) (Zhang et al., 2018b). It has been hypothesised that AM fungi can exude labile C to stimulate the activity of microbial decomposers and increase the decomposition of organic matter and mineralisation of nutrients (Bunn et al., 2019, Aristizábal et al., 2004).

2.2 Soil nutrient processes mediated by AM fungi

2.2.1 N Fixation

Biological N fixation is the largest natural source of N input into most terrestrial ecosystems (Galloway et al., 2004). During this process, symbiotic and free-living diazotrophic micro-organisms reduce atmospheric nitrogen gas (N_2) to reactive and biologically available forms. Multiple studies have observed that associated synergistic interactions between AM fungi and rhizobium lead to increased N fixation in legumes (Toro et al., 1998, Ibijbijen et al., 1996). AM fungal-mediated increases in plant acquisition of P and trace elements (e.g. Cu, Zn) are hypothesised to play a role in enhancing N fixation (Parniske, 2008), including via colonisation of root nodule tissue and nutrient transfer by AM fungi (Scheublin and van der Heijden (2006).

AM fungi can also affect N-fixing organisms that are free-living in the soil. Veresoglou et al. (2012b) observed that two AM fungal species, *Rhizophagus intraradices* and *Glomus margarita,* affected the presence of the *nifH* gene, encoding the nitrogenase iron protein, in the soil bacterial community. Moreover, it has been demonstrated that an AM fungus, *Gigaspora margarita* (BEG isolate 34), harbours a population of endobacteria (*Burkholderia*) in its cytoplasm throughout its life cycle and this bacterium possess genes encoding proteins for N fixation (Bianciotto et al., 1996).

2.2.2 Soil nutrient mineralisation

As mentioned above, AM fungi have weak capacity to decompose organic matter directly but can indirectly affect nutrient mineralisation via effects on saprotrophs in the microbial community. Recently, in a pot experiment, Bukovská et al. (2018) used isotopic ^{15}N-labeling to observe that when adding organic N (supplied as chitin) into a zone accessible solely to AM fungi hyphae, more than 20% of this organic N could be transferred to host plants. Similarly, Hestrin et al. (2019) observed a synergistic effect between AM fungi and soil microbes on plant (*Brachypodium distachyon*) organic N acquisition, with co-inoculation of AM fungi and soil microbes leading to a doubling of AM fungal acquisition of N from organic matter and a ten-fold increase in N acquisition compared with non-mycorrhizal plants grown in the absence of inoculated soil microbes. These studies suggest that mycorrhizal benefits can extend to free-living soil microbial communities, with enhanced rhizosphere C inputs increasing microbial mineralisation. Additional research is necessary in order to evaluate the net effect of AM-microbial synergies and associated soil N transformation on soil C and nutrient stocks.

2.2.3 Soil nitrification and denitrification

Nitrification is the enzymatic oxidation of NH_4^+ to NO_3^-, mediated by specific nitrifying micro-organisms such as ammonia-oxidizing bacteria (AOB), ammonia-oxidizing archaea (AOA), and nitrite-oxidizing bacteria. AOB and AOA are essential for the oxidation of ammonia via the expression of the ammonia monooxygenase gene (*amoA*) (Canfield et al., 2010). AM fungi could potentially compete with ammonia oxidisers under low concentrations of NH_4^+, thus reducing abundance of, and nitrification by, ammonia oxidisers and affecting AOB and AOA community composition. However, negative (Bukovská et al., 2018, Veresoglou et al., 2011), neutral (Cavagnaro et al., 2007) and positive (Teutscherova et al., 2019, Amora-Lazcano et al., 1998) correlations between AM fungal colonisation and the abundance of ammonia-oxidizing microbes have been reported.

Denitrification is a microbial-facilitated process where nitrate (NO_3^-) is reduced and ultimately produces N_2 through a series of intermediate gaseous nitrogen oxide products, including the greenhouse gas N_2O. Recent studies have observed that AM fungi can reduce N_2O fluxes and that the abundance of *nirK* genes (associated with N_2O production) and *nosZ* genes (associated with N_2O consumption) can be negatively and positively associated with AM fungal root colonisation, respectively (Bender et al., 2014). Additionally, AM fungi may affect rates of denitrification through the alteration of the physical soil environment, for example, through soil aggregation (Okiobe et al., 2019).

2.2.4 Spatial and temporal redistribution of nutrients in soil

Nutrients are distributed unevenly in soil, and studying how plants and AM fungi forage for and transport nutrients in light of their distributions can lead to improved understanding of nutrient use (Chen et al., 2016). Recently, Whiteside et al. (2019) used quantum dots (light-emitting particles) to track the movement of P through fungal hyphae and host roots in the presence of modified P distributions in space and observed that AM fungi transferred more P to the host when it was unevenly distributed in the *in vitro* study system. Movement of P was greater to parts of the system where it was initially scarce (Whiteside et al., 2019). More work is needed to investigate the effect that the movement of P, and possibly other nutrients, has on the rate of nutrient cycles in soil, as redistribution of nutrients and modifications to C exudation may also influence where those nutrient cycles are most active.

Little is known about the extent that AM fungal hyphae to act as a temporal storage pool for nutrients. AM fungal hyphae do break down slowly, with one observation observing that 40% of the starting length of hyphae remained after incubating samples of field soil for 150 days (Steinberg and Rillig, 2003). N stored in chitin is immobilised until hyphae die and microbial activity leads to mineralisation followed by competition for the newly available N by plants and microbes (Schimel and Bennett, 2004). P is stored in multiple pools in AM fungal hyphae, including phospholipids, DNA and polyphosphates, all of which are likely rapidly broken down when hyphae die (as observed for algae, see Gabrielson et al. (1983)).

2.2.5 Nutrient leaching

AM fungi have been observed to reduce N leaching but with inconsistent results among studies (Köhl and van der Heijden, 2016, Cavagnaro et al., 2015, Bender et al., 2015, Asghari and Cavagnaro, 2012, van der Heijden, 2010). Some studies observed that AM fungi resulted in large reductions of NH_4^+ or NO_3^- leaching (Asghari and Cavagnaro, 2012), while other studies observed no (Bender et al., 2015) or weak effects (van der Heijden, 2010). Observations of AM fungi reducing P leaching are more common but with variable effects in relation to soil P availability (van der Heijden, 2010), plant or AM fungi species identity (Köhl and van der Heijden, 2016), and ecosystem types (Bender et al., 2015).

2.2.6 The multifaceted effect of AM fungi on soil fertility

The extraradical hyphal mycelium of AM fungi can have positive effects on soil fertility, including chemical, physical and biological aspects of fertility (Abbott and Murphy, 2003). For chemical aspects of fertility, AM fungi have the capacity to increase nutrient input by facilitating nutrient mineralisation and N fixation and to reduce nutrient losses due to leaching and gaseous emissions (red pane

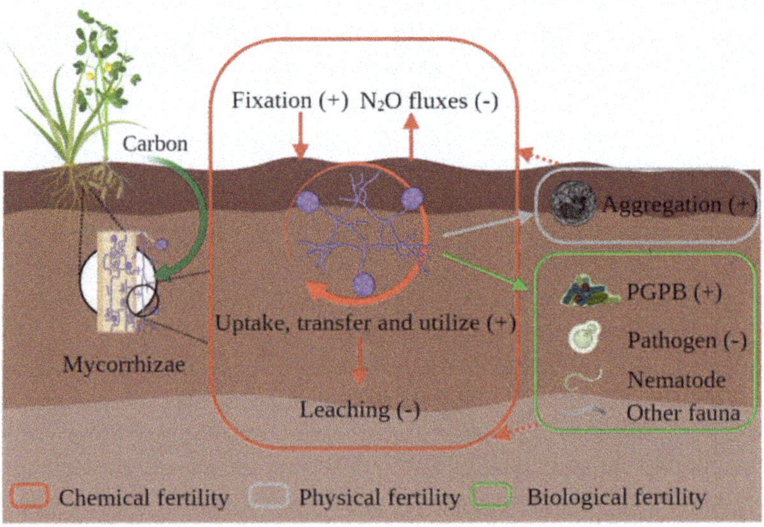

Figure 1 Schematic representation showing the various processes and mechanisms by which AM fungi can improve soil fertility. The red pane includes processes that are directly related to soil chemical fertility. The grey and green panes, respectively, indicate mechanisms by which AM fungi can indirectly influence fertility via effects on soil physical and biological properties. (PGPB: plant-growth promoting bacteria; +: positive effect; −: negative effect; others are unclear).

in Fig. 1). For physical aspects of fertility, AM fungal hyphae directly interact with soil particles to form aggregates that can contribute to the protection of soil organic matter (Clemmensen et al., 2015, Six and Paustian, 2014, Beare et al., 1997) and soil water retention (Mickan et al., 2016) (grey pane in Fig. 1). For biological aspects of fertility, AM fungi deliver root-derived C beyond the rhizosphere, with the potential to influence soil biota involved in the degradation of organic matter (Veresoglou et al., 2016) (green pane in Fig. 1). Effects on physical and biological fertility can result in positive feedback associated with chemical fertility.

3 AM fungal effects on soil fertility in an agronomic context

3.1 Interactions among agronomic practices and AM fungi

Several studies have observed the potential for AM fungi to improve soil fertility without increasing inorganic nutrient inputs and, therefore, to improve the nutrient-use efficiency of crop production systems. However, modern agriculture includes many practices that may be incompatible with AM fungi (Fig. 2), reducing the potential for AM fungi to contribute to gains in nutrient-use efficiency. High inputs of inorganic fertilisers to crop/pasture systems, together with mechanical tillage for weed control and other aspects of soil management, directly reduce AM fungal abundance (Ryan and Graham, 2018, Deng et al., 2017). AM-inducible plant transporters for nutrient uptake are well characterised but new functionalities have been recently unveiled, such as nutrient sensing. For example, Volpe et al. (2016) observed that the expression of *MtPT4* gene (AM-induced phosphorus transporter gene) was also found in the root tips of non-colonised plants and that these transporters may act as sensors of inorganic P and regulate arbuscule morphogenesis and control symbiotic associations (Yang et al., 2012, Xie et al., 2013). AM fungi can also be affected by other land management practices such as crop selection during rotation (Ryan and Graham, 2018, Ryan and Angus, 2003) and the use of cover crops (Bowles et al., 2017, Njeru et al., 2014, Lehman et al., 2012), while intensive agronomic management such as conventional tillage tends to disrupt AM fungal hyphal networks required to move nutrients to roots (Bowles et al., 2017).

Not all AM fungi are equally impacted by conventional agricultural practices and some are observed more frequently in intensively managed systems. Helgason et al. (1998) observed AM fungal communities associated with arable environments to be dominated almost entirely by *Funneliformis mosseae*, while Johnson (1993) observed that long-term inorganic fertilisation in a low-fertility grassland ecosystem led to increased frequency of *Rhizophagus intraradices* and reductions in other species. The AM fungi found in these intensively

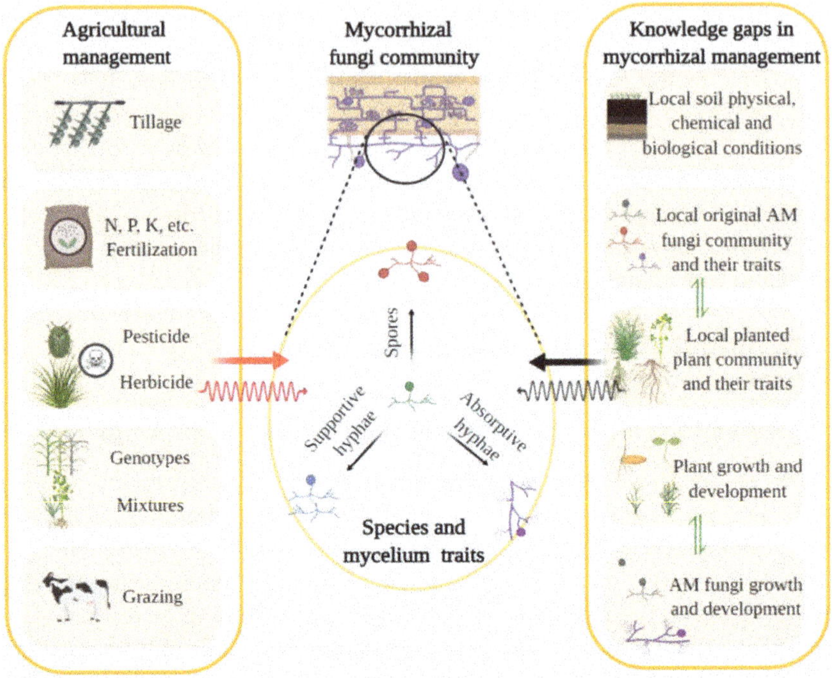

Figure 2 Conceptual figure representing a framework linking agricultural practices (left) and ecological interactions involving AM fungi (right) via attributes, or traits, of the fungi found in agricultural systems (middle). The framework is described as a way to understand the negative effects that practices have on AM fungi (straight red arrow) as well as the selective effects that lead to differences in the fungi that remain (wavy red arrow). The ecological interactions are also representative of knowledge gaps that need to be addressed to understand how AM fungi may be managed to be more abundant (straight black arrow) and to contain functional traits that are beneficial for nutrient-use efficiency (wavy black arrow). Some of the ecological interactions between crop plants and their associated AM fungi are explicitly linked (green equilibrium arrow).

managed environments may not be capable of providing the same nutritional benefits as those found under less-managed systems: AM fungi from long-term fertiliser plots did a poorer job of promoting growth of big bluestem (*Andropogon gerardii*) than did AM fungi from unfertilised plots (Johnson, 1993).

Despite these observations, a minimum level of N fertility may be required to support nutritional benefits provided by AM fungi as they may compete with plants for N under limiting conditions. In pot experiments with limited N, inoculation of grasses with AM fungi resulted in no increase in host biomass for *A. gerardii* (Johnson et al., 2015) and reductions in biomass and shoot N for *B. distachyon* (Riley et al., 2019) relative to un-inoculated controls. In the latter study, these negative effects of AM fungi under N limitation were particularly bad for the fast-growing genotype, exhibiting growth behaviour similar to that of cereal crops (Riley et al., 2019). Thus, careful use of fertilisers may result in compatible interactions with AM fungi, particularly if used in a way to avoid the negative effect that high rates of fertilisation have on AM fungi. When plants are grown at high levels of available P, AM colonisation can be extensively reduced (Balzergue et al., 2010, Breuillin et al., 2010), with Müller et al. (2019) identifying genes encoding CLAVATA3/ESR-related peptides that modulate root strigolactone production, a key molecule for AM fungi to recognise the host. However, Deng et al. (2017) observed that AM fungal colonisation was still observed following application of inorganic P at optimal rates for crop production, while reductions in root colonisation were much greater at rates that were above what was needed for crop production. Controlled-release P fertilisers may show promise for promoting crop production while limiting reductions in mycorrhizal-mediated nutrient-use efficiency. Van Geel et al. (2016) observed that three formulations of slow-release P fertilisers (mushroom compost, green compost and struvite) increased AM fungal diversity in an apple orchard relative to inorganic P and unfertilised control treatments. Further research is needed to evaluate whether new fertilisers that combine relatively insoluble forms of P with P-solubilising micro-organisms (Adesemoye and Kloepper, 2009) can lead to increased compatibility with AM fungi.

In the near future, global environmental change may interact with agricultural management and affect contributions of AM fungi to nutrient-use efficiency (Allen et al., 2003). For example, Martínez-García et al. (2017) and Bowles et al. (2018) both found that increasing rainfall amount and frequency significantly affected AM fungal-mediated effects on nutrient leaching, with reduction of leaching being improved after high rainfall events and with increased rainfall frequency. Our work (unpublished) has observed that, for fescue (*Festuca arundinacea*), AM fungal-mediated reductions of nutrient losses (including leaching and N_2O fluxes) may be lost under warming scenarios.

In addition, new developments in the ability to produce AM fungal inoculum and to support AM fungal growth under field conditions may be on the horizon. Until very recently, the state-of-the-art was that AM fungi relied entirely on carbohydrates from the host plant as their sole pathway of C acquisition (Smith and Read, 2010). Recent landmark studies indicate that AM fungi can also assimilate lipids from their plant hosts and that the fungi actually lack genes that would allow them to synthesise these lipids themselves (Jiang et al., 2017, Keymer et al., 2017, Luginbuehl et al., 2017). In a very recent development, Sugiura et al. (2019) observed that AM fungi can complete their life cycle in the absence of plant host when supplied with a variety of formulations of myristate (C:14), a common organic acid found in plant root exudates (Li et al., 2017). This latter finding in particular could be transformative when it comes to the cost-effective culturing of AM fungi for inoculation in agricultural systems or for maintaining the abundance of AM fungi in fields where non-hosts are included in rotation.

4 Future research: functional linkages between roots and AM fungi

4.1 Morphological and nutritional linkages between root and hyphal traits

Plants exhibit several adaptations associated with morphological root traits to enhance nutrient uptake including modifications to the density of root branching and the diameter of absorptive roots, as well as structural differences between fine roots responsible for absorbing nutrients and coarse roots responsible for transporting nutrients (Kong et al., 2019). For example, as soil nutrient availability decreases, plants tend to produce root systems with higher branching density to increase root absorptive area (Lambers et al., 2008). The extent to which AM fungal hyphae exhibit similar morphological responses to nutrient availability, either within or between species, is poorly understood. If they exhibit similar responses to plants, then high-input agricultural systems may be more likely to be dominated by AM fungi that exhibit mycelial traits that are less efficient for absorbing poorly available nutrients and/or transporting nutrients over long distances, both likely to be of key importance for improving nutrient-use efficiency. While currently untested, we can hypothesise that hyphal branching density is related to the ability of AM fungi to fully explore a volume of soil while hyphal diameter is related to the ability to transport nutrients over long distances (when large) and to be able to access nutrients within soil pores (when small). Resolving this hypothesis related to morphological linkages between root and hyphal functional traits may lead to answers regarding whether AM fungal communities can be managed to promote nutrient-use efficiency in agricultural systems, either via modification of agronomic practices or via inoculation with AM fungi that exhibit desirable functional traits.

Although the AM symbiosis is primarily a nutritional symbiosis, our knowledge about the chemical attributes of the fungal mycelium is surprisingly limited. For example, the ratios of nutritional elements in plant tissues are generally conserved but with deviations from these ratios observed under conditions where some nutrients are limited (Sterner and Elser, 2002). This ecological stoichiometry is an important aspect of plant behaviour since it can reflect differences among plants that are either selected to invest acquired resources into rapid growth or to conserve those resources for later use (Riley et al., 2019). AM fungi may be constrained to a different stoichiometric ratio compared with other non-fungal organisms due to the presence of chitin in cell walls and relatively high protein contents (Allen et al., 2003). This is consistent with observations that competition between AM fungi and host plants may be exacerbated by low N availability (Riley et al., 2019, Johnson et al., 2015), but little is known about how variable these behaviours are among AM fungal species and isolates and whether those AM fungi that are found in agricultural environments tend to occupy either end of this spectrum.

AM fungal species have been observed to exhibit differences in the C required from the host during symbiosis and the development of the extraradical hyphal network as well as in the nutritional benefits that are provided in return. For instance, AM fungi differed in the amount of C taken up from their host (Jakobsen and Rosendahl, 1990), P translocation to roots (Pearson and Jakobsen, 1993), C storage (Gavito and Olsson, 2003) and relative investment into extraradical versus intraradical biomass (Hart and Reader, 2002). Moreover, although variation exists, what drives the variation in the amount of nutrients that AM fungi keep for their own metabolism versus that which is transferred to the plant host is still unclear (Zhang et al., 2015), and although AM fungi can extract nutrients from recently deceased host plants (Johansen and Jensen, 1996), the extent that AM fungi vary in nutrient resorption from dead AM fungal hyphae is not known. This is an important ecological behaviour that varies among terrestrial plants (Aerts, 1996, Zhang et al., 2018a) and has the potential to influence the storage of nutrients in soil in agricultural systems where AM fungi are found.

Assessing these functional linkages between plants and fungi under field conditions will require observational approaches for characterizing AM fungi beyond those currently used in the field, typically microscopic analysis of root colonisation and sequencing of environmental DNA. Sufficient quantities of AM fungal extraradical hyphae for quantifying morphological and chemical traits (Fig. 3) can be obtained using hyphal ingrowth bags (Neumann and George, 2005) followed by techniques such as inductively coupled plasma mass spectrometry (ICP-MS) and nanoscale secondary ion mass spectrometry

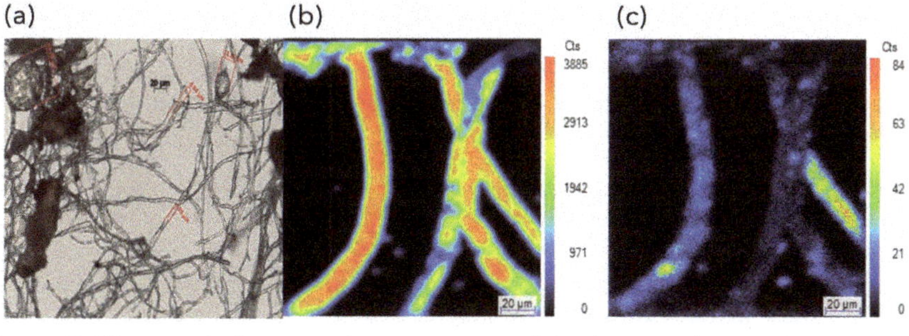

Figure 3 Microscopic images taken on extracted hyphae can be used to measure morphological (a) and chemical (b, c) traits of AM fungi. Light microscopy can be used to estimate hyphal diameter and branching (a) while secondary ion mass spectrometry can be used to estimate concentrations of carbon (b) and phosphorus (c), among other elements, in AM fungal hyphae (both images scaled at 150 × 150 µm).

(NanoSIMS) (Kaiser et al., 2015). Once technical aspects of these approaches are improved, this will lead to an enhanced understanding of how AM fungi interact with crop plants under field conditions and how these interactions respond to changing environmental conditions, leading to recommendations for how these interactions can be used positively to increase nutrient-use efficiency in agroecosystems.

5 Where to look for further information

A standard introduction to the subject of mycorrhizas is S. E. Smith and D. S. Read, *Mycorrhizal Symbiosis* (available through ScienceDirect). There are several literature reviews each offering different perspectives of the many roles that arbuscular mycorrhizas and other plant-associated fungi play in plant nutrition and soil nutrient dynamics. Some recent reviews include:

- Smith S. and Smith A. 2011. Roles of arbuscular mycorrhizas in plant nutrition and growth: new paradigms from cellular to ecosystem scales. *Annual Review of Plant Biology* 62, 227–250.
- Behie, S. 2017. Nutrient transfer in plant-fungal symbioses. *Trends in Plant Science* 19(11), 734–740.
- Cavagnaro, T. et al. 2015. The role of arbuscular mycorrhizas in reducing soil nutrient loss. *Trends in Plant Science* 20(5), 283–290.
- Aguilar-Trigueros, C. et al. 2014. Ecological understanding of root-infecting fungi using trait-based approaches. *Trends in Plant Science* 19(7), 432–438.

An extensive collection of information on mycorrhizas, images of fungi and roots, methodological descriptions and further resources can be found at Mycorrhizal associations: the web resources (https://mycorrhizas.info/index.html).

Multiple collections of AM fungal cultures exist, with each of these collections having information on these cultures and/or methods. A list of some of these is included here:

- International Culture Collection of VA Mycorrhizal Fungi (INVAM) (http://invam.wvu.edu/).
- International Bank for the Glomeromycota: a high-quality culture database of AM FUNGI (http://www.i-beg.eu).
- Glomeromycota *In vitro* Collection (GINCO) (http://www.mycorrhiza.be/ginco-bel/index.php).
- International Culture Collection of Glomeromycota (CICG) (http://www.furb.br/cicg/index.php?lang=EN).

Additional information regarding species of AM fungi has been compiled in the following databases:

- The Fun^{fun} database for fungal traits (https://github.com/traitecoevo/fungaltraits).
- UNITE database for molecular identification of fungi (http://unite.ut.ee).
- MycoDB, a global database of plant response to mycorrhizal fungi (https://www.ncbi.nlm.nih.gov/pmc/articles/PMC4862322/).

6 References

Abbott, L. K. and Murphy, D. V. 2003. *Soil Biological Fertility: A Key to Sustainable Land Use in Agriculture*, Springer Science & Business Media, Dordrecht, The Netherlands.

Adesemoye, A. O. and Kloepper, J. W. 2009. Plant-microbes interactions in enhanced fertilizer-use efficiency. *Applied Microbiology and Biotechnology* 85(1), 1-12.

Aerts, R. 1996. Nutrient resorption from senescing leaves of perennials: are there general patterns? *Journal of Ecology* 84(4), 597-608.

Allen, M. F., Swenson, W., Querejeta, J. I., Egerton-Warburton, L. M. and Treseder, K. K. 2003. Ecology of mycorrhizae: a conceptual framework for complex interactions among plants and fungi. *Annual Review of Phytopathology* 41, 271-303.

Amora-Lazcano, E., Vazquez, M. M. and Azcon, R. 1998. Response of nitrogen-transforming microorganisms to arbuscular mycorrhizal fungi. *Biology and Fertility of Soils* 27(1), 65-70.

Aristizábal, C., Rivera, E. L. and Janos, D. P. 2004. Arbuscular mycorrhizal fungi colonize decomposing leaves of Myrica parvifolia, M. pubescens and Paepalanthus sp. *Mycorrhiza* 14(4), 221-228.

Asghari, H. R. and Cavagnaro, T. R. 2012. Arbuscular mycorrhizas reduce nitrogen loss via leaching. *PLoS ONE* 7(1), e29825.

Balzergue, C., Puech-Pagès, V., Bécard, G. and Rochange, S. F. 2010. The regulation of arbuscular mycorrhizal symbiosis by phosphate in pea involves early and systemic signalling events. *Journal of Experimental Botany* 62(3), 1049-1060.

Beare, M. H., Reddy, M. V., Tian, G. and Srivastava, S. C. 1997. Agricultural intensification, soil biodiversity and agroecosystem function in the tropics: the role of decomposer biota. *Applied Soil Ecology* 6(1), 87-108.

Behie, S. W. and Bidochka, M. J. 2014. Nutrient transfer in plant-fungal symbioses. *Trends in Plant Science* 19(11), 734-740.

Bender, S. F., Conen, F. and van der Heijden, M. G. A. 2015. Mycorrhizal effects on nutrient cycling, nutrient leaching and N_2O production in experimental grassland. *Soil Biology and Biochemistry* 80, 283-292.

Bender, S. F., Plantenga, F., Neftel, A., Jocher, M., Oberholzer, H. R., Köhl, L., Giles, M., Daniell, T. J. and van der Heijden, M. G. 2014. Symbiotic relationships between soil fungi and plants reduce N_2O emissions from soil. *The ISME Journal* 8(6), 1336-1345.

Bianciotto, V., Bandi, C., Minerdi, D., Sironi, M., Tichy, H. V. and Bonfante, P. 1996. An obligately endosymbiotic mycorrhizal fungus itself harbors obligately intracellular bacteria. *Applied and Environmental Microbiology* 62(8), 3005-3010.

Bowles, T. M., Jackson, L. E. and Cavagnaro, T. R. 2018. Mycorrhizal fungi enhance plant nutrient acquisition and modulate nitrogen loss with variable water regimes. *Global Change Biology* 24(1), e171–e182.

Bowles, T. M., Jackson, L. E., Loeher, M. and Cavagnaro, T. R. 2017. Ecological intensification and arbuscular mycorrhizas: a meta-analysis of tillage and cover crop effects. *Journal of Applied Ecology* 54(6), 1785–1793.

Breuillin, F., Schramm, J., Hajirezaei, M., Ahkami, A., Favre, P., Druege, U., Hause, B., Bucher, M., Kretzschmar, T., Bossolini, E., Kuhlemeier, C., Martinoia, E., Franken, P., Scholz, U. and Reinhardt, D. 2010. Phosphate systemically inhibits development of arbuscular mycorrhiza in Petunia hybrida and represses genes involved in mycorrhizal functioning. *The Plant Journal: For Cell and Molecular Biology* 64(6), 1002–1017.

Bukovská, P., Bonkowski, M., Konvalinková, T., Beskid, O., Hujslová, M., Püschel, D., Řezáčová, V., Gutiérrez-Núñez, M. S., Gryndler, M. and Jansa, J. 2018. Utilization of organic nitrogen by arbuscular mycorrhizal fungi—is there a specific role for protists and ammonia oxidizers? *Mycorrhiza* 28(3), 269–283.

Bunn, R. A., Simpson, D. T., Bullington, L. S., Lekberg, Y. and Janos, D. P. 2019. Revisiting the 'direct mineral cycling' hypothesis: arbuscular mycorrhizal fungi colonize leaf litter, but why? *The ISME Journal* 13(8), 1891–1898.

Canfield, D. E., Glazer, A. N. and Falkowski, P. G. 2010. The evolution and future of Earth's nitrogen cycle. *Science* 330(6001), 192–196.

Cavagnaro, T. R., Bender, S. F., Asghari, H. R. and van der Heijden, M. G. A. V. 2015. The role of arbuscular mycorrhizas in reducing soil nutrient loss. *Trends in Plant Science* 20(5), 283–290.

Cavagnaro, T. R., Jackson, L. E., Scow, K. M. and Hristova, K. R. 2007. Effects of arbuscular mycorrhizas on ammonia oxidizing bacteria in an organic farm soil. *Microbial Ecology* 54(4), 618–626.

Chen, W., Koide, R. T., Adams, T. S., DeForest, J. L., Cheng, L. and Eissenstat, D. M. 2016. Root morphology and mycorrhizal symbioses together shape nutrient foraging strategies of temperate trees. *Proceedings of the National Academy of Sciences of the United States of America* 113(31), 8741–8746.

Clemmensen, K. E., Finlay, R. D., Dahlberg, A., Stenlid, J., Wardle, D. A. and Lindahl, B. D. 2015. Carbon sequestration is related to mycorrhizal fungal community shifts during long-term succession in boreal forests. *New Phytologist* 205(4), 1525–1536.

Coccina, A., Cavagnaro, T. R., Pellegrino, E., Ercoli, L., McLaughlin, M. J. and Watts-Williams, S. J. 2019. The mycorrhizal pathway of zinc uptake contributes to zinc accumulation in barley and wheat grain. *BMC Plant Biology* 19(1), 133.

Deng, Y., Feng, G., Chen, X. P. and Zou, C. Q. 2017. Arbuscular mycorrhizal fungal colonization is considerable at optimal Olsen-P levels for maximized yields in an intensive wheat-maize cropping system. *Field Crops Research* 209, 1–9.

Gabrielson, J. O., Birch, P. B. and Hamel, K. S. 1983. Decomposition of Cladophora II. In vitro studies of nitrogen and phosphorus regeneration. *Botanica Marina* 26(4), 173–180.

Galloway, J. N., Dentener, F. J., Capone, D. G., Boyer, E. W., Howarth, R. W., Seitzinger, S. P., Asner, G. P., Cleveland, C. C., Green, P. A., Holland, E. A., Karl, D. M., Michaels, A. F., Porter, J. H., Townsend, A. R. and Vörösmarty, C. J. 2004. Nitrogen cycles: past, present, and future. *Biogeochemistry* 70(2), 153–226.

Gavito, M. E. and Olsson, P. A. 2003. Allocation of plant carbon to foraging and storage in arbuscular mycorrhizal fungi. *FEMS Microbiology Ecology* 45(2), 181–187.

Harrison, M. J. 2005. Signaling in the arbuscular mycorrhizal symbiosis. *Annual Review of Microbiology* 59, 19–42.

Hart, M. M. and Reader, R. J. 2002. Taxonomic basis for variation in the colonization strategy of arbuscular mycorrhizal fungi. *New Phytologist* 153(2), 335–344.

Helgason, T., Daniell, T. J., Husband, R., Fitter, A. H. and Young, J. P. W. 1998. Ploughing up the wood-wide web? *Nature* 394(6692), 431–431.

Hestrin, R., Hammer, E. C., Mueller, C. W. and Lehmann, J. 2019. Synergies between mycorrhizal fungi and soil microbial communities increase plant nitrogen acquisition. *Communications Biology* 2, 233.

Hodge, A. and Fitter, A. H. 2010. Substantial nitrogen acquisition by arbuscular mycorrhizal fungi from organic material has implications for N cycling. *Proceedings of the National Academy of Sciences of the United States of America* 107(31), 13754–13759.

Ibijbijen, J., Urquiaga, S., Ismaili, M., Alves, B. J. R. and Boddey, R. M. 1996. Effect of arbuscular mycorrhizal fungi on growth, mineral nutrition and nitrogen fixation of three varieties of common beans (Phaseolus vulgaris). *New Phytologist* 134(2), 353–360.

Jakobsen, I. and Rosendahl, L. 1990. Carbon flow into soil and external hyphae from roots of mycorrhizal cucumber plants. *New Phytologist* 115(1), 77–83.

Jiang, Y., Wang, W., Xie, Q., Liu, N., Liu, L., Wang, D., Zhang, X., Yang, C., Chen, X., Tang, D. and Wang, E. 2017. Plants transfer lipids to sustain colonization by mutualistic mycorrhizal and parasitic fungi. *Science* 356(6343), 1172–1175.

Jin, H., Pfeffer, P. E., Douds, D. D., Piotrowski, E., Lammers, P. J. and Shachar-Hill, Y. 2005. The uptake, metabolism, transport and transfer of nitrogen in an arbuscular mycorrhizal symbiosis. *New Phytologist* 168(3), 687–696.

Johansen, A. and Jensen, E. S. 1996. Transfer of N and P from intact or decomposing roots of pea to barley interconnected by an arbuscular mycorrhizal fungus. *Soil Biology and Biochemistry* 28(1), 73–81.

Johnson, N. C. 1993. Can fertilization of soil select less mutualistic mycorrhizae? *Ecological Applications: A Publication of the Ecological Society of America* 3(4), 749–757.

Johnson, N. C., Wilson, G. W. T., Wilson, J. A., Miller, R. M. and Bowker, M. A. 2015. Mycorrhizal phenotypes and the Law of the Minimum. *New Phytologist* 205(4), 1473–1484.

Kaiser, C., Kilburn, M. R., Clode, P. L., Fuchslueger, L., Koranda, M., Cliff, J. B., Solaiman, Z. M. and Murphy, D. V. 2015. Exploring the transfer of recent plant photosynthates to soil microbes: mycorrhizal pathway vs direct root exudation. *New Phytologist* 205(4), 1537–1551.

Keymer, A., Pimprikar, P., Wewer, V., Huber, C., Brands, M., Bucerius, S. L., Delaux, P. M., Klingl, V., von Roepenack-Lahaye, E. V., Wang, T. L., Eisenreich, W., Dörmann, P., Parniske, M. and Gutjahr, C. 2017. Lipid transfer from plants to arbuscular mycorrhiza fungi. *eLife* 6, e29107.

Köhl, L. and van der Heijden, M. G. A. 2016. Arbuscular mycorrhizal fungal species differ in their effect on nutrient leaching. *Soil Biology and Biochemistry* 94, 191–199.

Kong, D., Wang, J., Wu, H., Valverde-Barrantes, O. J., Wang, R., Zeng, H., Kardol, P., Zhang, H. and Feng, Y. 2019. Nonlinearity of root trait relationships and the root economics spectrum. *Nature Communications* 10(1), 2203.

Lambers, H., Raven, J. A., Shaver, G. R. and Smith, S. E. 2008. Plant nutrient-acquisition strategies change with soil age. *Trends in Ecology and Evolution* 23(2), 95–103.

Lehman, R. M., Taheri, W. I., Osborne, S. L., Buyer, J. S. and Douds Jr., D. D. 2012. Fall cover cropping can increase arbuscular mycorrhizae in soils supporting intensive agricultural production. *Applied Soil Ecology* 61, 300–304.

Leigh, J., Hodge, A. and Fitter, A. H. 2009. Arbuscular mycorrhizal fungi can transfer substantial amounts of nitrogen to their host plant from organic material. *New Phytologist* 181(1), 199–207.

Li, S., Xu, C., Wang, J., Guo, B., Yang, L., Chen, J. and Ding, W. 2017. Cinnamic, myristic and fumaric acids in tobacco root exudates induce the infection of plants by Ralstonia solanacearum. *Plant and Soil* 412(1–2), 381–395.

Luginbuehl, L. H., Menard, G. N., Kurup, S., Van Erp, H., Radhakrishnan, G. V., Breakspear, A., Oldroyd, G. E. D. and Eastmond, P. J. 2017. Fatty acids in arbuscular mycorrhizal fungi are synthesized by the host plant. *Science* 356(6343), 1175–1178.

Martínez-García, L. B., De Deyn, G. B., Pugnaire, F. I., Kothamasi, D. and van der Heijden, M. G. A. 2017. Symbiotic soil fungi enhance ecosystem resilience to climate change. *Global Change Biology* 23(12), 5228–5236.

Mickan, B. S., Abbott, L. K., Stefanova, K. and Solaiman, Z. M. 2016. Interactions between biochar and mycorrhizal fungi in a water-stressed agricultural soil. *Mycorrhiza* 26(6), 565–574.

Müller, L. M., Flokova, K., Schnabel, E., Sun, X., Fei, Z., Frugoli, J., Bouwmeester, H. J. and Harrison, M. J. 2019. A CLE–SUNN module regulates strigolactone content and fungal colonization in arbuscular mycorrhiza. *Nature Plants* 5(9), 933–939.

Neumann, E. and George, E. 2005. Extraction of extraradical arbuscular mycorrhizal mycelium from compartments filled with soil and glass beads. *Mycorrhiza* 15(7), 533–537.

Njeru, E. M., Avio, L., Sbrana, C., Turrini, A., Bocci, G., Bàrberi, P. and Giovannetti, M. 2014. First evidence for a major cover crop effect on arbuscular mycorrhizal fungi and organic maize growth. *Agronomy for Sustainable Development* 34(4), 841–848.

Okiobe, S. T., Augustin, J., Mansour, I. and Veresoglou, S. D. 2019. Disentangling direct and indirect effects of mycorrhiza on nitrous oxide activity and denitrification. *Soil Biology and Biochemistry* 134, 142–151.

Parniske, M. 2008. Arbuscular mycorrhiza: the mother of plant root endosymbioses. *Nature Reviews. Microbiology* 6(10), 763–775.

Pearson, J. N. and Jakobsen, I. 1993. The relative contribution of hyphae and roots to phosphorus uptake by arbuscular mycorrhizal plants, measured by dual labelling with ^{32}P and ^{33}P. *New Phytologist* 124(3), 489–494.

Riley, R. C., Cavagnaro, T. R., Brien, C., Smith, F. A., Smith, S. E., Berger, B., Garnett, T., Stonor, R., Schilling, R. K., Chen, Z. H. and Powell, J. R. 2019. Resource allocation to growth or luxury consumption drives mycorrhizal responses. *Ecology Letters* 22(11), 1757–1766. DOI: 10.1111/ele.13353.

Rillig, M. C., Aguilar-Trigueros, C. A., Camenzind, T., Cavagnaro, T. R., Degrune, F., Hohmann, P., Lammel, D. R., Mansour, I., Roy, J., van der Heijden, M. G. A. and Yang, G. 2019. Why farmers should manage the arbuscular mycorrhizal symbiosis. *New Phytologist* 222(3), 1171–1175.

Rillig, M. C., Sosa-Hernández, M. A., Roy, J., Aguilar-Trigueros, C. A., Vályi, K. and Lehmann, A. 2016. Towards an integrated mycorrhizal technology: harnessing mycorrhiza for sustainable intensification in agriculture. *Frontiers in Plant Science* 7, 1625.

Ryan, M. H. and Angus, J. F. 2003. Arbuscular mycorrhizae in wheat and field pea crops on a low P soil: increased Zn-uptake but no increase in P-uptake or yield. *Plant and Soil* 250(2), 225–239.

Ryan, M. H. and Graham, J. H. 2018. Little evidence that farmers should consider abundance or diversity of arbuscular mycorrhizal fungi when managing crops. *New Phytologist* 220(4), 1092–1107.

Ryan, M. H., Graham, J. H., Morton, J. B. and Kirkegaard, J. A. 2019. Research must use a systems agronomy approach if management of the arbuscular mycorrhizal symbiosis is to contribute to sustainable intensification. *New Phytologist* 222(3), 1176–1178.

Schachtman, D. P., Reid, R. J. and Ayling, S. M. 1998. Phosphorus uptake by plants: from soil to cell. *Plant Physiology* 116(2), 447–453.

Scheublin, T. R. and van der Heijden, M. G. 2006. Arbuscular mycorrhizal fungi colonize nonfixing root nodules of several legume species. *New Phytologist* 172(4), 732–738.

Schimel, J. P. and Bennett, J. 2004. Nitrogen mineralization: challenges of a changing paradigm. *Ecology* 85(3), 591–602.

Six, J. and Paustian, K. 2014. Aggregate-associated soil organic matter as an ecosystem property and a measurement tool. *Soil Biology and Biochemistry* 68, A4–A9.

Smith, S. E., Jakobsen, I., Grønlund, M. and Smith, F. A. 2011. Roles of arbuscular mycorrhizas in plant phosphorus nutrition: interactions between pathways of phosphorus uptake in arbuscular mycorrhizal roots have important implications for understanding and manipulating plant phosphorus acquisition. *Plant Physiology* 156(3), 1050–1057.

Smith, S. E. and Read, D. J. 2010. *Mycorrhizal Symbiosis*, Academic Press, London.

Smith, S. E. and Smith, F. A. 2011. Roles of arbuscular mycorrhizas in plant nutrition and growth: new paradigms from cellular to ecosystem scales. *Annual Review of Plant Biology* 62, 227–250.

Steinberg, P. D. and Rillig, M. C. 2003. Differential decomposition of arbuscular mycorrhizal fungal hyphae and glomalin. *Soil Biology and Biochemistry* 35(1), 191–194.

Sterner, R. W. and Elser, J. J. 2002. *Ecological Stoichiometry: The Biology of Elements from Molecules to the Biosphere*, Princeton University Press, Princeton.

Sugiura, Y., Akiyama, R., Tanaka, S., Yano, K., Kameoka, H., Kawaguchi, M., Akiyama, K. and Saito, K. 2019. Myristate as a carbon and energy source for the asymbiotic growth of the arbuscular mycorrhizal fungus *Rhizophagus irregularis*. *bioRxiv*, 731489.

Talbot, J. M., Allison, S. D. and Treseder, K. K. 2008. Decomposers in disguise: mycorrhizal fungi as regulators of soil C dynamics in ecosystems under global change. *Functional Ecology* 22(6), 955–963.

Tanaka, Y. and Yano, K. 2005. Nitrogen delivery to maize via mycorrhizal hyphae depends on the form of N supplied. *Plant, Cell and Environment* 28(10), 1247–1254.

Teutscherova, N., Vazquez, E., Arango, J., Arevalo, A., Benito, M. and Pulleman, M. 2019. Native arbuscular mycorrhizal fungi increase the abundance of ammonia-oxidizing bacteria, but suppress nitrous oxide emissions shortly after urea application. *Geoderma* 338, 493–501.

Tian, C., Kasiborski, B., Koul, R., Lammers, P. J., Bücking, H. and Shachar-Hill, Y. 2010. Regulation of the nitrogen transfer pathway in the arbuscular mycorrhizal symbiosis: gene characterization and the coordination of expression with nitrogen flux. *Plant Physiology* 153(3), 1175–1187.

Toro, M., Azcón, R. and Barea, J. M. 1998. The use of isotopic dilution techniques to evaluate the interactive effects of Rhizobium genotype, mycorrhizal fungi, phosphate-solubilizing rhizobacteria and rock phosphate on nitrogen and phosphorus acquisition by Medicago sativa. *New Phytologist* 138(2), 265–273.

Van Der Heijden, M. G. A. 2010. Mycorrhizal fungi reduce nutrient loss from model grassland ecosystems. *Ecology* 91(4), 1163–1171.

Van Der Heijden, M. G. A., Klironomos, J. N., Ursic, M., Moutoglis, P., Streitwolf-Engel, R., Boller, T., Wiemken, A. and Sanders, I. R. 1998. Mycorrhizal fungal diversity determines plant biodiversity, ecosystem variability and productivity. *Nature* 396(6706), 69–72.

Van Geel, M., De Beenhouwer, M., Ceulemans, T., Caes, K., Ceustermans, A., Bylemans, D., Gomand, A., Lievens, B. and Honnay, O. 2016. Application of slow-release phosphorus fertilizers increases arbuscular mycorrhizal fungal diversity in the roots of apple trees. *Plant and Soil* 402(1–2), 291–301.

Veresoglou, S. D., Anderson, I. C., de Sousa, N. M., Hempel, S. and Rillig, M. C. 2016. Resilience of fungal communities to elevated CO_2. *Microbial Ecology* 72(2), 493–495.

Veresoglou, S. D., Chen, B. and Rillig, M. C. 2012a. Arbuscular mycorrhiza and soil nitrogen cycling. *Soil Biology and Biochemistry* 46, 53–62.

Veresoglou, S. D., Sen, R., Mamolos, A. P. and Veresoglou, D. S. 2011. Plant species identity and arbuscular mycorrhizal status modulate potential nitrification rates in nitrogen-limited grassland soils. *Journal of Ecology* 99(6), 1339–1349.

Veresoglou, S. D., Shaw, L. J., Hooker, J. E. and Sen, R. 2012b. Arbuscular mycorrhizal modulation of diazotrophic and denitrifying microbial communities in the (mycor) rhizosphere of Plantago lanceolata. *Soil Biology and Biochemistry* 53, 78–81.

Volpe, V., Giovannetti, M., Sun, X. G., Fiorilli, V. and Bonfante, P. 2016. The phosphate transporters LjPT4 and MtPT4 mediate early root responses to phosphate status in non mycorrhizal roots. *Plant, Cell and Environment* 39(3), 660–671.

Whiteside, M. D., Werner, G. D. A., Caldas, V. E. A., van't Padje, A., Dupin, S. E., Elbers, B., Bakker, M., Wyatt, G. A. K., Klein, M., Hink, M. A., Postma, M., Vaitla, B., Noë, R., Shimizu, T. S., West, S. A. and Kiers, E. T. 2019. Mycorrhizal fungi respond to resource inequality by moving phosphorus from rich to poor patches across networks. *Current Biology* 29(12), 2043–2050.e8.

Xie, X., Huang, W., Liu, F., Tang, N., Liu, Y., Lin, H. and Zhao, B. 2013. Functional analysis of the novel mycorrhiza-specific phosphate transporter AsPT1 and PHT1 family from Astragalus sinicus during the arbuscular mycorrhizal symbiosis. *New Phytologist* 198(3), 836–852.

Yang, S. Y., Grønlund, M., Jakobsen, I., Grotemeyer, M. S., Rentsch, D., Miyao, A., Hirochika, H., Kumar, C. S., Sundaresan, V., Salamin, N., Catausan, S., Mattes, N., Heuer, S. and Paszkowski, U. 2012. Nonredundant regulation of rice arbuscular mycorrhizal symbiosis by two members of the PHOSPHATE TRANSPORTER1 gene family. *Plant Cell* 24(10), 4236–4251.

Zhang, H., Ziegler, W., Han, X., Trumbore, S. and Hartmann, H. 2015. Plant carbon limitation does not reduce nitrogen transfer from arbuscular mycorrhizal fungi to Plantago lanceolata. *Plant and Soil* 396(1–2), 369–380.

Zhang, H. Y., Lü, X. T., Hartmann, H., Keller, A., Han, X. G., Trumbore, S. and Phillips, R. P. 2018a. Foliar nutrient resorption differs between arbuscular mycorrhizal and ectomycorrhizal trees at local and global scales. *Global Ecology and Biogeography* 27(7), 875–885.

Zhang, L., Feng, G. and Declerck, S. 2018b. Signal beyond nutrient, fructose, exuded by an arbuscular mycorrhizal fungus triggers phytate mineralization by a phosphate solubilizing bacterium. *The ISME Journal* 12(10), 2339–2351.

Part 2

Root response to biotic threats

Chapter 8

Advances in understanding plant root response to weedy root parasites

M. Fernández-Aparicio and D. Rubiales, CSIC, Spain

1 Introduction

Parasitic plants are plants that have evolved to feed on other plants. There are about 4500 species of such parasitic plants. Some of these have specialized to infect stems (i.e. *Cuscuta*, *Viscum*) and others to infect roots (i.e. *Orobanche*, *Phelipanche*, *Striga*, *Alectra*), but all have in common an infective organ, the haustorium, to transfer nutrients (Fig. 1). Most of these species are not economically important, infecting only wild plants in natural ecosystems, but a few of them have evolved to thrive in agricultural systems, becoming noxious weeds (Rubiales and Heide-Jørgensen, 2011; Parker, 2012).

This chapter deals with the root parasitic plants. The largest family is the Orobanchaceae, comprising about 1800 species with a wide degree of nutritional dependencies on the crop. This includes very damaging groups of species such as the hemiparasitic witchweeds (*Striga* spp.) and the holoparasitic broomrapes (*Orobanche* and *Phelipanche* spp.). The witchweeds are widespread in Africa and Asia. The most damaging witchweeds species are *Striga hermonthica* and *S. asiatica* that infect cereal crops, while *S. gesnerioides* is restricted to dicotyledonous crops in sub-Saharan Africa, mainly cowpea (Parker, 2012). The broomrapes are widespread in Mediterranean areas, Asia and Eastern Europe, attacking a wide range of dicotyledonous crops. Major broomrape weedy species are *Phelipanche ramosa* infecting vegetable crops (e.g. tomato, potato, cabbage) but also field crops like rapeseed, hemp, or

http://dx.doi.org/10.19103/AS.2020.0075.11

Figure 1 Parasitic plants. (a) *Orobanche* seedling penetrating a young pea root. (b) *Viscum* seedling penetrating the stem of an olive tree. (c) Two *Orobanche* flowering plants connected to a mature pea plant. (d) Multiple *Viscum* mature plants growing on the stems of an almond tree.

tobacco; *P. aegyptiaca* with a similar host range extending also to sunflower and legumes; *O. cumana* infecting sunflower only; *O. cernua* infecting solanaceous crops; *O. minor* infecting forage legumes; *O. crenata* infecting most grain and forage legumes as well as other crops including carrot or celery; and *O. foetida* infecting legumes (Parker, 2012; Rubiales and Fernández-Aparicio, 2012). In addition, some hemiparasitic species of *Alectra* and *Rhamphicarpa* genera are of economic importance locally.

2 Host-parasitic weed interaction

Orobanchaceae weeds interact with crop roots by means of haustorial connections to their vascular system. Their haustoria develop functional characteristics shared with common roots acting as organs for water and mineral acquisition from the host xylem (Westwood, 2013). In holoparasitic weeds the haustorium also acquires carbon from direct connections with the crop phloem. To invade the root and reach the conductive system for nutrients, the parasitic seedling quickly completes a series of independent life stages supported by nutrients stored in the tiny seeds (Joel et al., 2007). Seedling survival during the pre-attached stages depends on the timely detection of signals released by crop roots at a close distance.

2.1 Seed dispersal and germination

The chance of survival in an obligate parasitic seedling depends on the encounter with a host root. Any germination occurring away from a host root would lead to the death of the germinated seedling. To increase the likelihood of establishment on the roots of crops, the parasitic seedling develops a reproductive strategy that allows the formation of high numbers of seeds that are spread by wind, animals, or agricultural activity. Seed germination potential is preserved in the soil for decades by a mechanism that inhibits the germination in the absence of a host. This inhibition is relieved by a two-step dormancy-breaking signal. The first signal, a warm and humid conditioning of the seeds, announces the cropping season in which their preferred hosts grow. Then, conditioned seeds become prepared to detect germination stimulants exuded by host roots (Fernández-Aparicio et al., 2009; Lechat et al., 2012). If conditioned seeds are not challenged by proper signals from host roots, they become dormant until the next sequence of dry/wet seasons, carrying new opportunities to encounter a host.

Chemical signalling between the host and the parasite determines host specificity (Fernández-Aparicio et al., 2011a). Several plant metabolites are known to induce the seed germination of root parasitic weeds, mainly dihydrosorgoleone, sesquiterpene lactones and strigolactones (Yoneyama et al., 2009), but other metabolites have also been reported to contribute to host specificity (Brun et al., 2018). Strigolactones are a class of hormones that control plant architecture (Al-Babili & Bouwmeester, 2015). Under nutrient-deficient conditions, strigolactones are exuded by roots into the rhizosphere to promote symbiotic interactions (Akiyama et al., 2005). This strigolactone-mediated communication between plant root and symbionts is detected by seeds of parasitic weeds, causing them to germinate. Additional compounds with germination activity on parasitic weeds have been reported, such as dihydrosorgoleone, peagol, peagoldione, peapolyphenols A–C, soyasapogenol B, *trans*-22-dehydrocampesterol (Evidente et al., 2009, 2010, 2011), dehydrocostus lactone (Joel et al., 2011), or isothyocyanates (Auger et al., 2012). The germination response to different germination stimulants and root exudates differs between the species of parasitic weeds, allowing selective germination in the presence of the roots of their preferred hosts (Fernández-Aparicio et al., 2011b, 2014).

2.2 Host infection

Upon germination, a parasitic radicle emerges through the seed coat that grows to be a few millimetres long (Figs 1a and 2a,b). Once the radicle contacts a host root, it stops elongation and differentiates a terminal haustorium that attaches to and penetrates the root. Several root-derived host metabolites such as phenols, flavonoids, and quinones induce the formation of haustoria in hemiparasitic plants (Fig. 2c) (Bandaranayake and Yoder, 2013). Broomrapes were commonly believed to form haustoria with no need for any haustoria-inducing factors (Joel et al., 2007; Bandaranayake and Yoder, 2013). However, it has recently been reported that broomrapes radicles also respond to chemical factors such as sphaeropsidones and cytokinins to develop terminal haustoria (Figs 2d,e) (Fernández-Aparicio et al., 2016; Goyet et al., 2017).

The seedlings of facultative hemiparasites such as *Ramphicarpa* develop lateral haustoria (Fig. 2f). These seedlings differ from those of obligate parasitic weeds in the capacity for the radicle to resume growth if the lateral haustorium fails to contact the host root. Lateral haustoria in *Orobanche* seem to serve only for anchorage (Fig. 2g).

Haustorium penetration in the host root occurs as a result of a combination of mechanical forces, with enzymatic activity separating host cells without their lysis (Veronesi et al., 2005; Fig. 3a,b). The transfer of nutrients from host to parasite is performed through a continuous vascular system formed by the interconnection of both organisms (Fig. 3c,d). In addition to the xylem connections, *Orobanche* and *Phelipanche* species develop direct phloem connections (Dörr and Kollmann, 1995). The parasitic plant acts as a strong competitive sink for water and nutrients and develops a tubercle for nutrient storage above the surface of the host root, which, in turn, develops crown roots (Fig. 3c) carrying lateral haustoria with anchorage function (Fig. 2g). A shoot will also differentiate in the tubercle, which grows to emerge from the soil surface. Flowering and seed setting occur rapidly after shoot emergence from the soil (Fig. 4).

3 Host plant defence

Although parasitic weeds share some features with autotrophic weeds, yield losses by them occur by the extraction of water and nutrients from the host roots rather than by competition for them. As for any other type of disease, host plants have developed defence mechanisms against parasitic weed infection. The defence mechanisms may operate at any of the various steps in the infection process and can be incorporated by breeders into resistant cultivars

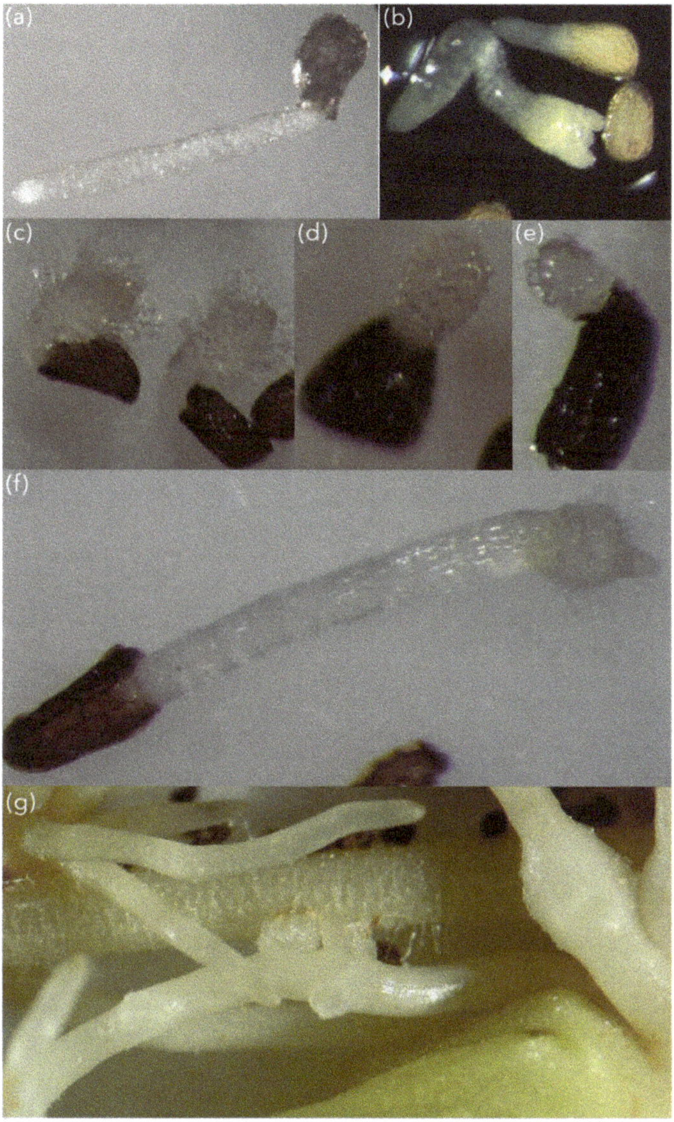

Figure 2 (a-f) Pre-attachment stages of obligated parasitic seedlings. (a) Seven-day-old *Orobanche crenata* seedling showing the maximum length of its radicle without haustorial induction. (b) Seedlings of *Striga gesnerioides* showing the cotyledons. (c) Terminal haustoria at the tip of *Striga hermonthica* seedlings. (d) Terminal haustoria at the tip of *O. crenata* seedlings. (e) Similar to D at the tip of *O. cumana* seedlings. (f) Haustoria formed at the tip of *Rhamphicarpa fistulosa* seedling resuming the longitudinal growth due to lack of host contact. (g) Penetration of vetch roots by lateral haustoria formed in the crown roots of *Phelipanche aegyptiaca* tubercles.

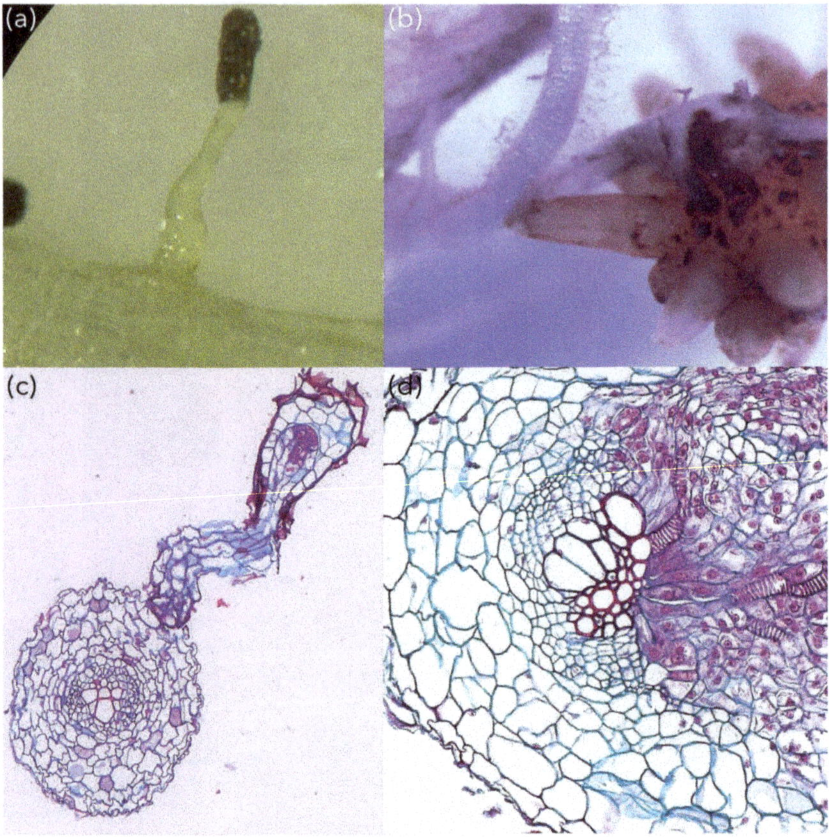

Figure 3 (a) *Orobanche* radicle attached to host root. (b) Young tubercle formed at the surface of the host showing young crown roots. (c) Longitudinal section of *Orobanche* radicle attached to the surface of the host root (shown in the transverse section). (d) Longitudinal section of *Orobanche* xylem vessels connected to the host xylem (shown in the transverse section).

individually or in combination to achieve cultivars that have durable resistance (Rubiales, 2003, 2018; Rispail et al., 2007).

3.1 Pre-penetration defence mechanisms

Parasitic weeds require host-derived molecules to stimulate the initiation of parasitism during both seed germination and haustorium initiation. Roots of crop genotypes inducing low levels of those signalling molecules are classical objectives in breeding programmes for parasitic weed resistance (Yoder and Scholes, 2010). Resistance based on the low activity of germination stimulants has been identified in sorghum against *S. hermonthica* (Ejeta, 2007), in chickpea, faba bean, and pea against *O. crenata* (Rubiales et al., 2004; Pérez-de-Luque

Figure 4 (a) Young *Striga* seedling with photosynthetic activity attached to a sorghum plant growing in vitro. (b) Young *Orobanche* seedling without photosynthetic activity attached to a pea plant growing in vitro. (c) Young *Orobanche* seedling without photosynthetic activity attached to roots of clover growing in the field. (d) A commercial field of red clover heavily parasitized by *Orobanche minor* plants.

et al., 2005a; Fernández-Aparicio et al., 2014; Pavan et al., 2016) and in tomato against *P. ramosa* (Dor et al., 2010).

This resistance has been shown in sorghum to be of simple inheritance, with molecular markers making it relatively simple to introgress into elite cultivars. Interestingly, mutation at the responsible site (*LGS1*) results in a change of the dominant strigolactone in root exudates, reducing levels of 5-deoxystrigol, but increasing orobanchol that has opposite stereochemistry. 5-deoxystrigol is a more active Striga germination stimulant than orobanchol. By this shift in the strigolactone balance, *Striga* germination is reduced, while retaining proper mycorrhizal colonization, favourable tillering, and root responsiveness to nutritional deficiencies (Gobena et al., 2017).

Some allelochemicals can inhibit seed germination or radicle growth (Serghini et al., 2001; El-Halmouch et al., 2006; Cimmino et al., 2015). Some genotypes have reduced production of haustorium-inducing factors but continue to exude normal levels of germination stimulants into the rhizosphere. Such genotypes prevent parasitism in the current crop and also reduce the parasitic seed bank by promoting suicidal germination of seedlings that have no ability to infect. Genotypes showing low induction of haustoria exist in crop species susceptible to *Striga* (Rich et al., 2004) but have not been found yet for *Orobanche* and *Phelipanche* species.

3.2 Post-penetration defence mechanisms

Once the haustorium is attached to the host root surface, the haustorial intrusive cells are stopped from penetrating resistant crops at three different levels. The first barriers are located in the cortex with reinforced cell walls mediated by either protein cross-linking or with the deposition of suberin or callose (Gurney et al., 2006; Pérez-de-Luque et al., 2006a,b; Amusan et al., 2008). In addition, the accumulation of phenolics at the host–parasite interface can be observed in some resistant crop genotypes (Goldwasser et al., 1999; Echevarría-Zomeño et al., 2006). Resistance to the penetration of the endodermis is mediated by the lignification of endodermal and pericycle cell walls (Pérez-de-Luque et al., 2005b). Once the parasite has reached the vascular tissues, resistance response can occur by the accumulation of phenolic compounds in host tissues surrounding the parasitic cells and nearby xylem vessels, inducing a toxic environment to the parasite, impeding further vascular connection (Lozano-Baena et al., 2007; Pérez-de-Luque et al., 2009). After successful connection of the parasite to the host vascular system, resistance can still operate by occlusion or sealing of host vessels by gel- or gum-like substances, peroxidase-related lignification, mucilage production, or haustorium disorganization, preventing the parasite's survival (Labrousse et al., 2001; Pérez-de-Luque et al., 2006a,b).

4 Future trends in research

Most of the parasitic plants are botanical wonders in natural ecosystems, but a few species have become noxious weeds posing a great risk to major agricultural crops. Their life cycle takes place underground, feeding on crop roots that hamper diagnosis and control. Huge efforts have been paid to develop management strategies, from quarantine, sanitation, chemigation, crop rotations, biological control, etc. Being flowering plants, parasitic weeds can be managed by a number of practices commonly used to control weeds. Particular efforts are needed on sanitation and quarantine strategies to limit the spread of seeds to areas still free of infestations and to prevent the expansion of new races. Seedbank demise can be efficiently achieved by fumigation or solarization; however, this is only economically feasible in high-value crops. Being plants, parasitic weeds can be managed with herbicides, mainly glyphosate (inhibitor of 5-enolpyruvylshikimate-3-phosphate synthase) and imidazolinones and sulfonylureas (inhibitors of acetolectate synthase) (Joel et al., 2007). Also, target site herbicide resistance might be a promising solution for controlling parasitic weeds by either classical breeding or mutagenesis of transgenic techniques (Gressel, 2013).

There are a number of alternative control strategies showing promise but are still under development, needing further refinement before they can be readily applicable at the commercial level. These include agronomic practices such as rotations with a trap or catch crops or intercropping susceptible crops with allelopathic crops (Rubiales et al., 2009; Fernández-Aparicio et al., 2011a; Pickett and Hooper, 2013); biological control with fungi, insects, or symbionts; or the use of environmental-friendly herbicidal actions such as aminoacids or fungal metabolites with orobanchicidal activity (Lendzemo et al., 2009; Mabrouk et al., 2007; Sauerborn et al., 2007; Vurro et al., 2009; Cimmino et al., 2014; Fernández-Aparicio et al., 2010, 2017).

Despite these efforts, parasitic weeds are increasing rather than being contained, extending their threats to new areas and new crops (Grenz and Sauerborn, 2007). They adapt genetically in response to changing management strategies and cropping systems. This is an active process in which non-weedy populations have developed weedy traits, or weedy populations have broadened their host range or have increased their virulence, developing new capabilities to attack the roots of resistant cultivars (Rubiales, 2018). Resistance, mostly under polygenic control, has been identified in several crops against parasitic weeds (Rubiales et al., 2016; Samejima et al., 2016; Rodenburg et al., 2017; Mbuvi et al., 2017). Monogenic resistance breeding has been successfully developed in sunflower against *O. cumana* and in cowpea against *S. gesnerioides* (Velasco et al., 2016; Omoigui et al., 2017) but, concurrently, the development of new races breaking that resistance has occurred (Botanga and Timko, 2006; Martín-Sanz et al., 2016). Resistance should be protected by complementary measures, reducing population size and preventing long-distance seed spread and reducing the chances of appearance, fixation, and spread of new genetic variants (Rubiales, 2018).

The integration of information obtained from QTL analysis with gene and protein expression analysis can aid conventional breeding or marker-assisted selection in identifying candidate genes (Castillejo et al., 2012; Dita et al., 2009; Swarbrick et al., 2007; Li et al., 2009; Yoder and Scholes, 2010; Louarn et al., 2016; Yang et al., 2017; Cvejić et al., 2020). Current progress on the genome assembly of both the parasites (Westwood et al., 2012; Gouzy et al., 2017) and the hosts (Duriez et al., 2019) will enable efficient studies of virulence and corresponding resistance genes. Developments in gene editing techniques promise useful applications, like the mutagenesis of the strigolactone-biosynthesis gene CCD8 to create *P. aegyptiaca*-resistant tomatoes (Bari et al., 2019).

5 Where to look for further information

The following special issues or newsletters can be visited for further information:

- *Parasitic Orobanchaceae*, D. M. Joel et al. (Eds.), Springer-Verlag, Berlin, Heidelberg, Germany, 2013. https://link.springer.com/book/10.1007/978-3-642-38146-1.
- *E-book Advances in Parasitic Weeds Research*. 2018. Rubiales et al. (Eds.), Frontiers Media, Lausanne. ISSN 1664-8714, doi: 10.3389/978-2-88945-633-8.
- *Haustorium, Parasitic Plants Newsletter*, Parker, C. (ed.), http://ww2.odu.edu/~lmusselm/haustorium/.
- Special Issue: Parasitic Weeds, 2009. *Weed Research* 49(S1), 1–87.
- Special Issue: Managing Parasitic Weeds, 2009. *Pest Management Science* 65(5), 451–614.

6 Acknowledgements

This work is funded by AEI (Spanish Research Agency) projects AGL2017-82907-R, AGL2017-87693-R, and RYC-2015-18961.

7 References

Akiyama, K., Matsuzaki, K. and Hayashi, H. 2005. Plant sesquiterpenes induce hyphal branching in arbuscular mycorrhizal fungi. *Nature* 435(7043), 824–827.

Al-Babili, S. and Bouwmeester, H. J. 2015. Strigolactones, a novel carotenoid-derived plant hormone. *Annual Review of Plant Biology* 66, 161–186.

Amusan, I. O., Rich, P. J., Menkir, A., Housley, T. and Ejeta, G. 2008. Resistance to *Striga hermonthica* in a maize inbred line derived from *Zea diploperennis*. *New Phytologist* 178(1), 157–166.

Auger, B., Pouvreau, J. B., Pouponneau, K., Yoneyama, K., Montiel, G., Le Bizec, B., Yoneyama, K., Delavault, P., Delourme, R. and Simier, P. 2012. Germination stimulants of *Phelipanche ramosa* in the rhizosphere of *Brassica napus* are derived from the glucosinolate pathway. *Molecular Plant-Microbe Interactions* 25(7), 993–1004.

Bandaranayake, P. C. G. and Yoder, J. I. 2013. Haustorium initiation and early development. In: Joel, D., Gressel, J. and Musselman, L. (Eds), *Parasitic Orobanchaceae*, Springer, Berlin, Heidelberg.

Bari, V. K., Nassar, J. A., Kheredin, S. M., Gal-On, A., Ron, M., Britt, A., Steele, D., Yoder, J. and Aly, R. 2019. CRISPR/Cas9-mediated mutagenesis of CAROTENOID CLEAVAGE DIOXYGENASE 8 in tomato provides resistance against the parasitic weed *Phelipanche aegyptiaca*. *Scientific Reports* 9(1), 11438.

Botanga, C. J. and Timko, M. P. 2006. Phenetic relationships among different races of *Striga gesnerioides* (Wild) Vatke from West Africa. *Genome* 49(11), 1351–1365.

Brun, G., Braem, L., Thoiron, S., Gevaert, K., Goormachting, S. and Delavault, P. 2018. Seed germination in parasitic plants: what insights can we expect from strigolactone research? *Journal of Experimental Botany* 69(9), 2265–2280.

Castillejo, M. Á, Fernández-Aparicio, M. and Rubiales, D. 2012. Proteomic analysis by two-dimensional differential in gel electrophoresis (2D DIGE) of the early response of *Pisum sativum* to *Orobanche crenata*. *Journal of Experimental Botany* 63(1), 107–119.

Cimmino, A., Fernández-Aparicio, M., Andolfi, A., Basso, S., Rubiales, D. and Evidente, A. 2014. Effect of fungal and plant metabolites on broomrapes (*Orobanche* and *Phelipanche* spp.) seed germination and radicle growth. *Journal of Agricultural and Food Chemistry* 62(43), 10485–10492.

Cimmino, A., Fernández-Aparicio, M., Avolio, F., Yoneyama, K., Rubiales, D. and Evidente, A. 2015. Ryecyanatines A and B and ryecarbonitrilines A and B, substituted cyanatophenol, cyanato-benzo[1,3] diole, and benzo[1,3]dioxolecarbonitriles from rye (*Secale cereale* L.) root exudates: new metabolites with allelophatic activity on *Orobanche* seed germination and radicle growth. *Phytochemistry* 109, 57–65. DOI: 10.1016/j.phytochem.2014.10.034.

Cvejić, S., Radanović, A., Dedić, B., Jocković, M., Jocić, S. and Miladinović, D. 2020. Genetic and genomic tools in sunflower breeding for broomrape resistance. *Genes* 11(2), 152.

Dita, M. A., Die, J. V., Román, B., Krajinski, F., Küster, H., Moreno, M. T., Cubero, J. I. and Rubiales, D. 2009. Gene expression profiling of *Medicago truncatula* roots in response to the parasitic plant *Orobanche crenata*. *Weed Research* 49, 66–80.

Dor, E., Alperin, B., Wininger, S., Ben-Dor, B., Somvanshi, V. S., Koltai, H., Kapulnik, Y. and Hershenhorn, J. 2010. Characterization of a novel tomato mutant resistant to *Orobanche* and *Phelipanche* spp. weedy parasites. *Euphytica* 171(3), 371–380.

Dörr, I. and Kollmann, R. 1995. Symplasmic sieve element continuity between Orobanche and its host. *Botanica Acta* 108(1), 47–55.

Duriez, P., Vautrin, S., Auriac, M. C., Bazerque, J., Boniface, M. C., Callot, C., Carrère, S., Cauet, S., Chabaud, M., Gentou, F., Lopez-Sendon, M., Paris, C., Pegot-Espagnet, P., Rousseaux, J. C., Pérez-Vich, B., Velasco, L., Bergès, H., Piquemal, J. and Muños, S. 2019. A receptor-like kinase enhances sunflower resistance to *Orobanche cumana*. *Nature Plants* 5(12), 1211–1215. DOI: 10.1038/s41477-019-0556-z.

Ejeta, G. 2007. Breeding for *Striga* resistance in sorghum: exploitation of an intricate host-parasite biology. *Crop Science* 47, S-216.

El-Halmouch, Y., Benharrat, H. and Thalouarn, P. 2006. Effect of root exudates from different tomato genotypes on broomrape (*O. aegyptiaca*) seed germination and tubercle development. *Crop Protection* 25(5), 501–507.

Evidente, A., Cimmino, A., Fernández-Aparicio, M., Andolfi, A., Rubiales, D. and Motta, A. 2010. Polyphenols, including the new peapolyphenols A-C, from root exudates stimulate *Orobanche foetida* seed germination. *Journal of Agricultural and Food Chemistry* 58(5), 2902–2907.

Evidente, A., Cimmino, A., Fernández-Aparicio, M., Rubiales, D., Andolfi, A. and Melck, D. 2011. Soyasapogenol B and *trans*-22-dehydrocampesterol from common vetch (*Vicia sativa* L.) root exudates stimulate broomrape seed germination. *Pest Management Science* 67(8), 1015–1022.

Evidente, A., Fernández-Aparicio, M., Cimmino, A., Rubiales, D., Andolfi, A.. and Motta, A. 2009. Peagol and peagoldione, two new strigolactone like metabolites isolated from pea root exudates. *Tetrahedron Letters* 50(50), 6955–6958.

Fernández-Aparicio, M., Bernard, A., Falchetto, L., Marget, P., Chauvel, B., Steinberg, C., Morris, C. E., Gibot-Leclerc, S., Boari, A., Vurro, M., Bohan, D. A., Sands, D. C. and Reboud, X. 2017. Investigation of amino acids as herbicides for control of *Orobanche minor* parasitism in red clover. *Frontiers in Plant Science* 8(842), 842.

Fernández-Aparicio, M., Flores, F. and Rubiales, D. 2009. Recognition of root exudates by seeds of broomrape (*Orobanche* and *Phelipanche*) species. *Annals of Botany* 103(3), 423–431.

Fernández-Aparicio, M., García-Garrido, J. M., Ocampo, J. A. and Rubiales, D. 2010. Colonization of field pea roots by arbuscular mycorrhizal fungi reduces *Orobanche* and *Phelipanche* species seed germination. *Weed Research* 50(3), 262–268.

Fernández-Aparicio, M., Kisugi, T., Xie, X., Rubiales, D. and Yoneyama, K. 2014. Low strigolactone root exudation: a novel mechanism of broomrape (*Orobanche* and *Phelipanche* spp.) resistance available for faba bean breeding. *Journal of Agricultural and Food Chemistry* 62(29), 7063–7071.

Fernández-Aparicio, M., Masi, M., Maddau, L., Cimmino, A., Evidente, M., Rubiales, D. and Evidente, A. 2016. Induction of haustorium development by sphaeropsidones in radicles of the parasitic weeds *Striga* and *Orobanche*. A structure-activity relationship study. *Journal of Agricultural and Food Chemistry* 64(25), 5188–5196.

Fernández-Aparicio, M., Westwood, J. H. and Rubiales, D. 2011a. Agronomic, breeding, and biotechnological approaches to parasitic plant management through manipulation of germination stimulant levels in agricultural soils. *Botany* 89(12), 813–826.

Fernández-Aparicio, M., Yoneyama, K. and Rubiales, D. 2011b. The role of strigolactones in host specificity of *Orobanche* and *Phelipanche* seed germination. *Seed Science Research* 21(1), 55–61.

Gobena, D., Shimels, M., Rich, P. J., Ruyter-Spira, C., Bouwmeester, H., Kanuganti, S., Mengiste, T. and Ejeta, G. 2017. Mutation in sorghum *LGS1* alters strigolactones. *Proceedings of the National Academy of Sciences of the United States of America* 114(17), 4471–4476. DOI: 10.1073/pnas.1618965114.

Goldwasser, Y., Hershenhorn, J., Plakhine, D., Kleifeld, Y. and Rubin, B. 1999. Biochemical factors involved in vetch resistance to *Orobanche aegyptiaca*. *Physiological and Molecular Plant Pathology* 54(3–4), 87–96.

Gouzy, J., Pouill, N., Boniface, M. C., Bouchez, O., Carrère, S., Catrice, O., Cauet, S., Claudel, C., Cottret, L., Faure, S., Calderón González, Á., Grand, X., Hu, L., Jéziorski, C., Lechat, M.-M., Legrand, L., Louarn, J., Marnade, W., Ribière, N., Sallet, E., Simier, P., Velasco Varo, L., Donnadieu, C., Jestin, C., Delavault, P., Bergès, H., Coque, M., Pérez-Vich, B. and Muños, S. 2017. The complete genome sequence of *Orobanche cumana* (sunflower broomrape). In: *Proceedings of the 14th World Congress on Parasitic Plants*, Pacific Grove, CA, USA, 25–30 June 2017.

Goyet, V., Billard, E., Pouvreau, J. B., Lechat, M. M., Pelletier, S., Bahut, M., Monteau, F., Spíchal, L., Delavault, P., Montiel, G. and Simier, P. 2017. Haustorium initiation in the obligate parasitic plant *Phelipanche ramosa* involves a host-exudated cytokinin signal. *Journal of Experimental Botany* 68(20), 5539-5552.

Grenz, J. H. and Sauerborn, J. 2007. Mechanisms limiting the geographical range of the parasitic weed *Orobanche crenata*. *Agriculture, Ecosystems and Environment* 122(3), 275-281.

Gressel, J. 2013. Biotechnologies for directly generating crops resistant to parasites. In: Joel, D.M., Gressel, J. and Musselman, L.J. (Eds), *Parasitic Orobanchaceae*. Springer, Berlin, Heidelberg.

Gurney, A. L., Slatem, J., Press, M. C. and Scholes, J. D. 2006. A novel form of resistance in rice to the angiosperm parasite *Striga hermonthica*. *New Phytologist* 169(1), 199-208.

Joel, D. M., Chaudhury, S. K., Plakhine, D., Ziadna, H. and Steffens, J. C. 2011. Dehydrocostus lactone is exuded from sunflower roots and stimulates germination of the root parasite *Orobanche cumana*. *Phytochemistry* 72(7), 624-634.

Joel, D. M., Hershenhorn, J., Eizenberg, H., Aly, R., Ejeta, G., Rich, P. J., Ransom, J. K., Sauerborn, J. and Rubiales, D. 2007. Biology and management of weedy root parasites. *Horticultural Reviews* 33, 267-349.

Labrousse, P., Arnaud, M. C., Serieys, H., Berville, A. and Thalouarn, P. 2001. Several mechanisms are involved in resistance of *Helianthus* to *Orobanche cumana* Wallr. *Annals of Botany* 88(5), 859-868.

Lechat, M. M., Pouvreau, J. B., Péron, T., Gauthier, M., Montiel, G., Veronesi, C., Todoroki, Y., Le Bizec, B., Monteau, F., Macherel, D., Simier, P., Thoiron, S. and Delavault, P. 2012. PrCYP707A1, an ABA catabolic gene, is a key component of *Phelipanche ramosa* seed germination in response to the strigolactone analogue GR24. *Journal of Experimental Botany* 63(14), 5311-5322.

Lendzemo, V., Kuyper, T. W. and Vierheilig, H. 2009. *Striga* seed-germination activity of root exudates and compounds present in stems of *Striga* host and nonhost (trap crop) plants is reduced due to root colonization by arbuscular mycorrhizal fungi. *Mycorrhiza* 19(5), 287-294.

Li, J., Lis, K. E. and Timko, M. P. 2009. Molecular genetics of race-specific resistance of cowpea to *Striga gesnerioides* (Willd.). *Pest Management Science* 65(5), 520-527.

Louarn, J., Boniface, M. C., Pouilly, N., Velasco, L., Pérez-Vich, B., Vincourt, P. and Muños, S. 2016. Sunflower resistance to broomrape (*Orobanche cumana*) is controlled by specific QTLs for different parasitism stages. *Frontiers in Plant Science* 7, 590.

Lozano-Baena, M. D., Prats, E., Moreno, M. T., Rubiales, D. and Pérez-de-Luque, A. 2007. *Medicago truncatula* as a model host for legumes-parasitic plants interactions: two phenotypes of resistance for one defensive mechanism. *Plant Physiology* 145(2), 437-449.

Mabrouk, Y., Zourgui, L., Sifi, B., Delavault, P., Simier, P. and Belhadj, O. 2007. Some compatible *Rhizobium leguminosarum* strains in peas decrease infections when parasitized by *Orobanche crenata*. *Weed Research* 47(1), 44-53.

Martín-Sanz, A., Malek, J., Fernández-Martínez, J. M., Pérez-Vich, B. and Velasco, L. 2016. Increased virulence in sunflower broomrape (*Orobanche cumana* Wallr.) populations from Southern Spain is associated with greater genetic diversity. *Frontiers in Plant Science* 7, 589.

Mbuvi, D. A., Masiga, C. W., Kuria, E., Masanga, J., Wamalwa, M., Mohamed, A., Odeny, D. A., Hamza, N., Timko, M. P. and Runo, S. 2017. Novel sources of witchweed (*Striga*) resistance from wild sorghum accessions. *Frontiers in Plant Science* 8, 116.

Omoigui, L. O., Kamara, A. Y., Alunyo, G. I., Bello, L. L., Oluoch, M., Timko, M. P. and Boukar, O. 2017. Identification of new sources of resistance to *Striga gesnerioides* in cowpea *Vigna unguiculata* accessions. *Genetic Resources and Crop Evolution* 64(5), 901-911.

Parker, C. 2012. Parasitic weeds: a world challenge. *Weed Science* 60(2), 269-276.

Pavan, S., Schiavulli, A., Marcotrigiano, A. R., Bardaro, N., Bracuto, V., Ricciardi, F., Charnikova, T., Lotti, C., Bouwmeester, H. and Ricciardi, L. 2016. Characterization of low-strigolactone germplasm in pea (*Pisum sativum* L.) resistant to crenate broomrape (*Orobanche crenata* Forsk.). *Molecular Plant-Microbe Interactions* 29(10), 743-749.

Pérez-de-Luque, A., Fondevilla, S., Pérez-Vich, B., Aly, R., Thoiron, S., Simier, P., Castillejo, M. A., Fernández-Martínez, J. M., Jorrín, J., Rubiales, D. and Delavault, P. 2009. Understanding *Orobanche* and *Phelipanche*-host plant interaction and developing resistance. *Weed Research* 49, 8-22.

Pérez-de-Luque, A., González-Verdejo, C. I., Lozano, M. D., Dita, M. A., Cubero, J. I., González-Melendi, P., Risueño, M. C. and Rubiales, D. 2006a. Protein cross-linking, peroxidase and β-1,3-endoglucanase involved in resistance of pea against *Orobanche crenata*. *Journal of Experimental Botany* 57(6), 1461-1469.

Pérez-de-Luque, A., Jorrín, J., Cubero, J. I. and Rubiales, D. 2005a. *Orobanche crenata* resistance and avoidance in pea (*Pisum* spp.) operate at different developmental stages of the parasite. *Weed Research* 45(5), 379-387.

Pérez-de-Luque, A., Lozano, M. D., Cubero, J. I., González-Melendi, P., Risueño, M. C. and Rubiales, D. 2006b. Mucilage production during the incompatible interaction between *Orobanche crenata* and *Vicia sativa*. *Journal of Experimental Botany* 57(4), 931-942.

Pérez-de-Luque, A., Rubiales, D., Cubero, J. I., Press, M. C., Scholes, J., Yoneyama, K., Takeuchi, Y., Plakhine, D. and Joel, D. M. 2005b. Interaction between *Orobanche crenata* and its host legumes: unsuccessful haustorial penetration and necrosis of the developing parasite. *Annals of Botany* 95(6), 935-942.

Pickett, J. A. and Hooper, A. M. 2013. Allelopathy. In: Joel, D. M., Gressel, J. and Musselman, L. J. (Eds), *Parasitic Orobanchaceae*. Springer-Verlag, Berlin Heidelberg, DOI: 10.1007/978-3-642-38146-1_25.

Rich, P. J., Grenier, C. and Ejeta, G. 2004. *Striga* resistance in the wild relatives of sorghum. *Crop Science* 44(6), 2221-2229.

Rispail, N., Dita, M. A., González-Verdejo, C., Pérez-de-Luque, A., Castillejo, M. A., Prats, E., Román, B., Jorrín, J. and Rubiales, D. 2007. Plant resistance to parasitic plants: molecular approaches to an old foe. *New Phytologist* 173(4), 703-712.

Rodenburg, J., Cissoko, M., Kayongo, N., Dieng, I., Bisikwa, J., Irakiza, R., Masoka, I., Midega, C. A. and Scholes, J. D. 2017. Genetic variation and host-parasite specificity of *Striga* resistance and tolerance in rice: the need for predictive breeding. *New Phytologist* 214(3), 1267-1280.

Rubiales, D. 2003. Parasitic plants, wild relatives and the nature of resistance. *New Phytologist* 160(3), 459-461.

Rubiales, D., Fernández-Aparicio, M., Wegmann, K. and Joel, D. M. 2009. Revisiting strategies for reducing the seedbank of *Orobanche* and *Phelipanche* spp. *Weed Research* 49, 23-33.

Rubiales, D. 2018. Can we breed for durable resistance to broomrapes? *Phytopathologia Mediterranea* 57(1), 170–185.

Rubiales, D., Alcántara, C. and Sillero, J. C. 2004. Variation in resistance to crenate broomrape (*Orobanche crenata*) in species of *Cicer*. *Weed Research* 44(1), 27–32.

Rubiales, D. and Fernández-Aparicio, M. 2012. Innovations in parasitic weeds management in legume crops. A review. *Agronomy for Sustainable Development* 32(2), 433–449.

Rubiales, D. and Heide-Jørgensen, H. S. 2011. Parasitic plants. In: *Encyclopedia of Life Sciences (ELS)*. John Wiley & Sons, Ltd, Chichester. DOI: 10.1002/9780470015902.a0021271.

Rubiales, D., Rojas-Molina, M. M. and Sillero, J. C. 2016. Characterization of resistance mechanisms in faba bean (*Vicia faba*) against broomrape species (*Orobanche* and *Phelipanche* spp.). *Frontiers in Plant Science* 7, 1747.

Samejima, H., Babiker, A. G., Mustafa, A. and Sugimoto, Y. 2016. Identification of *Striga hermonthica*-resistant upland rice varieties in Sudan and their resistance phenotypes. *Frontiers in Plant Science* 7, 634.

Sauerborn, J., Müller-Stöver, D. and Hershenhorn, J. 2007. The role of biological control in managing parasitic weeds. *Crop Protection* 26(3), 246–254.

Serghini, K., Pérez-de-Luque, A., Castejón-Muñoz, M., García-Torres, L. and Jorrín, J. V. 2001. Sunflower (*Helianthus annuus* L.) response to broomrape (*Orobanche cernua* Loefl.) parasitism: induced synthesis and excretion of 7-hydroxylated simple coumarins. *Journal of Experimental Botany* 52(364), 2227–2234.

Swarbrick, P. J., Huang, K., Liu, G., Slate, J., Press, M. C. and Scholes, J. D. 2007. Global patterns of gene expression in rice cultivars undergoing a susceptible or resistant interaction with the parasitic plant *Striga hermonthica*. *New Phytologist* 179(2), 515–529.

Velasco, L., Pérez-Vich, B. and Fernández-Martínez, J. M. 2016. Research on resistance to sunflower broomrape: an integrated vision. *OCL* 23(2). DOI: 10.1051/ocl/2016002.

Veronesi, C., Bonnin, E., Benharrat, H., Fer, A. and Thalouarn, P. 2005. Are pectinolytic activities of *Orobanche cumana* seedlings related to virulence towards sunflower? *Israeli Journal Plant Science* 53, 19–27.

Vurro, M., Boari, A., Evidente, A., Andolfi, A. and Zermane, N. 2009. Natural metabolites for parasitic weed management. *Pest Management Science* 65(5), 566–571.

Westwood, J. H. 2013. The physiology of the established parasite-host association. In: Joel, D. M., Gressel, J. and Musselman, L.(Eds), *Parasitic Orobanchaceae*, Pages 87–114. DOI: 10.10007/978-3-642-38146-1_4.

Westwood, J. H., dePamphilis, C. W., Das, M., Fernández-Aparicio, M., Honaas, L. A., Timko, M. P., Wafula, E. K., Wickett, N. J. and Yoder, J. I. 2012. The parasitic plant genome project: new tools for understanding the biology of Orobanche and Striga. *Weed Science* 60(2), 295–306. DOI: 10.1614/WS-D-11-00113.1.

Yang, C., Xu, L., Zhang, N., Islam, F., Song, W., Hu, L., Liu, D., Xie, X. and Zhou, W. 2017. iTRAQ-based proteomics of sunflower cultivars differing in resistance to parasitic weed *Orobanche cumana*. *Proteomics* 17(13–14), 28618117.

Yoder, J. I., Gunathilake, P., Wu, B., Tomilova, N. and Tomilov, A. A. 2009. Engineering host resistance against parasitic weeds with RNA interference. *Pest Management Science* 65(5), 460–466.

Yoder, J. I. and Scholes, J. D. 2010. Host plant resistance to parasitic weeds; recent progress and bottlenecks. *Current Opinion in Plant Biology* 13(4), 478–484.

Yoneyama, K., Xie, X., Yoneyama, K. and Takeuchi, Y. 2009. Strigolactones: structures and biological activities. *Pest Management Science* 65(5), 467–470.

Chapter 9

Advances in understanding plant root responses to root-feeding insects

Scott N. Johnson, Hawkesbury Institute for the Environment - Western Sydney University, Australia; and Ximena Cibils-Stewart, Hawkesbury Institute for the Environment - Western Sydney University, Australia and Instituto Nacional de Investigación Agropecuaria (INIA), Uruguay

1 Introduction

In terms of the number of animal and plant species that occupy our planet, around 26% are herbivorous insects (Strong et al., 1984). They represent enormous diversity, particularly in terms of how they consume and exploit plant resources. A comparatively small proportion of herbivorous insects, limited to six orders, feed belowground on plant roots (Brown and Gange, 1990). While some root herbivores spend their entire lifecycle belowground (e.g. grape phylloxera), it is likely that the majority of species have aboveground life stages (usually adult) and it is the juvenile (larval) life-stages that attack roots. Root-feeding insects differ markedly from shoot-feeding insects in terms of their ecology and life-history traits. While there are many exceptions to the rule, these differences can be broadly generalized (see Table 1) (Johnson et al., 2016b). In particular, root herbivores are probably represented in just 17% of families: they are predominantly chewers, have relatively long lifespans, live in highly aggregated populations and are in constant contact with the immense microbial communities found in the soil (Johnson et al., 2016b). As a group, there are still major gaps in our knowledge about how they interact

http://dx.doi.org/10.19103/AS.2020.0075.13

Table 1 Comparison of the life history traits and ecology of root and shoot herbivores, reproduced from Johnson et al. (2016b)

Insect life history traits and ecology	Root herbivores	Shoot herbivores
Diversity	Low Represented in 17% of herbivorous families[1]	High Represented in 81% of herbivorous families[1]
Feeding guild	Predominantly external chewers (c. 81%)[2]	Diverse feeding guilds
Lifespan	Long	Short
Interaction with abiotic environment	Limited variation (soil buffering) but tightly linked to patterns of herbivory	Highly variable; usually less influential
Exposure to the microbial environment	Constant; rhizopheric microbial community immense and highly diverse	Variable; mostly with pathogenic or endophytic fungi associated with the plant
Connectivity with host plant	Unattached and able to graze unconnected tissues	Often attached or feed sequentially on connected tissues
Distribution and mobility	Highly aggregated	Less aggregated and more mobile in many taxa

[1] Based on 257 North American insect herbivore families (Rasmann and Agrawal, 2008).
[2] Based on the experimental observations in the meta-analysis of Zvereva and Kozlov (2012).

with roots compared with our understanding of aboveground insect herbivore interactions (Hunter, 2001).

Compared to nematode damage, which is estimated to cost upwards of US $80 billion a year (Nicol et al., 2011; Jones et al., 2013) in crop losses, damage inflicted by root-feeding insects is probably less serious. Nonetheless, a number of species inflict significant economic losses to important agricultural sectors at regional scales. Examples include western corn rootworm (WCR), *Diabrotica virgifera virgifera*, whose damage and control costs exceed US $1 billion annually in the United States (Gray et al., 2009), greyback canegrub, *Dermolepida albohirtum*, which cause losses of up to AUD $28 million annually in Australia (Chandler, 2002), Japanese beetle grubs (*Popillia japonica*) that cost the US turfgrass industry US $450 million in control costs (Coy et al., 2019) and wireworms, whose damage and control costs to the Canadian potato industry approximates CAN $6 million (Agriculture and Agri-Food Canada, 2020).

Root herbivory can be especially damaging to crops when combined with abiotic stresses, particularly drought, which is often exacerbated by root damage as water uptake is reduced even further (Zvereva and Kozlov, 2012; Erb and Lu, 2013). Moreover, plants may not be able to tolerate root herbivory to the same extent as they can tolerate shoot herbivory, not only because the damage caused is acute but also because many root-feeding pests are extremely persistent, with root damage lasting many months or even years (Johnson et al., 2016a). The persistence of root herbivores in the soil can be extremely costly if prime agricultural land has to be taken out of production (Blackshaw and Kerry, 2008). Moreover, the cryptic nature of root herbivory means that infestations can go unnoticed and extensive damage to crops then becomes inevitable. Management options are costly and potentially more impactful to the environment, particularly when insecticides are applied prophylactically (often unnecessarily) in an attempt to avoid possible losses (Blackshaw and Kerry, 2008). The use of many insecticides, particularly the neonicotinoids, for pest management is now becoming impractical because of legislation restricting their use (Jactel et al., 2019) so it is possible that some root-feeding insects may become more problematic in the future.

In this chapter, we present an overview of the interactions between plant roots and root-feeding insect herbivores, focussing on changes in growth and physiology and crucially how roots are defended against insect attack. Several reviews have covered the ecology (Johnson et al., 2016b) and management (Johnson et al., 2016a) of insect root herbivores, together with their interactions with the abiotic (Barnett and Johnson, 2013; Erb and Lu, 2013; Hiltpold et al., 2013) and biotic (Johnson and Rasmann, 2015) soil environment. Hence, we focus particularly on advances in our understanding of how plant mutualistic fungi may affect root herbivores. This is an emerging area of research, with many attendant knowledge gaps, but we argue that this is an important component of how plants resist attack by belowground insect herbivores.

2 Root-feeding insects: an evolving research path

The pest status of root-feeding insects meant that most of the earliest research took an applied perspective, mostly directed at controlling pest populations. Beyond developing resistant crop varieties, there was little emphasis on understanding the nature of their interactions with plants (Johnson and Rasmann, 2015). While still firmly rooted in the agricultural literature, researchers started to take a more ecological perspective towards the end of the 1980s with the publication of several important reviews (Andersen, 1987; Stanton, 1988; Brown and Gange, 1990; Villani and Wright, 1990). These reviews, and particularly the paper by Brown and Gange (1990), provided a comprehensive survey of the previously fragmented literature on root-feeding insects. They established that root-feeding insects could have significant impacts in both managed and natural ecosystems via their capacity to affect other, often spatially separated, organisms in the ecosystem. As Brown and Gange (1990) commented, however, this evidence was either anecdotal or incidental to the agronomic goal of the research (i.e. experiments were not specifically designed to explore ecological responses). Nonetheless, these reviews provided a framework for subsequent ecological studies which established that root herbivores often played important roles in driving the populations of plant communities (Blossey and Hunt-Joshi, 2003; Borgström et al., 2017; Vachová and Walmsley, 2017), soil microbes (Treonis et al., 2005; Dematheis et al., 2012; Ourry et al., 2018), other herbivores (Johnson et al., 2012; Lackner et al., 2020) and aboveground food webs (Soler et al., 2012; Johnson et al., 2013).

Rather than being considered purely as a plant pest, research also suggested that root-feeding insects also can be seen as an effective tool for suppressing certain weed species. Blossey and Hunt-Joshi (2003) reported that, while establishment rates (e.g. successful colonization of plants) were similar for aboveground and belowground insect herbivores (67.2% and 77.5%, respectively), root-feeding insects contributed substantially more to weed suppression compared to aboveground herbivores (53.7% versus 33.6%, respectively). Reports of root herbivores making weeds more competitive by increasing root exudates with harmful effects on native plants (e.g. Thelen et al., 2005; Oduor et al., 2017), however, suggest that root herbivores may not be universally useful in this respect. Another positive dimension to root herbivory is that, at low densities, root-feeding herbivores can increase soil aeration and mixing of nutrients because they also feed on soil organic matter (mostly during early larval stages).

Research into root herbivory has focussed increasingly on identifying patterns of plant defence belowground, especially in the context of plant defence theory (Rasmann and Agrawal, 2008; Huber et al., 2016; Moore and Johnson, 2017; Hervé and Erb, 2019). This growing area of research was, at least in part, stimulated by the high profile work of Rasmann et al. (2005) which identified how roots produced specific chemicals that recruited natural enemies (entomopathogenic nematodes, EPNs) to root-feeding WCR. This capacity was lost in some domesticated maize lines, but could be restored (Degenhardt et al., 2009). In recent times, there has been an interest in understanding how root herbivory relates to defence phytohormone signalling in plants (Erb et al., 2012; Lu et al., 2015; Johnson et al., 2018), which was traditionally understood purely in relation to aboveground herbivory (Howe and Jander, 2008).

3 Changes in plant growth and physiology

Functionally, roots not only absorb water and nutrients that are essential for plant growth and survival, but they fulfil a number of important functions including acting as plant storage organs and providing plant anchorage (Gregory, 2006). Shoots, on the other hand, photosynthesize, transpire and are responsible for plant sexual and asexual reproduction (Groff and Kaplan, 1988). Processes in the roots and shoots are, hence, intimately related to one another in order to facilitate plant growth and survival. In this section we will focus on changes in growth, morphology and physiology that accompany root damage caused by belowground insect pests. As might be expected, root herbivory causes changes in the patterns of growth and physiological processes in both roots and the shoots.

It is estimated that plants allocate 50–90% of their biomass to roots (Bazzaz et al., 1987) and that 50% of their daily net CO_2 fixation is allocated belowground (Farrar and Jones, 2003). Building and maintaining roots therefore comes at a high metabolic cost to plants. Similar to drought stress, root consumption by belowground herbivores results in losses of resources stored belowground and reductions of the root surface area, which directly disrupts water and nutrient flows from roots to the rest of the plant (Goldson et al., 1987; Gange and Brown, 1989). Consequently, the same mechanisms that decrease photosynthetic rates in plants suffering from drought might be observed in plants experiencing root damage. Likewise, increased supply of water and nutrients might mitigate the adverse impacts of root herbivores on plant growth (Ladd, 1988; Gange and Brown, 1989).

It is well recognized that following herbivory, plants often accelerate the growth of adventitious and lateral roots as a form of compensatory growth (Brown and Gange, 1990; Riedell and Reese, 1999; Rubio and Lynch, 2007). Compensatory growth leads to an increased flow of photosynthates (assimilates) and nitrogen to the roots, often reducing allocation to aboveground tissues (Detling et al., 1980; Steinger and Müller-Schärer, 1992; Borowicz, 2010) and potentially reducing overall plant reproduction. Hence, in combination with decreased photosynthetic rates, the increased sink demands of regrowing roots may be especially detrimental (Zvereva and Kozlov, 2012). In terms of costs and benefits, compensatory root growth following root herbivory seems inefficient although if the plant stays herbivore-free for a longer period, compensatory growth becomes more cost-effective (Zvereva and Kozlov, 2012).

Zvereva and Kozlov (2012) summarized how detrimental effects of root herbivory on plant growth and reproduction result from five processes that include: (i) decreases in water and nutrition uptake due to direct reductions in root biomass (mostly when fine roots are consumed) or disruption of water and nutrient flows (mostly when the main root is damaged); (ii) consumption of resources stored belowground; (iii) decreases in photosynthetic rates due to water deficits imposed by root damage; (iv) translocation of assimilates to roots for root regrowth; and (v) direct costs of inducible defences. Depending on the extent to which resources are limited, root herbivory imposes differential effects on overall plant fitness ranging from negligible impacts on growth to rapid plant death (Gange et al., 1991; Maron, 2001).

In their seminal meta-analysis that summarized 85 experiments and included 36 species of belowground herbivores along with 75 plant species, Zvereva and Kozlov (2012) reported that, on average, belowground herbivory led to 36.3% losses of root biomass, and was accompanied by a reduction in aboveground growth, photosynthesis and reproduction of −16.3%, −11.7%, and −15.5%, respectively. Interestingly, Zvereva and Kozlov (2012) concluded that the impacts of root herbivory on plants may exceed the effects of aboveground herbivores, since root damage often elicits equally strong responses in both leaves and roots (Kaplan et al., 2008). This may result in higher fitness costs (e.g. diverting more resources away from reproduction), as opposed to aboveground herbivory that elicits mostly responses on leaves.

4 Root defences against root herbivory

In general terms, plant defences can be divided into two components: resistance (e.g. physical or chemical features that deter and/or harm herbivores) and tolerance (i.e. the ability of plants to regrow and/or reproduce during and after herbivory) (Stowe et al., 2000). Undoubtedly, there is more information about these processes in aboveground tissues, but we now have some grasp of how plants respond in terms of root herbivory.

4.1 Tolerance of root herbivory

Plant traits that confer tolerance to herbivory essentially limit the injury caused to plants following infestation and can be evident before, during or following herbivore attack (Strauss and Agrawal, 1999; Stowe et al., 2000; Stout, 2013). Perhaps counter-intuitively, tolerance may be a more robust form of defence than resistance, because it does not harm herbivore fitness and is therefore less likely to impose a strong selection pressure on pests to overcome its effects (Weis and Franks, 2006; McNickle and Evans, 2018). Mechanisms of herbivore tolerance include changes in photosynthesis and growth, phenology and remobilization of stored reserves (Tiffin, 2000). For root-feeding pests, changes in resource allocation and root growth seem to have been most widely studied. Diversion of resources belowground following root attack can compensate or even over-compensate for root loss (Quinn and Hall, 1992; Thelen et al., 2005; Ryalls et al., 2013). Such resource reallocation is less widely reported for root pests compared to shoot herbivores (where resources are moved from the roots to shoots). For instance, a meta-analysis showed that compensatory growth occurs in about 17% of cases of root herbivore attack (Zvereva and Kozlov, 2012), whereas compensatory growth is achieved in 35–44% of shoot herbivore attacks (Hawkes and Sullivan, 2001). It is possible that this situation could be improved with intervention strategies. For example, Johnson et al. (2019) reported that silicon fertilizer application allowed wheat to tolerate damage inflicted by white grubs. Similarly, inoculation with plant growth-promoting rhizobacteria (PGPR) facilitated compensatory root growth, and therefore tolerance, in tall fescue and Bermuda grass attacked by Japanese beetles (Coy et al., 2019).

An alternative tolerance strategy might be to divert resources away from damaged roots towards tissues that are not under attack (leaves, stems, tubers or healthy roots). Such resource diversion, termed 'resource sequestration', has been reported extensively in response to aboveground herbivory (in this case moving resources to the roots) (Schultz et al., 2013), but there is increasing evidence for resource movement in the opposite direction (i.e. from roots to shoots) following root herbivory. In particular, this has been reported for knapweed (Newingham et al., 2007), tomato (Henkes et al., 2008), potato (Poveda et al., 2010), maize (Robert et al., 2014) and sugarcane (Frew et al., 2020). In the latter case, arbuscular mycorrhizal fungi both augmented plant tolerance by facilitating resource reallocation aboveground, but also increased resistance to the root-feeding insect (*Dermolepida albohirtum*) (Frew et al., 2020). Resource reallocation is mainly thought to facilitate future investment in root growth until the threat of attack has passed, a process that may contribute to tolerance of WCR in herbivore-tolerant maize plants (Robert et al., 2015).

The value of breeding and selecting crops that are tolerant to root herbivory by insects is now recognized and was recently reviewed by Johnson et al. (2016a). This review pointed to the use of non-invasive high throughput phenotyping to screen large numbers of plant phenotypes to identify those desirable root traits (e.g. vigour) that may help to develop tolerant crop phenotypes (Fahlgren et al., 2015). This approach, it was argued, might be particularly productive when used in conjunction with quantitative trait loci (QTL) to identify genetic markers for these traits (Goggin et al., 2015).

4.2 Direct resistance to root herbivory

Resistance to root herbivory operates through physical and chemical mechanisms in the plant, although evidence for physical defence is scarce (Johnson et al., 2016a). Structural macro-molecules and crystalline deposits such as lignin, hemicellulose, cellulose, callose, silicon and calcium oxalate underpin root toughness (Arnott, 1966; Genet et al., 2005; Leroux et al., 2011), which could prevent root herbivores from chewing roots (Johnson et al., 2010). In terms of physical defences in the roots, incorporation of silicon as biogenic silica has been recently reported to provide resistance against greyback canegrub in sugarcane (Frew et al., 2017a,b,c). In shoots, tissue silicification has anti-herbivore effects because of the abrasive nature of silicon-rich bodies (phytoliths) on the leaf surface (Hartley et al., 2015), and general strengthening of tissues that make them harder to penetrate or crush (Hunt et al., 2008). In addition to being a physical barrier to feeding, extraction of nutrients from silicified tissues by herbivores is substantially more difficult (Massey and Hartley, 2006, 2009). It is plausible that silicon could confer similar physical defences in the roots since it is known that silicon increases root strength (Hansen et al., 1976). Moreover, root-specific phytoliths have been found in roots and tubers (Chandler-Ezell et al., 2006; Moore and Johnson, 2017) so the abrasive properties of silicon may play a role in the roots as well.

Resistance via chemical mechanisms is altogether better studied than physical resistance mechanisms. Once herbivory of plant tissues begins, various plant- and insect-derived chemical elicitors are released followed by activation of genes that underpin the reconfiguration of the plant's chemical profile (Erb et al., 2012; Waterman et al., 2019). Plant secondary metabolites promote resistance to root herbivory via toxic, deterrent or anti-feedant effects. Although secondary metabolites with anti-herbivore properties can be present throughout the entire plant, there is evidence for tissue-localization in either roots and shoots in many species (Rasmann and Agrawal, 2008; Kabouw et al., 2010; Huber et al., 2015). Moreover, accumulation of secondary metabolites can be induced locally following herbivore attack (van Dam and Raaijmakers, 2006; Robert et al., 2012b), although, generally speaking, this inducibility tends to be lower in the roots compared to the shoots (Erb et al., 2012). The lower levels of induced root secondary metabolites might be explained by their high constitutive concentrations, for example, in the case of glucosinolates (van Dam et al., 2009) and benzoxazinoids (Robert et al., 2012c).

Defensive proteins represent a class of inducible metabolites that potentially provide an effective form of resistance against root herbivores (Johnson et al., 2016a). Erb et al. (2009) proposed that nitrogen-rich defences might be more common in roots rather than carbon-rich defences in the leaves since nitrogen acquisition may be achieved at a lower metabolic cost in the roots compared to the leaves. In essence, the site of acquisition (nitrogen uptake in roots and carbon fixation in leaves) may influence whether nitrogen or carbon-based defences are deployed. For example, plant proteinase inhibitors (PIs) were induced in the roots attacked by the southern corn rootworm (*Diabrotica undecimpunctata*) (Lawrence et al., 2012) and the WCR (Robert et al., 2012b), with the PIs having anti-feedant properties for adult WCR (Kim and Mullin, 2003). Similarly, strawberry plants transformed with the Cowpea trypsin inhibitor gene supported a lower abundance of root-feeding vine weevils (*Otiorhynchus sulcatus*) larvae (Graham et al., 2002).

Plant secondary metabolites that provide effective defence against shoot-feeding insects can conversely act as attractants or promote performance of root herbivores. For example, development of cabbage root fly (*Delia radicum*) and vine weevil was improved when they fed on plants with higher concentrations of glucosinolates (van Leur et al., 2008) and phenolic acids (Clark et al., 2011; Johnson et al., 2011), respectively. Similarly, WCR larvae were able to tolerate relatively high concentrations of benzoxazinoids in maize roots (Robert et al., 2012c). Moreover, there is evidence that WRC larvae can use benzoxazinoids to locate the most nutritious part of the root system (Robert et al., 2012c). In the case root herbivores performing better on phenolic-rich plants (see discussion in Johnson and Nielsen, 2012), one possible reason is that there is a potential trade-off between silicon and carbon-based defences with phenolic-rich leaves typically having lower levels of silicon defences (Cooke and Leishman, 2012). Frew et al. (2016) showed a similar negative correlation between silicon and phenolics in the roots of sugarcane and established that silicon was the more effective form of defence, resulting in canegrubs performing best on phenolic-rich plants.

Potential mechanisms for exploiting direct defences in the roots have been considered by Johnson et al. (2016a), who suggested that genomic and molecular breeding techniques are probably the most promising route. Using molecular markers and genetic mapping, for instance, specific alleles can be selected or deleted (Moose and Mumm, 2008). Johnson et al. (2016a) mention the example of how the plants were manipulated to express insecticidal *Bacillus thuringiensis* (Bt) toxins against WCR (for review see Hilder and Boulter, 1999). Bt toxins work by binding selectively to receptors of the epithelial surface of the larval midgut which causes pore formation, cell rupture and septicaemia (Vachon et al., 2012). This was not successful as a means of suppressing WCR, however, as they quickly developed resistance to Bt toxin (Gassmann et al., 2011; Meihls et al., 2011; Gassmann, 2012). We are not aware of programmes that genetically select or manipulate root direct defences specifically, but there has been extensive screening for root herbivore-resistant lines in a number of crops. For example, phenotypic screening for root herbivore-resistant varieties in maize (Tollefson, 2007; Bernklau et al., 2010), potato (Parker and Howard, 2001) and Brassicaceae (Ellis et al., 1999; Dosdall et al., 2000) have been undertaken. Two QTLs, RM-G8 and RM-G4, encoding for resistance against the root maggot were discovered in Brassicaceae (Ekuere et al., 2005) and while promising candidates for breeding of resistant varieties, to our knowledge this has not been developed further. As with any form of breeding for resistance factors, it is likely to be associated with physiological costs (e.g. trade-offs with other defences, primary metabolism, crop quality) and ecological consequences (e.g. untargeted effects, emergence of adapted herbivore species) that need to be carefully evaluated before release.

4.3 Indirect resistance via natural enemy recruitment

Plants experiencing herbivory often increase production of volatile organic compounds (VOCs) that can be perceived and used by natural enemies to locate their herbivore prey, in the case of predators, or herbivore host in the case of parasites (Poveda et al., 2010). Conventionally, this was seen as an aboveground process, but root herbivore damage has also been shown to activate the production of VOCs in the soil (Rasmann and Agrawal, 2008).

Exudation of root VOCs can provide information cues (e.g. identity and location of the root herbivore) for various soil-dwelling organisms such as bacteria, fungi and nematodes or other arthropod species (Johnson and Rasmann, 2015). Such indirect defence mechanisms, especially those involving EPNs, have been widely considered in biological control programmes against root pests. EPNs belong to two families (Heterorhabditidae and Steinernematidae) and include about 60 known species (Ivezić et al., 2009). EPN species differ considerably in behaviour and foraging strategies, but they are all obligate parasites that penetrate and occupy the bodies of arthropod hosts for successful development and reproduction. They move between hosts as infective juveniles, a resistant form that can survive adverse environmental conditions for several days to months, even when deprived of food (Kaya and Gaugler, 1993). Once inside the host's body, they release symbiotic bacteria, which multiply and produce a toxin that causes septicaemia that kills the insect in a matter of days. The insect cadaver then provides a food source for the nematodes.

EPNs predominantly use olfactory cues for successful foraging (Hallem et al., 2011; Rasmann et al., 2012). While inorganic gases such as CO_2 released by roots have been implicated in host location, EPNs can integrate other organic volatile root signals, such as caryophyllene in maize, or geijerene and pregeijerene in citrus plants, to forage more efficiently (Rasmann et al., 2005; Ali et al., 2011; Turlings et al., 2012). Efforts to bioengineer these compounds, and others belonging to the terpene family, as a management tool for root herbivores have been reported (Huang and Osbourn, 2019). Bioengineering approaches focus on targeted or semi-targeted overexpression of the compounds in the plant, and complement this with other approaches including plant breeding, intercropping to interfere with signalling and trap cropping (Guerrieri et al., 2019). However, engineering plants to produce VOCs may come at a cost to plants in terms of reduced germination, growth and yield (Robert et al., 2013). Moreover, there are other challenges including the fact that VOCs such as (*E*)-β-caryophyllene are also attractive to several root-feeding pests, including WCR and *Spodoptera littoralis* larvae (Robert et al., 2012a).

Using root exudation of VOCs to stimulate indirect defence (i.e. via recruitment of the insects natural enemies) clearly has some way to go, but EPN attraction to root VOCs has now been reported in a number of crops including potato (Laznik and Trdan, 2016a,b), brassicas (Jagodič et al., 2017), maize (Rasmann et al., 2005), citrus (Ali et al., 2011) and sugarcane (Tonelli et al., 2016), so could be a useful management tool.

5 Root defence via fungal partners

Many grasses associate simultaneously with arbuscular mycorrhizal (AM) fungi (Glomeromycota) and *Epichloë* endophytes (Clavicipitaceae) in roots and shoots, respectively. Both AM fungi and endophytes are true biotrophic fungi, with a high degree of specialization. They form tight-mutualistic relationships with their host plants. They have lost their ability to decompose dead organic matter, and instead rely on their host for their carbon supply (Jung et al., 2012; Kohler et al., 2015; Bastias et al., 2017a). Even though functional traits and ecological roles of these organisms might vary, both AM fungi and endophytes have been reported to alter both productivity rates (fungi receive sugars generated from photosynthesis) and plant–insect interactions of their host, with many examples of them conferring resistance against insect herbivore pests (Hartley and Gange, 2009; Kivlin et al., 2013).

5.1 Arbuscular mycorrhizae (AM) fungi

Mycorrhizal associations are the oldest (c. 450 million years) and most frequent forms of symbiotic associations found in nature (Jung et al., 2012). It is estimated that approximately 80% of land plants form root-associations with AM, but less than 2% of fungal species form mycorrhizal relationships with plants (van der Heijden et al., 2015; Brundrett and Tedersoo, 2018).

AM fungi belong to the phylum Mucoromycota, sub-phylum Glomeromycotina (Schüßler and Walker, 2010). As true carbon-sinks, it is estimated that AM fungi receive up to 20% of photosynthetically assimilated plant carbon (Pepe et al., 2017). In exchange for carbon, AM fungi enhance plant nutrient supply, mainly phosphorus, nitrogen, copper, iron, potassium and zinc, and improve soil structure through interactions with soil pores and aggregates (Pepe et al., 2017). Enhancement of nutrient supply by AM fungi to its host is achieved through their highly specialized intracellular structures called arbuscules, vesicles and through the AM fungi's extracellular hyphal network (extraradical mycelium) that spreads widely into the surrounding soil, thereby reaching outside nutrient depletion zones caused by the roots alone (Hoeksema et al., 2010; Lanfranco et al., 2018). Specifically, AM fungi improve the supply of inorganic nutrients, especially phosphate and nitrate (Jung et al., 2012). Apart from providing the plant with nutrients, growing evidence suggests that AM fungi also improve host resistance to biotic and abiotic stresses (Koricheva et al., 2009; Jung et al., 2012; Talaat and Shawky, 2012; Yang et al., 2016; Helander et al., 2018). Enhancement of host resistance is most likely mediated by (1) the stimulation of jasmonic acid (JA) production through mycorrhiza-induced resistance (M-IR), as well as (2) the promotion of growth in the host plant via improved uptake of phosphorus and nitrogen (Thirkell et al., 2016).

5.2 Mycorrhiza mediated-induced resistance (M-IR)

Mycorrhiza mediated-induced resistance (M-IR) is the suggested mechanism by which mycorrhizae stimulate JA-production (a key plant defence phytohormone for the formation of plant-secondary metabolites; see Section 4.2), and thus the host's immune system, and thereby increasing plant resistance to biotic and abiotic stresses (Pozo and Azcón-Aguilar, 2007). Under phosphate-limiting conditions, plant roots release strigolactones (carotenoid-derived compounds) that stimulate branching and elongation of AM fungal hyphae in the soil, thus, enabling further host-microbe encounters (Lanfranco et al., 2018). Upon germination, AM spores form appressorium-like structures. These structures resemble those of biotrophic pathogens (disease agents) and are involved in salicylic acid (SA) upregulation in plants as a defence response; SA is another key plant defence phytohormone. Unlike the SA-responses to pathogens, SA-responses to AM fungal colonization are local and transient, since they are followed by rapid AM–host plant recognition (high degree of host-symbiont coordination) (Jung et al., 2012). Upon this host plant–AM fungi recognition, SA levels decrease and compatible interactions among host and AM fungi become established. It is then that AM fungi are capable of colonizing the root cortex, and the 'symbiotic program is activated' (Pozo and Azcón-Aguilar, 2007). This 'programming' permits physical accommodation of AM fungal arbuscules within root cells, a process that is regulated by jasmonates. This in turn triggers an increase of endogenous JA in plants, the phenomenon often referred to as 'mycorrhiza-priming' (Jung et al., 2012) or 'defence priming' (Martinez-Medina et al., 2016).

Defence priming (initial steps of chemical pathway activation) is common in other plant beneficial microorganisms' interactions, and has fitness benefits compared to direct activation of defence (Frost et al., 2008). Martinez-Medina et al. (2016) proposed that upon a priming stimulus, such as AM fungal colonization, primed plants activated transient defence responses. This transient activation of defences and subsequent symbiotic programme activation is associated with a fitness reduction when comparing these primed plants to their un-primed counterparts, mainly due to energetic costs of defence activation. However, upon a subsequent biotic or abiotic stress event, mycorrhized primed plants show a faster, stronger and a more sustained response to stress than their non-primed counterparts (Martinez-Medina et al., 2016). Therefore, the fitness reduction observed at the onset of priming is counteracted by the faster defence response upon attack by the triggering stress stimulus (e.g. insect attack); primed plants therefore display higher fitness over their un-primed partners following biotic stress (Martinez-Medina et al., 2016).

5.3 Mycorrhiza and root herbivory

For arthropod herbivores, feeding guild seems to influence the interaction with mycorrhizae. Specifically, when plants are attacked by phloem-feeding aphids, evidence suggests that AM fungal colonization enhances aphid performance parameters (Simon et al., 2017; Maurya et al., 2018; Meier and Hunter, 2018a; Tomczak and Müller, 2018), with some negative (e.g. Babikova et al., 2014), and neutral (e.g. Wilkinson et al., 2019) reports. This enhancement of performance of phloem-feeding insects has been linked to the downregulation of the SA pathway in mycorrhized plants (Simon et al., 2017; Maurya et al., 2018; Meier and Hunter, 2018b; Tomczak and Müller, 2018). This phenomenon has also been suggested for nematodes, which are affected by the same SA-plant responses (Frew et al., 2018). Interestingly, AM fungi were also associated with a reduction in benzoxazinoid defence compounds produced by wheat roots, which are usually under the control of the JA pathway, which made plants more susceptible to plant-parasitic nematodes (Frew et al., 2018). In general, the effect of AM fungi on root nematodes varies according to the mode of parasitism (Hol and Cook, 2005). For example, mycorrhized plants are generally more resistant to endoparasites than ectoparasites and support fewer cyst nematodes than root-knot nematodes (both sedentary endoparasites) (Hol and Cook, 2005).

For chewing herbivores, evidence seems to suggest that AM fungi either do not affect, or enhance plant protection against generalist chewers, meanwhile increasing susceptibility of plants to specialist feeders (Koricheva et al., 2009; Meier and Hunter, 2018b). In terms of belowground herbivores, although less studied, AM fungi have been reported to increase plant resistance to belowground herbivory (Johnson and Rasmann, 2015; Frew et al., 2017c). Interestingly, similar to foliar chewers, specialist root feeders seem to be less affected by the presence of AM fungi than are generalist species (Currie et al., 2011), although there are too few studies to make a conclusive generalization. Additionally, Bennett et al. (2013) demonstrated that root herbivory during AM fungi establishment can have detrimental effects on AM fungal colonization. As noted, there is a need to further explore AM fungal interactions with root herbivores, since most of the research has focussed on interactions of AM with aboveground herbivores, despite the fact that belowground herbivores share the same parts of the host as AM (Johnson and Rasmann, 2015).

5.4 *Epichloë* endophytes

Symbiotic relationships between plants and fungi have a long evolutionary history with plant fossils containing fungal endophytes dating back 400-million-years (Krings et al., 2007). It is estimated that approximately 30% of cool-season grasses (C_3) in the Poaceae family of plants associate asymptomatically with *Epichloë* fungi, colloquially known as foliar endophytes, *Neotyphodium* (anamorphic-asexual form) and *Epichloë* (teleomorphic-sexual form), Ascomycota: Clavicipitaceae (Clay and Schardl, 2002). *Epichloë* belong to class

I endophytes (consecutive mutualists) because they exclusively infect grasses, reside intracellularly in aerial plant parts, and are vertically transmitted (asexually) via seed (Clay and Schardl, 2002; Cavazos et al., 2018). In contrast, class II endophytes are non-systemic inducible endophytes that are taxonomically diverse, horizontally transmitted and colonize almost all plants (Singh et al., 2011). *Epichloë* intracellular hyphae run parallel to the longitudinal side of the plant cells, producing no aerial or sexual structures; hyphae are located in apoplastic plant spaces (Spiering et al., 2005). *Epichloë* hyphae are dormant in seed tissue, specifically between aleurone cells, with hyphal growth further activated by seed germination. Upon activation, endophytic hyphae colonize early leaf primordia attaching to growing plant cells; leaf expansion therefore induces hyphal stretching and triggers their unique host-synchronized pattern of growth (Scott et al., 2018).

Epichloë endophyte-grass associations are advantageous for plants because in exchange for photosynthetic carbon, endophytes protect grasses from biotic and abiotic stress by releasing nitrogen-rich alkaloid compounds in response to stress. Specifically, these compounds deter herbivores (both mammalian and arthropod) feeding (Novas et al., 2011; Cavazos et al., 2018). Enzymes for alkaloid synthesis are encoded within the fungal genome, and to date, most of the biosynthetic routes have been elucidated (Bastias et al., 2017b). Specifically, *Epichloë* endophytes have genes accountable for the production of four major groups of alkaloids: ergots (ergovalines), aminopyrrolizidines (lolines), indolediterpenes (lolitrem B) and pyrrolopyrazines (peramines) (Clay and Schardl, 2002; Bastias et al., 2017b). Despite their high insecticidal activity, ergovalines and lolitrems have negative impacts on grazing mammals, being the main cause of mammalian fescue toxicoses and ryegrass staggers, and therefore, not desired for grazing pastures (Young et al., 2013). In contrast, peramines and lolines confer biotic (insecticidal) resistance to stresses without negative consequences for grazing mammals (Bastias et al., 2017b). At present, breeding programmes, such as that in AgResearch New Zealand, have selected 'novel' animal safe endophyte strains with genes accountable only for peramine and loline production, and transferred these strains into varieties of agricultural interest to increase insecticidal activity of grasses without compromising the health of grazing mammals.

Advantages of plant-novel endophyte association include tolerance to: water deficit (Gundel et al., 2016; Dastogeer, 2018; Rho et al., 2018), fungal pathogens (Perez et al., 2017), aboveground (Bastias et al., 2018a) and belowground (Omacini et al., 2012; Rostás et al., 2015) insect herbivores. The underlying mechanisms of these benefits have been mainly associated with alkaloid expression, promotion of host foliar and root growth, increased vigour, and modified water relations (Franzluebbers and Hill, 2005; Novas et al., 2009; Saikkonen et al., 2013; Arrieta et al., 2015).

5.5 Epichloë endophyte mediated-induced resistance (E-IR)

While most studies have focussed on enhanced defence against herbivory provided by endophytes through direct alkaloid production, recent evidence suggests that *Epichloë* endophytes also alter JA-pathways in their grass hosts (Bastias et al., 2017a; Bastias et al., 2018a,b). The mechanisms underlying endophyte promotion of plant endogenous JA are analogous to those of AM fungi (see Section 5.2 covering M-IR and Bastias et al., 2017a). Essentially, endophytes are first perceived by the plant as biotrophic pathogens, and after mutualistic recognition (during seed germination), JA-production is stimulated to enable the persistence of symbiosis. Although this phenomenon has been reported, specific mechanisms involved in endophytes-priming are yet to be elucidated.

Priming by endophytes upregulates JA and often downregulates SA due to phytohormonal cross-talk (Kunkel and Brooks, 2002), making the plant more susceptible to biotrophic pathogens and phloem-feeding insects (e.g. aphids). Endophyte-produced alkaloids in response to stress, however, may counteract this susceptibility. Resistance to chewing insects, on the other hand, might be achieved via both specific endophyte alkaloids (Bastias et al., 2017b) and plant JA-based pathways via endophyte priming (Bastias et al., 2017a).

5.6 Epichloë endophytes and root herbivory

Although endophytes reside in foliar host tissue, a few studies have focussed on endophytic effects on the belowground ecosystem, suggesting that endophytes also confer protection belowground and that this involves processes and mechanisms other than bioactive alkaloids (Omacini et al., 2012; Rostás et al., 2015). Omacini et al. (2012) conducted a meta-analysis that synthesized knowledge of the effect that grass-*Epichloë* symbiosis has on belowground components and processes. Overall, they found that endophytes reduced root biomass and AM fungal colonization, but stimulated root exudation. Endophytes had no significant effect on soil respiration or litter decomposition and there were few consistent patterns in relation to soil fauna. More recent research, however, suggests endophytes confer resistance against root-feeding insects. For example, using *Costelytra zealandica* (Australasian grass grub), Rostás et al. (2015) demonstrated that larvae were deterred from attacking endophytic grass roots because their roots emitted less VOCs (see Section 4.3 for more information on VOCs) than endophyte-free plants. Reduced production of VOCs in endophytic plants had previously been demonstrated aboveground (Fuchs and Krauss, 2019). Further, Rojas et al. (2016) found that the presence of endophyte altered soil microbial communities, increasing and decreasing the relative abundance of Glomeromycota and Ascomycota, respectively. The latter includes *Lecanicillium, Volutella, Lipomyces, Pochonia* and *Rhizoctonia* which include pathogenic species so this could affect soil health more generally. Moreover, *Epichloë* have also been shown to negatively affect the performance of other soil-dwelling insects such as: *Heteronychus arator* (Popay and Baltus, 2001; Barker et al., 2015), *Costelytra zealandica* (Popay et al., 2003), *Aploneura lentisci* (Pennell et al., 2005) and *Porina* spp. (Popay et al., 2012; Hennessy et al., 2016).

5.7 Case study: Uruguay

Recently, novel endophyte strains (specifically AR584) were inserted in elite varieties of Uruguayan tall grass fescue in collaborative efforts between the Instituto Nacional de Investigación Agropecuaria (National Institute of Agricultural Research; INIA) in Uruguay and Ag-Research New Zealand. These new lines were utilized in laboratory and greenhouse-feeding trials where their potential for grub control was assessed using *Diloboderus abderus*. *Diloboderus abderus* is the most destructive native grub in Southern South America. Detailed information about the life-cycle and biology of *D. abderus* is provided by Del Valle et al. (2018) and Cibils-Stewart and Zerbino (2020). Preliminary results of these trials confirm the potential use of novel endophyte for managing *D. abderus* outbreaks in fescue paddocks (see Fig. 1 and Table 2). AR584 insertion in Uruguay fescue lines reduced consumption and weight gain by larvae in no-choice tests (Table 2); additionally choice tests revealed that larvae had a strong preference for non-symbiotic plant roots over *Epichloë*-infected counterparts (Table 2, Fig. 1d). Field trials with the same lines are being deployed to test if these interactions are observed in more realistic field conditions.

5.8 Entomopathogenic fungi as endophytes

In this chapter, we focus on root responses and root resistance to belowground herbivores but, in addition to free-living forms, some fungal entomopathogens occur as endophytic forms, in other words, they live asymptomatically within the plant host (Lewis and Bing, 1991; Lewis et al., 2001). We therefore, briefly consider how these fungi may moderate root interactions with belowground herbivores. The Hypocreales order of fungi contains the largest number of fungal entomopathogens including two of the most widely studied organisms, *Beauveria bassiana* and *Metarhizium anisopliae* (Roy et al., 2006; Meyling and Eilenberg, 2007). Their traditional mode of infection involves spore deposition on the insect cuticle followed by the formation of a germ tube, which through enzymatic and mechanical action penetrates the cuticle. Once in the insect hemocoele, hyphal growth causes tissue damage and nutrient depletion – in some cases releasing metabolites (Hajek and St. Leger, 1994). It is well known that these entomopathogenic fungi occur universally in soil and have been used extensively as commercial inundative microbial insecticides. Some of

Figure 1 Case study. Experimental units used in (a–b) laboratory (Petri dished), and (c–d) greenhouse (pots) trials evaluating the effect of novel endophyte (AR584) on *Diloboderus abderus* larvae root consumption. Units were either (a) un-inoculated (controls) or (b) inoculated with *D. abderus* larvae and used to run either antibiosis (no-choice) or antixenosis (choice) trials. Antibiosis units comprised of experimental units with two roots fragments (a–b) or two plants (c–d) of the same plant material on each end (e.g. F-/F- or F+/F+), meanwhile antixenosis units, consisted of contrasting plant material at each end (e.g. F-/F+). A clear example of antixenosis results is depicted in (b and d) where root consumption by *D. abderus* was notably less on infected (F+) vs. uninfected (F-) roots. Source of pictures: *X. Cibils-Stewart*.

the problems when using these products include inconsistent control, poor persistence (sensitivity to UV radiation, reduced humidity, and excessive rainfall), erratic product quality (especially challenge of synchronizing the biocontrol agent with target pests), poor shelf life and elevated costs relative to chemical alternatives (Bamisile et al., 2018a; Jaber and Ownley, 2018). Endophytic forms of these entomopathogens might decrease these problems mentioned earlier since their host plants provide some protection from such environmental factors.

Endophytic entomopathogenic fungi not only benefit their host plant in terms of resistance to biotic stresses (e.g. arthropod, nematodes and disease pathogens), but also against abiotic stresses (e.g. heat, salt and drought), the latter primarily by enhancing host growth and nutrient acquisition (Bamisile et al., 2018b; Jaber and Ownley, 2018). Specifically, endophytic *B. bassiana* and *M. anisopliae* have been demonstrated to effectively control *Ostrinia nubilalis* while being non-pathogenic to their maize host (Lewis and Bing, 1991; Lewis et al., 2001) and are compatible with other management strategies such as Bt and insecticide applications (Lewis et al., 1996).

Table 2 Case study. Results of antibiosis (no-choice) and antixenosis (choice) trials in both laboratory and greenhouse conditions. *Diloboderus abderus* larvae were fed tall grass fescue cv. INIA Fortuna (F) with (F+) and without (F−) the insertion of the AR584 novel (animal safe) *Epichloë*-endophyte. Laboratory trials consisted of placing larvae in the centre of a Petri dish with two pieces of excised roots (ex-situ feeding), either of the same endophyte status (no-choice; e.g. F−/F−) or a different endophyte status (choice; e.g. F−/F+) (Fig 2a-b) for 1 week. Greenhouse experimental units followed the same design (no-choice; e.g. F−/F− / choice; e.g. F−/F+) but plants were deployed in pots and larvae were subject to in-situ feeding (Fig 2c-d) for 3 weeks. For all trials (a) larval weight gain and (b) root dry weight was calculated.

		Pot	Petri
(a) *D. abderus* weight gain (g ± SEM).			
Antibiosis (F−/F−)		0.24 ± 0.2 B	0.07 ± 0.01a
Antibiosis (F+/F+)		0.05 ± 0.1 C	0.03 ± 0.01b
Antixenosis (F+/F−)		0.61 ± 0.5 A	0.10 ± 0.02 a
P-value		**0.0291**	**0.0395**
(b) Root dry weight (g ± SEM).			
Antibiosis (F−/F−)	*D. abderus*	0.03 ± 0.01 C	0.02 ± 0.01 c
	No grub	0.28 ± 0.09 A	0.08 ± .012 a
Antibiosis (F+/F+)	*D. abderus*	0.24 ± 0.06 A	0.06± 0.01 a
	No grub	0.29 ± 0.09 A	0.09 ± 0.02 a
Antixenosis (F+/F−)	*D. abderus*	0.16 ± 0.05 B	0.03 ± 0.03 b
	No grub	0.30 ± 0.03 A	0.07 ± 0.01 a
P-value		**0.001**	**0.001**

Currently this novel biological control mechanism (endophytic growth within their host) has been reported in several host plant species (Vega et al., 2008). As endophytes, entomopathogenic fungi form germ tubes, which gradually become hyphae that enter their plant host through natural openings or directly through the epidermal cell walls with the help of enzymes and/or mechanical pressure. Upon entering the plant, the hyphae grow in the space between the parenchyma cells or even in the xylem vessels (Mantzoukas and Eliopoulos, 2020). Although the mode of action for endophytic entomopathogenic fungi remains poorly understood (Vega et al., 2008), studies have reported high mortality, as well as, fungal sporulation on insect cadavers upon feeding on endophytic *B. bassiana* plants (Mantzoukas and Eliopoulos, 2020). These findings are important because sporulation of an insect cadaver after death represents new sources of secondary inoculum (Ramos et al., 2020).

Future efforts in understanding best inoculation methods, improved consistency of plant colonization, as well as more studies in field conditions, need to be developed as this is a promising area of research that could lead to more sustainable control methods (Parsa et al., 2013; Bamisile et al., 2018a; Jaber and Ownley, 2018). Although, entomopathogenic fungi occur naturally in soil, much of the research has focussed on their efficacy against foliar-feeding herbivores and their interactions with belowground herbivores remain poorly understood (Fuller-Schaefer et al., 2005; Vega et al., 2008). Interestingly, Parsa et al. (2013) demonstrated that drench inoculations were effective in establishing endophytic root colonization, and Fuller-Schaefer et al. (2005) reported colonization of sugarbeet roots by the fungal entomopathogens, *B. bassiana* and *M. anisopliae*. Both studies demonstrate the potential colonization of plant roots with entomopathogenic fungi, although to our knowledge only one study has investigated the efficacy of this novel technique for a soil-living herbivore. This study found suppressive growth effects of endophytic *Metarhizium robertsii* on larval black cutworm (*Agrotis ipsilon*) in maize (Ahmad et al., 2020).

6 Conclusion and future trends

Plant roots have evolved many adaptations to overcome root herbivory including changes in growth and physiology (Section 3), tolerance and resistance defence mechanisms (see Section 4), as well as mutualistic associations (see Section 5). While low levels of root feeding possibly result in a more vigorously growing root system, high levels of root feeding have extremely negative impacts on plant growth, often resulting in plant death (Zvereva and Kozlov, 2012). Root feeding can change the primary and secondary chemistry of a host plant, in both aboveground and belowground tissues with diverse effects on other organisms interacting with the host plant. Likewise, changes in the quality, quantity and composition of the foliage, induced by root feeding, can lead to altered litter quality, quantity and composition (Masters, 2004). Regardless of the level of herbivory, or ultimately the insect species, factors such as soil nutrient and water content play a crucial role on the overall effect (Barnett and Johnson, 2013).

Infestations of belowground insect herbivores in pastures and crops can often be avoided or addressed early if some general knowledge about the biology and behaviour of the species is known. Nonetheless, the persistence of some root herbivores, their capacity to adapt to plant resistance and future limitations in the use of insecticides necessitate development of sustainable management practices (Johnson et al., 2016a). Moreover, damage is exacerbated in drought situations, periods in which the multiplication and survival of root herbivores are favoured, whereas their natural enemies decline (e.g. Preisser and Strong, 2004). Because larval movement is limited, and their life cycle is long, damage is seen in patches that enlarge from year to year. Monitoring soil insect populations becomes therefore an essential tool for sustainable management.

6.1 *Mutualistic fungi and silicon defence - can they work together?*

As we describe, many grasses associate simultaneously with AM fungi and endophytes in the roots and shoots, respectively. Although these symbionts occupy different structures and niches in their host plant, interactions between the two taxa are possible via the changes in plant chemistry and physiology they cause (Guo et al., 2016; Yang et al., 2016; Vignale et al., 2018); these might be antagonistic, neutral or synergistic. Larimer et al. (2010) suggest that synergism among mutualists 'is most likely to be observed when partners exert functionally different jobs'. Therefore, antagonism is expected when mutualists are functionally equivalent, for instance, nutrient provision in the case of AM fungi and nitrogen-fixing bacteria. Likewise, Larimer et al. (2010) hypothesize that synergistic interactions are expected between AM fungi and endophyte symbionts, given that the primary function of AM fungi is to improve host nutrition, and that of endophytes is to improve host plant defence via production of alkaloids (Novas et al., 2009). The literature for AM fungi-endophyte interactions in grasses indicates both synergistic (i.e. positive) (Novas et al., 2005; Novas et al., 2009) and antagonistic (i.e. negative) (Chu-Chou et al., 1992; Müller, 2003; Omacini et al., 2006; Mack and Rudgers, 2008) effects. Interestingly, artificial associations (e.g. via inoculation of mutualists into grasses) usually yield antagonistic AM fungi-endophyte interactions (Müller, 2003), whereas synergistic associations between AM fungi-endophyte result when both mutualists naturally have shared evolutionary histories (Chu-Chou et al., 1992; Mack and Rudgers, 2008). Although in a natural context this makes sense, these arguments need further investigation.

Incorporation of silicon into root tissues has emerged as an effective defence against root herbivores, either through resistance (Frew et al., 2016, 2017a) or tolerance mechanisms (Johnson et al., 2019). However, a preliminary study by Redmond and Potter (2006) found equivocal impacts on *Cyclocephala* spp., although this may be due to the fact that concentrations of silicon in the roots were unchanged following treatment. Studies in the last few years suggest that silicon uptake is facilitated by AM fungi (Frew et al., 2017b) and endophytes (Huitu et al., 2014), with Frew et al. (2017b) also showing this increased defence against root herbivores. In some respects, research into root-feeding insects has come full circle. The early emphasis on pest management using different management approaches gave way to more ecological studies about the interactions between root herbivores and other organisms. Understanding these ecological interactions might now provide the route to effective pest management.

7 Where to look for further information

The following reviews and books provide a good overview of the subject:

- Andersen, D. C. (1987) Belowground herbivory in natural communities - a review emphasizing fossorial animals. *Quarterly Review of Biology* 62: 261–286.
- Barnett, K. and Johnson, S. N. (2013) Living in the soil matrix: abiotic factors affecting root herbivores. *Advances in Insect Physiology* 45: 1–52.
- Benefer, C. M. and Blackshaw, R. P. (2013) Molecular approaches for studying root herbivores. *Advances in Insect Physiology* 45: 219–255.
- Brown, V. K. and Gange, A. C. (1990) Insect herbivory below ground. *Advances in Ecological Research* 20: 1–58.
- Erb, M. (2012) The role of roots in plant defence. In: JM Mérillon and KG Ramawat (Eds), *Plant Defence: Biological Control*. Berlin: Springer, pp. 291–309.
- Erb, M., Huber, M., Robert, C. A. M. et al. (2013) The role of plant primary and secondary metabolites in root-herbivore behaviour, nutrition and physiology. *Advances in Insect Physiology* 45: 53–95.
- Hiltpold, I., Bernklau, E. J., Bjostad, L. B. et al. (2013) Nature, evolution and characterisation of rhizospheric chemical exudates affecting root herbivores. *Advances in Insect Physiology* 45: 97–157.
- Johnson, S. N., Benefer, C. M. Frew, A. et al. (2016a) New frontiers in belowground ecology for plant protection from root-feeding insects. *Applied Soil Ecology* 108: 96–107.
- Johnson, S. N., Erb, M. and Hartley, S. E. (2016b) Roots under attack: contrasting plant responses to below- and aboveground insect herbivory. *New Phytologist* 210: 413–418.
- Johnson, S. N., Hiltpold, I. and Turlings, T. C. J. (Eds) (2013) *Behaviour and Physiology of Root Herbivores*. Advances in Insect Physiology. London, UK: Academic Press.
- Johnson, S. N. and Murray, P. J. (2008) *Root Feeders - An Ecosystem Perspective*. 1st ed. Wallingford, UK: CABI, 256pp.
- Johnson, S. N. and Rasmann, S. (2015) Root-feeding insects and their interactions with organisms in the rhizosphere. *Annual Review of Entomology* 60: 517–535.
- Powell, K. S., Cooper, P. D. and Forneck, A. (2013) The biology, physiology and host-plant interactions of grape phylloxera *Daktulosphaira vitifoliae*. *Advances in Insect Physiology* 45: 159–218.
- Rasmann, S. and Agrawal, A. A. (2008) In defense of roots: a research agenda for studying plant resistance to belowground herbivory. *Plant Physiology* 146: 875–880.
- Villani, M. G., Allee, L. L., Díaz, A. et al. (1999) Adaptive strategies of edaphic arthropods. *Annual Review of Entomology* 44: 233–256.
- Villani, M. G. and Wright, R. J. (1990) Environmental influences on soil macroarthropod behavior in agricultural systems. *Annual Review of Entomology* 35: 249–269.

Key research in this area can be found at the following departments and organizations:

- Biotic Interactions within the Institute of Plant Sciences, University of Bern: https://www.ips.unibe.ch/research/interactions/index_eng.html.
- Fundamental and applied research in chemical ecology, University of Neuchatel: https://www.unine.ch/centres-of-excellence/home/ecologie-chimique.html. https://www.unine.ch/farce/home/recherche.html.
- German Centre for Integrative Biodiversity Research: https://www.idiv.de/en/groups_and_people/core_groups/molecular_interaction_ecology.html.
- Department of Entomology, Pennsylvania State University: https://sites.psu.edu/interactions/. https://ento.psu.edu/.

8 References

Agriculture and Agri-Food Canada (2020). *New Pest Management Practices for the Control of Wireworm*. Available at: http://www.agr.gc.ca/eng/news/science-of-agricultural-innovation/new-pest-management-practices-for-the-control-of-wireworm/?id=1454519421585. Accessed: 29 March 2020.

Ahmad, I., Jiménez-Gasco, M. d. M., Luthe, D. S., Shakeel, S. N. and Barbercheck, M. E. (2020). Endophytic *Metarhizium robertsii* promotes maize growth, suppresses insect growth, and alters plant defense gene expression. *Biological Control* 144: 104167.

Ali, J. G., Alborn, H. T. and Stelinski, L. L. (2011). Constitutive and induced subterranean plant volatiles attract both entomopathogenic and plant parasitic nematodes. *Journal of Ecology* 99(1): 26–35.

Andersen, D. C. (1987). Below-ground herbivory in natural communities - a review emphasizing fossorial animals. *Quarterly Review of Biology* 62(3): 261–286.

Arnott, H. J. (1966). Studies of calcification in plants. In Fleisch, H., Blackwood, J. H. and Owen, M. (Eds), *Third European Symposium on Calcified Tissues*. New York: Springer-Verlag, pp. 152–157.

Arrieta, A. M., Iannone, L. J., Scervino, J. M., Vignale, M. V. and Novas, M. V. (2015). A foliar endophyte increases the diversity of phosphorus-solubilizing rhizospheric fungi and mycorrhizal colonization in the wild grass *Bromus auleticus*. *Fungal Ecology* 17: 146–154.

Babikova, Z., Gilbert, L., Bruce, T., Dewhirst, S. Y., Pickett, J. A. and Johnson, D. (2014). Arbuscular mycorrhizal fungi and aphids interact by changing host plant quality and volatile emission. *Functional Ecology* 28(2): 375–385.

Bamisile, B. S., Dash, C. K., Akutse, K. S., Keppanan, R., Afolabi, O. G., Hussain, M., Qasim, M. and Wang, L. (2018a). Prospects of endophytic fungal entomopathogens as biocontrol and plant growth promoting agents: an insight on how artificial inoculation methods affect endophytic colonization of host plants. *Microbiological Research* 217: 34–50.

Bamisile, B. S., Dash, C. K., Akutse, K. S., Keppanan, R. and Wang, L. (2018b). Fungal endophytes: beyond herbivore management. *Frontiers in Microbiology* 9: 544.

Barker, G. M., Patchett, B. J. and Cameron, N. E. (2015). *Epichloë uncinata* infection and loline content afford *Festulolium* grasses protection from black beetle (*Heteronychus arator*). *New Zealand Journal of Agricultural Research* 58(1): 35-56.

Barnett, K. and Johnson, S. N. (2013). Living in the soil matrix: abiotic factors affecting root herbivores. *Advances in Insect Physiology* 45: 1-52.

Bastias, D. A., Martínez-Ghersa, M. A., Ballaré, C. L. and Gundel, P. E. (2017a). *Epichloë* fungal endophytes and plant defenses: not just alkaloids. *Trends in Plant Science* 22(11): 939-948.

Bastias, D. A., Martínez-Ghersa, M. A., Newman, J. A., Card, S. D., Mace, W. J. and Gundel, P. E. (2018a). Jasmonic acid regulation of the anti-herbivory mechanism conferred by fungal endophytes in grasses. *Journal of Ecology* 106(6): 2365-2379.

Bastias, D. A., Martínez-Ghersa, M. A., Newman, J. A., Card, S. D., Mace, W. J. and Gundel, P. E. (2018b). The plant hormone salicylic acid interacts with the mechanism of anti-herbivory conferred by fungal endophytes in grasses. *Plant, Cell and Environment* 41(2): 395-405.

Bastias, D. A., Ueno, A. C., Assefh, C. R. M., Alvarez, A. E., Young, C. A. and Gundel, P. E. (2017b). Metabolism or behavior: explaining the performance of aphids on alkaloid-producing fungal endophytes in annual ryegrass (*Lolium multiflorum*). *Oecologia* 185(2): 245-256.

Bazzaz, F. A., Chiariello, N. R., Coley, P. D. and Pitelka, L. F. (1987). Allocating resources to reproduction and defense. *BioScience* 37(1): 58-67.

Bennett, A. E., Macrae, A. M., Moore, B. D., Caul, S. and Johnson, S. N. (2013). Early Root Herbivory Impairs Arbuscular Mycorrhizal Fungal Colonization and Shifts Defence Allocation in Establishing *Plantago lanceolata*. *PLoS ONE* 8(6): e66053.

Bernklau, E. J., Hibbard, B. E. and Bjostad, L. B. (2010). Antixenosis in maize reduces feeding by western corn rootworm larvae (Coleoptera: Chrysomelidae). *Journal of Economic Entomology* 103(6): 2052-2060.

Blackshaw, R. P. and Kerry, B. R. (2008). Root herbivory in agricultural ecosystems. In Johnson, S. N. and Murray, P. J. (Eds), *Root Feeders - An Ecosystem Perspective*. Wallingford, UK: CABI Publishing, pp. 35-53.

Blossey, B. and Hunt-Joshi, T. R. (2003). Belowground herbivory by insects: influence on plants and aboveground herbivores. *Annual Review of Entomology* 48: 521-547.

Borgström, P., Strengbom, J., Marini, L., Viketoft, M. and Bommarco, R. (2017). Above- and belowground insect herbivory modifies the response of a grassland plant community to nitrogen eutrophication. *Ecology* 98(2): 545-554.

Borowicz, V. A. (2010). The impact of arbuscular mycorrhizal fungi on strawberry tolerance to root damage and drought stress. *Pedobiologia* 53(4): 265-270.

Brown, V. K. and Gange, A. C. (1990). Insect herbivory below ground. *Advances in Ecological Research* 20: 1-58.

Brundrett, M. C. and Tedersoo, L. (2018). Evolutionary history of mycorrhizal symbioses and global host plant diversity. *New Phytologist* 220(4): 1108-1115.

Cavazos, B. R., Bohner, T. F., Donald, M. L., Sneck, M. E., Shadow, A., Omacini, M., Rudgers, J. A. and Miller, T. E. X. (2018). Testing the roles of vertical transmission and drought stress in the prevalence of heritable fungal endophytes in annual grass populations. *New Phytologist* 219(3): 1075-1084.

Chandler, K. J. (2002). *Strategies to Control Greyback Canegrub in Early Harvested Ratoon Crops. SRDC Final Report: Project IPB001*. Queensland, Australia: Sugar Research Australia.

Chandler-Ezell, K., Pearsall, D. M. and Zeidler, J. A. (2006). Root and tuber phytoliths and starch grains document manioc (*Manihot esculenta*), arrowroot (*Maranta arundinacea*), and llerén (*Calathea* sp.) at the Real Alto site, Ecuador. *Economic Botany* 60(2): 103–120.

Chu-Chou, M., Guo, B., An, Z.-Q., Hendrix, J. W., Ferriss, R. S., Siegel, M. R., Dougherty, C. T. and Burrus, P. B. (1992). Suppression of mycorrhizal fungi in fescue by the *Acremonium coenophialum* endophyte. *Soil Biology and Biochemistry* 24(7): 633–637.

Cibils-Stewart, X. and Zerbino, S. (2020). ISOCAS: la importancia de conocer su biología para optimizar su manejo. "White grubs: the importance of knowing its biology to optimize its management". *Revista INIA* 60: 51–55.

Clark, K. E., Hartley, S. E. and Johnson, S. N. (2011). Does mother know best? The preference-performance hypothesis and parent-offspring conflict in aboveground-belowground herbivore life cycles. *Ecological Entomology* 36(2): 117–124.

Clay, K. and Schardl, C. (2002). Evolutionary origins and ecological consequences of endophyte symbiosis with grasses. *The American Naturalist* 160(Suppl. 4): S99–S127.

Cooke, J. and Leishman, M. R. (2012). Tradeoffs between foliar silicon and carbon-based defences: evidence from vegetation communities of contrasting soil types. *Oikos* 121(12): 2052–2060.

Coy, R. M., Held, D. W. and Kloepper, J. W. (2019). Rhizobacterial treatments of tall fescue and bermudagrass increases tolerance to damage from white grubs. *Pest Management Science* 75(12): 3210–3217.

Currie, A. F., Murray, P. J. and Gange, A. C. (2011). Is a specialist root-feeding insect affected by arbuscular mycorrhizal fungi? *Applied Soil Ecology* 47(2): 77–83.

Dastogeer, K. M. G. (2018). Influence of fungal endophytes on plant physiology is more pronounced under stress than well-watered conditions: a meta-analysis. *Planta* 248(6): 1403–1416.

Degenhardt, J., Hiltpold, I., Köllner, T. G., Frey, M., Gierl, A., Gershenzon, J., Hibbard, B. E., Ellersieck, M. R. and Turlings, T. C. (2009). Restoring a maize root signal that attracts insect-killing nematodes to control a major pest. *Proceedings of the National Academy of Sciences of the United States of America* 106(32): 13213–13218.

Del Valle, E. E. D., Mayer, G. F., Mazuquín, G. A., Nari, P., Toffoli, G. D. and Imvinkelried, H. O. (2018). Determination of damage caused by different populations of *Diloboderus abderus* (Coleoptera: Scarabaeidae) larvae on wheat. *Australian Journal of Crop Science* 12(12): 1899–1901.

Dematheis, F., Zimmerling, U., Flocco, C., Kurtz, B., Vidal, S., Kropf, S. and Smalla, K. (2012). Multitrophic interaction in the rhizosphere of maize: root feeding of western corn rootworm larvae alters the microbial community composition. *PLOS ONE* 7(5): e37288.

Detling, J. K., Winn, D. T., Procter-Gregg, C. and Painter, E. L. (1980). Effects of simulated grazing by below-ground herbivores on growth, CO_2 exchange, and carbon allocation patterns of *Bouteloua gracilis*. *Journal of Applied Ecology* 17(3): 771–778.

Dosdall, L. M., Good, A., Keddie, B. A., Ekuere, U. and Stringam, G. (2000). Identification and evaluation of root maggot (*Delia* spp.) (Diptera: Anthomyiidae) resistance within Brassicaceae. *Crop Protection* 19(4): 247-253.

Ekuere, U. U., Dosdall, L. M., Hills, M., Keddie, A. B., Kott, L. and Good, A. (2005). Identification, mapping, and economic evaluation of QTLs encoding root maggot resistance in *Brassica*. *Crop Science* 45(1): 371-378.

Ellis, P. R., Pink, D. A. C., Barber, N. E. and Mead, A. (1999). Identification of high levels of resistance to cabbage root fly, *Delia radicum*, in wild *Brassica* species. *Euphytica* 110(3): 207-214.

Erb, M., Glauser, G. and Robert, C. A. M. (2012). Induced immunity against belowground insect herbivores - activation of defenses in the absence of a jasmonate burst. *Journal of Chemical Ecology* 38(6): 629-640.

Erb, M., Lenk, C., Degenhardt, J. and Turlings, T. C. (2009). The underestimated role of roots in defense against leaf attackers. *Trends in Plant Science* 14(12): 653-659.

Erb, M. and Lu, J. (2013). Soil abiotic factors influence interactions between belowground herbivores and plant roots. *Journal of Experimental Botany* 64(5): 1295-1303.

Fahlgren, N., Gehan, M. A. and Baxter, I. (2015). Lights, camera, action: high-throughput plant phenotyping is ready for a close-up. *Current Opinion in Plant Biology* 24: 93-99.

Farrar, J. B. and Jones, D. H. (2003). The control of carbon acquisition by the growth of roots. In de Kroon, H. and Visser, E. J. W. (Eds), *Root Ecology*. Berlin, Germany: Springer-Verlag, pp. 91-124.

Franzluebbers, A. J. and Hill, N. S. (2005). Soil carbon, nitrogen, and ergot alkaloids with short- and long-term exposure to endophyte-infected and endophyte-free tall fescue. *Soil Science Society of America Journal* 69(2): 404-412.

Frew, A., Allsopp, P. G., Gherlenda, A. N. and Johnson, S. N. (2017a). Increased herbivory under elevated atmospheric carbon dioxide concentrations is reversed by silicon-based plant defences. *Journal of Applied Ecology* 54(5): 1310-1319.

Frew, A., Powell, J. R., Allsopp, P. G., Sallam, N. and Johnson, S. N. (2017b). Arbuscular mycorrhizal fungi promote silicon accumulation in plant roots, reducing the impacts of root herbivory. *Plant and Soil* 419(1-2): 423-433.

Frew, A., Powell, J. R., Glauser, G., Bennett, A. E. and Johnson, S. N. (2018). Mycorrhizal fungi enhance nutrient uptake but disarm defences in plant roots, promoting plant-parasitic nematode populations. *Soil Biology and Biochemistry* 126: 123-132.

Frew, A., Powell, J. R., Hiltpold, I., Allsopp, P. G., Sallam, N. and Johnson, S. N. (2017c). Host plant colonisation by arbuscular mycorrhizal fungi stimulates immune function whereas high root silicon concentrations diminish growth in a soil-dwelling herbivore. *Soil Biology and Biochemistry* 112: 117-126.

Frew, A., Powell, J. R. and Johnson, S. N. (2020). Aboveground resource allocation in response to root herbivory is enhanced by the arbuscular mycorrhizal symbiosis. *Plant and Soil* 447: 463-473. doi:10.1007/s11104-11019-04399-x.

Frew, A., Powell, J. R., Sallam, N., Allsopp, P. G. and Johnson, S. N. (2016). Trade-offs between silicon and phenolic defences may explain enhanced performance of root herbivores on phenolic-rich plants. *Journal of Chemical Ecology* 42(8): 768-771.

Frost, C. J., Mescher, M. C., Carlson, J. E. and De Moraes, C. M. (2008). Plant defense priming against herbivores: getting ready for a different battle. *Plant Physiology* 146(3): 818-824.

Fuchs, B. and Krauss, J. (2019). Can *Epichloë* endophytes enhance direct and indirect plant defence? *Fungal Ecology* 38: 98–103.

Fuller-Schaefer, C., Jung, K. and Jaronski, S. (2005). Colonization of sugarbeet roots by entomopathogenic fungi. In *Proceedings of the 38th Annual Meeting of the Society for Invertebrate Pathology*, Anchorage, AK, p. 49.

Gange, A. C. and Brown, V. K. (1989). Effects of root herbivory by an insect on a foliar-feeding species, mediated through changes in the host plant. *Oecologia* 81(1): 38–42.

Gange, A. C., Brown, V. K. and Farmer, L. M. (1991). Mechanisms of seedling mortality by subterranean insect herbivores. *Oecologia* 88(2): 228–232.

Gassmann, A. J. (2012). Field-evolved resistance to Bt maize by western corn rootworm: predictions from the laboratory and effects in the field. *Journal of Invertebrate Pathology* 110(3): 287–293.

Gassmann, A. J., Petzold-Maxwell, J. L., Keweshan, R. S. and Dunbar, M. W. (2011). Field-evolved resistance to Bt maize by western corn rootworm. *PLOS ONE* 6(7): e22629.

Genet, M., Stokes, A., Salin, F., Mickovski, S. B., Fourcaud, T., Dumail, J.-F. and van Beek, R. (2005). The influence of cellulose content on tensile strength in tree roots. *Plant and Soil* 278(1–2): 1–9.

Goggin, F. L., Lorence, A. and Topp, C. N. (2015). Applying high-throughput phenotyping to plant-insect interactions: picturing more resistant crops. *Current Opinion in Insect Science* 9: 69–76.

Goldson, S. L., Bourdot, G. W. and Proffitt, J. R. (1987). A study of the effects of *Sitona discoideus* (Coleoptera, Curculionidae) larval feeding on the growth and development of lucerne (*Medicago sativa*). *Journal of Applied Ecology* 24(1): 153–161.

Graham, J., Gordon, S. C., Smith, K., McNicol, R. J. and McNicol, J. W. (2002). The effect of the Cowpea trypsin inhibitor in strawberry on damage by vine weevil under field conditions. *Journal of Horticultural Science and Biotechnology* 77(1): 33–40.

Gray, M. E., Sappington, T. W., Miller, N. J., Moeser, J. and Bohn, M. O. (2009). Adaptation and invasiveness of western corn rootworm: intensifying research on a worsening pest. *Annual Review of Entomology* 54: 303–321.

Gregory, P. J. (2006). *Plant Roots - Growth, Activity and Interaction with Soils*. Oxford, UK: Blackwell Publishing.

Groff, P. A. and Kaplan, D. R. (1988). The relation of root systems to shoot systems in vascular plants. *Botanical Review* 54(4): 387–422.

Guerrieri, A., Dong, L. M. and Bouwmeester, H. J. (2019). Role and exploitation of underground chemical signaling in plants. *Pest Management Science* 75(9): 2455–2463.

Gundel, P. E., Irisarri, J. G. N., Fazio, L., Casas, C. and Pérez, L. I. (2016). Inferring field performance from drought experiments can be misleading: the case of symbiosis between grasses and *Epichloë* fungal endophytes. *Journal of Arid Environments* 132: 60–62.

Guo, J., McCulley, R. L., Phillips, T. D. and McNear, D. H. (2016). Fungal endophyte and tall fescue cultivar interact to differentially affect bulk and rhizosphere soil processes governing C and N cycling. *Soil Biology and Biochemistry* 101: 165–174.

Hajek, A. E. and St. Leger, R. J. (1994). Interactions between fungal pathogens and insect hosts. *Annual Review of Entomology* 39(1): 293–322.

Hallem, E. A., Dillman, A. R., Hong, A. V., Zhang, Y., Yano, J. M., DeMarco, S. F. and Sternberg, P. W. (2011). A sensory code for host seeking in parasitic nematodes. *Current Biology* 21(5): 377-383.

Hansen, D. J., Dayanandan, P., Kaufman, P. B. and Brotherson, J. D. (1976). Ecological adaptations of salt-marsh grass, *Distichlis spictata* (Gramineae) and environmental factors affecting its growth and distribution. *American Journal of Botany* 63(5): 635-650.

Hartley, S. E., Fitt, R. N., McLarnon, E. L. and Wade, R. N. (2015). Defending the leaf surface: intra- and inter-specific differences in silicon deposition in grasses in response to damage and silicon supply. *Frontiers in Plant Science* 6: 35.

Hartley, S. E. and Gange, A. C. (2009). Impacts of plant symbiotic fungi on insect herbivores: mutualism in a multitrophic context. *Annual Review of Entomology* 54: 323-342.

Hawkes, C. V. and Sullivan, J. J. (2001). The impact of herbivory on plants in different resource conditions: a meta-analysis. *Ecology* 82(7): 2045-2058.

Helander, M., Saloniemi, I., Omacini, M., Druille, M., Salminen, J. P. and Saikkonen, K. (2018). Glyphosate decreases mycorrhizal colonization and affects plant-soil feedback. *Science of the Total Environment* 642: 285-291.

Henkes, G. J., Thorpe, M. R., Minchin, P. E. H., Schurr, U. and Röse, U. S. R. (2008). Jasmonic acid treatment to part of the root system is consistent with simulated leaf herbivory, diverting recently assimilated carbon towards untreated roots within an hour. *Plant, Cell and Environment* 31(9): 1229-1236.

Hennessy, L. M., Popay, A. J., Finch, S. C., Clearwater, M. J. and Cave, V. M. (2016). Temperature and plant genotype alter alkaloid concentrations in ryegrass infected with an *Epichloë* endophyte and this affects an insect herbivore. *Frontiers in Plant Science* 7: 1097.

Hervé, M. R. and Erb, M. (2019). Distinct defense strategies allow different grassland species to cope with root herbivore attack. *Oecologia* 191(1): 127-139.

Hilder, V. A. and Boulter, D. (1999). Genetic engineering of crop plants for insect resistance - a critical review. *Crop Protection* 18(3): 177-191.

Hiltpold, I., Bernklau, E. J., Bjostad, L. B., Alvarez, N., Miller-Struttmann, N. E., Lundgren, J. G. and Hibbard, B. E. (2013). Nature, evolution and characterisation of rhizospheric chemical exudates affecting root herbivores. *Advances in Insect Physiology* 45: 97-157.

Hoeksema, J. D., Chaudhary, V. B., Gehring, C. A., Johnson, N. C., Karst, J., Koide, R. T., Pringle, A., Zabinski, C., Bever, J. D., Moore, J. C., Wilson, G. W., Klironomos, J. N. and Umbanhowar, J. (2010). A meta-analysis of context-dependency in plant response to inoculation with mycorrhizal fungi. *Ecology Letters* 13(3): 394-407.

Hol, W. H. G. and Cook, R. (2005). An overview of arbuscular mycorrhizal fungi-nematode interactions. *Basic and Applied Ecology* 6(6): 489-503.

Howe, G. A. and Jander, G. (2008). Plant immunity to insect herbivores. *Annual Review of Plant Biology* 59: 41-66.

Huang, A. C. C. and Osbourn, A. (2019). Plant terpenes that mediate below-ground interactions: prospects for bioengineering terpenoids for plant protection. *Pest Management Science* 75(9): 2368-2377.

Huber, M., Bont, Z., Fricke, J., Brillatz, T., Aziz, Z., Gershenzon, J. and Erb, M. (2016). A below-ground herbivore shapes root defensive chemistry in natural plant populations. *Proceedings of the Royal Society B: Biological Sciences* 283(1827): 20160285.

Huber, M., Triebwasser-Freese, D., Reichelt, M., Heiling, S., Paetz, C., Chandran, J. N., Bartram, S., Schneider, B., Gershenzon, J. and Erb, M. (2015). Identification, quantification, spatiotemporal distribution and genetic variation of major latex secondary metabolites in the common dandelion (*Taraxacum officinale* agg.). *Phytochemistry* 115: 89–98.

Huitu, O., Forbes, K. M., Helander, M., Julkunen-Tiitto, R., Lambin, X., Saikkonen, K., Stuart, P., Sulkama, S. and Hartley, S. (2014). Silicon, endophytes and secondary metabolites as grass defenses against mammalian herbivores. *Frontiers in Plant Science* 5: 478.

Hunt, J. W., Dean, A. P., Webster, R. E., Johnson, G. N. and Ennos, A. R. (2008). A novel mechanism by which silica defends grasses against herbivory. *Annals of Botany* 102(4): 653–656.

Hunter, M. D. (2001). Out of sight, out of mind: the impacts of root-feeding insects in natural and managed systems. *Agricultural and Forest Entomology* 3(1): 3–9.

Ivezić, M., Raspudić, E., Brmež, M., Majić, I., Brkić, I., Tollefson, J. J., Bohn, M., Hibbard, B. E. and Šimić, D. (2009). A review of resistance breeding options targeting western corn rootworm (*Diabrotica virgifera virgifera* LeConte). *Agricultural and Forest Entomology* 11(3): 307–311.

Jaber, L. R. and Ownley, B. H. (2018). Can we use entomopathogenic fungi as endophytes for dual biological control of insect pests and plant pathogens? *Biological Control* 116: 36–45.

Jactel, H., Verheggen, F., Thiéry, D., Escobar-Gutiérrez, A. J., Gachet, E., Desneux, N. and Neonicotinoids Working Group (2019). Alternatives to neonicotinoids. *Environment International* 129: 423–429.

Jagodič, A., Ipavec, N., Trdan, S. and Laznik, Ž. (2017). Attraction behaviors: are synthetic volatiles, typically emitted by insect-damaged *Brassica nigra* roots, navigation signals for entomopathogenic nematodes (*Steinernema* and *Heterorhabditis*)? *BioControl* 62(4): 515–524.

Johnson, S. N., Barton, A. T., Clark, K. E., Gregory, P. J., McMenemy, L. S. and Hancock, R. D. (2011). Elevated atmospheric carbon dioxide impairs the performance of root-feeding vine weevils by modifying root growth and secondary metabolites. *Global Change Biology* 17(2): 688–695.

Johnson, S. N., Benefer, C. M., Frew, A., Griffiths, B. S., Hartley, S. E., Karley, A. J., Rasmann, S., Schumann, M., Sonnemann, I. and Robert, C. A. M. (2016a). New frontiers in belowground ecology for plant protection from root-feeding insects. *Applied Soil Ecology* 108: 96–107.

Johnson, S. N., Clark, K. E., Hartley, S. E., Jones, T. H., McKenzie, S. W. and Koricheva, J. (2012). Aboveground-belowground herbivore interactions: a meta-analysis. *Ecology* 93(10): 2208–2215.

Johnson, S. N., Erb, M. and Hartley, S. E. (2016b). Roots under attack: contrasting plant responses to below- and aboveground insect herbivory. *New Phytologist* 210(2): 413–418.

Johnson, S. N., Glauser, G., Hiltpold, I., Moore, B. D. and Ryalls, J. M. W. (2018). Root herbivore performance suppressed when feeding on a jasmonate-induced pasture grass. *Ecological Entomology* 43(4): 547–550.

Johnson, S. N., Hallett, P. D., Gillespie, T. L. and Halpin, C. (2010). Belowground herbivory and root toughness: a potential model system using lignin-modified tobacco. *Physiological Entomology* 35(2): 186–191.

Johnson, S. N., Mitchell, C., McNicol, J. W., Thompson, J. and Karley, A. J. (2013). Downstairs drivers – root herbivores shape communities of above-ground herbivores and natural enemies via changes in plant nutrients. *Journal of Animal Ecology* 82(5): 1021-1030.

Johnson, S. N. and Nielsen, U. N. (2012). Foraging in the dark - chemically mediated host plant location by belowground insect herbivores. *Journal of Chemical Ecology* 38(6): 604-614.

Johnson, S. N. and Rasmann, S. (2015). Root-feeding insects and their interactions with organisms in the rhizosphere. *Annual Review of Entomology* 60: 517-535.

Johnson, S. N., Reynolds, O. L., Gurr, G. M., Esveld, J. L., Moore, B. D., Tory, G. J. and Gherlenda, A. N. (2019). When resistance is futile, tolerate instead: silicon promotes plant compensatory growth when attacked by above- and belowground herbivores. *Biology Letters* 15(7): 20190361.

Jones, J. T., Haegeman, A., Danchin, E. G. J., Gaur, H. S., Helder, J., Jones, M. G., Kikuchi, T., Manzanilla-López, R., Palomares-Rius, J. E., Wesemael, W. M. and Perry, R. N. (2013). Top 10 plant-parasitic nematodes in molecular plant pathology. *Molecular Plant Pathology* 14(9): 946-961.

Jung, S. C., Martinez-Medina, A., Lopez-Raez, J. A. and Pozo, M. J. (2012). Mycorrhiza-induced resistance and priming of plant defenses. *Journal of Chemical Ecology* 38(6): 651-664.

Kabouw, P., Biere, A., van der Putten, W. H. and van Dam, N. M. (2010). Intra-specific differences in root and shoot glucosinolate profiles among white cabbage (*Brassica oleracea* var. *capitata*) cultivars. *Journal of Agricultural and Food Chemistry* 58(1): 411-417.

Kaplan, I., Halitschke, R., Kessler, A., Sardanelli, S. and Denno, R. F. (2008). Constitutive and induced defenses to herbivory in above- and belowground plant tissues. *Ecology* 89(2): 392-406.

Kaya, H. K. and Gaugler, R. (1993). Entomopathogenic nematodes. *Annual Review of Entomology* 38(1): 181-206.

Kim, J. H. and Mullin, C. A. (2003). Antifeedant effects of proteinase inhibitors on feeding behaviors of adult western corn rootworm (*Diabrotica virgifera virgifera*). *Journal of Chemical Ecology* 29(4): 795-810.

Kivlin, S. N., Emery, S. M. and Rudgers, J. A. (2013). Fungal symbionts alter plant responses to global change. *American Journal of Botany* 100(7): 1445-1457.

Kohler, A., Kuo, A., Nagy, L. G., Morin, E., Barry, K. W., Buscot, F., Canbäck, B., Choi, C., Cichocki, N., Clum, A., Colpaert, J., Copeland, A., Costa, M. D., Doré, J., Floudas, D., Gay, G., Girlanda, M., Henrissat, B., Herrmann, S., Hess, J., Högberg, N., Johansson, T., Khouja, H.-R., LaButti, K., Lahrmann, U., Levasseur, A., Lindquist, E. A., Lipzen, A., Marmeisse, R., Martino, E., Murat, C., Ngan, C. Y., Nehls, U., Plett, J. M., Pringle, A., Ohm, R. A., Perotto, S., Peter, M., Riley, R., Rineau, F., Ruytinx, J., Salamov, A., Shah, F., Sun, H., Tarkka, M., Tritt, A., Veneault-Fourrey, C., Zuccaro, A., Mycorrhizal Genomics Initiative Consortium, Tunlid, A., Grigoriev, I. V., Hibbett, D. S. and Martin, F. (2015). Convergent losses of decay mechanisms and rapid turnover of symbiosis genes in mycorrhizal mutualists. *Nature Genetics* 47(4): 410-415.

Koricheva, J., Gange, A. C. and Jones, T. (2009). Effects of mycorrhizal fungi on insect herbivores: a meta-analysis. *Ecology* 90(8): 2088-2097.

Krings, M., Taylor, T. N., Hass, H., Kerp, H., Dotzler, N. and Hermsen, E. J. (2007). Fungal endophytes in a 400-million-yr-old land plant: infection pathways, spatial distribution, and host responses. *New Phytologist* 174(3): 648-657.

Kunkel, B. N. and Brooks, D. M. (2002). Cross talk between signaling pathways in pathogen defense. *Current Opinion in Plant Biology* 5(4): 325–331.

Lackner, S., Lackus, N. D., Paetz, C., Köllner, T. G. and Unsicker, S. B. (2020). Aboveground phytochemical responses to belowground herbivory in poplar trees and the consequence for leaf herbivore preference. *Plant, Cell and Environment* 42(12): 3293–3307.

Ladd, T. L. Jr. (1988). Japanese beetle (Coleoptera, Scarabaeidae): influence of sugars on feeding response of larvae. *Journal of Economic Entomology* 81(5): 1390–1393.

Lanfranco, L., Fiorilli, V. and Gutjahr, C. (2018). Partner communication and role of nutrients in the arbuscular mycorrhizal symbiosis. *New Phytologist* 220(4): 1031–1046.

Larimer, A. L., Bever, J. D. and Clay, K. (2010). The interactive effects of plant microbial symbionts: a review and meta-analysis. *Symbiosis* 51(2): 139–148.

Lawrence, S. D., Novak, N. G., El Kayal, W. E., Ju, C. J. and Cooke, J. E. (2012). Root herbivory: molecular analysis of the maize transcriptome upon infestation by Southern corn rootworm, *Diabrotica undecimpunctata howardi*. *Physiologia Plantarum* 144(4): 303–319.

Laznik, Ž. and Trdan, S. (2016a). Attraction behaviors of entomopathogenic nematodes (Steinernematidae and Heterorhabditidae) to synthetic volatiles emitted by insect damaged potato tubers. *Journal of Chemical Ecology* 42(4): 314–322.

Laznik, Ž. and Trdan, S. (2016b). The role of volatile substances emitted by cultivated plant's roots in indirect defense against soil herbivores. In Trdan, S. (Ed.), *Insecticides Resistance*. Intech, pp. 393–419. Available at: https://www.intechopen.com/books/insecticides-resistance/the-role-of-volatile-substances-emitted-by-cultivated-plant-s-roots-in-indirect-defense-against-soil.

Leroux, O., Leroux, F., Bagniewska-Zadworna, A., Knox, J. P., Claeys, M., Bals, S. and Viane, R. L. L. (2011). Ultrastructure and composition of cell wall appositions in the roots of *Asplenium* (Polypodiales). *Micron* 42(8): 863–870.

Lewis, L. C., Berry, E. C., Obrycki, J. J. and Bling, L. A. (1996). Aptness of insecticides (*Bacillus thuringiensis* and carbofuran) with endophytic *Beauveria bassiana*, in suppressing larval populations of the European corn borer. *Agriculture, Ecosystems and Environment* 57(1): 27–34.

Lewis, L. C. and Bing, L. A. (1991). *Bacillus thuringiensis* Berliner and *Beauveria bassiana* (Balsamo) Vuillimen for European corn borer control – program for immediate and season-long suppression. *Canadian Entomologist* 123(2): 387–393.

Lewis, L. C., Bruck, D. J., Gunnarson, R. D. and Bidne, K. G. (2001). Assessment of plant pathogenicity of endophytic *Beauveria bassiana* in Bt transgenic and non-transgenic corn. *Crop Science* 41(5): 1395–1400.

Lu, J., Robert, C. A. M., Riemann, M., Cosme, M., Mène-Saffrané, L., Massana, J., Stout, M. J., Lou, Y., Gershenzon, J. and Erb, M. (2015). Induced jasmonate signaling leads to contrasting effects on root damage and herbivore performance. *Plant Physiology* 167(3): 1100–1116.

Mack, K. M. L. and Rudgers, J. A. (2008). Balancing multiple mutualists: asymmetric interactions among plants, arbuscular mycorrhizal fungi, and fungal endophytes. *Oikos* 117(2): 310–320.

Mantzoukas, S. and Eliopoulos, P. A. (2020). Endophytic entomopathogenic fungi: a valuable biological control tool against plant pests. *Applied Sciences* 10(1): 360.

Maron, J. L. (2001). Intraspecific competition and subterranean herbivory: individual and interactive effects on bush lupine. *Oikos* 92(1): 178–186.

Martinez-Medina, A., Flors, V., Heil, M., Mauch-Mani, B., Pieterse, C. M. J., Pozo, M. J., Ton, J., van Dam, N. M. and Conrath, U. (2016). Recognizing plant defense priming. *Trends in Plant Science* 21(10): 818–822.

Massey, F. P. and Hartley, S. E. (2006). Experimental demonstration of the antiherbivore effects of silica in grasses: impacts on foliage digestibility and vole growth rates. *Proceedings of the Royal Society B: Biological Sciences* 273(1599): 2299–2304.

Massey, F. P. and Hartley, S. E. (2009). Physical defences wear you down: progressive and irreversible impacts of silica on insect herbivores. *Journal of Animal Ecology* 78(1): 281–291.

Masters, G. J. (2004). Below-ground herbivores and ecosystem processes. In Weisser, W. W. and Siemann E. (Eds), *Insects and Ecosystem Function*. Berlin, Germany: Springer-Verlag, pp. 94–112.

Maurya, A. K., Kelly, M. P., Mahaney, S. M. and Gomez, S. K. (2018). Arbuscular mycorrhizal symbiosis alters plant gene expression and aphid weight in a tripartite interaction. *Journal of Plant Interactions* 13(1): 294–305.

McNickle, G. G. and Evans, W. D. (2018). Toleration games: compensatory growth by plants in response to enemy attack is an evolutionarily stable strategy. *AoB PLANTS* 10(4): ply035.

Meier, A. R. and Hunter, M. D. (2018a). Arbuscular mycorrhizal fungi mediate herbivore-induction of plant defenses differently above and belowground. *Oikos* 127(12): 1759–1775.

Meier, A. R. and Hunter, M. D. (2018b). Mycorrhizae alter toxin sequestration and performance of two specialist herbivores. *Frontiers in Ecology and Evolution* 6: 33.

Meihls, L. N., Higdon, M. L., Ellersieck, M. and Hibbard, B. E. (2011). Selection for resistance to mCry3A-expressing transgenic corn in western corn rootworm. *Journal of Economic Entomology* 104(3): 1045–1054.

Meyling, N. V. and Eilenberg, J. (2007). Ecology of the entomopathogenic fungi *Beauveria bassiana* and *Metarhizium anisopliae* in temperate agroecosystems: potential for conservation biological control. *Biological Control* 43(2): 145–155.

Moore, B. D. and Johnson, S. N. (2017). Get tough, get toxic, or get a bodyguard: identifying candidate traits conferring belowground resistance to herbivores in grasses. *Frontiers in Plant Science* 7: 1925.

Moose, S. P. and Mumm, R. H. (2008). Molecular plant breeding as the foundation for 21st century crop improvement. *Plant Physiology* 147(3): 969–977.

Müller, J. (2003). Artificial infection by endophytes affects growth and mycorrhizal colonisation of *Lolium perenne*. *Functional Plant Biology* 30(4): 419–424.

Newingham, B. A., Callaway, R. M. and BassiriRad, H. (2007). Allocating nitrogen away from a herbivore: a novel compensatory response to root herbivory. *Oecologia* 153(4): 913–920.

Nicol, J. M., Turner, S. J., Coyne, D. L., Nijs, L., Hockland, S. and Maafi, Z. T. (2011). Current nematode threats to world agriculture. In Jones, J. T., Gheysen, G. and Fenoll, C. (Eds), *Genomics and Molecular Genetics of Plant-Nematode Interactions*. Heidelberg, Germany: Springer, pp. 21–44.

Novas, M. V., Cabral, D. and Godeas, A. M. (2005). Interaction between grass endophytes and mycorrhizas in *Bromus setifolius* from Patagonia, Argentina. *Symbiosis* 40: 23–30.

Novas, M. V., Iannone, L. J., Godeas, A. M. and Cabral, D. (2009). Positive association between mycorrhiza and foliar endophytes in *Poa bonariensis*, a native grass. *Mycological Progress* 8(1): 75–81.

Novas, M. V., Iannone, L. J., Godeas, A. M. and Scervino, J. M. (2011). Evidence for leaf endophyte regulation of root symbionts: effect of *Neotyphodium* endophytes on the pre-infective state of mycorrhizal fungi. *Symbiosis* 55(1): 19–28.

Oduor, A. M. O., van Kleunen, M. and Stift, M. (2017). In the presence of specialist root and shoot herbivory, invasive-range *Brassica nigra* populations have stronger competitive effects than native-range populations. *Journal of Ecology* 105(6): 1679–1686.

Omacini, M., Eggers, T., Bonkowski, M., Gange, A. C. and Jones, T. H. (2006). Leaf endophytes affect mycorrhizal status and growth of co-infected and neighbouring plants. *Functional Ecology* 20(2): 226–232.

Omacini, M., Semmartin, M., Peréz, L. I. and Gundel, P. E. (2012). Grass-endophyte symbiosis: a neglected aboveground interaction with multiple belowground consequences. *Applied Soil Ecology* 61: 273–279.

Ourry, M., Lebreton, L., Chaminade, V., Guillerm-Erckelboudt, A.-Y., Hervé, M., Linglin, J., Marnet, N., Ourry, A., Paty, C., Poinsot, D., Cortesero, A.-M. and Mougel, C. (2018). Influence of belowground herbivory on the dynamics of root and rhizosphere microbial communities. *Frontiers in Ecology and Evolution* 6: 91.

Parker, W. E. and Howard, J. J. (2001). The biology and management of wireworms (*Agriotes* spp.) on potato with particular reference to the UK. *Agricultural and Forest Entomology* 3(2): 85–98.

Parsa, S., Ortiz, V. and Vega, F. E. (2013). Establishing fungal entomopathogens as endophytes: towards endophytic biological control. *Journal of Visualized Experiments* (74): 50360.

Pennell, C. G. L., Popay, A. J., Ball, O. J.-P., Hume, D. E. and Baird, D. B. (2005). Occurrence and impact of pasture mealybug (*Balanococcus poae*) and root aphid (*Aploneura lentisci*) on ryegrass (*Lolium* spp.) with and without infection by *Neotyphodium* fungal endophytes. *New Zealand Journal of Agricultural Research* 48(3): 329–337.

Pepe, A., Sbrana, C., Ferrol, N. and Giovannetti, M. (2017). An in vivo whole-plant experimental system for the analysis of gene expression in extraradical mycorrhizal mycelium. *Mycorrhiza* 27(7): 659–668.

Perez, L. I., Gundel, P. E., Marrero, H. J., Arzac, A. G. and Omacini, M. (2017). Symbiosis with systemic fungal endophytes promotes host escape from vector-borne disease. *Oecologia* 184(1): 237–245.

Popay, A. and Baltus, J. (2001). Black beetle damage to perennial ryegrass infected with AR1 endophyte. *Proceedings of the New Zealand Grassland Association* 63: 267–272.

Popay, A. J., Cotching, B., Moorhead, A., and Ferguson, C. M. (2012). AR37 endophyte effects on porina and root aphid populations and ryegrass damage in the field. *Proceedings of the New Zealand Grassland Association* 74: 165–169.

Popay, A. J., Townsend, R. J. and Fletcher, L. R. (2003). The effect of endophyte (*Neotyphodium uncinatum*) in meadow fescue on grass grub larvae. *New Zealand Plant Protection* 56: 123–128.

Poveda, K., Gómez Jiménez, M. I. and Kessler, A. (2010). The enemy as ally: herbivore-induced increase in crop yield. *Ecological Applications* 20(7): 1787–1793.

Pozo, M. J. and Azcón-Aguilar, C. (2007). Unraveling mycorrhiza-induced resistance. *Current Opinion in Plant Biology* 10(4): 393–398.

Preisser, E. L. and Strong, D. R. (2004). Climate affects predator control of an herbivore outbreak. *The American Naturalist* 163(5): 754–762.

Quinn, M. A. and Hall, M. H. (1992). Compensatory response of a legume root-nodule system to nodule herbivory by *Sitona hispidulus*. *Entomologia Experimentalis et Applicata* 64(2): 167–176.

Ramos, Y., Taibo, A. D., Jiménez, J. A. and Portal, O. (2020). Endophytic establishment of *Beauveria bassiana* and *Metarhizium anisopliae* in maize plants and its effect against *Spodoptera frugiperda* (J. E. Smith) (Lepidoptera: Noctuidae) larvae. *Egyptian Journal of Biological Pest Control* 30(1): 20.

Rasmann, S. and Agrawal, A. A. (2008). In defense of roots: a research agenda for studying plant resistance to belowground herbivory. *Plant Physiology* 146(3): 875–880.

Rasmann, S., Ali, J. G., Helder, J. and van der Putten, W. H. (2012). Ecology and evolution of soil nematode chemotaxis. *Journal of Chemical Ecology* 38(6): 615–628.

Rasmann, S., Köllner, T. G., Degenhardt, J., Hiltpold, I., Toepfer, S., Kuhlmann, U., Gershenzon, J. and Turlings, T. C. J. (2005). Recruitment of entomopathogenic nematodes by insect-damaged maize roots. *Nature* 434(7034): 732–737.

Redmond, C. T. and Potter, D. A. (2006). Silicon fertilization does not enhance creeping bentgrass resistance to cutworms and white grubs. *Applied Turfgrass Science* 3: 1–6.

Rho, H., Hsieh, M., Kandel, S. L., Cantillo, J., Doty, S. L. and Kim, S. H. (2018). Do endophytes promote growth of host plants under stress? A meta-analysis on plant stress mitigation by endophytes. *Microbial Ecology* 75(2): 407–418.

Riedell, W. E. and Reese, R. N. (1999). Maize morphology and shoot CO_2 assimilation after root damage by western corn rootworm larvae. *Crop Science* 39(5): 1332–1340.

Robert, C. A. M., Erb, M., Duployer, M., Zwahlen, C., Doyen, G. R. and Turlings, T. C. J. (2012a). Herbivore-induced plant volatiles mediate host selection by a root herbivore. *New Phytologist* 194(4): 1061–1069.

Robert, C. A. M., Erb, M., Hibbard, B. E., Wade French, B., Zwahlen, C. and Turlings, T. C. J. (2012b). A specialist root herbivore reduces plant resistance and uses an induced plant volatile to aggregate in a density-dependent manner. *Functional Ecology* 26(6): 1429–1440.

Robert, C. A. M., Erb, M., Hiltpold, I., Hibbard, B. E., Gaillard, M. D. P., Bilat, J., Degenhardt, J., Cambet-Petit-Jean, X., Turlings, T. C. J. and Zwahlen, C. (2013). Genetically engineered maize plants reveal distinct costs and benefits of constitutive volatile emissions in the field. *Plant Biotechnology Journal* 11(5): 628–639.

Robert, C. A. M., Ferrieri, R. A., Schirmer, S., Babst, B. A., Schueller, M. J., Machado, R. A. R., Arce, C. C. M., Hibbard, B. E., Gershenzon, J., Turlings, T. C. J. and Erb, M. (2014). Induced carbon reallocation and compensatory growth as root herbivore tolerance mechanisms. *Plant, Cell and Environment* 37(11): 2613–2622.

Robert, C. A. M., Schirmer, S., Barry, J., Wade French, B., Hibbard, B. E. and Gershenzon, J. (2015). Belowground herbivore tolerance involves delayed overcompensatory root regrowth in maize. *Entomologia Experimentalis et Applicata* 157(1): 113–120.

Robert, C. A. M., Veyrat, N., Glauser, G., Marti, G., Doyen, G. R., Villard, N., Gaillard, M. D. P., Köllner, T. G., Giron, D., Body, M., Babst, B. A., Ferrieri, R. A., Turlings, T. C. J. and Erb, M. (2012c). A specialist root herbivore exploits defensive metabolites to locate nutritious tissues. *Ecology Letters* 15(1): 55–64.

Rojas, X., Guo, J. Q., Leff, J. W., McNear, D. H., Fierer, N. and McCulley, R. L. (2016). Infection with a shoot-specific fungal endophyte (*Epichloë*) alters tall fescue soil microbial communities. *Microbial Ecology* 72(1): 197–206.

Rostás, M., Cripps, M. G. and Silcock, P. (2015). Aboveground endophyte affects root volatile emission and host plant selection of a belowground insect. *Oecologia* 177(2): 487–497.

Roy, H. E., Steinkraus, D. C., Eilenberg, J., Hajek, A. E. and Pell, J. K. (2006). Bizarre interactions and endgames: entomopathogenic fungi and their arthropod hosts. *Annual Review of Entomology* 51: 331–357.

Rubio, G. and Lynch, J. P. (2007). Compensation among root classes in *Phaseolus vulgaris* L. *Plant and Soil* 290(1–2): 307–321.

Ryalls, J. M. W., Riegler, M., Moore, B. D., Lopaticki, G. and Johnson, S. N. (2013). Effects of elevated temperature and CO_2 on aboveground-belowground systems: a case study with plants, their mutualistic bacteria and root/shoot herbivores. *Frontiers in Plant Science* 4: 445.

Saikkonen, K., Gundel, P. E. and Helander, M. (2013). Chemical ecology mediated by fungal endophytes in grasses. *Journal of Chemical Ecology* 39(7): 962–968.

Schüßler, A. and Walker, C. (2010). *The Glomeromycota – A Species List With New Families and New Genera*. The Royal Botanic Garden Edinburgh, The Royal Botanic Garden Kew, Botanische Staatssammlung Munich, and Oregon State University. ISBN-13: 978-1466388048. ISBN-10: 1466388048. Available at http://www.amf-phylogeny.com.

Schultz, J. C., Appel, H. M., Ferrieri, A. P. and Arnold, T. M. (2013). Flexible resource allocation during plant defense responses. *Frontiers in Plant Science* 4: 324.

Scott, B., Green, K. and Berry, D. (2018). The fine balance between mutualism and antagonism in the *Epichloë festucae*-grass symbiotic interaction. *Current Opinion in Plant Biology* 44: 32–38.

Simon, A. L., Wellham, P. A. D., Aradottir, G. I. and Gange, A. C. (2017). Unravelling mycorrhiza-induced wheat susceptibility to the English grain aphid *Sitobion avenae*. *Scientific Reports* 7: 46497.

Singh, L. P., Gill, S. S. and Tuteja, N. (2011). Unraveling the role of fungal symbionts in plant abiotic stress tolerance. *Plant Signaling and Behavior* 6(2): 175–191.

Soler, R., Van der Putten, W. H., Harvey, J. A., Vet, L. E., Dicke, M. and Bezemer, T. M. (2012). Root herbivore effects on aboveground multitrophic interactions: patterns, processes and mechanisms. *Journal of Chemical Ecology* 38(6): 755–767.

Spiering, M. J., Lane, G. A., Christensen, M. J. and Schmid, J. (2005). Distribution of the fungal endophyte *Neotyphodium lolii* is not a major determinant of the distribution of fungal alkaloids in *Lolium perenne* plants. *Phytochemistry* 66(2): 195–202.

Stanton, N. L. (1988). The underground in grasslands. *Annual Review of Ecology and Systematics* 19(1): 573–589.

Steinger, T. and Müller-Schärer, H. (1992). Physiological and growth responses of *Centaurea maculosa* (Asteraceae) to root herbivory under varying levels of interspecific plant competition and soil nitrogen availability. *Oecologia* 91(1): 141–149.

Stout, M. J. (2013). Reevaluating the conceptual framework for applied research on host-plant resistance. *Insect Science* 20(3): 263–272.

Stowe, K. A., Marquis, R. J., Hochwender, C. G. and Simms, E. L. (2000). The evolutionary ecology of tolerance to consumer damage. *Annual Review of Ecology and Systematics* 31(1): 565–595.

Strauss, S. Y. and Agrawal, A. A. (1999). The ecology and evolution of plant tolerance to herbivory. *Trends in Ecology and Evolution* 14(5): 179–185.

Strong, D. R., Lawton, J. H. and Southwood, T. R. E. (1984). *Insects on Plants. Community Patterns and Mechanisms*. Oxford, UK: Blackwell Scientific Publications.

Talaat, N. B. and Shawky, B. T. (2012). Influence of arbuscular mycorrhizae on root colonization, growth and productivity of two wheat cultivars under salt stress. *Archives of Agronomy and Soil Science* 58(1): 85–100.

Thelen, G. C., Vivanco, J. M., Newingham, B., Good, W., Bais, H. P., Landres, P., Caesar, A. and Callaway, R. M. (2005). Insect herbivory stimulates allelopathic exudation by an invasive plant and the suppression of natives. *Ecology Letters* 8(2): 209–217.

Thirkell, T. J., Cameron, D. D. and Hodge, A. (2016). Resolving the 'nitrogen paradox' of arbuscular mycorrhizas: fertilization with organic matter brings considerable benefits for plant nutrition and growth. *Plant, Cell and Environment* 39(8): 1683–1690.

Tiffin, P. (2000). Mechanisms of tolerance to herbivore damage: what do we know? *Evolutionary Ecology* 14(4–6): 523–536.

Tollefson, J. J. (2007). Evaluating maize for resistance to *Diabrotica virgifera virgifera* LeConte (Coleoptera: Chrysomelidae). *Maydica* 52: 311–318.

Tomczak, V. V. and Müller, C. (2018). Plant species, mycorrhiza, and aphid age influence the performance and behaviour of a generalist. *Ecological Entomology* 43(1): 37–46.

Tonelli, M., Peñaflor, M. F. G. V., Leite, L. G., Silva, W. D., Martins, F. and Bento, J. M. S. (2016). Attraction of entomopathogenic nematodes to sugarcane root volatiles under herbivory by a sap-sucking insect. *Chemoecology* 26(2): 59–66.

Treonis, A. M., Grayston, S. J., Murray, P. J. and Dawson, L. A. (2005). Effects of root feeding, cranefly larvae on soil microorganisms and the composition of rhizosphere solutions collected from grassland plants. *Applied Soil Ecology* 28(3): 203–215.

Turlings, T. C. J., Hiltpold, I. and Rasmann, S. (2012). The importance of root-produced volatiles as foraging cues for entomopathogenic nematodes. *Plant and Soil* 358(1–2): 51–60.

Vachon, V., Laprade, R. and Schwartz, J.-L. (2012). Current models of the mode of action of *Bacillus thuringiensis* insecticidal crystal proteins: a critical review. *Journal of Invertebrate Pathology* 111(1): 1–12.

Vachová, P. and Walmsley, A. (2017). Wireworms suppress spreading of the expansive weed *Calamagrostis epigejos* (L) Roth by feeding on its rhizomes. *European Journal of Soil Biology* 81: 19–24.

van Dam, N. M. and Raaijmakers, C. E. (2006). Local and systemic induced responses to cabbage root fly larvae (*Delia radicum*) in *Brassica nigra* and *B. oleracea*. *Chemoecology* 16(1): 17–24.

van Dam, N. M., Tytgat, T. O. G. and Kirkegaard, J. A. (2009). Root and shoot glucosinolates: a comparison of their diversity, function and interactions in natural and managed ecosystems. *Phytochemistry Reviews* 8(1): 171–186.

van der Heijden, M. G. A., Martin, F. M., Selosse, M. A. and Sanders, I. R. (2015). Mycorrhizal ecology and evolution: the past, the present, and the future. *New Phytologist* 205(4): 1406–1423.

van Leur, H., Raaijmakers, C. E. and van Dam, N. M. (2008). Reciprocal interactions between the cabbage root fly (*Delia radicum*) and two glucosinolate phenotypes of *Barbarea vulgaris*. *Entomologia Experimentalis et Applicata* 128(2): 312–322.

Vega, F. E., Posada, F., Aime, M. C., Pava-Ripoll, M., Infante, F. and Rehner, S. A. (2008). Entomopathogenic fungal endophytes. *Biological Control* 46(1): 72–82.

Vignale, M. V., Iannone, L. J., Scervino, J. M. and Novas, M. V. (2018). *Epichloë* exudates promote in vitro and in vivo arbuscular mycorrhizal fungi development and plant growth. *Plant and Soil* 422(1-2): 267-281.

Villani, M. G. and Wright, R. J. (1990). Environmental influences on soil macroarthropod behavior in agricultural systems. *Annual Review of Entomology* 35(1): 249-269.

Waterman, J. M., Cazzonelli, C. I., Hartley, S. E. and Johnson, S. N. (2019). Simulated herbivory: the key to disentangling plant defence responses. *Trends in Ecology and Evolution* 34(5): 447-458.

Weis, A. E. and Franks, S. J. (2006). Herbivory tolerance and coevolution: an alternative to the arms race? *New Phytologist* 170(3): 423-425.

Wilkinson, T. D. J., Ferrari, J., Hartley, S. E. and Hodge, A. (2019). Aphids can acquire the nitrogen delivered to plants by arbuscular mycorrhizal fungi. *Functional Ecology* 33(4): 576-586.

Yang, H. S., Xu, J. L., Guo, Y., Koide, R. T., Dai, Y., Xu, M., Bian, L., Bian, X. and Zhang, Q. (2016). Predicting plant response to arbuscular mycorrhizas: the role of host functional traits. *Fungal Ecology* 20: 79-83.

Young, C. A., Hume, D. E. and McCulley, R. L. (2013). Forages and Pastures Symposium: Fungal endophytes of tall fescue and perennial ryegrass: pasture friend or foe? *Journal of Animal Science* 91(5): 2379-2394.

Zvereva, E. L. and Kozlov, M. V. (2012). Sources of variation in plant responses to belowground insect herbivory: a meta-analysis. *Oecologia* 169(2): 441-452.

Chapter 10

Advances in understanding plant root response to nematode attack

Shahid Siddique, University of California-Davis, USA; and John T. Jones, The James Hutton Institute and University of St Andrews, UK

1 Introduction

Plant-parasitic nematodes (PPN) cause huge damage to crops across the globe and are a major constraint in achieving food security (reviewed by Jones et al., 2013). Calculating the impact of the damage caused by PPN is challenging, but is estimated as being in excess of $80 billion per annum (Nicol et al., 2011). PPN are microscopically small, soil-dwelling organisms and the symptoms of PPN infection are rarely diagnostic or specific; in most cases infected plants appear stunted or yellowed and these symptoms are readily confused with nutrient deficiencies. Many growers therefore remain unaware of the damage caused by PPN.

Plant-parasitic nematodes are primarily pathogens of roots although there are notable exceptions. For example, the pine wilt nematode, *Bursaphelenchus xylophilus*, is a migratory endoparasitic pathogen of pine trees. This nematode is vectored to the host by Monochaimid beetles and infects twigs through feeding wounds made by the beetles. The nematode subsequently feeds on parenchymal cells, reproduces and migrates through the vascular tissues disrupting vascular function eventually causing death of the host (reviewed

http://dx.doi.org/10.19103/AS.2020.0075.14

by Jones et al., 2008). The rice tip nematode, *Aphelenchoides besseyi*, is an economically important pathogen of rice that is seed transmitted and that feeds on the growing points of the seedling stems and leaves (Duncan and Moens, 2013). However, the overwhelming majority of economically important PPN are soil-borne and cause damage to the roots. The results of nematode infection of the roots vary hugely depending on the nematode. In many cases roots may simply show reduced growth or may be damaged due to secondary infections of bacteria or fungi that are facilitated by the physical damage caused by PPN as they enter or migrate through the roots (e.g. Manzanilla-Lopez and Starr, 2009). However, in some cases, PPN infection can cause large-scale remodelling of root anatomy and architecture (Fig. 1).

2 Modes of parasitism

The ability to infect plants has arisen independently on at least four occasions, with PPN present in four of the twelve clades within the Nematoda identified by van Megen et al. (2009). PPN infect and exploit plant roots using a variety of different strategies, ranging from simple, ectoparasitic feeding to complex biotrophic interactions.

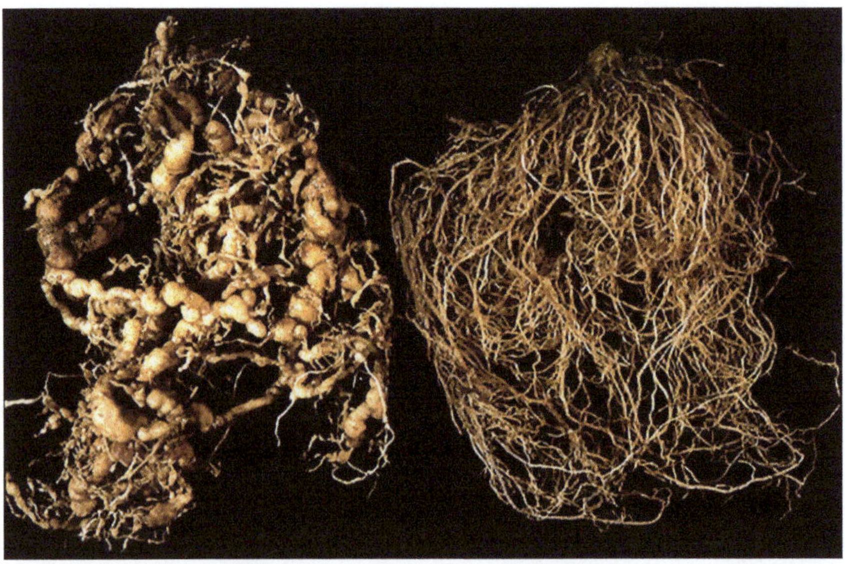

Figure 1 Structural changes in roots due to infection with root-knot nematodes *Meloidogyne* spp. Infected roots (left) show galling typically associated with root-knot nematode infection. Image courtesy of Nemapix (Eisenback and Zunke, 2000).

2.1 Migratory ectoparasites

The simplest interactions between PPN and plants are those seen between the migratory endoparasites and their hosts. In some cases, such as *Trichodorus* or *Paratrichodorus*, the interaction may be limited to the nematode locating the root, inserting the feeding apparatus and removing cell contents. However, even in these very basic interactions the nematode may produce a 'feeding tube' - a sheath formed from secretions produced by a specialized region of the pharynx. The feeding tube bridges the cell wall and may serve to protect the delicate feeding apparatus of the nematode (onchiostyle) while it removes cell contents (Eves-van den Akker et al., 2015). However, some ectoparasitic nematodes including *Xiphinema* and *Longidorus* species have more complex interactions with their hosts. For example, *Xiphinema index* feeds using its long hollow odontostyle. At sites other than the root tip, the nematode feeds from a column of cells in the root (Wyss, 1978) and induces necrosis in damaged tissues. However, when feeding at the root tip, large multinucleate, metabolically active cells, similar in appearance to the giant cells induced by root-knot nematodes (below), are formed (Wyss et al., 1980). Feeding behaviour and gall formation by *Longidorus elongatus* are similar, although this nematode feeds exclusively on the root tips of its hosts. However, this nematode does not induce multinucleate cells because both mitosis and cytokinesis occur when these feeding structures are induced (Robertson et al., 1984). Changes in nuclei and DNA levels within galls induced by *L. elongatus* have been recorded (Griffiths and Robertson, 1984). Migratory ectoparasitic nematodes feed from several different host plants during their life cycle and this has allowed them to become vectors of a range of economically damaging plant viruses.

2.2 Migratory endoparasites

Plants are also parasitized by a range of migratory endoparasitic nematodes (reviewed by Duncan and Moens, 2013). These nematodes feed and reproduce while migrating through plant tissues. Migratory endoparasites can be extremely damaging; other than the root-knot and cyst nematodes, *Pratylenchus* spp cause the greatest damage to crops on a worldwide basis (Castillo and Volvas, 2007). Migration can be extremely destructive where it occurs intracellularly, although in some cases nematodes may migrate between cells. In both cases wounding to plant tissues may well allow secondary infections of bacteria or fungi. Heavily infected root systems can become reduced in size or may show additional symptoms such as root swelling, stunted rootlets or 'witches broom'. Nematodes can reach huge numbers in highly susceptible plants, for example, *P. coffeae* on citrus, where population densities of 10 000 nematodes per gram of root have been reported (Duncan and Moens, 2013).

2.3 Sedentary endoparasites

Sedentary endoparasitic nematodes have complex, biotrophic interactions with their hosts. These nematodes enter their hosts, either fully or partially, and induce the formation of a long-lasting feeding structure on which they depend for all the nutrients required for development to the adult stage. The most economically important PPN, the cyst nematodes and root-knot nematodes, have prolonged biotrophic interactions with their hosts and maintain their feeding structures for up to six weeks, a period of biotrophy comparable to that of the classical biotrophic fungal pathogens. The feeding structures induced by PPN (nematode feeding sites - NFS) are extremely diverse in form and ontogeny and can be induced in a variety of root tissues. However, NFS have some conserved features. All provide a rich food source for the nematodes and show many of the characteristics of metabolically active plant tissues such as a cytoplasm enriched in subcellular organelles. Many NFS contain enlarged or multiple nuclei and may show signs of DNA replication. The range of NFS that can be induced by PPN has been reviewed by Wyss (2002). The simplest feeding structures are single cells that appear to be highly metabolically

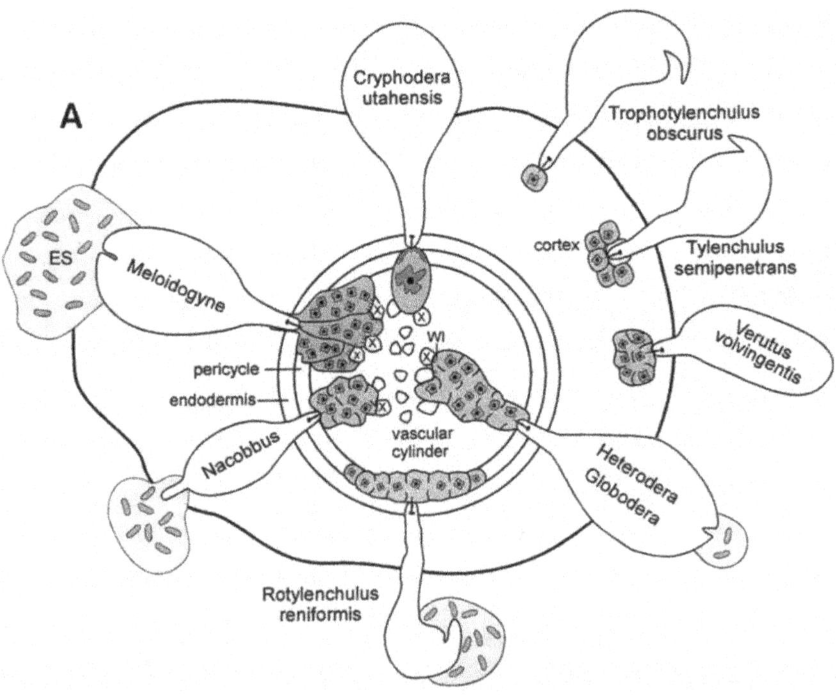

Figure 2 Schematic diagram of the variety of feeding structures induced by PPN. X - xylem, WI - cell wall ingrowths, ES - egg sac. Reproduced from Wyss (2002).

active such as those induced by *Trophotylenchus* spp. In some nematodes, such as *Cryphodera* spp., these single cells may become enlarged while other nematodes (e.g. *Tylenchus semipentrans*) induce clusters of such cells (Fig. 2). The most complex feeding structures fall into two broad categories: syncytia and giant cells. Both are large multinucleate cells, but have an entirely different ontogeny, and the ability to induce syncytia and giant cells seems to have evolved independently in the clade 12 PPN (Van Megen et al., 2009). Syncytia are formed by controlled breakdown of the cell walls between adjacent cells followed by fusion of the protoplasts to give rise to the mutinucleate condition (reviewed by Sobczak and Golinowski, 2011). Giant cells are formed as a result of repeated rounds of nuclear division in the absence of cytokinesis (reviewed by Jones and Goto, 2011). The detailed biology of these structures is discussed below.

3 Plant-nematode interactions

3.1 Invasion and migration

In all sedentary biotrophic species of PPN the second stage juvenile (J2) is the invasive stage. The moult from J1 to J2 occurs in the egg and the J2 hatches from the egg as the starting point of the parasitic phase of the life cycle. In host-specific PPN the J2 is a dormant stage and the hatch is induced by exposure of the eggs to host root exudates, providing the nematode with a means of coordinating the life cycle with the host. However, this requirement for the presence of host root diffusates to initiate hatch is not present in PPN that have a broad host range, such as root-knot nematodes, and that do not need to link their life cycle with the presence of a specific host plant (reviewed by Masler and Perry, 2018).

After finding the host roots, the migration behaviour of cyst nematodes and root-knot nematodes is markedly different. J2s of cyst nematodes invade the root by perforating the walls of multiple, adjacent epidermal cells using their stylet. As this process is repeated, these perforations merge, forming a slit through which the juvenile nematodes enter the root (Wyss and Grundler, 1992; Marhavy et al., 2019; Wyss and Zunke, 1986; Doncaster and Seymour, 1973). Entering the root through this process is physically demanding and requires the nematodes to use their entire body to exert sufficient pressure at the point of entry (Doncaster and Seymour, 1973). Once inside the root, the nematodes migrate towards the vascular tissue, moving intracellularly through different tissue layers by cutting slits and entering individual cells in a sequential manner, thereby causing extensive damage to the host tissue (Doncaster and Seymour, 1973; Marhavy et al., 2019) (Fig. 3). A cyst nematode takes between 1 and 15 min to breach a cell wall, with the speed of this process depending on factors such as cell wall rigidity and nematode vigour (Wyss and Zunke, 1986).

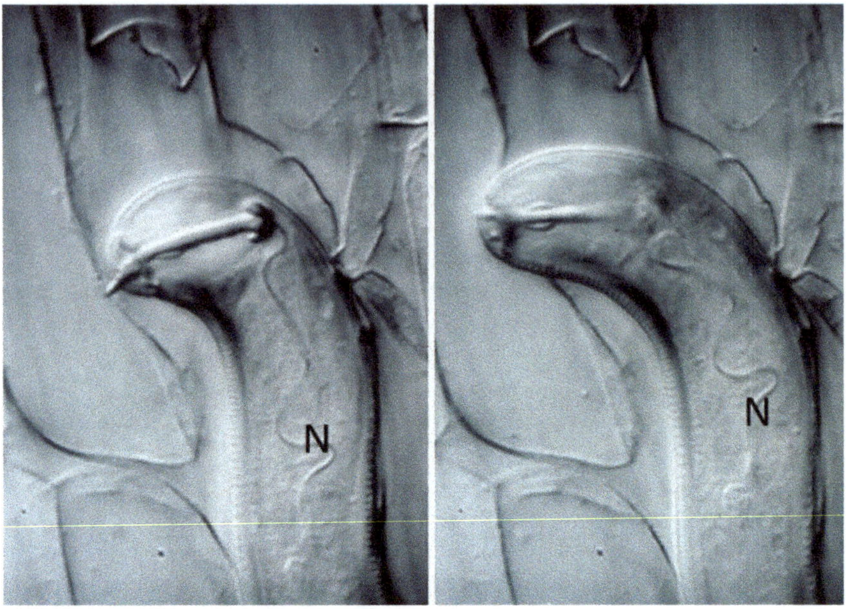

Figure 3 Migration through root cells by a cyst a nematode. A slit is made in cell wall (left) through which nematode enters (right) the neighbouring cell. N – nematode. Image courtesy of Nemapix (Wyss and Zunke, 1986).

By contrast, root-knot nematodes migrate through host roots intercellularly, thus avoiding the destruction of root cells seen during invasion and migration by cyst nematodes, and possibly avoiding eliciting some defence responses. In both cyst and root-knot nematodes, the process of migration is facilitated by the release of a cocktail of plant cell wall-degrading enzymes, which are thought to soften the plant cell wall. These are produced in the nematode's subventral oesophageal gland cells and secreted into the plant through the stylet. The range of cell wall-degrading enzymes produced by PPN includes cellulase, pectate lyases, polygalacturonases and arabinases (Smant et al., 1998; Popeijus et al., 2000; Danchin et al., 2010), as well as cell wall-modifying enzymes such as expansins (Qin et al., 2004). Remarkably, there is strong evidence to support the hypothesis that cell wall-degrading and cell wall-modifying enzymes present in PPN have been acquired by horizontal gene transfer from bacteria and fungi (see Section 3.1.1).

3.1.1 Case study: Horizontal gene transfer in plant-parasitic nematodes

Until 1998 it was assumed that animals were not able to produce enzymes that digest the plant cell wall. Where there was a requirement to break down the plant cell wall, for example, during digestion, a symbiotic relationship was present with bacteria which produced the enzymes required. For many years it had been recognized that the secretions of the oesophageal gland cells of sedentary endoparasitic nematodes were likely to play a critical role in the interaction between plant and nematode; electron microscopy studies had shown that the gland cells were packed with secretory granules and video microscopy suggested that materials produced in the gland cells were secreted into the plant (Hussey and Mimms, 1990; Wyss and Grundler, 1992). The first protein identified as being secreted from the subventral oesophageal gland cells had similarities to bacterial cellulases from glycosyl hydrolase family 5 (Smant et al., 1998). However, the gene encoding the protein had many features that showed it was an endogenous nematode gene. Subsequent work led to the identification of genes encoding many other plant cell wall-degrading enzymes, culminating in the analysis of the first genome sequences for PPN. The root-knot nematode *Meloidogyne incognita* was found to have 21 genes encoding cellulases, 6 xylanases, 2 polygalacturonases, 30 pectate lyases, 2 arabinases, 2 invertases as well as 20 putative expansins (Abad et al., 2008) while the potato cyst nematode *Globodera pallida* has 15 cellulases, 1 polygalacturonase, 2 arabinases, 7 pectate lyases and 9 expansins (Cotton et al., 2014). The presence of endogenous nematode genes showing such similarity to bacterial sequences led to the suggestion that these genes have been acquired by multiple horizontal gene transfers from bacteria, a hypothesis supported by detailed phylogenetic analysis of the sequences (Danchin et al., 2010).

Phylogenetic analysis of the phylum Nematoda suggests that the ability to parasitize plants has arisen independently on at least four occasions (Van Megen et al., 2009). Remarkably, analysis to date suggests that acquisition of cell wall-degrading enzymes through horizontal gene transfer has occurred independently in each of the phylogenetic groups of nematodes and that the genes have been acquired from different sources on each occasion. For example, while the cellulases present in clade 12 PPN, including the root-knot and cyst nematodes, are most similar to GHF5 cellulases from bacteria, the cellulases present in clade 10, *Bursaphelenchus* species, are most similar to GHF45 cellulases from fungi (Kikuchi et al., 2004) and the cellulases present in the clade 2 species, *Xiphinema index*, are similar to clade 12 cellulases from bacteria (Danchin et al., 2017). Horizontal gene transfer has most likely played a key role in the evolution of plant parasitism on multiple occasions within the Nematoda (Jones et al., 2005).

3.2 Cyst nematodes: morphological changes and syncytium development

Upon reaching the vascular tissue, cyst nematodes stop moving and their behaviours change. The nematode sequentially probes cells within its reach using gentle thrusts of the stylet. This process may continue for several hours until a suitable initial syncytial cell (ISC) is found (Marhavy et al., 2019). The factors involved in ISC selection are currently unknown, although it may be that the nematode rejects any potential ISC in which it detects elicitation of a defence response. Upon selecting an ISC, cyst nematodes become sedentary and enter the feeding-preparation phase, which lasts for 6–8 hours. During this time, the stylet remains inserted into the wall of the ISC, but the nematode does not release any secretions into the cell (Wyss, 1992; Wyss and Zunke, 1986). Upon completing the feeding-preparation phase, the stylet is withdrawn and reinserted at the same spot, a process facilitated by secretions from the nematode's oesophageal gland cells that are released into the cytoplasm of the ISC. Several hours after ISC selection, the stylet tip is covered with a callose-like material except for at its opening. The stylet is used to make a small puncture in the ISC plasmalemma through which nematode secretions are released and, following this, nematodes start feeding from the ISC (Sobczak, Golinowski and Grundler, 1999).

The ISC and its nucleus enlarge slightly within a few hours of the feeding-preparation phase. Approximately 24 hours after ISC selection, adjacent cells are fused with the ISC due to local dissolution of cell walls and the widening of pre-existing plasmodesmata. This controlled cell-wall degradation is mediated by the plants' own endogenous cell wall-degrading enzymes, which are specifically upregulated in the developing syncytium (Fudali et al., 2008; Wieczorek et al., 2006, 2008, 2014). Cells incorporated into the syncytium are hypertrophied and exhibit features typical of a syncytial cell within 3 days (Fig. 4). Over the next two weeks, the syncytium expands and the nematode undergoes three additional moults (to J3, J4 and adult) and develops into an adult. The males of cyst nematodes become mobile and leave the root in order to locate sedentary females for sexual reproduction. After mating, the body of the mature adult female is filled with eggs and forms a cyst structure that encloses the eggs after the female dies. There is evidence that sex is determined environmentally in cyst nematodes, with nematodes that induce a large feeding site and that access plentiful food supplies becoming females and nematodes that induce smaller syncytia that do not access the vascular tissues becoming males (Anjam et al., 2018; Sobczak and Golinowski, 2011; Molz, 1920; Ellenby, 1954; Grundler, Betka and Wyss, 1991; Den Ouden, 1960; Trudgill, 1967). Cyst nematodes withdraw food from their feeding site in cycles, with each cycle consisting of three distinct stages and lasting for a few hours. During stage I, the nematode's stylet is inserted into the syncytium to withdraw food. In stage II, the stylet is withdrawn and then inserted back at the same spot. Finally, in stage III, the nematode injects secretions into the syncytium (Steinbach, 1972). Some of these secretions form a feeding tube (Wyss and Zunke, 1986). The precise role of the feeding tube is uncertain but it may act

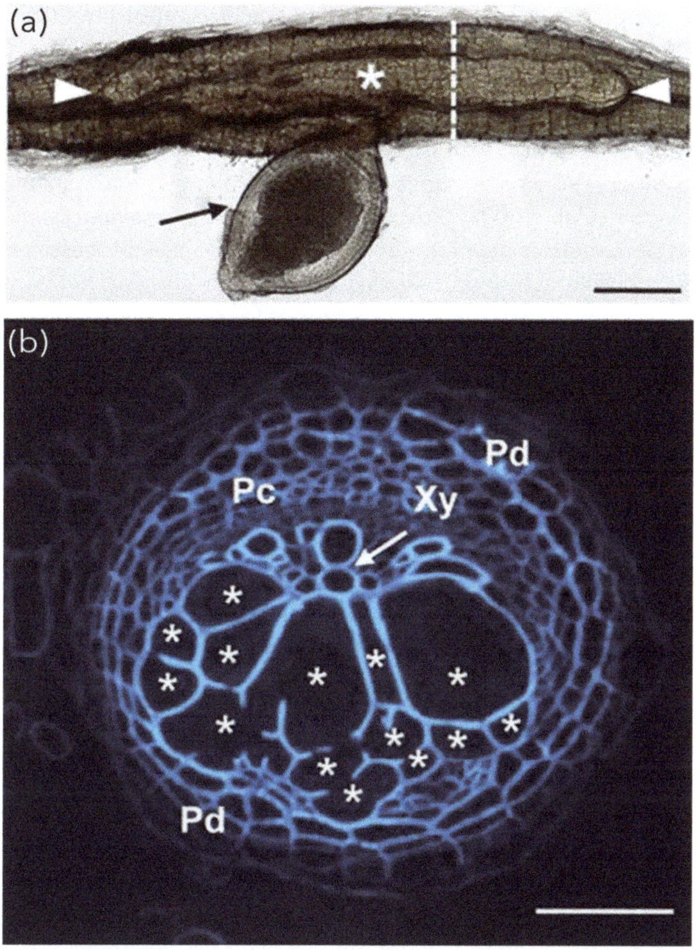

Figure 4 Structure of syncytia formed by a cyst nematode in *Arabidopsis* roots. (a) Syncytium induced by a female cyst nematode at day 14 post infection. Dotted line indicates the location of a typical transverse section as shown in Fig. 4b; arrowheads indicate the boundaries of the feeding site; asterisk indicates syncytium; arrow indicates adult female cyst nematode. (b) Calcofluor-white staining of a transverse section through a nematode-infected root section. Asterisks indicate syncytium; arrow indicates location of xylem vessels; Pc – procambium, Pd – periderm, Xy – xylem. Scale bars: (a) 20 μm; (b) 50 μm (reproduced from Davies et al., 2012).

as a molecular sieve, helping to ensure that the syncytium is not destroyed as a result of the nematode feeding (Eves-van den Akker et al., 2015).

Development of the syncytium is accompanied by a massive reprogramming of infected root cells. The cytoplasm is dense and rich in organelles such as ribosomes, endoplasmic reticulum (ER), mitochondria and plastids. Several smaller vacuoles are formed de novo via widening of ER cisternae in place of the usual large central vacuole seen in unparasitized root cells (Baranowski, et al., 2019). Nuclei are enlarged due to repeated cycles of endoreduplication (replication of nuclear genome without mitosis), and although mitotic activity is absent within the syncytium (Endo and Veech, 1970; Endo, 1964; de Almeida Engler et al., 1999), it is reactivated in cells surrounding it, leading to their division and subsequent incorporation into the syncytium (Niebel et al., 1996; Goverse et al., 2000a; Siddique et al., 2015a). Cytokinin is involved in regulation of numerous developmental and physiological processes in plants, including cell cycle activation and sink tissue establishment. In line with this, cytokinin has been shown to provide a trigger for reactivation of the cell cycle during syncytium development. Remarkably, nematodes also produce their own cytokinin, which is released into the host and significantly contributes to cell cycle reactivation and syncytium expansion (Siddique et al., 2015a).

Auxin (indole-3-acetic acid or IAA) is a key plant regulator involved in almost every aspect of plant growth and development including cell cycle activation, cell wall modifications and hypertrophy. Therefore, it is not surprising that initiation and expansion of the syncytium is associated with local accumulation of auxin (Karczmarek et al., 2004). Interestingly, the auxin response is more pronounced in the syncytium during early stages of infection. However, it becomes less specific to the syncytium and is more pronounced at the periphery of syncytium during later stages of infection (Grunewald et al., 2009; Goverse et al., 2000b). There is clear evidence showing that accumulation of auxin is required for normal syncytium development. For example, an auxin-insensitive tomato mutant (*dgt*) is almost completely resistant to cyst nematodes (Goverse et al., 2000b). Similarly, the *Arabidopsis axr2/iaa7* mutant, which is impaired in auxin signalling, supports a significantly reduced development of cyst nematodes (Goverse et al., 2000b). Most auxin in plants is produced in the shoot apical meristem, young leaves and flower buds, from where it is transported throughout the plant via the phloem. In addition to phloem transport, auxin is also distributed basipetally from cell to cell via polar auxin transport involving influx carriers AUXIN RESISTANT 1 (AUX1) and LIKE AUX1 (LAX) as well as efflux transporters (PIN FORMED (PINs)). Cyst nematodes manipulate polar auxin transport by secreting an effector (19C07) that targets an auxin influx carrier (LAX3), presumably increasing its activity and thus enhancing the influx of auxin into the syncytium (Lee et al., 2011). Moreover, expression of the efflux transporter PIN1 is downregulated in the syncytium, preventing auxin from being pumped out. Consequently, auxin accumulates in the syncytium. Remarkably, the polarity of another PIN protein, PIN3, changes from basal to lateral plasma membranes within the syncytium. Such a shift in polarity facilitates integration of neighbouring cells and radial expansion of the syncytium (Grunewald et al., 2009). This result also explains the pronounced auxin response towards the periphery of syncytium. In addition to the manipulation of auxin transport, accumulation of auxin could also be due to secretion by nematodes or local de novo biosynthesis. Indeed, auxin, mainly in conjugated form, is detected from secretions of cyst and root-knot nematodes (De Meutter et al., 2005). However, the physiological relevance of nematode-derived auxin in formation and functioning of the syncytium remains obscure.

3.3 Root-knot nematodes: giant cells

Most biotrophic plant pathogens have a relatively limited host range; however, root-knot nematodes (*Meloidogyne* spp.) are unusual in that their potential host range includes most species of flowering plants (Trudgill and Blok, 2001). The infective J2s of root-knot nematodes enter the host root in the elongation zone and move between cells until they reach the vascular tissue where they induce the formation of 5–7 giant cells (Holbein et al., 2019). Upon establishing these giant cells, the nematodes become immobile, begin feeding and moult three times before reaching the adult stage. Giant cells are metabolically hyperactive, providing nutrients for the rapidly developing nematodes. The formation of giant cells is accompanied by hyperplasia and hypertrophy of the cells surrounding the developing juvenile (Niebel et al., 1996; Goverse et al., 2000a; Siddique et al., 2015a), giving rise to the typical gall-like structure, which is the characteristic of root-knot nematode infection (Fig. 5).

Giant cells are large and may grow up to 100 times the size of normal root vascular parenchyma cells (Krusberg and Nielsen, 1958). Their cytoplasm, like that of syncytia, is condensed and rich in organelles, including ribosomes, mitochondria, Golgi apparatus, plastids and smooth ER. Instead of the large central vacuole present in uninfected root cells, giant cells contain several smaller vacuoles. Numerous ingrowths are formed at the cell wall interface with xylem cells, which are thought to increase the surface area for absorption of water and nutrients. Although the giant cells induced by root-knot nematodes share some structural and functional similarities with syncytia, they are

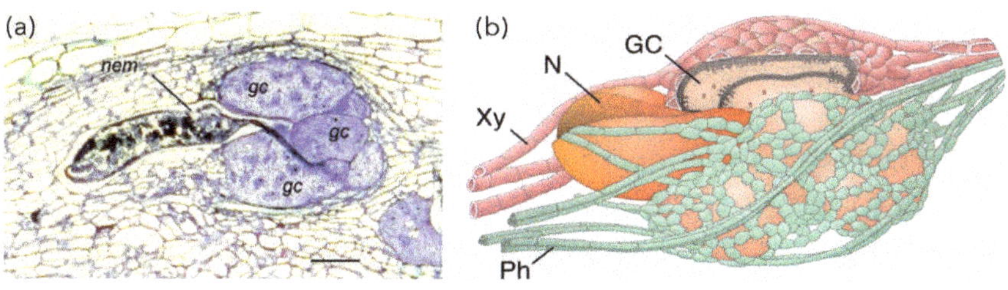

Figure 5 (a) Longitudinal section of a pepper root showing three giant cells (GC) induced by the nematode (nem) 14 days after inoculation. Scale bar = 50 μm (reproduced from Castagnone-Sereno and Danchin, 2014). (b) Schematic representation of the vascular tissue surrounding GC at the nutrient acquisition stage. Xy - xylem, Ph - phloem, N - nematode, and GC - giant cells (reproduced from Bartlem, Jones and Hammes, 2014).

formed by entirely different means. Giant cells undergo numerous cycles of acytokinetic mitosis, that is, nuclear division without cell division (Niebel et al., 1996; Goverse et al., 2000a; Siddique et al., 2015a, de Almeida Engler et al., 1999) and contain as many as 60 nuclei. These nuclei are enlarged and often grouped, presenting peculiar shapes and a variety of different sizes (Vieira, Engler and de Almeida Engler, 2012). Furthermore, the nuclei of giant cells are polyploid, most likely due to repeated cycles of endoreduplication (Wiggers, Starr and Price, 1990; Starr, 1993). As with syncytia, the involvement of cytokinin in cell cycle reactivation has been demonstrated. However, differences have also been observed in regulation of cytokinin signalling as well as biosynthesis and catabolism between giant cell and syncytium, suggesting that cyst and root-knot nematodes manipulate the cytokinin pathways differently (Dowd et al., 2017). For example, a cytokinin-biosynthesis gene has been identified in cyst nematodes; however, identification of such a gene in root-knot nematode remains elusive (Siddique and Grundler, 2018).

Auxin plays a key role in giant cell formation and functioning. Similar to syncytia, a strong activation of auxin response is found within young giant cells (Karczmarek et al., 2004). However, this response is transient and shifts to neighbouring cells during later stages (Absmanner, Stadler and Hammes, 2013). Root-knot nematodes also affect the polar transport of auxin to initiate and maintain giant cells. In young galls, expression of auxin influx carriers, AUX1 and LAX3, is induced, whereas expression of an efflux carrier, *PIN3*, is detected in the neighbouring cells but not within giant cells. At later stages, expression of *AUX1* and *LAX3* is absent in galls, but *PIN3* expression is present in all giant and neighbouring cells. Interestingly, expression of *PIN4* is enhanced throughout giant cell development. These observations suggest that auxin is accumulated in young giant cells due to a coordinated expression of *AUX1*, *LAX3* and *PIN3* (Kyndt et al., 2016). This auxin accumulation is vital for initiation and functioning of giant cells. For example, an auxin-insensitive tomato mutant (*dgt*) is unable to support root-knot nematode development due to an arrest in giant cell formation (Richardson and Price, 1984). Similarly, various loss-of-function auxin transport mutants show a significant reduction in the number and sizes of galls (Kyndt et al., 2016). As in cyst nematodes, auxin has been detected in the secretions of root-knot nematodes, although its relevance to giant cell formation remains unknown.

3.4 Changes in the cytoskeleton in syncytia and giant cells

The cytoskeleton is a network of interconnected protein polymers that helps cells maintain their shape and provides mechanical support to the cytoplasmic constituents. Changes in cell shape are therefore highly dependent on cytoskeleton reorganization. The cytoskeleton in plants consists of two main components: microtubules and microfilaments. Microtubules are made up of α- and β-tubulin dimers that are polymerized into long, hollow cylinders. Most microtubules are arranged in a highly ordered array over the cell cortex beneath the plasma membrane (cortical microtubule) and regulate the targeting of the cellulose synthase complex to the plasmalemma. During preprophase, a transient band consisting of dense cortical microtubule is formed, defining the plane of cell division (Van Damme, Vanstraelen and Geelen, 2007). Upon entry into mitosis, all cortical microtubules disappear and instead two successive microtubule arrays are formed: the mitotic spindle and phragmoplast (Goddard et al., 1994; de Almeida Engler and Favery, 2011). The microfilaments are long filamentous chains (F-actin) of globular actin molecules (G-actin) that are intertwined, giving the appearance of a double helix. In the interphase of plant cells, microfilaments are organized into three arrays: subcortical microtubule bundles, short cortical non-bundled microtubules and a perinuclear microfilament. During mitosis, microfilaments are observed in the preprophase band and phragmoplast and can contribute to the assembly of microtubules in these structures (de Almeida Engler and Favery, 2011). Both microtubule and microfilament cytoskeletons are associated with many different proteins that contribute to their organization and dynamics. For example, profilins are actin-depolymerizing factors that bind to G-actin and play an important role in the turnover and restructuring of microfilaments.

Both microfilaments and microtubules are continuously rearranged (polymerized and depolymerized) in response to various local and external stimuli such as pathogen attack. As syncytium and giant cell formation and expansion involve massive changes in cell structure, the role of the cytoskeleton in both types of feeding cells has been examined. Expression of actin genes has been shown to be upregulated in syncytia and galls, particularly during early stages of infection. Interestingly, vascular tissues surrounding feeding sites have also shown an increased expression of actin genes, suggesting that the cytoskeleton is also reorganized in neighbouring cells (de Almeida Engler et al., 2004). In comparison to actin, high expression of α- and β-tubulin has only been detected in giant and neighbouring cells, which agrees with the observation that giant cells undergo mitoses. The expression of tubulin genes is lower in syncytia, which lack mitotic activity. However, a higher tubulin expression has been observed in cells surrounding the syncytium, which undergo division prior to their incorporation into the syncytium (de Almeida Engler et al., 2004). Higher expression of both actin and tubulin genes in and around the feeding site is thought to allow the assembly of a new cytoskeleton in a rapidly expanding feeding site. Surprisingly, however, both microfilaments and microtubules appear disorganized and disrupted throughout NFS development. Nonetheless, the difference in the extent of depolymerization between giant cells and syncytium is notable. In the case of syncytia, no functional mitotic microtubule arrays (pre-prophase bands, spindles and phragmoplast) are observed, and both actin and tubulin networks seem completely disrupted. In contrast, microfilaments and microtubules, although disrupted, are still visible in giant cells. Furthermore, a functional mitotic apparatus consisting of multiple spindles and phragmoplasts has been observed throughout the giant cell development. However, no preprophase band of cortical microtubule is observed in giant cells. Importantly, three microtubule arrays (PPBs, spindles and phragmoplasts) are present in cells surrounding both types of NFS. Chemical stabilization of the microtubular cytoskeleton completely blocks the initiation of syncytia and giant cells. In contrast, nematodes are able to complete their life cycles in the presence of drugs that causes actin or tubulin cytoskeleton depolymerization (de Almeida Engler et al., 2004; de Almeida Engler and Favery, 2011). Taken together, formation of the nematode feeding site (NFS) appears to induce cytoskeleton rearrangement and depolymerization, which is essential for successful development of nematodes. A root-knot nematode effector, profilin (*MiPFN3*), has recently been shown to cause actin depolymerization in an in vitro assay, reducing the filamentous actin network when expressed in protoplasts (Leelarasamee, Zhang and Gleason, 2018).

3.5 Metabolism and nutrient delivery in syncytia and giant cells

As obligate biotrophs, cyst and root-knot nematodes are entirely dependent on plant-derived nutrients and solutes to fulfil their energy requirements throughout their life cycle. Thus, both types of feeding sites are sink tissues that cater to the needs of the rapidly developing nematode. A series of detailed transcriptomic analyses performed over the last decade have shown that the expression of a large number of plant genes is either upregulated or downregulated in NFS as compared to uninfected roots. For example, out of a total of 21 138 genes in *Arabidopsis*, 7231 genes (34.2%) are differentially expressed in syncytia, as compared with control roots (Szakasits et al., 2009). A gene ontology (GO) analysis of differentially expressed genes showed that categories related to high metabolic activity are preferentially upregulated, whereas those related to defense responses are downregulated (Szakasits et al., 2009; Hofmann et al., 2010; Hutten et al., 2015; Portillo et al., 2013; Barcala et al., 2010; Escobar, Brown and Mitchum, 2011; Ithal et al., 2007a,b; Jammes et al., 2005). Interestingly, the transcriptome of NFSs is more closely related to that of seeds or pollen than to uninfected roots, probably because seeds and pollen also exhibit high metabolic activity (Szakasits et al., 2009). Although details of the mechanisms that lead to such large-scale transcriptome changes are not clearly understood, recent studies point to the importance of host microRNA (*miRNA*) pathways in modulating gene expression in NFS. For example, *miR858* has been shown to facilitate cyst nematode parasitism by modulating the expression of a transcription factor *MYB83* in syncytia. Transcriptome analysis of lines overexpressing MYB83 showed that approximately 17% of the syncytial transcriptome is modulated by *MYB83*, suggesting a role for *mir858-MYB83* regulatory module in transcriptional reprogramming of syncytia (Piya et al., 2017). Metabolite profiling of syncytia and giant cells revealed a pronounced increase in the concentrations of various sugars, organic acids and amino acids suggesting a functional similarity between these two types of feeding sites despite their different ontogeny (Hofmann et al., 2010; Baldacci-Cresp et al., 2012; Siddique and Grundler, 2015).

Syncytia and giant cells, as metabolic sinks, need to have a strong connection to the phloem to ensure a continuous supply of nutrients. Indeed, de novo formation of phloem occurs around both syncytia and giant cells. Phloem tissue consists of two different cell types: sieve elements (SEs) and companion cells (CCs). The SEs are major conducting cells connected through pores in their cell wall. The CCs are nucleated having a large number of mitochondria. Since SEs lack a nucleus or any other organelles, they rely on CCs for their functions. Both SEs and CCs are derived from the same parent cell by mitosis and remain connected with each other through an extensive network of plasmodesmata. A large number of studies conducted over the past two decades have shown that syncytia are symplasmically isolated during the initial stage of development, which may facilitate cellular and ultrastructural reorganization. Solute transport during this time takes place through enhanced activity of membrane-localized transporters. Nevertheless, secondary plasmodesmata between the syncytium and sieve elements become active as the nematode and syncytium grow, thus fulfilling the high demand for assimilates (Hofmann et al., 2007, 2009; Hoth et al., 2008; Absmanner et al., 2013). By contrast, giant cells are symplasmically isolated and rely on membrane-localized transporters to obtain nutrients from phloem (Absmanner et al., 2013; Hoth et al., 2008).

4 Activation and suppression of plant defence responses

The function and evolution of the plant immune system has been summarized in the zigzag model (Jones and Dangl, 2006). Plants recognize the presence of potential pathogens through the detection by pattern recognition receptors of highly conserved pathogen-associated molecular patterns (PAMPs) or breakdown products of the plant cell wall produced as a result of pathogen degradation of this structure (damage-associated molecular patterns (DAMPs)). This activates a first layer of plant defence known as pattern-triggered immunity (PTI). Adapted pathogens secrete effectors into host tissues that suppress PTI and allow infection. However, in a second layer of plant immunity (effector-triggered immunity (ETI)) the presence of these effectors is recognized by the products of R genes. Many R genes encode nucleotide-binding site leucine-rich repeat (NB-LRR) proteins. Although the zigzag model was originally developed to describe the immune system in relation to bacterial, oomycete and fungal pathogens, it is clear that the same principles apply to plant-nematode interactions. Numerous effectors have been identified from PPN that suppress defence responses, and several NB-LRR type R proteins have been cloned that provide resistance against PPN (reviewed by Williamson and Kumar, 2006), as well as one avirulence effector recognized by the *Gpa2* R gene (Sacco et al., 2009). Defence responses in roots can clearly be underpinned by the same molecular processes that are seen in aerial parts of the plant. However, the requirement to maintain a feeding structure has provided plants with the opportunity to evolve other 'resistant' responses that do not operate through NB-LRR proteins but prevent normal feeding site development.

4.1 Pattern triggered immunity (PTI)

Plant PRRs are primarily membrane-localized receptor-like kinases (RKs) or receptor-like proteins (RPs) that are categorized based on the structural features of their extracellular domains (ECDs), which act as a ligand-binding site for diverse extracellular signals. Major classes of RKs are leucine-rich repeat (LRR)-RKs, lysin-motif (LysM)-related proteins, pathogenesis-related (PR5)-RLs (PR5-RKs), and lectin-RKs (LecRKs). Upon detection of PAMPs or DAMPs, PRRs initiate a broad range of downstream responses that restrict pathogen growth and development, including the activation of calcium- and mitogen-activated protein kinases, bursts of calcium and reactive oxygen species (ROS) release, the reprogramming of host defence-related transcripts, and cell wall reinforcement (Couto and Zipfel, 2016; Macho and Zipfel, 2014; Smakowska-Luzan et al., 2018; Zipfel and Oldroyd, 2017).

Nematode root invasion and migration within root tissue activate PTI responses. Historically, the study of such responses at a molecular level has been impeded by three major challenges: (i) synchronization of the early infection processes, (ii) isolation of infected plant material during the early stages of nematode infection and (iii) differentiation of general DAMP-associated host responses from specific responses associates with nematode recognition. However, several publications provide insights into the molecular mechanisms underlying the perception of nematodes by the host plant (Teixeira, Wei and Kaloshian, 2016; Mendy et al., 2017; Peng and Kaloshian, 2014; Manosalva et al., 2015; Teixeira et al., 2019). For example, PAMPs have been identified from PPN (Manosalva et al., 2015) as well as a PRR, which is involved in initiation of PTI against PPN (Mendy et al., 2017). Cyst nematodes migrate intracellularly and cause severe cell damage, which is likely to produce DAMPS and activate PTI responses (Shah et al., 2017; Lozano-Torres et al., 2019; Marhavy et al., 2019). In comparison to cyst nematodes, there is no clear evidence of DAMP-induced PTI by root-knot nematodes, which migrate intercellularly thus causing relatively little damage (Shah et al., 2017; Teixeira et al., 2016).

4.2 Suppression of host defences

All biotrophic plant pathogens need to suppress PTI in order to be able to feed from living host tissues. A wide range of PPN effectors, produced in the oesophageal gland cells and secreted into the host plant through the stylet, have been identified that suppress host defence responses. For example, a calreticulin was identified as a secreted effector from the root-knot nematode *M. incognita* (Jaouannet et al., 2013). This calreticulin is produced in the oesophageal gland cells and secreted into the root apoplast. The presence of the calreticulin in the apoplast, but not in the cytoplasm, suppresses several PTI responses normally observed in the presence of PAMPs including callose deposition and the upregulation of marker genes of defence responses. An effector consisting of a secreted ubiquitin extension protein, composed of a ubiquitin protein fused to a 12 amino acid C terminal extension domain, has been identified in a range of cyst nematodes that suppress PTI induced by exposure to the bacterial FLG22 PAMP (Chen, Chronis and Wang, 2013). Venom allergen proteins from *G. rostochiensis* also suppress plant immunity mediated by cell-surface receptors through targeting of the extracellular Rcr3 proteinase (Lozano-Torres et al., 2014 – below). An effector from *M. javanica* that is similar to transthyretins suppresses host defence responses through its interaction with ferredoxin:thioredoxinreductase catalytic subunit, a key component of host antioxidant system. This interaction results in increased host active oxygen scavenging activity, and thus in a reduction of host defence efficacy (Lin et al., 2015).

Effectors have also been identified from PPN that can suppress ETI. For example, a large family of secreted proteins that have a SPRY domain (SPRYSECs) have been identified from cyst nematodes (Mei et al., 2015) that suppress cell death responses induced by the activation of a variety of different R genes (Postma et al., 2012; Mei et al., 2018). The molecular mechanism by which this is achieved remains unknown. Further details on the role of ETI in plant-nematode interactions is provided in the next section.

5 Resistance

5.1 Control of PPN

The range of tools available to growers for control of PPN is limited. Chemical control is an effective method for managing nematode damage, but the number of active ingredients that are permitted to be used is decreasing due to changes in legislation. Biological control is still only suitable for use in a small number of specific situations and although genetic modification for control of PPN is, in principle, feasible – as has been demonstrated for a range of root-knot nematodes (e.g. Huang et al., 2006; Yang et al., 2013; Dinh et al., 2014) – there are still no commercial crops available on the market that have been engineered to provide resistance against PPN. Natural resistance is the most environmental-friendly and cost-effective method of controlling PPN, or indeed any other pathogen. However, natural resistance needs to be present in crop cultivars that have other commercially required characteristics if it is to be useful to growers. In addition, careful stewardship is required in order to prevent virulent populations of pathogens from emerging in response to selection from resistance. This selection can be seen in PPN even though most species undergo a relatively small number of reproductive cycles in a growing season compared to other plant pathogens. Repeated use of potato cultivars in the UK containing the H1 R gene against *Globodera rostochiensis* has led to the emergence of a second species, *Globodera pallida*, against which H1 is ineffective (Minnis et al., 2002).

5.2 Resistance against root-knot nematodes

Natural resistance against root-knot nematodes has been identified and major resistance loci have been mapped in various crop plants including tomato (*Mi*-1, Kaloshian et al., 1998), (*Mi*-3, Yaghoobi, Yates and Williamson, 2005), (*Mi*-9, Ammiraju et al., 2003), carrot (*Mj*-1, Boiteux et al., 2000), peach (*Mi* and *Mij*; Lu et al., 1999) (*Mia,* Yamamoto and Hayashi, 2002) and peanut (*Mae* and *Mag*, Garcia et al., 1996). However, *Mi-1* is the only root-knot nematode R gene to have been cloned. It encodes an NB-LRR protein with an N-terminal domain containing a coiled-coil (CC) motif (Milligan et al., 1998; Hwang and Williamson, 2003). In tomato plants containing the *Mi-1* gene, nematodes can invade the root and migrate towards the vascular tissue as usual. However, once inside the vascular cylinder, the nematodes do not induce a feeding site. Instead, necrosis of host tissues occurs near the head of the invading nematodes (Dropkin, 1969; Paulson and Webster, 1972), accompanied by an accumulation of ROS, leading to nematode death (Melillo et al., 2006; Zhou et al., 2018). Subcellular localization has revealed that ROS accumulation occurs primarily in the exterior of the infected root cells, including the plasma membranes, cell walls and intercellular spaces. A heat shock-factor gene promotes the production of ROS by binding to the promoter region of membrane-localized respiratory burst oxidase homolog (Rboh) gene (Zhou et al., 2018).

Mi-1 confers resistance to three commonly occurring and damaging species of root-knot nematodes (*Meloidogyne arenaria*, *M. incognita* and *M. javanica*) in tomato. However, it does not provide resistance to *M. hapla*, *M. enterolobii* or *M. floridensis*, which can be present in geographically overlapping regions (Roberts and Thomason, 1989). In addition to nematodes, *Mi-1* also confers resistance to potato aphid (*Macrosiphum euphorbiae*) and whitefly (*Bemisia tabaci*) (Rossi et al., 1998; Nombela, Williamson and Muniz, 2003). Subsequent studies have identified common components (*RME1,* resistance to Meloidogyne; SGT1, suppressor of G2 allele of *skp1; HSP90,* heat shock protein 90), required for *Mi-1*-mediated resistance against these three different pests (De Ilarduya, Moore and Kaloshian, 2001; Bhattarai, Bishnoi and Kaloshian, 2007; Xie et al., 2004). Although avirulence effectors that are recognized by Mi-1 have not yet been identified, it is likely that Mi-1 operates by sensing perturbations in RME1. Upon detection of RME1 modifications, Mi-1 recruits SGT1 and HSP90 to activate defense responses (Bhattarai et al., 2007).

5.3 Resistance against cyst nematodes

Resistance against cyst nematodes manifests itself in different ways. Very few cyst nematode R genes seem to provide an extremely rapid response against the invading nematode or the initial syncytial cell. One exception is the resistance present in the *Sinapsis alba* cultivar Maxi against *H. schachtii.* In this interaction a necrotic response is seen in cells surrounding the nematode during migration and nematodes are frequently unable to induce a syncytium (Soliman et al., 2005). However, in most other cases the resistance response is delayed and the initial syncytial cell develops as would be expected in susceptible plants for a few days (Sobczak and Golinowski, 2011). After this time the syncytium itself may show signs of degradation or may fail to expand as expected and become surrounded by necrotic cells that prevent it from fully developing. In such plants the syncytia may allow development of adult males while preventing development of females, which have greater nutritional requirements. For example, in potato plants containing the H1 gene against *G. rostochiensis,* nematodes invade and migrate as normal and induce syncytia in cortical cells. However, the incorporation of procambial cells seen in susceptible plants fails and the syncytium does not interface effectively with the vascular tissues (Rice et al., 1985). Similarly, syncytia in plants carrying the *Hero* gene expand in roots but fail to incorporate procambial cells effectively and are separated from vascular tissues by necrotized cells (Sobczak et al., 2005). By contrast, the delayed resistance response against *G. pallida* in potato carrying the *Gpa5* resistance source derived from *Solanum vernei* is targeted directly at the syncytium. Syncytia in these plants develop normally for up to seven days but then appear vacuolated and show signs of cytoplasmic degeneration (Rice et al., 1987).

To date three NB-LRR genes that confer resistance against cyst nematodes have been identified at a molecular level. One, *Gro1-4*, encodes a NB-LRR protein with a Toll-interleukin receptor (TIR) domain at the N terminus (Paal et al., 2004) while two others (*Hero* - Ernst et al. (2002) and *Gpa2* - Van der Vossen et al. (2000)) encode CC-NB-LRR proteins, which have a CC domain at the N terminus. In addition, two proteins that encode extracellular receptors that provide resistance against cyst nematodes have been identified. The *Hs1*$^{pro\text{-}1}$ gene provides resistance against *H. schachtii* in sugarbeet (Cai et al., 1997). However, the details of how this resistance functions remain unclear as a subsequent mapping study showed that the resistance phenotype was not correlated with the presence of the *Hs1*$^{pro\text{-}1}$ gene (Sandal et al., 1997). The *Cf-2* gene confers resistance against *G. rostochiensis* and encodes an extracellular receptor-like protein with a leucine-rich repeat domain (Lozano-Torres et al., 2012). *Cf-2* also confers resistance against the fungal pathogen *Cladosporum fulvum* and is known to operate by sensing attempts by the pathogen to interfere with the apoplastic cysteine proteinase Rcr3, mediated in *C. fulvum* by the *Avr2* effector (Rooney et al., 2005) and in *G. rostochiensis* by the VAP-1 effector. The targeting of the Rcr3 proteinase by both fungal and nematode pathogens shows that pathogens independently target the same potential host proteins and that by guarding such targets, plants can expand the range of pathogens recognized by their immune receptors. It is likely that the broad resistance spectrum provided by *Mi* is underpinned by similar evolutionary pressures.

5.4 Resistance against PPN conferred by novel mechanisms

The sources of resistance described above are 'classical' NB-LRR type R genes. However, the unique biology of the NFS has allowed plants to evolve different ways of achieving resistance against PPNs. Most notable are the two resistance sources, *rhg1* and *rhg4*, against the soybean cyst nematode (*Heterodera glycines*) in soybean (*Glycine max*). Although the phenotype - failure of parasitism and degradation of the feeding structure - of these two sources is the same, the underlying mechanism of resistance is different and is based on a lethal toxic effect within the feeding structure.

The *rhg1* source derived from the line PI88788 is widely used in soybean grown across the United States, and this source is present in 90% of cultivars grown in some regions (Cook et al., 2012). Initial studies showed that resistance was conferred by the presence of multiple copies of a segment of DNA encoding several different proteins; an amino acid transporter, an alpha-SNAP protein and a protein containing a wound-inducible domain. No NB-LRR type gene was present in the segment of DNA associated with inheritance of resistance and while overexpression of any individual gene did not confer resistance, overexpression of all three sequences gave the resistant phenotype (Cook et al., 2012). Subsequent analysis revealed that the αSNAP protein encoded at the *rhg1* locus has differences in the amino acid sequence that cause a failure of its normal, essential function (Bayless et al., 2016). This mutation is compensated for in resistant soybean plants by other 'wild type' αSNAP proteins that are present due to ancient genome duplications in this species. However, the cytotoxic form of the protein accumulates specifically in the NFS, causing cell death and resulting in resistance (Bayless et al., 2016). *Rhg4* is a dominant resistance QTL against *H. glycines* located on Chromosome 8. Positional cloning allowed a serine hydroxy methyl transferase (SHMT) to be identified as the causal sequence underpinning this resistance response, with a number of polymorphisms identified as being critical to the resistance response (Liu et al., 2012). The role of this protein in the resistance response was confirmed by silencing the expression of the gene encoding this protein, which led to an increase in susceptibility of the resistant soybean to *H. glycines*. Similarly, transfer of the SHMT allele from a resistant line to a susceptible plant conferred resistance. Further functional studies suggested that the biochemical properties of the SHMT enzyme present in resistant plants were different to those from susceptible lines and that these differences may impact the biology of the developing feeding site and thus cause it to fail to provide nutrients to the developing nematode. These unusual resistance mechanisms show that plants have been able to evolve novel ways of combatting PPN based on the reliance of these pathogens on the feeding structure.

5.5 Contemporary resistance breeding

Over the past two decades the cost of DNA sequencing has declined dramatically and at the same time the technology behind DNA sequencing has evolved to allow huge amounts of DNA sequence data to be generated. Sequencing of crops and their pathogens is now a routine and good-quality genome sequences are available for all major crops as well as comprehensive transcriptome sequences (reviewed by Marri et al., 2018). The availability of these resources has facilitated the development of powerful new tools for identification and high-resolution mapping of R genes. In particular, the development of enrichment sequencing specifically targeted at R genes (resistance gene enrichment sequencing (RenSeq)) has allowed improved annotation of R genes (Jupe et al., 2013) as well as mapping and identification of R genes (Witek et al., 2016; Chen et al., 2018). The potential for this technology to expedite identification of R genes against PPN was recently illustrated by the fine mapping of the H2 gene in potato, which provides control of the Pa1 population of *G. pallida* (Strachan et al., 2019). A combination of RenSeq and another enrichment sequencing technique based on 1800 single copy genes distributed across the genome (GenSeq) allowed mapping of H2 to an interval of less than 5 Mb on the genome. Critically, these techniques also allow the development of KASP markers that can be used for further refinement of the mapping of the gene, but that are also compatible with current marker-assisted breeding programmes. Marker development for breeding, as well as identification of R genes, is greatly expedited by the availability of these tools.

6 Future trends

As with almost all areas of contemporary biological science, the continuing developments in sequencing technologies and associated bioinformatic analyses are likely to drive progress in many areas of this field. High-quality genome sequences are likely to become available for all major PPN and their crops, enabling studies on the interactions between these organisms. In spite of this, the enormous range of biological tools and resources available for *Arabidopsis* mean that fundamental studies are still likely to focus in this area rather than on crops in the first instance. In terms of such fundamental studies we still need to understand how biotrophic PPN induce feeding structures in their hosts and how and why they deploy many hundreds of effectors to induce these structures and suppress host defences. Root-knot nematodes offer an opportunity to understand how a complex biotrophic pathogen can have a host range of unprecedented breadth. Many of these fundamental areas would benefit hugely from development of a system for genetic transformation of PPN, something that has eluded the field for several decades. We also need to understand how resistance works for all plant pathogens, including PPN. The advent of new genetic mapping techniques as described above offers the prospect of rapid development of markers for crop breeding.

7 Where to look for further information

7.1 Books

Several books are available that summarize the interactions between plants and nematodes. A good introduction to the topic of PPN is available in *Plant Nematology 2nd Edition* (edited by R. N. Perry & M. Moens and published by CABI – ISBN 978-1-78064-153-9). A broad overview of many different areas of the field can be found in *Genomics and molecular genetics of plant-nematode interactions* (edited by J. T. Jones, G. Gheysen and C. Fenoll and published by Springer – ISBN 978-94-007-0433-6). More detailed information on the two most economically important groups of PPN can be found in *Cyst Nematodes* (edited by R. N. Perry, M. Moens & J. T. Jones and published by CAB – ISBN 978-1-78639-083-7) and *Root Knot Nematodes* (edited by R. N. Perry, M. Moens & J. L. Starr and published by CABI – ISBN 978-1-84593-492-7).

7.2 Economic impact

Articles on the economic importance of PPN include Jones et al., (2013) and Coyne et al. (2018).

7.3 Video resources

A series of high resolution videos showing infection of plant roots by cyst and root-knot nematodes are available at: https://www.youtube.com/channel/UCfre06aTpLn2qa6r1KSX2xg.

7.4 Research centres

Many major universities and research institutes across the world host research programmes that focus on PPN. Major centres include Wageningen University in the Netherlands (https://www.wur.nl/en/Research-Results/Chair-groups/Plant-Sciences/Laboratory-of-Nematology.htm), The James Hutton Institute in the UK (https://www.hutton.ac.uk/research/groups/cell-and-molecular-sciences), various INRA centres in France and several US Universities including UC Davis (http://entomology.ucdavis.edu/) and Riverside (https://nematology.ucr.edu/), Iowa State University (www.iastate.edu), NC State University (www.ncsu.edu), University of Florida (www.entnemdept.ufl.edu) and The University of Idaho (www.uidaho.edu). The International Institute for Tropical Agriculture (IITA - https://www.iita.org/) hosts one of the main research programmes on PPN important to subsistence farmers in sub-Saharan Africa.

8 Acknowledgement

The James Hutton Institute receives funding from the Scottish Government Rural and Environment Science and Analytical Services (RESAS) division.

9 References

Abad, P., Gouzy, J., Aury, J. M., Castagnone-Sereno, P., Danchin, E. G. J., Deleury, E., Perfus-Barbeoch, L., Anthouard, V., Artiguenave, F., Blok, V. C., Caillaud, M. C., Coutinho, P. M., Dasilva, C., De Luca, F., Deau, F., Esquibet, M., Flutre, T., Goldstone, J. V., Hamamouch, N., Hewezi, T., Jaillon, O., Jubin, C., Leonetti, P., Magliano, M., Maier, T. R., Markov, G. V., McVeigh, P., Pesole, G., Poulain, J., Robinson-Rechavi, M., Sallet, E., Ségurens, B., Steinbach, D., Tytgat, T., Ugarte, E., van Ghelder, C., Veronico, P., Baum, T. J., Blaxter, M., Bleve-Zacheo, T., Davis, E. L., Ewbank, J. J., Favery, B., Grenier, E., Henrissat, B., Jones, J. T., Laudet, V., Maule, A. G., Quesneville, H., Rosso, M. N., Schiex, T., Smant, G., Weissenbach, J. and Wincker, P. (2008). Genome sequence of the metazoan plant-parasitic nematode *Meloidogyne incognita*. *Nature Biotechnology* 26(8), 909–915.

Absmanner, B., Stadler, R. and Hammes, U. Z. (2013). Phloem development in nematode-induced feeding sites: the implications of auxin and cytokinin. *Frontiers in Plant Science* 4, 241.

Ammiraju, J. S. S., Veremis, J. C., Huang, X., Roberts, P. A. and Kaloshian, I. (2003). The heat-stable root-knot nematode resistance gene Mi-9 from *Lycopersicon peruvianum* is localized on the short arm of chromosome 6. *Theoretical and Applied Genetics* 106, 478–484.

Anjam, M. S., Shah, S. J., Matera, C., Różańska, E., Sobczak, M., Siddique, S. and Grundler, F. M. W. (2018). Host factors influence the sex of nematodes parasitizing roots of *Arabidopsis thaliana*. *bioRxiv*, 273391.

Baldacci-Cresp, F., Chang, C., Maucourt, M., Deborde, C., Hopkins, J., Lecomte, P., Bernillon, S., Brouquisse, R., Moing, A., Abad, P., Hérouart, D., Puppo, A., Favery, B. and Frendo, P. (2012). (Homo)glutathione deficiency impairs root-knot nematode development in *Medicago truncatula*. *PLoS Pathogens* 8(1), e1002471.

Baranowski, Ł., Rozanska, E., Sanko-Sawczenko, I., Matuszkiewicz, M., Znojek, E., Filipecki, M., Grundler, F. M. W. and Sobczak, M. (2019). Arabidopsis tonoplast intrinsic protein and vacuolar H+-adenosinetriphosphatase reflect vacuole dynamics during development of syncytia induced by the beet cyst nematode *Heterodera schachtii*. *Protoplasma* 256(2), 419–429.

Barcala, M., Garcia, A., Cabrera, J., Casson, S., Lindsey, K., Favery, B., Garcia-Casado, G., Solano, R., Fenoll, C. and Escobar, C. (2010). Early transcriptomic events in microdissected Arabidopsis nematode-induced giant cells. *Plant Journal* 61(4), 698–712.

Bartlem, D. G., Jones, M. G. K. and Hammes, U. Z. (2014). Vascularization and nutrient delivery at root-knot nematode feeding sites in host roots. *Journal of Experimental Botany* 65(7), 1789–1798.

Bayless, A. M., Smith, J. M., Song, J., McMinn, P. H., Teillet, A., August, B. K. and Bent, A. F. (2016). Disease resistance by impairment of α-SNAP. *Proceedings of the National Academy of Sciences of the United States of America* 113(47), E7375–E7382.

Bhattarai, K. K., Bishnoi, U. and Kaloshian, I. (2007). Root-knot nematode/tomato interactions: transcriptome profiling and defense signaling pathways. *Journal of Nematology* 39, 73.

Boiteux, L. S., Belter, J. G., Roberts, P. A. and Simon, P. W. (2000). RAPD linkage map of the genomic region encompassing the root-knot nematode (*Meloidogyne javanica*) resistance locus in carrot. *Theoretical and Applied Genetics* 100(3–4), 439–446.

Cai, D., Kleine, M., Kifle, S., Harloff, H. J., Sandal, N. N., Marcker, K. A., Klein-Lankhorst, R. M., Salentijn, E. M. J., Lange, W., Stiekema, W. J., Wyss, U., Grundler, F. M. W. and Jung, C. (1997). Positional cloning of a gene for nematode resistance in sugar beet. *Science* 275(5301), 832–834.

Castagnone-Sereno, P. and Danchin, E. G. J. (2014). Parasitic success without sex - the nematode experience. *Journal of Evolutionary Biology* 27(7), 1323–1333.

Castillo, P. and Volvas, N. (2007). *Pratylenchus (Nematoda; Pratylenchidae): Diagnosis, Biology, Pathogenicity and Management*. *Nematology Monographs and Perspectives* 6 (Series Editors Hunt, D. J. and Perry, R. N.). Leiden: Brill.

Chen, S., Chronis, D. and Wang, X. (2013). The novel GrCEP12 peptide from the plant-parasitic nematode *Globodera rostochiensis* suppresses flg22 mediated PTI. *Plant Signaling and Behavior* 8(9), e25359.

Chen, X., Lewandowska, D., Armstrong, M. R., Baker, K., Lim, T. Y., Bayer, M., Harrower, B., McLean, K., Jupe, F., Witek, K., Lees, A. K., Jones, J. D., Bryan, G. J. and Hein, I. (2018). Identification and rapid mapping of a gene conferring broad-spectrum late blight resistance in the diploid potato species *Solanum verrucosum* through DNA capture technologies. *Theoretical and Applied Genetics* 131(6), 1287–1297.

Cook, D. E., Lee, T. G., Guo, X., Melito, S., Wang, K., Bayless, A. M., Wang, J., Hughes, T. J., Willis, D. K., Clemente, T. E., Diers, B. W., Jiang, J., Hudson, M. E. and Bent, A. F. (2012). Copy number variation of multiple genes at *rhg1* mediates nematode resistance in soybean. *Science* 338(6111), 1206–1209.

Cotton, J. A., Lilley, C. J., Jones, L. M., Kikuchi, T., Reid, A. J., Thorpe, P., Tsai, I. J., Beasley, H., Blok, V., Cock, P. J., Eves-van den Akker, S., Holroyd, N., Hunt, M., Mantelin, S., Naghra, H., Pain, A., Palomares-Rius, J. E., Zarowiecki, M., Berriman, M., Jones, J. T. and Urwin, P. E. (2014). The genome and life-stage specific transcriptomes of *Globodera pallida* elucidate key aspects of plant parasitism by a cyst nematode. *Genome Biology* 15(3), R43.

Couto, D. and Zipfel, C. (2016). Regulation of pattern recognition receptor signalling in plants. *Nature Reviews. Immunology* 16(9), 537–552.

Coyne, D. L., Cortada, L., Dalzell, J. J., Claudius-Cole, A. O., Haukeland, S., Luambano, N., and Talwana, H. (2018). Plant-parasitic nematodes and food security in sub-Saharan Africa. *Annual Review of Phytopathology* 56, 381–403.

Danchin, E. G. J., Rosso, M. N., Vieira, P., de Almeida Engler, J., Coutinho, P. M., Henrissat, B. and Abad, P. (2010). Multiple lateral gene transfers and duplications have promoted plant parasitism ability in nematodes. *Proceedings of the National Academy of Sciences of the United States of America* 107(41), 17651–17656.

Danchin, E. G. J., Thorpe, P., Perfus-Barbeoch, L., Rancurel, C., Da Rocha, M., Labadie, K., Guy, J., Da Silva, C., Bajew, S., Neilsen, R., Guzeeva, E., Esmenjaud, D., Helder, J., Jones, J. T. and Eves-van den Akker, S. (2017). The transcriptomes of *Xiphinema index* and *Longidorus elongatus* reveal independent acquisition of parasitism genes by horizontal gene transfer in early-branching plant-parasitic nematodes. *Genes* 8, 287.

Davies, L. J., Lilley, C. J., Paul Knox, J. and Urwin, P. E. (2012). Syncytia formed by adult female Heterodera schachtii in *Arabidopsis thaliana* roots have a distinct cell wall molecular architecture. *New Phytologist* 196(1), 238–246.

de Almeida Engler, J., Van Poucke, K., Karimi, M., De Groodt, R., Gheysen, G., Engler, G. and Gheysen, G. (2004). Dynamic cytoskeleton rearrangements in giant cells and syncytia of nematode-infected roots. *Plant Journal* 38(1), 12–26.

de Almeida Engler, J., De Vleesschauwer, V., Burssens, S., Celenza, J. L., Inze, D., Van Montagu, M., Engler, G. and Gheysen, G. (1999). Molecular markers and cell cycle inhibitors show the importance of cell cycle progression in nematode-induced galls and syncytia. *Plant Cell* 11(5), 793–808.

de Almeida Engler, J. and Favery, B. (2011). The plant cytoskeleton remodelling in nematode induced feeding sites. In: *Genomics and Molecular Genetics of Plant-Nematode Interactions*, Jones, J., Gheysen, G. and Fenoll, C. (Eds), 369–394. Dordrecht, The Netherlands: Springer.

De Ilarduya, O. M., Moore, A. E. and Kaloshian, I. (2001). The tomato *Rme1* locus is required for *Mi-1*-mediated resistance to root-knot nematodes and the potato aphid. *Plant Journal* 27(5), 417–425.

De Meutter, J., Robertson, L. , Parcy, F., Mena, M., Fenoll, C. and Gheysen, G. (2005). Differential activation of ABI3 and LEA genes upon plant parasitic nematode infection. *Molecular Plant Pathology* 6(3), 321–325.

Den Ouden, H. (1960). A note on parthenogenesis and sex determination in *Heterodera rostoschiensis* Woll. *Nematologica* 5(3), 215–216.

Dinh, P. T. Y., Brown, C. R. and Elling, A. A. (2014). RNA interference of an effector gene *Mc16D10L* confers resistance against *Meloidogyne chitwoodi* in *Arabidopsis* and potato. *Phytopathology* 104(10), 1098–1106.

Doncaster, C. and Seymour, M. K. (1973). Defaecation behaviour of *Aphelenchoides blastophthorus*. *Nematologica* 18(4), 463–468.

Dowd, C. D., Chronis, D., Radakovic, Z. S., Siddique, S., Schmulling, T., Werner, T., Kakimoto, T., Grundler, F. M. W. and Mitchum, M. G. (2017). Divergent expression of cytokinin biosynthesis, signaling and catabolism genes underlying differences in feeding sites induced by cyst and root-knot nematodes. *Plant Journal* 92(2), 211–228.

Dropkin, V. H. (1969). Cellular responses of plants to nematode infections. *Annual Review of Phytopathology* 7(1), 101–122.

Duncan, L. W. and Moens, M. (2013). Migratory endoparasitic nematodes. In: *Plant Nematology*, 2nd Edition, Perry, R. N. and Moens, M. (Eds), 144–178. Wallingford, UK: CAB International.

Eisenback, J. and Zunke, U. (2000). *Nemapix Vol. 1 – A Journal of Nematological Images*. Mactode Publications.

Ellenby, C. (1954). Environmental determination of the sex ratio of a plant parasitic nematode. *Nature* 174(4439), 1016–1017.

Endo, B. Y. (1964). Penetration and development of *Heterodera glycines* in soybean roots and related anatomical changes. *Phytopathology* 54, 79.

Endo, B. Y. and Veech, J. A. (1970). Morphology and histochemistry of soybean roots infected with *Heterodera glycines*. *Phytopathology* 60(10), 1493–1497.

Ernst, K., Kumar, A., Kriseleit, D., Kloos, D. U., Phillips, M. S. and Ganal, M. W. (2002). The broad-spectrum potato cyst nematode resistance gene (*Hero*) from tomato is the only member of a large gene family of NBS-LRR genes with an unusual amino acid repeat in the LRR region. *Plant Journal* 31(2), 127–136.

Escobar, C., Brown, S. and Mitchum, M. G. (2011). Transcriptomic and proteomic analysis of the plant response to nematode infection. *Genomics and Molecular Genetics of Plant-Nematode Interactions*, Jones, J., Gheysen, G. and Fenoll, C. (Eds), 157–173. Dordrecht, The Netherlands: Springer.

Eves-van den Akker, S., Lilley, C. J., Jones, J. T. and Urwin, P. E. (2015). Feeding tubes and plugs: new perspectives. *Nematology* 17, 1–9.

Fudali, S., Sobczak, M., Janakowski, S., Griesser, M., Grundler, F. M. W. and Golinowski, W. (2008). Expansins are among plant cell wall modifying agents specifically expressed during development of nematode-induced syncytia. *Plant Signaling and Behavior* 3(11), 969–971.

Garcia, G. M., Stalker, H. T., Shroeder, E. and Kochert, G. (1996). Identification of RAPD, SCAR, and RFLP markers tightly linked to nematode resistance genes introgressed from *Arachis cardenasii* into *Arachis hypogaea*. *Genome* 39(5), 836–845.

Goddard, R. H., Wick, S. M., Silflow, C. D. and Snustad, D. P. (1994). Microtubule components of the plant cell cytoskeleton. *Plant Physiology* 104(1), 1–6.

Goverse, A., de Engler, J. A., Verhees, J., van der Krol, S., Helder, J. H. and Gheysen, G. (2000a). Cell cycle activation by plant parasitic nematodes. *Plant Molecular Biology* 43(5-6), 747-761.

Goverse, A., Overmars, H., Engelbertink, J., Schots, A., Bakker, J. and Helder, J. (2000b). Both induction and morphogenesis of cyst nematode feeding cells are mediated by auxin. *Molecular Plant-Microbe Interactions* 13(10), 1121-1129.

Griffiths, B. S. and Robertson, W. M. (1984). Morphological and histochemical changes occurring during the life-span of root-tip galls on *Lolium perenne* induced by *Longidorus elongatus*. *Journal of Nematology* 16(3), 223-229.

Grundler, F. M. W., Betka, M. and Wyss, U. (1991). Influence of changes in the nurse cell system (syncytium) on sex determination and development of the cyst nematode *Heterodera schachtii* - Total amounts of proteins and amino-Acids. *Phytopathology* 81(1), 70-74.

Grunewald, W., van Noorden, G., van Isterdael, G., Beeckman, T., Gheysen, G. and Mathesius, U. (2009). Manipulation of auxin transport in plant roots during rhizobium symbiosis and nematode parasitism. *Plant Cell* 21(9), 2553-2562.

Hofmann, J., El Ashry, A. N., Anwar, S., Erban, A., Kopka, J. and Grundler, F. M. W. (2010). Metabolic profiling reveals local and systemic responses of host plants to nematode parasitism. *Plant Journal* 62(6), 1058-1071.

Hofmann, J., Hess, P. H., Szakasits, D., Blochl, A., Wieczorek, K., Daxbock-Horvath, S., Bohlmann, H., van Bel, A. J. E. and Grundler, F. M. W. (2009). Diversity and activity of sugar transporters in nematode-induced root syncytia. *Journal of Experimental Botany* 60(11), 3085-3095.

Hofmann, J., Wieczorek, K., Blochl, A. and Grundler, F. M. W. (2007). Sucrose supply to nematode-induced syncytia depends on the apoplasmic and symplasmic pathways. *Journal of Experimental Botany* 58(7), 1591-1601.

Holbein, J., Franke, R. B., Marhavý, P., Fujita, S., Górecka, M., Sobczak, M., Geldner, N., Schreiber, L., Grundler, F. M. W. and Siddique, S. (2019). Root endodermal barrier system contributes to defence against plant-parasitic cyst and root-knot nematodes. *Plant Journal* 100(2), 221-236.

Hoth, S., Stadler, R., Sauer, N. and Hammes, U. Z. (2008). Differential vascularization of nematode-induced feeding sites. *Proceedings of the National Academy of Sciences of the United States of America* 105(34), 12617-12622.

Huang, G., Allen, R., Davis, E. L., Baum, T. J. and Hussey, R. S. (2006). Engineering broad root-knot resistance in transgenic plants by RNAi silencing of a conserved and essential root-knot nematode parasitism gene. *Proceedings of the National Academy of Sciences of the United States of America* 103(39), 14302-14306.

Hussey, R. S. and Mimms, C. W. (1990). Ultrastructure of esophogeal glands and their secretory granules in the root-knot nematode *Meloidogyne incognita*. *Protoplasma* 162, 99-107.

Hutten, M., Geukes, M., Misas-Villamil, J. C., van der Hoorn, R. A. L., Grundler, F. M. W. and Siddique, S. (2015). Activity profiling reveals changes in the diversity and activity of proteins in Arabidopsis roots in response to nematode infection. *Plant Physiology and Biochemistry* 97, 36-43.

Hwang, C. F. and Williamson, V. M. (2003). Leucine-rich repeat-mediated intramolecular interactions in nematode recognition and cell death signaling by the tomato resistance protein *Mi*. *Plant Journal* 34(5), 585-593.

Ithal, N., Recknor, J., Nettleton, D., Hearne, L., Maier, T., Baum, T. J. and Mitchum, M. G. (2007a). Parallel genome-wide expression profiling of host and pathogen during soybean cyst nematode infection of soybean. *Molecular Plant-Microbe Interactions* 20(3), 293–305.

Ithal, N., Recknor, J., Nettleton, D., Maier, T., Baum, T. J. and Mitchum, M. G. (2007b). Developmental transcript profiling of cyst nematode feeding cells in soybean roots. *Molecular Plant-Microbe Interactions* 20(5), 510–525.

Jammes, F., Lecomte, P., de Almeida Engler, J., Bitton, F., Martin-Magniette, M. L., Renou, J. P., Abad, P. and Favery, B. (2005). Genome-wide expression profiling of the host response to root-knot nematode infection in Arabidopsis. *Plant Journal* 44(3), 447–458.

Jaouannet, M., Magliano, M., Arguel, M. J., Gourges, M., Evangelisti, E., Abad, P. and Rosso, M. N. (2013). The root-knot nematode calreticulin Mi-CRT is a key effector in plant defense suppression. *Molecular Plant-Microbe Interactions* 26(1), 97–105.

Jones, J. D. G. and Dangl, J. L. (2006). The plant immune system. *Nature* 444(7117), 323–329.

Jones, J. T., Furlanetto, C. and Kikuchi, T. (2005). Horizontal gene transfer from bacteria and fungi as a driving force in the evolution of plant parasitism in nematodes. *Nematology* 7(5), 641–646.

Jones, J. T., Haegeman, A., Danchin, E. G. J., Gaur, H. S., Helder, J., Jones, M. G. K., Kikuchi, T., Manzanilla-Lopez, R., Palomares-Rius, J. E., Wesemael, W. M. L. and Perry, R. N. (2013). Top 10 plant parasitic nematodes in molecular plant pathology. *Molecular Plant Pathology* 14(9), 946–961.

Jones, J. T., Li, H., Moens, M., Mota, M. and Kikuchi, T. (2008). *Bursaphelenchus xylophilus*: opportunities in comparative genomics and molecular host-parasite interactions. *Molecular Plant Pathology* 9(3), 357–368.

Jones, M. G. K. and Goto, D. B. (2011). Root-knot nematodes and giant cells. In: *Genomics and Molecular Genetics of Plant-Nematode Interactions*, Jones, J. T., Gheysen, G. and Fenoll, C. (Eds), 83–100. Dordrecht, The Netherlands: Springer.

Jupe, F., Witek, K., Verweij, W., Śliwka, J., Pritchard, L., Etherington, G. J., Maclean, D., Cock, P. J., Leggett, R. M., Bryan, G. J., Cardle, L., Hein, I. and Jones, J. D. G. (2013). Resistance gene enrichment sequencing (RenSeq) enables reannotation of the NB-LRR gene family from sequenced plant genomes and rapid mapping of resistance loci in segregating populations. *Plant Journal* 76(3), 530–544.

Kaloshian, I., Yaghoobi, J., Liharska, T., Hontelez, J., Hanson, D., Hogan, P., Jesse, T., Wijbrandi, J., Simons, G., Vos, P., Zabel, P. and Williamson, V. M. (1998). Genetic and physical localization of the root-knot nematode resistance locus *Mi* in tomato. *Molecular and General Genetics* 257(3), 376–385.

Karczmarek, A., Overmars, H., Helder, J. and Goverse, A. (2004). Feeding cell development by cyst and root-knot nematodes involves a similar early, local and transient activation of a specific auxin-inducible promoter element. *Molecular Plant Pathology* 5(4), 343–346.

Kikuchi, T., Jones, J. T., Aikawa, T., Kosaka, H. and Ogura, N. (2004). A family of GHF45 cellulases from the pine wood nematode *Bursaphelenchus xylophilus*. *FEBS Letters* 572(1–3), 201–205.

Krusberg, L. R. and Nielsen, L. W. (1958). Pathogenesis of root-knot nematodes to the Porto Rico variety of sweet potato. *Phytopathology* 48, 30–38.

Kyndt, T., Goverse, A., Haegeman, A., Warmerdam, S., Wanjau, C., Jahani, M., Engler, G., de Almeida Engler, J. and Gheysen, G. (2016). Redirection of auxin flow in *Arabidopsis thaliana* roots after infection by root-knot nematodes. *Journal of Experimental Botany* 67(15), 4559–4570.

Lee, C., Chronis, D., Kenning, C., Peret, B., Hewezi, T., Davis, E. L., Baum, T. J., Hussey, R., Bennett, M. and Mitchum, M. G. (2011). The novel cyst Nematode effector protein 19C07 interacts with the Arabidopsis auxin influx transporter LAX3 to control feeding site development. *Plant Physiology* 155(2), 866–880.

Leelarasamee, N., Zhang, L. and Gleason, C. (2018). The root-knot nematode effector *MiPFN3* disrupts plant actin filaments and promotes parasitism. *PLOS Pathogens* 14(3), e1006947.

Lin, B., Zhuo, K., Chen, S., Hu, L., Sun, L., Wang, X., Zhang, L.-H. and Liao, J. (2015). A novel nematode effector suppresses plant immunity by activating host reactive oxygen species-scavenging system. *New Phytologist* 3, 1159–1173.

Liu, S., Kandoth, P. K., Warren, S. D., Yeckel, G., Heinz, R., Alden, J., Yang, C., Jamai, A., El-Mellouki, T., Juvale, P. S., Hill, J., Baum, T. J., Cianzio, S., Whitham, S. A., Korkin, D., Mitchum, M. G. and Meksem, K. (2012). A soybean cyst nematode resistance gene points to a new mechanism of plant resistance to pathogens. *Nature* 492(7428), 256–260.

Lozano-Torres, J., Willig, J. J., Warmerdam, S., Van Schaik, C. C., Bakker, J., Goverse, A. and Smant, G. (2019). Proline-rich extensin-like receptor kinases mediate damage-triggered immune responses to nematode infections. *Molecular Plant-Microbe Interactions* 32, 233–233.

Lozano-Torres, J. L., Wilbers, R. H. P., Gawronski, P., Boshoven, J. C., Finkers-Tomczak, A., Cordewener, J. H., America, A. H., Overmars, H. A., Van 't Klooster, J. W., Baranowski, L., Sobczak, M., Ilyas, M., van der Hoorn, R. A., Schots, A., de Wit, P. J., Bakker, J., Goverse, A. and Smant, G. (2012). Dual disease resistance mediated by the immune receptor Cf-2 in tomato requires a common virulence target of a fungus and a nematode. *Proceedings of the National Academy of Sciences of the United States of America* 109(25), 10119–10124.

Lozano-Torres, J. L., Wilbers, R. H. P., Warmerdam, S., Finkers-Tomczak, A., Diaz-Granados, A., van Schaik, C. C., Helder, J., Bakker, J., Goverse, A., Schots, A. and Smant, G. (2014). Apoplastic venom allergen-like proteins of cyst nematodes modulate the activation of basal plant innate immunity by cell surface receptors. *PLoS Pathogens* 10(12), e1004569.

Lu, Z.-X., Sossey-Alaoui, K., Reighard, G. L., Baird, W. V. and Abbott, A. G. (1999). Development and characterization of a codominant marker linked to root-knot nematode resistance, and its application to peach rootstock breeding. *Theoretical and Applied Genetics* 99(1–2), 115–122.

Macho, A. P. and Zipfel, C. (2014). Plant PRRs and the activation of innate immune signaling. *Molecular Cell* 54(2), 263–272.

Manosalva, P., Manohar, M., von Reuss, S. H., Chen, S., Koch, A., Kaplan, F., Choe, A., Micikas, R. J., Wang, X., Kogel, K. H., Sternberg, P. W., Williamson, V. M., Schroeder, F. C. and Klessig, D. F. (2015). Conserved nematode signalling molecules elicit plant defenses and pathogen resistance. *Nature Communications* 6, Article 7795.

Manzanilla-Lopez, R. H. and Starr, J. L. (2009). Interactions with other pathogens. In: *Root-Knot Nematodes* Perry, R. N., Moens, M. and Starr, J. L. (Eds), 223–245. Wallingford, UK: CAB International.

Marhavy, P., Kurenda, A., Siddique, S. Tendon, V. D., Zhou, F., Holbein, J., Hasan, M. S., Grundler, F. M. W., Farmer, E. E. and Geldner, N. (2019). Single-cell damage elicits regional, nematode-restricting ethylene responses in roots. *EMBO Journal* 38, e100972.

Marri, P. R., Ye, L., Jia, Y., Jiang, K. and Rounsley, S. D. (2018). Advances in sequencing and resequencing in crop plants. In: *Plant Genetics and Molecular Biology. Advances in Biochemical Engineering/Biotechnology*, Varshney, R., Pandey, M. and Chitikineni, A. (Eds), vol. 164. Cham: Springer.

Masler, E. P. and Perry, R. N. (2018). Hatch, survival and sensory perception. In: *Cyst Nematodes*, Perry, R. N., Moens, M. and Jones, J. T. (Eds), 44–73. Wallingford, UK: CABI Publishing.

Mei, Y., Thorpe, P., Guzha, A., Haegeman, A., Blok, V. C., MacKenzie, K., Gheysen, G., Jones, J. T. and Mantelin, S. (2015). Only a small subset of the SPRY domain gene family in *Globodera pallida* is likely to encode effectors, two of which suppress host defences induced by the potato resistance gene *Gpa2*. *Nematology* 17(4), 409–424.

Mei, Y., Wright, K. M., Haegeman, A., Bauters, L., Diaz-Granados, A., Goverse, A., Gheysen, G., Jones, J. T. and Mantelin, S. (2018). The *Globodera pallida* SPRYSEC effector *Gp*SPRY-414-2 that suppresses plant defenses targets a regulatory component of the dynamic microtubule network. *Frontiers in Plant Science* 9, 1019.

Melillo, M. T., Leonetti, P., Bongiovanni, M., Castagnone-Sereno, P. and Bleve-Zacheo, T. (2006). Modulation of reactive oxygen species activities and H2O2 accumulation during compatible and incompatible tomato-root-knot nematode interactions. *New Phytologist* 170(3), 501–512.

Mendy, B., Wang'ombe, M. W., Radakovic, A. S., Holbein, J., Ilyas, M., Chopra, D., Holton, N., Zipfel, C., Grundler, F. M. W. and Siddique, S. (2017). Arabadopsis leucine-rich receptor kinase NILR1 is required for inducation of innate immunity to parasitic nematodes. *PLoS Pathogens* 13(4), e1006284.

Milligan, S. B., Bodeau, J., Yaghoobi, J., Kaloshian, I., Zabel, P. and Williamson, V. M. (1998). The root knot nematode resistance gene Mi from tomato is a member of the leucine zipper, nucleotide binding, leucine-rich repeat family of plant genes. *Plant Cell* 10(8), 1307–1319.

Minnis, S. T., Haydock, P. P. J., Ibrahim, S. K., Grove, I. G., Evans, K. and Russell, M. D. (2002). Potato cyst nematodes in England and Wales-occurrence and distribution. *Annals of Applied Biology* 140(2), 187–195.

Molz, E. (1920). Versuche zur Ermittlung des Einflusses aüsserer Faktoren auf des Geschlechtsverhältnis des Rübennematoden (Heterodera schachtii A. Schmidt). *Landwirthschaftliche Jahrbücher* 54, 769–791.

Nicol, J. M., Turner, S. J., Coyne, D. L., den Nijs, L., Hockland, S. and Maafi, Z. T. (2011). Current nematode threats to world agriculture. In: *Genomics and Molecular Genetics of Plant-Nematode Interactions*, Jones, J. T., Gheysen, G. and Fenoll, C. (Eds), 21–44. Dordrecht, The Netherlands: Springer.

Niebel, A., de Almeida Engler, J., Hemerly, A., Ferreira, P., Inze, D., VanMontagu, M. and Gheysen, G. (1996). Induction of cdc2a and cyc1At expression in *Arabidopsis thaliana* during early phases of nematode-induced feeding cell formation. *Plant Journal* 10(6), 1037–1043.

Nombela, G., Williamson, V. M. and Muniz, M. (2003). The root-knot nematode resistance gene Mi-1.2 of tomato is responsible for resistance against the whitefly *Bemisia tabaci*. *Molecular Plant-Microbe Interactions* 16(7), 645–649.

Paal, J., Henselewski, H., Muth, J., Meksem, K., Menendez, C. M., Salamini, F., Ballvora, A. and Gebhardt, C. (2004). Molecular cloning of the potato *Gro1-4* gene conferring

resistance to pathotype Ro1 of the root cyst nematode *Globodera rostochiensis*, based on a candidate gene approach. *Plant Journal* 38(2), 285–297.
Paulson, R. E. and Webster, J. M. (1972). Ultrastructure of the hypersensitive reaction in roots of tomato, *Lycopersicon esculentum* L. to infection by the root-knot nematode. *Physiological Plant Pathology* 2(3), 227–234.
Peng, H. C. and Kaloshian, I. (2014). The Tomato leucine-Rich Repeat Receptor-Like Kinases SlSERK3A and SlSERK3B have overlapping functions in bacterial and nematode innate immunity. *PLoS ONE* 9(3), e93302.
Piya, S., Kihm, C., Rice, J. H., Baum, T. J. and Hewezi, T. (2017). Cooperative regulatory functions of *miR858* and *MYB83* during cyst nematode parasitism. *Plant Physiology* 174(3), 1897–1912.
Popeijus, H., Overmars, H., Jones, J., Blok, V., Goverse, A., Helder, J., Schots, A., Bakker, J. and Smant, G. (2000). Enzymology – degradation of plant cell walls by a nematode. *Nature* 406(6791), 36–37.
Portillo, M., Cabrera, J., Lindsey, K., Topping, J., Andres, M. F., Emiliozzi, M., Oliveros, J. C., Garcia-Casado, G., Solano, R., Koltai, H., Resnick, N., Fenoll, C. and Escobar, C. (2013). Distinct and conserved transcriptomic changes during nematode-induced giant cell development in tomato compared with Arabidopsis: a functional role for gene repression. *New Phytologist* 197(4), 1276–1290.
Postma, W. J., Slootweg, E. J., Rehman, S., Finkers-Tomczak, A., Tytgat, T. O. G., van Gelderen, K., Lozano-Torres, J. L., Roosien, J., Pomp, R., van Schaik, C., Bakker, J., Goverse, A. and Smant, G. (2012). The effector SPRYSEC-19 of *Globodera rostochiensis* suppresses CC-NB-LRR mediated disease resistance in plants. *Plant Physiology* 160(2), 944–954.
Qin, L., Kudla, U., Roze, E. H. A., Goverse, A., Popeijus, H., Nieuwland, J., Overmars, H., Jones, J. T., Schots, A., Smant, G., Bakker, J. and Helder, J. (2004). Plant degradation: a nematode expansin acting on plants. *Nature* 427(6969), 30–30.
Rice, S. L., Leadbeater, B. S. C. and Stone, A. R. (1985). Changes in cell structure in roots of resistant potatoes parasitized by potato cyst nematodes. I. Potatoes with resistance gene H1 derived from *Solanum tuberosum ssp. Andigena*. *Physiological Plant Pathology* 27(2), 219–234.
Rice, S. L., Stone, A. R. and Leadbeater, B. S. C. (1987). Changes in cell structure in roots of resistant potatoes parasitized by potato cyst nematodes. 2. Potatoes with resistance derived from *Solanum vernei*. *Physiological and Molecular Plant Pathology* 31(1), 1–14.
Richardson, L. and Price, N. S. (1984). Observations on the biology of *Meloidogyne incognita* and the Diageotropica tomato mutant. *Revue Nematology* 7, 97–99.
Roberts, P. A. and Thomason, I. J. (1989). A review of variability in four *Meloidogyne* spp. measured by reproduction on several hosts including *Lycopersicon*. *Agricultural Zoology Reviews* 2, 225–252.
Robertson, W. M., Griffiths, B. S. and Trudgill, D. L. (1984). Feeding of *Longidorus elongatus* and *L. leptocephalus* on root-tip galls of perennial ryegrass (*Lolium perenne*). *Nematologica* 30(2), 222–229.
Rooney, H. C., Van't Klooster, J. W., van der Hoorn, R. A., Joosten, M. H., Jones, J. D. and de Wit, P. J. (2005). *Cladosporium* Avr2 inhibits tomato Rcr3 protease required for *Cf-2*-dependent disease resistance. *Science* 308(5729), 1783–1786.
Rossi, M., Goggin, F. L., Milligan, S. B., Kaloshian, I., Ullman, D. E. and Williamson, V. M. (1998). The nematode resistance gene Mi of tomato confers resistance against the

potato aphid. *Proceedings of the National Academy of Sciences of the United States of America* 95(17), 9750–9754.

Sacco, M. A., Koropacka, K., Grenier, E., Jaubert, M. J., Blanchard, A., Goverse, A., Smant, G. and Moffett, P. (2009). The cyst nematode SPRYSEC protein RBP-1 elicits Gpa2- and RanGAP2-dependent plant cell death. *PLoS Pathogens* 5(8), e1000564.

Sandal, N. N., Salentijn, E. M. J., Kleine, M., Cai, D., Arens-de Reuver, M., van Druten, M., de Bock, T. S. M., Lange, W., Steen, P., Jung, C., Marker, K., Stiekema, W. J. and Klein-Lankhorst, R. M. (1997). Backcrossing of nematode-resistant sugar beet: a second nematode resistance gene at the locus containing Hs1pro-1. *Molecular Breeding* 3(6), 471–480.

Shah, S. J., Anjam, M. S., Mendy, B., Anwer, M. A. S., Habash, S. S., Lozano-Torres, J. L., Grundler, F. M. W. and Siddique, S. (2017). Damage-associated responses of the host contribute to defence against cyst nematodes but not root-knot nematodes. *Journal of Experimental Botany* 68(21–22), 5949–5960.

Siddique, S. and Grundler, F. M. W. (2015). Metabolism in nematode feeding sites. *Advances in Botanical Research* 73, 119–138.

Siddique, S. and Grundler, F. M. W. (2018). Parasitic nematodes manipulate plant development to establish feeding sites. *Current Opinion in Microbiology* 46, 102–108.

Siddique, S., Radakovic, Z. S., De La Torre, C. M., Chronis, D., Novák, O., Ramireddy, E., Holbein, J., Matera, C., Hütten, M., Gutbrod, P., Anjam, M. S., Rozanska, E., Habash, S., Elashry, A., Sobczak, M., Kakimoto, T., Strnad, M., Schmülling, T., Mitchum, M. G. and Grundler, F. M. W. (2015a). A parasitic nematode releases cytokinin that controls cell division and orchestrates feeding site formation in host plants. *Proceedings of the National Academy of Sciences* 112(41), 12669–12674.

Smakowska-Luzan, E., Mott, G. A., Parys, K., Stegmann, M., Howton, T. C., Layeghifard, M., Neuhold, J., Lehner, A., Kong, J., Grünwald, K., Weinberger, N., Satbhai, S. B., Mayer, D., Busch, W., Madalinski, M., Stolt-Bergner, P., Provart, N. J., Mukhtar, M. S., Zipfel, C., Desveaux, D., Guttman, D. S. and Belkhadir, Y. (2018). An extracellular network of Arabidopsis leucine-rich repeat receptor kinases. *Nature* 553(7688), 342–346.

Smant, G., Stokkermans, J. P., Yan, Y., de Boer, J. M., Baum, T. J., Wang, X., Hussey, R. S., Gommers, F. J., Henrissat, B., Davis, E. L., Helder, J., Schots, A. and Bakker, J. (1998). Endogenous cellulases in animals: isolation of beta-1, 4-endoglucanase genes from two species of plant-parasitic cyst nematodes. *Proceedings of the National Academy of Sciences of the United States of America* 95(9), 4906–4911.

Sobczak, M., Avrova, A., Jupowicz, J., Phillips, M. S., Ernst, K. and Kumar, A. (2005). Characterization of susceptibility and resistance responses to potato cyst nematode (*Globodera* spp.) infection of tomato lines in the absence and presence of the broad-spectrum nematode resistance *Hero* gene. *Molecular Plant-Microbe Interactions* 18(2), 158–168.

Sobczak, M., Golinowski, W. and Grundler, F. M. W. (1999). Ultrastructure of feeding plugs and feeding tubes formed by *Heterodera schachtii*. *Nematology* 1(4), 363–374.

Sobczak, M. and Golinowski, W. (2011). Cyst nematodes and syncytia. In: *Genomics and Molecular Genetics of Plant-Nematode Interactions*, Jones, J. T., Gheysen, G. and Fenoll, C. (Eds), 61–82. Dordrecht: Springer.

Soliman, A.H., Sobczak, M. and Golinowski, W. (2005). Defence responses of white mustard, *Sinapsis alba*, to infection with the cyst nematode *Heterodera schachtii*. *Nematology* 7(6), 881–889.

Starr, J. L. (1993). Dynamics of the nuclear complement of giant cells induced by *Meloidogyne incognita*. *Journal of Nematology* 25(3), 416–421.

Steinbach, P. (1972). Untersuchungen über das Verhalten von Larven des Kartoffelzystenälchens (*Heterodera rostochiensis* Wollenweber, 1923) an und in Wurzeln der Wirtspflanze *Lycopersicon esculentum* Mill. I. Zur Anlockwirkung von Tomatensämlingswurzeln gegenüber Kartoffelnematodenlarven. *Biologisches Zentralblatt* 91, 493-500.

Strachan, S. M., Armstrong, M. R., Kaur, A., Wright, K. M., Tze-Yin Lim, J., Baker, K., Jones, J. T., Bryan, G. J., Blok, V. C. and Hein, I. (2019). Mapping the H2 resistance effective against *Globodera pallida* pathotype Pa1 in tetraploid potato. *Theoretical and Applied Genetics* 132, 1283–1294.

Szakasits, D., Heinen, P., Wieczorek, K., Hofmann, J., Wagner, F., Kreil, D. P., Sykacek, P., Grundler, F. M. W. and Bohlmann, H. (2009). The transcriptome of syncytia induced by the cyst nematode *Heterodera schachtii* in Arabidopsis roots. *Plant Journal* 57(5), 771–784.

Teixeira, M. A., Wei, L. H. and Kaloshian, I. (2016). Root-knot nematodes induce pattern-triggered immunity in *Arabidopsis thaliana* roots. *New Phytologist* 211(1), 276–287.

Teixeira, M. A., Zhou, D., He, J., Godinez-Vidal, D., Van Norman, J. and Kaloshian, I. (2019). Softening the breaks: releasing a key negative immune regulator results in enhanced resistance to nematodes. *Molecular Plant-Microbe Interactions* 32, 153–153.

Trudgill, D. L. (1967). The effect of environment on sex determination in *Heterodera rostochiensis*. *Nematologica* 13(2), 263–272.

Trudgill, D. L. and Blok, V. C. (2001). Apomictic, polyphagous root-knot nematodes: exceptionally successful and damaging biotrophic root pathogens. *Annual Review of Phytopathology* 39, 53–77.

Van Damme, D., Vanstraelen, M. and Geelen, D. (2007). Cortical division zone establishment in plant cells. *Trends in Plant Science* 12(10), 458–464.

Van der Vossen, E. A. G., Rouppe van der Voort, J. N. A. M., Kanyuka, K., Bendahmane, A., Sandbrink, H., Baulcombe, D. C., Bakker, J., Stiekema, W. J. and Klein-Lankhorst, R. M. (2000). Homologues of a single resistance-gene cluster in potato confer resistance to distinct pathogens: a virus and a nematode. *Plant Journal* 23(5), 567–576.

Van Megen, H., van den Elsen, S., Holterman, M., Karssen, G., Mooyman, P., Bongers, T., Holovachov, O., Bakker, J. and Helder, J. (2009). A phylogenetic tree of nematodes based on about 1200 full-length small subunit ribosomal DNA sequences. *Nematology* 11(6), 927–950.

Vieira, P., Engler, G. and de Almeida Engler, J. (2012). Whole-mount confocal imaging of nuclei in giant feeding cells induced by root-knot nematodes in Arabidopsis. *New Phytologist* 195(2), 488–496.

Wieczorek, K., Elashry, A., Quentin, M., Grundler, F. M. W., Favery, B., Seifert, G. J. and Bohlmann, H. (2014). A distinct role of pectate lyases in the formation of feeding structures induced by cyst and root-knot nematodes. *Molecular Plant-Microbe Interactions* 27(9), 901–912.

Wieczorek, K., Golecki, B., Gerdes, L., Heinen, P., Szakasits, D., Durachko, D. M., Cosgrove, D. J., Kreil, D. P., Puzio, P. S., Bohlmann, H. and Grundler, F. M. W. (2006). Expansins are involved in the formation of nematode-induced syncytia in roots of *Arabidopsis thaliana*. *Plant Journal* 48(1), 98–112.

Wieczorek, K., Hofmann, J., Blochl, A., Szakasits, D., Bohlmann, H. and Grundler, F. M. W. (2008). Arabidopsis endo-1,4-beta-glucanases are involved in the formation of root syncytia induced by *Heterodera schachtii*. *Plant Journal* 53(2), 336–351.

Wiggers, R. J., Starr, J. L. and Price, H. J. (1990). DNA content and variation in chromosome number in plant-cells affected by *Meloidogyne incognita* and *M. arenaria*. *Phytopathology* 80(12), 1391–1395.

Williamson, V. M. and Kumar, A. (2006). Nematode resistance in plants: the battle underground. *Trends in Genetics* 22(7), 396–403.

Witek, K., Jupe, F., Witek, A. I., Baker, D., Clark, M. D. and Jones, J. D. G. (2016). Accelerated cloning of a potato late blight-resistance gene using RenSeq and SMRT sequencing. *Nature Biotechnology* 34(6), 656–660.

Wyss, U. (1978). Root and cell response to feeding by *Xiphinema index*. *Nematologica* 24(2), 159–166.

Wyss, U. (1992). Observations on the feeding behaviour of *Heterodera schachtii* throughout development, including events during moulting. *Fundamental and Applied Nematology* 15, 75–89.

Wyss, U. (2002). Feeding mechanisms of plant parasitic nematodes. In: *The Biology of Nematodes*, Lee, D. L. (Ed), 233–259. London: Taylor & Francis.

Wyss, U. and Grundler, F. M. W. (1992). Feeding behaviour of sedentary plant parasitic nematodes. *Netherlands Journal of Plant Pathology* 98(S2), 165–173.

Wyss, U. and Zunke, U. (1986). Observations on the behaviour of second stage juveniles of *Heterodera schachtii* inside host roots. *Revue Nematology* 9, 153–165.

Wyss, U., Lehmann, H. and Jank-Ladwig, R. (1980). Ultrastructure of modified root-tip cells in *Ficus carica*, induced by the ectoparasitic nematode *Xiphinema index*. *Journal of Cell Science* 41, 193–208.

Xie, Q. G., Li, Q., Sergeant, D., Smith-Becker, J. and Kaloshian, I. (2004). Underlying pathways in the Mi-l-mediated resistance to nematode and aphid. *Phytopathology* 94, S112–S112.

Yaghoobi, J., Yates, J. L. and Williamson, V. M. (2005). Fine mapping of the nematode resistance gene Mi-3 in *Solanum peruvianum* and construction of a S-lycopersicum DNA contig spanning the locus. *Molecular Genetics and Genomics* 274(1), 60–69.

Yamamoto, T. and Hayashi, T. (2002). New root-knot nematode resistance genes and their STS markers in peach. *Scientia Horticulturae* 96(1–4), 81–90.

Yang, Y. A., Jittayasothorn, Y., Chronis, D., Wang, X. H., Cousins, P. and Zhong, G. Y. (2013). Molecular characteristics and efficacy of *16D10* siRNAs in inhibiting root-knot nematode infection in transgenic grape hairy roots. *PLoS ONE* 8(7), e69463.

Zhou, J., Xu, X. C., Cao, J. J., Yin, L. L., Xia, X. J., Shi, K., Zhou, Y. H. and Yu, J. Q. (2018). Heat shock Factor HsfA1a Is essential for R gene-mediated nematode resistance and triggers H2O2 production1. *Plant Physiology* 176(3), 2456–2471.

Zipfel, C. and Oldroyd, G. E. D. (2017). Plant signalling in symbiosis and immunity. *Nature* 543(7645), 328–336.

Part 3

Root uptake of nutrients and water

Chapter 11

Advances in the understanding of nitrogen (N) uptake by plant roots

Malcolm J. Hawkesford and William R. Whalley, Rothamsted Research, UK

1 Introduction

Efficient use of nitrogen (N) by plants and particularly crops is of global importance. In agriculture, high crop yields and protein content depend upon extensive N-inputs, but fertiliser N is costly to the farmer, and overuse can be damaging to the environment. A critical component of optimised usage is efficient capture by crop root systems. Globally the recovery of N by grain crops in terms of N harvested in the grain as a percentage of fertiliser inputs has been estimated to be as low as 33% when soil and rainfall inputs are taken into consideration (Raun and Johnson, 1999). In natural systems, complex cycles that exist in the soil and uptake processes may be quite different from those in cropping systems. Specifically pools of free N are relatively low as N is robustly recycled and is a resource heavily competed for. To succeed, plants have evolved highly efficient uptake mechanisms. These systems both enable high affinity for the substrates and are specialised in scavenging a wide range of N-compounds. Losses from these ecosystems are typically low.

http://dx.doi.org/10.19103/AS.2020.0075.15

All N is acquired from the soil via the roots and may be taken up in a variety of both inorganic and organic forms. This N may vary both spatially and temporally in the soil profile, and root systems need to be adaptable to optimise uptake. A measure of the effectiveness of capture and use is nitrogen use efficiency (NUE), often only considered in terms of yield per unit N applied, but is the product of both uptake efficiency into the plant (NUpE) and utilisation efficiency in terms of production or yield (NUtE). Roots have a major role contributing specifically to NUpE. While yield and NUE are important considerations, N recovery may be used as a measure of efficiency and is a particularly important concept in developing means of damaging environmental effects. The quoted global efficiency of 33% suggests that uptake efficiency is inefficient for crops. Improving this figure is a major challenge for agronomy and crop breeding in order to optimise the use of this important resource. Targets for crop improvement include optimising root structure and proliferation, as well as function, for example, manipulation of key membrane transporters involved in the transfer of N-compounds from the soil into roots cells. An important consideration is the interaction of the root with the soil, and physical factors such as soil strength and water availability will influence these interactions and impact the functioning of the acquisition systems.

2 Nitrogen in the soil

Nitrogen in the soil can go through many chemical transformations catalysed by the soil microbial community and plant root systems in the soil (Jackson et al., 2008). Although fertilisers may be applied in many forms including nitrate, ammonium and urea and all, in some cases following chemical transformations, provide nitrogen which is available to crops, nitrate is the major form taken up. Ammonium can be acquired directly by plant roots, but it is usually converted into nitrate in most non-acidic aerobic soils. Transporter systems for both nitrate and ammonium occur in plant roots, optimised for both low and high concentrations of these inorganic ions. Except after specific inputs such as fertilisers, free N-pool sizes are mostly quite low in soils and high-affinity transport systems are likely to be the key mechanism for capture. Nitrate and other forms of N in the soil diffuse towards the roots down the concentration gradients induced by plant uptake or move towards the root as a consequence of the mass flow of water driven by the transpiration of the crop (Matimati et al., 2014). Interception will depend on root architecture and proliferation and uptake by the transporters themselves.

Nitrate is very mobile in the soil and may be readily leached through the soil profile, contributing to the poor uptake efficiency when large fertiliser inputs occur. Targeted application in terms of trickle feeding or multiple partial applications based on demand, or spatially adjusted inputs, such as found in precision farming, are the major mechanisms for agronomic improvement. However, to best exploit these approaches, suitable germplasm with high NUE attributes must be employed.

In some agricultural situations, and in many natural ecosystems, organic forms of nitrogen may be more important for plant growth. Turnover of organic matter such as protein, in the soil and rhizosphere, will result in oligopeptide and free amino acid proliferation, both of which could be exploited by the root. Soil organic matter may originate predominantly from plant materials, either exudates or tissue decay, but turnover depends on microbial activity and roots must continually compete with the microbial population for N for de novo growth. In these circumstances, plant uptake systems for amino acids and small oligopeptides will be important.

3 Nitrogen uptake and NUE in agriculture

NUE can be considered in a number of ways and comprises terms relating to uptake and conversion into biomass. Roots are the principal site for uptake, and therefore the roots and processes in the root are clearly directly responsible for nitrogen uptake and capture efficiency. However, uptake also depends on sink strength, and hence a larger canopy has a greater demand for N, resulting in greater N-uptake. Factors contributing to N-uptake and NUE in a typical crop such as wheat are summarised in Fig. 1.

In agronomic terms, NUpE is the amount of nitrogen taken up into the crop biomass as a fraction of all N available to the crop (including fertiliser N, soil N and N from atmospheric deposition). The product of NUpE and NUtE (the harvested biomass, for example, grain as a function of the N taken up into the crop) gives overall grain NUE (grain yield per unit of available N) (Moll et al., 1982; Barraclough et al., 2010). An alternative way to consider NUE is to consider the recovery of applied N in the whole crop, irrespective of biomass yield. This approach is termed as the fertiliser recovery efficiency or alternatively the output/input ratio (N-retention). It is the amount recovered in the crop divided by that available (Zhang et al., 2015). Improving any of these measures in crops will result in better use of N and decreased environmental losses.

The efficiency of uptake by the root system depends upon intrinsic properties of the roots, including root depth, degree of branching, overall or local proliferation, and root angle as an indicator of root spread, as well as the activity of membrane nitrogen transporters. These traits, which are fundamentally genetically controlled, are also influenced by the form and concentration of nitrogen, fertiliser inputs in the case of agricultural systems, water availability, and physical properties of the soil. These factors will be discussed in subsequent sections.

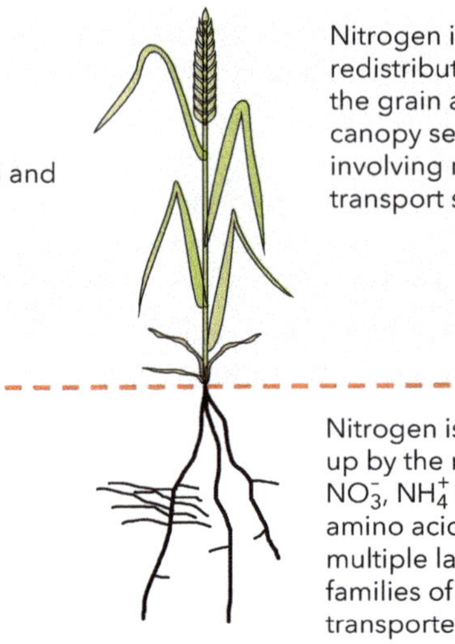

Figure 1 Diagrammatic representation of traits important for the uptake of N using wheat as an example.

4 Genetic variation in NUE and roots

Variation in rooting has a strong genetic basis and is dependent on the local conditions of water and nutrient availability. Across different ecosystems considerable variation occurs, for example, in rooting depth, between ecosystems, for species within these ecosystems, and for individuals within a species (Fig. 2). Within crop species there is diversity, which may be harnessed for crop improvement. The observed variation in NUE can be attributed to the various components of NUE and not just to the roots, but root architecture and proliferation should be expected generally to be a major determinant of N-uptake. Root growth and hence N-capture is required for different crop types at different specific growth stages, depending on whether vegetation or seeds are the final product (Lammerts van Bueren and Struik, 2017). Crop improvement strategies need to accommodate this variation in N requirement, which is dependent on the growth pattern.

Many studies have examined the genetic diversity of NUE in plant species. Substantial natural variation in NUE amongst 18 accessions of *Arabidopsis thaliana* was reported to be independent of N-supply but had the greatest variation at high N-supply (Chardon et al., 2010). This variation appeared to be mainly due to uptake and was strongly influenced by shoot biomass, indicating a strong control by sink demand.

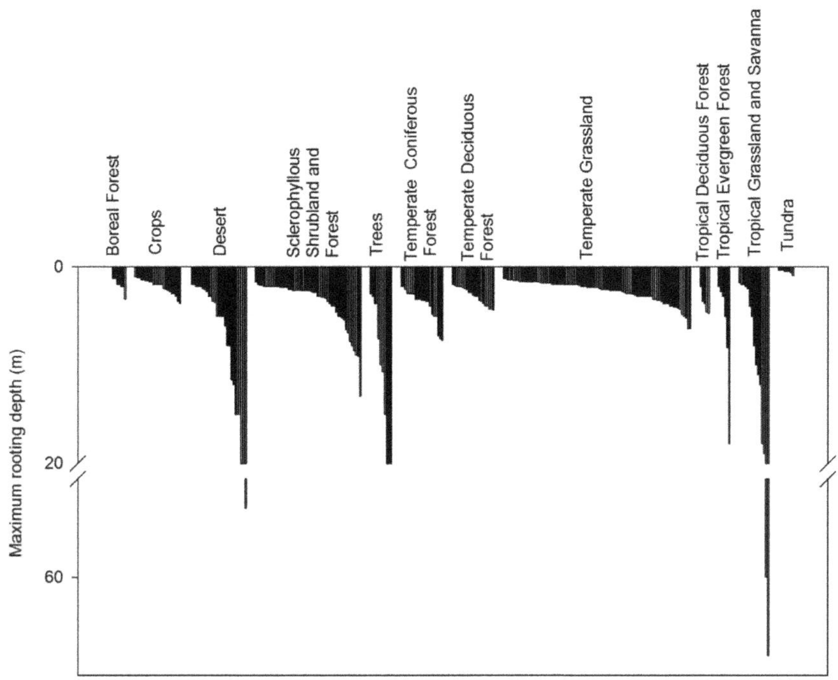

Figure 2 Maximum rooting depths for different biomes (from Canadell et al., 1996).

Genetic variation in NUE and component traits for wheat have been reported (Ortiz-Monasterio et al., 1997; Barraclough et al., 2010; Gaju et al., 2011; Allard et al., 2013), and overall diversity and prospects for improvement in wheat have been reviewed (Hawkesford and Griffiths, 2019). NUpE was reported to be the most important component of NUE for wheat grown at low N availability (Ortiz-Monasterio et al., 1997; Gaju et al., 2011). Variation in nitrogen uptake capability is mainly related to biomass (Barraclough et al., 2010), similar to the situation for *Arabidopsis* described earlier (Chardon et al., 2010). A similar conclusion was made by Allard et al. (2013). Although architectural traits such as root length and distribution are genetically controlled and can affect N-acquisition, the degree of tillering, which is genetically and nutritionally controlled, directly influences root production (Allard et al., 2013). These authors also suggest that root N re-distribution is not a major factor in NUE parameters. However, genetic variability in seedling root architecture was related to field NUpE in a study using a large wheat diversity panel (Kenobi et al., 2017).

While NUpE is mostly related only to root architectural traits, function can also be a major influence on efficiency. For example, in rice, genetic diversity in N-use between rice sub-species (*indica* and *japonica*) may be specifically attributed to variation in a single nitrate transporter, NRT1.1B (Hu et al., 2015).

5 Transporters involved in nitrogen uptake

Nutrients in the soil are taken up by root cells via membrane transporters. The initial site of cellular uptake may be at the root hairs, in the exodermis or at inner cell layers, and, subsequently, nutrients need to be transferred to the vascular tissue for further transport to the shoot tissues. Within the roots, the movement of nutrients may be via a number of alternative routes and involve symplastic or apoplastic mechanisms or a coupled trans-cellular pathway involving repeated uptake and efflux within the root, prior to loading of xylem tissues and transportation to the shoot (Barberon et al., 2016). The multiple routes for lateral movements are partly the result of a diffusion barrier in the apoplast, the Casparian strip, present to control the diffusion of nutrients and water and to protect against pathogens (Geldner, 2013).

Transport across cellular membranes includes sites at the initial location of uptake into the plant within the roots, as well as multiple locations of transfer to the various sink tissues throughout the plant. Multiple transporters are responsible for these movements for all of the different N-compounds found in the soil, which are utilisable by the plant. The transporters are mostly encoded by large gene families, with individual members being expressed in different tissues, responding differently to developmental and environmental cues, or even having different substrate specificities (Fan et al., 2017). There may be a direct transfer of nutrients from the soil through the root to other parts of the plant, or compounds may undergo chemical conversion within the plant. For example, the assimilation of nitrate into glutamine and other amino acids may also occur within the root tissues. Subsequent transfer steps around the plant require related transporters for nitrate or assimilation products.

Given the importance of nitrate as an N source in many environments, nitrate transporters have been the focus of extensive study. Two major types of nitrate transporters with contrasting substrate affinities exist in plants to cope with the perhaps four orders (from μM to mM) of magnitude difference in nitrate concentrations which can occur in soils: NPF (nitrate transporter 1/peptide family, formerly known as NRT1 transporters) or NRT2 (nitrate transporter 2) transporters (Tsay et al., 2007; Fan et al., 2017). Both forms are encoded by large gene families, reflecting the complexity of managing the movement of nitrogen within a plant. There may be some redundancy of function among the many homologues, but the major significance of these large gene families is differential spatial expression, different substrate affinities and substrate specificity. In some instances, individual transporters may be involved in the transport of compounds other than nitrate.

The NPF family transporters are mostly low-affinity nitrate transporters, but some also transport peptides and other anions. The gene family consists of 53 members in *Arabidopsis* (Tsay et al., 2007), and multiple NPF genes also exist in crop species such as wheat with up to 77 examples (Buchner and Hawkesford, 2014; Bajgain et al., 2018). The NPF transporters are able to catalyse the transport of many substrates including nitrate, amino acids and auxin. Spatial variation in the expression in different tissues, in response to development and nutrient availability, has been shown in wheat (Buchner and Hawkesford, 2014). At least one NPF transporter, AtNRT1.1, has a dual affinity for nitrate modulated by the phosphorylation of a threonine residue, switching this transporter from a low- to high-affinity mode; this is induced by external nitrate concentration (Liu and Tsay, 2003). This also clearly demonstrates a potential role in sensing nitrate, facilitating the plant response to nitrate in the environment (Fan et al., 2016; Wang et al., 2019).

The NRT2 family of high-affinity nitrate transporters is smaller in number, with seven members in *Arabidopsis* (Tsay et al., 2007). An extensive study in rice indicates spatial expression and regulation by N and C (Feng et al., 2011). These high-affinity transporters are critical for nitrate acquisition in many agricultural and natural environments.

In many agricultural and natural systems, particularly anaerobic or acidic soils, ammonium may be an important pool of N, and roots have specific ammonium transporters (AMT) to capture this N resource. AMT transporters are also encoded by a gene family, with 6 reported members for *Arabidopsis* (Yuan et al., 2007) and at least 23 in wheat (Li et al., 2017; Bajgain et al., 2018). Of the six AMT transporters in *Arabidopsis*, four are highly expressed in roots and up-regulated at the level of gene expression under nitrogen-limiting conditions. These transporters function as ammonium uniporters or proton/ammonia co-transporters or, in one case, as an ammonia channel (Yuan et al., 2007; Pantoja, 2012). Within the root, the differential spatial expression is optimised for efficient uptake from the soil solution and transfer to the xylem involving both symplastic and apoplastic routes (Yuan et al., 2007).

Nitrogen may be acquired from the soil in various organic forms. Transporters for organic N-compounds include members of the nitrate transporter family (Tsay et al., 2007), specific amino acid transporters which may transport single or multiple amino acids (AAT comprising AAAP and APC subgroups) (Wan et al., 2017) and peptide transporters such as Usually Multiple Acids Move In and out Transporters (UMAMIT) (Denancé et al., 2014). As with all transporter families, specific amino acid transporters from any of these families may be involved in primary uptake, cell-to-cell transfer, xylem loading or sink-tissue delivery. Complex spatiotemporal patterns of expression of individual isoforms and differing amino acid specificities aid in the plant management of nitrogen compounds, which, apart from being sources of N, are key metabolites for organ-to-organ N-transport, N-storage and amino acid metabolism, in general.

Developmental cues, coupled with N availability, impact expression patterns of most N transporters. For example, during plant development, demand for N fluctuates to facilitate canopy growth or storage organ development, and the expression of transporters is modulated at the level of gene expression (Buchner and Hawkesford, 2014; Garnett et al., 2013). This is generally not a rapid response, and, in addition, post-translational control may occur to effect rapid responses to fluctuating N availability in the environment, thereby maximising capture (Garnett et al., 2013).

6 Root responses to nitrogen availability

Nitrogen availability in the soil may vary in form, show considerable spatial variability and fluctuate temporally in both the short and long term. Plants and crops adapt to this variability with many levels of response, involving changes in the allocation of resources to roots, in architecture, and in biochemical and physiological changes to function (reviewed in Nacry et al., 2013). Responses to N availability involve molecular signalling pathways that respond to internal nutritional signals modulated by metabolism and to external signals indicating soil availability (reviewed by Jia and von Wirén, 2020).

A primary response to N imbalances between supply and demand is a change in the shoot/root ratio, with lower resource allocation to roots when N is plentiful, and greater allocation initially to enhance root proliferation under N-limitation (e.g. Agren and Franklin, 2003). The plasticity of root responses also includes the ability to respond to spatial variation in nitrogen availability. For example, the localised proliferation of roots is seen in response to the localised supply of nitrogen, either as nitrate or as ammonium (Drew et al., 1973; Drew, 1975). The specific primary signal appears to be local nitrate concentrations, but systemic signalling involving auxins is also involved. The components of the molecular signalling pathway of this response include a MADS-box transcription factor (ANR1) (Zhang and Forde, 1998; Gan et al., 2005). The NRT1.1 transporter is linked to this response, transporting auxin away from lateral root tips under low N conditions and thus relieving auxin repression of growth. At higher N-levels, this transporter is involved in nitrate uptake (Wang et al., 2019; Jia and von Wirén, 2020). Manipulating this pathway offers another target for crop improvement.

7 Effects of soil conditions and structure on root N-uptake and proliferation

Root proliferation is a major factor in the capacity for plant nutrient uptake, and in the field it is affected by many factors which vary in space and time, and which can interact with one another. Establishing which factors are limiting root growth is not easy. Soil compaction is associated with both decreased nitrogen uptake and yield (Alakukku and Elonen, 1995). Poor root growth in compacted soils is commonly used as at least a partial explanation. It is widely assumed that the effects of compact soil are simply to limit nitrogen availability by either decreased capture by roots or denitrification in waterlogged soils (e.g. Chamen et al., 2015). In the field roots experience a combination of physical and nutrient stresses. Root proliferation with depth depends on the combination of physical and nutrient stresses and the effects of their interaction. In a review of rooting depths, Canadell et al. (1996) showed that the enormous range of possible rooting depths (Fig. 2) is from less than 1 m up to 68 m. Some of the smallest rooting depths were in crops. The depth of roots is determined by genetics and soil conditions. At present there is limited evidence that the genetic selection within a crop species will provide deep rooting genotypes with any reliability (Bai et al., 2019). Soil conditions such as water and nutrient availability have a significant impact on both rooting depth and proliferation.

Penetrometer resistance is of considerable value because it is a good predictor of the ease with which roots can penetrate soil by deformation, provided all other stresses are not limiting. At deeper depths penetrometer resistance is greater due to the effects of the self-weight of soil or overburden pressure and internal soil friction (Richards and Greacen, 1986; Whitmore et al., 2011; Gao et al., 2016a,b). This is illustrated in Fig. 3, which compares the penetrometer resistance profile measured in a field with that estimated from laboratory measurements made on core extracted from different depths. Soil strength increases with depth, simply due to the weight of soil,

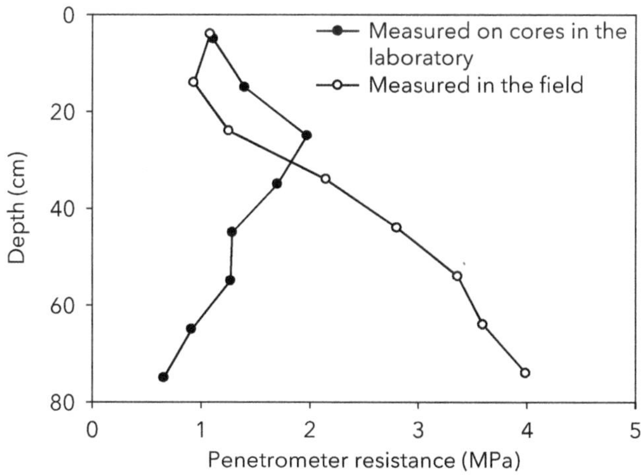

Figure 3 A comparison of penetrometer resistance measured in the field with that measured in the laboratory on cores taken in the field. In the field, penetrometer resistance increases with depth due to the effect of the increasing overburden pressure with depth. When cores are extracted, the overburden pressure is removed, and the soil has a much lower penetrometer resistance at depth. Irrespective of soil management, the soil penetrometer resistance increases with depth, see Zhou et al., (2020).

and is sufficiently great to limit root elongation by soil deformation. Typically, penetrometer resistances greater than 2.5 MPa seriously restrict root elongation in structureless soil (Bengough and Mullins, 1991). In deeper layers where soil penetrometer resistance is high, roots are confined to existing pore networks (Atkinson et al., 2019; White and Kirkegaard, 2010; Hodgkinson et al., 2017). Unfortunately, roots in pores are less effective at extracting water and nutrients from the soil because their hydraulic connection with the soil is poor compared to roots which have deformed soil in the surface layers. Both root hairs and mucilage exuded by roots improve soil-root contact and water uptake (Carminati et al., 2017; Zarebanadkouki et al., 2019). Mucilage has also been shown to improve nutrient diffusion in the rhizosphere (Zarebanadkouki et al., 2019). However, in deeper layers, due to a combination of low root density and poor soil-root contact, roots are unable to access all the available soil water. In Northern Europe, even in dry summer, the soil at depths of 0.5 m or more can be relatively wet because water is not accessed by roots (Whalley et al., 2006, 2008). A further issue is that the roots tend to be clumped so that several roots may occupy a single pore (White and Kirkegaard, 2010). While water and nitrate around clumped roots may be extracted more effectively than if the pore were to be occupied by a single root, the negative effect of clumping is to produce a non-optimal root distribution at depth for resource capture (Passioura, 1991).

In the surface layers, root proliferation is related to both water and nutrient availability. Figure 4 shows the total root length of wheat in a pot experiment (Ge et al., 2019) as a function of nitrate concentration of the soil water in well-watered and water-limited conditions. In common with many studies (e.g. Roycewicz and Malany, 2012), at higher nitrate levels, there is inhibition of

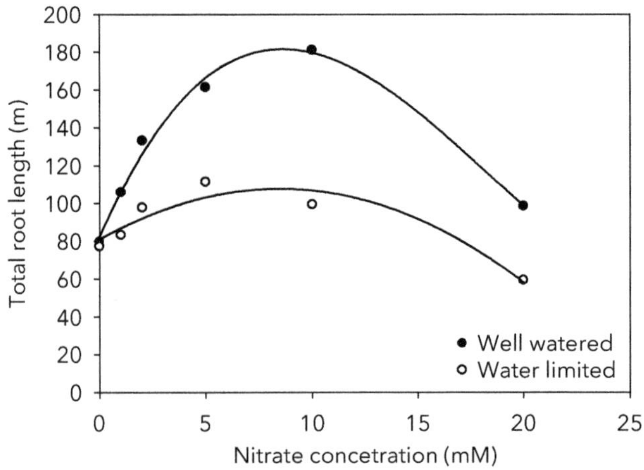

Figure 4 Total root length of wheat in a pot experiment as a function of nitrate concentration of the soil water in well-watered and water-limited conditions. Data is for wheat (cv. Cadenza) plotted from table 3 of Ge et al. (2019).

root elongation. Root diameter increases with root impedance (Atwell, 1990), and root diameter also increases with nitrate concentration. Shoot nitrate concentration is correlated with an increased mean root diameter (Darwent et al., 2003), which is consistent with the increased production of fine roots under low N-supply. Models designed to optimise the root ideotype in maize (Postma et al., 2014) predict that the genotypes with fewer branches are adapted to low-nitrate environments while the genotypes with more branches are adapted to low-phosphorus environments. This is consistent with experimental evidence (Ge et al., 2019) for wheat where fewer root axis with thinner roots have been observed under low-nitrate levels. Thinner roots maximised the soil exploration for a given carbon input (Postma et al., 2014). Thicker roots associated with higher nitrate levels are also associated with improved penetration of hard layers (Clark et al., 2008).

8 Future trends for crop improvement

Given the observed natural variation within populations for root traits and for nitrogen capture, there is a clear potential for breeding to optimise nitrogen capture in different environments. Many of the target traits are summarised in Fig. 1. The approaches would include exploiting genetic variability in transport and architectural traits to both enhance nutrient uptake efficiency and cope with difficult soils. Manipulating root systems will have consequences for all nutrient uptakes and may negatively impact required resources for roots to the detriment of shoot allocation, possibly requiring a holistic approach. Alternatively, a more targeted approach would be to focus on specific transporters and pathways associated with N-capture and delivery to sites of assimilation. In an agricultural context, tailoring the expression of capture mechanisms to cope with the fluctuations of availability due to transient changes induced by fertiliser applications would be beneficial.

Improvements in NUE would be beneficial to agricultural production and for alleviating the damaging environmental aspect of overuse of N. Improvements to N-capture would be directly beneficial for improving fertiliser recovery and might be not only obtained through improved agronomy but also aided by improved varieties, with roots optimised for architectural and functional traits relating to capture. There is also a need to separate yield and biomass from uptake and to manipulate these independently.

Substantial variation has been shown for root proliferation and root distribution in many studies (e.g. Atkinson et al., 2015; Bai et al., 2019), as well as specific genetic variability for individual gene expression and function, notably of transporters such as NRT1.1B (Hu et al., 2015). Expression of this allele, as that occurs in *Indica* rice, in lines such as *Japonica*, results in higher N-uptake. In some instances, increased expression of a specific transporter does not directly influence uptake from the soil solution but is the result of impacts of upstream processes. For example, a nitrate transporter which affects phloem pH regulation has a downstream effect on general nutrient acquisition in the roots and transfer to the shoot (Fan et al., 2016). Overexpression of this nitrate transporter, OsNRT2.3b, was shown to increase nutrient (N, Fe and P) uptake and crop yields and enhance overall NUE by 40%.

A novel approach to improving NUE is the overexpression of downstream elements involved in N-assimilation, effectively to increase sink strength. An example of this is the overexpression of alanine amino transferase under the control of a root-specific promoter, which serves to increase the flux of N to assimilation products, thereby increasing a local sink demand. This increased N flux induced an increased nitrate uptake mediated by the increased expression of high-affinity nitrate transporters (Good et al., 2007). The overall result in the field trials of *Brassica napus* was increased yield at lower N-levels, although little effect at high N-inputs.

Current evidence suggests that the soil structure may be at least as important as genetics in determining root proliferation, especially at depth. There is a clear role for improved soil management as one of the tools to increase NUE. Any management practice that increases connected porosity to depth will be of value. Furthermore, compaction of soil should be avoided because increased soil strength stunts shoot growth irrespective of N availability (Ge et al., 2019). Eventually the development of new nutrient-efficient genotypes needs to be coupled with screening in a wide variety of soil types and conditions.

In summary, both genetics and agronomy need to be combined to produce better crop varieties and farming practices to improve the current global inefficiencies of N-capture in agriculture.

9 Acknowledgements

Rothamsted Research receives grant-aided support from the Biotechnology and Biological Sciences Research Council (BBSRC) through the Designing Future Wheat programme [BB/P016855/1].

10 Where to look for further information

10.1 Further reading

- A general guide to nutrient acquisition is *Marschners Mineral Nutrition of Higher Plants* 3rd Edition 2012 ed. by Petra Marschner, Elsevier Academic Press.
- *Roots the Hidden Half* (Eds Amram Eshel and Tom Beeckman) Fourth Edition. 2013 ISBN 9781439846483.

- *A Journal of Experimental Botany Special Issue* (Volume 68, Issue 10, Issue 10, May 2017: Nitrogen Nutrition in Plants: Rapid Progress and New Challenges, contains many reviews on the current status of nitrogen acquisition and metabolism.

10.2 Key projects and websites

- The Defra-sponsored Wheat Genetic Improvement Network (WGIN) includes an extensive analysis of nitrogen use efficiency in wheat germplasm: http://www.wgin.org.uk.
- The Wheat Initiative has many wheat-orientated resources and specifically an expert working group on nutrient use efficiency and many other resources for wheat research: https://www.wheatinitiative.org.
- Designing Future Wheat is a UK BBSRC major programme coordinating wheat genetic resources and includes the aspects of root biology and efficient nutrient use: https://designingfuturewheat.org.uk.
- The use of 3D X-ray CT scanners for examining root stricture is the focus of the Nottingham Hounsfield facility: https://www.nottingham.ac.uk/microct/.

11 References

Agren, G. I. and Franklin, O. (2003). Root: shoot ratios, optimization and nitrogen productivity, *Ann. Bot.* 92(6): 795–800.

Alakukku, L. and Elonen, P. (1995). Long-term effects of a single compaction by heavy field traffic on yield and nitrogen uptake of annual crops, *Soil Till. Res.* 36(3–4): 141–152.

Allard, V., Martre, P. and Le Gouis, J. (2013). Genetic variability in biomass allocation to roots in wheat is mainly related to crop tillering dynamics and nitrogen status, *Eur. J. Agron.* 46: 68–76.

Atkinson, J. A., Wingen, L. U., Griffiths, M., Pound, M. P., Gaju, O., Foulkes, M. J., Le Gouis, J., Griffiths, S., Bennett, M. J., King, J. and Wells, D. M. (2015). Phenotyping pipeline reveals major seedling root growth QTL in hexaploid wheat, *J. Exp. Bot.* 66(8): 2283–2292.

Atkinson, J. A., Hawkesford, M. J., Whalley, W. R., Hu, Z. and Mooney, S. J. (2019). Soil strength influences wheat root interactions with soil macropores, *Plant Cell Environ.* 43(1): 235–245.

Atwell, B. J. (1990). The effect of soil compaction on wheat during early tillering. I. Growth, development and root structure, *New Phytol.* 115(1): 29–35.

Bai, C., Ge, Y., Ashton, R. W., Evans, J., Milne, A., Hawkesford, M. J., Whalley, W. R., Parry, M. A. J., Melichar, J., Feuerhelm, D., Bansept Basler, P. B. and Bartsch, M. (2019). The relationships between seedling root screens, root growth in the field and grain yield for wheat, *Plant Soil* 440(1–2): 311–326.

Bajgain, P., Russell, B. and Mohammadi, M. (2018). Phylogenetic analyses and in-seedling expression of ammonium and nitrate transporters in wheat, *Sci. Rep.* 8(1): 7082.

Barberon, M., Vermeer, J. E. M., De Bellis, D., Wang, P., Naseer, S., Andersen, T. G., Humbel, B. M., Nawrath, C., Takano, J., Salt, D. E. and Geldner, N. (2016). Adaptation of root function by nutrient-induced plasticity of endodermal differentiation, *Cell* 164(3): 447–459.

Barraclough, P. B., Howarth, J. R., Jones, J., Lopez-Bellido, R., Parmar, S., Shepherd, C. E. and Hawkesford, M. J. (2010). Nitrogen efficiency of wheat: genotypic and environmental variation and prospects for improvement, *Eur. J. Agron.* 33(1): 1–11.

Bengough, A. G. and Mullins, C. E. (1991). Penetrometer resistance, root penetration resistance and root elongation rate in two sandy loam soils, *Plant Soil* 131(1): 59–66.

Buchner, P. and Hawkesford, M. J. (2014). Complex phylogeny and gene expression patterns of members of the nitrate transporter 1/peptide transporter family (NPF) in wheat, *J. Exp. Bot.* 65(19): 5697–5710.

Canadell, J., Jackson, R. B., Ehleringer, J. B., Mooney, H. A., Sala, O. E. and Schulze, E. D. (1996). Maximum rooting depth of vegetation types at the global scale, *Oecologia* 108(4): 583–595.

Carminati, A., Passioura, J. B., Zarebanadkouki, M., Ahmed, M. A., Ryan, P. R., Watt, M. and Delhaize, E. (2017). Root hairs enable high transpiration rates in drying soils, *New Phytol.* 216(3): 771–781.

Chamen, W. C., Moxey, A. P., Towers, W., Balana, B. and Hallett, P. D. (2015). Mitigating arable soil compaction: a review and analysis of available cost and benefit data, *Soil Till. Res.* 146: 10–25.

Chardon, F., Barthelemy, J., Daniel-Vedele, F. and Masclaux-Daubresse, C. (2010). Natural variation of nitrate uptake and nitrogen use efficiency in *Arabidopsis thaliana* cultivated with limiting and ample nitrogen supply, *J. Exp. Bot.* 61(9): 2293–2302.

Clark, L. J., Ferraris, S., Price, A. H. and Whalley, W. R. (2008). A gradual rather than abrupt increase in soil strength gives better root penetration of strong layers, *Plant Soil* 307(1–2): 235–242.

Darwent, M. J., Paterson, E., McDonald, A. J. S. and Tomos, A. D. (2003). Biosensor reporting of root exudation from *Hordeum vulgare* in relation to shoot nitrate concentration, *J. Exp. Bot.* 54(381): 325–334.

Denancé, N., Szurek, B. and Noël, L. D. (2014). Emerging functions of nodulin-like proteins in non-nodulating plant species, *Plant Cell Physiol.* 55(3): 469–474.

Drew, M. C. (1975). Comparison of effects of a localized supply of phosphate, nitrate, ammonium and potassium on growth of seminal root system, and shoot, in barley, *New Phytol.* 75(3): 479–490.

Drew, M. C., Saker, L. R. and Ashley, T. W. (1973). Nutrient supply and the growth of the seminal root system in barley I. The effect of nitrate concentration on the growth of axes and laterals. *J. Exp. Bot.* 24, 1189.

Fan, X., Tang, Z., Tan, Y., Zhang, Y., Luo, B., Yang, M., Lian, X., Shen, Q., Miller, A. J. and Xu, G. (2016). Overexpression of a pH-sensitive nitrate transporter in rice increases crop yields, *Proc. Natl. Acad. Sci. U.S.A.* 113(26): 7118–7123.

Fan, X., Naz, M., Fan, X., Xuan, W., Miller, A. J. and Xu, G. (2017). Plant nitrate transporters: from gene function to application, *J. Exp. Bot.* 68(10): 2463–2475.

Feng, H., Yan, M., Fan, X., Li, B., Shen, Q., Miller, A. J. and Xu, G. (2011). Spatial expression and regulation of rice high-affinity nitrate transporters by nitrogen and carbon status, *J. Exp. Bot.* 62(7): 2319–2332.

Gaju, O., Allard, V., Martre, P., Snape, J. W., Heumez, E., LeGouis, J., Moreau, D., Bogard, M., Griffiths, S., Orford, S., Hubbart, S. and Foulkes, M. J. (2011). Identification of traits to improve the nitrogen use efficiency of wheat genotypes, *Field Crops Res.* 123(2): 139–152.

Gan, Y., Filleur, S., Rahman, A., Gotensparre, S. and Forde, B. G. (2005). Nutritional regulation of ANR1 and other root-expressed MADS-box genes in *Arabidopsis thaliana*, *Planta* 222(4): 730–742.

Gao, W., Whalley, W. R., Tian, Z., Liu, J. and Ren, T. (2016a). A simple model to predict soil penetrometer resistance as a function of density, drying and depth in the field, *Soil Till. Res.* 155: 190–198.

Gao, W., Hodgkinson, L., Jin, K., Watts, C. W., Ashton, R. W., Shen, J., Ren, T., Dodd, I. C., Binley, A., Phillips, A. L., Hedden, P., Hawkesford, M. J. and Whalley, W. R. (2016b). Deep roots and soil structure, *Plant Cell Environ.* 39(8): 1662 –1668.

Garnett, T., Conn, V., Plett, D., Conn, S., Zanghellini, J., Mackenzie, N., Enju, A., Francis, K., Holtham, L., Roessner, U., Boughton, B., Bacic, A., Shirley, N., Rafalski, A., Dhugga, K., Tester, M. and Kaiser, B. N. (2013). The response of the maize nitrate transport system to nitrogen demand and supply across the life cycle, *New Phytol.* 198: 829–894.

Ge, Y., Hawkesford, M. J., Rosolem, C. A., Mooney, S. J., Ashton, R. W., Evans, J. and Whalley, W. R. (2019). Multiple abiotic stress, nitrate availability and the growth of wheat, *Soil Till. Res.* 191: 171–184.

Geldner, N. (2013). The endodermis, *Annu. Rev. Plant Biol.* 64(1): 531–558.

Good, A. G., Johnson, S. J., De Pauw, M., Carroll, R. T., Savidov, N., Vidmar, J., Lu, Z., Taylor, G. and Stroeher, V. (2007). Engineering nitrogen use efficiency with alanine aminotransferase, *Can. J. Bot.* 85(3): 252–262.

Hawkesford, M. J. and Griffiths, S. (2019). Exploiting genetic variation in nitrogen use efficiency for cereal crop improvement, *Curr. Opin. Plant Biol.* 49: 35–42.

Hodgkinson, L., Dodd, I. C., Binley, A., Ashton, R. W., White, R. P., Watts, C. W. and Whalley, W. R. (2017). Root growth in field-grown winter wheat: some effects of soil conditions, season and genotype, *Eur. J. Agron.* 91: 74–83.

Hu, B., Wang, W., Ou, S., Tang, J., Li, H., Che, R., Zhang, Z., Chai, X., Wang, H., Wang, Y., Liang, C., Liu, L., Piao, Z., Deng, Q., Deng, K., Xu, C., Liang, Y., Zhang, L., Li, L. and Chu, C. (2015). Variation in NRT1.1B contributes to nitrate-use divergence between rice subspecies, *Nat. Genet.* 47(7): 834–838.

Jackson, L. E., Burger, M. and Cavagnaro, T. R. (2008). Roots, nitrogen transformations, and ecosystem services, *Annu. Rev. Plant Biol.* 59: 341–363.

Jia, Z. and von Wirén, N. (2020). Signaling pathways underlying nitrogen-dependent changes in root system architecture: from model to crop species, *J. Exp. Bot.* 71(15): 4393–4404. https://doi.org/10.1093/jxb/eraa033.

Kenobi, K., Atkinson, J. A., Wells, D. M., Gaju, O., De Silva, J. G., Foulkes, M. J., Dryden, I. L., Wood, A. T. A. and Bennett, M. J. (2017). Linear discriminant analysis reveals differences in root architecture in wheat seedlings related to nitrogen uptake efficiency, *J. Exp. Bot.* 68(17): 4969–4981.

Lammerts van Bueren, E. T. and Struik, P. C. (2017). Diverse concepts of breeding for nitrogen use efficiency. A review, *Agron. Sustain. Dev.* 37(5): 50.

Li, T. Y., Liao, K., Xu, X. F., Gao, Y., Wang, Z. Y., Zhu, X. F., Jia, B. L. and Xuan, Y. H. (2017). Wheat Ammonium Transporter (AMT) Gene family: diversity and possible role in host-pathogen interaction with stem rust, *Front. Plant Sci.* 8: 1637.

Liu, K. H. and Tsay, Y. F. (2003). Switching between the two action modes of the dual-affinity nitrate transporter CHL1 by phosphorylation, *EMBO J.* 22(5): 1005–1013.

Matimati, I., Verboom, G. A. and Cramer, M. D. (2014). Nitrogen regulation of transpiration controls mass-flow acquisition of nutrients, *J. Exp. Bot.* 65(1): 159–168.

Moll, R. H., Kamprath, E. J. and Jackson, W. A. (1982). Analysis and interpretation of factors which contribute to efficiency of nitrogen-utilization, *Agron. J.* 74(3): 562–564.

Nacry, P., Bouguyon, E. and Gojon, A. (2013). Nitrogen acquisition by roots: physiological and developmental mechanisms ensuring plant adaption to a fluctuating resource, *Plant Soil* 370(1–2): 1–29.

Ortiz-Monasterio, J. I., Sayre, K. D., Rajaram, S. and McMahon, M. (1997). Genetic progress in wheat yield and nitrogen use efficiency under four nitrogen rates, *Crop Sci.* 37(3): 898–904.

Pantoja, A. (2012). High affinity ammonium transporters; molecular mechanism of action, *Front. Plant Sci.* 3: 34.

Passioura, J. B. (1991). Soil structure and plant growth, *Soil Res.* 29(6): 717–728.

Postma, J. A., Dathe, A. and Lynch, J. P. (2014). The optimal lateral root branching density for maize depends on nitrogen and phosphorus availability, *Plant Physiol.* 166(2): 590–602.

Raun, W. R. and Johnson, G. V. (1999). Improving nitrogen use efficiency for cereal production, *Agron J.* 91(3): 357–363.

Richards, B. G. and Greacen, E. L. (1986). Mechanical stresses on an expanding cylindrical root analogue in granular media, *Aust. J. Soil Res.* 24: 393–404.

Roycewicz, P. and Malany, J. E. (2012). Dissecting the effects on nitrate, sucrose and osmotic potential on Arabidopsis root and shoot system growth in laboratory assays, *Philos. Trans. R. Soc. Lond. B Biol. Sci.* 367(1595): 1489–1500.

Tsay, Y. F., Chiu, C. C., Tsai, C. B., Ho, C. H. and Hsu, P. K. (2007). Nitrate transporters and peptide transporters, *FEBS Lett.* 581(12): 2290–2300.

Wan, Y., King, R., Mitchell, R. A. C., Hassani-Pak, K. and Hawkesford, M. J. (2017). Spatiotemporal expression patterns of wheat amino acid transporters reveal their putative roles in nitrogen transport and responses to abiotic stress, *Sci. Rep.* 7(1): 5461.

Wang, W., Hu, B., Li, A. and Chu, C. (2019). NRT1.1s in plants: functions beyond nitrate transport, *J. Exp. Bot.* 71(15): 4373–4379. https://doi.org/10.1093/jxb/erz554.

Whalley, W. R., Clark, L. J., Gowing, D. J. G., Cope, R. E., Lodge, R. J. and Leeds-Harrison, P. B. (2006). Does soil strength play a role in wheat yield losses caused by soil drying?, *Plant Soil* 280(1–2): 279–290.

Whalley, W. R., Watts, C. W., Gregory, A. S., Mooney, S. J., Clark, L. J. and Whitmore, A. P. (2008). The effect of soil strength on the yield of wheat, *Plant Soil* 306(1–2): 237–247.

White, R. G. and Kirkegaard, J. A. (2010). The distribution and abundance of wheat roots in a dense, structured subsoil – implications for water uptake, *Plant Cell Environ.* 33(2): 133–148.

Whitmore, A. P., Whalley, W. R., Bird, N. R. A., Watts, C. W. and Gregory, A. S. (2011). Estimating soil strength in the rooting zone of wheat, *Plant Soil* 339(1–2): 363–375.

Yuan, L., Loqué, D., Kojima, S., Rauch, S., Ishiyama, K., Inoue, E., Takahashi, H. and von Wirén, N. (2007). The organization of high-affinity ammonium uptake in Arabidopsis roots depends on the spatial arrangement and biochemical properties of AMT1-type transporters, *Plant Cell* 19(8): 2636–2652.

Zarebanadkouki, M., Fink, T., Benard, P. and Banfield, C. C. (2019). Mucilage facilitates nutrient diffusion in the drying rhizosphere, *Vadose Zone J.* 18(1): 1–13. https://doi.org/10.2136/vzj2019.02.0021

Zhang, H. M. and Forde, B. G. (1998). An Arabidopsis MADS box gene that controls nutrient-induced changes in root architecture, *Science* 279(5349): 407–409.

Zhang, X., Davidson, E. A., Mauzerall, D. L., Searchinger, T. D., Dumas, P. and Shen, Y. (2015) Managing nitrogen for sustainable development, *Nature* 528(7580): 51–59. Epub 2015/11/26.

Zhou, H., Whalley, E. R., Hawkesford, M. J., Ashton, R. W., Atkinson, B., Atkinson, J. A., Sturrock, C. J., Bennett, M. J. and Mooney, S. J. (2020). The interaction between wheat roots and soil pores in structured field soil, *J. Exp. Bot.* https://doi.org/10.1093/jxb/eraa475.

Chapter 12

Advances in understanding plant root uptake of phosphorus

Jiayin Pang, The University of Western Australia, Australia; Zhihui Wen, The University of Western Australia, Australia and China Agricultural University, China; Daniel Kidd and Megan H. Ryan, The University of Western Australia, Australia; Rui-Peng Yu, Long Li and Wen-Feng Cong, China Agricultural University, China; Kadambot H. M. Siddique, The University of Western Australia, Australia; and Hans Lambers, The University of Western Australia, Australia and China Agricultural University, China

1 Introduction

Phosphorus (P) is an important structural component of nucleic acids, coenzymes, phospholipids and sugar phosphates (Veneklaas et al., 2012). Deficiency of P is a major constraint to crop production, and therefore, sustaining food production for a growing world population requires a large input of P fertilisers, which are

http://dx.doi.org/10.19103/AS.2020.0075.16

manufactured from non-renewable resources that are expected to diminish dramatically in the next centuries (Fixen and Johnston, 2012). With the decline of rock phosphate, phosphate fertiliser will inevitably become a scarcer and more costly resource (Cordell et al., 2009). Phosphorus resources are mainly restricted to a few countries, with most rock phosphate reserves found in Morocco, China and the United States (Cordell et al., 2009). At the global scale, P-deficient soils comprise almost 30% of the global cropland area; on the other hand, P has also been overused by applying P fertiliser in excess of crop requirements in many other regions around the world, especially in some developed and rapidly developing countries (MacDonald et al., 2011, Vitousek et al., 2009). High P-fertiliser application is often associated with areas of relatively low P-use efficiency (55% of the global cropland area) (MacDonald et al., 2011). Where long-term applications of P fertiliser have occurred, a legacy soil P bank has built-up which is, however, largely unavailable to most crop plants (Menezes-Blackburn et al., 2018). For soils that exceed the critical values for P (usually defined as the soil P level corresponding to 90% or 95% of maximum yield), there is a need to adopt P maintenance practices to improve financial and environmental outcomes (Weaver and Wong, 2011). Syer et al. (2008) suggested that once critical P values are reached, maintenance rates of P can be based on replacing the amount of P removed in harvested products. Improved crop P-acquisition efficiency would allow lower target critical soil P values and savings in P fertiliser (Simpson et al., 2011), as relatively more P is lost through run-off, leaching and/or soil sorption when P-fertiliser applications are high (MacDonald et al., 2011).

Efficient P management in agriculture will be key to maximise crop productivity in the short- and long-term. Such systems must be constructed to reduce fertiliser-input costs, but without compromising yields, or achieving higher yields with the current rates of fertiliser. This could occur through (1) genetic improvement to increase crop capacity to explore soil for soluble P (Bovill et al., 2013); and/or (2) improved plant capacity to access the large pools of sorbed P (Menezes-Blackburn et al., 2018).

The aim of this chapter is to review our current understanding of crop P acquisition, and to promote future research of the best management of farming systems with a large amount of legacy P, and to suggest key future research directions.

2 Root architecture and morphology associated with phosphorus (P) uptake

Plants are able to respond to varying P availability by changing their root architectural and root morphological specialisations (Lambers et al., 2006, Shen et al., 2011, Lynch, 2019). Compared with plants with an optimum P supply, most P-deficient plants often show an increased root mass fraction, lateral root branching, root hair length and density, and specific root length/surface area (Lynch and Brown, 2008, Kidd et al., 2015, Wen et al., 2019). Conversely, the formation of specialised roots such as cluster roots only occurs in a limited number of species (e.g. in most Proteaceae and some Fabaceae) (Shen et al., 2003, Shane et al., 2006, Lambers et al., 2011). More and more architectural and morphological root traits associated with enhanced P acquisition (in the following sections) have been explored systematically among various species and genotypes within a species (Zhu et al., 2010, Brown et al., 2013, Haling et al., 2018, Sun et al., 2018), and the potential mechanisms underlying their differential P-acquisition efficiency have been discussed.

Large intraspecific variation in root architecture and morphology associated with enhanced P acquisition has been found among genotypes/cultivars within various crop/pasture species: for example, maize (*Zea mays*) (Postma and Lynch, 2011), common bean (*Phaseolus vulgaris*) (Strock et al., 2018), barley (*Hordeum vulgare*) (Schneider and Lynch, 2018) and subterranean clover (*Trifolium subterraneum*) (Jeffery et al., 2017b). For instance, maize genotypes with shallow root growth angles through greater production of crown roots (Bayuelo-Jiménez et al., 2011), with greater lateral root density (Zhu and Lynch, 2004, Postma et al., 2014, Jia et al., 2018), and with longer/denser root hairs (Zhu et al., 2010, Miguel et al., 2015) usually exhibit greater P uptake and plant performance in low-P soils. Moreover, maize genotypes with more root cortical aerenchyma formation (Postma and Lynch, 2011), with greater root cortical senescence resulting in reduced root cortical construction costs (Schneider and Lynch, 2018, Galindo-Castaneda et al., 2019) and common bean genotypes with reduced root secondary growth (Strock et al., 2018) all had advantages in root growth, shoot biomass and P acquisition under limiting soil P availability. Overall, these findings highlight that: (1) root traits that increase soil exploration, especially topsoil-foraging capacity (toward topsoil-foraging ideotypes), enhance P capture, because of the low mobility of P in most soils and its greatest bioavailability in the topsoil layers; (2) root traits that reduce the metabolic cost of soil exploration with lower root-construction costs enable resources to be allocated to other plant functions such as new root growth, P-mobilising exudates and mycorrhizas, which may facilitate overall resource capture and plant performance.

Wide interspecific variation in root architecture and morphology related to contrasting P-acquisition strategies has been demonstrated among different species (Hill et al., 2010, Kidd et al., 2015, Waddell et al., 2015, Yang et al., 2015b, Haling et al., 2016, Lyu et al., 2016, Wen et al., 2019). For example, some species of annual pasture legumes show more than three-fold variation in their critical external P requirement (i.e. P required for 90% of maximum yield) (Haling et al., 2016). Phosphorus-efficient species tend to possess relatively long root hairs, high specific root length and relatively high root length density, and consequently a large root hair cylinder volume (a function of root length, root hair length and average root diameter) at low soil P fertility. This increases soil exploration and thus is associated with a lower critical external P requirement (Kidd et al., 2015, Waddell et al., 2015, Yang et al., 2015b, Haling et al., 2016). Moreover, in response to varying P availability, plants also showed diverse responses (magnitudes and directions) in root architectural and morphological traits to maximise their P capture (Haling et al., 2016, Lyu et al., 2016, Wen et al., 2019). For instance, species with thin roots such as wheat (*Triticum aestivum*) and maize show stronger responses in root branching, first-order root length and specific root length in response to decreasing P availability, while species with thick roots such as faba bean (*Vicia faba*) and chickpea (*Cicer arietinum*) display a limited root morphological response (Lyu et al., 2016, Wen et al., 2019). The grass species *Lolium perenne* and *Rytidosperma richardsonii* increased specific root length in response to suboptimal soil P availability, but different strategies were expressed, with *L. perenne* tending to reduce root tissue density, while *R. richardsonii* tended to increase the proportion of fine roots (Waddell et al., 2015). Collectively, these results showed: (1) wide interspecific differences in root architectural and morphological traits that underpin differences in P-acquisition efficiency and critical P requirement; (2) species-specific variation that fine-tune root architectural and morphological trait(s) in response to varying P availability, with a large variation in root morphological plasticity among species.

Plant-available P shows considerable spatial and temporal variability in natural soils, because it is easily sorbed by soil particles and has an extremely low mobility in soil (Hodge, 2004, York et al., 2016). In response to heterogeneous P distribution (patches) in soil, plant roots exhibit varying extents of proliferation (e.g. lateral root numbers, root growth rate and retention time in the patches), and hence will differ in the extent to which they increase P acquisition from patches (Drew, 1975, Hodge, 2004, Li et al., 2014a, Liu et al., 2015). In general, species with thin roots and/or with fast growth rates showed a greater root morphological plasticity to local P supply than species with thick roots and/or slow growth rates (Fitter, 1994, Li et al., 2014a, Liu et al., 2015, Zhang et al., 2016, Zhang et al., 2019a). This suggests two contrasting resource-acquisition strategies: an acquisitive strategy versus a conservative strategy. However, the mechanisms of P-dependent changes in root morphology in response to local P supply are not fully understood, and thus merit further study.

Although various adjustments in root architecture and morphology play key roles in enhancing P acquisition and plant growth, no single root morphological trait has been identified that universally increases the ability of the root system to explore the soil for P (Yang et al., 2015b, Waddell et al., 2015, Haling et al., 2016). Ideal root traits that enhance P capture often incur trade-offs for the capture of other resources such as nitrate and water (Bishopp and Lynch, 2015, Lynch, 2015, 2019). Therefore we should consider optimised root morphological traits/trait combinations associated with P acquisition under specific environmental conditions.

In summary, plant roots exhibit great variation in root architecture and morphology associated with P acquisition among species and genotypes within a species. Developing and adopting crop genotypes and species with desirable root traits for improved P-acquisition efficiency would reduce P input and dependence on rock phosphate reserves.

3 Root biochemistry associated with P uptake

3.1 Role of carboxylates in P uptake

The exudation of carboxylates (the anion component of organic acids) into the rhizosphere can give some plants a competitive advantage in P-limiting environments (Jones, 1998). Carboxylates interact with the soil which provides a number of ways to enhance P supply to roots. For example, carboxylates have a strong affinity for the sorption sites occupied by P, thus allowing P to move into the soil solution. They also have the capacity for dissolution of strongly bound P (i.e. P associated with oxides and hydroxides of Fe and Al or with Ca) and the mobilisation of organic P, creating a source of P that was previously unavailable to the plant. Moreover, carboxylates also act as a substrate for micro-organisms and can promote the activity of P-solubilising bacteria in the rhizosphere.

The impact of carboxylates on P acquisition is most evident in plant species that occupy severely P-impoverished soils such as members of the Proteaceae (Shane and Lambers, 2005). However, the finding of significant carboxylate exudation in crop species such as lupins (e.g. *Lupinus albus*) and chickpeas has sparked suggestions about utilising carboxylates to develop crop genotypes with a lower external critical P requirement (Wouterlood et al., 2004) or to enhance plant growth in P-sorbing soils. The most widely studied crop species with respect to carboxylate release are related to the formation of cluster roots in *L. albus*. The mechanism of P mobilisation in the rhizosheath of *L. albus* was investigated under controlled conditions (Gardner et al., 1982a,b, 1983a,b, Gardner and Boundy, 1983). Despite having a root system architecture that was considered unfavourable for efficient nutrient capture (i.e. thick lateral roots and no mycorrhizas), this species was able to survive in P-impoverished soils due to the formation of cluster roots, which exude significant quantities of citrate into the rhizosphere with strongly bound inorganic P ($Ca_3(PO_4)_2$, $FePO_4$, $AlPO_4$). By a process of reduction of Fe^{3+}, acidification and dissolution of phosphate complexes and an increased net uptake of P, the plants maintained growth in P-deficient soils which demonstrates that root physiological activities may be more important than root morphological traits for P acquisition in P-impoverished soils (Lyu et al., 2016).

A number of studies show a role for carboxylates in P mobilisation, with a range of crop species studied with regard to P-fertiliser responses in pots and under field conditions using a wide range of P supplies (Brennan and Bolland, 2001, Bolland et al., 1999, 2000, Bolland and Brennan, 2008, Bolland, 1995, 1997). Figure 1, adapted from Bolland (1997), shows that the two lupin species that produce cluster roots (*L. albus* and *L. cosentinii*) perform better in a field experiment at a low rate of P-fertiliser application than the two species that lack cluster roots (*L. angustifolius* and *L. luteus*), even though these also release carboxylates (Hocking and Jeffery, 2004). Yet, in a review, Lynch (2019) concluded that variation in the production of P-mobilising carboxylates is associated with P mobilisation in vitro, but that rigorous analyses have failed to display the benefits of such variation for P acquisition in a range of soils in the field. These rigorous analyses are obviously available from papers by Bolland and co-workers.

There are few field studies that conclusively demonstrate the benefits of P mobilisation by species such as lupin on subsequent crops, because these fields are often managed with relatively high levels of P fertiliser (Angus et al., 2015). Profitable agricultural practices in developed countries rarely operate in the realms of P deficiency where P mobilisation from carboxylates has proven most effective. For example, significant dry-matter improvements from carboxylate exudation in wheat were reported in solution culture by Ryan et al. (2014), but increases attributed to citrate efflux could only be apportioned to two of six pot studies (and only under severe P deficiency), and there was no yield benefit under field conditions. Citrate is a substrate for microbes and it seems that the released citrate was rapidly consumed in the rhizosphere (Martin et al., 2016). A further complication is that there are nutrient deficiencies, other than P, that can drive changes in root exudation. For example, excessive cation uptake has been linked to increases in malate production and proton release by plants (Yan et al., 1992), and rapid rates of nitrate uptake were positively correlated with increased synthesis of organic anions (Haynes, 1990), particularly citrate and malate (Kirkby and Mengel, 1967). Deficiency of molybdenum, iron and manganese may also cause an imbalance in the ratio of cations to anions which can stimulate an increase in carboxylate exudation.

Despite other potential nutritional factors, P deficiency remains the major driver for carboxylate research. There is no doubt that plants that release far more carboxylates than wheat does, in particular those that produce cluster roots, show enhanced P acquisition and growth when soil P is limiting.

3.2 Finding practical applications for root exudate research

Our knowledge of the role and function of root exudates is continuously advancing, yet there remains a distinct lack of sampling techniques for

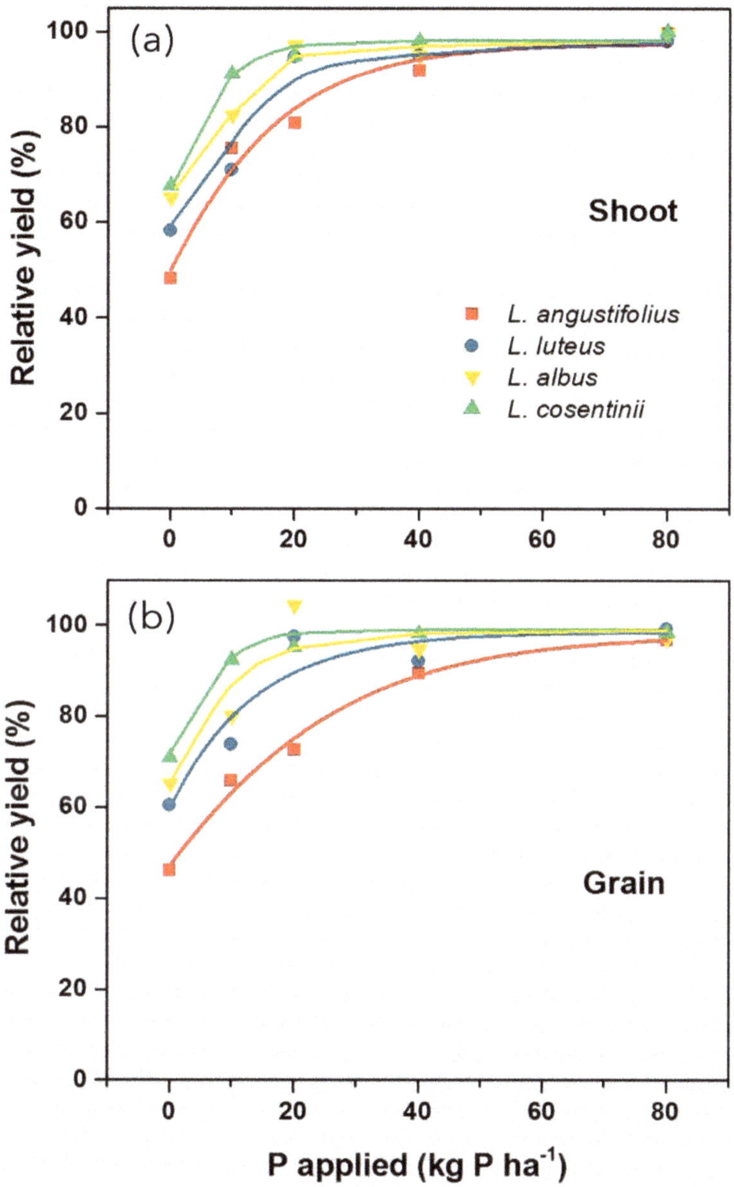

Figure 1 Relationship between relative yield in (a) shoot and (b) grain and the level of phosphorus (P) applied. Lines are fitted using the Mitscherlich equation. The dots and lines in red and blue represent two species without cluster roots (*Lupinus angustifolius*, *L. luteus*), while yellow and green represent two species with cluster roots (*L. albus*, *L. cosentinii*). Adapted from Bolland (1997).

field-grown plants. The relative benefits and shortcomings of the various root exudate collection techniques have been discussed previously (Jones et al., 2003, Oburger et al., 2013, Oburger and Jones, 2018). Generally, growing plants in solution culture alleviates the effects that microbial degradation and soil adsorption can have on plant-derived carboxylates, but results in potentially large sample volumes and precludes the typical root architectural responses to soil. Soil-grown plants often require destructive sampling techniques, which cause root damage and the potential release of carboxylates from root cells, and results vary with soil type (Clark, 1967, Oburger et al., 2009) and soil moisture (Fan et al., 2015). Moreover, the microbial populations in soil not only have the ability to utilise exudates, but can also produce them (Rózycki and Strzelczyk, 1986). Thus, for soil-grown plants, the presence of soil microbes may influence the composition and amount of rhizosphere exudates (Barber and Lynch, 1977). There are few studies of root exudates carried out in the field due to the destructive sampling techniques required. However, rhizosheath sampling is one method that has been used successfully, and despite its destructive nature, the results are surprisingly consistent and repeatable. This method involves excavating plants, removing the bulk soil and rinsing off the rhizosheath (Schefe et al., 2008, Veneklaas et al., 2003, Wang et al., 2017b). It provides information on the root exudates at the time of sampling, and reflects the total of plant and microbial processes. Therefore, field analysis of carboxylates is possible, but results do come with some limitations.

When measuring carboxylates in controlled conditions, careful consideration must be given to the supplied nutrients and pH. For solution culture experiments, the result is a homogenous nutrient concentration where roots and root hairs grow unimpeded, and plants lack the capacity to significantly alter root architecture. The ionic strength of the solution can also have significant impacts (Blamey et al., 1983). While soil studies are a step closer to field-relevant outcomes with regard to carboxylate exudation, the effects of microbes and pathogens can influence root growth, root architecture and hence root exudation. Thus, discussions around soil pasteurisation to minimise their influence must be considered. Often, substrates such as river sand, which have reportedly less microbial activity and fewer adsorption sites, can offer a useful medium for carboxylate measurements (Ryan et al., 2012). However, once added, nutrient concentrations are usually homogenously distributed in the profile, and there is no spatial heterogeneity with regards to soil pH and nutrients, particularly the stratified positioning of P and other nutrients in the soil profile. Repacking the soil to emulate the field soil profile or stratifying soil nutrients significantly alters root architecture (Haling et al., 2016), and hence the distribution of root exudates in the soil profile. The verification of findings from controlled conditions to field experimentation must occur, if results are to have any agricultural or ecological relevance.

There are other abiotic factors that affect carboxylate exudation which are often not controlled in pot studies or are simply too difficult to replicate. For example, temperature is often set at a constant day and night limit, and rarely reflects the daily fluctuations that occur. Equally, light intensity varies seasonally and diurnally and can be highly variable throughout the day due to cloud cover, yet is often set at a constant intensity and tends not to reach the high levels of natural sunlight. Both factors affect photosynthesis and root exudation (Watt and Evans, 1999). Light intensity can also be influenced by shading. While many crop plants often have an inter-row spacing that allows some light penetration to lower levels in the canopy, forage plants are naturally grown in dense swards, where the competitive effects of shading influence plant growth. Stems and petioles become etiolated when exposed to light with a low-red:far-red ratio, and root architecture and root exudation are significantly altered as a result. Experiments that incorporate reflective sleeves around pots that are continually raised to canopy height have been used in forage crop pot trials to emulate field conditions (Jeffery et al., 2017a), and may significantly increase root exudation (>10-fold increase per unit of root length), compared with unconstrained plant growth. Significant advances in controlled-environment technology (i.e. LED lighting) provide the capacity to simulate dawn and dusk lighting to make day/night transitions less abrupt, and enable light intensities approaching that of uninterrupted sunlight. Therefore, some of these abiotic factors may be better simulated in future experiments.

Due to the vast array of sampling techniques used in this field, there are also a number of ways of standardising the data, and this creates difficulties in comparing experimental outcomes (Oburger and Jones, 2018). Although root exudation occurs primarily at the root tips, it is routinely recommended that data are presented as micromoles of root exudate per gram of root dry mass ($\mu mol\ g^{-1}$ root DM). This kind of standardised measurement allows direct comparison with other research, and is very useful when comparing root exudation from members of the same species. However, when comparing a range of species with different root architectural types, this method may not adequately differentiate the species. For example, in Kidd et al. (2018), comparing a species with a comparatively large root diameter and low specific root length (i.e. chickpea) with a number of species with very small root diameters and high-specific root length (i.e. *Lotus* and *Ornithopus* spp.), standardising data by root DM gave distinctly higher values for species with long, thin roots, suggesting their carboxylate exudation was similar to that of chickpea. However, once the root exudation was standardised by root length, the results clearly showed that chickpea had greater concentrations in the

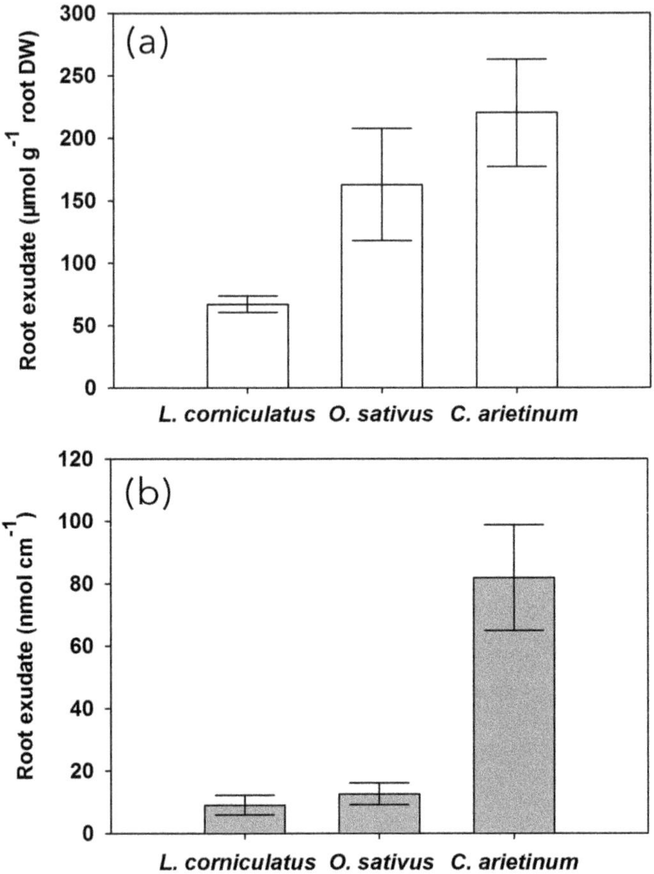

Figure 2 Root exudates collected from the rhizosheath of birdsfoot trefoil (*Lotus corniculatus*), which has a medium specific root length (SRL), French serradella (*Ornithopus sativus*), which has a high SRL, and chickpea (*Cicer arietinum*), which has a low SRL. Amounts of root exudates were not significantly different for chickpea and French serradella when expressed per unit of root dry mass (DM) (a), but they were when root length was used to express the data (b). Adapted from (Kidd et al., 2018).

rhizosheath due to its lower total root length (Fig. 2). This suggests that for comparisons across species, carboxylate amount is best expressed per unit of root length, because these units better reflects differences in root exploration capacity and subsequent interaction between the root and rhizosheath.

It seems that for carboxylate research there is no single method that perfectly addresses the complexity of interactions in field-grown plants. However, if results are repeatable and adequately benchmarked or referenced against previous research, then the use of hydroponics, pot studies or field measurements will be valuable in advancing this research area.

3.3 Role of carboxylates in intercropping systems

Plant species vary widely in their capacity to mobilise sorbed inorganic and organic P (Pearse et al., 2006). We consider plant species that release carboxylates, protons or phosphatases into the rhizosphere, and readily convert unavailable soil P into an available form, P-mobilising species. In contrast, species with a weak or no capacity to access poorly available soil P are non-P-mobilising species (Lambers et al., 2008). Phosphorus-mobilising species may increase the P content and biomass of neighbouring non-P-mobilising species via increased P availability in the rhizosphere, that is, rhizosphere P facilitation (Li et al., 2014b, Waddell et al., 2015, Lambers et al., 2018). The mechanisms underlying interspecific facilitation result in greater P acquisition from both inorganic and organic P sources. Species like faba bean, lupin and *Banksia* spp. release a large amount of protons and carboxylates (e.g. citrate and malate), and thus mobilise sorbed inorganic P (e.g. P bound to oxides and hydroxides of aluminium or iron in acid soils or to calcium in alkaline soils) into plant-available forms that they take up, and this facilitates the growth of their neighbours (Li et al., 2007, Waddell et al., 2015, Dissanayaka et al., 2015, Lambers et al., 2018). Species like chickpea and white lupin that efficiently access organic P because of exudation of phosphatases into the rhizosphere which hydrolyse organic P into inorganic P also benefit the neighbours when growing in close proximity (Li et al., 2004, Li et al., 2003, Waddell et al., 2015, Dissanayaka et al., 2015).

Root barriers are used to measure interspecific rhizosphere effects in glasshouse (Li et al., 2007, Li et al., 2004, Muler et al., 2014, Teste et al., 2014) and field experiments (Li et al., 1999, 2007). A 37-μm nylon mesh barrier blocks root intermingling, but permits exchange of root exudates and hyphal penetration, and a solid barrier eliminates any root interaction by preventing root exudates moving between root zones of two species (Li et al., 2007). If there is greater performance of non-P-mobilising species with a mesh barrier (with root interaction) than with a solid barrier (no root interaction), this provides direct evidence of positive interaction via rhizosphere processes. Interspecific facilitation by P-mobilising species leading to greater P content and performance on P-deficient soils has been observed in several intercropping systems/natural species mixtures in the field and in glasshouse pot experiments: for example, wheat/white lupin (mobilising crop species) (Gardner and Boundy, 1983), wheat/chickpea (mobilising crop species) (Li et al., 2003), maize/faba bean (mobilising crop species) (Li et al., 2007), maize/chickpea (mobilising crop species) (Li et al., 2004), maize/white lupin (mobilising crop species) (Dissanayaka et al., 2015), *Scholtzia involucrata*/*Banksia attenuata* (mobilising wild species) (Muler et al., 2014) and *Cleistogenes squarrosa*/*Melissilus ruthenicus* (mobilising wild species) (Yu et al., 2020).

In cropping systems, there have been only a few studies showing the facilitative effects of plants exuding large amounts of carboxylates on neighbouring plants that release little carboxylates. For example, in a maize/faba bean intercropping system under field conditions, the yield of intercropped maize and faba bean was 43% and 26% greater, respectively, than that in monoculture (Li et al., 2007). Another example is intercropping of durum wheat and common bean (Li et al., 2008). More often, intercropping results in a yield improvement of one crop species to the detriment of the companion species as for white lupin with wheat or maize (Li et al., 2010) or faba bean with maize (Li et al., 2010). In some instances, there are yield penalties for both plants, as for intercropped maize and peanut (Inal et al., 2007).

There are still a number of questions related to the benefits of using P-mobilising species for the purposes of P acquisition. Plants grown under P deficiency will often exude greater quantities of carboxylates (Grierson, 1992, Lipton et al., 1987), and for *L. albus*, increasing P supply will lead to a decrease in the formation of cluster roots and subsequent P mobilisation due to citrate efflux. Thus, for a companion species to receive the P-derived benefits from lupin, the soil would need to be relatively P deficient. The ultimate goal of such research is for intercropping of these P-mobilising species with crop species that lack this trait. Through this association, organic P or insoluble inorganic P can be made available to the crop plant through release of carboxylates, protons and phosphatases (Li et al., 2014b). In some cases, the improved yield of intercropping of two species in low-P soil is more likely due to 'complementarity' or 'niche differentiation', with the companion species exploring different soil horizons and/or different soil P sources (Becquer et al., 2017). So, the question of in what kind of agricultural system would this method of P acquisition be functional or even profitable deserves further study.

There are some examples where a low-input agricultural system might benefit from P mobilisation through intercropping. These include areas of sub-Saharan Africa, where P fertilisers are largely unavailable or where phosphate buffering of the soil is so high that the rates of P fertiliser required to obtain a growth response are cost prohibitive. In developed countries, perhaps a more pertinent system lies in a resurgent organic agricultural industry, where there are no synthetic fertiliser inputs at all, and soil organic P inputs are insufficient. However, some of the other functional traits associated with root exudates (including carboxylates) will also be important such as the promotion of beneficial interactions with plant-growth promoting micro-organisms (Wang and Lambers, 2019). The control of rhizosphere micro-organisms by the plant through root exudates is likely the next frontier in rhizosphere research. Therefore, further studies into the production of not only carboxylates, but also other root exudates in various plant species will prove increasingly valuable. The selection of appropriate plant species to match soil chemical properties with plant root exudates that facilitate nutrient acquisition might greatly benefit agricultural production.

3.4 Leaf manganese (Mn) as a proxy of belowground carboxylates for plant breeding

Belowground functional traits, such as exudation of carboxylates, have become increasingly interesting to plant scientists; however, as discussed above, the measurement of carboxylates in the rhizosphere is often laborious (Lambers et al., 2015a). Shane and Lambers (2005) found that the Mn concentration in old leaves of *Hakea prostrata* (a species with cluster roots native to Australia) increased with an increasing percentage of root mass invested in cluster roots, attributed to the increased release of carboxylates and protons. In a natural ecosystem across a 2-million-year-old dune chronosequence in Jurien Bay in Australia, Hayes et al. (2014) observed that the leaf Mn concentration in non-mycorrhizal species known to release carboxylates was higher than that of co-occurring mycorrhizal species, regardless of soil age. The non-mycorrhizal carboxylate-releasing species belonged to a range of functional types: for example, species producing cluster roots, dauciform roots and sand-binding roots. In *Lupinus albus*, which also forms cluster roots, Gardner et al. (1982a,b) found that leaf Mn concentration was highly correlated with the weight of cluster roots, which are the main site for the reduction of MnO_2 in the rhizosphere.

Further to the findings in woody plants and lupin with specialised roots, Pang et al. (2018) found that the Mn concentration in mature leaves on the main stems was highly correlated with the amount of carboxylates relative to root dry weight (DW) in the rhizosheath, particularly malonate, in 100 chickpea genotypes with diverse genetic background (Fig. 3). The importance of carboxylates for Mn acquisition in chickpea significantly extends the findings in plant species with cluster roots, showing a significant contribution of carboxylate exudation to Mn uptake in chickpea plants with no cluster roots or functionally equivalent roots. Therefore, the study by Pang et al. (2018) provides strong evidence that Mn concentration in mature leaves can be used as an easily measurable proxy for carboxylates in the rhizosheath. However, the study was undertaken when chickpea was grown in a low-P environment, and whether this result would apply to plants grown at higher P levels and in soil with a high P-buffer index is still unknown and worth further research. The applicability of the relationship in other crop species needs to be verified and studies are underway in *Glycine max* and *Brassica napus* in research groups in China. Leaf Mn concentration as a proxy for belowground carboxylates offers an important breeding trait, and might provide a valuable tool for the screening of a large number of plants in breeding programmes.

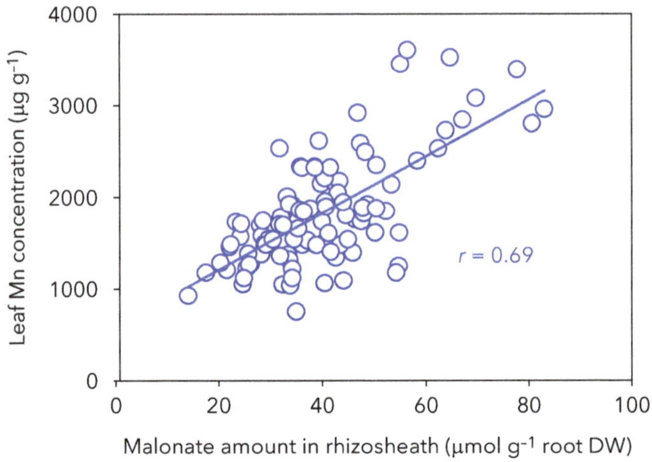

Figure 3 Correlation between manganese (Mn) concentration in mature leaves on the main stems and the amount of malonate in the rhizosheath soil per root dry weight for 100 chickpea genotypes with diverse genetic background, after growing for 7 weeks in washed river sand supplied with 10 µg P g^{-1} dry soil as $FePO_4$, which was growth-limiting. Adapted from Pang et al. (2018).

Given that the species in a mixture share the consequences of rhizosphere modification, it is difficult to separate the effect of each species; however, leaf manganese concentrations ([Mn]) can be used as a proxy for interspecific facilitation via rhizosphere carboxylate release (Lambers et al., 2015a, Pang et al., 2018, Wang and Lambers, 2019). If there is a greater leaf [Mn] of non-P-mobilising species in intercropping/mixtures than that in monoculture, it would provide evidence for facilitation via carboxylate release of P-mobilising species. For example, greater leaf [Mn] of wheat was observed when grown with white lupin, which increased with increasing density of lupin (Fig. 4; Gardner and Boundy, 1983). When grown with *Carex korshinskyi*, a forb that releases a large amount of carboxylates, neighbouring species such as *Leymus chinensis* and *Stipa grandis* usually had a greater leaf [Mn] and aboveground biomass (R.-P. Yu, unpublished data), providing compelling evidence that interspecific facilitation via carboxylate release prevails on P-impoverished soils.

3.5 Advances in phosphatases in P acquisition

Soil organic P represents 40–80% of the total P pool in soils (Menezes-Blackburn et al., 2013). Plants cannot take up organic P directly, but rather, organic P is first hydrolysed to orthophosphate by root or microbial-released phosphatases (Tarafdar and Jungk, 1987, Richardson et al., 2011). Different enzyme types are involved in mineralisation of organic P, and these have considerable differences in hydrolysis efficiency (Menezes-Blackburn et al.,

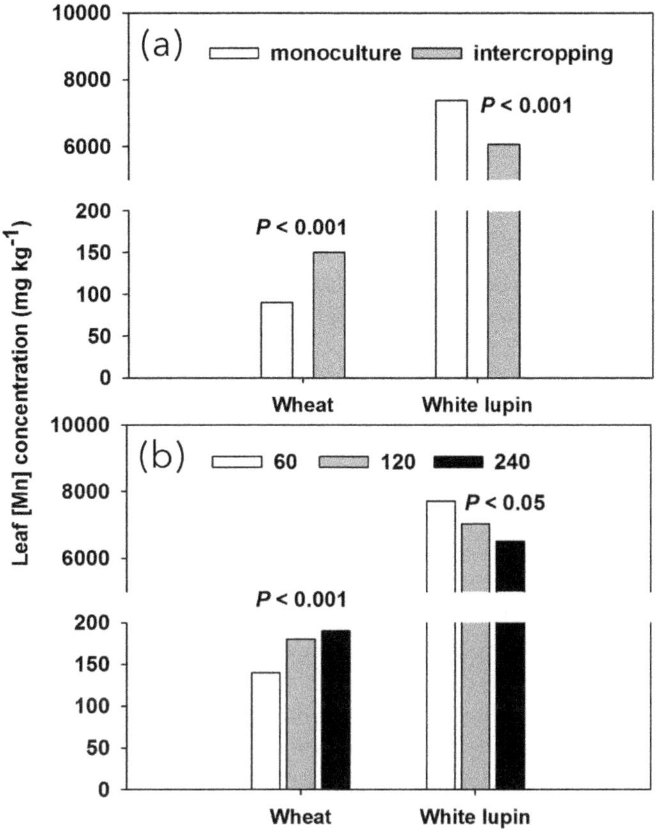

Figure 4 Leaf manganese concentrations ([Mn]) of wheat and white lupin in (a) monoculture (sowing rate: wheat 85 kg ha^{-1}, white lupin 180 kg ha^{-1}) and intercropping (sowing rate: wheat 85 kg ha^{-1}, white lupin 135 kg ha^{-1}), and (b) intercropping with different sowing rate of white lupin (60, 120 and 240 kg ha^{-1}). Bars represent mean values (n=4). P value indicates difference between (a) monoculture and intercropping, and (b) different sowing rate of white lupin (LSD). Adapted from Gardner and Boundy (1983).

2013). Some phosphatases and carboxylates have a synergistic effect on organic P mineralisation. Clarholm et al. (2015) summarizes that citrate or oxalate chelates and removes metal anions from the supramolecule aggregate or soil organic matter molecules first, and then phosphatases (e.g., acid phosphatase) hydrolyse organic P into inorganic P from newly exposed materials. Among phosphatase types, phytases are capable of initiating the cleavage of phytate molecules, making P plant-available (George et al., 2007a). Rapid utilisation of phytate-P in soil also needs the synergistic effects of carboxylates to mobilise phytate (Dissanayaka et al., 2015, Giles et al., 2017, 2018, Maruyama and Wasaki, 2017).

Soil zymography, an in situ non-destructive method combined with fluorescent substrates and ^{14}C imaging, has been used to investigate the two-dimensional distribution of phosphatases (acid and alkaline) and the interaction with microbes in the rhizosheath (Spohn and Kuzyakov, 2013, Razavi et al., 2019). The results show a spatial separation among acid and alkaline phosphatase activity and hotspots of bacteria in the rhizosheath in response to different soil P levels (Spohn et al., 2015), P sources (Ma et al., 2019), root exudate composition (Zhang et al., 2019d), growth stages (Ma et al., 2019) and plant species (Razavi et al., 2016) in rhizobox studies. This method reveals the linkages between spatial and temporal dynamics of enzymes and microbes for organic P mobilisation (Kuzyakov and Razavi, 2019). A unified visualisation of phosphatase activity released by roots or microbes, coupled with other imaging techniques and analytical methods shines light on hotspot and hot moments in the rhizosphere (Kuzyakov and Razavi, 2019, Razavi et al., 2019).

Genetically modified (GM) organisms will likely enhance the capacity of plants to use organic P. For example, greater acid phosphatase activity and P content were observed when *LASAP2* was overexpressed in *Nicotiana tabacum*, tobacco (Wasaki et al., 2009). Overexpression of phytase-encoding genes such as *OsPHY1* (Li et al., 2012) or *GmPAP4* (Kong et al., 2014) in plants enhanced plant growth and P acquisition when supplied with phytate as a P source. Enhanced biomass and P content in *Arabidopsis* were observed by transgenic expression of *MtPAP1* from *Medicago truncatula* (Xiao et al., 2006). Similarly, introduction of a phytase gene, *MtPHY*, in *M. sativa* improved plant performance when supplied with phytate (Ma et al., 2012). However, most studies on GM plants have been conducted in agar or in pots in short-term experiments, rather than field conditions (George et al., 2005, 2007b). The efficiency of GM plants to mobilise organic P depends on the P source and amount of the expressed phosphatase or released carboxylates, as well as whether the exudates are close to site of organic P mineralisation. This is because of the low mobility of phosphatase, and the required spatial and temporal vicinity needs further study (Giles et al., 2016, 2017, 2018).

3.6 Enhancing storage of P in root cells to improve fertiliser-use efficiency

For crops, P fertiliser is often applied as a band below the seed at sowing (McLaughlin et al., 2011). However, the proportion of fertiliser P taken up by the crop can be low due to losses prior to plant uptake, because of reaction with soil components to form poorly soluble inorganic compounds (especially in acid soils), uptake by microbes, and, in high-rainfall areas with sandy soils, run-off and leaching (McLaughlin et al., 2011, Simpson et al., 2015). Faster uptake by roots would reduce these losses. However, a factor limiting the uptake of

P by seedling roots is, presumably, high-tissue P concentrations. Phosphorus is stored in seeds primarily in the form of phytate (Raboy, 2009), an organic macromolecule, which allows safe storage of high concentrations of P without impacting cell osmotic pressure. After germination, the seedling quickly (within days) converts phytate-P in the seed back to inorganic P, while also commencing P uptake from the soil (Ryan et al., 2019b, Nadeem et al., 2011, 2012). Hence, young seedlings high in leaf P take up P from soil including fertiliser P, but are no longer able to store P as phytate. If fertiliser P could be absorbed faster by a seedling (and safely stored for future use), then less fertiliser P would be lost prior to root uptake, and hence fertiliser rates could be reduced. However, this requires seedlings with high-tissue P concentrations to quickly absorb and safely store large amounts of P.

The mechanism of P regulation in plant and seedling tissues (to avoid deficiency or toxicity) is storage of P in cell vacuoles: this has been assumed to be the only significant mechanism (Lee and Ratcliffe, 1993, Liu et al., 2016b). However, there is a second mechanism for P storage in roots that involves plastids (Ryan et al., 2019b), validating a suggestion by Brodelius and Vogel (1985). Ryan and co-workers showed that roots of seedlings of a pasture legume, subterranean clover, contained globular structures, similar to the globoids that contain phytate in seeds (Lott et al., 1978), and that these exhibit an extraordinarily high P concentration (up to and > 3000 mmol kg^{-1}). These do not contain phytate, but precipitated/crystallised salts of P. When roots were exposed to additional P, they stored it in these structures. Molecular analyses at whole transcriptome resolution combined with high-resolution electron microscopy indicated that the globular structures were plastids (Ryan et al., 2019b). This work led to the hypothesis that regulating the key genes controlling P movement through the root into the plastids could enhance uptake of P from P fertiliser, and thereby improve the P-uptake efficiency from P fertiliser and reduce fertiliser application rates (Fig. 5). Further work in this area is warranted.

4 Arbuscular mycorrhizal fungi (AMF) and fine root endophytes

The many ways in which the symbiosis between crops and root-colonising fungi, AMF, can impact crop growth are reviewed elsewhere in this volume. The literature pertaining to the need to manage occurrence of AMF in agricultural systems with a view to maximise crop yield was reviewed by Ryan and Graham (2018). Several of their key points are reiterated here. One was, that much of the literature lacks applicability to farming systems due to weaknesses with the design and interpretation of glasshouse experiments (see also Section 7 below) and a lack of understanding of the commercial agricultural systems to

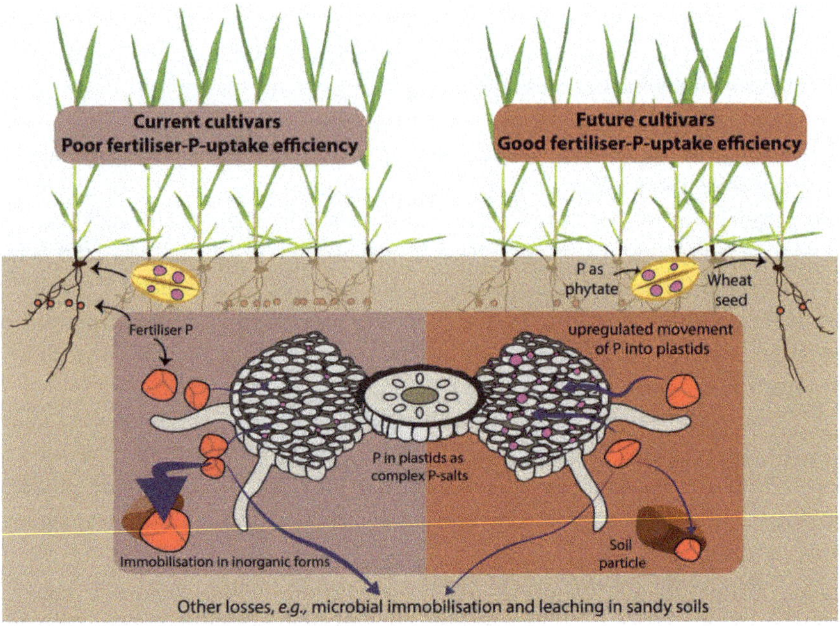

Figure 5 Schematic drawing showing how a cultivar with enhanced capacity to store phosphorus (P) in root plastids could reduce losses of fertiliser P, and hence reduce P-fertiliser requirements for wheat growing in an acid soil.

which results were extrapolated, and this extends to poor understanding of basic agronomy field techniques and science. While controversial (see Rillig et al., 2019), the failure of a very large body of literature to influence crop breeding or farm management suggests that there is a need to consider their arguments. A second key point made by Ryan and Graham (2018) and reinforced by Ryan et al. (2019a) is the need to apply a systems agronomy approach, that is, the benchmarking of credible physiological crop production limits in order to understand and remove the constraints in achieving them with the aim to develop regionally adapted, resource-efficient agronomic practices. Hence, research on AMF and crop-P nutrition requires the contribution of AMF towards P uptake and crop yield to be explored within the framework of the crop critical external P requirement. Research in this context is scarce. However, two excellent examples came from China (Deng et al., 2017, Mai et al., 2018). Deng et al. (2017) grew a wheat-maize rotation over two years with six rates of P addition. They determined the external critical soil P concentration, and found that this corresponded to a relatively high level of colonisation by AMF (Fig. 6). This work suggests that for a farming system with a target soil P sufficient to meet the crop external critical P requirement, there may be no need to alter management to enhance colonisation by AMF. This is a useful finding as

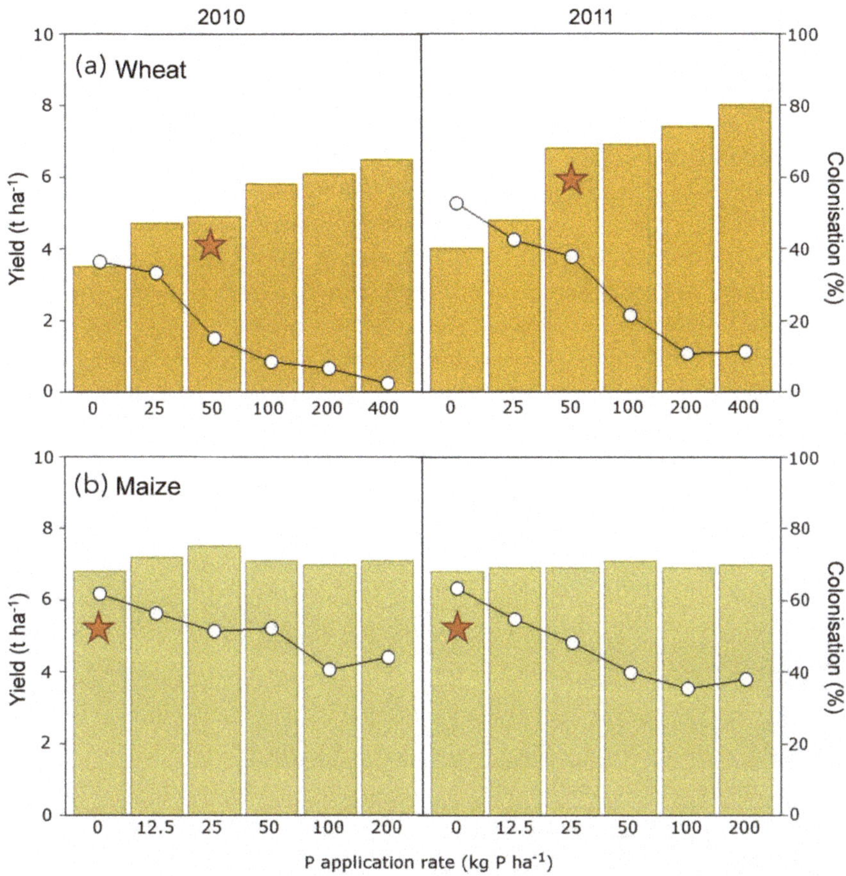

Figure 6 An example of investigating the interaction between crop yield, phosphorus (P) supply and colonisation by AMF in a farming systems context: (a) wheat; (b) maize. Adapted from Deng et al. (2017).

commercial inoculants are not widely used (Ryan and Graham, 2018, Hart et al., 2018).

Suggestions that crop breeders should consider how to enhance the level of colonisation and the mycorrhizal contribution to P uptake (e.g. Martin-Robles et al., 2018) need to be critically appraised in the context of the need to have a target soil P that meets crop external critical P requirements. In addition, it should be considered that higher colonisation may have unintended negative consequences in some circumstances such as reduced yield (Ryan et al., 2005) or promotion of plant parasitic nematodes (Frew et al., 2018). Moreover, attempts to change other root traits to enhance P uptake may require consideration of AMF. For instance, while, functionally, AMF were found by Jakobsen et al. (2005) to substitute for root hairs on mutant barley plants lacking root hairs, they were suggested to possibly override the P-acquisition benefits of selecting for longer root hairs in white clover (Caradus, 2012). The trade-offs that can occur at the genotype or individual plant level due to the carbon costs of colonisation by AMF and other roots traits are discussed in Section 5.

An interesting finding, whose significance for root uptake of P is yet to be explored, is that by Orchard et al. (2017a) that the 'fine root endophyte' previously thought a minor species of AMF, are in fact members of a different subphylum than the AMF, the Mucoromycotina. Orchard et al. (2017b) suggested that they belong to the Densosporaceae family within the Endogonales. Hence, it seems that plants contain two groups of arbuscule-forming endophytes that may be involved in plant P nutrition. Fine root endophytes (FRE) are relatively little researched, but are prolific in some agricultural systems including pastures in New Zealand and dryland cereal crops and pastures subjected to seasonal waterlogging in Australia (Orchard et al., 2017b). They may also enhance host

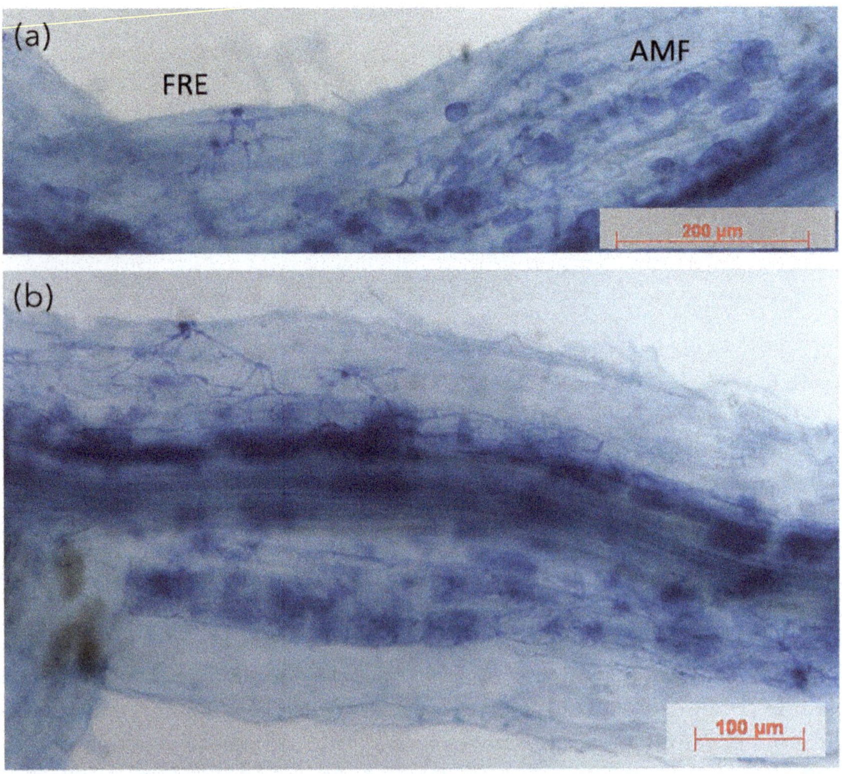

Figure 7 The arbuscule-forming root symbiont, fine root endophyte (FRE), colonises roots either in close association with arbuscular mycorrhizal fungi (AMF) (a) or by itself (b); its role in plant root phosphorus (P) uptake remains to be explored. Photograph by Jeremy Bougoure.

plant P uptake (e.g. Crush, 1973) and respond negatively to increasing P supply in a way similar to AMF (Jeffery et al., 2018). It is likely that the literature on AMF in agricultural systems has been confused by researchers either assuming FRE were AMF when scoring colonisation using light microscopy or missed their presence when using molecular techniques to characterise populations. FRE and AMF often colonise the same plant and even co-habitat the same sections of roots (Fig. 7), untangling their relative contribution to plant root P uptake will likely prove challenging. Beyond FRE, there is also an increasingly complex range of interactions between AMF and other soil biota (Ryan and Graham, 2018) including those closely associated with AMF, both internally (Salvioli et al., 2016) and externally (Zhang et al., 2014b, 2019b). However, harnessing such interactions for agricultural purposes will be a long-term endeavour (Ryan and Graham, 2018).

5 Trade-offs in traits

Roots exhibit diverse functional traits, which allow them to acclimate to low-P soils and to increase soil P acquisition including changes in root mass allocation, root morphology, P-mining exudates and mycorrhizal symbioses (Vance et al., 2003, Lambers et al., 2006, Shen et al., 2011). All these strategies pertain to different soil P sources in contrasting ways, presumably with different costs (carbon and other resources) and benefits (Lynch and Ho, 2005, Ryan et al., 2012, Raven et al., 2018). Thus, a plausible hypothesis is that the cost of each adjustment may limit the ability of plants to express all strategies simultaneously (Ryan et al., 2012, Nazeri et al., 2014). Indeed, plant species seem to show contrasting ways to express these root functional traits (Lambers et al., 2008, Ryan et al., 2016, Lyu et al., 2016). For instance, crops such as maize and wheat exhibit stronger root morphological responses than physiological responses to P deficiency (Lyu et al., 2016, Deng et al., 2014, Wen et al., 2017). Other crops such as chickpea and faba bean show a limited root morphological response to varying P supply, but a significant modification in root exudation (Liu et al., 2016a, Lyu et al., 2016, Zhang et al., 2016). Plant species that are adapted to either relatively fertile soils (e.g. Brassicaceae) or severely P-impoverished soils (e.g. Proteaceae), tend to be non-mycorrhizal (Lambers et al., 2008, 2015b, Brundrett, 2009), but species that are adapted to less extreme environments exhibit all three groups of root functional traits (i.e. root morphology, P-mobilising exudates and mycorrhizal symbioses) with an intermediate strategy. An interesting question is how plants coordinate the different root functional traits to enhance soil P acquisition under contrasting soil P availability.

Great progress has been made in revealing the evidence associated with coordination and trade-offs among root functional traits within or among species, linked with their nutrient-acquisition strategies, environmental gradients and evolutionary history (Reinhart et al., 2012, Waddell et al., 2015, Ryan et al., 2012, Roumet et al., 2016, Li et al., 2017, Ma et al., 2018, Wen et al., 2019). Generally, thin roots with a relatively high-specific root length, are associated with rapid growth rates and low root tissue density, but short life span, indicating a resource-acquisitive strategy. In contrast, thick roots with a relatively low specific root length are linked with slow growth rates and high root tissue density, but long life span, suggesting a resource-conservative strategy (Valenzuela-Estrada et al., 2008, McCormack et al., 2015, Freschet and Roumet, 2017). At the species level, this is particularly prevalent in natural forest and grassland ecosystems, and species with thinner roots (thin-root species) always have lower colonisation by AMF, but higher root branching, whereas species with thicker roots (thick-root species) generally exhibit higher colonisation by AMF, but consistently lower root branching (Liu et al., 2015, McCormack et al., 2015, Chen et al., 2016, Li et al., 2017). Wen et al. (2019) considered root exudation together with root morphology and mycorrhizal symbioses in shaping P-acquisition strategies of 16 crops, with a range of trade-offs among these root-functional traits. Taken together, in response to a low P availability, species with thinner roots mainly depended on their root morphology (e.g. high root-branching intensity, high specific root length) to enhance nutrient (including P) acquisition by increasing soil exploration. Conversely, species with thicker roots relied more on higher colonisation by AMF to compensate for a low root-absorptive surface and/or more P-mobilising exudates (e.g. significant carboxylate exudation) to mine sparingly soluble P in the rhizosheath (Kidd et al., 2015, Wen et al., 2019).

According to the principle of carbon economy, optimisation of internal resource allocation should occur, underpinning the adjustments of various root functional traits at the individual plant level, towards a balance between costs and benefits (Lynch and Ho, 2005). Supporting this idea, Ryan et al. (2012) found for 10 *Kennedia* species that under low-P treatment, inoculation with AMF decreased the amount of carboxylates in their rhizosheath that likely resulted in inoculated plants accessing a greater proportion of soil P from a labile inorganic P pool, rather than sparingly available P sources. Compared with plants without inoculation, association with AMF can also modify root architecture and morphology (Hetrick et al., 1991, Atkinson et al., 1994, Berta et al., 1995, Yang et al., 2015a). However, the association between root architecture and mycorrhizal growth response remains contentious (Brundrett, 2002, Maherali, 2014, Smith and Read, 2008). Generally, colonisation by AMF decreases the root mass fraction which may contribute to more carbon diversion to mycorrhizal symbionts, and thereby less carbon being allocated to root growth (Berta et al., 1995, Johnson et al., 1997, Graham and Abbott, 2000, Ryan et al., 2016). It also decreases specific root length (Berta et al., 1995, Mendoza, 2001, Sun and Tang, 2013), but increases root tissue density

(Zangaro et al., 2007), which suggests the root morphology of inoculated plants becomes more conserved. Not all the effects are due to improved P nutrition, because similar results can be found in high-P conditions (Berta et al., 1995, Zangaro et al., 2007, Maherali, 2014).

Overall, the findings show that, at the species level and at individual plant level, root morphology, P-mobilising exudates and mycorrhizal symbioses together shape P-acquisition strategies, with a range of trade-offs among these root functional traits.

6 Microbially-mediated efficient P acquisition in species-diverse systems

Soil micro-organisms play a pivotal role as P sinks and sources, and mediate key soil P processes such as P mineralisation and immobilisation (Oberson and Joner, 2005). The density and diversity of soil micro-organisms are particularly great in the rhizosphere. Plants can affect rhizosphere micro-organisms to mobilise soil P resources, and thus enhance P acquisition through exuding C-rich and diverse rhizodeposits (Badri et al., 2009, Drogue et al., 2012). Yet, it remains unclear whether and how soil/rhizosphere micro-organisms can mediate plant P acquisition in species-diverse systems such as natural grasslands and intercropped agroecosystems.

Soil microbial P pools represent an important source of P for plants (Hinsinger et al., 2015, Bünemann et al., 2013), because most of the microbial P is located in labile intracellular compounds, and the turnover of microbial P is often fast, for example, in grassland soils (Oberson and Joner, 2005). In natural grasslands, plant diversity increases soil microbial activity (Chung et al., 2007, Eisenhauer et al., 2010, Zak et al., 2003). This greatly contributes to the increase in overall plant production (Zak et al., 2003, Hooper et al., 2005), which generally increases with increasing number of plant species grown together. This suggests that systems where multiple species are grown together may enhance soil microbial P pools and turnover, and thus facilitate plant P acquisition. In agroecosystems, intercropping alters soil microbial biomass and composition compared with monocropping systems (Song et al., 2007, Tang et al., 2014, Wang et al., 2007). For example, Tang et al. (2014) showed that both durum wheat/chickpea and durum wheat/lentil intercropping promoted P storage in soil microbial biomass compared with the respective monocrops. In addition to microbial biomass, cereal-legume intercropping (e.g. wheat/soybean) increased root fungal diversity of intercropped wheat which was correlated with shoot P content (Bargaz et al., 2017). However, increased soil microbial P pools in species-diverse systems do not always translate into enhanced plant P acquisition and productivity (Li et al., 2010), suggesting that other mechanisms may also determine the soil P turnover and subsequent P acquisition.

Phosphate-solubilising micro-organisms including bacteria, fungi and actinomycetes promote plant growth by transforming sparingly soluble P into soluble P forms in different production systems (Khan et al., 2014). Intercropping rubber (*Hevea brasiliensis*) with banana (*Musa acuminata*) or cassava (*Manihot esculenta*) significantly enhanced the diversity of phosphate-solubilising bacteria, which was associated with increased P solubilisation (Sungthongwises, 2016). Enhanced activity of phosphate-solubilising bacteria can be attributed to the exudation of lectins by crops which positively influence plant growth-promoting rhizobacteria (Schelud'ko et al., 2009). In addition, phosphate-solubilising bacteria such as Gram-negative *Pseudomonas fluorescens* may exude antibiotics (Yang et al., 2008, Taurian et al., 2010), providing protection for plants against soil-borne pathogens, which could influence plant health and subsequent plant P acquisition (Singh et al., 2010).

Future work should not only investigate microbial biomass or coarse-scale microbial composition in species-diverse systems, but also identify specific microbial functional groups, and their role and interactions with plant species. In addition, bacteria-feeding nematodes may increase soil phosphatase activity and soil P availability through grazing bacteria (Djigal et al., 2004). It might be worthwhile to examine how intercropping affects soil fauna such as earthworms and nematodes, and their role in soil P cycling.

7 Farming-management practice in P acquisition at the intensive agricultural system level

There are many reviews and summaries on soil- and crop P-management strategies (Shen et al., 2013, Withers et al., 2014, Zhang et al., 2019c). The beneficial management practices include developing new fertilisers, shaping root-microbe interactions and rhizosphere management (Table 1). Here, we emphasise the role of root functional traits in improving P acquisition in the context of intensive agricultural systems, especially under high P input and high plant-density conditions.

The importance of root systems in supporting plant growth and P uptake has been extensively studied (Teng et al., 2013, Wen et al., 2017, Jeffery et al., 2017b, Haling et al., 2018, Pang et al., 2018). However, many of these studies were performed under P-deficient conditions and/or on individual plants (Pearse et al., 2007, Lyu et al., 2016, Wen et al., 2017, Pang et al., 2018). Few have considered the context of intensive agricultural systems, which are characterised by high resource input (including P fertiliser) and high plant density (Shen et al., 2013, Withers et al., 2014, Mi et al., 2016). Under such a condition, with a large quantity of fertiliser applied, crop growth is typically not limited by nutrients (Withers et al., 2014). Moreover, high plant density potentially leads to high root-root competition, so that many root traits that

Table 1 Examples of phosphorus (P)-management practices that have the potential for development and application in cropping systems

Management practices	Description	Comments	References
Novel/smart P fertilisers	Considering nutrient-release rate matches crop P demand, soil properties and climatic feature	Integrate knowledge in plant science, agrochemicals and materials; the high cost often limits the application in grain crops	Calabi-Floody et al. (2018); Weeks and Hettiarachchi (2019)
Microbial fertiliser (bio-fertiliser, bioformulation fertiliser)	Shaping beneficial root-microbial relationships to promote plant growth and P uptake	The functions involved in production and regulation of phytohormones, release of nutrients (P) to plants and control of pathogens; the cost may be high and the beneficial effects may be unstable under changing soil conditions	Richardson et al. (2011); Nadeem et al. (2014); Calabi-Floody et al. (2018)
Intercropping (mixed cropping)	Two or more crop species or genotypes with niche complementarity, growing together and coexisting for a prolonged time	Improve plant diversity; suitable for small-hold farming and low-input/resource-limited agricultural systems, but may not be suitable for intensive cropping systems and mechanisation	Li et al. (2014b); Brooker et al. (2015)
'4R' nutrient stewardship	Applying the right source, at the right rate, right time and in the right place	It supports multiple ecosystem services; the challenge for 4R stewardship is to develop an effective management package that works within a site-specific, dynamic, and complex system	Bruulsema et al. (2019); Grant and Flaten (2019)
Soil-based P-management strategies	Based on soil P test and yield response; considering a nutrient balance of P fertiliser input and plant uptake	Applying fertilisers to build-up and maintain an 'insurance' soil P level for maximum crop growth and yield	Li et al. (2011); Withers et al. (2014)
Plant/rhizosphere-based P-management strategies	Maximising root/rhizosphere efficiency to enhance soil P acquisition	Enhance the utilisation of legacy soil P and increase fertiliser-P recovery by rhizosphere management	Li et al. (2011); Shen et al. (2013); Zhang et al. (2019c)

are advantageous for P uptake in individual plants, may not be suitable for a population in intensive cropping systems (Weiner et al., 2010, York et al., 2015, Mi et al., 2016, Zhu et al., 2019). So, for sustainable cropping systems and P management, we should reconsider the roles of root functional traits in enhancing P acquisition to combine the root functional traits of high P input (high legacy soil P reserves) and high plant density.

The improvement of crop productivity has relied heavily on the input of chemical fertilisers including P (Li et al., 2011, Shen et al., 2011, Withers et al., 2014, Zhang et al., 2019c). The continuous overuse of P not only resulted in a significant increase in plant-available P concentrations, but also led to the accumulation of a largely unavailable soil P pool (soil P reserves) (MacDonald et al., 2011, Sattari et al., 2012). How to make more efficient use of both applied P and the soil P reserves, and reduce target soil P levels has become increasingly urgent for the development of sustainable cropping systems. Soil P availability significantly affects the response intensity and the effectiveness of root functional traits. Generally, adjustments of root morphology, release of P-mobilising exudates and colonisation by AMF are enhanced by a low shoot P status and low soil P availability, and suppressed by increasing plant P status and soil P availability (Deng et al., 2014, Wang et al., 2016, Wen et al., 2017). Supporting this view, Teng et al. (2013) found that the optimal P supply for maximum grain yield of wheat coincided with the critical P concentration for P-starvation marker gene responses in roots, and was at or near the change points for downregulating root responses to P supply. At an optimal soil P level, grain yield was highest, but also maintained a relatively high response intensity of root traits and a relatively high expression level of genes encoding P transporters and phosphatases (Teng et al., 2013). Similar results were found in maize in response to increasing P supply in the field (Deng et al., 2014). Overall, in the context of intensive agricultural systems, it is an important strategy to reduce or withhold P applications, while using the soil P reserves for crop growth, until soil P concentrations decline to the optimum level for maximum crop productivity (critical soil P level). That means, improving P-use efficiency and crop production can be achieved through enhancing root/rhizosphere efficiency with lower external P input.

Increasing plant density is one of the most important ways to increase yield per hectare (Duvick, 2005, York et al., 2015). When plant density is too low, each individual plant may perform at its maximum capacity, but there are not enough plants to reach the maximum yield per hectare (Shao et al., 2018). In contrast, if the plant density is too high, plants may compete with each other, and the performance of individual plants becomes a limiting factor for maximum crop yield (Xu et al., 2017, Shao et al., 2018). Thus, it is important to use appropriate plant densities to reach a high yield. Understanding root responses, and thereby P uptake, to high planting density will provide important information for better crop management in the context of intensive agricultural systems. In response to increasing plant density, plants exhibit significant modifications in root growth and spatial distribution such as a marked reduction in overall root size, root mass and total root length. These changes have an indirect effect on plant nutrient and water uptake (Chen et al., 2013, York et al., 2015, Shao et al., 2018). Different genotypes show contrasting tolerance to high plant density, opening up the possibility of selecting elite genotypes with high yield performance under high plant density (York et al., 2015, Mi et al., 2016). Comparing eight maize hybrids under three plant densities, Shao et al. (2019) found that genotypes with less variation in root size, medium root size, medium broad-root system and more roots distributed in the inter-row space reduced root competition between neighbouring plants. These hybrids tended to achieve higher yield at high plant density. In summary, plants may show shifts in P-acquisition strategies at the individual plant level and at the community level. Aiming to optimise yield performance and nutrient-acquisition efficiency at the community level, we should consider root functional traits to improve P acquisition in the context of high plant density.

Screening genotypes for root traits may be done best under high-density conditions that replicate those of the target crop. Glasshouse experiments in spaced pots where the canopy can expand beyond the edge of the pot or spaced single plants or spaced rows in the field may not properly replicate these conditions. For instance, when Jeffery et al. (2017a) grew annual pasture legumes in a glasshouse with and without reflective sleeves to mimic sward conditions, the addition of the sleeves reduced shoot dry mass by 10–28%. However, their impact on roots was much greater, being a 39–59% decrease in root dry mass, and an increase in specific root length. Most notably, there was a 3.5–12-fold increase in the amount of rhizosheath carboxylates expressed relative to unit root length.

8 Breeding to improve P acquisition

A review by Wang et al. (2019) summarised the parental lines, types and sizes of mapping populations used to identify quantitative trait loci (QTLs) associated with P efficiency in various crops including rice, maize, wheat, common bean, soybean and oilseed rape. With the rapid development of high-throughput sequencing technology, and the construction of high-density genetic maps, genome-wide association studies (GWAS) has been increasingly used for the mapping of P-efficient genes, for example, in soybean (Ning et al., 2016) and oilseed rape (Wang et al., 2017a). Phosphorus-efficiency traits used for QTL mapping studies primarily involve four groups: (1) root traits improving soil inorganic P availability such as acidification of the rhizosphere and exudation of carboxylates, hydrolytic enzymes, for example, acid phosphatases, ribonuclease, phytase; (2) traits improving P uptake such as root architecture and morphology, AMF and beneficial rhizosphere micro-organisms; (3) traits improving the physiological P-utilisation efficiency; and (4) traits associated with yield production (Wang et al., 2019).

Though numerous QTLs affecting P efficiency have been identified in key crops, few causal genes have been identified, and the progress in the breeding of P-efficient crops using marker-assisted selection (MAS) has been limited (Wang et al., 2019). However, there is a very successful example in rice. One major QTL cluster, *Pup1* (phosphorus uptake1) has been identified and introgressed into several rice varieties using a marker-assisted backcrossing approach, leading to a dramatic increase in P-acquisition efficiency and enhanced yield of those lines when grown under a suboptimal P supply (Chin et al., 2011). Functional characterisation of a *Pup1* QTL in rice identified a protein kinase gene (*PSTOL1*, phosphorus starvation tolerance 1), which increased adventitious root development under low P availability and conferred low-P tolerance (Gamuyao et al., 2012). Cloning of *PSTOL1* has been a major breakthrough in a P-efficient rice breeding programme. Further to the findings in rice, several *PSTOL1* genes were identified in sorghum (*Sorghum bicolor*) which were associated with P efficiency, attributed to larger root systems with longer and finer roots and more laterals (Hufnagel et al., 2014). Leiser et al. (2014) found that *PSTOL1* homologs consistently enhanced sorghum P efficiency between two different association panels in Africa and Brazil with different genetic backgrounds and environments, suggesting the potential for using markers linked to *PSTOL* genes for molecular breeding to improve P efficiency in sorghum. Another interesting example, in soybean, is the identification of a QTL, both in GWAS and linkage mapping for a candidate gene-encoding acid phosphatase, *GmACP1* (Zhang et al., 2014a). As P acquisition is a complex trait due to the significant plant-soil interaction, the lack of understanding of P uptake is a bottleneck for both conventional and GM approaches. The availability of complete genome sequences for more crops, and the combination of conventional linkage mapping, association mapping, QTL-sequencing, transcriptomics and gene-editing technologies will speed up the identification of genes underlying QTLs associated with P efficiency. Further progress is required using gene-linked markers to improve nutrient-acquisition efficiency and crop yield. The identification of these genes would provide the opportunity to improve P efficiency through MAS or gene manipulation.

For the successful breeding of P-efficient crops, high-throughput selection of genotypes is required. The development of high-throughput plant phenomics facilities has enabled accurate and standardised investigation of P-efficiency-related traits (Furbank and Tester, 2011, Ubbens and Stavness, 2017). So far, much of the reported work comes from glasshouse or hydroponic studies. Although field assessments would be the ultimate test of the superiority of new varieties, controlled environments reduce environmental variation, thus revealing traits associated with P efficiency. Assessment of P efficiency in the field is often problematic due to the large spatial heterogeneity of soil P, masking genetic effects leading to rather low heritability (Van De Wiel et al., 2016). As plant growth and P uptake rely on a complex interaction of plant, soil and rhizosphere (Bovill et al., 2013), traits for P improvement identified under controlled conditions often fail to show advantages in the field, with significant variation among phenotypes in different culture systems. For example, using three culture systems (i.e. hydroponics, vermiculite and paper rolls) for the phenotyping of eight root traits in a maize recombinant inbred-line population at seedling stage, Liu et al. (2017) found that the root traits in hydroponics and vermiculite were weakly correlated with P-uptake efficiency in the field, while there was no correlation for plants grown in paper rolls. High-throughput screening is desirable, but if the results do not correlate with those obtained in the field their application will be limited.

In many crops, selection for P efficiency has been slow, partly due to inconsistent definitions of P efficiency. Often, different terms are used, even though the efficiency is calculated in the same way, leading to different conclusions about genetic differences in nutrient efficiency depending on the criterion of efficiency used (Bovill et al., 2013). More consistent use of terminology and definitions for improving P efficiency are needed. Both single traits and relative traits (the ratio of the value of a trait for plants grown at a reduced P supply to that of plants grown with optimal P supply) have been used for the identification of QTLs for the effects of P supply on yield, rhizosphere acidification and biomass-related traits. While single traits are often commonly used for QTL associated with P-efficiency traits, relative traits indicate a genotype's tolerance to lower P supply. For example, in barley Gong et al. (2016) reported that a QTL for P responsiveness was mapped to the same locus as a QTL for yield at 30 kg P ha^{-1}, suggesting that P responsiveness and grain yield could be improved simultaneously under high-input cropping systems. Therefore, it would be more useful to identify the co-located QTLs, both for a single trait and for a relative trait in breeding programmes (Wang et al., 2019).

9 Case study: mobilisation of phosphorus and manganese in cluster roots of *Lupinus albus*

While working towards his PhD, Bill Gardner was the first to show that cluster roots of *Lupinus albus*, a non-mycorrhizal species, release citrate, providing a mechanism for their highly-efficient P mobilisation (Gardner et al., 1981). The authors also showed that these cluster roots were equally efficient at mobilising Mn, and followed up their laboratory studies with field experiments, demonstrating that intercropped wheat benefitted from the P-mobilisation capacity of *L. albus* and showed higher leaf [Mn] (Gardner and Boundy, 1983).

Much later, leaf [Mn] was proposed as a proxy for carboxylate concentrations in the rhizosheath (Lambers et al., 2015a), and as a tool to screen for carboxylate release in *Cicer arietinum* (Pang et al., 2018). Mycorrhizal strategies are effective at low soil P availability, but when the P availability is very low, for example, in strongly P-sorbing soils, a strategy based on carboxylate release is more effective (Lambers et al., 2018, Raven et al., 2018, Parfitt, 1979). That would account for lupins being a successful crop on both sandy soils in Australia (Gladstones, 1970, Lambers et al., 2013) and strongly P-sorbing volcanic soils in Chile (Lambers et al., 2013, Baer, 2006).

As far as we know, all *Lupinus* species are non-mycorrhizal, but only a small number produce cluster roots, and the ones that do not, likely all release carboxylates (Lambers et al., 2013). Bolland (1997) compared four *Lupinus* species in a field experiment, showing that the two with cluster roots (*L. albus* and *L. cosentinii*) grow better and produce more grain at a low P supply than the two without cluster roots (*L. angustifolius*, *L. luteus*). Bolland and Brennan (2008) compared *L. angustifolius* with two crop species that do not release large amounts of carboxylates in a pot experiment using the top 0.10 m of a sandy soil that was collected under remnant vegetation that had never been fertilised and used for agriculture. To produce 90% of the maximum grain yield, *L. angustifolius* required ~67% less P than wheat, and canola required ~75% more P than *L. angustifolius*.

Cluster roots in *L. albus* are suppressed at high P supply (Gardner et al., 1983a, Keerthisinghe et al., 1998), as a result of a systemic signal associated with high leaf P concentrations (Marschner et al., 1987, Shane et al., 2003). When plants relied on symbiotic N_2 fixation, which increased their demand for P, their investment in cluster roots was greater than when they were being fed with nitrate (Wang et al., 2018b).

There is overwhelming evidence from field and laboratory studies that cluster roots are a very effective strategy to acquire P from soils with a very low P availability, both in natural and managed systems (Dinkelaker et al., 1989, Shane and Lambers, 2005). Species that released carboxylates without specialised structures were not quite as effective as those with cluster roots, but more effective than mycorrhizal species that did not release significant amounts of carboxylates (Pearse et al., 2006, Veneklaas et al., 2003). We therefore strongly disagree with a statement by Lynch (2019) that variation for the production of carboxylates is associated with P mobilisation in vitro only, and cited numerous references above on crops and native species showing that rigorous analyses have convincingly shown a benefit of such variation for P acquisition in a range of soils in the field.

Pot experiments have shown a significant P benefit of white lupin on the subsequent wheat crop (Nuruzzaman et al., 2005a,b). There is insufficient information to conclude that this effect also happens under field conditions, partly because the subsequent wheat crop was grown with an adequate P supply (Angus et al., 2015). However, in a field experiment on two acid soils of southern Cameroon, significant P benefits were shown for maize grown after some soybean cultivars, but not after others (Jemo et al., 2006). Positive effects would be expected only on soils with a low P availability, especially when these soils contain high levels of total P. Future field work should target such soils, rather than soils that have been fertilised for a very long time.

10 Conclusion

There are two contrasting soil P levels that need our urgent attention: low levels in low-input agricultural system such as many parts of Africa, and accumulation of P due to use of P fertiliser in excess of crop requirement in many developed and rapidly developing countries. It is necessary to improve the acquisition of soluble P under both scenarios, while also increasing a crop's capacity to access large pools of insoluble P. To improve P acquisition, further understanding of root traits is required including (1) improving P availability through exudation of carboxylates, protons and phosphatases; and (2) improving P uptake through changes in root architecture and morphology, AMF and beneficial rhizosphere micro-organisms. Improved crop P-acquisition efficiency would lower critical P requirements, and, therefore, reduce P-fertiliser input, while minimising soil P loss and environmental pollution.

GWAS is increasingly used for mapping P-efficient genes in a number of crops, but few causal genes have been identified. More progress is required in using gene-linked markers to improve P-acquisition efficiency and crop yield. Along with the availability of complete genome sequences for more crops, the combination of conventional linkage mapping, association mapping, QTL-sequencing, transcriptomics and gene-editing technologies would need to be combined to speed up the identification of genes underlying QTLs associated with P efficiency. Identification of these genes would provide the opportunity to improve P efficiency through MAS or gene manipulation. Although yield and P acquisition under field conditions is the ultimate aim, due to the large environmental effects on plant performance and P acquisition, trait-based selection using high-throughput phenotyping platforms with accurate and standardised protocols will be useful for the identification of P-acquisition-related traits. In addition, given the challenges of belowground sampling and observations, the development of easily measurable aboveground proxies for belowground traits will be highly desirable and ideal for the high-throughput phenotyping of large numbers of genotypes. Accompanying the genetic improvement in P acquisition, improvement in the design of cropping systems and rhizosphere management will also be required such as intercropping, rotations and modification of rhizosphere processes.

11 Future trends in research

Figure 8 shows challenges in improving P acquisition and future efforts towards this. We suggest the following trends:

i Many studies on P efficiency have been undertaken on severely P-deficient plants or soils. In many developing countries, P deficiency is a limiting factor for crop production. However, in many cereal-based systems in North America, Europe, Australia and China, there is a long history of P-fertiliser applications so that severe P deficiency is rare. However, improving P efficiency in those situations is still needed, as P recovery and P-fertiliser efficiency may be low. So far, most attempts to

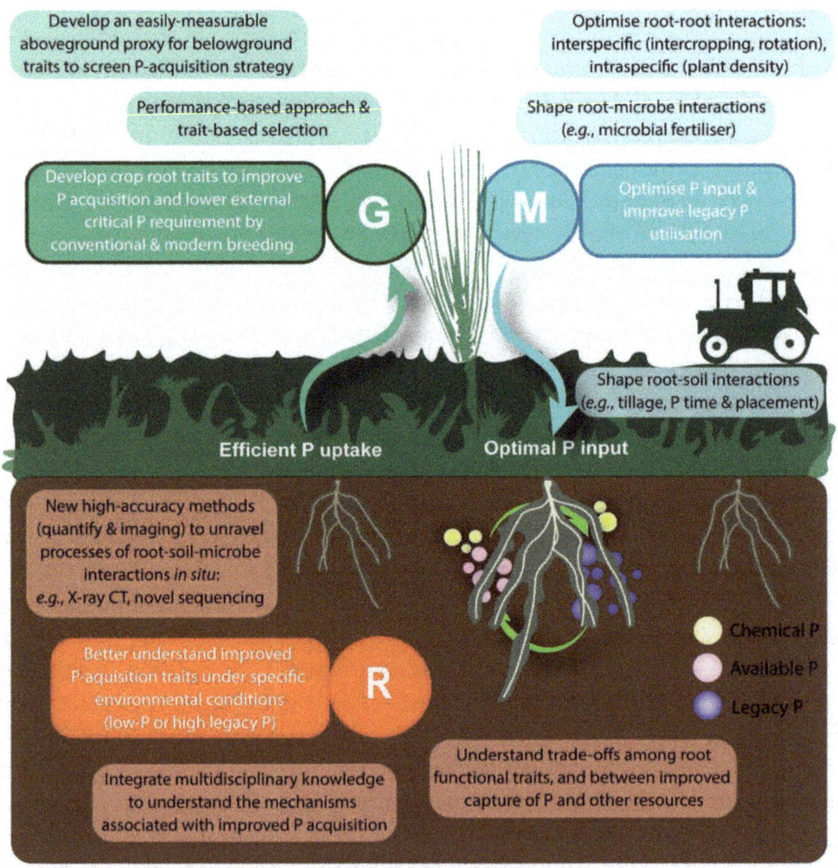

Figure 8 A summary diagram showing the challenges in improving phosphorus (P) acquisition and future efforts towards this. G, development of GENOTYPES with improved P acquisition; M, optimisation of P MANAGEMENT to increase P acquisition; R, better RESEARCH on belowground traits associated with P acquisition.

improve P efficiency have been made in low-P environment, but efforts to improve P efficiency within the context of farming systems with higher P levels must also be considered. On farms with reasonable soil P levels, selecting for P uptake under severely P-limited conditions is unrealistic; instead, systematic evaluation of putative traits for P-acquisition efficiency at different P levels is required. This would make clear whether characteristics contributing to enhanced P uptake and growth under severe P stress would still be useful at higher soil P levels. In the context of the need to maximise yields, there should be a focus on the traits that confer a decrease in the critical external P requirement of crops, and how the target critical P should be determined in a diverse system (rotated or intercropping crops) and how fertiliser practices can be best modified to ensure yield and P-use efficiency in such a system (Deng et al., 2017). Such an approach requires a good understanding of the target commercial agricultural systems, and failure to acquire and apply such knowledge can result in irrelevant research as discussed for AMF by Ryan and Graham (2018).

ii In addition to a traditional performance-based approach for yield and/or nutrient uptake, the trait-based selection to improve nutrient uptake and yield would be useful to overcome the problem of having low heritability in the field due to often large environmental effects. With the rapid advances in molecular technology, reliable phenotyping will be key to the identification of QTLs and genes associated with P-acquisition efficiency. The development of automatic high-throughput phenotyping platforms with accurate and standardised investigation of P-acquisition-related traits will be required, as the final aim will be to improve yield and P acquisition in field. It is imperative that glasshouse screening is conducted under conditions as relevant to the field as logistically possible, with an important preliminary step being determination of the field conditions that should be replicated. For instance, presence of AMF or other soil organisms, stratification of P, environmental conditions and fluctuations, plant age or phenological stage(s) at assessment and plant density/sward conditions (Ryan and Graham, 2018, Jeffery et al., 2017a, Haling et al., 2016, 2018). Ultimately, field assessment under commercially realistic management conditions will be required.

iii Due to the difficulty of studying roots in soil under field conditions, the development of easily measurable aboveground proxies for belowground traits is highly desirable and ideal for high-throughput phenotyping of large numbers of genotypes. For example, the use of leaf [Mn] as a proxy for rhizosheath carboxylates in chickpea (Pang et al., 2018), and blumenols as shoot markers of root symbiosis with AMF (Wang et al., 2018a). More work like this on different crops and in different environments is needed to make the high-throughput phenotyping of belowground traits more accessible. In addition, new methods for root studies would be required for future root study. For example, soil zymography (Spohn and Kuzyakov, 2013) is an in-situ technique for localisation and quantification of enzyme activity in soil using thin gels with embedded substrates. It is non-destructive for soil structure and displays enzyme activities more realistically than traditional enzyme assays, which involve the destruction of soil structure.

iv There is a need to adopt better agronomic-management practices such as the design of smart farming system in intercropping and rotation to maximise P acquisition of crops. The concept of 'building up and maintenance' by Zhang et al. (2019c) suggests limiting P input in fields with high existing P levels by maintaining plant-available P in an optimal range for the yield of the target plants. This will improve the P-use efficiency of applied fertiliser, while also reducing the sorption of P fertiliser in soil. Limiting P input will also be a very effective strategy to reduce soil P loss causing environmental pollution. By optimising cropping systems and rhizosphere management such as intercropping, rotations and modification of rhizosphere processes, soil legacy P could be used more efficiently allowing a decrease of plant-available soil P to a critical level, without decreasing crop yields.

v Because P acquisition involves a number of complex traits, there is a need to develop collaboration among the disciplines of agronomy, physiology, modelling, genetics, genomics, metabolomics, proteomics, transcriptomics and bioinformatics.

12 Where to look for further information

- The Global Phosphorus Research initiative (GPRI), a collaboration between six independent research institutes in Europe, Australia and North America. http://phosphorusfutures.net.
- The Sustainable Phosphorus Alliance hosts an Annual Phosphorus Forum on topics related to phosphorus sustainability. The organisation also hosts technical webinars and workshops on phosphorus management issues. https://phosphorusalliance.org.
- International Society of Root Research (ISRR) aims to promote cooperation and communication between root researchers around the world. The society holds an international symposium every three years. https://www.rootresearch.org/.
- Rhizosphere Conferences have been held every four years around the world, providing a multidisciplinary forum for researchers in the area of the rhizosphere, to understand its complexity, and role in both natural

and agricultural ecosystem processes. https://www.rhizo5.org/home/index.html.

12.1 Review articles

- Baker, A., Ceaser, S. A., Palmer, A. J., Paterson, J. B., Qi, W., Muench, S. P. and Baldwin, S. A. (2015). Replace, reuse, recycle: improving the sustainable use of phosphorus by plants. *Journal of Experimental Botany* 66: 3523–3540.
- Cordell, D. and White, S. (2014). Life's bottleneck: sustaining the world's phosphorus for a food secure future. *Annual Review of Environment and Resources* 39: 161–188.
- Menezes-Blackburn, D., Giles, C., Darch, T., George, T. S., Blackwell, M., Stutter, M., Shand, C., Lumsdon, D., Cooper, P., Wendler, R., et al. (2018). Opportunities for mobilizing recalcitrant phosphorus from agricultural soils: a review. *Plant and Soil* 427: 5–16.

13 Acknowledgement

Zhihui Wen was supported by the China Scholarship Council (CSC). Daniel Kidd was funded by Meat and Livestock Australia (MLA) and Australian Wool Industry (AWI). We acknowledge OOid scientific for the preparation of Figs 5 and 8.

14 References

Angus, J. F., Kirkegaard, J. A., Hunt, J. R., Ryan, M. H., Ohlander, L. and Peoples, M. B. 2015. Break crops and rotations for wheat. *Crop and Pasture Science* 66(6), 523–552.

Atkinson, D., Berta, G. and Hooker, J. E. 1994. Impact of mycorrhizal colonisation on root architecture, root longevity and the formation of growth regulators. In: Gianinazzi, S. and Schüepp, H. (Eds) *Impact of Arbuscular Mycorrhizas on Sustainable Agriculture and Natural Ecosystems*. Basel: Birkhäuser Basel. pp. 89–99.

Badri, D. V., Weir, T. L., Van Der Lelie, D. and Vivanco, J. M. 2009. Rhizosphere chemical dialogues: plant-microbe interactions. *Current Opinion in Biotechnology* 20(6), 642–650.

Baer, E. V. 2006. Relevant points for the production and use of sweet lupin in Chile. México, where old and new world lupins meet. Proceedings of the 11th International Lupin Conference, Guadalajara, Jalisco, Mexico, 4–9 May 2005. International Lupin Association. pp. 116–119.

Barber, D. A. and Lynch, J. M. 1977. Microbial growth in the rhizosphere. *Soil Biology and Biochemistry* 9(5), 305–308.

Bargaz, A., Noyce, G. L., Fulthorpe, R., Carlsson, G., Furze, J. R., Jensen, E. S., Dhiba, D. and Isaac, M. E. 2017. Species interactions enhance root allocation, microbial diversity and P acquisition in intercropped wheat and soybean under P deficiency. *Applied Soil Ecology* 120, 179–188.

Bayuelo-Jiménez, J. S., Gallardo-Valdéz, M., Pérez-Decelis, V. A., Magdaleno-Armas, L., Ochoa, I. and Lynch, J. P. 2011. Genotypic variation for root traits of maize (*Zea mays* L.)

from the Purhepecha Plateau under contrasting phosphorus availability. *Field Crops Research* 121(3), 350–362.

Becquer, A., Haling, R., Stefanski, A., Richardson, A. and Simpson, R. 2017. Complementary phosphorus acquisition strategies of interplanted subterranean clover and white lupin increase sward yield in a low phosphorus soil. *Doing More with Less*. Proceedings of the 18th Australian Agronomy Conference 2017, Ballarat, Victoria, Australia. Australian Society of Agronomy Inc. pp. 1–4.

Berta, G., Trotta, A., Fusconi, A., Hooker, J. E., Munro, M., Atkinson, D., Giovannetti, M., Morini, S., Fortuna, P., Tisserant, B., Gianinazzi-Pearson, V. and Gianinazzi, S. 1995. Arbuscular mycorrhizal induced changes to plant growth and root system morphology in *Prunus cerasifera*. *Tree Physiology* 15(5), 281–293.

Bishopp, A. and Lynch, J. P. 2015. The hidden half of crop yields. *Nature Plants* 1, 15117.

Blamey, F. P. C., Edwards, D. G. and Asher, C. J. 1983. Effects of aluminum, OH-Al and P-Al molar ratios, and ionic-strength on soybean root elongation in solution culture. *Soil Science* 136(4), 197–207.

Bolland, M. D. A. 1995. *Lupinus cosentinii* more effectively utilizes low levels of phosphorus from superphosphate than *Lupinus angustifolius*. *Journal of Plant Nutrition* 18(3), 421–435.

Bolland, M. D. A. 1997. Comparative phosphorus requirement of four lupin species. *Journal of Plant Nutrition* 20(10), 1239–1253.

Bolland, M. D. A. and Brennan, R. F. 2008. Comparing the phosphorus requirements of wheat, lupin, and canola. *Australian Journal of Agricultural Research* 59(11), 983–998.

Bolland, M. D. A., Siddique, K. H. M., Loss, S. P. and Baker, M. J. 1999. Comparing responses of grain legumes, wheat and canola to applications of superphosphate. *Nutrient Cycling in Agroecosystems* 53(2), 157–175.

Bolland, M. D. A., Sweetingham, M. W. and Jarvis, R. J. 2000. Effect of applied phosphorus on the growth of *Lupinus luteus*, L. *angustifolius* and *L. albus* in acidic soils in the south-west of Western Australia. *Australian Journal of Experimental Agriculture* 40(1), 79–92.

Bovill, W. D., Huang, C. Y. and McDonald, G. K. 2013. Genetic approaches to enhancing phosphorus-use efficiency (PUE) in crops: challenges and directions. *Crop and Pasture Science* 64(3), 179–198.

Brennan, R. F. and Bolland, M. D. A. 2001. Comparing fertilizer phosphorus requirements of canola, lupin, and wheat. *Journal of Plant Nutrition* 24(12), 1885–1900.

Brodelius, P. and Vogel, H. J. 1985. A phosphorus-31 nuclear magnetic resonance study of phosphate uptake and storage in cultured *Catharanthus roseus* and *Daucus carota* plant cells. *Journal of Biological Chemistry* 260(6), 3556–3560.

Brooker, R. W., Bennett, A. E., Cong, W. F., Daniell, T. J., George, T. S., Hallett, P. D., Hawes, C., Iannetta, P. P. M., Jones, H. G., Karley, A. J., Li, L., McKenzie, B. M., Pakeman, R. J., Paterson, E., Schöb, C., Shen, J., Squire, G., Watson, C. A., Zhang, C., Zhang, F., Zhang, J. and White, P. J. 2015. Improving intercropping: a synthesis of research in agronomy, plant physiology and ecology. *New Phytologist* 206(1), 107–117.

Brown, L. K., George, T. S., Barrett, G. E., Hubbard, S. F. and White, P. J. 2013. Interactions between root hair length and arbuscular mycorrhizal colonisation in phosphorus deficient barley (*Hordeum vulgare*). *Plant and Soil* 372(1–2), 195–205.

Brundrett, M. C. 2002. Coevolution of roots and mycorrhizas of land plants. *New Phytologist* 154(2), 275–304.

Brundrett, M. C. 2009. Mycorrhizal associations and other means of nutrition of vascular plants: understanding the global diversity of host plants by resolving conflicting information and developing reliable means of diagnosis. *Plant and Soil* 320(1-2), 37-77.

Bruulsema, T. W., Peterson, H. M. and Prochnow, L. I. 2019. The science of 4R nutrient stewardship for phosphorus management across latitudes. *Journal of Environmental Quality* 48(5), 1295-1299.

Bünemann, E. K., Keller, B., Hoop, D., Jud, K., Boivin, P. and Frossard, E. 2013. Increased availability of phosphorus after drying and rewetting of a grassland soil: processes and plant use. *Plant and Soil* 370(1-2), 511-526.

Calabi-Floody, M., Medina, J., Rumpel, C., Condron, L. M., Hernandez, M., Dumont, M. and Mora, M. D. L. L. 2018. Smart fertilizers as a strategy for sustainable agriculture. *Advances in Agronomy* 147, 119-157.

Caradus, J. R. 2012. Effect of root hair length on white clover growth over a range of soil phosphorus levels. *New Zealand Journal of Agricultural Research* 24(3-4), 353-358.

Chen, W., Koide, R. T., Adams, T. S., Deforest, J. L., Cheng, L. and Eissenstat, D. M. 2016. Root morphology and mycorrhizal symbioses together shape nutrient foraging strategies of temperate trees. *Proceedings of the National Academy of Sciences of the United States of America* 113(31), 8741-8746.

Chen, X., Chen, F., Chen, Y., Gao, Q., Yang, X., Yuan, L., Zhang, F. and Mi, G. 2013. Modern maize hybrids in Northeast China exhibit increased yield potential and resource use efficiency despite adverse climate change. *Global Change Biology* 19(3), 923-936.

Chin, J. H., Gamuyao, R., Dalid, C., Bustamam, M., Prasetiyono, J., Moeljopawiro, S., Wissuwa, M. and Heuer, S. 2011. Developing rice with high yield under phosphorus deficiency: *Pup1* sequence to application. *Plant Physiology* 156(3), 1202-1216.

Chung, H., Zak, D. R., Reich, P. B. and Ellsworth, D. S. 2007. Plant species richness, elevated CO_2, and atmospheric nitrogen deposition alter soil microbial community composition and function. *Global Change Biology* 13(5), 980-989.

Clarholm, M., Skyllberg, U. and Rosling, A. 2015. Organic acid induced release of nutrients from metal-stabilized soil organic matter - the unbutton model. *Soil Biology and Biochemistry* 84, 168-176.

Clark, F. E. 1967. Bacteria in soil. In: Burges, A. and Raw, F. (Eds) *Soil Biology*. London: Academic Press. pp. 15-49.

Cordell, D., Drangert, J. O. and White, S. 2009. The story of phosphorus: global food security and food for thought. *Global Environmental Change* 19(2), 292-305.

Crush, J. R. 1973. The effect of *Rhizophagus tenuis* mycorrhizas on ryegrass, cocksfoot and sweet vernal. *New Phytologist* 72(5), 965-973.

Deng, Y., Chen, K., Teng, W., Zhan, A., Tong, Y., Feng, G., Cui, Z., Zhang, F. and Chen, X. 2014. Is the inherent potential of maize roots efficient for soil phosphorus acquisition? *PLoS ONE* 9(3), e90287.

Deng, Y., Feng, G., Chen, X. and Zou, C. 2017. Arbuscular mycorrhizal fungal colonization is considerable at optimal Olsen-P levels for maximized yields in an intensive wheat-maize cropping system. *Field Crops Research* 209, 1-9.

Dinkelaker, B., Romheld, V. and Marschner, H. 1989. Citric acid excretion and precipitation of calcium citrate in the rhizosphere of white lupin (*Lupinus albus* L.). *Plant, Cell and Environment* 12(3), 285-292.

Dissanayaka, D. M. S., Maruyama, H., Masuda, G. and Wasaki, J. 2015. Interspecific facilitation of P acquisition in intercropping of maize with white lupin in two contrasting soils as influenced by different rates and forms of P supply. *Plant and Soil* 390(1-2), 223-236.

Djigal, D., Brauman, A., Diop, T. A., Chotte, J. L. and Villenave, C. 2004. Influence of bacterial-feeding nematodes (Cephalobidae) on soil microbial communities during maize growth. *Soil Biology and Biochemistry* 36(2), 323-331.

Drew, M. C. 1975. Comparison of the effects of a localised supply of phosphate, nitrate, ammonium and potassium on the growth of the seminal root system, and the shoot, in barley. *New Phytologist* 75(3), 479-490.

Drogue, B., Doré, H., Borland, S., Wisniewski-Dyé, F. and Prigent-Combaret, C. 2012. Which specificity in cooperation between phytostimulating rhizobacteria and plants? *Research in Microbiology* 163(8), 500-510.

Duvick, D. N. 2005. The contribution of breeding to yield advances in maize (*Zea mays* L.). *Advances in Agronomy* 86, 83-145.

Eisenhauer, N. Beßler, H., Engels, C., Gleixner, G., Habekost, M., Milcu, A., Partsch, S., Sabais, A. C. W., Scherber, C., Steinbeiss, S., Weigelt, A., Weisser, W. W. and Scheu, S. 2010. Plant diversity effects on soil microorganisms support the singular hypothesis. *Ecology* 91, 485-496.

Fan, J., Du, Y., Turner, N. C., Wang, B., Fang, Y., Xi, Y., Guo, X. and Li, F. 2015. Changes in root morphology and physiology to limited phosphorus and moisture in a locally-selected cultivar and an introduced cultivar of *Medicago sativa* L. growing in alkaline soil. *Plant and Soil* 392(1-2), 215-226.

Fitter, A. H. 1994. Architecture and biomass allocation as components of the plastic response of root systems to soil heterogeneity. In: Martyn, M. C. and Robert, W. P. (Eds) *Exploitation of Environmental Heterogeneity by Plants*. Boston, MA: Academic Press. pp. 305-323.

Fixen, P. E. and Johnston, A. M. 2012. World fertilizer nutrient reserves: a view to the future. *Journal of the Science of Food and Agriculture* 92(5), 1001-1005.

Freschet, G. T. and Roumet, C. 2017. Sampling roots to capture plant and soil functions. *Functional Ecology* 31(8), 1506-1518.

Frew, A., Powell, J. R., Glauser, G., Bennett, A. E. and Johnson, S. N. 2018. Mycorrhizal fungi enhance nutrient uptake but disarm defences in plant roots, promoting plant-parasitic nematode populations. *Soil Biology and Biochemistry* 126, 123-132.

Furbank, R. T. and Tester, M. 2011. Phenomics - technologies to relieve the phenotyping bottleneck. *Trends in Plant Science* 16(12), 635-644.

Galindo-Castaneda, T., Brown, K. M., Kuldau, G. A., Roth, G. W., Wenner, N. G., Ray, S., Schneider, H. and Lynch, J. P. 2019. Root cortical anatomy is associated with differential pathogenic and symbiotic fungal colonization in maize. *Plant, Cell and Environment* 42(11), 2999-3014.

Gamuyao, R., Chin, J. H., Pariasca-Tanaka, J., Pesaresi, P., Catausan, S., Dalid, C., Slamet-Loedin, I., Tecson-Mendoza, E. M., Wissuwa, M. and Heuer, S. 2012. The protein kinase Pstol1 from traditional rice confers tolerance of phosphorus deficiency. *Nature* 488(7412), 535-539.

Gardner, W. K., Barber, D. A. and Parbery, D. G. 1983a. The acquisition of phosphorus byLupinus albus L.: III. The probable mechanism by which phosphorus movement in the soil/root interface is enhanced. *Plant and Soil* 70(1), 107-124.

Gardner, W. K. and Boundy, K. A. 1983. The acquisition of phosphorus by *Lupinus albus* L. IV. The effect of interplanting wheat and white lupin on the growth and mineral composition of the two species. *Plant and Soil* 70(3), 391-402.

Gardner, W. K., Parbery, D. G. and Barber, D. A. 1981. Proteoid root morphology and function in *Lupinus albus*. *Plant and Soil* 60(1), 143–147.
Gardner, W. K., Parbery, D. G. and Barber, D. A. 1982a. The acquisition of phosphorus by *Lupinus albus* L. I. Some characteristics of the soil/root interface. *Plant and Soil* 68(1), 19–32.
Gardner, W. K., Parbery, D. G. and Barber, D. A. 1982b. The acquisition of phosphorus by *Lupinus albus* L. II. The effect of varying phosphorus supply and soil type on some characteristics of the soil/root interface. *Plant and Soil* 68(1), 33–41.
Gardner, W. K., Parbery, D. G., Barber, D. A., Swinden, L. J. P. and SOIL 1983b. The acquisition of phosphorus by *Lupinus albus* L. V. The diffusion of exudates away from roots: a computer simulation. *Plant and Soil* 72(1), 13–29.
George, T. S., Quiquampoix, H., Simpson, R. J. and Richardson, A. E. 2007a. Interactions between phytases and soil constituents: implications for the hydrolysis of inositol phosphates. In: Turner, B. L., Richardson, A. E. and Mullaney, E. J. (Eds) *Inositol Phosphates: Linking Agriculture and the Environment*. Wallingford: CAB International. pp. 221–241.
George, T. S., Richardson, A. E., Smith, J. B., Hadobas, P. A. and Simpson, R. J. 2005. Limitations to the potential of transgenic *Trifolium subterraneum* L. plants that exude phytase when grown in soils with a range of organic P content. *Plant and Soil* 278(1–2), 263–274.
George, T. S., Simpson, R. J., Hadobas, P. A., Marshall, D. J. and Richardson, A. E. 2007b. Accumulation and phosphatase-lability of organic phosphorus in fertilised pasture soils. *Australian Journal of Agricultural Research* 58(1), 47–55.
Giles, C. D., Brown, L. K., Adu, M. O., Mezeli, M. M., Sandral, G. A., Simpson, R. J., Wendler, R., Shand, C. A., Menezes-Blackburn, D., Darch, T., Stutter, M. I., Lumsdon, D. G., Zhang, H., Blackwell, M. S., Wearing, C., Cooper, P., Haygarth, P. M. and George, T. S. 2017. Response-based selection of barley cultivars and legume species for complementarity: root morphology and exudation in relation to nutrient source. *Plant Science* 255, 12–28.
Giles, C. D., George, T. S., Brown, L. K., Mezeli, M. M., Richardson, A. E., Shand, C. A., Wendler, R., Darch, T., Menezes-Blackburn, D., Cooper, P., Stutter, M. I., Lumsdon, D. G., Blackwell, M. S. A., Wearing, C., Zhang, H. and Haygarth, P. M. 2016. Does the combination of citrate and phytase exudation in *Nicotiana tabacum* promote the acquisition of endogenous soil organic phosphorus? *Plant and Soil* 412(1–2), 43–59.
Giles, C. D., Richardson, A. E., Cade-Menun, B. J., Mezeli, M. M., Brown, L. K., Menezes-Blackburn, D., Darch, T., Blackwell, M. S., Shand, C. A., Stutter, M. I., Wendler, R., Cooper, P., Lumsdon, D. G., Wearing, C., Zhang, H., Haygarth, P. M. and George, T. S. 2018. Phosphorus acquisition by citrate- and phytase-exuding *Nicotiana tabacum* plant mixtures depends on soil phosphorus availability and root intermingling. *Physiologia Plantarum* 163, 356–371.
Gladstones, J. 1970. Lupins as crop plants. *Field Crop Abstracts* 23, 123–148.
Gong, X., Wheeler, R., Bovill, W. D. and McDonald, G. K. 2016. QTL mapping of grain yield and phosphorus efficiency in barley in a Mediterranean-like environment. *Theoretical and Applied Genetics* 129(9), 1657–1672.
Graham, J. H. and Abbott, L. K. 2000. Wheat responses to aggressive and non-aggressive arbuscular mycorrhizal fungi. *Plant and Soil* 220(1/2), 207–218.
Grant, C. A. and Flaten, D. N. 2019. 4R management of phosphorus fertilizer in the Northern Great Plains. *Journal of Environmental Quality* 48(5), 1356–1369.

Grierson, P. F. 1992. Organic acids in the rhizosphere of *Banksia integrifolia* L.f. *Plant and Soil* 144(2), 259–265.

Haling, R. E., Brown, L. K., Stefanski, A., Kidd, D. R., Ryan, M. H., Sandral, G. A., George, T. S., Lambers, H. and Simpson, R. J. 2018. Differences in nutrient foraging among *Trifolium subterraneum* cultivars deliver improved P-acquisition efficiency. *Plant and Soil* 424(1–2), 539–554.

Haling, R. E., Yang, Z., Shadwell, N., Culvenor, R. A., Stefanski, A., Ryan, M. H., Sandral, G. A., Kidd, D. R., Lambers, H. and Simpson, R. J. 2016. Growth and root dry matter allocation by pasture legumes and a grass with contrasting external critical phosphorus requirements. *Plant and Soil* 407(1–2), 67–79.

Hart, M. M., Antunes, P. M., Chaudhary, V. B., Abbott, L. K. and Field, K. 2018. Fungal inoculants in the field: is the reward greater than the risk? *Functional Ecology* 32(1), 126–135.

Hayes, P., Turner, B. L., Lambers, H. and Laliberté, E. 2014. Foliar nutrient concentrations and resorption efficiency in plants of contrasting nutrient-acquisition strategies along a 2-million-year dune chronosequence. *Journal of Ecology* 102(2), 396–410.

Haynes, R. J. 1990. Active ion uptake and maintenance of cation-anion balance: A critical examination of their role in regulating rhizosphere pH. *Plant and Soil* 126(2), 247–264.

Hetrick, B. A. D., Wilson, G. W. T. and Leslie, J. F. 1991. Root architecture of warm- and cool-season grasses: relationship to mycorrhizal dependence. *Canadian Journal of Botany* 69(1), 112–118.

Hill, J. O., Simpson, R. J., Ryan, M. H. and Chapman, D. F. 2010. Root hair morphology and mycorrhizal colonisation of pasture species in response to phosphorus and nitrogen nutrition. *Crop and Pasture Science* 61(2), 122–131.

Hinsinger, P., Herrmann, L., Lesueur, D., Robin, A., Trap, J., Waithaisong, K. and Plassard, C. 2015. Impact of roots, microorganisms and microfauna on the fate of soil phosphorus in the rhizosphere. *Annual Plant Reviews* 48, 377–408.

Hocking, P. J. and Jeffery, S. 2004. Cluster-root production and organic anion exudation in a group of old-world lupins and a new-world lupin. *Plant and Soil* 258(1), 135–150.

Hodge, A. 2004. The plastic plant: root responses to heterogeneous supplies of nutrients. *New Phytologist* 162(1), 9–24.

Hooper, D. U., Chapin III, F. S., Ewel, J. J., Hector, A., Inchausti, P., Lavorel, S., Lawton, J. H., Lodge, D. M., Loreau, M., Naeem, S., Schmid, B., Setälä, H., Symstad, A. J., Vandermeer, J. and Wardle, D. A. 2005. Effects of biodiversity on ecosystem functioning: A consensus of current knowledge. *Ecological Monographs* 75(1), 3–35.

Hufnagel, B., De Sousa, S. M., Assis, L., Guimaraes, C. T., Leiser, W., Azevedo, G. C., Negri, B., Larson, B. G., Shaff, J. E., Pastina, M. M., Barros, B. A., Weltzien, E., Rattunde, H. F. W., Viana, J. H., Clark, R. T., Falcão, A., Gazaffi, R., Garcia, A. A. F., Schaffert, R. E., Kochian, L. V. and Magalhaes, J. V. 2014. Duplicate and conquer: multiple homologs of *PHOSPHORUS-STARVATION TOLERANCE1* enhance phosphorus acquisition and sorghum performance on low-phosphorus soils. *Plant Physiology* 166(2), 659–677.

Inal, A., Gunes, A., Zhang, F. and Cakmak, I. 2007. Peanut/maize intercropping induced changes in rhizosphere and nutrient concentrations in shoots. *Plant Physiology and Biochemistry* 45(5), 350–356.

Jakobsen, I., Chen, B., Munkvold, L., Lundsgaard, T. and Zhu, Y.-G. 2005. Contrasting phosphate acquisition of mycorrhizal fungi with that of root hairs using the root hairless barley mutant. *Plant, Cell and Environment* 28(7), 928–938.

Jeffery, R. P., Simpson, R. J., Lambers, H., Kidd, D. R. and Ryan, M. H. 2017a. Plants in constrained canopy micro-swards compensate for decreased root biomass and soil exploration with increased amounts of rhizosphere carboxylates. *Functional Plant Biology* 44(5), 552–562.

Jeffery, R. P., Simpson, R. J., Lambers, H., Kidd, D. R. and Ryan, M. H. 2017b. Root morphology acclimation to phosphorus supply by six cultivars of *Trifolium subterraneum* L. *Plant and Soil* 412(1–2), 21–34.

Jeffery, R. P., Simpson, R. J., Lambers, H., Orchard, S., Kidd, D. R., Haling, R. E. and Ryan, M. H. 2018. Contrasting communities of arbuscule-forming root symbionts change external critical phosphorus requirements of some annual pasture legumes. *Applied Soil Ecology* 126, 88–97.

Jemo, M., Abaidoo, R. C., Nolte, C., Tchienkoua, M., Sanginga, N. and Horst, W. J. 2006. Phosphorus benefits from grain-legume crops to subsequent maize grown on acid soils of southern Cameroon. *Plant and Soil* 284(1–2), 385–397.

Jia, X., Liu, P. and Lynch, J. P. 2018. Greater lateral root branching density in maize improves phosphorus acquisition from low phosphorus soil. *Journal of Experimental Botany* 69(20), 4961–4970.

Johnson, N. C., Graham, J. H. and Smith, F. A. 1997. Functioning of mycorrhizal associations along the mutualism-parasitism continuum. *New Phytologist* 135(4), 575–585.

Jones, D. L. 1998. Organic acids in the rhizosphere. *Plant and Soil* 205(1), 25–44.

Jones, D. L., Dennis, P. G., Owen, A. G. and Van Hees, P. A. W. 2003. Organic acid behavior in soils – misconceptions and knowledge gaps. *Plant and Soil* 248(1/2), 31–41.

Keerthisinghe, G., Hocking, P. J., Ryan, P. R. and Delhaize, E. 1998. Effect of phosphorus supply on the formation and function of proteoid roots of white lupin (*Lupinus albus* L.). *Plant, Cell and Environment* 21(5), 467–478.

Khan, M. S., Zaidi, A. and Musarrad, J. 2014. *Phosphate Solubilizing Microorganisms, Principles and Application of Microphos Technology*. New Delhi: Springer International Publishing.

Kidd, D. R., Ryan, M. H., Hahne, D., Haling, R. E., Lambers, H., Sandral, G. A., Simpson, R. J. and Cawthray, G. R. 2018. The carboxylate composition of rhizosheath and root exudates from twelve species of grassland and crop legumes with special reference to the occurrence of citramalate. *Plant and Soil* 424(1–2), 389–403.

Kidd, D. R., Ryan, M. H., Haling, R. E., Lambers, H., Sandral, G. A., Yang, Z., Culvenor, R. A., Cawthray, G. R., Stefanski, A. and Simpson, R. J. 2015. Rhizosphere carboxylates and morphological root traits in pasture legumes and grasses. *Plant and Soil* 402(1–2), 77–89.

Kirkby, E. A. and Mengel, K. 1967. Ionic balance in different tissues of the tomato plant in relation to nitrate, urea, or ammonium nutrition. *Plant Physiology* 42(1), 6–14.

Kong, Y., Li, X., Ma, J., Li, W., Yan, G. and Zhang, C. 2014. *GmPAP4*, a novel purple acid phosphatase gene isolated from soybean (*Glycine max*), enhanced extracellular phytate utilization in *Arabidopsis thaliana*. *Plant Cell Reports* 33(4), 655–667.

Kuzyakov, Y. and Razavi, B. S. 2019. Rhizosphere size and shape: temporal dynamics and spatial stationarity. *Soil Biology and Biochemistry* 135, 343–360.

Lambers, H., Albornoz, F., Kotula, L., Laliberté, E., Ranathunge, K., Teste, F. P. and Zemunik, G. 2018. How belowground interactions contribute to the coexistence of mycorrhizal and non-mycorrhizal species in severely phosphorus-impoverished hyperdiverse ecosystems. *Plant and Soil* 424(1–2), 11–33.

Lambers, H., Clements, J. C. and Nelson, M. N. 2013. How a phosphorus-acquisition strategy based on carboxylate exudation powers the success and agronomic potential of lupines (*Lupinus*, Fabaceae). *American Journal of Botany* 100(2), 263–288.

Lambers, H., Finnegan, P. M., Laliberte, E., Pearse, S. J., Ryan, M. H., Shane, M. W. and Veneklaas, E. J. 2011. Update on phosphorus nutrition in Proteaceae. Phosphorus nutrition of Proteaceae in severely phosphorus-impoverished soils: are there lessons to be learned for future crops? *Plant Physiology* 156(3), 1058–1066.

Lambers, H., Hayes, P. E., Laliberté, E., Oliveira, R. S. and Turner, B. L. 2015a. Leaf manganese accumulation and phosphorus-acquisition efficiency. *Trends in Plant Science* 20(2), 83–90.

Lambers, H., Martinoia, E. and Renton, M. 2015b. Plant adaptations to severely phosphorus-impoverished soils. *Current Opinion in Plant Biology* 25, 23–31.

Lambers, H., Raven, J. A., Shaver, G. R. and Smith, S. E. 2008. Plant nutrient-acquisition strategies change with soil age. *Trends in Ecology and Evolution* 23(2), 95–103.

Lambers, H., Shane, M. W., Cramer, M. D., Pearse, S. J. and Veneklaas, E. J. 2006. Root structure and functioning for efficient acquisition of phosphorus: matching morphological and physiological traits. *Annals of Botany* 98(4), 693–713.

Lee, R. B. and Ratcliffe, R. G. 1993. Subcellular distribution of inorganic phosphate, and levels of nucleoside triphosphate, in mature maize roots at low external phosphate concentrations: measurements with ^{31}P-NMR. *Journal of Experimental Botany* 44(3), 587–598.

Leiser, W. L., Rattunde, H. F. W., Weltzien, E., Cisse, N., Abdou, M., Diallo, A., Tourè, A. O., Magalhaes, J. V. and Haussmann, B. I. G. 2014. Two in one sweep: aluminum tolerance and grain yield in P-limited soils are associated to the same genomic region in West African sorghum. *BMC Plant Biology* 14, 206–206.

Li, H., Huang, G., Meng, Q., Ma, L., Yuan, L., Wang, F., Zhang, W., Cui, Z., Shen, J., Chen, X., Jiang, R. and Zhang, F. 2011. Integrated soil and plant phosphorus management for crop and environment in China: a review. *Plant and Soil* 349(1–2), 157–167.

Li, H., Liu, B., Mccormack, M. L., Ma, Z. and Guo, D. 2017. Diverse belowground resource strategies underlie plant species coexistence and spatial distribution in three grasslands along a precipitation gradient. *New Phytologist* 216(4), 1140–1150.

Li, H., Shen, J., Zhang, F., Clairotte, M., Drevon, J. J., Le Cadre, E. and Hinsinger, P. 2008. Dynamics of phosphorus fractions in the rhizosphere of common bean (*Phaseolus vulgaris* L.) and durum wheat (*Triticum turgidum durum* L.) grown in monocropping and intercropping systems. *Plant and Soil* 312(1–2), 139–150.

Li, H., Shen, J., Zhang, F., Marschner, P., Cawthray, G. and Rengel, Z. 2010. Phosphorus uptake and rhizosphere properties of intercropped and monocropped maize, faba bean, and white lupin in acidic soil. *Biology and Fertility of Soils* 46(2), 79–91.

Li, H., Ma, Q., Li, H., Zhang, F., Rengel, Z. and Shen, J. 2014a. Root morphological responses to localized nutrient supply differ among crop species with contrasting root traits. *Plant and Soil* 376(1–2), 151–163.

Li, L., Li, S. M., Sun, J. H., Zhou, L. L., Bao, X. G., Zhang, H. G. and Zhang, F. S. 2007. Diversity enhances agricultural productivity via rhizosphere phosphorus facilitation on phosphorus-deficient soils. *Proceedings of the National Academy of Sciences of the United States of America* 104(27), 11192-11196.

Li, L., Tang, C., Rengel, Z. and Zhang, F. 2003. Chickpea facilitates phosphorus uptake by intercropped wheat from an organic phosphorus source. *Plant and Soil* 248(1/2), 297–303.

Li, L., Tilman, D., Lambers, H. and Zhang, F. S. 2014b. Plant diversity and overyielding: insights from belowground facilitation of intercropping in agriculture. *New Phytologist* 203(1), 63-69.
Li, L., Yang, S., Li, X., Zhang, F. and Christie, P. 1999. Interspecific complementary and competitive interactions between intercropped maize and faba bean. *Plant and Soil* 212(2), 105-114.
Li, R., Lu, W., Guo, C., Li, X., Gu, J. and Xiao, K. 2012. Molecular characterization and functional analysis of OsPHY1, a purple acid phosphatase (PAP)-type phytase gene in rice (*Oryza sativa* L.). *Journal of Integrative Agriculture* 11(8), 1217-1226.
Li, S. M., Li, L., Zhang, F. S. and Tang, C. 2004. Acid phosphatase role in chickpea/maize intercropping. *Annals of Botany* 94(2), 297-303.
Lipton, D. S., Blanchar, R. W. and Blevins, D. G. 1987. Citrate, malate, and succinate concentration in exudates from P-sufficient and P-stressed *Medicago sativa* L. seedlings. *Plant Physiology* 85(2), 315-317.
Liu, B., Li, H., Zhu, B., Koide, R. T., Eissenstat, D. M. and Guo, D. 2015. Complementarity in nutrient foraging strategies of absorptive fine roots and arbuscular mycorrhizal fungi across 14 coexisting subtropical tree species. *New Phytologist* 208(1), 125-136.
Liu, H. T., White, P. J. and Li, C. J. 2016a. Biomass partitioning and rhizosphere responses of maize and faba bean to phosphorus deficiency. *Crop and Pasture Science* 67(8), 847-856.
Liu, T. Y., Huang, T. K., Yang, S. Y., Hong, Y. T., Huang, S. M., Wang, F. N., Chiang, S. F., Tsai, S. Y., Lu, W. C. and Chiou, T. J. 2016b. Identification of plant vacuolar transporters mediating phosphate storage. *Nature Communications* 7, 11095.
Liu, Z., Gao, K., Shan, S., Gu, R., Wang, Z., Craft, E. J., Mi, G., Yuan, L. and Chen, F. 2017. Comparative analysis of root traits and the associated QTLs for maize seedlings grown in paper roll, hydroponics and vermiculite culture system. *Frontiers in Plant Science* 8, 436.
Lott, J. N., Greenwood, J. S., Vollmer, C. M. and Buttrose, M. S. 1978. Energy-dispersive X-ray analysis of phosphorus, potassium, magnesium, and calcium in globoid crystals in protein bodies from different regions of *Cucurbita maxima* embryos. *Plant Physiology* 61(6), 984-988.
Lynch, J. P. 2015. Root phenes that reduce the metabolic costs of soil exploration: opportunities for 21st century agriculture. *Plant, Cell and Environment* 38(9), 1775-1784.
Lynch, J. P. 2019. Root phenotypes for improved nutrient capture: an underexploited opportunity for global agriculture. *New Phytologist* 223(2), 548-564.
Lynch, J. P. and Brown, K. M. 2008. Root strategies for phosphorus acquisition. In: White, P. J. and Hammond, J. P. (Eds) *The Ecophysiology of Plant-Phosphorus Interactions*. Netherlands: Springer. pp. 83-116.
Lynch, J. P. and Ho, M. D. 2005. Rhizoeconomics: carbon costs of phosphorus acquisition. *Plant and Soil* 269(1-2), 45-56.
Lyu, Y., Tang, H., Li, H., Zhang, F., Rengel, Z., Whalley, W. R. and Shen, J. 2016. Major crop species show differential balance between root morphological and physiological responses to variable phosphorus supply. *Frontiers in Plant Science* 7, 1939.
Ma, X. F., Tudor, S., Butler, T., Ge, Y., Xi, Y., Bouton, J., Harrison, M. and Wang, Z. Y. 2012. Transgenic expression of phytase and acid phosphatase genes in alfalfa (*Medicago sativa*) leads to improved phosphate uptake in natural soils. *Molecular Breeding* 30(1), 377-391.

Ma, X. M., Mason-Jones, K., Liu, Y., Blagodatskaya, E., Kuzyakov, Y., Guber, A., Dippold, M. A. and Razavi, B. S. 2019. Coupling zymography with pH mapping reveals a shift in lupine phosphorus acquisition strategy driven by cluster roots. *Soil Biology and Biochemistry* 135, 420–428.

Ma, Z., Guo, D., Xu, X., Lu, M., Bardgett, R. D., Eissenstat, D. M., McCormack, M. L. and Hedin, L. O. 2018. Evolutionary history resolves global organization of root functional traits. *Nature* 555(7694), 94–97.

MacDonald, G. K., Bennett, E. M., Potter, P. A. and Ramankutty, N. 2011. Agronomic phosphorus imbalances across the world's croplands. *Proceedings of the National Academy of Sciences of the United States of America* 108(7), 3086–3091.

Maherali, H. 2014. Is there an association between root architecture and mycorrhizal growth response? *New Phytologist* 204(1), 192–200.

Mai, W., Xue, X., Feng, G., Yang, R. and Tian, C. 2018. Can optimization of phosphorus input lead to high productivity and high phosphorus use efficiency of cotton through maximization of root/mycorrhizal efficiency in phosphorus acquisition? *Field Crops Research* 216, 100–108.

Marschner, H., Romheld, V. and Cakmak, I. 1987. Root-induced changes of nutrient availability in the rhizosphere. *Journal of Plant Nutrition* 10(9), 1175–1184.

Martin-Robles, N., Lehmann, A., Seco, E., Aroca, R., Rillig, M. C. and Milla, R. 2018. Impacts of domestication on the arbuscular mycorrhizal symbiosis of 27 crop species. *New Phytologist* 218(1), 322–334.

Martin, B. C., George, S. J., Price, C. A., Shahsavari, E., Ball, A. S., Tibbett, M. and Ryan, M. H. 2016. Citrate and malonate increase microbial activity and alter microbial community composition in uncontaminated and diesel-contaminated soil microcosms. *SOIL* 2(3), 487–498.

Maruyama, H. and Wasaki, J. 2017. Transgenic approaches for improving phosphorus use efficiency in plants. In: Hossain, M. A., Kamiya, T., Burritt, D. J., Tran, L.-S. P. and Fujiwara, T. (Eds) *Plant Macronutrient Use Efficiency*. London: Academic Press. pp. 323–338.

McCormack, M. L., Dickie, I. A., Eissenstat, D. M., Fahey, T. J., Fernandez, C. W., Guo, D., Helmisaari, H. S., Hobbie, E. A., Iversen, C. M., Jackson, R. B., Leppalammi-Kujansuu, J., Norby, R. J., Phillips, R. P., Pregitzer, K. S., Pritchard, S. G., Rewald, B. and Zadworny, M. 2015. Redefining fine roots improves understanding of below-ground contributions to terrestrial biosphere processes. *New Phytologist* 207(3), 505–518.

McLaughlin, M. J., Mcbeath, T. M., Smernik, R., Stacey, S. P., Ajiboye, B. and Guppy, C. 2011. The chemical nature of P accumulation in agricultural soils–implications for fertiliser management and design: an Australian perspective. *Plant and Soil* 349(1-2), 69–87.

Mendoza, R. 2001. Phosphorus nutrition and mycorrhizal growth response of broadleaf and narrowleaf birdsfoot trefoils. *Journal of Plant Nutrition* 24(1), 203–214.

Menezes-Blackburn, D., Giles, C., Darch, T., George, T. S., Blackwell, M., Stutter, M., Shand, C., Lumsdon, D., Cooper, P., Wendler, R., Brown, L., Almeida, D. S., Wearing, C., Zhang, H. and Haygarth, P. M. 2018. Opportunities for mobilizing recalcitrant phosphorus from agricultural soils: a review. *Plant and Soil* 427(1), 5–16.

Menezes-Blackburn, D., Jorquera, M. A., Greiner, R., Gianfreda, L. and De La Luz Mora, M. J. C. R. I. E. S. 2013. Phytases and phytase-labile organic phosphorus in manures and soils. *Critical Reviews in Environmental Science and Technology* 43(9), 916–954.

Mi, G., Chen, F., Yuan, L. and Zhang, F. 2016. Ideotype root system architecture for maize to achieve high yield and resource use efficiency in intensive cropping systems. *Advances in Agronomy* 139, 73–97.

Miguel, M. A., Postma, J. A. and Lynch, J. P. 2015. Phene synergism between root hair length and basal root growth angle for phosphorus acquisition. *Plant Physiology* 167(4), 1430–1439.

Muler, A. L., Oliveira, R. S., Lambers, H. and Veneklaas, E. J. 2014. Does cluster-root activity benefit nutrient uptake and growth of co-existing species? *Oecologia* 174(1), 23–31.

Nadeem, M., Mollier, A., Morel, C., Vives, A., Prud'homme, L. and Pellerin, S. 2011. Relative contribution of seed phosphorus reserves and exogenous phosphorus uptake to maize (*Zea mays* L.) nutrition during early growth stages. *Plant and Soil* 346(1–2), 231–244.

Nadeem, M., Mollier, A., Morel, C., Vives, A., Prud'homme, L. and Pellerin, S. 2012. Maize (*Zea mays* L.) endogenous seed phosphorus remobilization is not influenced by exogenous phosphorus availability during germination and early growth stages. *Plant and Soil* 357(1–2), 13–24.

Nadeem, S. M., Ahmad, M., Zahir, Z. A., Javaid, A. and Ashraf, M. 2014. The role of mycorrhizae and plant growth promoting rhizobacteria (PGPR) in improving crop productivity under stressful environments. *Biotechnology Advances* 32(2), 429–448.

Nazeri, N. K., Lambers, H., Tibbett, M. and Ryan, M. H. 2014. Moderating mycorrhizas: arbuscular mycorrhizas modify rhizosphere chemistry and maintain plant phosphorus status within narrow boundaries. *Plant, Cell and Environment* 37(4), 911–921.

Ning, L., Kan, G., Du, W., Guo, S., Wang, Q., Zhang, G., Cheng, H. and Yu, D. 2016. Association analysis for detecting significant single nucleotide polymorphisms for phosphorus-deficiency tolerance at the seedling stage in soybean [*Glycine max* (L) Merr.]. *Breeding Science* 66(2), 191–203.

Nuruzzaman, M., Lambers, H., Bolland, M. D. and Veneklaas, E. J. 2005a. Phosphorus benefits of different legume crops to subsequent wheat grown in different soils of Western Australia. *Plant and Soil* 271(1–2), 175–187.

Nuruzzaman, M., Lambers, H., Bolland, M. D. A. and Veneklaas, E. J. 2005b. Phosphorus uptake by grain legumes and subsequently grown wheat at different levels of residual phosphorus fertiliser. *Australian Journal of Agricultural Research* 56(10), 1041–1047.

Oberson, A. and Joner, E. J. 2005. Microbial turnover of phosphorus in soil. In: Turner, B., Frossard, E. and Baldwin, D. (Eds) *Organic Phosphorus in the Environment*. Wallingford: CABI Publishing. pp. 133–164.

Oburger, E., Dell'Mour, M., Hann, S., Wieshammer, G., Puschenreiter, M. and Wenzel, W. W. 2013. Evaluation of a novel tool for sampling root exudates from soil-grown plants compared to conventional techniques. *Environmental and Experimental Botany* 87, 235–247.

Oburger, E. and Jones, D. L. 2018. Sampling root exudates-mission impossible? *Rhizosphere* 6, 116–133.

Oburger, E., Kirk, G. J. D., Wenzel, W. W., Puschenreiter, M. and Jones, D. L. 2009. Interactive effects of organic acids in the rhizosphere. *Soil Biology and Biochemistry* 41(3), 449–457.

Orchard, S., Hilton, S., Bending, G. D., Dickie, I. A., Standish, R. J., Gleeson, D. B., Jeffery, R. P., Powell, J. R., Walker, C., Bass, D., Monk, J., Simonin, A. and Ryan, M. H. 2017a. Fine endophytes (*Glomus tenue*) are related to Mucoromycotina, not Glomeromycota. *New Phytologist* 213(2), 481–486.

Orchard, S., Standish, R. J., Dickie, I. A., Renton, M., Walker, C., Moot, D. and Ryan, M. H. 2017b. Fine root endophytes under scrutiny: a review of the literature on arbuscule-producing fungi recently suggested to belong to the Mucoromycotina. *Mycorrhiza* 27(7), 619–638.

Pang, J., Bansal, R., Zhao, H., Bohuon, E., Lambers, H., Ryan, M. H., Ranathunge, K. and Siddique, K. H. M. 2018. The carboxylate-releasing phosphorus-mobilizing strategy can be proxied by foliar manganese concentration in a large set of chickpea germplasm under low phosphorus supply. *New Phytologist* 219(2), 518–529.

Parfitt, R. L. 1979. The availability of P from phosphate-goethite bridging complexes. Desorption and uptake by ryegrass. *Plant and Soil* 53(1–2), 55–65.

Pearse, S. J., Veneklaas, E. J., Cawthray, G., Bolland, M. D. and Lambers, H. 2007. Carboxylate composition of root exudates does not relate consistently to a crop species' ability to use phosphorus from aluminium, iron or calcium phosphate sources. *New Phytologist* 173(1), 181–190.

Pearse, S. J., Veneklaas, E. J., Cawthray, G. R., Bolland, M. D. A. and Lambers, H. 2006. Carboxylate release of wheat, canola and 11 grain legume species as affected by phosphorus status. *Plant and Soil* 288(1–2), 127–139.

Postma, J. A., Dathe, A. and Lynch, J. P. 2014. The optimal lateral root branching density for maize depends on nitrogen and phosphorus availability. *Plant Physiology* 166(2), 590–602.

Postma, J. A. and Lynch, J. P. 2011. Root cortical aerenchyma enhances the growth of maize on soils with suboptimal availability of nitrogen, phosphorus, and potassium. *Plant Physiology* 156(3), 1190–1201.

Raboy, V. 2009. Approaches and challenges to engineering seed phytate and total phosphorus. *Plant Science* 177(4), 281–296.

Raven, J. A., Lambers, H., Smith, S. E. and Westoby, M. 2018. Costs of acquiring phosphorus by vascular land plants: patterns and implications for plant coexistence. *New Phytologist* 217(4), 1420–1427.

Razavi, B. S., Zarebanadkouki, M., Blagodatskaya, E. and Kuzyakov, Y. 2016. Rhizosphere shape of lentil and maize: spatial distribution of enzyme activities. *Soil Biology and Biochemistry* 96, 229–237.

Razavi, B. S., Zhang, X., Bilyera, N., Guber, A. and Zarebanadkouki, M. 2019. Soil zymography: simple and reliable? Review of current knowledge and optimization of the method. *Rhizosphere* 11, 100161.

Reinhart, K. O., Wilson, G. W. and Rinella, M. J. 2012. Predicting plant responses to mycorrhizae: integrating evolutionary history and plant traits. *Ecology Letters* 15(7), 689–695.

Richardson, A. E., Lynch, J. P., Ryan, P. R., Delhaize, E., Smith, F. A., Smith, S. E., Harvey, P. R., Ryan, M. H., Veneklaas, E. J., Lambers, H., Oberson, A., Culvenor, R. A. and Simpson, R. J. 2011. Plant and microbial strategies to improve the phosphorus efficiency of agriculture. *Plant and Soil* 349(1–2), 121–156.

Rillig, M. C., Aguilar-Trigueros, C. A., Camenzind, T., Cavagnaro, T. R., Degrune, F., Hohmann, P., Lammel, D. R., Mansour, I., Roy, J., Van Der Heijden, M. G. A. and Yang, G. 2019. Why farmers should manage the arbuscular mycorrhizal symbiosis. *New Phytologist* 222(3), 1171–1175.

Roumet, C., Birouste, M., Picon-Cochard, C., Ghestem, M., Osman, N., Vrignon-Brenas, S., Cao, K. F. and Stokes, A. 2016. Root structure-function relationships in 74 species: evidence of a root economics spectrum related to carbon economy. *New Phytologist* 210(3), 815–826.

Rózycki, H. and Strzelczyk, E. 1986. Organic acids production by *Streptomyces* spp. isolated from soil, rhizosphere and mycorrhizosphere of pine (*Pinus sylvestris* L.). *Plant and Soil* 96(3), 337–345.

Ryan, M. H. and Graham, J. H. 2018. Little evidence that farmers should consider abundance or diversity of arbuscular mycorrhizal fungi when managing crops. *New Phytologist* 220(4), 1092–1107.

Ryan, M. H., Graham, J. H., Morton, J. B. and Kirkegaard, J. A. 2019a. Research must use a systems agronomy approach if management of the arbuscular mycorrhizal symbiosis is to contribute to sustainable intensification. *New Phytologist* 222(3), 1176–1178.

Ryan, M. H., Herwaarden, A. F. V., Angus, J. F. and Kirkegaard, J. A. 2005. Reduced growth of autumn-sown wheat in a low-P soil is associated with high colonisation by arbuscular mycorrhizal fungi. *Plant and Soil* 270(1), 275–286.

Ryan, M. H., Kaur, P., Nazeri, N. K., Clode, P. L., Keeble-Gagnere, G., Doolette, A. L., Smernik, R. J., Van Aken, O., Nicol, D., Maruyama, H., Ezawa, T., Lambers, H., Millar, A. H. and Appels, R. 2019b. Globular structures in roots accumulate phosphorus to extremely high concentrations following phosphorus addition. *Plant, Cell and Environment* 42(6), 1987–2002.

Ryan, M. H., Kidd, D. R., Sandral, G. A., Yang, Z., Lambers, H., Culvenor, R. A., Stefanski, A., Nichols, P. G. H., Haling, R. E. and Simpson, R. J. 2016. High variation in the percentage of root length colonised by arbuscular mycorrhizal fungi among 139 lines representing the species subterranean clover (*Trifolium subterraneum*). *Applied Soil Ecology* 98, 221–232.

Ryan, M. H., Tibbett, M., Edmonds-Tibbett, T., Suriyagoda, L. D., Lambers, H., Cawthray, G. R. and Pang, J. 2012. Carbon trading for phosphorus gain: the balance between rhizosphere carboxylates and arbuscular mycorrhizal symbiosis in plant phosphorus acquisition. *Plant, Cell and Environment* 35(12), 2170–2180.

Ryan, P. R., James, R. A., Weligama, C., Delhaize, E., Rattey, A., Lewis, D. C., Bovill, W. D., Mcdonald, G., Rathjen, T. M., Wang, E., Fettell, N. A. and Richardson, A. E. 2014. Can citrate efflux from roots improve phosphorus uptake by plants? Testing the hypothesis with near-isogenic lines of wheat. *Physiologia Plantarum* 151(3), 230–242.

Salvioli, A., Ghignone, S., Novero, M., Navazio, L., Venice, F., Bagnaresi, P. and Bonfante, P. 2016. Symbiosis with an endobacterium increases the fitness of a mycorrhizal fungus, raising its bioenergetic potential. *The ISME Journal* 10(1), 130–144.

Sattari, S. Z., Bouwman, A. F., Giller, K. E. and Van Ittersum, M. K. 2012. Residual soil phosphorus as the missing piece in the global phosphorus crisis puzzle. *Proceedings of the National Academy of Sciences of the United States of America* 109(16), 6348–6353.

Schefe, C. R., Watt, M., Slattery, W. J. and Mele, P. M. 2008. Organic anions in the rhizosphere of Al-tolerant and Al-sensitive wheat lines grown in an acid soil in controlled and field environments. *Soil Research* 46(3), 257–264.

Schelud'ko, A. V., Makrushin, K. V., Tugarova, A. V., Krestinenko, V. A., Panasenko, V. I., Antonyuk, L. P. and Katsy, E. I. 2009. Changes in motility of the rhizobacterium *Azospirillum brasilense* in the presence of plant lectins. *Microbiological Research* 164(2), 149–156.

Schneider, H. M. and Lynch, J. P. 2018. Functional implications of root cortical senescence for soil resource capture. *Plant and Soil* 423(1-2), 13-26.
Shane, M. W., Cawthray, G. R., Cramer, M. D., Kuo, J. and Lambers, H. 2006. Specialized 'dauciform' roots of Cyperaceae are structurally distinct, but functionally analogous with 'cluster' roots. *Plant, Cell and Environment* 29(10), 1989-1999.
Shane, M. W., De Vos, M., De Roock, S. and Lambers, H. 2003. Shoot P status regulates cluster-root growth and citrate exudation in Lupinus albus grown with a divided root system. *Plant, Cell and Environment* 26(2), 265-273.
Shane, M. W. and Lambers, H. 2005. Cluster roots: a curiosity in context. *Plant and Soil* 274(1-2), 101-125.
Shao, H., Shi, D. F., Shi, W. J., Ban, X. B., Chen, Y. C., Ren, W., Chen, F. J. and Mi, G. H. 2019. Genotypic difference in the plasticity of root system architecture of field-grown maize in response to plant density. *Plant and Soil* 439(1-2), 201-217.
Shao, H., Xia, T., Wu, D., Chen, F. and Mi, G. 2018. Root growth and root system architecture of field-grown maize in response to high planting density. *Plant and Soil* 430(1-2), 395-411.
Shen, J., Li, C., Mi, G., Li, L., Yuan, L., Jiang, R. and Zhang, F. 2013. Maximizing root/rhizosphere efficiency to improve crop productivity and nutrient use efficiency in intensive agriculture of China. *Journal of Experimental Botany* 64(5), 1181-1192.
Shen, J., Rengel, Z., Tang, C. and Zhang, F. 2003. Role of phosphorus nutrition in development of cluster roots and release of carboxylates in soil-grown *Lupinus albus*. *Plant and Soil* 248(1/2), 199-206.
Shen, J., Yuan, L., Zhang, J., Li, H., Bai, Z., Chen, X., Zhang, W. and Zhang, F. 2011. Phosphorus dynamics: from soil to plant. *Plant Physiology* 156(3), 997-1005.
Simpson, R. J., Oberson, A., Culvenor, R. A., Ryan, M. H., Veneklaas, E. J., Lambers, H., Lynch, J. P., Ryan, P. R., Delhaize, E., Smith, F. A., Smith, S. E., Harvey, P. R. and Richardson, A. E. 2011. Strategies and agronomic interventions to improve the phosphorus-use efficiency of farming systems. *Plant and Soil* 349(1-2), 89-120.
Simpson, R. J., Stefanski, A., Marshall, D. J., Moore, A. D. and Richardson, A. E. 2015. Management of soil phosphorus fertility determines the phosphorus budget of a temperate grazing system and is the key to improving phosphorus efficiency. *Agriculture, Ecosystems and Environment* 212, 263-277.
Singh, N., Kumar, S., Bajpai, V. K., Dubey, R. C., Maheshwari, D. K. and Kang, S. C. 2010. Biological control of *Macrophomina phaseolina* by chemotactic fluorescent *Pseudomonas aeruginosa* PN1 and its plant growth promotory activity in chir-pine. *Crop Protection* 29(10), 1142-1147.
Smith, S. and Read, D. (Eds.) 2008. *Mycorrhizal Symbiosis*. London: Academic Press.
Song, Y. N., Zhang, F. S., Marschner, P., Fan, F. L., Gao, H. M., Bao, X. G., Sun, J. H. and Li, L. 2007. Effect of intercropping on crop yield and chemical and microbiological properties in rhizosphere of wheat (*Triticum aestivum* L.), maize (*Zea mays* L.), and faba bean (*Vicia faba* L.). *Biology and Fertility of Soils* 43(5), 565-574.
Spohn, M. and Kuzyakov, Y. 2013. Distribution of microbial- and root-derived phosphatase activities in the rhizosphere depending on P availability and C allocation - coupling soil zymography with ^{14}C imaging. *Soil Biology and Biochemistry* 67, 106-113.
Spohn, M., Treichel, N. S., Cormann, M., Schloter, M. and Fischer, D. 2015. Distribution of phosphatase activity and various bacterial phyla in the rhizosphere of Hordeum vulgare L. depending on P availability. *Soil Biology and Biochemistry* 89, 44-51.

Strock, C. F., Morrow De La Riva, L. and Lynch, J. P. 2018. Reduction in root secondary growth as a strategy for phosphorus acquisition. *Plant Physiology* 176(1), 691-703.

Sun, B., Gao, Y. and Lynch, J. P. 2018. Large crown root number improves topsoil foraging and phosphorus acquisition. *Plant Physiology* 177(1), 90-104.

Sun, X. and Tang, M. 2013. Effect of arbuscular mycorrhizal fungi inoculation on root traits and root volatile organic compound emissions of *Sorghum bicolor*. *South African Journal of Botany* 88, 373-379.

Sungthongwises, K. 2016. Diversity of phosphate solubilizing bacteria under rubber intercropping. *Asian Journal of Plant Sciences* 15(3), 75-80.

Syer, J., Johnston, A. and Curtin, D. 2008. Efficiency of soil and fertilizer phosphorus use - reconciling changing concepts of soil phosphorus behaviour with agronomic information. FAO Fertilizer and Plant Nutrition Bulletin 18. Rome.

Tang, X., Bernard, L., Brauman, A., Daufresne, T., Deleporte, P., Desclaux, D., Souche, G., Placella, S. A. and Hinsinger, P. 2014. Increase in microbial biomass and phosphorus availability in the rhizosphere of intercropped cereal and legumes under field conditions. *Soil Biology and Biochemistry* 75, 86-93.

Tarafdar, J. C. and Jungk, A. 1987. Phosphatase activity in the rhizosphere and its relation to the depletion of soil organic phosphorus. *Biology and Fertility of Soils* 3(4), 199-204.

Taurian, T., Anzuay, M. S., Angelini, J. G., Tonelli, M. L., Ludueña, L., Pena, D., Ibáñez, F. and Fabra, A. 2010. Phosphate-solubilizing peanut associated bacteria: screening for plant growth-promoting activities. *Plant and Soil* 329(1-2), 421-431.

Teng, W., Deng, Y., Chen, X. P., Xu, X. F., Chen, R. Y., Lv, Y., Zhao, Y. Y., Zhao, X. Q., He, X., Li, B., Tong, Y. P., Zhang, F. S. and Li, Z. S. 2013. Characterization of root response to phosphorus supply from morphology to gene analysis in field-grown wheat. *Journal of Experimental Botany* 64(5), 1403-1411.

Teste, F. P., Veneklaas, E. J., Dixon, K. W. and Lambers, H. 2014. Complementary plant nutrient-acquisition strategies promote growth of neighbour species. *Functional Ecology* 28(4), 819-828.

Ubbens, J. R. and Stavness, I. 2017. Deep plant phenomics: a deep learning platform for complex plant phenotyping tasks. *Frontiers in Plant Science* 8, 1190.

Valenzuela-Estrada, L. R., Vera-Caraballo, V., Ruth, L. E. and Eissenstat, D. M. 2008. Root anatomy, morphology, and longevity among root orders in *Vaccinium corymbosum* (Ericaceae). *American Journal of Botany* 95(12), 1506-1514.

Van De Wiel, C. C. M., Van Der Linden, C. G. and Scholten, O. E. 2016. Improving phosphorus use efficiency in agriculture: opportunities for breeding. *Euphytica* 207(1), 1-22.

Vance, C. P., Uhde-Stone, C. and Allan, D. L. 2003. Phosphorus acquisition and use: critical adaptations by plants for securing a nonrenewable resource. *New Phytologist* 157(3), 423-447.

Veneklaas, E. J., Lambers, H., Bragg, J., Finnegan, P. M., Lovelock, C. E., Plaxton, W. C., Price, C. A., Scheible, W. R., Shane, M. W., White, P. J. and Raven, J. A. 2012. Opportunities for improving phosphorus-use efficiency in crop plants. *New Phytologist* 195(2), 306-320.

Veneklaas, E. J., Stevens, J., Cawthray, G. R., Turner, S., Grigg, A. M. and Lambers, H. 2003. Chickpea and white lupin rhizosphere carboxylates vary with soil properties and enhance phosphorus uptake. *Plant and Soil* 248(1/2), 187-197.

Vitousek, P. M., Naylor, R., Crews, T., David, M. B., Drinkwater, L. E., Holland, E., Johnes, P. J., Katzenberger, J., Martinelli, L. A., Matson, P. A., Nziguheba, G., Ojima, D., Palm, C.

A., Robertson, G. P., Sanchez, P. A., Townsend, A. R. and Zhang, F. S. 2009. Nutrient imbalances in agricultural development. *Science* 324(5934), 1519–1520.

Waddell, H., Simpson, R., Henderson, B., Ryan, M., Lambers, H., Garden, D. and Richardson, A. 2015. Differential growth response of *Rytidosperma* species (wallaby grass) to phosphorus application and its implications for grassland management. *Grass and Forage Science* 17, 245–258.

Wang, C., White, P. J. and Li, C. 2016. Colonization and community structure of arbuscular mycorrhizal fungi in maize roots at different depths in the soil profile respond differently to phosphorus inputs on a long-term experimental site. *Mycorrhiza* 27(4), 369–381.

Wang, D., Marschner, P., Solaiman, Z. and Rengel, Z. 2007. Growth, P uptake and rhizosphere properties of intercropped wheat and chickpea in soil amended with iron phosphate or phytate. *Soil Biology and Biochemistry* 39(1), 249–256.

Wang, M., Schafer, M., Li, D., Halitschke, R., Dong, C., Mcgale, E., Paetz, C., Song, Y., Li, S., Dong, J., Heiling, S., Groten, K., Franken, P., Bitterlich, M., Harrison, M. J., Paszkowski, U. and Baldwin, I. T. 2018a. Blumenols as shoot markers of root symbiosis with arbuscular mycorrhizal fungi. *eLife* 7, e37093.

Wang, W., Ding, G. D., White, P. J., Wang, X. H., Jin, K. M., Xu, F. S. and Shi, L. 2019. Mapping and cloning of quantitative trait loci for phosphorus efficiency in crops: opportunities and challenges. *Plant and Soil* 439(1–2), 91–112.

Wang, X., Chen, Y., Thomas, C. L., Ding, G., Xu, P., Shi, D., Grandke, F., Jin, K., Cai, H., Xu, F., Yi, B., Broadley, M. R. and Shi, L. 2017a. Genetic variants associated with the root system architecture of oilseed rape (*Brassica napus* L.) under contrasting phosphate supply. *DNA Research* 24(4), 407–417.

Wang, X., Ding, W. and Lambers, H. 2018b. Nodulation promotes cluster-root formation in Lupinus albus under low phosphorus conditions. *Plant and Soil* 439(1–2), 233–242.

Wang, Y., Krogstad, T., Clarke, N., Øgaard, A. F. and Clarke, J. L. 2017b. Impact of phosphorus on rhizosphere organic anions of wheat at different growth stages under field conditions. *AoB Plants* 9(2), plx008.

Wang, Y. and Lambers, H. 2019. Root-released organic anions in response to low phosphorus availability: recent progress, challenges and future perspectives. *Plant and Soil* 447(1–2), 135–156, https://doi.org/10.1007/s11104-019-03972-8.

Wasaki, J., Maruyama, H., Tanaka, M., Yamamura, T., Dateki, H., Shinano, T., Ito, S. and Osaki, M. 2009. Overexpression of the *LASAP2* gene for secretory acid phosphatase in white lupin improves the phosphorus uptake and growth of tobacco plants. *Soil Science and Plant Nutrition* 55(1), 107–113.

Watt, M. and Evans, J. R. 1999. Linking development and determinacy with organic acid efflux from proteoid roots of white lupin grown with low phosphorus and ambient or elevated atmospheric CO_2 concentration. *Plant Physiology* 120(3), 705–716.

Weaver, D. M. and Wong, M. T. F. 2011. Scope to improve phosphorus (P) management and balance efficiency of crop and pasture soils with contrasting P status and buffering indices. *Plant and Soil* 349(1–2), 37–54.

Weeks, J. J. and Hettiarachchi, G. M. 2019. A review of the latest in phosphorus fertilizer technology: possibilities and pragmatism. *Journal of Environmental Quality* 48(5), 1300–1313.

Weiner, J., Andersen, S. B., Wille, W. K. M., Griepentrog, H. W. and Olsen, J. M. 2010. Evolutionary agroecology: the potential for cooperative, high density, weed-suppressing cereals. *Evolutionary Applications* 3(5–6), 473–479.

Wen, Z., Li, H., Shen, Q., Tang, X., Xiong, C., Li, H., Pang, J., Ryan, M. H., Lambers, H. and Shen, J. 2019. Tradeoffs among root morphology, exudation and mycorrhizal symbioses for phosphorus-acquisition strategies of 16 crop species. *New Phytologist* 223(2), 882–895.

Wen, Z. H., Li, H. G., Shen, J. B. and Rengel, Z. 2017. Maize responds to low shoot P concentration by altering root morphology rather than increasing root exudation. *Plant and Soil* 416(1-2), 377-389.

Withers, P. J., Sylvester-Bradley, R., Jones, D. L., Healey, J. R. and Talboys, P. J. 2014. Feed the crop not the soil: rethinking phosphorus management in the food chain. *Environmental Science and Technology* 48(12), 6523-6530.

Wouterlood, M., Cawthray, G. R., Scanlon, T. T., Lambers, H. and Veneklaas, E. J. 2004. Carboxylate concentrations in the rhizosphere of lateral roots of chickpea (*Cicer arietinum*) increase during plant development, but are not correlated with phosphorus status of soil or plants. *New Phytologist* 162(3), 745–753.

Xiao, K., Katagi, H., Harrison, M. and Wang, Z.-Y. 2006. Improved phosphorus acquisition and biomass production in *Arabidopsis* by transgenic expression of a purple acid phosphatase gene from *M. truncatula*. *Plant Science* 170(2), 191–202.

Xu, C. L., Huang, S. B., Tian, B. J., Ren, J. H., Meng, Q. F. and Wang, P. 2017. Manipulating planting density and nitrogen fertilizer application to improve yield and reduce environmental impact in Chinese maize production. *Frontiers in Plant Science* 8, 1234.

Yan, F., Schubert, S. and Mengel, K. 1992. Effect of low root medium pH on net proton release, root respiration, and root growth of corn (*Zea mays* L.) and broad bean (*Vicia faba* L.). *Plant Physiology* 99(2), 415–421.

Yang, H., Zhang, Q., Dai, Y., Liu, Q., Tang, J., Bian, X. and Chen, X. 2015a. Effects of arbuscular mycorrhizal fungi on plant growth depend on root system: a meta-analysis. *Plant and Soil* 389(1–2), 361–374.

Yang, L., Xie, J., Jiang, D., Fu, Y., Li, G. and Lin, F. 2008. Antifungal substances produced by *Penicillium oxalicum* strain PY-1—potential antibiotics against plant pathogenic fungi. *World Journal of Microbiology and Biotechnology* 24(7), 909–915.

Yang, Z., Culvenor, R. A., Haling, R. E., Stefanski, A., Ryan, M. H., Sandral, G. A., Kidd, D. R., Lambers, H. and Simpson, R. J. 2015b. Variation in root traits associated with nutrient foraging among temperate pasture legumes and grasses. *Grass and Forage Science* 72(1), 93–103.

York, L. M., Carminati, A., Mooney, S. J., Ritz, K. and Bennett, M. J. 2016. The holistic rhizosphere: integrating zones, processes, and semantics in the soil influenced by roots. *Journal of Experimental Botany* 67(12), 3629-3643.

York, L. M., Galindo-Castaneda, T., Schussler, J. R. and Lynch, J. P. 2015. Evolution of US maize (*Zea mays* L.) root architectural and anatomical phenes over the past 100 years corresponds to increased tolerance of nitrogen stress. *Journal of Experimental Botany* 66(8), 2347-2358.

Yu, R. P., Li, X. X., Xiao, Z. H., Lambers, H. and Li, L. 2020. Phosphorus facilitation and covariation of root traits in steppe species. *New Phytology* 226, 1285–1298.

Zak, D. R., Holmes, W. E., White, D. C., Peacock, A. D. and Tilman, D. 2003. Plant diversity, soil microbial communities, and ecosystem function: are there any links? *Ecology* 84(8), 2042–2050.

Zangaro, W., Nishidate, F. R., Vandresen, J., Andrade, G. and Nogueira, M. A. 2007. Root mycorrhizal colonization and plant responsiveness are related to root plasticity, soil fertility and successional status of native woody species in southern Brazil. *Journal of Tropical Ecology* 23(1), 53-62.

Zhang, D., Song, H., Cheng, H., Hao, D., Wang, H., Kan, G., Jin, H. and Yu, D. 2014a. The acid phosphatase-encoding gene *GmACP1* contributes to soybean tolerance to low-phosphorus stress. *PLoS Genetics* 10(1), e1004061.

Zhang, D., Wang, Y., Tang, X., Zhang, A., Li, H. and Rengel, Z. 2019a. Early priority effects of occupying a nutrient patch do not influence final maize growth in intensive cropping systems. *Plant and Soil* 442(1-2), 285-298.

Zhang, D., Zhang, C., Tang, X., Li, H., Zhang, F., Rengel, Z., Whalley, W. R., Davies, W. J. and Shen, J. 2016. Increased soil phosphorus availability induced by faba bean root exudation stimulates root growth and phosphorus uptake in neighbouring maize. *New Phytologist* 209(2), 823-831.

Zhang, L., Fan, J., Ding, X., He, X., Zhang, F. and Feng, G. 2014b. Hyphosphere interactions between an arbuscular mycorrhizal fungus and a phosphate solubilizing bacterium promote phytate mineralization in soil. *Soil Biology and Biochemistry* 74, 177-183.

Zhang, L., Fan, J., Feng, G. and Declerck, S. 2019b. The arbuscular mycorrhizal fungus *Rhizophagus irregularis* MUCL 43194 induces the gene expression of citrate synthase in the tricarboxylic acid cycle of the phosphate-solubilizing bacterium *Rahnella aquatilis* HX2. *Mycorrhiza* 29(1), 69-75.

Zhang, W., Tang, X., Feng, X., Wang, E., Li, H., Shen, J. and Zhang, F. 2019c. Management strategies to optimize soil phosphorus utilization and alleviate environmental risk in China. *Journal of Environmental Quality* 48(5), 1167-1175.

Zhang, X. C., Dippold, M. A., Kuzyakov, Y. and Razavi, B. S. 2019d. Spatial pattern of enzyme activities depends on root exudate composition. *Soil Biology and Biochemistry* 133, 83-93.

Zhu, J. M. and Lynch, J. P. 2004. The contribution of lateral rooting to phosphorus acquisition efficiency in maize (*Zea mays*) seedlings. *Functional Plant Biology* 31(10), 949-958.

Zhu, J. M., Zhang, C. C. and Lynch, J. P. 2010. The utility of phenotypic plasticity of root hair length for phosphorus acquisition. *Functional Plant Biology* 37(4), 313-322.

Zhu, Y. H., Weiner, J., Yu, M. X. and Li, F. M. 2019. Evolutionary agroecology: trends in root architecture during wheat breeding. *Evolutionary Applications* 12(4), 733-743.

Chapter 13

Advances in understanding plant root water uptake

Mutez Ali Ahmed, University of Bayreuth, Germany; Doris Vetterlein, Helmholtz Centre for Environmental Research-UFZ, Germany; and Andrea Carminati, University of Bayreuth, Germany

1 Introduction

1.1 The agronomic context of root water uptake and drought tolerance

Water deficit, defined as insufficient soil water availability to meet the demand of a crop at a particular growth stage, is one of the primary limitations to crop production worldwide (Vadez, 2014; Tardieu et al., 2018; Lynch, 2018). This is not surprising since plants transpire large volumes of water, which causes their water potential to largely oscillate within a few hours (Tardieu & Parent, 2017). Therefore, it is of primary importance to understand the mechanisms that confer drought tolerance to crops. The definition of 'drought tolerance' in agriculture is context-dependent (Passioura, 2007). In an agricultural context, it is defined as the ability of a crop to maintain yield during gradual and moderate soil water deficits (Tardieu et al., 2018). Drought tolerance of crops depends on the environment (atmospheric and soil) and is the result of different strategies, which include: increasing extractability of soil water, saving water for critical growth stages, or completing the development cycle, thereby producing seeds, in a short period of time after rainfall. The success of these contrasting

http://dx.doi.org/10.19103/AS.2020.0075.17

strategies depends on environmental conditions (Tardieu, 2012; Vadez, 2014; Ahmed et al., 2018b; Tardieu et al., 2018).

Because soil water is taken up by roots, a pertinent question is what root and rhizosphere properties confer drought tolerance? In this context, researchers focussed on deep and profuse root systems, which were hypothesized to increase the volume of water available to roots (Wasson et al., 2012). While deep roots were shown to increase yield in specific scenarios and environments (Lilley & Kirkegaard, 2008), they do not contribute to drought adaptation in all stress conditions (Zaman-Allah et al., 2011). This is simply because the location and magnitude of root water uptake is a result of complex interactions between the distribution of roots, their hydraulic properties and the hydraulic properties of the soil (Draye et al., 2010). Tardieu and Parent (2017) showed that rapid and deep growth of roots often results in a negative effect on yield, as this is associated with faster plant growth and faster depletion of water resources. Furthermore, water uptake during critical stages (reproduction and grain filling) was shown to be more important than the total water extraction during the complete vegetative period (Zaman-Allah et al., 2011; Vadez, 2014). Root architecture would certainly have an effect on water uptake during these critical stages, but it was suggested to be mainly determined by their hydraulic properties, rather than simply by the size of the root system (Vadez, 2014), which would also have the considerable metabolic cost of respiration (Rachmilevitch et al., 2006). This by no means underestimates the importance of deep roots, but their role for improving yield under water deficit has to be contextualized.

Vadez (2014) proposed that root hydraulics, more than root depth or general root distribution, is the appropriate framework to investigate the role of roots in linking soil drying to the water-use economy of crops. The idea is that hydraulic limitation, for instance, due to a loss of soil-plant hydraulic conductance during soil drying, leads to a drop in leaf water potential that triggers stomatal closure. Stomatal closure can take place over progressive soil drying (over periods of weeks), but also due to diurnal variations in vapour pressure deficit (over periods of minutes to hours). The cumulative result of short-term stomatal regulation over several weeks might confer advantages in climates where water resources are limited, helping plants to reach the flowering stage with still sufficient water (Ahmed et al., 2018b).

In this chapter, we follow a similar hydraulic approach to link below- and above-ground hydraulics to stomatal regulation. We review the principles of water flow in soil and roots and its link to stomatal regulation (Fig. 1). We will mainly focus on below-ground processes, but we will also make the necessary connection to stomatal regulation, and finally to the economy of water use by crops.

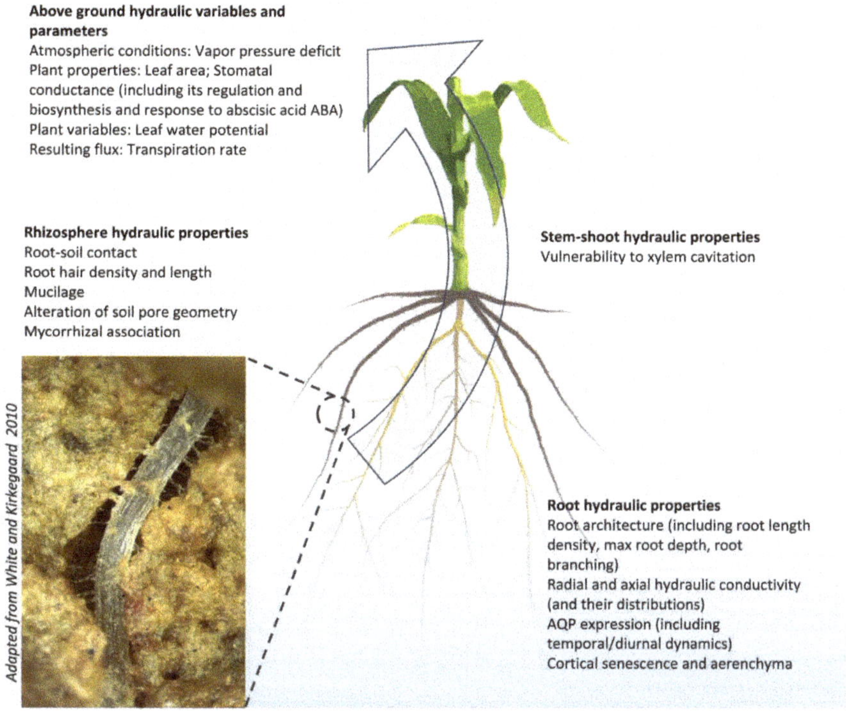

Figure 1 Traits and processes impacting water flow across the soil-plant atmosphere continuum.

2 Principles of water flow in soil and plants

2.1 Transpiration, stomatal conductance and leaf water potential

The flux of water across the soil-plant-atmosphere continuum is controlled by leaf area, stomatal conductance (*gs*) and atmospheric demand. Water evaporation at the leaf surface decreases the leaf water potential (ψ_{leaf}) and creates a tension in the leaf that propagates through the xylem vessels down to the roots and the soil. The leaf water potential depends on the soil water potential and on the hydraulic conductivities of the elements composing the soil-plant continuum. Although transpiration is set by atmospheric conditions and stomatal conductance, the soil-plant hydraulic conductance (including the contribution of the root system and xylem vessels of stem and leaves) affects the leaf water potential and thus the stomatal conductance. A review of stomatal response to decreases in leaf water potential showed a nonlinear decrease in stomatal conductance at a critical leaf water potential (Anderegg et al., 2017). Despite the variations across species, the general behaviour of this relationship was similar.

There is sufficient evidence that stomatal conductance is closely related to leaf water potential (Tardieu & Davies, 1992; Anderegg et al., 2017). Stomata respond to leaf water potential via passive and active mechanisms (Brodribb & McAdam, 2011; Buckley, 2019). In angiosperms, the closure of stomata is through an active mechanism, in which the biosynthesis and response to abscisic acid (ABA) is crucial (McAdam & Brodribb, 2015). Despite the consensus on the pivotal role of ABA in stomatal regulation, the location of ABA production is still a matter of debate. According to McAdam and Brodribb (2018), the leaf mesophyll is the main location of ABA, and the fact that ABA is also found in the roots is due to its transport from the leaves through the phloem. However, there is also a body of evidence supporting the hypothesis that ABA is produced in the roots when the soil water potential decreases and it is then transported to the leaves where it triggers stomatal closure (Davies & Zhang, 1991; Dodd, 2005). The hypothesis that stomatal regulation does not solely depend on the leaf water potential, but is also an explicit function of the soil water potential, is supported by experiments with pressurized plants (Gollan et al., 1986) in which stomata were found to close despite the leaves being turgid. This explicit dependence on soil water potential has profound consequence on the decrease of transpiration during soil drying. The idea is that stomata close when a part of the root-zone dries. As the soil dries and the drying front moves downwards, a progressively larger fraction of the root system is in dry soils and might send sufficient ABA to the leaves closing stomata. It could be expected that the more heterogeneous the soil water potential in the root zone, the stronger is the effect of the local production of ABA on transpiration regulation. This concept has been utilized to reduce stomatal conductance and increase water-use efficiency during partial root zone irrigation (Dodd et al., 2006).

2.2 Plant hydraulic conductance and its components

Transpiration generates a tension in the xylem that is transported down along the xylem vessels to the roots. The leaf water potential depends on the transpiration rate (*E*), the soil-plant hydraulic conductance and the soil water potential. Despite the complexity of the flow path in soils and plants, a simple relation can be found between transpiration and leaf water potential:

$$E = K_{tot}(\psi_{s,eq} - \psi_l) \quad (1)$$

where K_{tot} is the total conductance of the soil-plant system and $\psi_{s,eq}$ is the equivalent soil water potential experienced by the plant. Couvreur et al. (2012) showed that $\psi_{s,eq}$ is an average potential weighted by the local water uptake. The more water is extracted from a layer, the more strongly the plant reflects the water potential from this soil layer. Experimentally, $\psi_{s,eq}$ is measured just before dawn, when transpiration is still negligible and water is assumed to have reached an equilibrium between soil and plant tissues.

Plants can control the leaf water potential by closing stomata or by altering the hydraulic conductance. Here key questions are: what is the element of the soil-plant continuum that more strongly affect the total soil-plant conductance? To what extent do plants control these conductivities?

The overall hydraulic conductance of a plant depends on the hydraulic properties of roots, xylem vessels and leaves (Maurel et al., 2008). The xylem, typically has a higher hydraulic conductance than that of living tissues, and is considered as a connecting organ between roots and leaves (Angeles, 2004). In the root, water flows radially from the root surface to the xylem vessels crossing several cellular tissues and compartments: the cortex, where water can flow in the apoplast and across cells, and the endodermis, where apoplastic flow is impeded and is the main resistance to the radial flow.

It has been shown that plants control the hydraulic resistance within the root via the development of impermeable barriers (Kreszies et al., 2018) as well as through the activity of water channels (aquaporins (AQP)). For instance, Caldeira et al. (2014) showed that AQPs follow a circadian rhythm and proposed that they affected root hydraulic conductance, increasing its value at night and early morning, when the transpiration rate is minimal, while decreasing it in the afternoon when transpiration demand is maximal. They formulated the hypothesis that the reduction of root conductance, before transpiration peaks, functionally triggers a partial closure of stomata that avoids the rhizosphere becoming water depleted (Caldeira et al., 2014). As soil processes are non-linear and hysteretic (Passioura, 1980), a moderate reduction of water fluxes in the soil would result in a significant increase in plant water potential.

In the stele, the number, size and arrangement of xylem vessels affect the axial movement of water (Richards & Passioura, 1989). The number and size of xylem vessels increase during maturation. In contrast to barriers, AQP and root-soil interface traits (see Section 3), and the impact of xylem on yield has been tested in the field. Richards and Passioura (1989) conducted a breeding programme to reduce the diameter of the central xylem vessels in seminal axes of wheat and produced lines with approximately 15% narrower xylem vessels. The selected lines yielded about 5% more in dry years. In addition to these permanent features (i.e. xylem number and diameter), it has been shown that rapid reduction in axial conductance could occur as a result of xylem vessel embolism, or cavitation (Sperry & Pockman, 1993). Cavitation is a key process in understanding tree mortality, but it might also play a role in crops. For instance, Skelton et al. (2017) showed that cavitation occurs in roots, stems and leaves of herbaceous species (tomato). Furthermore, different tissues exhibited similar vulnerability to embolism. In maize, McCully et al. (1998)

observed a daily occurrence of embolism and refilling in roots of field-grown plants. Li et al. (2009) measured the vulnerability of xylem to cavitation and the hydraulic conductance of maize and found, surprisingly, that a drought-resistant variety did not exhibit the trade-off between resistance to cavitation and water-transport efficiency observed previously in woody species (Tyree et al., 1994).

In summary, plants continuously adapt to variable atmospheric and soil conditions by altering the hydraulic conductivity of key elements below and above ground, but our understanding of this hydraulic acclimatisation is still incomplete. Similarly, the primary constraint to water flow across the soil-root-stem-leaf continuum is not clear. It could be the soil, its interface with the roots, the root themselves (with their radial and axial components), or the cavitation of the xylem along the water path. A big challenge to increase this understanding is the difficulty of measuring the hydraulic conductance of key elements of the system. A recent experimental study with olive trees (Rodriguez-Dominguez & Brodribb, 2020) illustrates a method to measure and partition key hydraulic components. This is a major advancement. The authors found that the drop in hydraulic conductance of the root-soil interface is the primary constraint to water flow from the soil to the leaves during soil drying and that this decline in hydraulic conductance is concomitant with stomatal closure. This study calls for a more comprehensive investigation of below- and above-ground plant hydraulics, in particular, concerning the link between soil drying and stomatal closure.

2.3 Root water uptake and soil drying

Soil drying due to root water uptake has different spatial and temporal scales. The soil profile dries over several weeks. Typically, the drying front moves downwards, with the upper soil layers being dried first because of (1) the greater root length density, and (2) the greater suction in the more proximal parts of the root system (Ahmed et al., 2016b). The spatial and temporal dynamics of soil moisture distribution over the root zone depend on both, root length distribution, root hydraulic properties and soil hydraulic properties. Recent developments in modelling root water uptake in three dimensions with explicit description of water flow in the root system combined with soil water redistribution have shed light on disentangling these factors (Doussan et al., 2003; Javaux et al., 2008). For instance, Javaux et al. (2008) showed that the locations of root water uptake are not proportional to the root length density and that the dissipation of water potential along the root system plays a role in the locations of soil water extraction. In general, the higher the axial conductivity, the more uniform becomes the root water uptake along the root axis (Zwieniecki et al., 2002).

Simulation studies with root architecture models have shown that different root architecture can lead to different water uses during soil drying and that the advantages of a deep and profuse root system depend on the atmospheric conditions (Leitner et al., 2014). A limit of such models is the challenge to assign to the root system the correct hydraulic properties. These complex inverse problems have been solved for some specific examples, such as lupins and maize in wet soils (Meunier et al., 2018), but parameterizing of root architecture models remains a big challenge.

2.4 Radial flow of water from the soil to the root surface

Although the soil dries over weeks, millimetre-scale gradients of soil moisture develop around roots over minutes when the transpiration demand cannot be matched by the flow of water through the soil (Carminati et al., 2017b). Such gradients in water potential around single roots have been investigated since the pioneering work by Gardner (1960). Gardner (1960) found that in wet soils the gradients in soil water potential around roots are negligible and that large gradients develop only close to the wilting point. However, more accurate simulations of water flow around the roots showed more marked gradients (Fig. 2). These gradients are the result of the nonlinearity of the soil hydraulic conductivity, which decreases by several orders as the soil matric potential drops, and of the radial geometry of the flow problem, with the flux of water (cm s^{-1}) increasing with decreasing distance from the root centre. The radial flow model predicts that for any imposed flux at the root-soil interface, there is a minimum soil water potential for which this flux can be sustained.

Despite its simplifications, the radial flow model is capable of reproducing experimental observations of leaf water potential and transpiration rates with a limited number of parameters (Passioura, 1980; Carminati et al., 2017b; Ahmed et al., 2018d; Hayat et al., 2020; Cai et al., 2020). Central to this model are the soil hydraulic properties, the total root length active in water uptake and the root radius. All these parameters require a discussion.

The soil hydraulic properties could be measured from soil samples collected from the root zone. The assumption is that the structure of the soil, its hydraulic properties and its general flow geometry are representative of the radial flow of water towards the root surface. The validity of this assumption is questionable. Several studies showed that the structure and hydraulic properties of the soil near the roots, the rhizosphere, differ from the bulk soil (Carminati et al., 2010; Moradi et al., 2011; Carminati & Vetterlein, 2013; Ahmed et al., 2016a). Additionally, the average distance for radial flow of water, which can be approximated as $\sqrt{1/\pi L}$ with L being the active root length, is typically in the range from mm to cm, and in a structured and aggregate soils, such as the rhizosphere, it is questionable whether 1 cm is sufficient to represent the

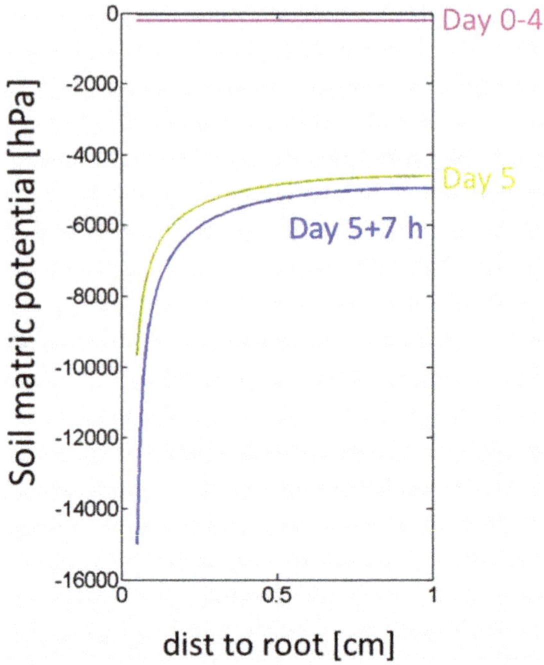

Figure 2 Gradients in soil matric potential around a single root taking up water at constant rate from a drying soil. As the soil water potential decreases, large gradients in matric potential develop near the root surface, ultimately limiting water flow to the root surface (lines corresponding to different times).

effective hydraulic conductivity that is measured in soil samples. Similarly, the length of the active roots is an empirical fitting parameter that is not easy to predict. Root architecture models (such as those introduced in the previous paragraph) as well as measurements of water fluxes into the root system (Ahmed et al., 2016b, 2018e) suggest that the radial flow of water into crop roots varies along the root system by orders of magnitude. Therefore, the definition of a root length with a constant water flux is poorly defined and has to be considered as an empirical parameter. Passioura (1980) used a root length equal to 30% of the measured one to fit the observed leaf water potential during soil drying. A similar consideration can be made for the root radius. A reasonable assumption is to use the radius of fine roots, which are supposed to be active in water uptake (Ahmed et al., 2016b). However, the question is whether the length of root hairs should be added to the radius of the root. Carminati et al. (2017b) showed that root hairs decrease the matric potential towards the root for a given transpiration rate. These experimental results could be simulated by adding the length of root hairs to the root radius. However, as for the active root length, this effective root radius has to be considered as a fitting parameter.

Another challenge for such radial flow models is to include the fact that the contact between roots and soil is not constant and might decrease as the soil dries. De Willigen et al. (2018) calculated the effects of a partial contact between root and soils. However, estimating the effects of the root-soil contact in a dynamic way is not simple. Roots shrink in response to low water potential, thereby creating air-filled gaps that break the hydraulic continuity between the soil and the root surface (Huck et al., 1970; Nobel & Cui, 1992). Carminati et al. (2013) investigated the effect of air-filled gaps on root water uptake. They found that air-filled gaps were formed after the transpiration rate had already fallen and concluded that gaps are a consequence and not the initial cause of limited water uptake (Carminati et al., 2013). Their conclusion might have been affected by the lack of spatial resolution, and gaps might have formed earlier than reported and might form concomitant (and not subsequent) to stomatal closure. An image of the root-soil interface at higher spatial resolution is shown in Fig. 3. The picture shows a seminal root of maize in a loamy soil that was severely dried. The picture clearly shows that the root occupies only a fraction of the original pore and is in partial contact with the soil matrix. The picture also shows low water saturation of the root cortex. It is likely that both the drying of the cortex and the loss of contact with the soil matrix affected the radial flow of water to the root surface. Recently, Rodriguez-Dominguez and Brodribb (2020) showed that upon soil drying the hydraulic resistance of the root-soil interface increased to ca. 90% of the total soil-plant hydraulic resistance and thus it represented the main constraint to transpiration. They speculated that the drop in hydraulic conductance of the root-soil interface was caused by

Figure 3 Syncrotron X-ray CT of a maize root in a relatively dry loamy soil showing an air-filled gap between the root and the soil matrix (field of view ca. 1 mm at a pixel size of 0.65 × 0.65 μm^2; air is black, soil is light grey and root is dark grey).

the loss of contact between the root surface and the soil. However, it could also be that a decrease in hydraulic conductance of the cortex might have affected the total conductance of the system. These considerations should also be extended to roots growing in the field, which are typically growing in continuous macropores, with only partial contact with the soil matrix (White & Kirkegaard, 2010).

Plants have developed ways of coping with the gradients of water potential at the root-soil interface. For instance, root hairs may help to bridge the gaps between roots and soil, thereby extending the effective root radius and avoiding increasingly steep gradients in water potential close to the root surface (Carminati et al., 2017b). Similarly, mucilage secreted by roots was postulated to maintain the continuity of water flow across the soil-root interface (Carminati et al., 2017a). Additionally, plant roots and microorganisms can modify the pore space in the rhizosphere in a dynamic way (Koebernick et al., 2017; Bitterlich et al., 2018; Helliwell et al., 2019; Lucas et al., 2019a,b). In the forthcoming paragraphs we review the rhizosphere mechanisms that are likely to affect the ability of roots to extract water from the soil.

3 Rhizosphere processes relevant for root water uptake

3.1 Root hairs

Root hairs bridge the gaps between roots and soil, thereby extending the effective root radius and avoiding increasingly steep gradients in water potential close to the root-soil surface. The mechanism by which root hairs bridge gaps between roots and soil is illustrated in Fig. 4a, which shows that root hairs increase the surface of the root extracting water from the soil and thereby decreases the velocity of water passing across the root-soil interface. This can be particularly important in structured and aggregated soils, where root hairs can increase contact with the soil structures that are still hydraulically conductive in dry conditions, but that are not in direct contact with the root epidermis. The result is a reduction in the decrease of water potential towards the root surface and thus a facilitation of water extraction from the soil.

The experimental evidence for the role of root hairs on water uptake is limited. Recent advances in imaging methods allow the visualization of hairs in a three-dimensional soil matrix (Keyes et al., 2013; Koebernick et al., 2017). These images have been used to parameterize models of nutrient uptake (Keyes et al., 2013). An example of high-resolution imaging of the root-soil interface including a root hair connecting the root surface to a soil aggregate is shown in Fig. 4b. These high-resolution images support the role of hairs in water and nutrient uptake, but do not prove it. For this, measurements of leaf water potential and transpiration are needed. Experiments with mutants lacking root hairs provided new insight on this topic. Experiments with a root pressure

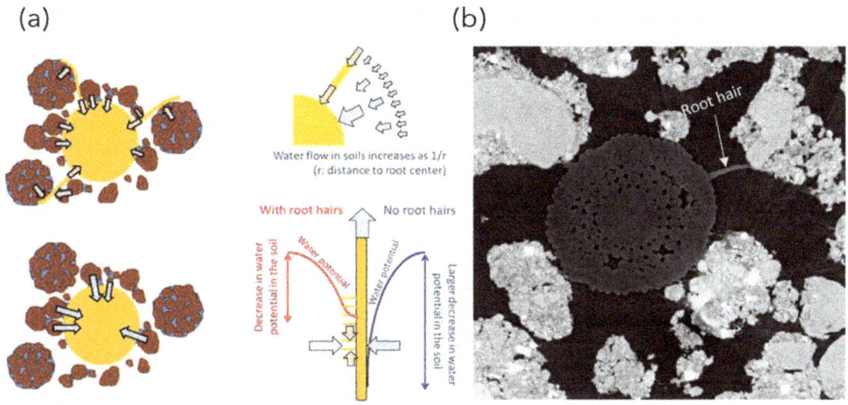

Figure 4 Root hairs bridge the gaps between roots and soil. (a) Root hairs increase the surface of the root extracting water from the soil and thus it decreases the velocity of the water across the root-soil interface, which could be particularly important in structured and aggregated soils, where the root hairs can largely increase the contact with the soil structures that are still hydraulically conductive in dry conditions. (b) Syncrotron X-ray CT of a maize root in a relatively dry loamy soil showing a root hair connecting the root surface to a soil aggregate (field of view ca. 1 mm at a pixel size of 0.65×0.65 μm^2; air is black, soil is light grey and root is dark grey).

chamber showed that wild-type barley plants (WT) possessing root hairs have a less negative leaf water potential compared to their hairless mutant. However, this effect was only visible at high transpiration rates in relatively dry soils. These experimental observations support the mechanism illustrated in Fig. 4a, thereby providing the first evidence for the ability of root hairs to take up water from drying soil (Carminati et al., 2017b). However, this limited evidence is not unequivocal. Dodd and Diatloff (2016) did not find a significant capacity for roots hairs to extract water from the soil; the increased root length of the hairless mutants compensated for the lack of hairs. In maize, mycorrhiza might also compensate for lack of, or reduced, root hairs. In summary, there is limited evidence that root hairs allow crops to sustain high transpiration rates in drying soils, as root and rhizosphere plasticity might compensate for the lack of hairs.

3.2 Mucilage

Mucilage, a polymeric gel exuded at the root tip and capable of absorbing large volumes of water, is another potential strategy that plants use to dynamically alter the gradients in water potential at the root-soil interface. The premise is that mucilage keeps the rhizosphere wet, connected to the root surface and hydraulically well conductive, especially in drying soil (Young, 1995; Ahmed et al., 2014; Carminati et al., 2016; Benard et al., 2019). Carminati et al. (2011) demonstrated that the enhanced water retention in the rhizosphere buffers the decrease in hydraulic conductivity in drying soil. They postulated that mucilage facilitates the extraction of water from drying soils by attenuating the drop in water potential across the rhizosphere. The physical explanation of this effect is the result of the interplay between water adsorption, surface tension and viscosity (Carminati et al., 2017a). Mucilage decreases the surface tension of the soil solution and increases its viscosity. The increasing importance of viscous forces prevents the break-up of liquid bridges between soil particles, leading to the formation of one-dimensional threads and two-dimensional surfaces that span through multiple pores (Benard et al., 2019). The filaments are significantly thinner than root hairs, and have a two-dimensional shape, as hollow cylinders. The result of the high viscosity of mucilage is that the liquid phase remains connected during soil drying. Simplified experiments with maize mucilage in soils of different texture supported this concept and showed that mucilage reduces the decrease in soil hydraulic conductivity during soil drying.

Despite these advances in our understanding of the effect of mucilage on soil hydraulic properties, its role on root water uptake is not yet clear, and the hypothesis that mucilage helps plants to extract water from the soil remains largely speculative. Most likely, mucilage is crucial in maintaining the contact between the root surface and the soil matrix, particularly during root shrinkage. This might explain why, despite root shrinkage, a significant fraction of the root surface remains in contact with the soil matrix, as illustrated in Fig. 3.

One aspect of soil-mucilage-plant relations that is relatively clear is the role of mucilage on rhizosphere rewetting. Mucilage turns water repellent upon drying (Ahmed et al., 2015; Zickenrott et al., 2016; Ahmed et al., 2016a, 2018d) and delays the rewetting of the rhizosphere (Carminati et al., 2010). The effect is greater in soils with coarse texture (Benard et al., 2018). In sandy soil, it was shown that rhizosphere water repellency limits the flux to roots upon soil rewetting (Zarebanadkouki et al., 2018). The effect is persistent for a few hours. However, the implications for plant performance are unclear. An interesting hypothesis is that rhizosphere water-repellency isolates roots from the driest soil layers, limiting the efflux of water from the roots during periods of low transpiration demand (i.e. at night).

Additionally, care is required in generalizing these concepts to all plant species and soil types. Mucilage from different species was found to have different properties (Naveed et al., 2019; Zickenrott et al., 2016). For instance, barley mucilage is much less viscous than maize and its effect on altering the water retention curve and hydraulic conductivity might be the opposite than that of maize mucilage (i.e. the effect of the reduced surface tension, and thus reduced capillary forces, might be lower than the effect of water adsorption and increased viscosity). Again, these results are likely to vary among soil types, but the general soil physical mechanisms have been understood.

A crucial open question is the degradation rate of mucilage in different soils and for different soil moistures. This question affects the length of the root system where mucilage plays a role. Ahmed et al. (2018a,c) used a natural abundance approach to measure the effect of soil moisture on maize mucilage degradation. They found that, after two weeks, mucilage was nearly completely decomposed at high water content (80% WHC). Mucilage degradation was significantly suppressed under drought (30% WHC). The effects of mucilage are therefore strongly dynamic in space and time and still to be fully understood.

3.3 Root-induced changes in pore size distribution

As roots grow and tips push through the soil they compact the soil in their immediate vicinity (Dexter, 1987; Vollsnes et al., 2010). Compaction is synonymous with alteration of pore size distribution. Hence, soil water retention curve, as well as hydraulic conductivity, differs between the rhizosphere and the bulk soil. The extent of compaction is a function of root diameter (Lucas et al., 2019b). Whether compaction occurs depends also on texture (Helliwell et al., 2019), the initial bulk density and the presence of connected macropores which enable roots to grow without the necessity to create new macropores (Lucas et al., 2019a,b). The presence of root hairs may contribute to an alteration of compaction profiles around roots (Koebernick et al., 2017), although this effect was not consistently observed (Koebernick et al., 2019). If compaction does occur the consequences for water uptake are expected to be similar to those described for mucilage during a drying event, that is, the decrease in hydraulic conductivity occurs at lower matric potentials. Changes in rhizosphere soil bulk density are expected to be long-lasting as they have been observed in the field for root-derived biopores accumulating over time in an agricultural chronosequence (Lucas et al., 2019a). Little is known about the extent to which biopores are reused and what this means for maintenance of soil-root hydraulic contact in general, and more specifically during periods of limited water availability.

3.4 Mycorrhiza

The contribution of arbuscular mycorrhizal fungi (AMF) to plant phosphorus (P) uptake is well documented and the direct and indirect mechanisms have been investigated in detail (Smith & Read, 2008). Much less is known about the contribution of AMF to plant drought stress tolerance or more specifically to plant water uptake (Augé et al., 2015). While it was frequently observed that mycorrhizal plants are able to extract more water in general, the underlying mechanism is difficult to decipher as mycorrhizal and non-mycorrhizal plants in most studies differ in shoot size and root development. A substantial contribution of hyphal water transport to plant water uptake would require unrealistically high flow rates within hyphae and has not been observed (George et al., 1992). Bitterlich et al. (2018) investigated a hypothesis which is analogous to the processes discussed above for the contribution of mucilage, root hairs and the alteration of soil structure by roots per se. They demonstrated that the soil water retention curve is altered by the presence of hyphae and that soil hydraulic conductivity for a given matric potential is increased. They argued that their observations can be explained by alterations in soil structure and wettability related to the ability of hyphae to bridge large pores, and contribute to soil aggregate formation and the rearrangement of pore size distribution. A direct observation of such structural changes related to hyphal growth is urgently needed but so far hindered by the trade-off between sample size required for establishment of mycorrhizal symbioses allowing hyphae to extend several centimetres from the root surface into the soil, and the sample size which enables a resolution as it is seen in Fig. 4b.

4 Conclusion and future trends

We have reviewed the root-rhizosphere physical processes that affect the ability of plant roots to extract water from the soil. Prominent aspects of the rhizosphere hydraulic properties altering water uptake are: root shrinkage, alteration of pore geometry in the rhizosphere, effect of mucilage on water retention, conductivity and water repellency, root hairs and mycorrhiza connecting the root surface to the soil matrix. All these factors are strongly dynamic, changing over time and with soil moisture conditions. Although our understanding of the mechanisms related to these factors has advanced significantly in the last ten years, the relative importance of these rhizosphere processes for the ability of crops to extract water from the soil and better tolerate drought is still largely unclear, even more so when results from homogenized soils in pot experiments are extrapolated to structured soils in the field. We do not know the plasticity of these rhizosphere processes in relation to environmental changes and have just started to explore the genetic variability associate with some of these mechanisms.

The agronomic and environmental context is fundamental to link rhizosphere processes to their consequences for crop tolerance to drought. Rhizosphere processes facilitating water extraction might be favourable in moderately dry conditions, but in more severe and dry conditions processes the same processes that reduce water extractability, such as root shrinkage or rhizosphere water repellency, might save water and eventually decrease drought stress.

In summary, we have shown that roots modify the hydraulic properties of the rhizosphere in multiple ways and that these modifications are strongly dynamic. There is increasing evidence that the rhizosphere is a vital region where plants can alter the interactions with soil water and their access to it. The next research step is to investigate the implications of these rhizosphere processes on crop growth and water-use economy and then use this knowledge to grow more resilient crops that match to their environment.

5 Acknowledgements

The authors acknowledge the Deutsche Forschungsgemeinschaft (DFG, German Research Foundation) for funding of the priority programme 2089, project numbers 403670197 *"Emerging effects of root hairs and mucilage on plant scale soil water relations"*, 403640522 *"Spatial organization of the liquid phase in the rhizosphere"* and 403640293 *"Relevance of root growth and related soil structure formation for spatiotemporal patterns of chemical and biological properties and emergent system functions"* and 403801423 *"Coordination funds"* to MAA, DV and AC. We would like to acknowledge the Paul Scherrer Institute, Villigen, Switzerland, for provision of synchrotron radiation beamtime at the TOMCAT beamline X02DA of the SLS and would like to thank Dr. Goran Lovric for assistance.

6 References

Ahmed, M. A., Banfield, C. C., Sanaullah, M., Gunina, A. and Dippold, M. A. 2018a. Utilisation of mucilage C by microbial communities under drought. *Biology and Fertility of Soils* 54(1): 83–94.

Ahmed, M. A., Holz, M., Woche, S. K., Bachmann, J. and Carminati, A. 2015. Effect of soil drying on mucilage exudation and its water repellency: a new method to collect mucilage. *Journal of Plant Nutrition and Soil Science* 178(6): 821–824.

Ahmed, M. A., Kroener, E., Benard, P., Zarebanadkouki, M., Kaestner, A. and Carminati, A. 2016a. Drying of mucilage causes water repellency in the rhizosphere of maize: measurements and modelling. *Plant and Soil* 407(1–2): 161–171.

Ahmed, M. A., Kroener, E., Holz, M., Zarebanadkouki, M. and Carminati, A. 2014. Mucilage exudation facilitates root water uptake in dry soils. *Functional Plant Biology* 41(11): 1129–1137.

Ahmed, M. A., Passioura, J. and Carminati, A. 2018b. Hydraulic processes in roots and the rhizosphere pertinent to increasing yield of water-limited grain crops: a critical review. *Journal of Experimental Botany* 69(13): 3255–3265.

Ahmed, M. A., Sanaullah, M., Blagodatskaya, E., Mason-Jones, K., Jawad, H., Kuzyakov, Y. and Dippold, M. A. 2018c. Soil microorganisms exhibit enzymatic and priming response to root mucilage under drought. *Soil Biology and Biochemistry* 116: 410–418.

Ahmed, M. A., Zarebanadkouki, M., Ahmadi, K., Kroener, E., Kostka, S., Kaestner, A. and Carminati, A. 2018d. Engineering rhizosphere hydraulics: pathways to improve plant adaptation to drought. *Vadose Zone Journal* 17(1): 1–12.

Ahmed, M. A., Zarebanadkouki, M., Kaestner, A. and Carminati, A. 2016b. Measurements of water uptake of maize roots: the key function of lateral roots. *Plant and Soil* 398(1–2): 59–77.

Ahmed, M. A., Zarebanadkouki, M., Meunier, F., Javaux, M., Kaestner, A. and Carminati, A. 2018e. Root type matters: measurement of water uptake by seminal, crown, and lateral roots in maize. *Journal of Experimental Botany* 69(5): 1199–1206.

Anderegg, W. R. L., Wolf, A., Arango-Velez, A., Choat, B., Chmura, D. J., Jansen, S., Kolb, T., Li, S., Meinzer, F., Pita, P., Resco de Dios, V., Sperry, J. S., Wolfe, B. T. and Pacala, S. 2017. Plant water potential improves prediction of empirical stomatal models. *PLoS ONE* 12(10): e0185481.

Angeles, G. 2004. The cohesion-tension theory. *New Phytologist* 163(3): 451–452.

Augé, R. M., Toler, H. D. and Saxton, A. M. 2015. Arbuscular mycorrhizal symbiosis alters stomatal conductance of host plants more under drought than under amply watered conditions: a meta-analysis. *Mycorrhiza* 25(1): 13–24.

Benard, P., Zarebanadkouki, M., Brax, M., Kaltenbach, R., Jerjen, I., Marone, F., Couradeau, E., Felde, V. J. M. N. L., Kaestner, A. and Carminati, A. 2019. Microhydrological niches in soils: how mucilage and EPS alter the biophysical properties of the rhizosphere and other biological hotspots. *Vadose Zone Journal* 18(1): 1–10.

Benard, P., Zarebanadkouki, M., Hedwig, C., Holz, M., Ahmed, M. A. and Carminati, A. 2018. Pore-scale distribution of mucilage affecting water repellency in the rhizosphere. *Vadose Zone Journal* 17(1): 1–9.

Bitterlich, M., Franken, P. and Graefe, J. 2018. Arbuscular mycorrhiza improves substrate hydraulic conductivity in the plant available moisture range under root growth exclusion. *Frontiers in Plant Science* 9: 301.

Brodribb, T. J. and McAdam, S. A. M. 2011. Passive origins of stomatal control in vascular plants. *Science* 331(6017): 582–585.

Buckley, T. N. 2019. How do stomata respond to water status? *New Phytologist* 224(1): 21–36.

Cai, G., Ahmed, M. A., Dippold, M. A., Zarebanadkouki, M. and Carminati, A. 2020. Linear relation between leaf xylem water potential and transpiration in pearl millet during soil drying. *Plant and Soil* 447(1–2): 565–578.

Caldeira, C. F., Jeanguenin, L., Chaumont, F. and Tardieu, F. 2014. Circadian rhythms of hydraulic conductance and growth are enhanced by drought and improve plant performance. *Nature Communications* 5: 5365.

Carminati, A., Benard, P., Ahmed, M. A. and Zarebanadkouki, M. 2017a. Liquid bridges at the root-soil interface. *Plant and Soil* 417(1–2): 1–15.

Carminati, A., Moradi, A. B., Vetterlein, D., Vontobel, P., Lehmann, E., Weller, U., Vogel, H.-J. and Oswald, S. E. 2010. Dynamics of soil water content in the rhizosphere. *Plant and Soil* 332(1–2): 163–176.

Carminati, A., Passioura, J. B., Zarebanadkouki, M., Ahmed, M. A., Ryan, P. R., Watt, M. and Delhaize, E. 2017b. Root hairs enable high transpiration rates in drying soils. *New Phytologist* 216(3): 771–781.

Carminati, A., Schneider, C. L., Moradi, A. B., Zarebanadkouki, M., Vetterlein, D., Vogel, H.-J., Hildebrandt, A., Weller, U., Schüler, L. and Oswald, S. E. 2011. How the rhizosphere may favor water availability to roots. *Vadose Zone Journal* 10(3): 988–998.

Carminati, A. and Vetterlein, D. 2013. Plasticity of rhizosphere hydraulic properties as a key for efficient utilization of scarce resources. *Annals of Botany* 112(2): 277–290.

Carminati, A., Vetterlein, D., Koebernick, N., Blaser, S., Weller, U. and Vogel, H.-J. 2013. Do roots mind the gap? *Plant and Soil* 367(1–2): 651–661.

Carminati, A., Zarebanadkouki, M., Kroener, E., Ahmed, M. A. and Holz, M. 2016. Biophysical rhizosphere processes affecting root water uptake. *Annals of Botany* 118(4): 561–571.
Couvreur, V., Vanderborght, J. and Javaux, M. 2012. A simple three-dimensional macroscopic root water uptake model based on the hydraulic architecture approach. *Hydrology and Earth System Sciences* 16(8): 2957–2971.
Davies, W. J. and Zhang, J. 1991. Root signals and the regulation of growth and development of plants in drying soil. *Annual Review of Plant Physiology and Plant Molecular Biology* 42(1): 55–76.
Dexter, A. R. 1987. Compression of soil around roots. *Plant and Soil* 97(3): 401–406.
de Willigen, P., Heinen, M. and van Noordwijk, M. 2018. Roots partially in contact with soil: analytical solutions and approximation in models of nutrient and water uptake. *Vadose Zone Journal* 17(1): 1–16.
Dodd, I. C. 2005. Root-to-shoot signalling: assessing the roles of 'up' in the up and down world of long-distance signalling *in planta*. *Plant and Soil* 274(1–2): 251–270.
Dodd, I. C. and Diatloff, E. 2016. Enhanced root growth of the brb (bald root barley) mutant in drying soil allows similar shoot physiological responses to soil water deficit as wild-type plants. *Functional Plant Biology* 43(2): 199–206.
Dodd, I. C., Theobald, J. C., Bacon, M. A. and Davies, W. J. 2006. Alternation of wet and dry sides during partial rootzone drying irrigation alters root-to-shoot signalling of abscisic acid. *Functional Plant Biology* 33(12): 1081–1089.
Doussan, C., Pagès, L. and Pierret, A. 2003. Soil exploration and resource acquisition by plant roots: an architectural and modelling point of view. *Agronomie* 23(5–6): 419–431.
Draye, X., Kim, Y., Lobet, G. and Javaux, M. 2010. Model-assisted integration of physiological and environmental constraints affecting the dynamic and spatial patterns of root water uptake from soils. *Journal of Experimental Botany* 61(8): 2145–2155.
Gardner, W. R. 1960. Dynamic aspects of water availability to plants. *Soil Science* 89(2): 63–73.
George, E., Häussler, K.-U., Vetterlein, D., Gorgus, E. and Marschner, H. 1992. Water and nutrient translocation by hyphae of *Glomus mosseae*. *Canadian Journal of Botany* 70(11): 2130–2137.
Gollan, T., Richards, R. A., Rawson, H. M., Passioura, J. B., Johnson, D. A. and Munns, R. 1986. Soil water status affects the stomata. *Functional Plant Biology* 13(4): 459–464.
Hayat, F., Ahmed, M. A., Zarebanadkouki, M., Javaux, M., Cai, G. and Carminati, A. 2020. Transpiration reduction in maize (*Zea mays* L) in response to soil drying. *Frontiers in Plant Science* 10: 1695.
Helliwell, J. R., Sturrock, C. J., Miller, A. J., Whalley, W. R. and Mooney, S. J. 2019. The role of plant species and soil condition in the structural development of the rhizosphere. *Plant, Cell and Environment* 42(6): 1974–1986.
Huck, M. G., Klepper, B. and Taylor, H. M. 1970. Diurnal variations in root diameter. *Plant Physiology* 45(4): 529–530.
Javaux, M., Schröder, T., Vanderborght, J. and Vereecken, H. 2008. Use of a three-dimensional detailed modeling approach for predicting root water uptake. *Vadose Zone Journal* 7(3): 1079–1088.
Keyes, S. D., Daly, K. R., Gostling, N. J., Jones, D. L., Talboys, P., Pinzer, B. R., Boardman, R., Sinclair, I., Marchant, A. and Roose, T. 2013. High resolution synchrotron imaging of

wheat root hairs growing in soil and image based modelling of phosphate uptake. *New Phytologist* 198(4): 1023–1029.

Koebernick, N., Daly, K. R., Keyes, S. D., Bengough, A. G., Brown, L. K., Cooper, L. J., George, T. S., Hallett, P. D., Naveed, M., Raffan, A. and Roose, T. 2019. Imaging microstructure of the barley rhizosphere: particle packing and root hair influences. *New Phytologist* 221(4): 1878–1889.

Koebernick, N., Daly, K. R., Keyes, S. D., George, T. S., Brown, L. K., Raffan, A., Cooper, L. J., Naveed, M., Bengough, A. G., Sinclair, I., Hallett, P. D. and Roose, T. 2017. High-resolution synchrotron imaging shows that root hairs influence rhizosphere soil structure formation. *New Phytologist* 216(1): 124–135.

Kreszies, T., Schreiber, L. and Ranathunge, K. 2018. Suberized transport barriers in *Arabidopsis*, barley and rice roots: from the model plant to crop species. *Journal of Plant Physiology* 227: 75–83.

Leitner, D., Meunier, F., Bodner, G., Javaux, M. and Schnepf, A. 2014. Impact of contrasted maize root traits at flowering on water stress tolerance – a simulation study. *Field Crops Research* 165: 125–137.

Li, Y., Sperry, J. S. and Shao, M. 2009. Hydraulic conductance and vulnerability to cavitation in corn (*Zea mays* L.) hybrids of differing drought resistance. *Environmental and Experimental Botany* 66(2): 341–346.

Lilley, J. M. and Kirkegaard, J. A. 2008. Seasonal variation in the value of subsoil water to wheat: simulation studies in southern New South Wales. *Australian Journal of Agricultural Research* 58(12): 1115–1128.

Lucas, M., Schlüter, S., Vogel, H.-J. and Vetterlein, D. 2019a. Soil structure formation along an agricultural chronosequence. *Geoderma* 350: 61–72.

Lucas, M., Schlüter, S., Vogel, H.-J. and Vetterlein, D. 2019b. Roots compact the surrounding soil depending on the structures they encounter. *Scientific Reports* 9(1): 16236.

Lynch, J. P. 2018. Rightsizing root phenotypes for drought resistance. *Journal of Experimental Botany* 69(13): 3279–3292.

Maurel, C., Verdoucq, L., Luu, D.-T. and Santoni, V. 2008. Plant aquaporins: membrane channels with multiple integrated functions. *Annual Review of Plant Biology* 59: 595–624.

McAdam, S. A. M. and Brodribb, T. J. 2015. The evolution of mechanisms driving the stomatal response to vapor pressure deficit. *Plant Physiology* 167(3): 833–843.

McAdam, S. A. M. and Brodribb, T. J. 2018. Mesophyll cells are the main site of abscisic acid biosynthesis in water-stressed leaves. *Plant Physiology* 177(3): 911–917.

McCully, M. E., Huang, C. X. and Ling, L. E. C. 1998. Daily embolism and refilling of xylem vessels in the roots of field-grown maize. *New Phytologist* 138(2): 327–342.

Meunier, F., Zarebanadkouki, M., Ahmed, M. A., Carminati, A., Couvreur, V. and Javaux, M. 2018. Hydraulic conductivity of soil-grown lupine and maize unbranched roots and maize root-shoot junctions. *Journal of Plant Physiology* 227: 31–44.

Moradi, A. B., Carminati, A., Vetterlein, D., Vontobel, P., Lehmann, E., Weller, U., Hopmans, J. W., Vogel, H.-J. and Oswald, S. E. 2011. Three-dimensional visualization and quantification of water content in the rhizosphere. *New Phytologist* 192(3): 653–663.

Naveed, M., Ahmed, M. A., Benard, P., Brown, L. K., George, T. S., Bengough, A. G., Roose, T., Koebernick, N. and Hallett, P. D. 2019. Surface tension, rheology and hydrophobicity of rhizodeposits and seed mucilage influence soil water retention and hysteresis. *Plant and Soil* 437(1): 65–81.

Nobel, P. S. and Cui, M. 1992. Hydraulic conductances of the soil, the root-soil air gap, and the root: changes for desert succulents in drying soil. *Journal of Experimental Botany* 43(3): 319-326.

Passioura, J. B.. 1980. The transport of water from soil to shoot in wheat seedlings. *Journal of Experimental Botany* 31(1): 333-345.

Passioura, J. 2007. The drought environment: physical, biological and agricultural perspectives. *Journal of Experimental Botany* 58(2): 113-117.

Rachmilevitch, S., Lambers, H. and Huang, B. 2006. Root respiratory characteristics associated with plant adaptation to high soil temperature for geothermal and turf-type *Agrostis* species. *Journal of Experimental Botany* 57(3): 623-631.

Richards, R. and Passioura, J. 1989. A breeding program to reduce the diameter of the major xylem vessel in the seminal roots of wheat and its effect on grain yield in rain-fed environments. *Australian Journal of Agricultural Research* 40(5): 943-950.

Rodriguez-Dominguez, C. M. and Brodribb, T. J. 2020. Declining root water transport drives stomatal closure in olive under moderate water stress. *New Phytologist* 225(1): 126-134.

Skelton, R. P., Brodribb, T. J. and Choat, B. 2017. Casting light on xylem vulnerability in an herbaceous species reveals a lack of segmentation. *New Phytologist* 214(2): 561-569.

Smith, S. and Read, D. 2008. *Mycorrhizal Symbiosis*. Amsterdam: Academic Press.

Sperry, J. S. and Pockman, W. T. 1993. Limitation of transpiration by hydraulic conductance and xylem cavitation in *Betula occidentalis*. *Plant, Cell and Environment* 16(3): 279-287.

Tardieu, F. 2012. Any trait or trait-related allele can confer drought tolerance: just design the right drought scenario. *Journal of Experimental Botany* 63(1): 25-31.

Tardieu, F. and Davies, W. J. 1992. Stomatal response to abscisic acid is a function of current plant water status. *Plant Physiology* 98(2): 540-545.

Tardieu, F. and Parent, B. 2017. Predictable 'meta-mechanisms' emerge from feedbacks between transpiration and plant growth and cannot be simply deduced from short-term mechanisms. *Plant, Cell and Environment* 40(6): 846-857.

Tardieu, F., Simonneau, T. and Muller, B. 2018. The physiological basis of drought tolerance in crop plants: a scenario-dependent probabilistic approach. *Annual Review of Plant Biology* 69: 733-759.

Tyree, M. T., Davis, S. D. and Cochard, H. 1994. Biophysical perspectives of xylem evolution: is there a tradeoff of hydraulic efficiency for vulnerability to dysfunction? *IAWA Journal* 15(4): 335-360.

Vadez, V. 2014. Root hydraulics: the forgotten side of roots in drought adaptation. *Field Crops Research* 165: 15-24.

Vollsnes, A. V., Futsaether, C. M. and Bengough, A. G. 2010. Quantifying rhizosphere particle movement around mutant maize roots using time-lapse imaging and particle image velocimetry. *European Journal of Soil Science* 61(6): 926-939.

Wasson, A. P., Richards, R. A., Chatrath, R., Misra, S. C., Prasad, S. V. S., Rebetzke, G. J., Kirkegaard, J. A., Christopher, J. and Watt, M. 2012. Traits and selection strategies to improve root systems and water uptake in water-limited wheat crops. *Journal of Experimental Botany* 63(9): 3485-3498.

White, R. G. and Kirkegaard, J. A. 2010. The distribution and abundance of wheat roots in a dense, structured subsoil - implications for water uptake. *Plant, Cell and Environment* 33(2): 133–148.

Young, I. M. 1995. Variation in moisture contents between bulk soil and the rhizosheath of wheat (*Triticum aestivum* L. cv. Wembley). *New Phytologist* 130(1): 135–139.

Zaman-Allah, M., Jenkinson, D. M. and Vadez, V. 2011. A conservative pattern of water use, rather than deep or profuse rooting, is critical for the terminal drought tolerance of chickpea. *Journal of Experimental Botany* 62(12): 4239–4252.

Zarebanadkouki, M., Ahmed, M., Hedwig, C., Benard, P., Kostka, S. J., Kastner, A. and Carminati, A. 2018. Rhizosphere hydrophobicity limits root water uptake after drying and subsequent rewetting. *Plant and Soil* 428(1-2): 265–277.

Zickenrott, I.-M., Woche, S. K., Bachmann, J., Ahmed, M. A. and Vetterlein, D. 2016. An efficient method for the collection of root mucilage from different plant species—a case study on the effect of mucilage on soil water repellency. *Journal of Plant Nutrition and Soil Science* 179(2): 294–302.

Zwieniecki, M. A., Thompson, M. V. and Holbrook, N. M. 2002. Understanding the hydraulics of porous pipes: tradeoffs between water uptake and root length utilization. *Journal of Plant Growth Regulation* 21(4): 315–323.

Part 4

Improving root function

Chapter 14

Understanding and exploiting the genetics of plant root traits

Roberto Tuberosa, Elisabetta Frascaroli, Marco Maccaferri and Silvio Salvi, University of Bologna, Italy

1 Introduction

Major increases in crop productivity are urgently needed to meet the growing demand in food, feed and biofuel required by a burgeoning human population while minimizing the environmental footprint of agriculture and ensuring its long-term sustainability. Achieving this formidable goal will require a paradigm shift in breeding practices based on genomics-assisted approaches (Huang et al., 2015; Jung and McCouch, 2013; Kole et al., 2015; Langridge and Reynolds, 2015; Stadlmeier et al., 2018; Varshney et al., 2018a), faster selection cycles (Hickey et al., 2019; Watson et al., 2018) and increased attention to yield-related proxies with a simpler genetic architecture and more heritable than yield itself (Kuijken et al., 2015; Li et al., 2019b; Reynolds and Tuberosa, 2008). Among yield-related proxies, root traits have long been recognized for their pivotal role in plant growth and crop productivity through their mechanical and functional roles, particularly those related to water and nutrient uptake (Blum, 1988; Hasanuzzaman et al., 2017; King et al., 2003; Liu et al., 2020; Lynch, 1995, 2007, 2013; Reynolds and Tuberosa, 2008; Richards et al., 2010; Xu and Takahashi, 2020). Collectively, root traits determine the overall architecture and physiology of the root system, hence defining its spatial configuration in the soil and describing its morphology and shape (Gregory et al., 2009).

http://dx.doi.org/10.19103/AS.2020.0075.21

The variability observed in root traits of field-grown plants is determined by a plethora of quantitative trait loci (QTLs) and their interaction with the environment and management practices (G × E × M) adopted by farmers. This, notwithstanding, until recently our current understanding of the genes and molecular mechanisms that shape root architecture and control its functions were mainly contributed by studies conducted on mutants. Comprehensive reviews on the genetic control of RSA morphology in cereal crops are available for rice (Meng et al., 2019), barley (Rossini et al., 2018), sorghum (Zheng et al., 2020) and maize (Hochholdinger et al., 2018).

Ample genotypic variability for root traits and their plasticity under different environmental conditions have been documented in most crops (Ayele et al., 2001; Clarke et al., 2017; Fromm, 2019; Gregory et al., 2009; Ryan et al., 2016; Salvi, 2017; Sandhu et al., 2016; Schoppach et al., 2017; Tuberosa et al., 2003; Wu et al., 2018b; Yan et al., 2004; Yu et al., 2019). Leveraging the genetic variability of root traits, also defined as root phenes (Lynch, 2011), to maximize crop productivity while enhancing water- and nutrient-use efficiency requires a deep understanding of the genetic make-up of root system architecture (RSA) traits and their functions (Del Bianco and Kepinski, 2018; Gregory et al., 2009; Oyiga et al., 2020; Shahan and Benfey, 2020; Varshney et al., 2013; Wachsman and Benfey, 2020; Ye et al., 2018a). Based on such knowledge, genomics-assisted crop improvement provides breeders novel opportunities to accelerate the rate of genetic gain in crop yield while advancing our understanding of the genetic and functional basis of RSA traits underlying the adaptive response of crops to abiotic and biotic stresses (Roorkiwal et al., 2020; Varshney et al., 2018b; Voss-Fels et al., 2018b).

This chapter illustrates how genomics and other -omics approaches coupled with new-generation sequencing (NGS) platforms have been deployed to dissect the genetic make-up of RSA traits and better understand their functions, particularly under environmentally constrained conditions that commonly occur in most farmed soils. The major emphasis is devoted to studies during the past two decades in crops, and only occasional reference is provided to the vast literature from RSA studies conducted in *Arabidopsis* and other model plants. The chapter also provides examples of how, in some cases, this knowledge is already benefiting farmers and how it can help in reducing the environmental impact of agriculture worldwide.

2 Mapping loci for root traits

The genetic dissection of RSA traits is an essential prerequisite to (i) implement marker-assisted selection (MAS) in order to tailor root systems matching an ideotype perceived as optimal for the release of improved cultivars (Ahmadi et al., 2014; Li et al., 2019a; Lynch, 2018; Voss-Fels et al., 2018b; Wissuwa et al.,

2016) and, once a sufficiently high genetic resolution is achieved, (ii) attempt the cloning of the relevant loci (Arai-Sanoh et al., 2014; Kitomi et al., 2018; Salvi et al., 2016; Salvi and Tuberosa, 2007; Uga et al., 2013; Voss-Fels et al., 2018a). Since the late 1980s, molecular markers have streamlined the assembly of genetic maps for mapping mutants and QTLs influencing RSA traits and eventually cloning the causative sequences underlying their effects (Uga et al., 2013; Voss-Fels et al., 2018a; Xu et al., 2018). Collectively, these studies have contributed a wealth of information as reported in dozens of scientific reports (see Tables 1–7) and hereafter. This notwithstanding, so far only a handful of studies have completed comprehensive meta-analysis for RSA traits in order to compare and cross-link the results reported in multiple genetic backgrounds (Courtois et al., 2009; Guo et al., 2018; Jia et al., 2019b; Soriano and Alvaro, 2019; Zhang et al., 2016) and/or between different species (Tuberosa et al., 2003; Zheng et al., 2020), an important goal to identify the most appropriate loci to be prioritized for further studies and, most importantly, for breeding purposes (Collins et al., 2008; Maccaferri et al., 2016; Ramstein et al., 2019; Schaefer et al., 2018).

This section will first address genetic studies based on Mendelian loci, mainly mutants, a powerful approach for unveiling the genetic landscape of RSA or any other trait (Hirochika, 2010; Jiang and Ramachandran, 2010; Penna and Jain, 2017; Tsuchida-Mayama et al., 2010). The second part of this section will deal with the search and characterization of the QTLs that through their interaction with the environment and management practices (G × E × M) control the variability that breeders routinely deal with in their nurseries (Bray and Topp, 2018; Nadeau and Frankel, 2000). An important outcome of the studies conducted during the past two decades some of which have been reported hereafter clearly shows that there are no clear boundaries between Mendelian and quantitative genetics, in as much as a number of studies have shown the co-location between mutants and QTLs that underline the same trait including also RSA loci (Bray and Topp, 2018; Salvi et al., 2016), hence supporting Robertson's hypothesis (Robertson, 1985).

2.1 Mutants

The high selection pressure applied in crop breeding since their domestication thousands of years ago has drastically narrowed their genetic variability (Avni et al., 2017; Maccaferri et al., 2019; Steiner et al., 2019; Tanksley and McCouch, 1997; Vaughan et al., 2007), a major shortcoming that can be partially overcome with the creation of new alleles through mutagenesis (Herbert et al., 2020; Till et al., 2018; Uauy et al., 2017; Viana et al., 2019). A technique increasingly adopted in the past decade to identify mutants in a more targeted and systematic fashion is TILLING (targeting induced local lesions in genomes),

Table 1 References on studies addressing the genetic dissection and/or characterization of the root system in *Arabidopsis*

Title	Authors
MicroRNA directs mRNA cleavage of the transcription factor NAC1 to downregulate auxin signals for *Arabidopsis* lateral root development	Guo et al. (2005)
Whole-genome analysis of the SHORT-ROOT developmental pathway in *Arabidopsis*	Levesque et al. (2006)
PHENOPSIS, an automated platform for reproducible phenotyping of plant responses to soil water deficit in *Arabidopsis thaliana*, permitted the identification of accession with low sensitivity to soil water deficit	Granier et al. (2006)
Identification of QTL controlling root growth response to phosphate starvation in *Arabidopsis thaliana*	Reymond et al. (2006)
Periodic lateral root priming: what makes it tick?	Laskowski and ten Tusscher (2012)
Root patterning: it takes two to tangle	Scheres and Laskowski (2016)
Marker-aided incorporation of *Saltol*, a major QTL-associated with seedling stage salt tolerance, into *Oryza sativa* 'Pusa Basmati 1121'	Babu et al. (2016)
Robust root growth in altered hydrotropic response 1 (*ahr1*) mutant of *Arabidopsis* is maintained by a high rate of cell production at low water potential gradient	Salazar-Blas et al. (2017)
Transcription factors PRE3 and WOX11 are involved in the formation of new lateral roots from secondary growth taproot in *A. thaliana*	Baesso et al. (2018)
Root branching toward water involves post-translational modification of transcription factor ARF7	Orosa-Puente et al. (2018)
Lateral root formation in *Arabidopsis*: A well-ordered LRexit	Banda et al. (2019)
Root system depth in *Arabidopsis* is shaped by *EXOCYST70A3* via the dynamic modulation of auxin transport	Ogura et al. (2019)
PIN-FORMED and PIN-LIKES auxin transport facilitators	Suaer et al. (2019)
The microRNA397b-LACCASE2 module regulates root lignification under water and phosphate deficiency (1)	Khandal et al. (2020)
Lateral root development differs between main and secondary roots and depends on the ecotype	Perotti et al. (2020)
Overexpression of the*NMig1*gene encoding a NudC domain protein enhances root growth and abiotic stress tolerance in *Arabidopsis thaliana*	Velinov et al. (2020)
Early developmental plasticity of lateral roots in response to asymmetric water availability	von Wangenheim et al. (2020)
Lateral root initiation: the emergence of new primordia following cell death	Wachsman and Benfey (2020)
AtDRO1 is nuclear localized in root tips under native conditions and impacts auxin localization	Waite et al. (2020)
RGF1 controls root meristem size through ROS signaling	Yamada et al. (2020)
Novel regulators of sugar-mediated lateral root development in *Arabidopsis thaliana*	Li et al. (2000)

Table 2 References on studies addressing the genetic dissection and/or characterization of the root system in rice

Title	Authors
Locating genes associated with root morphology and drought avoidance in rice via linkage to molecular markers	Champoux et al. (1995)
Mapping genes controlling root morphology and root distribution in a doubled-haploid population of rice	Yadav et al. (1997)
Mapping QTLs associated with drought avoidance in upland rice	Courtois et al. (2000)
Deep rooting conferred by *DEEPER ROOTING 1* enhances rice yield in paddy fields	Arai-Sanoh et al. (2014)
Physiological mechanisms contributing to the QTL-combination effects on improved performance of IR64 rice NILs under drought	Henry et al. (2015)
QTLs underlying natural variation of root growth angle among rice cultivars with the same functional allele of *DEEPER ROOTING 1*	Kitomi et al. (2015)
qRT9, a quantitative trait locus controlling root thickness and root length in upland rice	Li et al. (2015a)
Transcriptome profiling of drought-responsive noncoding RNAs and their target genes in rice	Chung et al. (2016)
Identification of candidate network hubs involved in metabolic adjustments of rice under drought stress by integrating transcriptome data and genome-scale metabolic network	Mohanty et al. (2016)
Discovery of QTLs for water mining and water-use efficiency traits in rice under water-limited condition through association mapping	Raju et al. (2016)
Mapping quantitative trait loci associated with toot traits using sequencing-based genotyping chromosome segment substitution lines derived from 9311 and Nipponbare in rice (*Oryza sativa* L.)	Zhou et al. (2016)
Marker-aided incorporation of *Saltol*, a major QTL-associated with seedling stage salt tolerance, into *Oryza sativa* 'Pusa Basmati 1121'	Babu et al. (2017a)
Genetic loci governing grain yield and root development under variable rice cultivation conditions	Catolos et al. (2017)
Near-isogenic lines of IR64 (*Oryza sativa* subsp *indica* cv.) introgressed with *DEEPER ROOTING 1* and *STELE TRANSVERSAL AREA 1* improve rice yield formation over the background parent across three water management regimes	Deshmukh et al. (2017)
RSL class I genes positively regulate root hair development in *Oryza sativa*	Kim et al. (2017)
Genetic control of the root system in rice under normal and drought stress conditions by genome-wide association study	Li et al. (2017)
Genetic factors underlying boron toxicity tolerance in rice: genome-wide association study and transcriptomic analysis	Neto et al. (2017)
Association between root growth angle and root length density of a near-isogenic line of IR64 rice with *DEEPER ROOTING 1* under different levels of soil compaction	Ramalingam et al. (2017)
Genetic analysis of yield and agronomic traits under reproductive-stage drought stress in rice using a high-resolution linkage map	Bhattarai and Subudhi (2018)

(*Continued*)

Table 2 (Continued)

Title	Authors
Genome-wide association studies of image traits reveal genetic architecture of drought resistance in rice	Guo et al. (2018b)
Fine mapping of *QUICK ROOTING 1* and *2*, quantitative trait loci increasing root length in rice	Kitomi et al. (2018)
Root plasticity under fluctuating soil moisture stress exhibited by backcross inbred line of a rice variety, Nipponbare carrying introgressed segments from KDML105 and detection of the associated QTLs	Owusu-Nketia et al. (2018a)
OsARD4 encoding an acireductone dioxygenase improves root architecture in rice by promoting development of secondary roots	Ramalingam et al. (2018)
Transcriptome analysis of two rice varieties contrasting for nitrogen-use efficiency under chronic N starvation reveals differences in chloroplast and starch metabolism-related genes	Subodh et al. (2018)
OsNOX3, encoding a NADPH oxidase, regulates root hair initiation and elongation in rice	Wang et al. (2018)
Detection of QTL for exudation rate at ripening stage in rice and its contribution to hydraulic conductance	Yamamoto et al. (2018)
Regulation of root traits for internal aeration and tolerance to soil waterlogging-flooding stress	Yamauchi et al. (2018)
Tuning a timing device that regulates lateral root development in rice	Acevedo et al. (2020)
Introgression of root and water-use efficiency traits enhances water productivity: an evidence for physiological breeding in rice (*Oryza sativa* L.)	Dharmappa et al. (2019)
RSL Class II transcription factors guide the nuclear localization of RHL1 to regulate root hair development	Moon et al. (2019b)
The interaction of genes controlling root traits is required for the developmental acquisition of deep and thick root traits and improving root architecture in response to low water or nitrogen content in rice (*Oryza sativa* L.)	Nada et al. (2019)
Genetic mapping identifies consistent quantitative trait loci for yield traits of rice under greenhouse drought conditions	Baisakh et al. (2020)

a method that allows directed identification of mutant alleles in a specific gene. TILLING was introduced in 2000 using the model plant *Arabidopsis* (McCallum et al., 2000) and has since been used as a reverse-genetics method in crops and also as a breeding technique for the release of novel cultivars (Kashtwari et al., 2019; Till et al., 2003, 2007, 2018).

Studies conducted with root mutants have been instrumental in elucidating the genetic, physiological and biochemical basis of RSA phenes and functions particularly in *Arabidopsis* and other model species. Table 1 lists some of the studies in *Arabidopsis* that have contributed to shed light on

Table 3 References on studies addressing the genetic dissection and/or characterization of the root system in barley

Title	Authors
Mapping of chromosome regions conferring boron toxicity tolerance in barley (*Hordeum vulgare* L.)	Jefferies et al. (1999)
Phenotype/genotype associations for yield and salt tolerance in a mapping population segregating for two dwarfing genes	Ellis et al. (2002)
Boron-toxicity tolerance in barley arising from efflux transporter amplification	Sutton et al. (2007)
Boron toxicity tolerance in wheat and barley: Australian perspectives	Schnurbuschn et al. (2010)
Assessment of the agronomic value of QTL on chromosomes 2H and 4H linked to tolerance to boron toxicity in barley (*Hordeum vulgare* L.)	McDonald et al. (2010)
QTL mapping for salt tolerance in barley at seedling growth stage	Ahmadi-Ochtapeh et al. (2015)
Genomic regions influencing seminal root traits in barley	Robinson et al. (2016)
Detection of QTLs for seedling characteristics in barley (*Hordeum vulgare* L.) grown under hydroponic culture condition	Wang et al. (2017)
A major QTL on chromosome 7HS controls the response of barley seedling to salt stress in the Nure × Tremois population	Xue et al. (2017)
Root architectural traits and yield: exploring the relationship in barley breeding trials	Robinson et al. (2018)
VERNALIZATION1 modulates root system architecture in wheat and barley	Voss-Fels et al. (2018a)
Genetic dissection of root system architectural traits in spring barley	Jia et al. (2019b)
Genetic components of root architecture and anatomy adjustments to water-deficit stress in spring barley	Oyiga et al. (2020)

the genes shaping RSA in land plants. The residual synteny found between *Arabidopsis* and dicot crops is valuable for comparative studies aimed at leveraging the wealth of information available from the most studied model plant (Chen et al., 2018; Guo et al., 2018; Moon et al., 2018). Regrettably, the extensive level of rearrangements which occurred following the separation between dicots and monocots makes *Arabidopsis* unsuitable for meaningful synteny-based application to cereal crops (van Buuren et al., 2002). Although *Arabidopsis* differs from cereals in both overall architecture and the anatomy of individual roots, genes cloned in *Arabidopsis* could in some cases provide interesting leads for the identification of candidate genes for root-related QTLs in cereals, particularly for those functional and morphological features

Table 4 References on studies addressing the genetic dissection and/or characterization of the root system in durum wheat

Title	Authors
Association mapping for root architectural traits in durum wheat seedlings as related to agronomic performance	Canè et al. (2014)
A consensus framework map of durum wheat (*Triticum durum* Desf.) suitable for linkage disequilibrium analysis and genome-wide association mapping	Maccaferri et al. (2014)
Genetic analysis of root morphological traits in wheat	Petrarulo et al. (2015)
Prioritizing quantitative trait loci for root system architecture in tetraploid wheat	Maccaferri et al. (2016)
Genetic variation for aerenchyma and other root anatomical traits in durum wheat (*Triticum durum* Desf.).	Nazemi et al. (2016)
Identification and characterization of durum wheat microRNAs in leaf and root tissues	Fileccia et al. (2017)
Mapping QTL for root and shoot morphological traits in a durum wheat × *T. dicoccum* segregating population at seedling stage	Iannucci et al. (2017)
Ancestral QTL alleles from wild emmer wheat enhance root development under drought in modern wheat	Merchuk-Ovnat et al. (2017)
Genetic markers associated to arbuscular mycorrhizal colonization in durum wheat.	De Vita et al. (2018)
Root system architecture and its association with yield under different water regimes in durum wheat	El Hassouni et al. (2018)
Yield of chromosomally engineered durum wheat-*Thinopyrum ponticum* recombinant lines in a range of contrasting rain-fed environments.	Kuzmanovic et al. (2018)
A major root architecture QTL responding to water limitation in durum wheat	Alahmad et al. (2019)
Genetic dissection of the seminal root system architecture in Mediterranean durum wheat landraces by genome-wide association study	Rosello et al. (2019)
Discovering consensus genomic regions in wheat for root-related traits by QTL meta-analysis	Soriano et al. (2019)

of root development that may have been conserved to a greater extent from an evolutionary standpoint. Therefore, root mutants in *Arabidopsis* and the related database are a treasure trove for identifying candidate genes at the peak region of loci influencing RSA traits in cereals and evaluating their effects (Kulkarni et al., 2017).

In cereals, the move from structural to functional genomics is now possible due to the availability of large collections of mutants used to implement TILLING (Caldwell et al., 2004; Szurman-Zubrzycka et al., 2018; Talamè et al., 2008; Till et al., 2007, 2018), sequence database (King et al., 2015; Schreiber

Table 5 References on studies addressing the genetic dissection and/or characterization of the root system in bread wheat

Title	Authors
Mapping QTLs for nitrogen uptake in relation to the early growth of wheat	An et al. (2006)
Fine mapping and targeted SNP survey using rice-wheat gene collinearity in the region of the *Bo1* boron toxicity tolerance locus of bread wheat	Schnurbusch et al. (2007)
Detection of two major grain yield QTL in bread wheat (*Triticum aestivum* L.) under heat, drought and high yield potential environments	Bennett et al. (2012)
Tolerance of wheat to vegetative stage soil waterlogging is conditioned by both constitutive and adaptive QTL	Ballesteros et al. (2015)
Meta-analysis of wheat QTL regions associated with adaptation to drought and heat stress	Acuna-Galindo et al. (2015)
Phenotyping pipeline reveals major seedling root growth QTL in hexaploid wheat	Atkinson et al. (2015)
Marker-assisted breeding for *TaALMT1*, a major gene-conferring aluminum tolerance to wheat	Soto-Cerda et al. (2015)
High-resolution mapping of traits related to whole-plant transpiration under increasing evaporative demand in wheat	Schoppach et al. (2016)
Mapping of quantitative trait loci for root hair length in wheat identifies loci that co-locate with loci for yield components	Horn et al. (2016)
Identification and validation of root length QTLs for water stress resistance in hexaploid wheat (*Triticum aestivum* L.)	Ayalew et al. (2017)
Drought response in wheat: key genes and regulatory mechanisms controlling root system architecture and transpiration efficiency	Kulkarni et al. (2017)
Analysis of aneuploid lines of bread wheat to map chromosomal locations of genes controlling root hair length	Liu et al. (2017)
Genetics of mycorrhizal symbiosis in winter wheat (*Triticum aestivum*)	Lehnert et al. (2017)
Quantifying rooting at depth in a wheat doubled haploid population with introgression from wild emmer	Clarke et al. (2017)
Genome-wide association studies reveal genomic regions associated with the response of wheat (*Triticum aestivum* L.) to mycorrhizae under drought stress conditions	Lehnert et al. (2018)
Genomic regions responsible for seminal and crown root lengths identified by 2D and 3D root system image analysis	Uga et al. (2018)
QTL mapping for seedling morphology under drought stress in wheat cross synthetic (W7984)/Opata	Khalid et al. (2018)
Three genomes differentially contribute to the seedling lateral root number in allohexaploid wheat: evidence from phenotype evolution and gene expression	Wang et al. (2018)
Mapping of novel salt tolerance QTL in an Excalibur × Kukri doubled haploid wheat population	Asif et al. (2018)
Selection in early generations to shift allele frequency for seminal root angle in wheat	Richard et al. (2018)
Understanding nitrate uptake, signaling and remobilization for improving plant nitrogen use efficiency	Kant et al. (2018)

(*Continued*)

Table 5 (Continued)

Title	Authors
Comparative analysis of root transcriptome profiles between drought-tolerant and susceptible wheat genotypes in response to water stress	Ling et al. (2018)
Association study reveals novel genomic regions for grain yield and yield-related traits in drought-stressed synthetic hexaploid wheat	Bhatta et al. (2018)
QTL for stay-green traits in wheat in well-watered and water-limited environments	Christopher et al. (2018)
Selection in early generations to shift allele frequency for seminal root angle in wheat	Richard et al. (2018)
Do longer root hairs improve phosphorus uptake? Testing the hypothesis with transgenic *Brachypodium distachyon* lines overexpressing endogenous RSL genes	Zhang et al. (2018)
Detection of *T. urartu* introgressions in wheat and development of a panel of interspecific introgression lines	Grewal et al. (2018)
Cytokinin dehydrogenase: a genetic target for yield improvement in wheat	Chen et al. (2019a)
Evolution of *Deeper Rooting 1*-like homoeologs in wheat entails the C-terminus mutations as well as gain and loss of auxin response elements	Ashraf et al. (2019)
A wheat/rye polymorphism affects seminal root length and yield across different irrigation regimes	Howell et al. (2019)
A novel QTL *QTrl.saw-2D.2* associated with the total root length identified by linkage and association analyses in wheat (*Triticum aestivum* L.)	Zheng et al. (2019)
Comparative metabolomic profiling in the roots and leaves in contrasting genotypes reveals complex mechanisms involved in post-anthesis drought tolerance in wheat	Kang et al. (2019)
Loci and candidate genes controlling root traits in wheat seedlings - a wheat root GWAS	Beyer et al. (2019)
Transcriptome analysis of osmotic-responsive genes in ABA-dependent and ABA-independent pathways in wheat (*Triticum aestivum* L.) roots	Li et al. (2019c)
Transcriptome analysis in roots and leaves of wheat seedlings in response to low-phosphorus stress	Wang et al. (2019b)
Molecular signal communication during arbuscular mycorrhizal formation induces significant transcriptional reprogramming of wheat (*Triticum aestivum*) roots	Tian et al. (2019)
Early root transcriptomic changes in wheat seedlings colonized by *Trichoderma harzianum* under different inorganic nitrogen supplies	Rubio et al. (2019)
Quantitative trait loci for seeding root traits and the relationships between root and agronomic traits in common wheat	Li et al. (2020b)
The *LATERAL ROOT DENSITY* gene regulates root growth during water stress in wheat. (Article; Early Access)	Placido et al. (2020)
Mapping QTL for agronomic and root traits in the Kukri/RAC875 wheat (*Triticum aestivum* L.) population under drought stress conditions	Salarpour et al. (2020)
Exploring the genetic architecture of root-related traits in Mediterranean bread wheat landraces by genome-wide association analysis	Rufo et al. (2020)

Table 6 References on studies addressing the genetic dissection and/or characterization of the root system in maize and sorghum

Title	Authors
Genetic variation for nitrogen-use efficiency in a set of recombinant inbred lines II - QTL detection and coincidences	Bertin and Gallais (2001)
Lateral roots affect the proteome of the primary root of maize (*Zea mays* L.)	Hochholdinger et al. (2004a)
Detection of quantitative trait loci for seminal root traits in maize (*Zea mays* L.) seedlings grown under differential phosphorus levels	Zhu et al. (2006a)
Localization and quantification of plasma membrane aquaporin expression in maize primary root: A clue to understanding their role as cellular plumbers	Hachez et al. (2006)
Transcriptomic and proteomic analyses of pericycle cells of the maize primary root	Dembinsky et al. (2007)
Microarray analysis reveals early responsive genes possibly involved in localized nitrate stimulation of lateral root development in maize (*Zea mays* L.)	Liu et al. (2008)
Genetic variation for N-remobilization and post-silking N-uptake in a set of maize recombinant inbred lines. 3. QTL detection and coincidences	Coque et al. (2008)
Differential expression of miRNAs in response to salt stress in maize roots	Ding et al. (2009)
Two functionally distinct members of the MATE (multi-drug and toxic compound extrusion) family of transporters potentially underlie two major aluminum tolerance QTLs in maize	Maron et al. (2010)
QTL mapping and phenotypic variation for root architectural traits in maize (*Zea mays* L.)	Burton et al. (2014)
Multiple interval QTL mapping and searching for *PSTOL1* homologs associated with root morphology, biomass accumulation and phosphorus content in maize seedlings under low-P	Azevedo et al. (2015)
Association analysis of genes involved in (*Zea mays* L.) root development of maize with seedling and agronomic traits under contrasting nitrogen levels	Abdel-Ghani et al. (2015)
A genetic relationship between nitrogen-use efficiency and seedling root traits in maize as revealed by QTL analysis.	Li et al. (2015b)
Quantitative trait locus mapping reveals regions of the maize genome controlling root system architecture	Zurek et al. (2015)
Genetic dissection of maize seedling root system architecture traits using an ultra-high density bin-map and a recombinant inbred line population	Song et al. (2016)
Two major quantitative trait loci controlling the number of seminal roots in maize co-map with the root developmental genes *rtcs* and *rum1*	Salvi et al. (2016)
Synergy between root hydrotropic response and root biomass in maize (*Zea mays* L.) enhances drought avoidance	Eapen et al. (2017)
Reduced root cortical burden improves growth and grain yield under low phosphorus availability in maize	Galindo-Castaneda et al. (2018)
Predicting water and nitrogen requirements for maize under semi-arid conditions using the CSM-CERES-Maize model	Hammad et al. (2018)
Meta-QTL analysis and identification of candidate genes related to root traits in maize	Guo et al. (2018a)
QTL-by-environment interaction in the response of maize root and shoot traits to different water regimes	Li et al. (2018c)

(*Continued*)

Table 6 (Continued)

Title	Authors
Integrating GWAS and gene expression analysis identifies candidate genes for root morphology traits in maize at the seedling stage	Wang et al. (2019a)
Natural variation of *ZmHKT1* affects root morphology in maize at the seedling stage	Li et al. (2019d)
Maize lateral root developmental plasticity induced by mild water stress. I: Genotypic variation across a high-resolution series of water potentials	Dowd et al. (2019)
Carbon isotope composition, water-use efficiency, and drought sensitivity are controlled by a common genomic segment in maize	Avramova et al. (2019)
Cell type-specific transcriptomics of lateral root formation and plasticity	Kortz et al. (2019)
Plasticity of lateral root branching in maize	Yu et al. (2019)
Transcriptome analysis of maize seedling roots in response to nitrogen-, phosphorus-, and potassium-deficiency	Ma et al. (2020)
Nucleotide diversity and association analysis of *ZmMADS60* with root length in the maize seedling stage	Li et al. (2020a)
Genome-scale mining of root-preferential genes from maize and characterization of their promoter activity	Li et al. (2020b)
A comprehensive transcriptomics analysis reveals long non-coding RNA to be involved in the key metabolic pathway in response to waterlogging stress in maize	Yu et al. (2020)
Multi-environment QTL mapping of crown root traits in a maize RIL population	Li et al. (2020a)
Sorghum	
QTL for nodal root angle in sorghum (*Sorghum bicolor* L. Moench) co-locate with QTL for traits associated with drought adaptation	Mace et al. (2012)
Studying the genetic basis of drought tolerance in sorghum by managed stress trials and adjustments for phenological and plant height differences	Sabadin et al. (2012)
Drought adaptation of stay-green sorghum is associated with canopy development, leaf anatomy, root growth and water uptake	Borrell et al. (2014a)
Duplicate and conquer: Multiple homologs of *PHOSPHORUS1 STARVATION TOLERANCE1* enhances phosphorus acquisition and sorghum performance on low-phosphorus soils.	Hufnagel et al. (2014)
Validation of QTL mapping and transcriptome profiling for identification of candidate genes associated with nitrogen stress tolerance in sorghum	Gelli et al. (2017)
QTLs associated with crown root angle, stomatal conductance and maturity in sorghum	Lopez et al. (2017)
The genetic architecture of phosphorus efficiency in sorghum involves pleiotropic QTL for root morphology and grain yield under low phosphorus availability in the soil	Bernardino et al. (2019)
Repeat variants for the *SbMATE* transporter protect sorghum roots from aluminum toxicity by transcriptional interplay in cis and trans	Melo et al. (2019)
Root-knot nematode resistance is identified from a South African sweet sorghum line	Harris-Shultz et al. (2019)
Transcriptome analysis reveals candidate genes related to phosphorus starvation tolerance in sorghum	Zhang et al. (2019)
Shared genetic control of root system architecture between *Zea mays* and *Sorghum bicolor*	Zheng et al. (2020)

Table 7 References on studies addressing the genetic dissection and/or characterization of the root system in pulses and other dicot crops

Title	Authors
Pulses	
QTL mapping of root hair and acid exudation traits and their relationship to phosphorus uptake in common bean	Yan et al. (2004)
Quantitative trait loci for root architecture traits correlated with phosphorus acquisition in common bean	Beebe et al. (2006)
QTL analysis of adventitious root formation in common bean under contrasting phosphorus availability	Ochoa et al. (2006)
Quantitative trait loci for rooting pattern traits of common beans grown under drought stress versus non-stress conditions	Asfaw and Blair (2012)
Identification QTLs controlling genes for Se uptake in lentil seeds	Ates et al. (2016)
Identification of QTL controlling root and shoot traits associated with drought tolerance in a lentil (*Lens culinaris* Medik.) recombinant inbred line population	Idrissi et al. (2016)
Advancements in breeding, genetics and genomics for resistance to three nematode species in soybean	Kim et al. (2016)
Genome-wide association mapping and agronomic impact of cowpea root architecture	Burridge et al. (2017)
Root traits, nodulation and root distribution in soil for five wild lentil species and *Lens culinaris* (Medik.) grown under well-watered conditions	Gorim et al. (2017)
Genome-wide analysis of expansin superfamily in wild *Arachis* discloses a stress-responsive expansin-like B gene	Guimaraes et al. (2017)
Genetic diversity and genomic strategies for improving drought and waterlogging tolerance in soybeans	Valliyodan et al. (2017)
Comparative genome-wide-association mapping identifies common loci controlling root system architecture and resistance to *Aphanomyces euteiches* in pea	Desgroux et al. (2018)
Genetic diversity of root system architecture in response to drought stress in grain legumes	Ye et al. (2018a)
A major natural genetic variation associated with root system architecture and plasticity improves waterlogging tolerance and yield in soybean	Ye et al. (2018b)
Transcriptome analysis of waterlogging-sensitive and tolerant Zombi pea (*Vigna vexillata*) reveals energy conservation and root plasticity controlling waterlogging tolerance	Butsayawarapat et al. (2019)
Comparative genome-wide-association mapping identifies common loci controlling root system architecture and resistance to *Aphanomyces euteiches* in pea	Lo et al. (2019)
Understanding genetic control of root system architecture in soybean: Insights into the genetic basis of lateral root number	Prince et al. (2019)
Genetic diversity linked to haplotype variation in the world core collection of *Trifolium subterraneum* for boron toxicity tolerance provides valuable markers for pasture breeding	Tahghighi et al. (2019)

(*Continued*)

Table 7 (Continued)

Title	Authors
Root and shoot variation in relation to potential intermittent drought adaptation of Mesoamerican wild common bean (*Phaseolus vulgaris* L.)	Teran et al. (2019)
Identification and mapping of two independent recessive loci for the root hairless mutant phenotype in soybean	Yang et al. (2019a)
Genetic analysis and fine mapping of *phosphorus efficiency locus 1* (*PE1*) in soybean	Yang et al. (2019b)
LAZY3 plays a pivotal role in positive root gravitropism in *Lotus japonicus*	Chen et al. (2020)
Dissecting the genetic architecture of Aphanomyces root rot resistance in lentil by QTL mapping and genome-wide association study	Ma et al. (2020)
A co-opted regulator of lateral root development controls nodule organogenesis in *Lotus*	Shahan and Benfey (2020)
Other dicot crops	
Advanced backcross QTL analysis in tomato. I. Identification of QTLs for traits of agronomic importance from *Lycopersicon hirsutum*	Bernacchi et al. (1998)
Implications of the genetics of root structure in melon breeding	Fita et al. (2006)
Genetics of root system architecture using near-isogenic lines of melon	Fita et al. (2008)
QTL meta-analysis of root traits in *Brassica napus* under contrasting phosphorus supply in two growth systems	Zhang et al. (2016)
Boron-demanding tissues of *Brassica napus* express specific sets of functional Nodulin26-like intrinsic proteins and BOR1 transporters	Diehn et al. (2019)
Species-wide variation in shoot nitrate concentration, and genetic loci controlling nitrate, phosphorus and potassium accumulation in *Brassica napus* L.	Alcock et al. (2018)
A comparison of lateral root patterning among dicot and monocot plants	Chen et al. (2018)
Multiple genomic regions influence root morphology and seedling growth in cultivated sunflower (*Helianthus annuus* L.) under well-watered and water-limited conditions	Masalia et al. (2018)
The major-effect quantitative trait locus *CsARN6.1* encodes an AAA ATPase domain-containing protein that is associated with waterlogging stress tolerance by promoting adventitious root formation	Xu et al. (2018)
Transcriptomic responses to water deficit and nematode infection in mycorrhizal tomato roots	Balestrini et al. (2019)
Genetic analysis of tomato root colonization by arbuscular mycorrhizal fungi	Plouznikoff et al. (2019)
Downregulation and loss of auxin response factor 4 function using CRISPR/Cas9 alters plant growth, stomatal function and improves tomato tolerance to salinity and osmotic stress	Bouzroud et al. (2020)
Functional genomics by integrated analysis of transcriptome of sweet potato (*Ipomoea batatas* (L.) Lam.) during root formation	Kim et al. (2020)
BrEXLB1, a *Brassica rapa* expansin-like B1 gene is associated with root development, drought stress response and seed germination	Muthusami et al. (2020)

et al., 2019; Taheri et al., 2017) and more recently, targeted mutagenesis via multiple genome editing able to target many loci simultaneously (Najera et al., 2019; Zhu et al., 2017). The results of the studies listed in Table 1, and those summarized hereafter highlight the valuable role played by mutants to dissect the genetic basis of RSA traits and to elucidate the function of candidate genes.

Among RSA traits, root hairs have been frequently targeted with mutagenesis studies aimed at identifying the genes involved in root hairs ontogeny. Root hairs are important for the absorption of nutrients and water from the rhizosphere (Cai et al., 2020; Carminati et al., 2017; Ruiz et al., 2020). The density and/or length of root hairs are thought to have a major role in the plant's performance under stressful conditions (Zhang et al., 2018). Accordingly, improving a crop's root system will require a detailed understanding of the processes of root hairs differentiation. Marzec et al. (2015) reviewed the progress and shortcomings in uncovering the molecular basis of root epidermis specialization in the grasses with particular emphasis on the role of mutants in the quest for the genes underlying root hair ontogeny. Root hairs develop as tubular outgrowths from a subset of epidermal cells called trichoblasts, which are present in several billion per plant and greatly expand the root surface area in contact with soil (Dolan, 2017; Marzec et al., 2015). Root hair ontogeny entails three distinct phases: the establishment of root epidermal patterning, the formation of the bulge on the root surface, and finally, root hair growth. Approximately 50% of the epidermal cells develop root hairs in Poaceae, so that root hair cells and hairless epidermal cells (atrichoblasts) alternate along each epidermal cell file of the mature root (Marzec et al., 2015).

In barley, a number of mutant collections have been phenotyped for root phenes (Bovina et al., 2011; Gregory et al., 2009; Szurman-Zubrzycka et al., 2018; Talamè et al., 2008). A forward-genetics analysis was performed on a portion of TILLMore, a chemically mutagenized population of barley cv. 'Morex', to identify root morphology alterations by comparison with 'Morex' wild-type (Bovina et al., 2011). For this purpose, a simple paper-roll approach was used to identify phenotypic variants at the seedling stage. The analysis of approximately 1 000 M_4 families allowed for the identification of 66 M_4 lines with a range of altered root morphology. Figure 1 shows four of the most common mutants. Pictures of the root system of the other M_4 lines with altered root morphology are available at: https://site.unibo.it/plantgenomics/en/collections/tillmore.

Janiak and Szarejko (2007) mapped four barley root hair mutants identified after chemical mutagenesis of spring cultivar 'Karat': the root hairless mutant *rhl1.b*, mutant *rhp1.b* with root hair development blocked at the initial bulge formation, mutant *rhi1*.a with an irregular pattern of sparsely located root hairs and mutant *rhs1*.a with very short root hairs. Fine mapping narrowed the interval encompassing *rhl.b* to a 577-kb region on the barley physical map

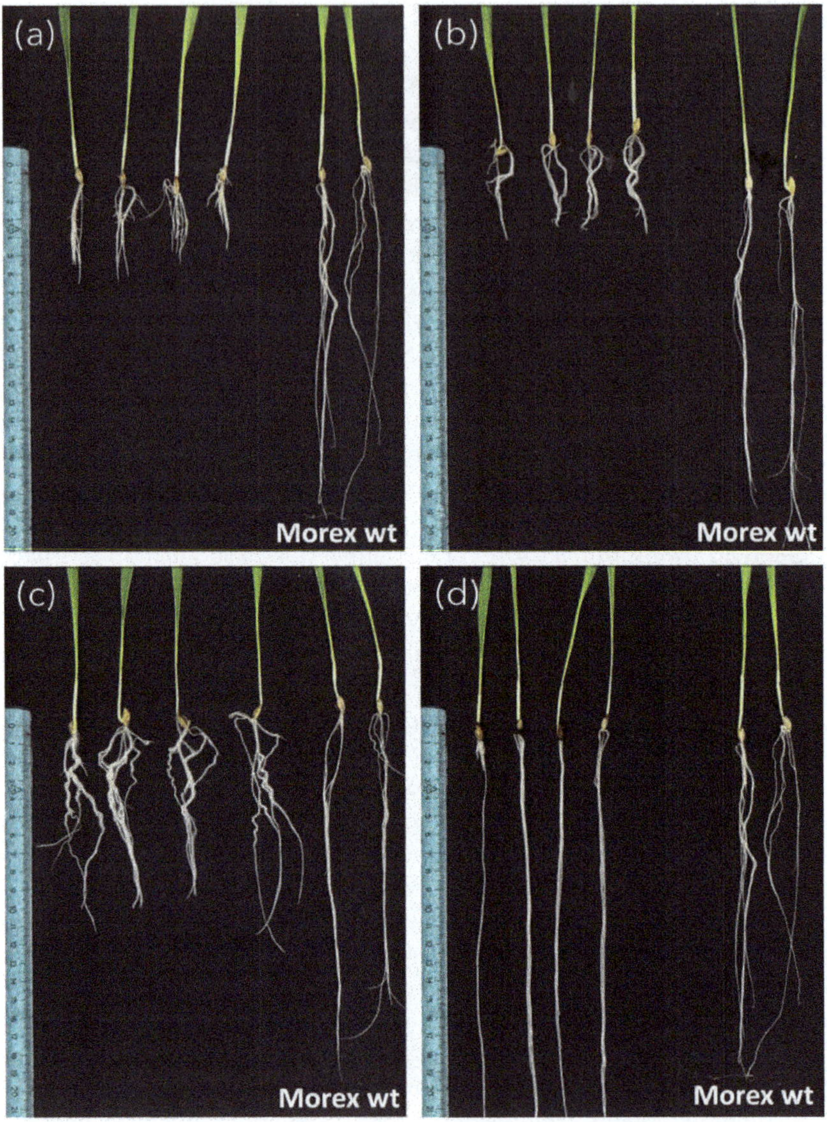

Figure 1 Root mutants in the TILLMore collection in the background of cv. Morex. Examples of mutations for root length (a and b), circumnutation (c) and hypergravitropism (d).

of chromosome 7HS of cv. 'Morex'. Among the five high-confidence genes within this region, sequencing identified an A > T mutation in one candidate, *HORVU7Hr1G030250* (MLOC_38567), which resulted in the introduction of premature STOP codon in mRNA. *HORVU7Hr1G030250* encodes a bHLH transcription factor with LRL domain and was suggested as the best candidate responsible for the early stages of rhizodermis differentiation in barley (Gajewska et al., 2018).

The root hair defective-six like (RSL) Class II family of transcription factors is expressed preferentially in root hairs and has a conserved role in root hair development in land plants. In rice, (Moon et al., 2019b) functionally characterized the seven members of the RSL Class II subfamily in root hairs. Six of these genes were preferentially expressed, and four were strongly expressed. Phenotypic analysis of each mutant revealed that *Os07g39940* plays a major role in root hair formation, based on observations of a short root hair phenotype in those mutants. Overexpression for each of four family members in rice resulted in an increased density and length of root hairs. These four members contain a transcription activation domain and are targeted to the nucleus. They interact with rice *Root Hairless1* (*OsRHL1*), a key regulator of root hair development. Transcriptome analysis using *Os07g39940*-OX plants revealed that 86 genes, including Class III peroxidases, were highly upregulated. Furthermore, reactive oxygen species (ROS) levels in the root hairs were increased in *Os07g39940*-OX plants but were drastically reduced in the *os07g39940* and *rhl1* mutants. These results clearly demonstrate that RSL Class II members function as essential regulators of root hair development in rice (Moon et al., 2019b).

Moon and co-workers have used root hair mutants to elucidate the ontogeny of these tubular-shaped protuberances of root epidermal cells found in nearly all vascular plants (Moon et al., 2018, 2019a,b). Root hair cell elongation requires the coordinated expression of a number of root hair genes. *OsRHC*, a rice root hair-specific putative cellulase gene is specifically expressed in elongating root hairs, and the OsRHC targets the plasma membrane. Moon et al. (2018) showed that the mutation of the *OsRHC* gene caused by a T-DNA knock-out and CRISPR-Cas9 system severely reduced root hair length and were able to demonstrate that the OsRHC protein interacts with a root hair-specific cellulose synthase protein (OsCSLD1) in the plasma membrane. Additionally, they observed a moderate reduction of cellulose content in the *osrhc* mutant. Altogether, these results suggested that the plasma membrane-localized OsRHC plays a critical role in cell wall remodeling during root hair extension.

In wheat, a high-throughput screen for variation in seminal root number was conducted using an exome-sequenced mutant population derived from the hexaploid wheat cultivar Cadenza (Shorinola et al., 2019). The screen identified seven independent mutants with homozygous and stably altered seminal root number phenotypes. One mutant, Cadenza0900, displayed a recessive extra seminal root number phenotype, while six mutants (Cadenza0062, Cadenza0369, Cadenza0393, Cadenza0465, Cadenza0818 and Cadenza1273) showed lower seminal root number phenotypes most likely originating from defects in the formation and activation of seminal root primordia. Segregation analysis in F_2 populations suggested that the phenotype of Cadenza0900 is controlled by multiple loci, whereas the Cadenza0062 phenotype fits a 3:1 mutant:wild-type segregation ratio characteristic of dominant single gene action. The work of Shorinola et al. (2019) highlights the potential of using the sequenced wheat mutant population as a forward-genetics resource to uncover novel variation in agronomic traits, such as seminal root architecture.

In rice, Kota et al. (2020) generated more than 12 000 activation-tagged mutants in a high-yielding *indica* rice variety, 'BPT 5204', employing transposable elements based on a maize *Ac/Ds* system. The germinal transposition frequency of *Ds* element in different T_2 progeny of rice plants was found to be about 18.0%. Different stable tagged mutants exhibited marked increases in root length, plant height, number of tillers, leaf size, panicle size, seed size and number of grains per plant. Notably, when compared to control plants, the *Ds*-RFP (V6-*Ds*-RFP35)- and *Ds-bar* (D1-*Ds*-bar2)-tagged mutant plants revealed a 1.52- and 1.50-fold increase in root length, respectively. Additionally, overexpression of the rice root-specific gene *RCc3* was found to improve the development of RSA, plant growth and salt stress tolerance (Li et al., 2018b). Overall, the results indicate that the genes associated with these traits are upregulated by the enhancer element in activation-tagged mutants. As such, the various tagged mutant lines provide a valuable genetic resource for the identification of key genes modulating the expression of agronomic traits of interest in rice.

In maize, eight genes responsible for major checkpoints of root development have been cloned starting from corresponding morphological mutants (Hochholdinger et al., 2004, 2018; Woll et al., 2005). Molecular cloning of these genes highlighted that key elements of auxin signal transduction, such as LOB domain (LBD) and Aux/IAA proteins, are instrumental for seminal, shoot-borne and lateral root initiation. Moreover, genetic analyses have demonstrated that genes related to exocytotic vesicle docking, cell wall loosening and cellulose synthesis and organization control root hair elongation (Hochholdinger et al., 2018; Kortz et al., 2019).

2.2 Dissecting the QTLome for root traits

The variability that breeders deal with in their programs reflects the action of multiple native alleles at the QTLs that govern the variability of any quantitative trait, including those that shape RSA (Fig. 2). Each QTL defines a portion of chromosome varying from approximately 5 to 30 centiMorgan (cM) depending on the level of genetic resolution achieved (Xu and Crouch, 2008). The presence of a QTL is determined through an appropriate statistical analysis of phenotypic and molecular data of an artificial mapping population (e.g. linkage mapping) or a collection of unrelated genotypes (e.g. association mapping).

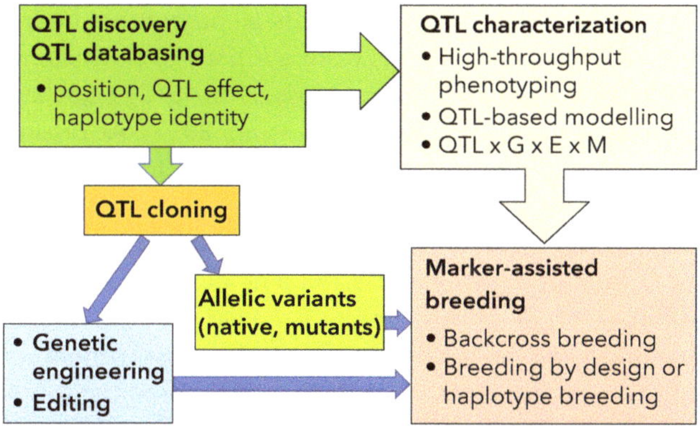

Figure 2 Leveraging the root system QTLome (Salvi and Tuberosa, 2015).

Collectively, all the experimentally supported QTLs that contribute to the native variability of a specific trait in one species concur to define the trait QTLome (Salvi and Tuberosa, 2015). A major shortcoming of any QTL study is the low accuracy in detecting the real number of QTLs affecting the genetic variation of the investigated traits (Xu, 2003). In a simulation study applied to experimental data, Beavis (1994) clearly showed that populations of approximately 100–200 progenies only captured a modest fraction of QTLs; furthermore, the effect of each single QTL was usually overestimated. Another study showed that with fewer than 500 progenies, and independently from marker density, it is very difficult to detect QTLs of small effect (Beavis, 1994).

2.2.1 Linkage mapping with biparental populations

The genetic dissection of the RSA QTLome through linkage mapping requires the evaluation of suitable mapping populations of recombinant inbred lines (RILs) more commonly developed by selfing the progeny of a biparental cross between two lines differing for the trait of interest (Champoux et al., 1995; Lafitte et al., 2004; Price et al., 2002b; Uga et al., 2013). The main shortcoming of biparental linkage mapping, namely the low genetic resolution, can be partially overcome using multiple RIL populations like the nested association mapping (NAM) approach (Atkinson et al., 2020; Yu et al., 2008) which has been adopted in soybean to identify QTLs for root length and yield under drought conditions (Khan et al., 2018) and in wheat to identify QTLs for dry root weight and root/shoot ratio as related to nitrogen-deficiency tolerance at the seedling stage (Ren et al., 2018). However, the limited number of RILs evaluated in these two studies did not allow for taking full advantage of the lower linkage disequilibrium of the NAM approach as compared to studies based upon a single biparental cross. More recently, multi-parental crosses between four and more (up to 16) lines where all parents are inter-mated prior to proceeding with repeated selfings were adopted to produce the so-called multi-parent advanced generation intercrosses (MAGIC) populations (Huang et al., 2015; Milner et al., 2016; Raghavan et al., 2017). At present, no study has been reported on their use to map QTLs for RSA traits, most likely due to the large number of RILs commonly comprising a MAGIC population. Future studies conducted with NAM and/or MAGIC populations will provide valuable opportunities for further unraveling the QTLome of RSA traits.

The assembly of a genetic map is based on the level of linkage disequilibrium (LD, i.e. the level of non-random assortment of alleles at different loci) among adjacent markers on the same chromosome. Accordingly, mapping the QTLs of the target trait is also based on the LD between the QTL and nearby markers (Darvasi et al., 1993; Tanksley, 1993; Tanksley et al., 1989). For breeding purposes, a density of one marker every 5–10 cM is sufficient for most applications. However, backcrossing a particular QTL from a wild or distant relative of the crop will require high resolution (< 1 cM) in order to avoid the negative effects (linkage drag) caused by undesirable effects of wild alleles closely linked to the QTL being introgressed into the elite line.

A notable example specifically related to the effects of a major RSA QTL in bread wheat is provided by the work of Voss-Fels et al. (2017) who demonstrated that linkage drag in European bread wheat, driven by strong selection for a haplotype variant controlling heading date, has eliminated a specific combination of two flanking, highly conserved haplotype variants whose interaction confers increased root biomass. The authors suggested that reversing this inadvertent consequence of selection could recover beneficial root diversity that may prove to be essential for future food production in fluctuating environments (Voss-Fels et al., 2017). Additionally, highly conserved synteny to rice across this chromosome segment suggested that adaptive selection has shaped the diversity landscape of this locus across different, globally important cereal crops. By mining wheat gene expression data, the authors identified root-expressed genes within the region of interest that could help breeders to select positive variants adapted to specific target soil environments (Voss-Fels et al., 2017). A much higher level of genetic resolution (< 0.1 cM) is necessary when the final goal is the isolation of the sequence governing the expression of the target QTL, which requires the screening of several thousand plants (Salvi and Tuberosa, 2007). The recent availability of extremely dense single nucleotide polymorphism (SNP)-based maps provides excellent opportunities for accurate mapping and positional cloning (Avni et al., 2014; Maccaferri et al., 2015; Song et al., 2016; Zhou et al., 2017).

A representative, albeit incomplete list of the most recent QTL studies on RSA traits conducted in different crops is reported in Tables 2–5. The early QTL work of Ray et al. (1996) and Price and co-workers in rice (Price and Courtois, 1999; Price et al., 2000, 2002b,d; Price and Tomos, 1997), maize (Giuliani et al., 2005b; Landi et al., 2002a; Mano et al., 2006; Tuberosa et al., 2002b; Zhu et al., 2005), sorghum (Borrell et al., 2014b; Christopher et al., 2013; Harris et al., 2007), barley (Chloupek et al., 2006), wheat (Kubo et al., 2007; Navakode et al., 2009; Zwart et al., 2010), bean (Beaver and Osorno, 2009; Lopez-Marin et al., 2009; Ochoa et al., 2006; Yan et al., 2004) and chickpea (Varshney et al., 2013) targeted RSA traits in studies conducted either under controlled conditions or in field conditions and, only in a few cases, in both conditions (Landi et al., 2002b; Tuberosa et al., 2002b). Notable examples of field-based studies are provided by the aforementioned work of Price and co-workers, Steele and co-workers (Steele et al., 2006, 2007) and Uga and co-workers in rice (Arai-Sanoh et al., 2014; Hanzawa et al., 2013; Uga et al., 2008, 2011, 2013), by Lynch and co-workers in bean (Beebe et al., 2006; Ochoa et al., 2006; Yan et al., 2004) and maize (Zhu et al., 2005), and by Varshney and colleagues in chickpea (Sivasakthi et al., 2018; Varshney et al., 2013). A representative, albeit incomplete, list of the most recent QTL studies on RSA traits conducted in different crops is reported in Tables 2–7.

Among the different RSA phenes targeted in QTL studies, root growth angle (RGA) is one of the most frequently investigated, most likely due to the possibility to measure it at the seedling stage in controlled, non-soil conditions and owing to its high heritability as shown in wheat (Alahmad et al., 2019; Bengough et al., 2004; Christopher et al., 2018; Maccaferri et al., 2016; Richard et al., 2015, 2018; Voss-Fels et al., 2018a), barley (Jia et al., 2019b; Voss-Fels et al., 2018a), rice (Uga et al., 2015a) and sorghum (Christopher et al., 2013). In durum wheat, seminal RGA variability was significantly correlated with the variability observed in field conditions at maturity, an important prerequisite for breeding purposes. Accordingly, divergent RGA phenotypes observed by seminal root screening were validated by root phenotyping of field-grown adult plants (Maccaferri et al., 2016). The same authors reported concurrent effects of RGA QTLs on grain yield across a broad range of soil moisture with the most notable effects on chromosome 6AL. In rice, an ideotype characterized by narrow RGA and deep root ideotype for enhancing drought resistance has been successfully pursued in two different breeding programs based on direct field observation of root distribution (Steele et al., 2013; Uga et al., 2013) or RGA measurements in rhizotrons (Kitomi et al., 2015; Uga et al., 2015a,b). In sorghum under drought conditions, stay-green genotypes have contributed additional evidence for the positive role on yield of QTL alleles for narrow RGA (Borrell et al., 2014a,b). Additionally, RGA is also of paramount importance for the acquisition of phosphorus, a low-mobility nutrient abundant in the upper soil layer as shown by extensive work in bean (Burridge et al., 2019; Jia et al., 2018; Miguel et al., 2015).

A widely investigated RSA phene is deep rooting, a key trait that helps plants to extract water from deeper soil layers, hence allowing crops to avoid the negative effects of drought (Blum, 2011; Li et al., 2019a; Richards, 2006; Tracy et al., 2020). Among cereal crops, upland rice provides some of the best examples of the beneficial effects of QTL alleles for deeper roots on grain yield. Following the seminal work of Price and co-workers (Price, 2002; Price et al., 1997, 2000, 2002a,c,d; Price and Tomos, 1997), which led to the discovery of four major QTLs governing root depth, the beneficial haplotypes for deeper rooting at three QTLs were introgressed by marker-assisted backcrossing (MABC) into Kalinga III, a popular improved upland rice variety in eastern India, a drought-prone area. Pyramiding (stacking) of the QTL alleles into the Kalinga III background contributed to the selection of Birsa Vikas Dhan 111, the first drought-tolerant cultivar released through marker-assisted backcrossing (MABC) for an RSA trait (Steele et al., 2006, 2007, 2013).

Deep rooting has also been shown to have a positive effect in drought resistance in sorghum (Borrell et al., 2014a; Boyles et al., 2019) and maize (Burton et al., 2014; Lynch, 2013; Lynch et al., 2014; Lynch and Wojciechowski, 2015). Sorghum production is prevalently located in semiarid environments, which makes drought resistance a target trait for genetic dissection and improvement. Genotypes with post-flowering drought resistance in sorghum exhibit higher chlorophyll and photosynthetic capability under drought compared with wild-type lines, a phenomenon commonly recognized as 'stay-green', a phene that significantly improves sorghum yield under water stress during grain fill (Borrell et al., 2000, 2014a). The grain yield advantage of the stay-green phenotype under drought was related to reduced vegetative biomass and water uptake during the pre-flowering growth stages (Borrell et al., 2014a). Under artificial conditions, sorghum root length during the seedling stage was found to be a major factor in drought tolerance (Bibi et al., 2012). Notably, Mace et al. (2012) found a relationship between drought adaptation and nodal root angle to further support the role of belowground biomass traits in sorghum production under water stress.

In maize, a major QTL (*root-ABA1*) on bin 2.04 near *csu133* originally shown to consistently affect leaf ABA concentration (Sanguineti et al., 1999; Tuberosa et al., 1998) was shown to affect root depth and root mass (Giuliani et al., 2005b; Landi et al., 2005, 2007). Marker-assisted selection was used for backcrossing the parental QTL regions in both parental lines, allowing the assembly of pairs of near-isogenic lines (NILs) contrasted for the *root-ABA1* QTL region.

Figure 1 shows the effects of *root-ABA1* on root architecture in plants at the flowering stage. Interestingly, similar experimental evidence on the influence of the region near *csu133* on root architecture and L-ABA was also gathered from Polj17 × F2 (Lebreton et al., 1995). More recently, the same QTL region was found to affect yield in different soil moisture scenarios across Europe in multiple seasons (Millet et al., 2016). A field evaluation conducted under well-watered and water-stressed conditions during two consecutive seasons indicated that each pair of *root-ABA1* NILs differed significantly for root lodging, root mass, brace root angle and grain yield (Giuliani et al., 2005b; Landi et al., 2005).

A major QTL for RSA in maize was also identified on bin 1.06 in a biparental (Lo964 × Lo1016) RIL population tested in hydroponics and also in the field for vertical root pulling resistance and grain yield under both well-watered and drought conditions (Landi et al., 2001, 2002b; Tuberosa et al., 2002b). These results indicate the feasibility of using hydroponics in maize to identify QTL regions controlling root traits at an early growth stage and also influencing GY in the field. A comparative analysis of the QTL regions with those described in previous studies investigating root traits in different maize populations revealed a number of QTLs in common, with a particularly notable effect of RSA and yield of the QTL near *csu61b* on bin 1.06 (Tuberosa et al., 2002a). Starting from different F_4 families, two pairs of NILs were developed for the QTL near *csu61b* named *root-yield-1.06* due to its concurrent effects on RSA and yield. The NILs *per se* and their crosses with the parental lines (Lo964 and Lo1016) were field-tested in two years under well-watered (WW) and water-stressed (WS) conditions. *Root-yield-1.06* showed a major constitutive effect on root size, plant vigor and productivity across water regimes, genetic backgrounds and inbreeding levels, hence indicating its suitability for cloning the sequence underlying its effects and for marker-assisted selection to improve yield stability in maize (Landi et al., 2010).

More recently, Uga and co-workers mapped and cloned *DEEPER ROOTING 1* (*DRO1*), a major QTL for root angle and deep rooting in rice (Ahmadi et al., 2014; Arai-Sanoh et al., 2014; Uga et al., 2013). The *DRO1*-introgressed lines exhibited deeper roots and higher yield owing to enhanced photosynthetic rates and grain filling, which were associated to higher expression of *DRO1* and deeper roots as compared to the recurrent, drought-susceptible cultivars (Kitomi et al., 2018; Tao et al., 2019; Tracy et al., 2020; Uga et al., 2018; Wallace et al., 2018; Yoshino et al., 2019). Notably, Guseman et al. (2017) found that *DRO1* and *DRO1*-related genes are present across diverse plant phyla, and fall within the IGT gene family. The IGT family also includes TAC1 and LAZY1, which are known to affect the orientation of lateral shoots. Consistent with a potential role in root development, *DRO1* homologs in *Arabidopsis* and peach showed root-specific expression. Promoter-reporter constructs revealed that *AtDRO1* was predominantly expressed in both the root vasculature and root tips, in a distinct developmental pattern while mutation of *AtDRO1* led to more horizontal lateral root angles (Guseman et al., 2017). Overexpression of *AtDRO1* under a constitutive promoter resulted in steeper lateral root angles, as well as shoot phenotypes including upward leaf curling, shortened siliques and narrow lateral branch angles. Overexpression of *PpeDRO1* in *Prunus domestica* (plum) led to deeper-rooting phenotypes. Based on these data, Guseman et al. (2017) suggested a potential application for *DRO1*-related genes to alter root architecture for drought avoidance and improved resource use.

In winter bread wheat (*Triticum aestivum* L., 2n = 6x = 42, AABBDD genomes), Kan et al. (2020) described *QYld.osu-1BS*, a major QTL for grain yield in a population of 260 doubled haploid (DH) lines derived from the cross of Duster and Billings, two important cultivars in the Southern Great Plains for their outstanding yielding ability. In two years of field trials, *QYld.osu-1BS* explained 13.9 and 23.5% of the total phenotypic variation for yield. Notably, no crossover was observed in the region from the telomere to 18.4 Mb when 4146 individual plants within $F_{2:4}$ lines derived from Duster × Billings were screened. The 1BL-1RS translocation was not observed in the region with the abnormal recombination rate in Duster or Billings. Duster is a unique haplotype in the whole exome capture dataset, compared with 57 cultivars and breeding lines with different genetic backgrounds. Unique sequences of the *QYld.osu-1BS* allele for the higher grain yield in Duster were used to design competitive allele-specific PCR (KASP) markers for the unique sequences suitable to implement MAS to develop novel cultivars with increased grain yield (Kan et al., 2020).

The introgression of a small segment chromosome arm 1BS of bread wheat in the distal region of the rye 1RS-BL arm translocation in wheat (henceforth 1RS(RW)) was previously associated with reduced grain yield, carbon isotope discrimination and stomatal conductance, suggesting reduced access to soil moisture (Howell et al., 2019). The authors showed that lines with the normal 1RS arm have longer roots than lines with the 1RS(RW) arm in both field and hydroponic experiments. In the 1RS(RW) lines, differences in seminal root length were associated with a developmentally regulated arrest of the root apical meristem (RAM). Approximately 10 days after germination, the seminal roots of the 1RS(RW) plants showed a gradual reduction in elongation rate, and stopped growing a week later. Seventeen days after germination, the roots of the 1RS(RW) plants showed altered gradients of ROS and the emergence of lateral roots close to the RAM, suggesting changes in the root meristem. The 1RS(RW) lines also showed reduced biomass and grain yield relative to the 1RS lines, with larger differences under reduced or excessive irrigation than under normal irrigation which suggests that this genetic variation could be useful to modulate root architecture (Howell et al., 2019).

An additional and more systematic opportunity for the dissection of the RSA QTLome in a biparental background is made possible by the assembly of a collection of introgression library lines (ca. 80–100) obtained by subsequent backcrosses of a recurrent parent (usually an elite cultivar) with a donor parent, usually a line highly diversified from the recurrent parent for one or more traits (Frary et al., 2004; Tanksley et al., 1996; Zamir, 2001). Depending on the number of backcrosses, each introgression library line (ILL) carries a fragment (from ca. 20 to 40 cM) of the donor genome different from that carried by the recurrent line. Collectively, the fragments of all ILLs cover the entire genome with limited overlap among lines. Isogenic library lines have been instrumental for mapping major adaptive QTLs in maize (Salvi et al., 2011; Zhan et al., 2018), rice (Balakrishnan et al., 2019; Dharmappa et al., 2019; Mei et al., 2006) and wheat (Grewal et al., 2018; Merchuk-Ovnat et al., 2017; Salvi et al., 2011; Zhan et al., 2018). Once two or more major QTLs for the target traits are identified in ILLs, testing for epistasis becomes feasible using a small number of genotypes, unlike with mapping populations, where testing for epistasis would require the evaluation of at least 200 RIL families (Lippman et al., 2007; Sannemann et al., 2018; Wang et al., 2012; Yadav et al., 2019).

In maize, the 75 ILLs that were originally assembled by Salvi et al. (2011) to dissect the flowering time QTLome were also used to genetically dissect RSA traits at an early stage (Salvi et al., 2016). Unlike other cereals, in maize, limited attention has been devoted to the genetic analysis of RSA diversity at an early stage. The difference in embryonic (including seminal and primary) root architecture between the maize reference line B73 which mostly develops three seminal roots and the landrace Gaspé Flint (with virtually no seminal roots) was genetically dissected using a collection of ILLs grown in paper rolls and pots. Three QTLs were identified for seminal root number (SRN) on chromosome bins 1.02, 3.07 and 8.04–8.05, which collectively explained 66% of the phenotypic variation for SRN. In all three cases, Gaspé Flint contributed the allele for lower SRN. Primary root dry weight was negatively correlated with SRN ($r = -0.52$), and QTLs for primary root size co-mapped with SRN QTLs, suggesting a concurrent effect of SRN QTLs on the primary root, most probably caused by competition for seed resources. Notably, two out of three SRN QTLs co-mapped with *rtcs* and *rum1*, the only two known maize genes affecting seminal root number. The strong additive effect of the three QTLs and the development of near-isogenic lines for each QTL in the elite B73 background provide unique opportunities to functionally characterize the genes involved in root development and to evaluate how root architecture affects seedling establishment, early development and eventually yield in maize.

2.2.2 Association mapping with diversity panels

In the past decade, as an alternative to linkage mapping with biparental populations of RILs, NILs and/or ILLs, association mapping, also known as genome-wide association (GWA) mapping, based on the evaluation of panels of unrelated accessions (typically 200 or more) has increasingly been adopted for the dissection of the RSA QTLome (Beyer et al., 2019; He et al., 2019; Sertse et al., 2019; To et al., 2019). The assumption underlying the use of GWA studies (GWAS) to detect the presence of loci influencing the target trait is that (i) alleles at two closely linked loci share a historical ancestor and (ii) this original co-occurrence will gradually decay in the population due to recombination events that occurred during subsequent meiosis. Consequently, the relative allele distributions of an unknown gene and that of a closely linked marker will be nonrandom because the two are in linkage disequilibrium (LD). A major factor to be accounted for in GWAS is the presence of population structure, which will significantly bias the results and inflate the detection of spurious marker-trait associations, hence resulting in false positives (Bernardo and Yu, 2007).

As compared to linkage mapping with biparental populations, an important advantage of association mapping is that the linkage reflects a much larger number of historic meiosis, hence providing a considerably lower LD and higher genetic resolution (Flint-Garcia et al., 2005; Yu and Buckler, 2006). Another important advantage is that the allelic variability explored by a panel of unrelated accessions is much larger than that present in a segregating population derived from two parental lines, hence two alleles per locus. However, GWAS do not allow for the detection of the effect that a rare, but otherwise agronomically valuable, allele may have on the target trait. In fact, the statistical procedures adopted for revealing the effects associated with a particular locus/haplotype usually disregard rare alleles with a frequency lower than 2–10% in the entire population, depending on the number of accessions being considered (Abdel-Ghani et al., 2019; Varshney et al., 2007; Xu et al., 2017). As compared to biparental mapping populations, common features of GWAS are the higher genetic resolution, the higher number of sampled alleles and the identification of candidate genes facilitated by the low LD, hence genetic resolution and the increasing availability of sequencing information. A number of GWAS targeting RSA traits in crops are listed in Tables 2–7. In wheat, association mapping has been adopted to dissect the genetic basis of root growth angle, a trait of great interest from a breeding standpoint owing to its high heritability (Alahmad et al., 2019; Canè et al., 2014; Sanguineti et al., 2007).

3 Fine mapping and cloning of loci for root traits

The identification of major QTLs for RSA traits is the first step of a longer process aimed at identifying and isolating the underlying molecular cause at the sequence level (i.e. polymorphism) of the functional variation revealed by QTL analysis. The cloning of the sequence underpinning the QTL is an essential prerequisite for a targeted manipulation of the trait through genetic engineering and/or gene editing via CRISPR/Cas9 (Salvi and Tuberosa,

2015). Importantly, QTL cloning allows breeders to access the native (natural) genetic variability present in locally adapted materials (e.g. landraces) and/or ancestral relatives in order to identify novel and often rare beneficial alleles lost during either modern breeding and/or domestication bottlenecks (Eshed and Zamir, 1994; Zamir, 2001). This aspect is of great practical importance in view of the legislative boundaries presently limiting the cultivation of edited crops in European countries (Borrill et al., 2019; Herbert et al., 2020; Ming et al., 2020).

The first step in sorting through the plethora of QTL information in order to identify those most valuable for cloning comes from QTL meta-analysis (Sosnowski et al., 2012), a statistical framework to project QTLs on a consensus map and to identify coincident QTLs among independent experiments. The resulting meta-QTLs are also expected to better define the boundaries of the causative genomic intervals. QTL meta-analyses are used to (i) summarize QTL information for one or more traits (Soriano and Alvaro, 2019; Zhang et al., 2016), (ii) locally verify the co-location of QTLs between different populations as the first step toward QTL validation (Landi et al., 2010; Maccaferri et al., 2016) and/or (iii) prioritize candidate genes (Guo et al., 2018; Zhang et al., 2014). This notwithstanding, current meta-analysis has notable shortcomings such as lack of integration of QTL haplotype information and no obvious protocol able to integrate QTL results from experimental populations with those from GWAS (Salvi and Tuberosa, 2015).

Once identified, the target QTL is fine mapped to reduce the interval, most likely harboring the polymorphic sequence (coding or non-coding) affecting the phenotype. This step requires the identification of a great number of SNPs in the chromosome segment being fine mapped. For this purpose, the availability of NILs greatly facilitates the characterization of the QTL region with multi-omics approaches (Bernardino et al., 2019; Giuliani et al., 2005a; Kang et al., 2019; Khandal et al., 2020; Torabi et al., 2009; Yoshino et al., 2019), a rather costly undertaking and usually not an option with RIL populations, unless bulked samples of the two tails of the population are considered (Li et al., 2018a, 2019a; Rebetzke et al., 2017). The availability of -omics data allows for the identification of differentially expressed genes (DEG) at candidate regions (Gelli et al., 2017), hence reducing the number of candidate genes to be validated in order to pinpoint the functionally polymorphic DNA responsible for the target phene (Grosskinsky et al., 2018).

In maize, residual heterozygosity (RH) in a collection of 2196 heterogeneous inbred family (HIF) lines derived from 12 RIL populations was used (Liu et al., 2018) to facilitate fine mapping at regions where RH was present. A total of 18 615 unique RH intervals were identified, ranging from 505 to 2095 intervals per population, with average maize genome coverage of 94.8%. A given RH region was present in an average of five different individuals within a population.

The potential use of this HIF library for the fine mapping of QTLs was assessed based on publicly available QTL information, achieving a <= 1 Mb resolution on average, hence providing a useful resource for the fine mapping and cloning of QTLs for RSA or other important agronomic traits in maize (Liu et al., 2018).

Transcriptome profiling provides valuable clues toward the identification of polymorphisms influencing the level of expression of genes in the target region and possibly of the gene/s underlying the QTL (Butsayawarapat et al., 2019; Hochholdinger et al., 2018; To et al., 2019; Wu et al., 2018a; Yoshino et al., 2019). Following the fine mapping, the markers more tightly linked to the QTL are used to anchor the genetic region underscoring the QTL peak to a physical chromosome region. Once the crop genome sequence becomes available, the anchoring will be extended to the complete genome sequence, thus making positional cloning of Mendelian loci more of a routine practice and strongly facilitating QTL cloning (Salvi et al., 2007; Tuberosa et al., 2007).

An element of complexity inherent to the organization of cereals genome that should be duly considered for its practical implications is the lack of micro-colinearity between the sequence of different lines of the same species, a phenomenon particularly widespread in maize (Messing and Dooner, 2006). This notwithstanding, as an increasing number of genome sequences become available in different crops, there is a growing understanding that the genome of a single genotype is insufficient to represent the gene diversity within a whole species. Many studies examine the sequence diversity within genes, and this allelic variation is an important source of phenotypic variation. However, the significant gene presence/absence variation present within species and its impact on traits is being addressed only now in greater detail to define the so-called pangenome, namely the sum of the genes for a species, and the determination and characterization of the pangenome is an important prerequisite to understand the functional variation within a species (Bolger et al., 2019; Golicz et al., 2016, 2020; Khan et al., 2020).

Polymorphic genes or genomic sequences that are found to be completely linked with the QTL are then functionally tested with a number of different approaches (e.g. genetic engineering, identification of knockouts, editing, TILLING, etc.) to validate their role (Pariasca-Tanaka et al., 2020; Uauy et al., 2017; Wang et al., 2019a). In barley, knockout of the *HvCKX1* or *HvCKX3* gene by RNA-guided Cas9 nuclease affected the regulation of cytokinin metabolism and root morphology (Gasparis et al., 2019).

The availability of complete genome sequences for more crops, and the combination of conventional linkage mapping, association mapping (also known as genome-wide association mapping), QTL-seq, transcriptomics and gene-editing technologies can accelerate the cloning and confirmation of genes underlying QTL affecting the target RSA trait and its functions. The application of new generation sequencing (NGS) at increasingly lower costs has been a game-changer in the way we can fine map target loci and identify gene functions and how the expression of candidate genes within the confidence interval on the locus is modulated by both growth stage and the signaling cascade triggered by abiotic (Ajmera et al., 2019; Khandal et al., 2020) and biotic stresses as well as their interactions (Chen et al., 2019c; Shahan and Benfey, 2020; Yamada et al., 2020).

Machine-learning methodologies can be applied readily to biological problems within the context of predicting mRNA expression levels from whole-genome DNA sequence data and are applicable across biological organisms (Washburn et al., 2019), hence underlining the value of *Arabidopsis* (Gautam and Gupta, 2020; Gupta et al., 2020; Jia et al., 2019a; Wachsman and Benfey, 2020; Yamada et al., 2020; Yang et al., 2019) and other model species (Chen et al., 2019a,c; Shahan and Benfey, 2020; Sun et al., 2019; Wang et al., 2019b) to streamline and accelerate the identification of the most plausible candidates. Additionally, machine learning is an effective method for identifying true SNPs in polyploid plants such as peanut (Korani et al., 2019).

In wheat, the development of high-throughput SNP genotyping platforms such as the Axiom® HD wheat genotyping array (Winfield et al., 2016) and the wheat 660K SNP array (Sun et al., 2020) in hexaploid wheat has facilitated the assembly of dense genetic maps to streamline fine mapping while advancing genetics and breeding efforts in wheat. Similar efforts have allowed for the assembly of the 'Axiom®CicerSNP array' in chickpea (Roorkiwal et al., 2018), hence promoting genomics-based studies and breeding applications in pulse crops. In rice, the use of an ultra-high density SNP map allowed for the identification of 21 QTLs for nitrogen and phosphorus uptake-related traits some of which were clustered in hotspots linked to variability in RSA (Fu et al., 2019).

Currently, *DRO1* and *PSTOL1* are the only two RSA QTLs that were both cloned in rice with a map-based (positional) cloning approach. *DRO1* enhances root gravitropism to increase root growth angle, which maintained higher photosynthesis and yield performance under severe water-stressed conditions; the beneficial functional variant of *DRO1* was identified in a drought-tolerant cultivar 'Kinandang Patong' from the Philippines (Uga et al., 2013). *PSTOL1* acts as an enhancer of early root growth and promotes root growth under both high- and low-phosphorus conditions; the beneficial functional variant was found in the low phosphorus-tolerant landrace 'Kasalath' (Gamuyao et al., 2012). These two genes were identified in landrace germplasm rather than elite breeding lines.

In wheat and barley, Voss-Fels et al. (2018a) demonstrated that *VERNALIZATION1* (*VRN1*), a key regulator of flowering time in cereals (Deng et al., 2015) also modulates root architecture.

The authors initially mapped a highly significant QTL for nodal root angle on *Triticum aestivum* chromosome 5B using a diversity panel of 219 hexaploid winter wheat accessions. Additionally, the authors reported that the QTL includes six SNP markers in strong linkage disequilibrium which span the B-subgenome homoeolog of *VERNALIZATION1* (*VRN1*), one of the most important developmental genes in wheat and barley and known to encode for a MADS-box transcription factor which regulates the expression of genes influencing reproductive plant organs and flowering (Deng et al., 2015). Based on this and other experimental evidence, Voss-Fels et al. (2018a) suggested that selection for *VRN-H1* variants that simultaneously induce early flowering and maintain the 'steep, cheap and deep' root systems ideotype advocated by Lynch (2013) provides a dual mechanism imparting flowering-mediated drought escape coupled with improved water or nutrient acquisition. More recently, Lynch (2018) proposed that reduced root development would be advantageous for drought resistance in high-input agroecosystems. Selection regimes for crop ancestors and landraces include multiple stresses, intense competition and variable resource distribution, which favored prolific root production, developmental plasticity in response to resource availability and maintenance of unspecialized root tissues. Accordingly, Lynch (2018) speculated that root phenotypes that focus on water capture at the expense of ancestral adaptations would be better suited to high-input agroecosystems. Parsimonious architectural phenotypes include fewer axial roots, reduced density of lateral roots, reduced growth responsiveness to local resource availability, and greater loss of roots that do not contribute to water capture while parsimonious anatomical phenotypes include a reduced number of cortical cell files, greater loss of cortical parenchyma to aerenchyma and senescence, and larger cortical cell size (Lynch, 2018). Parsimonious root phenotypes may be less useful in low-input agroecosystems, which are characterized by multiple challenges and trade-offs for root function in addition to water capture (Lynch, 2018).

In wheat, in order to investigate the phenotypic architecture of shoot- and root-related traits and quantify the impact of primary and secondary domestication, Gioia et al. (2015) examined a collection of 36 wheat genotypes under optimal and nitrogen-starvation conditions. These represented three taxa that correspond to key steps in the recent evolution of tetraploid wheat (i.e. wild emmer, emmer and durum wheat). Overall, nitrogen starvation reduced the shoot growth of all genotypes, while it induced the opposite effect on root traits (Gioia et al., 2015). An overall increase in all of the shoot and root growth traits was observed from wild emmer to durum wheat, while emmer was generally very similar to wild emmer but intermediate between these two subspecies. While the differences in phenotypic diversity due to the effects of primary domestication were not significant, the secondary domestication transition from emmer to durum wheat was marked by a large and significant decrease in the coefficient of additive genetic variation (Gioia et al., 2015). Notably, this reduction was very strong under the optimal condition and less intense under nitrogen starvation. Moreover, although under the optimal condition both root and shoot traits showed significantly reduced diversity due to secondary domestication, under nitrogen starvation the reduced diversity was significant only for shoot traits. Overall, a considerable amount of phenotypic variation was observed in wild emmer and emmer, which could be exploited for the development of pre-breeding strategies (Gioia et al., 2015).

More recently, Golan et al. (2018) showed that upon domestication the shoot root number (SRN) in wheat increased from three to five due to the activation of two root primordia that are suppressed in wild wheat, a feature controlled by loci expressed in the germinating embryo. Suppression of root primordia did not limit water uptake, indicating that three seminal roots adequately maintain growth during seedling development. The persistence of roots at their primordial state promoted seedling recovery from water stress through reactivation of suppressed primordia upon rehydration. Based on these results, Golan et al. (2018) suggested that under well-watered conditions, SRN is not a limiting factor, and an excessive number of roots may be costly and maladaptive. Conversely, the authors further suggested that upon water stress lack of substantial root system suppresses growth and rapid recovery of the root system, an essential feature for seedling recovery. While this study underscored the role of SRN as a key adaptive trait reshaped upon domestication, the maintenance of roots at their primordial state during seedling development may be regarded as a seedling protective mechanism against water stress (Golan et al., 2018).

In maize grown in hydroponics, several QTLs for RSA in hydroponics significantly affected P-uptake efficiency (PupE) and PUE in a biparental cross between two parental lines differing for RSA traits and PUE (Gu et al., 2016). In particular, one region on chromosome 3 (bin 3.04) was shown to harbor two QTLs for PUE, three for PupE and three for RSA that were assigned to two adjacent but distinct QTL clusters, *Cl-bin3.04a* and *Cl-bin3.04b*, where the favorable effects from the alleles contributed by the large-rooted and high-PupE parent. Notably, nine lines obtained from MAS backcrossing for the beneficial alleles at *Cl-bin3.04a* or *Cl-bin3.04b* displayed mean increases of 22–26% in PUE in low-P field conditions. Furthermore, a line L224 pyramiding *Cl-bin3.04a* and *Cl-bin3.04b* showed enhanced PupE, relying mainly on changes in root morphology, rather than root physiology (e.g. P-uptake efficiency), under both hydroponic and field conditions. These results highlight the contributions of RSA to maize PupE and provide a successful example of how QTL-based selection can contribute to develop P-efficient crops (Gu et al., 2016).

4 Genomics-assisted breeding for root traits

With only a few notable exceptions, breeders have indirectly and mostly unwillingly selected for root traits based on yield and/or aboveground proxy traits such as stay-green in sorghum (Borrell et al., 2014a; Pinto and Reynolds, 2015) and leaf temperature in wheat (Pinto and Reynolds, 2015), both cases being related to the capacity of the plant to capture soil moisture. In many studies (see Tables 2–7), root traits and functions have been investigated with the long-term goal to provide breeders with suitable markers to implement marker-assisted selection (MAS) to enhance and stabilize yield, particularly in adverse soil conditions such as an excess of salinity (Hazzouri et al., 2018; Munns et al., 2020; Nounjan et al., 2018), aluminum (Inostroza-Blancheteau et al., 2010; Ma et al., 2006; Melo et al., 2019) and boron (Neto et al., 2017; Pallotta et al., 2014; Reid, 2010; Rodda et al., 2018; Sutton et al., 2007; Yoshinari and Takano, 2017) or limited availability of nitrogen (Bahrman et al., 2005; Blummel et al., 2015; Hirel et al., 2001, 2007; Kichey et al., 2007; Li et al., 2015), phosphorus (Burton et al., 2014; Li et al., 2009; Ochoa et al., 2006; Yan et al., 2004) and soil moisture (Deshmukh et al., 2017; Kulkarni et al., 2017; Lynch, 2018; Maccaferri et al., 2016; Oyiga et al., 2020; Purushothaman et al., 2017; Tardieu et al., 2017; Uga et al., 2015a), the most prevalent and devastating abiotic stress curtailing crop yield on a global scale.

Notwithstanding the abundant literature reporting on the genomics of RSA traits and functions in crops (see Tables 2–7), the number of successful cases in which such information eventually delivered a new cultivar is limited to a handful of examples reviewed hereafter. A number of reviews have critically addressed the main reasons accounting for the difficulty to (i) effectively incorporate genomics-assisted crop improvement into extant breeding programs and (ii) leverage genomics-based knowledge of complex traits; such are those that govern the shape and functions of roots, to deliver new, profitable cultivars to farmers (Bernardo, 2016, 2020; Blum, 2014; Collins et al., 2008; Tardieu et al., 2018; Turner et al., 2014; Varshney et al., 2018c).

4.1 Rice

Among all crops, rice is the one showing the most remarkable examples of how genomics-assisted crop improvement for RSA traits and functions can deliver new cultivars to farmers. In upland rice, following the detection of major QTLs for root depth in a RIL population derived from a Bala × Azucena cross (Price et al., 2002a,c,d; Steele et al., 2006), a marker-assisted back-crossing (MABC) breeding program was implemented by Steele et al. (2006) to improve root depth, and thereby drought tolerance, of the Indian upland rice variety, Kalinga III. This variety, the recurrent parent in the MABC, had not previously been used for QTL mapping. The donor parent for the alleles contributing longer roots was Azucena, an upland japonica variety from the Philippines. Among the five segments on different chromosomes targeted for introgression, four carried QTLs on chromosomes 2, 7, 9 and 11 for improved root morphological traits (root length and thickness) and the fifth on chromosome 8 carried a recessive QTL for aroma. Selection made in three backcross (BC) generations and two further crosses between BC_3 lines allowed Steele et al. (2006) to pyramid (stack) all five target segments. Pyramids with four root QTLs were obtained in eight generations, completed in six years using 3000 marker assays in a total of 323 lines. Twenty-two near-isogenic lines (NILs) were evaluated for root traits in five field experiments in Bangalore, India. The target segment on chromosome 9 significantly increased root length under both irrigated and drought stress treatments, confirming that this root length QTL from Azucena was the most effective one to function in a novel genetic background. Notably, the other four targets showed no significant effects on root length (Steele et al., 2006) clearly indicating a strong QTL × genetic background interaction, the most common shortcoming when attempting MABC (Collins et al., 2008). The introgressed lines and the recurrent parent were grown for 6 years by resource-poor farmers in upland sites in Eastern India and yields recorded. In combination the QTLs significantly increased yield by 1 $t\,ha^{-1}$ under relatively favorable field conditions. In less-favorable trials, the QTL effects were not detected due to greater heterogeneity in soil-water availability in very low-yielding environments and consequent yield variability. Root studies under controlled conditions showed that lines with the introgressions had longer roots throughout tillering than the recurrent parent (14 cm longer 2 weeks after sowing). Therefore, both longer roots and increased yield can be attributed to the introgression of QTLs (Steele et al., 2013).

An additional, remarkable example of the important role played by RSA traits in regulating water use is provided by the work of Uga and co-workers on *DRO1,* a major QTL on chromosome 7 that governs root angle and deep rooting in rice (Ahmadi et al., 2014; Arai-Sanoh et al., 2014; Uga et al., 2013). Following the cloning of *DRO1,* a collection of rice cultivars with the functional *DRO1* allele were evaluated for RGA using different RIL populations in order to identify novel QTLs for RGA (Kitomi et al., 2015). The presence of novel QTLs was detected on chromosomes 2 (*DRO4*), 4 (two QTLs) and 6 (*DRO5*). Among them, the two QTLs on chromosome 4 were located near *DRO2*, which has been previously reported as a major QTL for RGA, whereas the two major QTLs for RGA on chromosomes 2 (*DRO4*) and 6 (*DRO5*) were novel. When the LOD threshold was reduced to 3.0, several minor QTLs for RGA were also detected in each one of the three populations, hence indicating that natural variation in RGA in rice cultivars carrying functional *DRO1* alleles may be controlled by a few major QTLs and by several additional minor QTLs (Kitomi et al.,

2015). More recently, Uga and co-workers applied a multi-omics approach to decipher the adaptive response of root development under various abiotic stresses such as drought, salinity and heat while assessing the status of root phenotyping technology and modeling in rice. This knowledge allowed Uga and co-workers to contemplate the possibility of applying an integrated multi-omics dataset from RSA to molecular breeding of climate-resilient rice (Yoshino et al., 2019).

Prolonged submergence is a major constraint to rice production, affecting over 15 million ha in South and Southeast Asia during the monsoon season, and causing an annual yield loss of over US$ 600 million because all the modern high-yielding rice varieties cultivated until recently are highly sensitive to complete submergence (Singh et al., 2009). In terms of the direct impact of MAS for root features, the best and most successful example is provided by the work in rice that followed the identification of *Sub1*, a major locus for submergence tolerance mapped on rice chromosome 9 (Xu and Mackill, 1996). Following fine mapping (Xu et al., 2000), positional cloning showed that *Sub1* effects were due to an ethylene-response-factor-like gene that confers submergence tolerance to rice (Xu et al., 2006). Shortly after, *Sub1* effects were evaluated in numerous field trials conducted with several elite backgrounds (Fujita et al., 2009; Neeraja et al., 2007; Sarkar et al., 2009; Singh et al., 2009) that fully confirmed the great value of *Sub1* in alleviating or even avoiding the devastating effects caused by the prolonged (> 14 days) submersion of rice. The most notable marker-assisted backcrossing of *Sub1* was into IR64 (Septiningsih et al., 2009), the most important and successful mega-variety of cultivated rice (Mackill and Khush, 2018). The rice submergence tolerance gene *SUB1* was introduced by backcrossing into IR64 and several other popular varieties.

A remarkable outcome of almost two decades of interdisciplinary genomics studies in rice is the positional cloning of *Pstol1*, a major QTL for phosphorus uptake that confers tolerance to phosphorus deficiency (Gamuyao et al., 2012). Rice is very sensitive to phosphorus (P) starvation as shown by stunted root establishment/development and negative effects on vegetative growth and reproductive stage development with a severe reduction in yield. Since P deficiency of soils is a major concern of rice production areas, it is necessary to develop new cultivars with enhanced P tolerance, particularly in southern states of India where rice cultivation is intensive. To this end, Chithrameenal et al. (2018) introgressed the P-starvation tolerance (*OsPSTOL1*) gene through marker-assisted backcross breeding into two popular local-varieties namely, ASD 16 and ADT 43, which harbor bacterial blight and blast resistance genes. The developed backcross inbred lines showed altered root architecture pattern and greater root surface area with increased P uptake, confirming their adaptability to P-deficient soil conditions. The enhanced root characteristics, therefore, enabled the plants to access and effectively absorb available nutrients from the soil.

Cloning of the P uptake1 (*Pup1*) locus allowed Neelam et al. (2017) to search for novel haplotypes at the phosphorus-starvation tolerance 1 (*PSTOL1*) gene in a germplasm collection of 182 accessions of the wild species rice *Oryza rufipogon*. The analysis showed that all of the *O. rufipogon* accessions had an insertion of 90 kb region, including Pup1-K46, a diagnostic marker for *PSTOL1*, and found to be absent among *O. sativa* cv. PR114, PR121 and PR122. The complete *PSTOL1* gene was also sequenced in 67 representative accessions of *O. rufipogon* and cv. Vandana as a positive control. From comparative sequence analysis, 53 mutations (52 SNPs and 1 nonsense mutation) were found in the *PSTOL1* coding region, which represent 17 haplotypes (Neelam et al., 2017). A major shared haplotype was observed among 44 accessions of *O. rufipogon* along with Vandana and Kasalath. Out of 17 haplotypes, accessions representing eight haplotypes were grown under phosphorus-deficient conditions in hydroponics for 60 days. Significant differences were observed in root length and weight among all the genotypes when grown under phosphorus-deficient conditions as compared to phosphorus-sufficient conditions. The *O. rufipogon* accession IRGC 106506 from Laos performed significantly better, with 2.5-fold higher root weight and P content as compared to the positive control Vandana. In terms of P-uptake efficiency, the *O. rufipogon* accessions IRGC 104639, 104712 and 105569 also showed nearly two times higher P content than Vandana. Based on these results, Neelam et al. (2017) suggested that these *O. rufipogon* accessions could be used as the potential donor for improving P-uptake efficiency of elite rice cultivars.

4.2 Wheat

In wheat, a number of conventional breeding programs have targeted root traits for either morphological features (e.g. root depth) and/or functional features influencing the uptake of nutrients and/or toxic minerals (e.g. boron, salt, aluminum, etc.). This notwithstanding, to date one of the most well-known and striking examples of selection-driven RSA manipulation in conventional wheat breeding is the introgression of the rye/wheat 1RS.1BL translocation, which is now used worldwide (Howell et al., 2014; Schlegel and Meinel, 1994). This QTL was first described by Schlegel and Meinel (1994) based on the evaluation of more than 200 wheat varieties from all over the world showing the presence of a 1RS.1BL wheat-rye chromosome translocation. Yield testing and analysis of yield components of F_{2-5} subpopulations with and without the translocation revealed that plants carrying the rye translocation significantly outyielded the control plants mainly due to increased spikelet fertility. Based on these results, Schlegel and Meinel (1994) postulated the presence on the short arm of rye chromosome 1R of a major QTL-affecting wheat performance under drought conditions. From a breeding standpoint, a major shortcoming of the 1RS.1BL wheat-rye chromosome translocation was the poor quality due to the negative effects of the rye genes. Following a complex work of chromosome engineering, near-isogenic lines were assembled with a much smaller rye introgression that still retained the beneficial rye QTL allele but otherwise flanked by wheat-derived regions conferring the high flour-quality genes (Howell et al., 2014). The introgression of a small segment of wheat chromosome arm 1BS in the distal region of the rye 1RS.1BL arm translocation in wheat (henceforth $1RS^{RW}$) was shown to be associated with reduced grain yield, carbon isotope discrimination and stomatal conductance, suggesting reduced access to soil moisture (Howell et al., 2014). Subsequently, Howell et al. (2019) showed that lines with the normal 1RS arm have longer roots than lines with the $1RS^{RW}$ arm in both field and hydroponic experiments. In the $1RS^{RW}$ lines, differences in seminal root length were associated with a developmentally regulated arrest of the root apical meristem. Approximately 10 days after germination, the seminal roots of the $1RS^{RW}$ plants showed a gradual reduction in elongation rate and stopped growing a week later. Seventeen days after germination, the roots of the $1RS^{RW}$ plants showed altered gradients of ROS and the emergence of lateral roots close to the RAM, suggesting changes in the root meristem. The $1RS^{RW}$ lines also showed reduced biomass and grain yield relative to the 1RS lines, with larger differences under reduced or excessive irrigation than under normal irrigation hence suggesting that this genetic variation could be useful to modulate root architecture (Howell et al., 2019). Based on the results reported by Howell and co-workers, the $1RS^{RW}$ appears a valuable target for MAS to enhance drought tolerance in wheat.

Root system depth has been associated with adaptation to heat and drought stress (Lopes et al., 2015; Lopes and Reynolds, 2010). Therefore, root and root-related traits can be integrated into wheat improvement programs to develop cultivars more resilient to the negative effects of climate change. Enhanced root development (based on root dry weight analysis) in wheat is negatively correlated with WSCs under drought-stress condition indicating extensive root development enhances drought tolerance by the accumulation of carbohydrates (Mkhabela et al., 2019; Pinto et al., 2010). QTLs for root and root-related traits have been mapped in both durum and bread wheat (Alahmad et al., 2020; Atkinson et al., 2015; Canè et al., 2014; Li et al., 2020b; Petrarulo et al., 2015; Soriano and Alvaro, 2019; Voss-Fels et al., 2018a; Zheng et al., 2019). In some cases, QTLs for root component traits were reported to be co-located with those controlling plant height, indicating that wheat genotypes with height-reducing genes (*Rht*) may have reduced root growth under water-stressed environments (Bai et al., 2013). The QTLs can be transferred to new or well-adapted cultivars to improve yield in wheat (Tshikunde et al., 2019).

Other notable examples for potential application of MAS in wheat are related to genes/QTLs conferring tolerance to (i) salinity through the action at the root level of membrane transporters acting on Na^+ or Cl^- exclusion (Huang et al., 2006; James et al., 2012; Munns et al., 2012; Munns and Tester, 2008) or (ii) aluminum toxicity in acid soils (Inostroza-Blancheteau et al., 2010; Raman et al., 2008; Zhou et al., 2007), a major limitation to crop production worldwide. In wheat, the Al-activated malate transporter (*TaALMT1*) gene located on chromosome 4DL co-segregates with a major QTL associated with malate efflux and Al-tolerance (Zhou et al., 2007). To introgress Al-tolerance from the breeding line CAR3911 into the high-yielding Al-sensitive cultivar Kumpa-INIA, Soto-Cerda et al. (2015) undertook phenotypic and molecular characterizations of gene/QTL underlying Al-tolerance in CAR3911 followed by marker-assisted backcrossing (MAS-BC). Al-tolerant backcross (BC) lines were selected using the functional marker *ALMT1-4* designed immediately upstream of the *TaALMT1* coding region. Foreground and background selections using ALMT1-4 and microsatellite markers were conducted. The MAS-BC strategy allowed (Soto-Cerda et al., 2015) to select Al-tolerant lines with the smallest introgressed region (6 cM) on chromosome 4D and the highest recurrent parent genome (RPG) (98%) covering 2194 cM of the wheat genome. The homozygous BC_3F_2 line named Kumpa-INIA-TaALMT1 expressed a 3-fold higher Al-tolerance than its isogenic line Kumpa-INIA. The MAS-BC strategy was successful for the introgression of the *TaALMT1* gene into Kumpa-INIA in only three BC generations, shortening the breeding cycle to 24 months, a promising opportunity to increase wheat production and achieve greater yield stability in the acid soils of Southern Chile (Soto-Cerda et al., 2015).

4.3 Pulses

4.3.1 Chickpea

Beside the cultivars released through MAS and MABC for RSA traits in rice as described in the previous section, Varshney and co-workers mapped and introgressed a *QTL-hotspot* that harbors QTLs for several drought-tolerance related traits including the root depth trait, into several elite genetic backgrounds of chickpea in South Asia and Sub-Saharan Africa (Varshney et al., 2013). In South Asia, especially in India, the *QTL-hotspot* was introgressed into elite chickpea cultivars like Pusa 362, Pusa 372, DCP 92-3 and JG 11 backgrounds. In the case of Sub-Saharan Africa the *QTL-hotspot* was introgressed into Ejere (in Ethiopia) and Chania Desi II, LDT 068 (in Kenya). As a result 20, 1, 21, 3, 7 and 19 molecular breeding lines with enhanced drought tolerance and yield were developed in the genetic background of Pusa 362, Pusa 372, DCP92-3, Ejere, LDT 068 and

Chania Desi II, respectively. In addition to developing molecular breeding lines, two high-yielding and drought-tolerant chickpea varieties (one each in India and Ethiopia) were released (for more details, see attached websites). In India, Pusa Chickpea 10216 (https://icar.org.in/content/development-two-superior-chickpea-varieties-genomics-assisted-breeding) that exhibited 16% yield advantage over recurrent parent Pusa 372 was released for commercial cultivation by the central varietal release committee. In Ethiopia, the root-enhanced variety Geletu (genetic background of JG 11) released by EIAR for mid- to high-altitude areas (1800–2700 m above sea level) in sub-moist agroecological zones (https://www.icrisat.org/first-ever-high-yielding-chickpea-variety-developed-using-marker-assisted-backcrossing-mabc-released-in-ethiopia/), showed a 15% higher yield over improved check variety Teketay and 78% more than the local check.

4.3.2 Soybean

In soybean, root QTLs play a key role in plant adaptation and productivity in nematode-infested soils. Advancements in genetics, genomics and bioinformatics shed light on the molecular and genetic mechanisms of nematode resistance and enabled the identification and cloning of *Rhg1* and *Rhg4*, two major soybean cyst nematode-resistant loci and the discovery of other novel resistance QTLs (Ha et al., 2007; Kim et al., 2016). Based on these discoveries, gene-specific DNA markers have been developed for marker-assisted selection of both *Rhg1* and *Rhg4* loci. The efficacy of soybean *Rhg1* recently shown by Butler et al. (2019) in transgenic *Arabidopsis* and potato demonstrates that the encoded mechanisms can function across plant and cyst nematode species and offers a possible avenue for engineered resistance in other species.

Genetic variation in waterlogging tolerance in soybean is controlled by multiple genes mapped as QTLs. Two QTLs associated with waterlogging tolerance were mapped by Ye et al. (2018b) from an elite/exotic soybean cross. The subsequent research was focused on a major QTL (*qWT_Gm03*) with the tolerant allele from the exotic parent. This QTL was isolated into near-isogenic backgrounds, and its effects on waterlogging tolerance were validated in multiple environments (Ye et al., 2018b). Fine mapping narrowed *qWT_Gm03* into a genomic region of < 380 kbp where the tolerant allele of *qWT_Gm03* promotes root growth under non-stress conditions and favorable root plasticity under waterlogging, resulting in improved waterlogging tolerance, yield and also drought tolerance-related traits, possibly through more efficient water/nutrient uptakes. Meanwhile, the involvement of auxin pathways was also identified in the regulation of waterlogging tolerance, as the genotypic differences of *qWT_Gm03* in waterlogging tolerance and formation of adventitious/aerial roots can be complemented by an exogenous auxin-biosynthesis inhibitor. These findings provided genetic resources to address the urgent demand of improving waterlogging tolerance in soybean and revealed the determinant roles of root architecture and plasticity in plant adaptation to waterlogging (Ye et al., 2018b). Major natural genetic variation/QTL to improve waterlogging tolerance was identified from an exotic soybean line. The tolerant allele of this natural genetic variation showed better root growth and more favorable root plasticity to facilitate adequate water and nutrient uptake under nonstress and waterlogging stress conditions, possibly through auxin pathways. In addition to its role in waterlogging tolerance, the tolerant allele improved yield and drought tolerance-related traits possibly due to better root growth as suggested by Ye et al., (2018b) who also summarized current progress in the genetic diversity in major legume crops, QTLs associated with RSA and the importance and applications of recent discoveries associated with beneficial root traits toward better RSA for enhanced drought tolerance and yield.

In soybean, screening for natural genetic variation in drought- and flooding tolerance-related traits, including RSA, water and nitrogen-fixation efficiency, and yield performance indices, has helped to identify the best resources for genetic studies and breeding purposes (Prince et al., 2019; Valliyodan et al., 2017, 2019). Developing crops with better root systems is a promising strategy to ensure productivity in both optimum and stress environments. Root system architectural traits in 397 soybean accessions were characterized, and a high-density SNP-based GWA study identified SNPs associated with root architectural traits specific to landraces and elite germplasm pools. Four loci were detected in landraces for lateral root number (LRN) and distribution of root thickness in diameter Class I with a major locus on chromosome 16. This major locus was detected in the coding region of an unknown protein, and subsequent analyses demonstrated that root traits are affected with mutated haplotypes of the gene. In an elite germplasm pool, three significant SNPs in alanine-glyoxalate aminotransferase, leucine-rich repeat receptor/no apical meristem and unknown functional genes were found to govern multiple traits, including root surface area and volume. However, no major loci were detected for LRN in elite germplasm. Nucleotide diversity analysis found evidence of selective sweeps around the landraces LRN gene. Soybean accessions with minor and mutated allelic variants of LRN gene were found to perform better in both water-limited and optimal field conditions (Prince et al., 2019).

5 Conclusion and future trends

The past two decades have witnessed tremendous progress in our capacity to dissect the genome of cereals and other crops in order to identify and clone the genes and QTLs tailoring the morphology of the root system and controlling its functions and adaptive response to environmental cues, pathogens and pests. Our capacity to sequence and annotate entire genomes of crops has increased exponentially, paralleled by a similar reduction of the costs to decipher an entire genome even in large polyploid genomes as exemplified in wild emmer (Avni et al., 2017), bread wheat (Appels et al., 2018) and durum wheat (Maccaferri et al., 2019).

Despite crop genomics coming of age even in the polyploid Triticeae (Tuberosa and Pozniak, 2014; Uauy, 2017) and the other important crops (Bohra et al., 2020; Dubey et al., 2019; Jha et al., 2020; Varshney et al., 2012, 2018a; Wallace et al., 2018) that feed mankind, major hurdles still remain in order to take full advantage of its potential to optimize RSA and its functions. A major limiting factor is the difficulty to phenotype RSA traits in a high-throughput fashion, particularly in field conditions where the destructive protocols commonly adopted are time-consuming, expensive and unable to reveal the fine details of the root system while it grows. Traditional destructive phenotyping contrasts with the flexibility and precision offered by nondestructive, noninvasive, multisensor and multitrait phenotyping discussed by Tracy et al. (2020). Importantly, traditional destructive methods do not allow combinations of developmental periods and dynamic environmental phenotypes, and new seed production is only possible by using the original genetic resources of the parental lines. Conversely, noninvasive phenotyping allows traits to be measured over time on the same plants of the large populations commonly generated from multiparent advanced generation intercross populations (Huang et al., 2015; Stadlmeier et al., 2018).

An additional hurdle in genomic RSA studies attempting to fill in the genotype-to-phenotype gap is related to the fact that root and shoot phenotypes are expressed differently depending on soil conditions (e.g. nitrogen, see Gioia et al., 2015; water, see Nagel et al., 2015; strength, see Nagel et al., 2015, Pfeifer et al., 2014) and also depending on the growth stage (Chen et al., 2014; Shi et al., 2018). This problem can be mitigated with the use of automated, high-throughput platforms that are able to move plants, using automated robotics, to imaging and weighing stations and sensors, and throughput is high enough to handle mapping populations and/or diversity panels to identify genetic variation and QTLs for further pre-breeding (Junker et al., 2015; Nagel et al., 2012; Pieruschka and Poorter, 2012; Shi et al., 2018; Wasson et al., 2020). Importantly, there is a need for clarification of terms and concepts in the root phenotyping community. Along this line, (Lobet et al., 2019) advocated the use of a reference naming system (ontologies) for roots and root phenes. Such uniformity would not only allow a better understanding of research results but would also enable a better sharing of data. Importantly, Lobet et al. (2019) and also Schneider and Lynch (2020) highlighted the need to incorporate in breeding programs the concept of plasticity as an essential component of root system development in heterogeneous environments. A critical issue is to what extent the phenotyping data collected with high-throughput platforms in controlled conditions are able to appropriately mimic and report on the dynamics of multi-scale QTL effects estimated in field conditions (Stahl et al., 2020; Tuberosa, 2012). In this respect, pot effects deserve particular attention, since insufficient soil volumes almost invariably decrease root growth potential and the root-shoot biomass ratio, enhance soil temperature, cause fluctuations in soil moisture, increase irrigation requirements due to elevated water extraction and potentially increase salinity (Poorter et al., 2012; Stahl et al., 2020).

On the genomics/sequencing side for the identification of candidate genes at target QTL or their validation, data storage, accessibility and processing are key factors that can create bottlenecks and require adequate computing infrastructures to process the burgeoning amount of genome sequences, genome maps, QTLs, gene expression atlas and other multi-omics database deposited on the web (Chen et al., 2019b; Jaganathan et al. 2020). New approaches that pave the way to fast gene discovery through high-resolution mapping based on new generation sequencing (NGS) technologies have recently been discussed in Jaganathan et al. (2020) and are herewith summarized:

High-density SNP arrays for faster mapping: The SNP arrays or high-density genotyping based on resequencing is being used to generate large-scale marker profiles for a number of individuals (Winfield et al., 2016). The SNP chip is preferred over other high-density genotyping platforms as the SNP data it generates are less computationally demanding.

In wheat, a high-throughput genotyping array (TaBW280K) containing 280 226 SNPs was developed and deployed for assessing germplasm diversity as well as high-density linkage mapping (Rimbert et al., 2018).

Bulked segregant RNA Seq (BSR-Seq) for gene discovery: Whole transcriptome sequencing of contrasting bulks is referred to as BSR-Seq. This approach is particularly effective for crops with large and complex genomes like wheat, where resequencing still remains cost-ineffective (Li et al., 2018a; Zou et al., 2016). Also, BSR-seq is suitable for fine mapping in crops that still lack a reference genome sequence. Similar to QTL-Seq, BSR-Seq relies on sequencing RNAs (cDNA) from extreme bulks for the trait of interest. It facilitates the identification of the target region and the development of markers near or within the gene of interest.

QTL-Seq for gene identification: QTL-Seq integrates traditional bulk segregant analysis (BSA) with sequencing methods. This approach has been established as highly efficient for rapid discovery of candidate genes for the trait of interest. By allowing placement of QTLs within a smaller genomic segment, it facilitates both detection of QTL and, at a stretch, its fine mapping.

The procedure involves the creation of bulks by selecting lines with extremely low and high values of a trait from a segregating population. The pools, along with one of the parental genotypes, are subsequently sequenced using NGS. A reference genome assembly for the parental genotype is developed based on sequencing data. Subsequently, SNP index for each bulk is computed based on the number of reads for an SNP that differs from the SNP in the parent reference assembly. Notably, in the case of chickpea (*Cicer arietinum*), Singh et al. (2016) refined a 7 Mb QTL region on linkage group (LG) CaLG04 (previously identified using non-NGS-based markers) to a ~1 Mb region for root and seed weight traits in chickpea (Singh et al., 2016).

Latest advances in sequencing technologies, providing millions of single nucleotide polymorphism (SNP) markers at a low cost, have revolutionized the field of plant genomics. A paradigm shift from marker-based to sequencing-based genotyping of breeders' germplasm panels has been observed in the post-genome sequencing era (Sehgal et al., 2020). It is expected that the genetic toolkits available in most crops coupled with non-destructive phenotyping platforms will facilitate the genetic dissection of RSA phenes by GWAS and predicting field performance by genomic selection (GS) will become increasingly feasible. Genomic selection is another widely used genomics-assisted approach in which genome-wide markers are used to predict the breeding value of individuals in a breeding population. It offers the potential to accelerate genetic gain by increasing selection accuracy and intensity and shortening the lengths of breeding cycles (Sehgal et al., 2020).

A topic that so far has received limited attention is the dissection of the genetic basis of the interaction of roots with arbuscular mycorrhizal fungi and the microbiome and how this knowledge can eventually be exploited to select genotypes more resource-use efficient and more resilient to adverse environmental conditions. Only a limited number of studies have addressed these issues and in some cases have reported QTLs that regulate the interaction between the plant and the rhizobiome (Barzana et al., 2014; Campos et al., 2019; Davidson et al., 2019; De Vita et al., 2018; Goswami and Deka, 2020; Kaushal, 2019; Lehnert et al., 2018; Yadav et al., 2020). Increasing attention will likely be devoted to elucidate the genetic basis of the complex dynamics of root-shoot signaling in response to fluctuating environmental cues (Fichman and Mittler, 2020; Peck and Mittler, 2020; Zandalinas et al., 2020) and how such signals affect the plasticity of root morphology, anatomy and functions (Duque and Villordon, 2019; Fromm, 2019; Lynch, 2018; von Wangenheim et al., 2020).

The inconsistency of effects on yield of most RSA QTLs according to where the crop is grown is probably the main reason for the overall modest impact from a breeding standpoint of genomics-assisted breeding based on RSA traits and other yield proxies as discussed in a number of reviews (Collins et al.,

2008; Tardieu, 2012; Tardieu and Tuberosa, 2010). Modeling field performance of crops including also the effects of RSA QTLs and their interaction with the environment and management practices to predict more accurately the yield of a specific genotype based on its haplotypes effects across a broad range of environments can partially mitigate this shortcoming (van Eeuwijk et al., 2019). Accordingly, Tardieu et al. (2018) recently proposed an approach that analyzed the genetic architecture of traits in phenotyping platforms and of yield in maize in tens of field experiments. A combination of modeling and genomic prediction was then used to estimate the comparative interests of combinations of alleles depending on drought scenarios. Drought tolerance was understood probabilistically by estimating the benefit and risk of each combination of alleles (Tardieu et al., 2018).

From a functional standpoint, the studies reported herewith clearly indicate the outstanding contribution of genetics and genomics toward a better understanding of the developmental mechanisms that govern root growth and branching into soil regions based on the presence of water, nutrients, metals and/or minerals able to affect root growth (Bao et al., 2014; Dinneny, 2019; Feng et al., 2016; Gupta et al., 2020; Lynch, 2018; Motte and Beeckman, 2019; Motte et al., 2019; Orosa-Puente et al., 2018; Robbins and Dinneny, 2015, 2018). Despite the limited number of novel cultivars released so far through genomics-assisted breeding for RSA traits, cloning of the loci governing root growth and functions will provide a much-needed impetus toward the use of targeted approaches via genome editing and/or leveraging native variability toward the cultivation of climate-resilient cultivars for an agriculture with a lower environmental footprint and able to sustainably feed a growing population.

6 Where to look for further information

To look for further information on the topic of the genetic control and genomics of RSA and its functions, and applications to crop improvement, the following websites are suggested.

The Plantstress website (http://www.plantstress.com) lists manuscripts on different abiotic stresses in plants. Several manuscripts report on the role and genetic control of RSA phenes in the adaptive response to different abiotic stresses. The purpose of this website originally conceived by the late Dr. Abraham Blum (see Varshney et al., 2018c) is to serve as a brokerage of information, a consultation facility and a source for the professional update on the most important issues of plant abiotic stress. The site is dynamic and also offers basic educational materials to newcomers into this area who wish to use the site for learning. The most important goal of this website is to promote interaction among stakeholders interested in mitigating the negative effects of environmental stress in agriculture.

The Benfey Lab website (https://sites.duke.edu/benfey/) investigates how root cells acquire their identities using a combination of genetics, molecular biology and genomics to identify and characterize the genes that regulate the formation of the root in the plant model system *Arabidopsis thaliana*. The choice of the root as a model was based on the simplicity of its organization and its stereotyped developmental program.

The overall research focus of the Roots Lab at Penn State (https://plantscience.psu.edu/research/labs/roots) is to understand the genetic, physiological and ecological basis of root architecture and anatomy on plant adaptation to drought and low soil fertility.

7 References

Abdel-Ghani, A. H., Sharma, R., Wabila, C., Dhanagond, S., Owais, S. J., Duwayri, M. A., Al-Dalain, S. A., Klukas, C., Chen, D. J., Lubberstedt, T., von Wiren, N., Graner, A., Kilian, B. and Neumann, K. (2019). Genome-wide association mapping in a diverse spring barley collection reveals the presence of QTL hotspots and candidate genes for root and shoot architecture traits at seedling stage. *BMC Plant Biology* 19(1), 216.

Acevedo, L. A., Korson, N. E., Williams, J. M. and Nicholson, L. K. (2020). Tuning a timing device that regulates lateral root development in rice. *Journal of Biomolecular NMR* 73(8–9), 493–507.

Ahmadi, N., Audebert, A., Bennett, M. J., Bishopp, A., de Oliveira, A. C., Courtois, B., Diedhiou, A., Dievart, A., Gantet, P., Ghesquiere, A., Guiderdoni, E., Henry, A., Inukai, Y., Kochian, L., Laplaze, L., Lucas, M., Luu, D. T., Manneh, B., Mo, X. R., Muthurajan, R., Perin, C., Price, A., Robin, S., Sentenac, H., Sine, B., Uga, Y., Very, A. A., Wissuwa, M., Wu, P. and Xu, J. (2014). The roots of future rice harvests. *Rice* 7(1), 29.

Ajmera, I., Hodgman, T. C. and Lu, C. G. (2019). An integrative systems perspective on plant phosphate research. *Genes* 10(2).

Alahmad, S., El Hassouni, K., Bassi, F. M., Dinglasan, E., Youssef, C., Quarry, G., Aksoy, A., Mazzucotelli, E., Juhasz, A., Able, J. A., Christopher, J., Voss-Fels, K. P. and Hickey, L. T. (2019). A major root architecture QTL responding to water limitation in durum wheat. *Frontiers in Plant Science* 10, 436.

Alahmad, S., Kang, Y. C., Dinglasan, E., Mazzucotelli, E., Voss-Fels, K. P., Able, J. A., Christopher, J., Bassi, F. M. and Hickey, L. T. (2020). Adaptive traits to improve durum wheat yield in drought and crown rot environments. *International Journal of Molecular Sciences* 21(15), 5260.

Arai-Sanoh, Y., Takai, T., Yoshinaga, S., Nakano, H., Kojima, M., Sakakibara, H., Kondo, M. and Uga, Y. (2014). Deep rooting conferred by *DEEPER ROOTING 1* enhances rice yield in paddy fields. *Scientific Reports* 4, 5563.

Atkinson, J. A., Hawkesford, M. J., Whalley, W. R., Zhou, H. and Mooney, S. J. (2020). Soil strength influences wheat root interactions with soil macropores. *Plant, Cell and Environment* 43(1), 235–245.

Atkinson, J. A., Wingen, L. U., Griffiths, M., Pound, M. P., Gaju, O., Foulkes, M. J., Le Gouis, J., Griffiths, S., Bennett, M. J., King, J. and Wells, D. M. (2015). Phenotyping pipeline reveals major seedling root growth QTL in hexaploid wheat. *Journal of Experimental Botany* 66(8), 2283–2292.

Avni, R., Nave, M., Barad, O., Baruch, K., Twardziok, S. O., Gundlach, H., Hale, I., Mascher, M., Spannagl, M., Wiebe, K., Jordan, K. W., Golan, G., Deek, J., Ben-Zvi, B., Ben-Zvi, G., Himmelbach, A., MacLachlan, R. P., Sharpe, A. G., Fritz, A., Ben-David, R., Budak, H., Fahima, T., Korol, A., Faris, J. D., Hernandez, A., Mikel, M. A., Levy, A. A., Steffenson, B., Maccaferri, M., Tuberosa, R., Cattivelli, L., Faccioli, P., Ceriotti, A., Kashkush, K., Pourkheirandish, M., Komatsuda, T., Eilam, T., Sela, H., Sharon, A., Ohad, N., Chamovitz, D. A., Mayer, K. F. X., Stein, N., Ronen, G., Peleg, Z., Pozniak, C. J., Akhunov, E. D. and Distelfeld, A. (2017). Wild emmer genome architecture and diversity elucidate wheat evolution and domestication. *Science* 357(6346), 93-97.

Avni, R., Nave, M., Eilam, T., Sela, H., Alekperov, C., Peleg, Z., Dvorak, J., Korol, A. and Distelfeld, A. (2014). Ultra-dense genetic map of durum wheat x wild emmer wheat developed using the 90K iSelect SNP genotyping assay. *Molecular Breeding* 34(4), 1549-1562.

Ayele, M., Blum, A. and Nguyen, H. T. (2001). Diversity for osmotic adjustment and root depth in TEF [*Eragrostis tef* (Zucc) Trotter]. *Euphytica* 121(3), 237-249.

Bahrman, N., Gouy, A., Devienne-Barret, F., Hirel, B., Vedele, F. and Le Gouis, J. L. (2005). Differential change in root protein patterns of two wheat varieties under high and low nitrogen nutrition levels. *Plant Science* 168(1), 81-87.

Bai, C., Liang, Y. and Hawkesford, M. J. (2013). Identification of QTLs associated with seedling root traits and their correlation with plant height in wheat. *Journal of Experimental Botany* 64(6), 1745-1753.

Balakrishnan, D., Surapaneni, M., Mesapogu, S. and Neelamraju, S. (2019). Development and use of chromosome segment substitution lines as a genetic resource for crop improvement. *Theoretical and Applied Genetics* 132(1), 1-25.

Bao, Y., Aggarwal, P., Robbins, N. E., Sturrock, C. J., Thompson, M. C., Tan, H. Q., Tham, C., Duan, L. N., Rodriguez, P. L., Vernoux, T., Mooney, S. J., Bennett, M. J. and Dinneny, J. R. (2014). Plant roots use a patterning mechanism to position lateral root branches toward available water. *Proceedings of the National Academy of Sciences of the United States of America* 111(25), 9319-9324.

Barzana, G., Aroca, R., Bienert, G. P., Chaumont, F. and Ruiz-Lozano, J. M. (2014). New insights into the regulation of aquaporins by the arbuscular mycorrhizal symbiosis in maize plants Under drought stress and possible implications for plant performance. *Molecular Plant-Microbe Interactions* 27(4), 349-363.

Beaver, J. S. and Osorno, J. M. (2009). Achievements and limitations of contemporary common bean breeding using conventional and molecular approaches. *Euphytica* 168(2), 145-175.

Beavis, W. (1994). The power and deceit of QTL experiments: lessons from comparative QTL studies. In: 49th Annual Corn Sorghum Research Conference (D. Wilkinson, Ed.), pp. 250-266. Chicago.

Beebe, S. E., Rojas-Pierce, M., Yan, X. L., Blair, M. W., Pedraza, F., Munoz, F., Tohme, J. and Lynch, J. P. (2006). Quantitative trait loci for root architecture traits correlated with phosphorus acquisition in common bean. *Crop Science* 46(1), 413-423.

Bengough, A. G., Gordon, D. C., Al-Menaie, H., Ellis, R. P., Allan, D., Keith, R., Thomas, W. T. B. and Forster, B. P. (2004). Gel observation chamber for rapid screening of root traits in cereal seedlings. *Plant and Soil* 262(1/2), 63-70.

Bernardino, K. C., Pastina, M. M., Menezes, C. B., de Sousa, S. M., Maciel, L. S., Carvalho, G., Guimaraes, C. T., Barros, B. A., Silva, L. D. E., Carneiro, P. C. S., Schaffert, R. E., Kochian, L. V. and Magalhaes, J. V. (2019). The genetic architecture of phosphorus efficiency in sorghum involves pleiotropic QTL for root morphology and grain yield under low phosphorus availability in the soil. *BMC Plant Biology* 19(1), 87.

Bernardo, R. (2016). Bandwagons I, too, have known. *Theoretical and Applied Genetics* 129(12), 2323–2332.

Bernardo, R. (2020). Reinventing quantitative genetics for plant breeding: something old, something new, something borrowed, something BLUE. *Heredity*, 125(6), 375–385.

Bernardo, R. and Yu, J. M. (2007). Prospects for genomewide selection for quantitative traits in maize. *Crop Science* 47(3), 1082–1090.

Beyer, S., Daba, S., Tyagi, P., Bockelman, H., Brown-Guedira, G. and Mohammadi, M. (2019). Loci and candidate genes controlling root traits in wheat seedling - A wheat root GWAS. *Functional and Integrative Genomics* 19(1), 91–107.

Bibi, A., Sadaqat, H. A., Tahir, M. H. N. and Akram, H. M. (2012). Screening of sorghum (*Sorghum bicolor* Var Moench) for drought tolerance at seedling stage in polyethylene glycol. *Journal of Animal and Plant Sciences* 22, 671–678.

Blum, A. (1988). *Plant Breeding for Stress Environments*. CRC Press, Boca Raton, FL. ISBN 978131589661, pp 231.

Blum, A. (2011). *Plant Breeding for Water-Limited Environments*. Springer, ISBN 978-1-4419-7491-4.

Blum, A. (2014). Genomics for drought resistance - getting down to earth. *Functional Plant Biology* 41(11), 1191–1198.

Blummel, M., Deshpande, S., Kholova, J. and Vadez, V. (2015). Introgression of staygreen QTLs for concomitant improvement of food and fodder traits in *Sorghum bicolor*. *Field Crops Research* 180, 228–237.

Bohra, A., Saxena, K. B., Varshney, R. K. and Saxena, R. K. (2020). Genomics-assisted breeding for pigeonpea improvement. *Theoretical and Applied Genetics* 133(5), 1721–1737.

Bolger, A. M., Poorter, H., Dumschott, K., Bolger, M. E., Arend, D., Osorio, S., Gundlach, H., Mayer, K. F. X., Lange, M., Scholz, U. and Usadel, B. (2019). Computational aspects underlying genome to phenome analysis in plants. *Plant Journal* 97(1), 182–198.

Borrell, A. K., Hammer, G. L. and Henzell, R. G. (2000). Does maintaining green leaf area in sorghum improve yield under drought? II. Dry matter production and yield. *Crop Science* 40(4), 1037–1048.

Borrell, A. K., Mullet, J. E., George-Jaeggli, B., van Oosterom, E. J., Hammer, G. L., Klein, P. E. and Jordan, D. R. (2014a). Drought adaptation of stay-green sorghum is associated with canopy development, leaf anatomy, root growth, and water uptake. *Journal of Experimental Botany* 65(21), 6251–6263.

Borrell, A. K., van Oosterom, E. J., Mullet, J. E., George-Jaeggli, B., Jordan, D. R., Klein, P. E. and Hammer, G. L. (2014b). Stay-green alleles individually enhance grain yield in sorghum under drought by modifying canopy development and water uptake patterns. *New Phytologist* 203(3), 817–830.

Borrill, P., Harrington, S. A. and Uauy, C. (2019). Applying the latest advances in genomics and phenomics for trait discovery in polyploid wheat. *Plant Journal* 97(1), 56–72.

Bovina, R., Talame, V., Ferri, M., Tuberosa, R., Chmielewska, B., Szarejko, I. and Sanguineti, M. C. (2011). Identification of root morphology mutants in barley. *Plant Genetic Resources* 9(2), 357–360.

Boyles, R. E., Brenton, Z. W. and Kresovich, S. (2019). Genetic and genomic resources of sorghum to connect genotype with phenotype in contrasting environments. *Plant Journal* 97(1), 19–39.

Bray, A. L. and Topp, C. N. (2018). The quantitative genetic control of root architecture in maize. *Plant and Cell Physiology* 59(10), 1919-1930.

Burridge, J. D., Findeis, J. L., Jochua, C. N., Miguel, M. A., Mubichi-Kut, F. M., Quinhentos, M. L., Xerinda, S. A. and Lynch, J. P. (2019). A case study on the efficacy of root phenotypic selection for edaphic stress tolerance in low-input agriculture: common bean breeding in Mozambique. *Field Crops Research* 244, 11.

Burton, A. L., Johnson, J. M., Foerster, J. M., Hirsch, C. N., Buell, C. R., Hanlon, M. T., Kaeppler, S. M., Brown, K. M. and Lynch, J. P. (2014). QTL mapping and phenotypic variation for root architectural traits in maize (*Zea mays* L.). *Theoretical and Applied Genetics* 127(11), 2293-2311.

Butler, K. J., Chen, S. Y., Smith, J. M., Wang, X. H. and Bent, A. F. (2019). Soybean resistance locus Rhg1 confers resistance to multiple cyst nematodes in diverse plant species. *Phytopathology* 109(12), 2107-2115.

Butsayawarapat, P., Juntawong, P., Khamsuk, O. and Somta, P. (2019). Comparative transcriptome analysis of waterlogging-sensitive and tolerant Zombi pea (*Vigna vexillata*) reveals energy conservation and root plasticity controlling waterlogging tolerance. *Plants* 8(8), 264.

Cai, G. C., Ahmed, M. A., Dippold, M. A., Zarebanadkouki, M. and Carminati, A. (2020). Linear relation between leaf xylem water potential and transpiration in pearl millet during soil drying. *Plant and Soil* 447(1-2), 565-578.

Caldwell, D. G., McCallum, N., Shaw, P., Muehlbauer, G. J., Marshall, D. F. and Waugh, R. (2004). A structured mutant population for forward and reverse genetics in barley (*Hordeum vulgare* L.). *Plant Journal* 40(1), 143-150.

Campos, C., Nobre, T., Goss, M. J., Faria, J., Barrulas, P. and Carvalho, M. (2019). Transcriptome analysis of wheat roots reveals a differential regulation of stress responses related to arbuscular mycorrhizal fungi and soil disturbance. *Biology-Basel* 8(4).

Canè, M. A., Maccaferri, M., Nazemi, G., Salvi, S., Francia, R., Colalongo, C. and Tuberosa, R. (2014). Association mapping for root architectural traits in durum wheat seedlings as related to agronomic performance. *Molecular Breeding* 34(4), 1629-1645.

Carminati, A., Passioura, J. B., Zarebanadkouki, M., Ahmed, M. A., Ryan, P. R., Watt, M. and Delhaize, E. (2017). Root hairs enable high transpiration rates in drying soils. *New Phytologist* 216(3), 771-781.

Champoux, M. C., Wang, G., Sarkarung, S., Mackill, D. J., O'Toole, J. C., Huang, N. and McCouch, S. R. (1995). Locating genes associated with root morphology and drought avoidance in rice via linkage to molecular markers. *Theoretical and Applied Genetics* 90(7-8), 969-981.

Chen, D. J., Neumann, K., Friedel, S., Kilian, B., Chen, M., Altmann, T. and Klukas, C. (2014). Dissecting the phenotypic components of crop plant growth and drought responses based on high-throughput image analysis. *Plant Cell* 26(12), 4636-4655.

Chen, L., Meng, J. and Luan, Y. S. (2019a). miR1916 plays a role as a negative regulator in drought stress resistance in tomato and tobacco. *Biochemical and Biophysical Research Communications* 508(2), 597-602.

Chen, T. W., Cabrera-Bosquet, L., Prado, S. A., Perez, R., Artzet, S., Pradal, C., Coupel-Ledru, A., Fournier, C. and Tardieu, F. (2019b). Genetic and environmental dissection of biomass accumulation in multi-genotype maize canopies. *Journal of Experimental Botany* 70(9), 2523-2534.

Chen, W. Q., Drapek, C., Li, D. X., Xu, Z. H., Benfey, P. N. and Bai, S. N. (2019c). Histone deacetylase HDA19 affects root cortical cell fate by interacting with SCARECROW. *Plant Physiology* 180(1), 276–288.

Chen, Y. Q., Xie, Y. M., Song, C. H., Zheng, L. L., Rong, X., Jia, L. T., Luo, L., Zhang, C., Qu, X. X. and Xuan, W. (2018). A comparison of lateral root patterning among dicot and monocot plants. *Plant Science* 274, 201–211.

Chithrameenal, K., Alagarasan, G., Raveendran, M., Robin, S., Meena, S., Ramanathan, A. and Ramalingam, J. (2018). Genetic enhancement of phosphorus starvation tolerance through marker assisted introgression of *OsPSTOL1* gene in rice genotypes harbouring bacterial blight and blast resistance. *PLoS ONE* 13(9), e0204144.

Chloupek, O., Forster, B. P. and Thomas, W. T. B. (2006). The effect of semi-dwarf genes on root system size in field-grown barley. *Theoretical and Applied Genetics* 112(5), 779–786.

Christopher, J., Christopher, M., Jennings, R., Jones, S., Fletcher, S., Borrell, A., Manschadi, A. M., Jordan, D., Mace, E. and Hammer, G. (2013). QTL for root angle and number in a population developed from bread wheats (*Triticum aestivum*) with contrasting adaptation to water-limited environments. *Theoretical and Applied Genetics* 126(6), 1563–1574.

Christopher, M., Chenu, K., Jennings, R., Fletcher, S., Butler, D., Borrell, A. and Christopher, J. (2018). QTL for stay-green traits in wheat in well-watered and water-limited environments. *Field Crops Research* 217, 32–44.

Clarke, C. K., Gregory, P. J., Lukac, M., Burridge, A. J., Allen, A. M., Edwards, K. J. and Gooding, M. J. (2017). Quantifying rooting at depth in a wheat doubled haploid population with introgression from wild emmer. *Annals of Botany* 120(3), 457–470.

Collins, N. C., Tardieu, F. and Tuberosa, R. (2008). Quantitative trait loci and crop performance under abiotic stress: where do we stand? *Plant Physiology* 147(2), 469–486.

Courtois, B., Ahmadi, N., Khowaja, F., Price, A. H., Rami, J. F., Frouin, J., Hamelin, C. and Ruiz, M. (2009). Rice root genetic architecture: meta-analysis from a drought QTL database. *Rice* 2(2–3), 115–128.

Darvasi, A., Weinreb, A., Minke, V., Weller, J. I. and Soller, M. (1993). Detecting marker-QTL linkage and estimating QTL gene effect and map location using a saturated genetic-map. *Genetics* 134(3), 943–951.

Davidson, H., Shrestha, R., Cornulier, T., Douglas, A., Travis, T., Johnson, D. and Price, A. H. (2019). Spatial effects and GWA mapping of root colonization assessed in the interaction between the rice diversity Panel 1 and an arbuscular mycorrhizal fungus. *Frontiers in Plant Science* 10, 633.

De Vita, P., Avio, L., Sbrana, C., Laido, G., Marone, D., Mastrangelo, A. M., Cattivelli, L. and Giovannetti, M. (2018). Genetic markers associated to arbuscular mycorrhizal colonization in durum wheat. *Scientific Reports* 8(1), 10612.

Del Bianco, M. and Kepinski, S. (2018). Building a future with root architecture. *Journal of Experimental Botany* 69(22), 5319–5323.

Deng, W. W., Casao, M. C., Wang, P. H., Sato, K., Hayes, P. M., Finnegan, E. J. and Trevaskis, B. (2015). Direct links between the vernalization response and other key traits of cereal crops. *Nature Communications* 6, 5882.

Deshmukh, V., Kamoshita, A., Norisada, M. and Uga, Y. (2017). Near-isogenic lines of IR64 (*Oryza sativa* subsp *indica* cv.) introgressed with *DEEPER ROOTING 1* and *STELE TRANSVERSAL AREA 1* improve rice yield formation over the background parent across three water management regimes. *Plant Production Science* 20(3), 249–261.

Dharmappa, P. M., Doddaraju, P., Malagondanahalli, M. V., Rangappa, R. B., Mallikarjuna, N. M., Rajendrareddy, S. H., Ramanjinappa, R., Mavinahalli, R. P., Prasad, T. G., Udayakumar, M. and Sheshshayee, S. M. (2019). Introgression of root and water use efficiency traits enhances water productivity: an evidence for physiological breeding in rice (*Oryza sativa* L.). *Rice* 12(1), 14.

Diehn, T. A., Bienert, M. D., Pommerrenig, B., Liu, Z. J., Spitzer, C., Bernhardt, N., Fuge, J., Bieber, A., Richet, N., Chaumont, F. and Bienert, G. P. (2019). Boron demanding tissues of *Brassica napus* express specific sets of functional Nodulin26-like intrinsic proteins and BOR1 transporters. *Plant Journal* 100(1), 68–82.

Dinneny, J. R. (2019). Developmental responses to water and salinity in root systems. In: *Annual Review of Cell and Developmental Biology*, Vol 35 (R. Lehmann, Ed.), pp. 239–257.

Dolan, L. (2017). Root hair development in grasses and cereals (Poaceae). *Current Opinion in Genetics and Development* 45, 76–81.

Dubey, A., Kumar, A., Abd Allah, E. F., Hashem, A. and Khan, M. L. (2019). Growing more with less: breeding and developing drought resilient soybean to improve food security. *Ecological Indicators* 105, 425–437.

Duque, L. O. and Villordon, A. (2019). Root branching and nutrient efficiency: status and way forward in root and tuber crops. *Frontiers in Plant Science* 10, 237.

Eshed, Y. and Zamir, D. (1994). A genomic library of *Lycopersicon pennellii* in *L. esculentum*: a tool for fine-mapping of genes. *Euphytica* 79(3), 175–179.

Feng, W., Lindner, H., Robbins, N. E. and Dinneny, J. R. (2016). Growing out of stress: the role of cell- and organ-scale growth control in plant water-stress responses. *Plant Cell* 28(8), 1769–1782.

Fichman, Y. and Mittler, R. (2020). Rapid systemic signaling during abiotic and biotic stresses: is the ROS wave master of all trades? *Plant Journal* 102(5), 887–896.

Flint-Garcia, S. A., Thuillet, A. C., Yu, J. M., Pressoir, G., Romero, S. M., Mitchell, S. E., Doebley, J., Kresovich, S., Goodman, M. M. and Buckler, E. S. (2005). Maize association population: a high-resolution platform for quantitative trait locus dissection. *Plant Journal* 44(6), 1054–1064.

Frary, A., Fulton, T. M., Zamir, D. and Tanksley, S. D. (2004). Advanced backcross QTL analysis of a *Lycopersicon esculentum* L. *pennellii* cross and identification of possible orthologs in the Solanaceae. *Theoretical and Applied Genetics* 108, 485–496.

Fromm, H. (2019). Root plasticity in the pursuit of water. *Plants-Basel* 8(7), 236.

Fu, Y. Q., Zhong, X. H., Pan, J. F., Liang, K. M., Liu, Y. Z., Peng, B. L., Hu, X. Y. and Huang, N. (2019). QTLs identification for nitrogen and phosphorus uptake-related traits using ultra-high density SNP linkage. *Plant Science* 288(110209).

Fujita, D., Santos, R. E., Ebron, L. A., Telebanco-Yanoria, M. J., Kato, H., Kobayashi, S., Uga, Y., Araki, E., Takai, T., Tsunematsu, H., Imbe, T., Khush, G. S., Brar, D. S., Fukuta, Y. and Kobayashi, N. (2009). Development of introgression lines of an Indica-type rice variety, IR64, for unique agronomic traits and detection of the responsible chromosomal regions. *Field Crops Research* 114(2), 244–254.

Gajewska, P., Janiak, A., Kwasniewski, M., Kedziorski, P. and Szarejko, I. (2018). Forward genetics approach reveals a mutation in bHLH transcription factor-encoding gene as the best candidate for the root hairless phenotype in barley. *Frontiers in Plant Science* 9, 1229.

Gamuyao, R., Chin, J. H., Pariasca-Tanaka, J., Pesaresi, P., Catausan, S., Dalid, C., Slamet-Loedin, I., Tecson-Mendoza, E. M., Wissuwa, M. and Heuer, S. (2012). The protein kinase Pstol1 from traditional rice confers tolerance of phosphorus deficiency. *Nature* 488(7412), 535-539.

Gasparis, S., Przyborowski, M., Kala, M. and Nadolska-Orczyk, A. (2019). Knockout of the *HvCKX1* or *HvCKX3* gene in barley (*Hordeum vulgare* L.) by RNA-guided Cas9 nuclease affects the regulation of cytokinin metabolism and root morphology. *Cells* 8(8), 782.

Gautam, T. and Gupta, P. K. (2020). Sequence variation in genes encoding miRNAs/targets and other related approaches for possible use in crop improvement. *Plant Breeding* 139(1), 28–41.

Gelli, M., Konda, A. R., Liu, K., Zhang, C., Clemente, T. E., Holding, D. R. and Dweikat, I. M. (2017). Validation of QTL mapping and transcriptome profiling for identification of candidate genes associated with nitrogen stress tolerance in sorghum. *BMC Plant Biology* 17(1), 123.

Gioia, T., Nagel, K. A., Beleggia, R., Fragasso, M., Ficco, D. B. M., Pieruschka, R., De Vita, P., Fiorani, F. and Papa, R. (2015). Impact of domestication on the phenotypic architecture of durum wheat under contrasting nitrogen fertilization. *Journal of Experimental Botany* 66(18), 5519–5530.

Giuliani, S., Clarke, J., Kreps, J. A., Sanguineti, M. C., Salvi, S., Landi, P., Zhu, T. and Tuberosa, R. (2005a). Microarray analysis of backcrossed-derived lines differing for *root-ABA1*, a major QTL controlling root characteristics and ABA concentration in maize. In: Proceedings of an International Congress "In the Wake of the Double Helix: From the Green Revolution to the Gene Revolution", 27–31 May 2003, Bologna, Italy, pp. 463–490. Avenue Media, Bologna, Italy.

Giuliani, S., Sanguineti, M. C., Tuberosa, R., Bellotti, M., Salvi, S. and Landi, P. (2005b). *Root-ABA1*, a major constitutive QTL, affects maize root architecture and leaf ABA concentration at different water regimes. *Journal of Experimental Botany* 56(422), 3061–3070.

Golan, G., Hendel, E., Espitia, G. E. M., Schwartz, N. and Peleg, Z. (2018). Activation of seminal root primordia during wheat domestication reveals underlying mechanisms of plant resilience. *Plant, Cell and Environment* 41(4), 755–766.

Golicz, A. A., Batley, J. and Edwards, D. (2016). Towards plant pangenomics. *Plant Biotechnology Journal* 14(4), 1099–1105.

Golicz, A. A., Bayer, P. E., Bhalla, P. L., Batley, J. and Edwards, D. (2020). Pangenomics comes of age: from bacteria to plant and animal applications. *Trends in Genetics* 36(2), 132–145.

Goswami, M. and Deka, S. (2020). Plant growth-promoting rhizobacteria-alleviators of abiotic stresses in soil: a review. *Pedosphere* 30(1), 40–61.

Gregory, P. J., Bengough, A. G., Grinev, D., Schmidt, S., Thomas, W. B. T. B., Wojciechowski, T. and Young, I. M. (2009). Root phenomics of crops: opportunities and challenges. *Functional Plant Biology* 36(11), 922–929.

Grewal, S., Hubbart-Edwards, S., Yang, C. Y., Scholefield, D., Ashling, S., Burridge, A., Wilkinson, P. A., King, I. P. and King, J. (2018). Detection of *T. urartu* introgressions in wheat and development of a panel of interspecific introgression lines. *Frontiers in Plant Science* 9, 1565.

Grosskinsky, D. K., Syaifullah, S. J. and Roitsch, T. (2018). Integration of multi-omics techniques and physiological phenotyping within a holistic phenomics approach to study senescence in model and crop plants. *Journal of Experimental Botany* 69(4), 825–844.

Gu, R. L., Chen, F. J., Long, L. Z., Cai, H. G., Liu, Z. G., Yang, J. B., Wang, L. F., Li, H. Y., Li, J. H., Liu, W. X., Mi, G. H., Zhang, F. S. and Yuan, L. X. (2016). Enhancing phosphorus uptake efficiency through QTL-based selection for root system architecture in maize. *Journal of Genetics and Genomics* 43(11), 663–672.

Guo, J., Chen, L., Li, Y. X., Shi, Y. S., Song, Y. C., Zhang, D. F., Li, Y., Wang, T. Y., Yang, D. G. and Li, C. H. (2018). Meta-QTL analysis and identification of candidate genes related to root traits in maize. *Euphytica* 214(12) 223.

Gupta, A., Rico-Medina, A. and Cano-Delgado, A. I. (2020). The physiology of plant responses to drought. *Science* 368(6488), 266–269.

Guseman, J. M., Webb, K., Srinivasan, C. and Dardick, C. (2017). *DRO1* influences root system architecture in Arabidopsis and Prunus species. *Plant Journal* 89(6), 1093–1105.

Ha, B. K., Hussey, R. S. and Boerma, H. R. (2007). Development of SNP assays for marker-assisted selection of two southern root-knot nematode resistance QTL in soybean. *Crop Science* 47, S73–S82.

Hanzawa, E., Sasaki, K., Nagai, S., Obara, M., Fukuta, Y., Uga, Y., Miyao, A., Hirochika, H., Higashitani, A., Maekawa, M. and Sato, T. (2013). Isolation of a novel mutant gene for soil-surface rooting in rice (*Oryza sativa* L.). *Rice* 6(1), 30.

Harris, K., Subudhi, P. K., Borrell, A., Jordan, D., Rosenow, D., Nguyen, H., Klein, P., Klein, R. and Mullet, J. (2007). Sorghum stay-green QTL individually reduce post-flowering drought-induced leaf senescence. *Journal of Experimental Botany* 58(2), 327–338.

Hasanuzzaman, M., Shabala, L., Brodribb, T. J., Zhou, M. X. and Shabala, S. (2017). Assessing the suitability of various screening methods as a proxy for drought tolerance in barley. *Functional Plant Biology* 44(2), 253–266.

Hazzouri, K. M., Khraiwesh, B., Amiri, K. M. A., Pauli, D., Blake, T., Shahid, M., Mullath, S. K., Nelson, D., Mansour, A. L., Salehi-Ashtiani, K., Purugganan, M. and Masmoudi, K. (2018). Mapping of *HKT1;5* gene in barley using GWAS approach and its implication in salt tolerance mechanism. *Frontiers in Plant Science* 9, 156.

He, Y. J., Hu, D. X., You, J. C., Wu, D. M., Cui, Y. X., Dong, H. L., Li, J. N. and Qian, W. (2019). Genome-wide association study and protein network analysis for understanding candidate genes involved in root development at the rapeseed seedling stage. *Plant Physiology and Biochemistry* 137, 42–52.

Herbert, L., Meunier, A. C., Bes, M., Vernet, A., Portefaix, M., Durandet, F., Michel, R., Chaine, C., This, P., Guiderdoni, E. and Perin, C. (2020). Beyond seek and destroy: how to generate allelic series using genome editing tools. *Rice* 13(1), 5.

Hickey, L. T., Hafeez, A. N., Robinson, H., Jackson, S. A., Leal-Bertioli, S. C. M., Tester, M., Gao, C. X., Godwin, I. D., Hayes, B. J. and Wulff, B. B. H. (2019). Breeding crops to feed 10 billion. *Nature Biotechnology* 37(7), 744–754.

Hirel, B., Le Gouis, J., Ney, B. and Gallais, A. (2007). The challenge of improving nitrogen use efficiency in crop plants: towards a more central role for genetic variability and quantitative genetics within integrated approaches. *Journal of Experimental Botany* 58(9), 2369–2387.

Hirel, B. X., Bertin, P., Quiller, I., Bourdoncle, W., Attagnant, C., Dellay, C., Gouy, A., Cadiou, S., Retailliau, C., Falque, M. and Gallais, A. (2001). Towards a better understanding of the genetic and physiological basis for nitrogen use efficiency in maize. *Plant Physiology* 125(3), 1258–1270.

Hirochika, H. (2010). Insertional mutagenesis with *Tos17* for functional analysis of rice genes. *Breeding Science* 60(5), 486–492.

Hochholdinger, F., Woll, K., Sauer, M. and Dembinsky, D. (2004). Genetic dissection of root formation in maize (*Zea mays*) reveals root-type specific developmental programmes. *Annals of Botany* 93(4), 359–368.
Hochholdinger, F., Yu, P. and Marcon, C. (2018). Genetic control of root system development in maize. *Trends in Plant Science* 23(1), 79–88.
Howell, T., Hale, I., Jankuloski, L., Bonafede, M., Gilbert, M. and Dubcovsky, J. (2014). Mapping a region within the 1RS.1BL translocation in common wheat affecting grain yield and canopy water status. *Theoretical and Applied Genetics* 127(12), 2695–2709.
Howell, T., Moriconi, J. I., Zhao, X. Q., Hegarty, J., Fahima, T., Santa-Maria, G. E. and Dubcovsky, J. (2019). A wheat/rye polymorphism affects seminal root length and yield across different irrigation regimes. *Journal of Experimental Botany* 70(15), 4027–4037.
Huang, B. E., Verbyla, K. L., Verbyla, A. P., Raghavan, C., Singh, V. K., Gaur, P., Leung, H., Varshney, R. K. and Cavanagh, C. R. (2015). Magic populations in crops: current status and future prospects. *Theoretical and Applied Genetics* 128(6), 999–1017.
Huang, S. B., Spielmeyer, W., Lagudah, E. S., James, R. A., Platten, J. D., Dennis, E. S. and Munns, R. (2006). A sodium transporter (HKT7) is a candidate for *Nax1*, a gene for salt tolerance in durum wheat. *Plant Physiology* 142(4), 1718–1727.
Inostroza-Blancheteau, C., Soto, B., Ibanez, C., Ulloa, P., Aquea, F., Arce-Johnson, P. and Reyes-Diaz, M. (2010). Mapping aluminum tolerance loci in cereals: A tool available for crop breeding. *Electronic Journal of Biotechnology* 13(4), 12.
International Wheat Genome Sequencing Consortium (IWGSC) (2018). Shifting the limits in wheat research and breeding using a fully annotated reference genome. *Science*, 2018, 361.6403.
Jaganathan, D., Bohra, A., Thudi, M. and Varshney, R. K. (2020). Fine mapping and gene cloning in the post-NGS era: advances and prospects. *Theoretical and Applied Genetics* 133(5), 1791–1810.
James, R. A., Blake, C., Zwart, A. B., Hare, R. A., Rathjen, A. J. and Munns, R. (2012). Impact of ancestral wheat sodium exclusion genes *Nax1* and *Nax2* on grain yield of durum wheat on saline soils. *Functional Plant Biology* 39(7), 609–618.
Janiak, A. and Szarejko, I. (2007). Molecular mapping of genes involved in root hair formation in barley. *Euphytica* 157(1–2), 95–111.
Jha, U. C., Bohra, A. and Nayyar, H. (2020). Advances in "omics" approaches to tackle drought stress in grain legumes. *Plant Breeding* 139(1), 1–27.
Jia, K. P., Dickinson, A. J., Mi, J. N., Cui, G. X., Xiao, T. T., Kharbatia, N. M., Guo, X. J., Sugiono, E., Aranda, M., Blilou, I., Rueping, M., Benfey, P. N. and Al-Babili, S. (2019a). Anchorene is a carotenoid-derived regulatory metabolite required for anchor root formation in Arabidopsis. *Science Advances* 5(11), eaaw6787.
Jia, X. C., Liu, P. and Lynch, J. P. (2018). Greater lateral root branching density in maize improves phosphorus acquisition from low phosphorus soil. *Journal of Experimental Botany* 69(20), 4961–4970.
Jia, Z. T., Liu, Y., Gruber, B. D., Neumann, K., Kilian, B., Graner, A. and von Wiren, N. (2019b). Genetic dissection of root system architectural traits in spring barley. *Frontiers in Plant Science* 10, 400.
Jiang, S. Y. and Ramachandran, S. (2010). Assigning biological functions to rice genes by genome annotation, expression analysis and mutagenesis. *Biotechnology Letters* 32(12), 1753–1763.

Jung, J. K. H. and McCouch, S. (2013). Getting to the roots of it: genetic and hormonal control of root architecture. *Frontiers in Plant Science* 4, 186.
Junker, A., Muraya, M. M., Weigelt-Fischer, K., Arana-Ceballos, F., Klukas, C., Melchinger, A. E., Meyer, R. C., Riewe, D. and Altmann, T. (2015). Optimizing experimental procedures for quantitative evaluation of crop plant performance in high throughput phenotyping systems. *Frontiers in Plant Science* 5, 770.
Kan, C. C., Jia, H. Y., Powers, C., Carver, B. F. and Yan, L. L. (2020). Genetic characterization and deployment of a major gene for grain yield on chromosome arm 1BS in winter wheat. *Molecular Breeding* 40(3), 26.
Kang, Z. Y., Babar, M. A., Khan, N., Guo, J., Khan, J., Islam, S., Shrestha, S. and Shahi, D. (2019). Comparative metabolomic profiling in the roots and leaves in contrasting genotypes reveals complex mechanisms involved in post-anthesis drought tolerance in wheat. *PLOS ONE* 14(3), e0213502.
Kashtwari, M., Wani, A. A. and Rather, R. N. (2019). TILLING: an alternative path for crop improvement. *Journal of Crop Improvement* 33(1), 83–109.
Kaushal, M. (2019). Microbes in cahoots with plants: MIST to hit the jackpot of agricultural productivity during drought. *International Journal of Molecular Sciences* 20(7), 1769.
Khan, A. W., Garg, V., Roorkiwal, M., Golicz, A. A., Edwards, D. and Varshney, R. K. (2020). Super-pangenome by integrating the wild side of a species for accelerated crop improvement. *Trends in Plant Science* 25(2), 148–158.
Khan, M. A., Tong, F., Wang, W. B., He, J. B., Zhao, T. J. and Gai, J. Y. (2018). Analysis of QTL-allele system conferring drought tolerance at seedling stage in a nested association mapping population of soybean *Glycine max* (L.) Merr. using a novel GWAS procedure. *Planta* 248(4), 947–962.
Khandal, H., Singh, A. P. and Chattopadhyay, D. (2020). The MicroRNA397b-LACCASE2 module regulates root lignification under water and phosphate deficiency(1). *Plant Physiology* 182, 1387–1403.
Kichey, T., Hirel, B., Heumez, E., Dubois, F. and Le Gouis, J. (2007). In winter wheat (*Triticum aestivum* L.), post-anthesis nitrogen uptake and remobilisation to the grain correlates with agronomic traits and nitrogen physiological markers. *Field Crops Research* 102(1), 22–32.
Kim, K. S., Vuong, T. D., Qiu, D., Robbins, R. T., Shannon, J. G., Li, Z. L. and Nguyen, H. T. (2016). Advancements in breeding, genetics, and genomics for resistance to three nematode species in soybean. *Theoretical and Applied Genetics* 129, 2295–2311.
King, J., Gay, A., Sylvester-Bradley, R., Bingham, I., Foulkes, J., Gregory, P., Robinson, D. (2003). Modelling cereal root systems for water and nitrogen capture: towards an economic optimum. *Annals of Botany* 91(3), 383–390.
King, R., Bird, N., Ramirez-Gonzalez, R., Coghill, J. A., Patil, A., Hassani-Pak, K., Uauy, C. and Phillips, A. L. (2015). Mutation scanning in wheat by exon capture and next-generation sequencing. *PLOS ONE* 10(9), e0137549.
Kitomi, Y., Kanno, N., Kawai, S., Mizubayashi, T., Fukuoka, S. and Uga, Y. (2015). QTLs underlying natural variation of root growth angle among rice cultivars with the same functional allele of *DEEPER ROOTING*. *Rice* 1, 8.
Kitomi, Y., Nakao, E., Kawai, S., Kanno, N., Ando, T., Fukuoka, S., Irie, K. and Uga, Y. (2018). Fine mapping of *QUICK ROOTING 1* and *2*, quantitative trait loci increasing root length in rice. *Genes Genomes Genetics* 8(2), 727–735.
Kole, C., Muthamilarasan, M., Henry, R., Edwards, D., Sharma, R., Abberton, M., Batley, J., Bentley, A., Blakeney, M., Bryant, J., Cai, H., Cakir, M., Cseke, L. J., Cockram, J.,

de Oliveira, A. C., De Pace, C., Dempewolf, H., Ellison, S., Gepts, P., Greenland, A., Hall, A., Hori, K., Hughes, S., Humphreys, M. W., Iorizzo, M., Ismail, A. M., Marshall, A., Mayes, S., Nguyen, H. T., Ogbonnaya, F. C., Ortiz, R., Paterson, A. H., Simon, P. W., Tohme, J., Tuberosa, R., Valliyodan, B., Varshney, R. K., Wullschleger, S. D., Yano, M. and Prasad, M. (2015). Application of genomics-assisted breeding for generation of climate resilient crops: progress and prospects. *Frontiers in Plant Science* 6, 563.

Korani, W., Clevenger, J. P., Chu, Y. and Ozias-Akins, P. (2019). Machine learning as an effective method for identifying true single nucleotide polymorphisms in polyploid plants. *Plant Genome* 12(1), 1–10.

Kortz, A., Hochholdinger, F. and Yu, P. (2019). Cell type-specific transcriptomics of latesral root formation and plasticity. *Frontiers in Plant Science* 10, 21.

Kota, V., Gundra, S. R., Vudem, D. R., Pulugurtha, B. K. and Khareedu, V. R. (2020). Development of a large population of activation-tagged mutants in an elite *indica* rice variety. *Plant Breeding* 139(2): 328–343.

Kubo, K., Elouafi, I., Watanabe, N., Nachit, M. M., Inagaki, M. N., Iwama, K. and Jitsuyama, Y. (2007). Quantitative trait loci for soil-penetrating ability of roots in durum wheat. *Plant Breeding* 126(4), 375–378.

Kuijken, R. C. P., van Eeuwijk, F. A., Marcelis, L. F. M. and Bouwmeester, H. J. (2015). Root phenotyping: from component trait in the lab to breeding. *Journal of Experimental Botany* 66(18), 5389–5401.

Kulkarni, M., Soolanayakanahally, R., Ogawa, S., Uga, Y., Selvaraj, M. G. and Kagale, S. (2017). Drought response in wheat: key genes and regulatory mechanisms controlling root system architecture and transpiration efficiency. *Frontiers in Chemistry* 5, 106.

Lafitte, H. R., Price, A. H. and Courtois, B. (2004). Yield response to water deficit in an upland rice mapping population: associations among traits and genetic markers. *Theoretical and Applied Genetics* 109(6), 1237–1246.

Landi, P., Giuliani, M. M., Darrah, L. L., Tuberosa, R., Conti, S. and Sanguineti, M. C. (2001). Variability for root and shoot traits in a maize population grown in hydroponics and in the field and their relationships with vertical root pulling resistance. *Maydica* 46, 177–182.

Landi, P., Giuliani, S., Salvi, S., Ferri, M., Tuberosa, R. and Sanguineti, M. C. (2010). Characterization of *root-yield-1.06*, a major constitutive QTL for root and agronomic traits in maize across water regimes. *Journal of Experimental Botany* 61(13), 3553–3562.

Landi, P., Sanguineti, M., Darrah, L., Giuliani, M., Salvi, S., Conti, S. and Tuberosa, R. (2002a). Detection of QTLs for vertical root pulling resistance in maize and overlap with QTLs for root traits in hydroponics and for grain yield under different water regimes. *Maydica* 47, 233–243.

Landi, P., Sanguineti, M. C., Salvi, S., Giuliani, S., Bellotti, M., Maccaferri, M., Conti, S. and Tuberosa, R. (2005). Validation and characterization of a major QTL affecting leaf ABA concentration in maize. *Molecular Breeding* 15(3), 291–303.

Landi, P., Sanguineti, M. C., Darrah, L. L., Giuliani, M. M., Salvi, S., Conti, S. and Tuberosa, R. (2002b). Detection of QTLs for vertical root pulling resistance in maize and overlap with QTLs for root traits in hydroponics and for grain yield under different water regimes. *Maydica* 47, 233–243.

Landi, P., Sanguineti, M. C., Liu, C., Li, Y., Wang, T. Y., Giuliani, S., Bellotti, M., Salvi, S. and Tuberosa, R. (2007). *Root-ABA1* QTL affects root lodging, grain yield, and other agronomic traits in maize grown under well-watered and water-stressed conditions. *Journal of Experimental Botany* 58(2), 319–326.

Langridge, P. and Reynolds, M. P. (2015). Genomic tools to assist breeding for drought tolerance. *Current Opinion in Biotechnology* 32, 130–135.

Lebreton, C., Lazic-Jancic, V., Steed, A., Pekic, S. and Quarrie, S. A. (1995). Identification of QTL for drought responses in maize and their use in testing causal relationships between traits. *Journal of Experimental Botany* 46, 853–865.

Lehnert, H., Serfling, A., Friedt, W. and Ordon, F. (2018). Genome-wide association studies reveal genomic regions associated with the response of wheat (*Triticum aestivum* L.) to mycorrhizae Under drought stress conditions. *Frontiers in Plant Science* 9, 1728.

Li, J. Z., Xie, Y., Dai, A. Y., Liu, L. F. and Li, Z. C. (2009). Root and shoot traits responses to phosphorus deficiency and QTL analysis at seedling stage using introgression lines of rice. *Journal of Genetics and Genomics* 36(3), 173–183.

Li, P. C., Chen, F. J., Cai, H. G., Liu, J. C., Pan, Q. C., Liu, Z. G., Gu, R. L., Mi, G. H., Zhang, F. S. and Yuan, L. X. (2015). A genetic relationship between nitrogen use efficiency and seedling root traits in maize as revealed by QTL analysis. *Journal of Experimental Botany* 66(11), 3175–3188.

Li, M. M., Li, B. B., Guo, G. H., Chen, Y. X., Xie, J. Z., Lu, P., Wu, Q. H., Zhang, D. Y., Zhang, H. Z., Yang, J., Zhang, P. P., Zhang, Y. and Liu, Z. Y. (2018a). Mapping a leaf senescence gene *els*1 by BSR-Seq in common wheat. *Crop Journal* 6(3), 236–243.

Li, X. X., Chen, R. R., Chu, Y. L., Huang, J. Y., Jin, L., Wang, G. X. and Huang, J. L. (2018b). Overexpression of *RCc3* improves root system architecture and enhances salt tolerance in rice. *Plant Physiology and Biochemistry* 130, 566–576.

Li, X. X., Ingvordsen, C. H., Weiss, M., Rebetzke, G. J., Condon, A. G., James, R. A. and Richards, R. A. (2019a). Deeper roots associated with cooler canopies, higher normalized difference vegetation index, and greater yield in three wheat populations grown on stored soil water. *Journal of Experimental Botany* 70(18), 4963–4974.

Li, L., Mao, X., Wang, J. Y., Chang, X. P., Reynolds, M. and Jing, R. L. (2019b). Genetic dissection of drought and heat-responsive agronomic traits in wheat. *Plant, Cell and Environment* 42(9), 2540–2553.

Li, P. C., Fan, Y. Y., Yin, S. Y., Wang, Y. Y., Wang, H. M., Xu, Y., Yang, Z. F. and Xu, C. W. (2020a). Multi-environment QTL mapping of crown root traits in a maize RIL population. *Crop Journal* 8(4), 645–654.

Li, T., Ma, J., Zou, Y. Y., Chen, G. D., Ding, P. Y., Zhang, H., Yang, C. C., Mu, Y., Tang, H. P., Liu, Y. X., Jiang, Q. T., Chen, G. Y., Qi, P. F., Wei, Y. M., Zheng, Y. L. and Lan, X. J. (2020b). Quantitative trait loci for seeding root traits and the relationships between root and agronomic traits in common wheat. *Genome* 63(1), 27–36.

Lippman, Z. B., Semel, Y. and Zamir, D. (2007). An integrated view of quantitative trait variation using tomato interspecific introgression lines. *Current Opinion in Genetics and Development* 17(6), 545–552.

Liu, B. H., Wu, J. Y., Yang, S. Q., Schiefelbein, J. and Gan, Y. B. (2020). Nitrate regulation of lateral root and root hair development in plants. *Journal of Experimental Botany* 71(15), 4405–4414.

Liu, N. N., Liu, J. X., Li, W. Q., Pan, Q. C., Liu, J., Yang, X. H., Yan, J. B. and Xiao, Y. J. (2018). Intraspecific variation of residual heterozygosity and its utility for quantitative genetic studies in maize. *BMC Plant Biology* 18(1), 66.

Lobet, G., Paez-Garcia, A., Schneider, H., Junker, A., Atkinson, J. A. and Tracy, S. (2019). Demystifying roots: a need for clarification and extended concepts in root phenotyping. *Plant Science* 282, 11–13.

Lopes, M. S., Dreisigacker, S., Pena, R. J., Sukumaran, S. and Reynolds, M. P. (2015). Genetic characterization of the wheat association mapping initiative (WAMI) panel for dissection of complex traits in spring wheat. *Theoretical and Applied Genetics* 128(3), 453–464.

Lopes, M. S. and Reynolds, M. P. (2010). Partitioning of assimilates to deeper roots is associated with cooler canopies and increased yield under drought in wheat. *Functional Plant Biology* 37(2), 147–156.

Lopez-Marin, H. D., Rao, I. M. and Blair, M. W. (2009). Quantitative trait loci for root morphology traits under aluminum stress in common bean (*Phaseolus vulgaris* L.). *Theoretical and Applied Genetics* 119(3), 449–458.

Lynch, J. (1995). Root architecture and plant productivity. *Plant Physiology* 109(1), 7–13.

Lynch, J. P. (2007). Roots of the second green revolution. *Australian Journal of Botany* 55(5), 493–512.

Lynch, J. P. (2011). Root phenes for enhanced soil exploration and phosphorus acquisition: tools for future crops. *Plant Physiology* 156(3), 1041–1049.

Lynch, J. P. (2013). Steep, cheap and deep: an ideotype to optimize water and N acquisition by maize root systems. *Annals of Botany* 112(2), 347–357.

Lynch, J. P. (2018). Rightsizing root phenotypes for drought resistance. *Journal of Experimental Botany* 69(13), 3279–3292.

Lynch, J. P., Chimungu, J. G. and Brown, K. M. (2014). Root anatomical phenes associated with water acquisition from drying soil: targets for crop improvement. *Journal of Experimental Botany* 65(21), 6155–6166.

Lynch, J. P. and Wojciechowski, T. (2015). Opportunities and challenges in the subsoil: pathways to deeper rooted crops. *Journal of Experimental Botany* 66(8), 2199–2210.

Ma, H. X., Bai, G. H. and Lu, W. Z. (2006). Quantitative trait loci for aluminum resistance in wheat cultivar Chinese Spring. *Plant and Soil* 283(1–2), 239–249.

Ma, N. N., Dong, L. N., Lu, W., Lu, J., Meng, Q. W. and Liu, P. (2020). Transcriptome analysis of maize seedling roots in response to nitrogen-, phosphorus-, and potassium deficiency. *Plant and Soil* 447, 637–658.

Maccaferri, M., El-Feki, W., Nazemi, G., Salvi, S., Cane, M. A., Colalongo, M. C., Stefanelli, S. and Tuberosa, R. (2016). Prioritizing quantitative trait loci for root system architecture in tetraploid wheat. *Journal of Experimental Botany* 67(4), 1161–1178.

Maccaferri, M., Harris, N. S., Twardziok, S. O., Pasam, R. K., Gundlach, H., Spannagl, M., Ormanbekova, D., Lux, T., Prade, V. M., Milner, S. G., Himmelbach, A., Mascher, M., Bagnaresi, P., Faccioli, P., Cozzi, P., Lauria, M., Lazzari, B., Stella, A., Manconi, A., Gnocchi, M., Moscatelli, M., Avni, R., Deek, J., Biyiklioglu, S., Frascaroli, E., Corneti, S., Salvi, S., Sonnante, G., Desiderio, F., Mare, C., Crosatti, C., Mica, E., Ozkan, H., Kilian, B., De Vita, P., Marone, D., Joukhadar, R., Mazzucotelli, E., Nigro, D., Gadaleta, A., Chao, S. M., Faris, J. D., Melo, A. T. O., Pumphrey, M., Pecchioni, N., Milanesi, L., Wiebe, K., Ens, J., MacLachlan, R. P., Clarke, J. M., Sharpe, A. G., Koh, C. S., Liang, K. Y. H., Taylor, G. J., Knox, R., Budak, H., Mastrangelo, A. M., Xu, S. S., Stein, N., Hale, I., Distelfeld, A., Hayden, M. J., Tuberosa, R., Walkowiak, S., Mayer, K. F. X., Ceriotti, A., Pozniak, C. J. and Cattivelli, L. (2019). Durum wheat genome highlights past domestication signatures and future improvement targets. *Nature Genetics* 51(5), 885–895.

Maccaferri, M., Ricci, A., Salvi, S., Milner, S. G., Noli, E., Martelli, P. L., Casadio, R., Akhunov, E., Scalabrin, S., Vendramin, V., Ammar, K., Blanco, A., Desiderio, F., Distelfeld, A., Dubcovsky, J., Fahima, T., Faris, J., Korol, A., Massi, A., Mastrangelo, A. M., Morgante, M., Pozniak, C., N'Diaye, A., Xu, S. and Tuberosa, R. (2015). A high-density, SNP-based consensus map of tetraploid wheat as a bridge to integrate durum and bread wheat genomics and breeding. *Plant Biotechnology Journal* 13(5), 648–663.

Mace, E. S., Singh, V., Van Oosterom, E. J., Hammer, G. L., Hunt, C. H. and Jordan, D. R. (2012). QTL for nodal root angle in sorghum (*Sorghum bicolor* L. Moench) co-locate with QTL for traits associated with drought adaptation. *Theoretical and Applied Genetics* 124(1), 97–109.

Mackill, D. J. and Khush, G. S. (2018). IR64: a high-quality and high-yielding mega variety. *Rice* 11(1), 18.

Mano, Y., Muraki, M. and Takamizo, T. (2006). Identification of QTL controlling flooding tolerance in reducing soil conditions in maize (*Zea mays* L.) seedlings. *Plant Production Science* 9(2), 176–181.

Marzec, M., Melzer, M. and Szarejko, I. (2015). Root hair development in the grasses: what we already know and what we still need to know. *Plant Physiology* 168(2), 407–414.

McCallum, C. M., Comai, L., Greene, E. A. and Henikoff, S. (2000). Targeting induced local lesions in genomes (TILLING) for plant functional genomics. *Plant Physiology* 123(2), 439–442.

Mei, H. W., Xu, J. L., Li, Z. K., Yu, X. Q., Guo, L. B., Wang, Y. P., Ying, C. S. and Luo, L. J. (2006). QTLs influencing panicle size detected in two reciprocal introgressive line (IL) populations in rice (*Oryza sativa* L.). *Theoretical and Applied Genetics* 112(4), 648–656.

Melo, J. O., Martins, L. G. C., Barros, B. A., Pimenta, M. R., Lana, U. G. P., Duarte, C. E. M., Pastina, M. M., Guimaraes, C. T., Schaffert, R. E., Kochian, L. V., Fontes, E. P. B. and Magalhaes, J. V. (2019). Repeat variants for the SbMATE transporter protect sorghum roots from aluminum toxicity by transcriptional interplay in cis and trans. *Proceedings of the National Academy of Sciences of the United States of America* 116(1), 313–318.

Meng, F. N., Xiang, D., Zhu, J. S., Li, Y. and Mao, C. Z. (2019). Molecular mechanisms of root development in rice. *Rice* 12(1), 10.

Merchuk-Ovnat, L., Fahima, T., Ephrath, J. E., Krugman, T. and Saranga, Y. (2017). Ancestral QTL alleles from wild emmer wheat enhance root development under drought in modern wheat. *Frontiers in Plant Science* 8, 703.

Messing, J. and Dooner, H. K. (2006). Organization and variability of the maize genome. *Current Opinion in Plant Biology* 9(2), 157–163.

Miguel, M. A., Postma, J. A. and Lynch, J. P. (2015). Phene synergism between root hair length and basal root growth angle for phosphorus acquisition. *Plant Physiology* 167(4), 1430–1439.

Millet, E. J., Welcker, C., Kruijer, W., Negro, S., Coupel-Ledru, A., Nicolas, S. D., Laborde, J., Bauland, C., Praud, S., Ranc, N., Presterl, T., Tuberosa, R., Bedo, Z., Draye, X., Usadel, B., Charcosset, A., Van Eeuwijk, F. and Tardieu, F. (2016). Genome-wide analysis of yield in Europe: allelic effects vary with drought and heat scenarios. *Plant Physiology* 172(2), 749–764.

Milner, S. G., Maccaferri, M., Huang, B. E., Mantovani, P., Massi, A., Frascaroli, E., Tuberosa, R. and Salvi, S. (2016). A multiparental cross population for mapping QTL for agronomic traits in durum wheat (*Triticum turgidum* ssp *durum*). *Plant Biotechnology Journal* 14(2), 735–748.

Ming, M. L., Ren, Q. R., Pan, C. T., He, Y., Zhang, Y. X., Liu, S. S., Zhong, Z. H., Wang, J. H., Malzahn, A. A., Wu, J., Zheng, X. L., Zhang, Y. and Qi, Y. P. (2020). CRISPR-Cas12b enables efficient plant genome engineering. *Nature Plants* 6(3), 202–208.

Mkhabela, S. S., Shimelis, H., Odindo, A. O. and Mashilo, J. (2019). Response of selected drought tolerant wheat (*Triticum aestivum* L.) genotypes for agronomic traits and biochemical markers under drought-stressed and non-stressed conditions. *Acta Agriculturae Scandinavica, Section B – Soil and Plant Science* 69(8), 674–689.

Moon, S., Chandran, A. K. N., An, G., Lee, C. and Jung, K. H. (2018). Genome-wide analysis of root hair-preferential genes in rice. *Rice* 11(1), 48.

Moon, S., Chandran, A. K. N., Kim, Y. J., Gho, Y., Hong, W. J., An, G., Lee, C. and Jung, K. H. (2019a). Rice *RHC* encoding a putative cellulase is essential for normal root hair elongation. *Journal of Plant Biology* 62(1), 82–91.

Moon, S., Cho, L. H., Kim, Y. J., Gho, Y. S., Jeong, H. Y., Hong, W. J., Lee, C., Park, H., Jwa, N. S., Dangol, S., Chen, Y., Park, H., Cho, H. S., An, G. and Jung, K. H. (2019b). RSL Class II transcription factors guide the nuclear localization of RHL1 to regulate root hair development. *Plant Physiology* 179(2), 558–568.

Motte, H. and Beeckman, T. (2019). The evolution of root branching: increasing the level of plasticity. *Journal of Experimental Botany* 70(3), 785–793.

Motte, H., Vanneste, S. and Beeckman, T. (2019). Molecular and environmental regulation of root development. *Annual Review of Plant Biology*, 70, 465–488.

Munns, R., Day, D. A., Fricke, W., Watt, M., Arsova, B., Barkla, B. J., Bose, J., Byrt, C. S., Chen, Z. H., Foster, K. J., Gilliham, M., Henderson, S. W., Jenkins, C. L. D., Kronzucker, H. J., Miklavcic, S. J., Plett, D., Roy, S. J., Shabala, S., Shelden, M. C., Soole, K. L., Taylor, N. L., Tester, M., Wege, S., Wegner, L. H. and Tyerman, S. D. (2020). Energy costs of salt tolerance in crop plants. *New Phytologist* 225(3), 1072–1090.

Munns, R., James, R. A., Xu, B., Athman, A., Conn, S. J., Jordans, C., Byrt, C. S., Hare, R. A., Tyerman, S. D., Tester, M., Plett, D. and Gilliham, M. (2012). Wheat grain yield on saline soils is improved by an ancestral Na^+ transporter gene. *Nature Biotechnology* 30(4), 360–364.

Munns, R. and Tester, M. (2008). Mechanisms of salinity tolerance. *Annual Review of Plant Biology* 59, 651–681.

Nadeau, J. H. and Frankel, W. N. (2000). The roads from phenotypic variation to gene discovery: mutagenesis versus QTLs. *Nature Genetics* 25(4), 381–384.

Nagel, K. A., Bonnett, D., Furbank, R., Walter, A., Schurr, U. and Watt, M. (2015). Simultaneous effects of leaf irradiance and soil moisture on growth and root system architecture of novel wheat genotypes: implications for phenotyping. *Journal of Experimental Botany* 66(18), 5441–5452.

Nagel, K. A., Putz, A., Gilmer, F., Heinz, K., Fischbach, A., Pfeifer, J., Faget, M., Blossfeld, S., Ernst, M., Dimaki, C., Kastenholz, B., Kleinert, A. K., Galinski, A., Scharr, H., Fiorani, F. and Schurr, U. (2012). GROWSCREEN-Rhizo is a novel phenotyping robot enabling simultaneous measurements of root and shoot growth for plants grown in soil-filled rhizotrons. *Functional Plant Biology* 39(11), 891–904.

Najera, V. A., Twyman, R. M., Christou, P. and Zhu, C. F. (2019). Applications of multiplex genome editing in higher plants. *Current Opinion in Biotechnology* 59, 93–102.

Navakode, S., Weidner, A., Lohwasser, U., Roder, M. S. and Borner, A. (2009). Molecular mapping of quantitative trait loci (QTLs) controlling aluminium tolerance in bread wheat. *Euphytica* 166(2), 283–290.

Neelam, K., Thakur, S., Neha, Yadav, I. S., Kumar, K., Dhaliwal, S. S. and Singh, K. (2017). Novel alleles of phosphorus-starvation tolerance 1 gene (*PSTOL1*) from *Oryza rufipogon* confers high phosphorus uptake efficiency. *Frontiers in Plant Science* 8, 509.

Neeraja, C. N., Maghirang-Rodriguez, R., Pamplona, A., Heuer, S., Collard, B. C. Y., Septiningsih, E. M., Vergara, G., Sanchez, D., Xu, K., Ismail, A. M. and Mackill, D. J. (2007). A marker-assisted backcross approach for developing submergence-tolerant rice cultivars. *Theoretical and Applied Genetics* 115(6), 767-776.

Neto, J. B., Hurtado-Perez, M. C., Wimmer, M. A. and Frei, M. (2017). Genetic factors underlying boron toxicity tolerance in rice: genome-wide association study and transcriptomic analysis. *Journal of Experimental Botany* 68(3), 687-700.

Nounjan, N., Chansongkrow, P., Charoensawan, V., Siangliw, J. L., Toojinda, T., Chadchawan, S. and Theerakulpisut, P. (2018). High performance of photosynthesis and osmotic adjustment are associated With salt tolerance ability in rice carrying drought tolerance QTL: physiological and co-expression network analysis. *Frontiers in Plant Science* 9, 1135.

Ochoa, I. E., Blair, M. W. and Lynch, J. P. (2006). QTL analysis of adventitious root formation in common bean under contrasting phosphorus availability. *Crop Science* 46(4), 1609-1621.

Orosa-Puente, B., Leftley, N., von Wangenheim, D., Banda, J., Srivastava, A. K., Hill, K., Truskina, J., Bhosale, R., Morris, E., Srivastava, M., Kumpers, B., Goh, T., Fukaki, H., Vermeer, J. E. M., Vernoux, T., Dinneny, J. R., French, A. P., Bishopp, A., Sadanandom, A. and Bennett, M. J. (2018). Root branching toward water involves posttranslational modification of transcription factor ARF7. *Science* 362(6421), 1407-1410.

Oyiga, B. C., Palczak, J., Wojciechowski, T., Lynch, J. P., Naz, A. A., Leon, J. and Ballvora, A. (2020). Genetic components of root architecture and anatomy adjustments to water-deficit stress in spring barley. *Plant, Cell and Environment* 43(3), 692-711.

Pallotta, M., Schnurbusch, T., Hayes, J., Hay, A., Baumann, U., Paull, J., Langridge, P. and Sutton, T. (2014). Molecular basis of adaptation to high soil boron in wheat landraces and elite cultivars. *Nature* 514(7520), 88-91.

Pariasca-Tanaka, J., Baertschi, C. and Wissuwa, M. (2020). Identification of loci through genome-wide association studies to improve tolerance to sulfur deficiency in rice. *Frontiers in Plant Science* 10, 1668.

Peck, S. and Mittler, R. (2020). Plant signaling in biotic and abiotic stress. *Journal of Experimental Botany* 71(5), 1649-1651.

Penna, S. and Jain, S. M. (2017). Mutant resources and mutagenomics in crop plants. *Emirates Journal of Food and Agriculture* 29, 651-657.

Perotti, M. F., Ariel, F. D. and Chan, R. L. (2020). Lateral root development differs between main and secondary roots and depends on the ecotype. *Plant Signaling and Behavior* 15(6), 1755504.

Petrarulo, M., Marone, D., Ferragonio, P., Cattivelli, L., Rubiales, D., De Vita, P. and Mastrangelo, A. M. (2015). Genetic analysis of root morphological traits in wheat. *Molecular Genetics and Genomics* 290(3), 785-806.

Pfeifer, J., Faget, M., Walter, A., Blossfeld, S., Fiorani, F., Schurr, U. and Nagel, K. A. (2014). Spring barley shows dynamic compensatory root and shoot growth responses when exposed to localised soil compaction and fertilisation. *Functional Plant Biology* 41(6), 581-597.

Pieruschka, R. and Poorter, H. (2012). Phenotyping plants: genes, phenes and machines. *Functional Plant Biology* 39(11), 813-820.

Pinto, R. S. and Reynolds, M. P. (2015). Common genetic basis for canopy temperature depression under heat and drought stress associated with optimized root distribution in bread wheat. *Theoretical and Applied Genetics* 128(4), 575-585.

Pinto, R. S., Reynolds, M. P., Mathews, K. L., McIntyre, C. L., Olivares-Villegas, J. J. and Chapman, S. C. (2010). Heat and drought adaptive QTL in a wheat population designed to minimize confounding agronomic effects. *Theoretical and Applied Genetics* 121(6), 1001–1021.

Poorter, H., Buhler, J., van Dusschoten, D., Climent, J. and Postma, J. A. (2012). Pot size matters: a meta-analysis of the effects of rooting volume on plant growth. *Functional Plant Biology* 39(11), 839–850.

Price, A. (2002). QTLs for rooth growth and drought resistance in rice. *Molecular Techniques in Crop Improvement*, 563-584.

Price, A. and Courtois, B. (1999). Mapping QTLs associated with drought resistance in rice: progress, problems and prospects. *Plant Growth Regulation* 29(1/2), 123–133.

Price, A. H., Cairns, J. E., Horton, P., Jones, H. G. and Griffiths, H. (2002a). Linking drought-resistance mechanisms to drought avoidance in upland rice using a QTL approach: progress and new opportunities to integrate stomatal and mesophyll responses. *Journal of Experimental Botany* 53(371), 989–1004.

Price, A. H., Steele, K. A., Gorham, J., Bridges, J. M., Moore, B. J., Evans, J. L., Richardson, P. and Jones, R. G. W. (2002b). Upland rice grown in soil-filled chambers and exposed to contrasting water-deficit regimes I. Root distribution, water use and plant water status. *Field Crops Research* 76(1), 11–24.

Price, A. H., Steele, K. A., Moore, B. J., Barraclough, P. P. and Clark, L. J. (2000). A combined RFLP and AFLP linkage map of upland rice (*Oryza sativa* L.) used to identify QTLs for root-penetration ability. *Theoretical and Applied Genetics* 100(1), 49–56.

Price, A. H., Steele, K. A., Moore, B. J. and Jones, R. G. W. (2002c). Upland rice grown in soil-filled chambers and exposed to contrasting water-deficit regimes II. Mapping quantitative trait loci for root morphology and distribution. *Field Crops Research* 76(1), 25–43.

Price, A. H. and Tomos, A. D. (1997). Genetic dissection of root growth in rice (*Oryza sativa* L.). II: Mapping quantitative trait loci using molecular markers. *Theoretical and Applied Genetics* 95(1–2), 143–152.

Price, A. H., Tomos, A. D. and Virk, D. S. (1997). Genetic dissection of root growth in rice (Oryza sativa L) .1. a hydrophonic screen. *Theoretical and Applied Genetics* 95(1–2), 132-142.

Price, A. H., Townend, J., Jones, M. P., Audebert, A. and Courtois, B. (2002d). Mapping QTLs associated with drought avoidance in upland rice grown in the Philippines and West Africa. *Plant Molecular Biology* 48(5–6), 683–695.

Prince, S. J., Valliyodan, B., Ye, H., Yang, M., Tai, S. S., Hu, W. S., Murphy, M., Durnell, L. A., Song, L., Joshi, T., Liu, Y., Van de Velde, J., Vandepoele, K., Shannon, J. G. and Nguyen, H. T. (2019). Understanding genetic control of root system architecture in soybean: insights into the genetic basis of lateral root number. *Plant, Cell and Environment* 42(1), 212–229.

Purushothaman, R., Krishnamurthy, L., Upadhyaya, H. D., Vadez, V. and Varshney, R. K. (2017). Genotypic variation in soil water use and root distribution and their implications for drought tolerance in chickpea. *Functional Plant Biology* 44(2), 235-252.

Raghavan, C., Mauleon, R., Lacorte, V., Jubay, M., Zaw, H., Bonifacio, J., Singh, R. K., Huang, B. E. and Leung, H. (2017). Approaches in characterizing genetic structure and mapping in a rice multiparental population. *G3-Genes Genomes Genetics* 7(6), 1721–1730.

Raman, H., Ryan, P. R., Raman, R., Stodart, B. J., Zhang, K., Martin, P., Wood, R., Sasaki, T., Yamamoto, Y., Mackay, M., Hebb, D. M. and Delhaize, E. (2008). Analysis of *TaALMT1* traces the transmission of aluminum resistance in cultivated common wheat (*Triticum aestivum* L.). *Theoretical and Applied Genetics* 116(3), 343-354.
Ramstein, G. P., Jensen, S. E. and Buckler, E. S. (2019). Breaking the curse of dimensionality to identify causal variants in breeding 4. *Theoretical and Applied Genetics* 132(3), 559-567.
Ray, J. D., Yu, L., McCouch, S. R., Champoux, M. C., Wang, G. and Nguyen, H. T. (1996). Mapping quantitative trait loci associated with root penetration ability in rice (Oryza sativa L). *Theoretical and Applied Genetics* 92(6), 627-636.
Rebetzke, G. J., Richards, R. A. and Holland, J. B. (2017). Population extremes for assessing trait value and correlated response of genetically complex traits. *Field Crops Research* 201, 122-132.
Reid, R. (2010). Can we really increase yields by making crop plants tolerant to boron toxicity? *Plant Science* 178(1), 9-11.
Ren, D. Q., Fang, X. J., Jiang, P., Zhang, G. X., Hu, J. M., Wang, X. Q., Meng, Q., Cui, W. A., Lan, S. J., Ma, X., Wang, H. W. and Kong, L. R. (2018). Genetic architecture of nitrogen-deficiency tolerance in wheat seedlings based on a Nested Association Mapping (NAM) population. *Frontiers in Plant Science* 9, 845.
Reynolds, M. and Tuberosa, R. (2008). Translational research impacting on crop productivity in drought-prone environments. *Current Opinion in Plant Biology* 11(2), 171-179.
Richard, C., Christopher, J., Chenu, K., Borrell, A., Christopher, M. and Hickey, L. (2018). Selection in early generations to shift allele frequency for seminal root angle in wheat. *Plant Genome* 11(2), 1-12.
Richard, C. A. I., Hickey, L. T., Fletcher, S., Jennings, R., Chenu, K. and Christopher, J. T. (2015). High-throughput phenotyping of seminal root traits in wheat. *Plant Methods* 11, 13.
Richards, R. A. (2006). Physiological traits used in the breeding of new cultivars for water-scarce environments. *Agricultural Water Management* 80(1-3), 197-211.
Richards, R. A., Rebetzke, G. J., Watt, M., Condon, A. G., Spielmeyer, W. and Dolferus, R. (2010). Breeding for improved water productivity in temperate cereals: phenotyping, quantitative trait loci, markers and the selection environment. *Functional Plant Biology* 37(2), 85-97.
Rimbert, H., Darrier, B., Navarro, J., Kitt, J., Choulet, F., Leveugle, M., Duarte, J., Riviere, N., Eversole, K., Le Gouis, J., Davassi, A., Balfourier, F., Le Paslier, M. C., Berard, A., Brunel, D., Feuillet, C., Poncet, C., Sourdille, P., Paux, E., Int Wheat Genome Sequencing, C., and BreedWheat, C. (2018). High throughput SNP discovery and genotyping in hexaploid wheat. *PLoS ONE* 13(1), e0186329.
Robbins, N. E. and Dinneny, J. R. (2015). The divining root: moisture-driven responses of roots at the micro- and macro-scale. *Journal of Experimental Botany* 66(8), 2145-2154.
Robbins, N. E. and Dinneny, J. R. (2018). Growth is required for perception of water availability to pattern root branches in plants. *Proceedings of the National Academy of Sciences of the United States of America* 115(4), E822-E831.
Robertson, D. S. (1985). A possible technique for isolating genic DNA for quantitative traits in plants. *Journal of Theoretical Biology* 117(1), 1-10.
Rodda, M. S., Sudheesh, S., Javid, M., Noy, D., Gnanasambandam, A., Slater, A. T., Rosewarne, G. M. and Kaur, S. (2018). Breeding for boron tolerance in lentil (*Lens*

culinaris Medik.) using a high-throughput phenotypic assay and molecular markers. *Plant Breeding* 137(4), 492–501.

Roorkiwal, M., Bharadwaj, C., Barmukh, R., Dixit, G. P., Thudi, M., Gaur, P. M., Chaturvedi, S. K., Fikre, A., Hamwieh, A., Kumar, S., Sachdeva, S., Ojiewo, C. O., Tar'an, B., Wordofa, N. G., Singh, N. P., Siddique, K. H. M. and Varshney, R. K. (2020). Integrating genomics for chickpea improvement: achievements and opportunities. *Theoretical and Applied Genetics* 133(5), 1703–1720.

Roorkiwal, M., Jain, A., Kale, S. M., Doddamani, D., Chitikineni, A., Thudi, M. and Varshney, R. K. (2018). Development and evaluation of high-density Axiom (R) CicerSNP Array for high-resolution genetic mapping and breeding applications in chickpea. *Plant Biotechnology Journal* 16(4), 890–901.

Rossini, L., Muehlbauer G. J., Okagaki R., Salvi, S. and von Korff, M. (2018). Genetics of whole plant morphology and architecture. In Stein, N. and Muehlbauer, G. (eds), *The Barley Genome. Compendium of Plant Genomes.* Springer, Cham. https://doi.org/10.1007/978-3-319-92528-8_13.

Ruiz, S., Koebernick, N., Duncan, S., Fletcher, D. M., Scotson, C., Boghi, A., Marin, M., Bengough, A. G., George, T. S., Brown, L. K., Hallett, P. D. and Roose, T. (2020). Significance of root hairs at the field scale – modelling root water and phosphorus uptake under different field conditions. *Plant and Soil* 447(1), 281–304.

Ryan, P. R., Delhaize, E., Watt, M. and Richardson, A. E. (2016). Plant roots: understanding structure and function in an ocean of complexity. *Annals of Botany* 118(4), 555–559.

Salarpour, M., Pakniyat, H., Abdolshahi, R., Heidari, B., Razi, H. and Afzali, R. (2020). Mapping QTL for agronomic and root traits in the Kukri/RAC875 wheat (*Triticum aestivum* L.) population under drought stress conditions. *Euphytica* 216(7), 105.

Salvi, S. (2017). An evo-devo perspective on root genetic variation in cereals. *Journal of Experimental Botany* 68(3), 351–354.

Salvi, S., Corneti, S., Bellotti, M., Carraro, N., Sanguineti, M. C., Castelletti, S. and Tuberosa, R. (2011). Genetic dissection of maize phenology using an intraspecific introgression library. *BMC Plant Biology* 11, 4.

Salvi, S., Giuliani, S., Ricciolini, C., Carraro, N., Maccaferri, M., Presterl, T., Ouzunova, M. and Tuberosa, R. (2016). Two major quantitative trait loci controlling the number of seminal roots in maize co-map with the root developmental genes rtcs and rum1. *Journal of Experimental Botany* 67(4), 1149–1159.

Salvi, S., Sponza, G., Morgante, M., Tomes, D., Niu, X., Fengler, K. A., Meeley, R., Ananiev, E. V., Svitashev, S., Bruggemann, E., Li, B., Hainey, C. F., Radovic, S., Zaina, G., Rafalski, J. A., Tingey, S. V., Miao, G. H., Phillips, R. L. and Tuberosa, R. (2007). Conserved noncoding genomic sequences associated with a flowering-time quantitative trait locus m maize. *Proceedings of the National Academy of Sciences of the United States of America* 104(27), 11376–11381.

Salvi, S. and Tuberosa, R. (2007). Cloning QTLs in plants. In: *Genomics-Assisted Crop Improvement-Vol 1: Genomics Approaches and Platforms* (R. K. Varshney and R. Tuberosa, Eds), pp. 207–226. Springer, New York.

Salvi, S. and Tuberosa, R. (2015). The crop QTLome comes of age. *Current Opinion in Biotechnology* 32, 179–185.

Sandhu, N., Raman, K. A., Torres, R. O., Audebert, A., Dardou, A., Kumar, A. and Henry, A. (2016). Rice root architectural plasticity traits and genetic regions for adaptability to variable cultivation and stress conditions. *Plant Physiology* 171(4), 2562–2576.

Sanguineti, M. C., Tuberosa, R., Landi, P., Salvi, S., Maccaferri, M., Casarini, E. and Conti, S. (1999). QTL analysis of drought related traits and grain yield in relation to genetic variation for leaf abscisic acid concentration in field-grown maize. *Journal of Experimental Botany* 50(337), 1289–1297.

Sanguineti, M. C., Li, S., Maccaferri, M., Corneti, S., Rotondo, F., Chiari, T. and Tuberosa, R. (2007). Genetic dissection of seminal root architecture in elite durum wheat germplasm. *Annals of Applied Biology* 151(3), 291–305.

Sannemann, W., Lisker, A., Maurer, A., Leon, J., Kazman, E., Coster, H., Holzapfel, J., Kempf, H., Korzun, V., Ebmeyer, E. and Pillen, K. (2018). Adaptive selection of founder segments and epistatic control of plant height in the Magic winter wheat population WM-800. *BMC Genomics* 19(1), 559.

Sarkar, R. K., Panda, D., Reddy, J. N., Patnaik, S. S. C., Mackill, D. J. and Ismail, A. M. (2009). Performance of submergence tolerant rice (*Oryza sativa*) genotypes carrying the *Sub1* quantitative trait locus under stressed and non-stressed natural field conditions. *Indian Journal of Agricultural Sciences* 79, 876–883.

Schaefer, R. J., Michno, J. M., Jeffers, J., Hoekenga, O., Dilkes, B., Baxter, I. and Myers, C. L. (2018). Integrating coexpression networks with GWAS to prioritize causal genes in maize. *Plant Cell* 30(12), 2922–2942.

Schlegel, R. and Meinel, A. (1994). A quantitative trait locus (QTL) on chromosome arm 1RS of rye and its effect on yield performance of hexaploid wheat. *Cereal Research Communications* 22, 7–13.

Schneider, H. M. and Lynch, J. P. (2020). Should root plasticity be a crop breeding target? *Frontiers in Plant Science* 11, 546.

Schoppach, R., Fleury, D., Sinclair, T. R. and Sadok, W. (2017). Transpiration sensitivity to evaporative demand Across 120 years of breeding of Australian wheat cultivars. *Journal of Agronomy and Crop Science* 203(3), 219–226.

Schreiber, M., Barakate, A., Uzrek, N., Macaulay, M., Sourdille, A., Morris, J., Hedley, P. E., Ramsay, L. and Waugh, R. (2019). A highly mutagenised barley (cv. Golden Promise) TILLING population coupled with strategies for screening-by-sequencing. *Plant Methods* 15, 99.

Sehgal, D., Rosyara, U., Mondal, S., Singh, R., Poland, J. and Dreisigacker, S. (2020). Incorporating genome-wide association mapping results into genomic prediction models for grain yield and yield stability in CIMMYT spring bread wheat. *Frontiers in Plant Science* 11, 197.

Septiningsih, E. M., Pamplona, A. M., Sanchez, D. L., Neeraja, C. N., Vergara, G. V., Heuer, S., Ismail, A. M. and Mackill, D. J. (2009). Development of submergence-tolerant rice cultivars: the Sub1 locus and beyond. *Annals of Botany* 103(2), 151–160.

Sertse, D., You, F. M., Ravichandran, S. and Cloutier, S. (2019). The complex genetic architecture of early root and shoot traits in flax revealed by Genome-Wide Association Analyses. *Frontiers in Plant Science* 10, 1483.

Shahan, R. and Benfey, P. N. (2020). A co-opted regulator of lateral root development controls nodule organogenesis in lotus. *Developmental Cell* 52(1), 6–7.

Shi, R. L., Junker, A., Seiler, C. and Altmann, T. (2018). Phenotyping roots in darkness: disturbance-free root imaging with near infrared illumination. *Functional Plant Biology* 45(4), 400–411.

Shorinola, O., Kaye, R., Golan, G., Peleg, Z., Kepinski, S. and Uauy, C. (2019). Genetic screening for mutants with altered seminal root numbers in hexaploid wheat using a high-throughput root phenotyping platform. *G3-Genes Genomes Genetics* 9(9), 2799–2809.

Singh, S., Mackill, D. J. and Ismail, A. M. (2009). Responses of *SUB1* rice introgression lines to submergence in the field: yield and grain quality. *Field Crops Research* 113(1), 12–23.

Singh, V. K., Khan, A. W., Jaganathan, D., Thudi, M., Roorkiwal, M., Takagi, H., Garg, V., Kumar, V., Chitikineni, A., Gaur, P. M., Sutton, T., Terauchi, R. and Varshney, R. K. (2016). QTL-seq for rapid identification of candidate genes for 100-seed weight and root/total plant dry weight ratio under rainfed conditions in chickpea. *Plant Biotechnology Journal* 14(11), 2110–2119.

Sivasakthi, K., Thudi, M., Tharanya, M., Kale, S. M., Kholova, J., Halime, M. H., Jaganathan, D., Baddam, R., Thirunalasundari, T., Gaur, P. M., Varshney, R. K. and Vadez, V. (2018). Plant vigour QTLs co-map with an earlier reported QTL hotspot for drought tolerance while water saving QTLs map in other regions of the chickpea genome. *BMC Plant Biology* 18(1), 29.

Song, W. B., Wang, B. B., Hauck, A. L., Dong, X. M., Li, J. P. and Lai, J. S. (2016). Genetic dissection of maize seedling root system architecture traits using an ultra-high density bin-map and a recombinant inbred line population. *Journal of Integrative Plant Biology* 58(3), 266–279.

Soriano, J. M. and Alvaro, F. (2019). Discovering consensus genomic regions in wheat for root-related traits by QTL meta-analysis. *Scientific Reports* 9(1), 10537.

Sosnowski, O., Charcosset, A. and Joets, J. (2012). BioMercator V3: an upgrade of genetic map compilation and quantitative trait loci meta-analysis algorithms. *Bioinformatics* 28(15), 2082–2083.

Soto-Cerda, B. J., Inostroza-Blancheteau, C., Mathias, M., Penaloza, E., Zuniga, J., Munoz, G., Rengel, Z. and Salvo-Garrido, H. (2015). Marker-assisted breeding for *TaALMT1*, a major gene conferring aluminium tolerance to wheat. *Biologia Plantarum* 59(1), 83–91.

Stadlmeier, M., Hartl, L. and Mohler, V. (2018). Usefulness of a multiparent advanced generation intercross population with a greatly reduced mating design for genetic studies in winter wheat. *Frontiers in Plant Science* 9, 1825.

Stahl, A., Wittkop, B. and Snowdon, R. J. (2020). High-resolution digital phenotyping of water uptake and transpiration efficiency. *Trends in Plant Science* 25(5), 429–433.

Steele, K. A., Price, A. H., Shashidhar, H. E. and Witcombe, J. R. (2006). Marker-assisted selection to introgress rice QTLs controlling root traits into an Indian upland rice variety. *Theoretical and Applied Genetics* 112(2), 208–221.

Steele, K. A., Price, A. H., Witcombe, J. R., Shrestha, R., Singh, B. N., Gibbons, J. M. and Virk, D. S. (2013). QTLs associated with root traits increase yield in upland rice when transferred through marker-assisted selection. *Theoretical and Applied Genetics* 126(1), 101–108.

Steele, K. A., Virk, D. S., Kumar, R., Prasad, S. C. and Witcombe, J. R. (2007). Field evaluation of upland rice lines selected for QTLs controlling root traits. *Field Crops Research* 101(2), 180–186.

Steiner, B., Michel, S., Maccaferri, M., Lemmens, M., Tuberosa, R. and Buerstmayr, H. (2019). Exploring and exploiting the genetic variation of Fusarium head blight resistance for genomic-assisted breeding in the elite durum wheat gene pool. *Theoretical and Applied Genetics* 132(4), 969–988.

Sun, C. W., Dong, Z. D., Zhao, L., Ren, Y., Zhang, N. and Chen, F. (2020). The Wheat 660K SNP array demonstrates great potential for marker-assisted selection in polyploid wheat. *Plant Biotechnology Journal* 18(6), 1354–1360.

Sun, J., Luu, N. S., Chen, Z. H., Chen, B., Cui, X. A., Wu, J. X., Zhang, Z. G. and Lu, T. G. (2019). Generation and characterization of a foxtail millet (*Setaria italica*) mutant library. *Frontiers in Plant Science* 10, 369.

Sutton, T., Baumann, U., Hayes, J., Collins, N. C., Shi, B. J., Schnurbusch, T., Hay, A., Mayo, G., Pallotta, M., Tester, M. and Langridge, P. (2007). Boron-toxicity tolerance in barley arising from efflux transporter amplification. *Science* 318(5855), 1446–1449.

Szurman-Zubrzycka, M. E., Zbieszczyk, J., Marzec, M., Jelonek, J., Chmielewska, B., Kurowska, M. M., Krok, M., Daszkowska-Golec, A., Guzy-Wrobelska, J., Gruszka, D., Gajecka, M., Gajewska, P., Stolarek, M., Tylec, P., Sega, P., Lip, S., Kudelko, M., Lorek, M., Gorniak-Walas, M., Malolepszy, A., Podsiadlo, N., Szyrajew, K. P., Keisa, A., Mbambo, Z., Todorowska, E., Gaj, M., Nita, Z., Orlowska-Job, W., Maluszynski, M. and Szarejko, I. (2018). *Hor*TILLUS-A rich and renewable source of induced mutations for forward/reverse genetics and pre-breeding programs in barley (*Hordeum vulgare* L.). *Frontiers in Plant Science* 9, 216.

Taheri, S., Abdullah, T. L., Jain, S. M., Sahebi, M. and Azizi, P. (2017). TILLING, high-resolution melting (HRM), and next-generation sequencing (NGS) techniques in plant mutation breeding. *Molecular Breeding* 37(3), 40.

Tahghighi, H., Erskine, W., Bennett, R. G., Bayer, P. E., Pazos-Navarro, M. and Kaur, P. (2019). Genetic diversity linked to haplotype variation in the world core collection of *Trifolium subterraneum* for boron toxicity tolerance provides valuable markers for pasture breeding. *Frontiers in Plant Science* 10, 1043.

Talamè, V., Bovina, R., Sanguineti, M. C., Tuberosa, R., Lundqvist, U. and Salvi, S. (2008). TILLMore, a resource for the discovery of chemically induced mutants in barley. *Plant Biotechnology Journal* 6(5), 477–485.

Tanksley, S. D. (1993). Mapping polygenes. *Annual Review of Genetics* 27, 205–233.

Tanksley, S. D., Grandillo, S., Fulton, T. M., Zamir, D., Eshed, Y., Petiard, V., Lopez, J. and BeckBunn, T. (1996). Advanced backcross QTL analysis in a cross between an elite processing line of tomato and its wild relative *L. pimpinellifolium*. *Theoretical and Applied Genetics* 92(2), 213–224.

Tanksley, S. D. and McCouch, S. R. (1997). Seed banks and molecular maps: unlocking genetic potential from the wild. *Science* 277(5329), 1063–1066.

Tanksley, S. D., Young, N. D., Paterson, A. H. and Bonierbale, M. W. (1989). RFLP mapping in plant-breeding – new tools for an old science. *Bio/Technology* 7, 257–264.

Tao, Y. F., Zhao, X. R., Mace, E., Henry, R. and Jordan, D. (2019). Exploring and exploiting pan-genomics for crop improvement. *Molecular Plant* 12(2), 156–169.

Tardieu, F. (2012). Any trait or trait-related allele can confer drought tolerance: just design the right drought scenario. *Journal of Experimental Botany* 63(1), 25–31.

Tardieu, F., Simonneau, T. and Muller, B. (2018). The physiological basis of drought tolerance in crop plants: A scenario-dependent probabilistic approach. In: *Annual Review of Plant Biology*, Vol 69 (S. S. Merchant, Ed.), 733–759.

Tardieu, F. and Tuberosa, R. (2010). Dissection and modelling of abiotic stress tolerance in plants. *Current Opinion in Plant Biology* 13(2), 206–212.

Tardieu, F., Varshney, R. K. and Tuberosa, R. (2017). Improving crop performance under drought – cross-fertilization of disciplines. *Journal of Experimental Botany* 68(7), 1393–1398.

Till, B. J., Cooper, J., Tai, T. H., Colowit, P., Greene, E. A., Henikoff, S. and Comai, L. (2007). Discovery of chemically induced mutations in rice by TILLING. *BMC Plant Biology* 7, 12.

Till, B. J., Datta, S. and Jankowicz-Cieslak, J. (2018). TILLING: the next generation. In: *Plant Genetics and Molecular Biology*, Vol 164 (R. K. Varshney, M. K. Pandey and A. Chitikineni, Eds), pp. 139–160. Springer International Publishing, Cham.

Till, B. J., Reynolds, S. H., Greene, E. A., Codomo, C. A., Enns, L. C., Johnson, J. E., Burtner, C., Odden, A. R., Young, K., Taylor, N. E., Henikoff, J. G., Comai, L. and Henikoff, S. (2003). Large-scale discovery of induced point mutations with high-throughput TILLING. *Genome Research* 13(3), 524–530.

To, H. T. M., Nguyen, H. T., Dang, N. T. M., Nguyen, N. H., Bui, T. X., Lavarenne, J., Phung, N. T. P., Gantet, P., Lebrun, M., Bellafiore, S. and Champion, A. (2019). Unraveling the genetic elements involved in shoot and root growth regulation by jasmonate in rice using a genome-wide association study. *Rice* 12(1), 69.

Torabi, S., Wissuwa, M., Heidari, M., Naghavi, M. R., Gilany, K., Hajirezaei, M. R., Omidi, M., Yazdi-Samadi, B., Ismail, A. M. and Salekdeh, G. H. (2009). A comparative proteome approach to decipher the mechanism of rice adaptation to phosphorous deficiency. *Proteomics* 9(1), 159–170.

Tracy, S. R., Nagel, K. A., Postma, J. A., Fassbender, H., Wasson, A. and Watt, M. (2020). Crop improvement from phenotyping roots: highlights reveal expanding opportunities. *Trends in Plant Science* 25(1), 105–118.

Tshikunde, N. M., Mashilo, J., Shimelis, H. and Odindo, A. (2019). Agronomic and physiological traits, and associated quantitative trait loci (QTL) affecting yield response in wheat (*Triticum aestivum* L.): a review. *Frontiers in Plant Science* 10, 1428.

Tsuchida-Mayama, T., Nakamura, H., Hakata, M. and Ichikawa, H. (2010). Rice transgenic resources with gain-of-function phenotypes. *Breeding Science* 60(5), 493–501.

Tuberosa, R. (2012). Phenotyping for drought tolerance of crops in the genomics era. *Frontiers in Physiology* 3, 347.

Tuberosa, R. and Pozniak, C. (2014). Durum wheat genomics comes of age Introduction to the special issue on durum wheat genomics. *Molecular Breeding* 34(4), 1527–1530.

Tuberosa, R., Salvi, S., Giuliani, S., Sanguineti, M. C., Bellotti, M., Conti, S. and Landi, P. (2007). Genome-wide approaches to investigate and improve maize response to drought. *Crop Science* 47, S120–S141.

Tuberosa, R., Salvi, S., Sanguineti, M. C., Landi, P., MacCaferri, M. and Conti, S. (2002a). Mapping QTLs regulating morpho-physiological traits and yield: case studies, shortcomings and perspectives in drought-stressed maize. *Annals of Botany* 89, 941-963.

Tuberosa, R., Salvi, S., Sanguineti, M. C., Maccaferri, M., Giuliani, S. and Landi, P. (2003). Searching for quantitative trait loci controlling root traits in maize: a critical appraisal. *Plant and Soil* 255(1), 35–54.

Tuberosa, R., Sanguineti, M. C., Landi, P., Michela Giuliani, M. M., Salvi, S. and Conti, S. (2002b). Identification of QTLs for root characteristics in maize grown in hydroponics and analysis of their overlap with QTLs for grain yield in the field at two water regimes. *Plant Molecular Biology* 48(5–6), 697–712.

Tuberosa, R., Sanguineti, M. C., Landi, P., Salvi, S., Casarini, E. and Conti, S. (1998). RFLP mapping of quantitative trait loci controlling abscisic acid concentration in leaves of drought-stressed maize (*Zea mays* L.). *Theoretical and Applied Genetics* 97(5–6), 744–755.

Turner, N. C., Blum, A., Cakir, M., Steduto, P., Tuberosa, R. and Young, N. (2014). Strategies to increase the yield and yield stability of crops under drought - are we making progress? *Functional Plant Biology* 41(11), 1199–1206.

Uauy, C. (2017). Plant genomics: unlocking the genome of wheat's progenitor. *Current Biology* 27(20), R1122–R1124.

Uauy, C., Wulff, B. B. H. and Dubcovsky, J. (2017). Combining traditional mutagenesis with new high-throughput sequencing and genome editing to reveal hidden variation in polyploid wheat. *Annual Review of Genetics* 51(51), 435–454.

Uga, Y., Assaranurak, I., Kitomi, Y., Larson, B. G., Craft, E. J., Shaff, J. E., McCouch, S. R. and Kochian, L. V. (2018). Genomic regions responsible for seminal and crown root lengths identified by 2D & 3D root system image analysis. *BMC Genomics* 19(1), 273.

Uga, Y., Kitomi, Y., Ishikawa, S. and Yano, M. (2015a). Genetic improvement for root growth angle to enhance crop production. *Breeding Science* 65(2), 111–119.

Uga, Y., Kitomi, Y., Yamamoto, E., Kanno, N., Kawai, S., Mizubayashi, T. and Fukuoka, S. (2015b). A QTL for root growth angle on rice chromosome 7 is involved in the genetic pathway of *DEEPER ROOTING 1*. *Rice* 8, 8.

Uga, Y., Okuno, K. and Yano, M. (2008). QTLs underlying natural variation in stele and xylem structures of rice root. *Breeding Science* 58(1), 7–14.

Uga, Y., Okuno, K. and Yano, M. (2011). Dro1, a major QTL involved in deep rooting of rice under upland field conditions. *Journal of Experimental Botany* 62(8), 2485–2494.

Uga, Y., Sugimoto, K., Ogawa, S., Rane, J., Ishitani, M., Hara, N., Kitomi, Y., Inukai, Y., Ono, K., Kanno, N., Inoue, H., Takehisa, H., Motoyama, R., Nagamura, Y., Wu, J., Matsumoto, T., Takai, T., Okuno, K. and Yano, M. (2013). Control of root system architecture by *DEEPER ROOTING 1* increases rice yield under drought conditions. *Nature Genetics* 45(9), 1097–1102.

Valliyodan, B., Cannon, S. B., Bayer, P. E., Shu, S. Q., Brown, A. V., Ren, L. H., Jenkins, J., Chung, C. Y. L., Chan, T. F., Daum, C. G., Plott, C., Hastie, A., Baruch, K., Barry, K. W., Huang, W., Patil, G., Varshney, R. K., Hu, H. F., Batley, J., Yuan, Y. X., Song, Q. J., Stupar, R. M., Goodstein, D. M., Stacey, G., Lam, H. M., Jackson, S. A., Schmutz, J., Grimwood, J., Edwards, D. and Nguyen, H. T. (2019). Construction and comparison of three reference-quality genome assemblies for soybean. *Plant Journal* 100(5), 1066–1082.

Valliyodan, B., Ye, H., Song, L., Murphy, M., Shannon, J. G. and Nguyen, H. T. (2017). Genetic diversity and genomic strategies for improving drought and waterlogging tolerance in soybeans. *Journal of Experimental Botany* 68(8), 1835–1849.

van Buuren, M. L., Salvi, S., Morgante, M., Serhani, B. and Tuberosa, R. (2002). Comparative genomic mapping between a 754 kb region flanking *DREB1A* in *Arabidopsis thaliana* and maize. *Plant Molecular Biology* 48(5–6), 741–750.

van Eeuwijk, F. A., Bustos-Korts, D., Millet, E. J., Boer, M. P., Kruijer, W., Thompson, A., Malosetti, M., Iwata, H., Quiroz, R., Kuppe, C., Muller, O., Blazakis, K. N., Yu, K., Tardieu, F. and Chapman, S. C. (2019). Modelling strategies for assessing and increasing the effectiveness of new phenotyping techniques in plant breeding. *Plant Science* 282, 23–39.

Varshney, R. K., Gaur, P. M., Chamarthi, S. K., Krishnamurthy, L., Tripathi, S., Kashiwagi, J., Samineni, S., Singh, V. K., Thudi, M. and Jaganathan, D. (2013). Fast-track introgression of "QTL-hotspot" for root traits and other drought tolerance traits in JG 11, an elite and leading variety of chickpea. *Plant Genome* 6(3).

Varshney, R. K., Langridge, P. and Graner, A. (2007). Application of genomics to molecular breeding of wheat and barley. *Advances in Genetics* 58, 121–155.

Varshney, R. K., Ribaut, J. M., Buckler, E. S., Tuberosa, R., Rafalski, J. A. and Langridge, P. (2012). Can genomics boost productivity of orphan crops? *Nature Biotechnology* 30(12), 1172–1176.

Varshney, R. K., Singh, V. K., Kumar, A., Powell, W. and Sorrells, M. E. (2018a). Can genomics deliver climate-change ready crops? *Current Opinion in Plant Biology* 45(B), 205–211.

Varshney, R. K., Thudi, M., Pandey, M. K., Tardieu, F., Ojiewo, C., Vadez, V., Whitbread, A. M., Siddique, K. H. M., Nguyen, H. T., Carberry, P. S. and Bergvinson, D. (2018b). Accelerating genetic gains in legumes for the development of prosperous smallholder agriculture: integrating genomics, phenotyping, systems modelling and agronomy. *Journal of Experimental Botany* 69(13), 3293–3312.

Varshney, R. K., Tuberosa, R. and Tardieu, F. (2018c). Progress in understanding drought tolerance: from alleles to cropping systems. *Journal of Experimental Botany* 69(13), 3175–3179.

Vaughan, D. A., Balazs, E. and Heslop-Harrison, J. S. (2007). From crop domestication to super-domestication. *Annals of Botany* 100(5), 893–901.

Velinov, V., Vaseva, I., Zehirov, G., Zhiponova, M., Georgieva, M., Vangheluwe, N., Beeckman, T. and Vassileva, V. (2020). Overexpression of the *NMig1* gene encoding a NudC domain protein enhances root growth and abiotic stress tolerance in *Arabidopsis thaliana*. *Frontiers in Plant Science* 11, 815.

Viana, V. E., Pegoraro, C., Busanello, C. and de Oliveira, A. C. (2019). Mutagenesis in rice: the basis for breeding a new super plant. *Frontiers in Plant Science* 10, 1326.

von Wangenheim, D., Banda, J., Schmitz, A., Boland, J., Bishopp, A., Maizel, A., Stelzer, E. H. K. and Bennett, M. (2020). Early developmental plasticity of lateral roots in response to asymmetric water availability. *Nature Plants* 6(2), 73–77.

Voss-Fels, K. P., Qian, L. W., Parra-Londono, S., Uptmoor, R., Frisch, M., Keeble-Gagnere, G., Appels, R. and Snowdon, R. J. (2017). Linkage drag constrains the roots of modern wheat. *Plant, Cell and Environment* 40(5), 717–725.

Voss-Fels, K. P., Robinson, H., Mudge, S. R., Richard, C., Newman, S., Wittkop, B., Stahl, A., Friedt, W., Frisch, M., Gabur, I., Miller-Cooper, A., Campbell, B. C., Kelly, A., Fox, G., Christopher, J., Christopher, M., Chenu, K., Franckowiak, J., Mace, E. S., Borrell, A. K., Eagles, H., Jordan, D. R., Botella, J. R., Hammer, G., Godwin, I. D., Trevaskis, B., Snowdon, R. J. and Hickey, L. T. (2018a). *VERNALIZATION1* modulates root system architecture in wheat and barley. *Molecular Plant* 11(1), 226–229.

Voss-Fels, K. P., Snowdon, R. J. and Hickey, L. T. (2018b). Designer roots for future crops. *Trends in Plant Science* 23(11), 957–960.

Wachsman, G. and Benfey, P. N. (2020). Lateral root initiation: the emergence of new primordia following cell death. *Current Biology* 30(3), R121–R122.

Wallace, J. G., Rodgers-Melnick, E. and Buckler, E. S. (2018). On the road to breeding 4.0: unraveling the good, the bad, and the boring of crop quantitative genomics. In: *Annual Review of Genetics*, Vol 52 (N. M. Bonini, Ed.), pp. 421–444. Annual Reviews, Palo Alto.

Wang, W., Ding, G. D., White, P. J., Wang, X. H., Jin, K. M., Xu, F. S. and Shi, L. (2019a). Mapping and cloning of quantitative trait loci for phosphorus efficiency in crops: opportunities and challenges. *Plant and Soil* 439(1–2), 91–112.

Wang, Z. F., Cheng, J. P., Chen, Z. W., Huang, J., Bao, Y. M., Wang, J. F. and Zhang, H. S. (2012). Identification of QTLs with main, epistatic and QTL x environment interaction effects for salt tolerance in rice seedlings under different salinity conditions. *Theoretical and Applied Genetics* 125(4), 807–815.

Wang, Z. L., Wang, J., Peng, J. X., Du, X. F., Jiang, M. S., Li, Y. F., Han, F., Du, G. H., Yang, H. Q., Lian, S. C., Yong, J. P., Cai, W., Cui, J. D., Han, K. N., Yuan, F., Chang, F., Yuan, G. B., Zhang, W. N., Zhang, L. Y., Peng, S. Z., Zou, H. F. and Guo, E. H. (2019b). QTL mapping for 11 agronomic traits based on a genome-wide Bin-map in a large F_2 population of foxtail millet (*Setaria italica* (L.) P. Beauv). *Molecular Breeding* 39, 13.

Washburn, J. D., Mejia-Guerra, M. K., Ramstein, G., Kremling, K. A., Valluru, R., Buckler, E. S. and Wang, H. (2019). Evolutionarily informed deep learning methods for predicting relative transcript abundance from DNA sequence. *Proceedings of the National Academy of Sciences of the United States of America* 116(12), 5542–5549.

Wasson, A. P., Nagel, K. A., Tracy, S. and Watt, M. (2020). Beyond digging: noninvasive root and rhizosphere phenotyping. *Trends in Plant Science* 25(1), 119–120.

Watson, A., Ghosh, S., Williams, M. J., Cuddy, W. S., Simmonds, J., Rey, M. D., Hatta, M. A. M., Hinchliffe, A., Steed, A., Reynolds, D., Adamski, N. M., Breakspear, A., Korolev, A., Rayner, T., Dixon, L. E., Riaz, A., Martin, W., Ryan, M., Edwards, D., Batley, J., Raman, H., Carter, J., Rogers, C., Domoney, C., Moore, G., Harwood, W., Nicholson, P., Dieters, M. J., DeLacy, I. H., Zhou, J., Uauy, C., Boden, S. A., Park, R. F., Wulff, B. B. H. and Hickey, L. T. (2018). Speed breeding is a powerful tool to accelerate crop research and breeding. *Nature Plants* 4(1), 23–29.

Winfield, M. O., Allen, A. M., Burridge, A. J., Barker, G. L. A., Benbow, H. R., Wilkinson, P. A., Coghill, J., Waterfall, C., Davassi, A., Scopes, G., Pirani, A., Webster, T., Brew, F., Bloor, C., King, J., West, C., Griffiths, S., King, I., Bentley, A. R. and Edwards, K. J. (2016). High-density SNP genotyping array for hexaploid wheat and its secondary and tertiary gene pool. *Plant Biotechnology Journal* 14(5), 1195–1206.

Wissuwa, M., Kretzschmar, T. and Rose, T. J. (2016). From promise to application: root traits for enhanced nutrient capture in rice breeding. *Journal of Experimental Botany* 67(12), 3605–3615.

Woll, K., Borsuk, L. A., Stransky, H., Nettleton, D., Schnable, P. S. and Hochholdinger, F. (2005). Isolation, characterization, and pericycle-specific transcriptome analyses of the novel maize lateral and seminal root initiation mutant rum1. *Plant Physiology* 139(3), 1255–1267.

Wu, L. Y., Yu, J. H., Shen, Q. F., Huang, L., Wu, D. Z. and Zhang, G. P. (2018a). Identification of microRNAs in response to aluminum stress in the roots of Tibetan wild barley and cultivated barley. *BMC Genomics* 19(1), 560.

Wu, W., Ma, B. L. and Whalen, J. K. (2018b). Enhancing rapeseed tolerance to heat and drought stresses in a changing climate: perspectives for stress adaptation from root system architecture. *Advances in Agronomy* 151(151), 87–157.

Xu, G. H. and Takahashi, H. (2020). Improving nitrogen use efficiency: from cells to plant systems. *Journal of Experimental Botany* 71(15), 4359–4364.

Xu, K., Xu, X., Fukao, T., Canlas, P., Maghirang-Rodriguez, R., Heuer, S., Ismail, A. M., Bailey-Serres, J., Ronald, P. C. and Mackill, D. J. (2006). *Sub1A* is an ethylene-response-factor-like gene that confers submergence tolerance to rice. *Nature* 442(7103), 705–708.

Xu, K. N. and Mackill, D. J. (1996). A major locus for submergence tolerance mapped on rice chromosome 9. *Molecular Breeding* 2(3), 219–224.

Xu, S. Z. (2003). Theoretical basis of the Beavis effect. *Genetics* 165(4), 2259–2268.

Xu, W. W., Subudhi, P. K., Crasta, O. R., Rosenow, D. T., Mullet, J. E. and Nguyen, H. T. (2000). Molecular mapping of QTLs conferring stay-green in grain sorghum (*Sorghum bicolor* L. Moench). *Genome* 43, 461–469.

Xu, X. W., Ji, J., Xu, Q., Qi, X. H., Weng, Y. Q. and Chen, X. H. (2018). The major-effect quantitative trait locus *CsARN6.1* encodes an AAA ATPase domain-containing protein that is associated with waterlogging stress tolerance by promoting adventitious root formation. *Plant Journal* 93(5), 917–930.

Xu, Y. B. and Crouch, J. H. (2008). Marker-assisted selection in plant breeding: From publications to practice. *Crop Science* 48(2), 391–407.

Xu, Y. B., Li, P., Zou, C., Lu, Y. L., Xie, C. X., Zhang, X. C., Prasanna, B. M. and Olsen, M. S. (2017). Enhancing genetic gain in the era of molecular breeding. *Journal of Experimental Botany* 68(11), 2641–2666.

Yadav, S., Sandhu, N., Majumder, R. R., Dixit, S., Kumar, S., Singh, S. P., Mande, N. P., Das, S. P., Yadaw, R. B., Singh, V. K., Sinha, P., Varshney, R. K. and Kumar, A. (2019). Epistatic interactions of major effect drought QTLs with genetic background loci determine grain yield of rice under drought stress. *Scientific Reports* 9(1), 2616.

Yadav, V. K., Raghav, M., Sharma, S. K. and Bhagat, N. (2020). Rhizobacteriome: promising candidate for conferring drought tolerance in crops. *Journal of Pure and Applied Microbiology* 14(1), 73–92.

Yamada, M., Han, X. W. and Benfey, P. N. (2020). *RGF1* controls root meristem size through ROS signalling. *Nature* 577(7788), 85–88.

Yan, X. L., Liao, H., Beebe, S. E., Blair, M. W. and Lynch, J. P. (2004). QTL mapping of root hair and acid exudation traits and their relationship to phosphorus uptake in common bean. *Plant and Soil* 265(1–2), 17–29.

Yang, Z. Y., Liu, J. H., Poree, F., Schaeufele, R., Helmke, H., Frackenpohl, J., Lehr, S., von Koskull-Doring, P., Christmann, A., Schnyder, H., Schmidhalter, U. and Grill, E. (2019). Abscisic acid receptors and coreceptors modulate plant water use efficiency and water productivity. *Plant Physiology* 180(2), 1066–1080.

Ye, H., Roorkiwal, M., Valliyodan, B., Zhou, L. J., Chen, P. Y., Varshney, R. K. and Nguyen, H. T. (2018a). Genetic diversity of root system architecture in response to drought stress in grain legumes. *Journal of Experimental Botany* 69(13), 3267–3277.

Ye, H., Song, L., Chen, H. T., Valliyodan, B., Cheng, P., Ali, L., Vuong, T., Wu, C. J., Orlowski, J., Buckley, B., Chen, P. Y., Shannon, J. G. and Nguyen, H. T. (2018b). A major natural genetic variation associated with root system architecture and plasticity improves waterlogging tolerance and yield in soybean. *Plant, Cell and Environment* 41, 2169–2182.

Yoshinari, A. and Takano, J. (2017). Insights into the mechanisms underlying boron homeostasis in plants. *Frontiers in Plant Science* 8, 1951.

Yoshino, K., Numajiri, Y., Teramoto, S., Kawachi, N., Tanabata, T., Tanaka, T., Hayashi, T., Kawakatsu, T. and Uga, Y. (2019). Towards a deeper integrated multi-omics approach in the root system to develop climate-resilient rice. *Molecular Breeding* 39(12), 165.

Yu, J., Holland, J. B., McMullen, M. D. and Buckler, E. S. (2008). Genetic design and statistical power of nested association mapping in maize. *Genetics* 178(1), 539–551.

Yu, J. M. and Buckler, E. S. (2006). Genetic association mapping and genome organization of maize. *Current Opinion in Biotechnology* 17(2), 155–160.

Yu, P., Hochholdinger, F. and Li, C. J. (2019). Plasticity of lateral root branching in maize. *Frontiers in Plant Science* 10, 363.

Zamir, D. (2001). Improving plant breeding with exotic genetic libraries. *Nature Reviews. Genetics* 2(12), 983–989.

Zandalinas, S. I., Fritschi, F. B. and Mittler, R. (2020). Signal transduction networks during stress combination. *Journal of Experimental Botany* 71(5), 1734–1741.

Zhan, J. J., Wang, F., Xing, W. H., Liu, J., Fan, Z. Y. and Tao, Y. S. (2018). Fine mapping and candidate gene prediction of a major QTL for kernel number per ear in maize. *Molecular Breeding* 38(3).

Zhang, C. Y., Simpson, R. J., Kim, C. M., Warthmann, N., Delhaize, E., Dolan, L., Byrne, M. E., Wu, Y. and Ryan, P. R. (2018). Do longer root hairs improve phosphorus uptake? Testing the hypothesis with transgenic *Brachypodium distachyon* lines over-expressing endogenous RSL genes. *New Phytologist* 217(4), 1654–1666.

Zhang, H., Uddin, M. S., Zou, C., Xie, C., Xu, Y. and Li, W. X. (2014). Meta-analysis and candidate gene mining of low-phosphorus tolerance in maize. *Journal of Integrative Plant Biology* 56(3), 262–270.

Zhang, Y., Thomas, C. L., Xiang, J. X., Long, Y., Wang, X. H., Zou, J., Luo, Z. L., Ding, G. D., Cai, H. M., Graham, N. S., Hammond, J. P., King, G. J., White, P. J., Xu, F. S., Broadley, M. R., Shi, L. and Meng, J. L. (2016). QTL meta-analysis of root traits in *Brassica napus* under contrasting phosphorus supply in two growth systems. *Scientific Reports* 6, 33113.

Zheng, X. W., Wen, X. J., Qiao, L., Zhao, J. J., Zhang, X. J., Li, X., Zhang, S. W., Yang, Z. J., Chang, Z. J., Chen, J. L. and Zheng, J. (2019). A novel QTL *QTrl.saw-2D.2* associated with the total root length identified by linkage and association analyses in wheat (*Triticum aestivum* L.). *Planta* 250(1), 129–143.

Zheng, Z. H., Hey, S., Jubery, T., Liu, H. Y., Yang, Y., Coffey, L., Miao, C. Y., Sigmon, B., Schnable, J. C., Hochholdinger, F., Ganapathysubramanian, B. and Schnable, P. S. (2020). Shared genetic control of root system architecture between *Zea mays* and *Sorghum bicolor*. *Plant Physiology* 182(2), 977–991.

Zhou, L. L., Bai, G. H., Ma, H. X. and Carver, B. F. (2007). Quantitative trait loci for aluminum resistance in wheat. *Molecular Breeding* 19(2), 153–161.

Zhou, Y., Tao, Y. J., Tang, D. N., Wang, J., Zhong, J., Wang, Y., Yuan, Q. M., Yu, X. F., Zhang, Y., Wang, Y. L., Liang, G. H. and Dong, G. C. (2017). Identification of QTL associated with nitrogen uptake and nitrogen use efficiency using high throughput genotyped CSSLs in rice (*Oryza sativa* L.). *Frontiers in Plant Science* 8, 1166.

Zhu, C. F., Bortesi, L., Baysal, C., Twyman, R. M., Fischer, R., Capell, T., Schillberg, S. and Christou, P. (2017). Characteristics of genome editing mutations in cereal crops. *Trends in Plant Science* 22(1), 38–52.

Zhu, J. M., Kaeppler, S. M. and Lynch, J. P. (2005). Mapping of QTL controlling root hair length in maize (Zea mays L.) under phosphorus deficiency. *Plant and Soil* 270(1), 299–310.

Zou, C., Wang, P. X. and Xu, Y. B. (2016). Bulked sample analysis in genetics, genomics and crop improvement. *Plant Biotechnology Journal* 14(10), 1941–1955.

Zwart, R. S., Thompson, J. P., Milgate, A. W., Bansal, U. K., Williamson, P. M., Raman, H. and Bariana, H. S. (2010). QTL mapping of multiple foliar disease and root-lesion nematode resistances in wheat. *Molecular Breeding* 26(1), 107–124.

Chapter 15

The use of plant growth-promoting rhizobacteria (PGPR) to improve root function and crop nutrient use efficiency

Melissa M. Larrabee and Louise M. Nelson, University of British Columbia, Canada

1 Introduction

With the world population expected to reach over 9 billion by 2050, there is an urgent need for increased food production (Alexandratos and Bruinsma, 2012). In the first green revolution over 50 years ago, this need was met by the development of high-yielding crop varieties that were dependent on intensive management and the use of synthetic chemical fertilizers and pesticides. This approach, which contributed significantly to feeding the world population, has led to unintended detrimental impacts on the environment, including soil degradation, soil and water pollution and increased greenhouse gas emissions. These issues, along with the expected damaging effects associated with the climate change, necessitate the move to a more 'sustainable intensification of agriculture', defined as increasing yields without adverse environmental impact and without cultivation of more land (Royal Society, 2009). Bacteria that colonize plant roots and enhance plant growth and development, plant growth-promoting rhizobacteria (PGPR), can contribute to this more sustainable approach by decreasing the use of synthetic fertilizers and enhancing nutrient uptake and efficiency (Adesemoye and Kloepper, 2009; Dodd and Ruiz-Lozano, 2012; Vacheron et al., 2013).

http://dx.doi.org/10.19103/AS.2020.0075.22

PGPR are found in the soil under the influence of the root (rhizosphere) and within the root (endosphere) and are taxonomically highly diverse (Reed and Glick, 2013; Carro and Nouioui, 2017; Backer et al., 2018). They are a component of the root microbiome which comprises all of the microorganisms (bacteria, archaea, fungi, algae, protozoa, viruses) influenced by the plant root and its exudates. PGPR, in turn, exert their beneficial effects on plant growth and health through a variety of mechanisms, both direct and indirect. These have been the subject of extensive research, and several recent reviews are cited here. Direct mechanisms include facilitating nutrient acquisition, modulating phytohormones and enhancing abiotic stress tolerance (Glick, 2012; Parnell et al., 2016; Ilangumaran and Smith, 2017; Majeed et al., 2018). Indirect mechanisms of plant growth promotion include controlling plant pathogens via the production of antibiotics, lytic enzymes and siderophores; the induction of systemic resistance; and competition (Glick, 2012; Rosier et al., 2018). This chapter will focus on the use of PGPR as biofertilizers and, hence, on the enhancement of plant growth via direct mechanisms.

The focus of much early PGPR research has been on the culture-based isolation and identification and characterization of individual PGPR species and strains from the roots of crop plants and plants in natural ecosystems. This has been followed by the determination of their potential mechanisms of action in pure culture; testing in greenhouses, often using model host plants; and, finally, testing as single-strain inoculants under field conditions (Finkel et al., 2017; Rosier et al., 2018). Beneficial effects observed in the lab and greenhouse are not always evident or consistent in the field (Nelson, 2004; Backer et al., 2018; Sessitsch et al., 2019). There are challenges with specificity for the host and the environment, as well as in formulating, storing and applying these inoculants within existing agricultural management practices, and this has deterred their wide adoption in developed countries (Glick, 2012; Parnell et al., 2016; Timmusk et al., 2017; Backer et al., 2018). Although considerable progress has been made in identifying mechanisms using mutants of both PGPR and host plants, there is still much that is unknown about how these microorganisms function *in situ*, interact with other members of the root microbiome and respond to varying abiotic and biotic stresses.

The advent of culture-independent approaches and the increasing accessibility and affordability of new 'omics' tools are revolutionizing our ability to understand and exploit the complex interactions of PGPR and their crop hosts (Finkel et al., 2017; Paterson et al., 2017; Backer et al., 2018; Rosier et al., 2018). Whole genome sequencing and comparison of individual PGPR strains and species are providing insight into conserved and unique PGPR-associated traits (Bruto et al., 2014; Xie et al., 2016). Metagenomic sequencing of the root microbiome of crop species from different environments is revealing the diversity of the microbial communities, while transcriptomic and metabolomic analyses are providing insight into microbial and plant gene expression and function as influenced by their interactions (Finkel et al., 2017; Garcia and Kao-Kniffen, 2018; Rosier et al., 2018). This information contributes to our understanding of the concepts of the holobiont and hologenome (Rosenberg and Zilber-Rosenberg, 2016) and the core microbiome (Toju et al., 2018), the construction of synthetic microbiomes (Grosskopf and Soyer, 2014) and the use of network analysis to dissect these interactions (Faust and Raes, 2012; Garcia and Kao-Kniffen, 2018; Toju et al., 2018). These novel approaches will facilitate the development and application of the more effective use of PGPR for sustainable agriculture.

In this chapter we will focus on recent advances in the use of PGPR to enhance root function and nutrient use with emphasis on studies that move from the lab to the field and the insight these studies provide with respect to potential mechanisms. In addition, we will review some of the novel approaches being developed to overcome the challenges of practical adoption and use of PGPR in commercial agriculture.

2 The effect of plant growth-promoting rhizobacteria on root function

In this section we summarize research on the role of PGPR in enhancing the root function, with emphasis on their effects on root architecture, metabolism and adaptation to abiotic stress.

2.1 Root architecture

It is now well established that PGPR can affect root architecture. As reviewed by Vacheron et al. (2013) and Verbon and Liberman (2016), commonly observed phenotypes include an inhibition of the primary root growth and an increase in the number and/or length of lateral roots and root hairs, leading to an increase in the uptake of mineral nutrients and water and greater plant biomass. Studies in *Arabidopsis thaliana* show that PGPR affect cell division and differentiation at the primary root meristem and at the sites of lateral root formation by modulating plant hormonal balance and signalling pathways. Auxins such as indole-3-acetic acid (IAA), cytokinins, ethylene, reactive oxygen species (ROS), quorum sensing molecules and bacterial volatiles have been shown to be involved in altering this balance (reviewed by Vacheron et al., 2013; Verbon and Liberman, 2016; Sharifi and Ryu, 2018; Ortiz-Castro and Lopez-Bucio, 2019).

Much of this research has been carried out with model plants such as *Arabidopsis* under gnotobiotic conditions using agar plate systems and bacterial and plant mutants to enable the determination of the molecular mechanisms involved (see Asari et al., 2017; Wang et al., 2018; Zhao et al., 2018 as examples). However, these growth conditions are far from those encountered by PGPR in the field. Klikno and Kutschera (2017) partially addressed this issue by showing that root hair elongation of *A. thaliana* grown in the presence of a soil extract containing microorganisms was promoted compared to gnotobiotic seedlings grown on agar. This effect on root development was mimicked by the addition of cytokinins, or ethylene, or by the addition of *Methylobacterium* species, PGPR that produce and secrete cytokinins. Other studies have observed similar effects in crop plants grown under greenhouse and/or field conditions.

Vacheron et al. (2018) compared the effects of a wild-type PGPR, *Pseudomonas kilonensis* F113, four deletion mutants impaired in traits associated with plant growth promotion, and a quadruple mutant, impaired in all four traits, on the root system architecture of *A. thaliana* Col-0 grown on agar plates and of *Zea mays* grown in pots in non-sterile agricultural soil. Inoculation with the wild-type F113 strain modified the root architecture of both *Arabidopsis* and maize seedlings, inhibiting the primary root length and increasing the root hair length of *Arabidopsis*, while enhancing the root length and surface area of maize. The relative effect of each mutation differed, but mutants deficient in 1-aminocyclopropane-1-carboxylate (ACC) deaminase or nitrite reductase had a significant effect on root architecture in both plant species under *in vitro* and *in vivo* growth conditions. Mutations in these two traits also impaired the ability of these strains to colonize maize roots over 20 days in non-sterile soil.

Lally et al. (2017) conducted greenhouse and field studies of oilseed rape, *Brassica napus*, inoculated with *Pseudomonas fluorescens* strain F113, three other endophytic *P. fluorescens* isolates from Ireland and a consortium of these three isolates and seven other PGPR isolates. Plant biomass above ground was increased with the inoculation of three of the four strains in the greenhouse, but effects on root biomass were not reported. In the field pre-harvest, only the 10-strain consortium significantly increased above-ground dry weight, and the consortium and a single strain, L321, increased the total plant length (a measure of root and stem length). Gomes et al. (2018) observed that inoculation with *Azospirillum brasilense* Ab-V5 increased the root biomass of *B. napus* grown in a greenhouse in sterile soil and root length was increased following inoculation with *Rhizobium* strain 8121.

Azospirillum sp. has probably been the most studied PGPR and the most widely applied inoculum under field conditions, particularly in South America and Asia (reviewed by Diaz-Zorita et al., 2015; Cassan and Diaz-Zorita, 2016). Meta-analyses of diverse field experiments with maize, cereals and legumes in mainly semi-arid environments showed inconsistent responses to inoculation; however, positive responses to inoculation were observed about 70% of the time. In maize increases in crop productivity following inoculation with *Azospirillum* were usually correlated with increased root surface area, leading to greater exploration of the soil volume during the early growth of the crop and were often associated with abiotic stresses during the early growth (Cassan and Diaz-Zorita, 2016). Lade et al. (2018) examined the effect of inoculation with *A. brasilense* Sp7 on the early root growth of four barley cultivars in a growth chamber in a sterile potting substrate and in an *in vitro* system in plastic bags. Inoculation with Sp7 increased the root length, surface area and number of root tips in both growth systems, but the magnitude of effects varied with the cultivar.

These studies suggest that by altering root architecture and increasing root surface area PGPR can have a significant beneficial effect on nutrient acquisition and overall crop productivity, but further research is required to understand these interactions and optimize conditions in the field to achieve this benefit consistently. The alteration of root architecture can also lead to changes in root and whole plant metabolism.

2.2 Plant metabolism

Vacheron et al. (2013) suggest that the ability of PGPR to modify whole plant physiology and metabolism at sites distant from the site of colonization occurs via two major mechanisms. By altering root architecture to enhance the nutrient uptake, whole plant nutrition and metabolism are stimulated, leading to enhanced growth. PGPR may also affect endogenous hormonal and signalling pathways (Verbon and Liberman, 2016), leading to modifications in plant metabolism and functioning. The accessibility of transcriptomic and metabolomic tools to analyse the effects of PGPR inoculation on plant gene expression and metabolites, respectively, is providing new perspectives in this area. It is not the intent of this review to provide an exhaustive listing of this research, but a few examples are discussed in the following text to illustrate the wealth of new information becoming available.

Kwon et al. (2016) investigated the root and shoot responses of *Arabidopsis* to inoculation with the PGPR strain *Paenibacillus polymyxa* E681 using proteome, transcriptome and metabolome analyses. Plants sampled 7 days after PGPR inoculation showed 17 protein spots in the root and 24 protein spots in the shoots that were differentially expressed compared to the control. These were classified into seven different protein groups using bioinformatic tools. Thirty-six of these proteins related to antioxidants, defence and stress, amino acid metabolism, carbohydrate metabolism, photosynthesis and plant hormone functions were up-regulated and five proteins related to carbohydrate metabolism and amino acid metabolism were down-regulated. Metabolite analysis showed increased levels of tryptophan (a precursor of both IAA and camalexin), indole-3-acetonitrile, IAA and camalexin, a defence-related protein in response to inoculation.

Hao et al. (2016) used transcriptomic analysis to determine the response of *Arabidopsis* at two different growth stages to volatiles emitted by the PGPR strain

Bacillus amyloliquefaciens. At the seedling stage, 1507 and 820 differentially expressed genes were identified in leaves and roots, respectively, and many of these were enriched for genes related to plant hormone biosynthesis and signalling pathways, cell wall modifications and protection against stress, but differences with tissue and growth stage were evident. These data provide support for the role of volatiles produced by PGPR in promoting plant growth and enhancing plant resistance to abiotic and biotic stresses.

Ibort et al. (2018) used proteomic analysis to determine the role of ethylene perception in the interaction of two PGPR strains, *Bacillus megaterium* and *Enterobacter* sp. C7, with tomato, by comparing differences in the root proteomic profiles of an ethylene-insensitive tomato line and its isogenic wild type. PGPR inoculation affected the proteomic profile in a strain- and genotype-specific manner, modifying the levels of stress-related and interaction proteins and showing inoculation effects on antioxidant content and phosphorus acquisition capacity. *B. megaterium* promoted growth only in wild-type plants, and this was associated with increased levels of reduced glutathione, which has been shown to increase plant tolerance to stress. *Enterobacter* inoculation promoted the growth of both tomato genotypes and improved phosphorus nutrition independent of ethylene sensitivity.

Scotti et al. (2019) examined the early effects of inoculation with PGPR, *P. fluorescens* CREA-C16 (Pf-16), on the tomato root transcriptome. Pf-16 strongly repressed defence response in roots within 48 h of inoculation by down-regulating resistance genes, biotic stress-related genes, transcription factors involved in salicylic and jasmonic acid metabolism, ethylene biosynthesis/signalling and protease-dependent mechanisms. Between 48 and 72 h Pf-16 up-regulated genes involved in plant cell wall modification, leading to partial cell wall disassembly which may be important for bacterial colonization. After 72 h Pf-16 up-regulated genes involved in plant growth promotion, including those facilitating the transport of nutrients and solutes.

Gamez et al. (2019) reported on the effects of inoculation of two PGPR with differing colonization patterns, *B. amyloliquefaciens* Bs006 and *P. fluorescens* Ps006, on the banana (*Musa acuminata*) transcriptome. Both strains significantly increased plant height, leaf area and fresh and dry weight of shoots and roots following 55 days of growth, but *B. amyloliquefaciens* colonized the surface of banana roots via biofilms while *P. fluorescens* colonized endophytically within the roots. Transcriptome profiles obtained 1 h, 48 h and 96 h post-inoculation showed differential gene expression in response to inoculation with increasing numbers of differentially expressed genes over time observed in response to *P. fluorescens* inoculation and decreasing numbers in response to *B. amyloliquefaciens* inoculation. The authors observed the differential expression of several thousand genes over the 96-hour period post-inoculation but focused on 22 experimentally validated genes involved in plant growth promotion, defence and biotic or abiotic stress. Their expression varied temporally and between the two PGPR strains.

These few examples illustrate the complexity of information made available using 'omics' tools and demonstrate the profound, rapid and diverse effect the inoculation of PGPR, or their metabolites, has on plant gene expression, metabolism and functioning.

2.3 Abiotic stress

PGPR promote plant growth under stress conditions by modifying the root system, enhancing mobilization and increasing the uptake of essential nutrients (N, P and K) (Egamberdieva et al., 2017; Parray et al., 2016). Many PGPR act as biofertilizers through the action of several important enzymes and metabolites. Bacterial ACC deaminase cleaves the precursor to the plant hormone ethylene, which is responsible for plant stress responses that decrease growth (Eida et al., 2018; El-Esawi et al., 2018b; Etesami and Maheshwari, 2018; Ilangumaran and Smith, 2017). Many PGPR produce plant growth regulators such as IAA, cytokinins, gibberellic acid and ethylene that may directly affect plant growth or modulate endogenous plant hormonal pathways (Vacheron et al., 2013). Some bacteria also modulate the abscisic acid (ABA) pathways in plants to enhance plant growth (Ilangumaran and Smith, 2017). PGPR are effective candidates for the amelioration of salinity, drought, heavy metal toxicity and nutrient imbalance in abiotically stressed plants (Etesami and Maheshwari, 2018). This section on the response of root function, growth and/or crop yield to PGPR inoculation under saline and drought conditions.

2.3.1 Salinity

Soil salinity is a leading cause of economic losses in agriculture in arid and semi-arid regions (Egamberdieva et al., 2017). High soil salinity severely affects the physiological responses of plants, such as lipid metabolism, protein synthesis and photosynthesis, thereby decreasing crop growth and performance (El-Esawi et al., 2018b; Parida and Das, 2005). Due to the osmotic effects of Na^+ and Cl^-, water uptake by roots is reduced, causing a build-up of toxic ROS which causes oxidative damage to lipids, proteins and nucleic acids (El-Esawi et al., 2018a; Ilangumaran and Smith, 2017; Parida and Das, 2005). Plant cell division and elongation are repressed in high salinity environments, ultimately affecting root growth (Egamberdieva et al., 2017; El-Esawi et al., 2018a,b; Ilangumaran and Smith, 2017). After initial exposure to salinity, root elongation may recover; however, over extended periods of high salinity, root architecture changes via the repression of lateral root formation (Ilangumaran and Smith, 2017).

PGPR mediate salt tolerance in plants through a variety of mechanisms that are not all well understood (Ilangumaran and Smith, 2017). Evidence suggests that PGPR function in plant osmotic balance, ion homeostasis and metabolic/phytohormone signalling (Ilangumaran and Smith, 2017). In brief, bacteria exposed to hyperosmotic environments accumulate osmolytes (proline, trehalose, glycine) in their cytosol, which are then absorbed by plants through their roots to aid in maintaining the osmotic balance. Bacteria trap toxic salt cations in their exopolysaccharide matrix, alter root structure and regulate the expression of plant ion transporters. Lastly, PGPR alter phytohormone signalling through the production of enzymes such as ACC deaminase and the phytohormone IAA (Ilangumaran and Smith, 2017). Several PGPR genera, including *Serratia*, *Pseudomonas*, *Bacillus* and *Bradyrhizobium*, have been shown to retain and/or enhance crop yield of plants in saline soils without harming the environment (Table 1) (El-Esawi et al., 2018a; Etesami and Maheshwari, 2018).

El-Esawi et al. (2018b) showed that *Serratia liquefaciens* KM4 was able to enhance maize growth and biomass yield compared to non-inoculated plants when under salinity stress in growth chambers. Under saline conditions, KM4 inoculation mitigated the expected uptake of Na^+ and Cl^- ions and increased accumulation of K^+ and Ca^{2+} ions inside of maize cells compared to the control. The authors suggested that KM4 decreased the content of the toxic salt ions by sequestering them into vacuoles, exporting them out of the cells and circulating them to the root. KM4 was also able to increase the levels of osmoprotectants and antioxidants in the leaves under saline conditions; these changes were corroborated by the up-regulation of stress-related genes. By improving growth and calcium and potassium acquisition under saline conditions, KM4 shows a promising potential for ameliorating salt stress in maize (El-Esawi et al., 2018b).

Sapre et al. (2018) reported that *Klebsiella* sp. successfully ameliorated salt stress in oat seedlings (*Avena sativa*) by increasing root length and dry weight under high salinity conditions in a hydroponic growth chamber. In another growth chamber study, *Bacillus firmus* SW5 was able to promote the root length and volume of soybean plants in saline environments compared to non-inoculated controls (El-Esawi et al., 2018a). Inoculating with SW5 also increased shoot growth and biomass in salt-stressed soybeans compared to the controls. Inoculation enhanced phosphorus and nitrogen content under both control (not salt-stressed) and salt-stressed conditions, induced nodulation in soybean and enhanced chlorophyll content in the salt-stressed plants (El-Esawi et al., 2018a). Typically, salt stress reduces chlorophyll content by inducing the pigment-degrading enzyme chlorophyllase (Abd_Allah et al., 2017; El-Esawi et al., 2018a). El-Esawi et al. (2018a) suggest that SW5 was able to enhance chlorophyll content by increasing nutrient uptake, which in turn induced pigment biosynthesis.

Table 1 Summary of selected recent studies of PGPR-induced root architecture changes and crop growth in stressed environments

Type of stress	Inoculation method, experiment type	Crop	PGPR	Effect	Reference
Salinity	Seed, sterilized soil growth chamber	Maize	*Serratia liquefaciens* KM4	Increased plant growth and biomass	El-Esawi et al. (2018b)
	Seed, hydroponic pots	Oat	*Klebsiella* sp.	Increased root length and dry weight	Sapre et al. (2018)
	Seed, hydroponic pots	Soybean	*Bacillus firmus* SW5	Increased root length, volume and nodulation	El-Esawi et al. (2018a)
	Seed, hydroponic pots	Soybean	Consortium: *Bradyrhizobium japonicum* *Pseudomonas putida*	Increased root length, volume, surface and project area, as well as nodulation	Egamberdieva et al. (2017)
	Seed, field	Wheat	Consortium: *Bacillus cereus* strain Y5, *Bacillus* sp. Y14 and *Bacillus subtilis* strain Y16	Increased plant growth and yield	Shahzad et al. (2017)
Drought	Soil, microcosm	Finger millet	Single: *Pseudomonas* sp. strain DPB13, DPB15, DPB16	Increased root dry weight and elongation	Chandra et al. (2018)
	Soil, microcosm	French lavender and sage	Single: *Bacillus thuringiensis* *Bacillus megaterium*	Increased shoot and root dry weight	Armada et al. (2014)
	Seed, jars	Wheat	*Rhizobium leguminosarum* (LR-30)	Increased root dry weight and length	Hussain et al. (2014)
	Seed, greenhouse	Lentil	*Methylobacterium oryzae*	Increased root and shoot growth	Jorge et al. (2019)

PGPR that were able to enhance plant growth and resistance to stressed environments in the laboratory sometimes fail to do so in field conditions (Etesami and Maheshwari, 2018). It is thought that the ability of PGPR to succeed in field conditions relies mainly on their ability to compete with the native microbiota for colonization and nutrients and their ability to survive the stressful environment (Etesami and Maheshwari, 2018; Shahzad et al., 2017). Single-strain inoculants often have poor root colonization in environmental field conditions (Shahzad et al., 2017). To improve field effectiveness, multi-strain inoculants (consortia) can be applied with synergistic effects due to the various plant-growth-promoting mechanisms specific to each strain (Egamberdieva et al., 2017; Shahzad et al., 2017). Consortia can stimulate the formation of biofilms that protect against degradation as well as enhance communication between microbes (Shahzad et al., 2017).

A hydroponic laboratory study conducted by Egamberdieva et al. (2017) showed that the combined inoculation of soybean with the symbiont *Bradyrhizobium japonicum* USDA 110 and the root-colonizing salt-tolerant *Pseudomonas putida* TSAU1 significantly improved plant growth, root length and volume, root surface and projected area, as well as nodulation compared to inoculation with *B. japonicum* alone. Alteration of root architecture as a result of PGPR colonization may enhance the ability of a plant to acquire more nutrients under salt stress (Egamberdieva et al., 2017). Shahzad et al. (2017) showed that a bacterial consortium of three PGPR strains (*Bacillus cereus* Y5, *Bacillus* sp. Y14 and *Bacillus subtilis* Y16) was able to substantially increase wheat growth and yield in salt-stressed soils in the field.

2.3.2 Drought

As in salt stress, drought conditions result in plant cell dehydration and increased ROS, which ultimately decrease plant growth. Water deficits cause stomatal closure and decreases in chlorophyll content, thereby reducing photosynthesis (Kaushal, 2019). PGPR enhance the drought tolerance of crops by similar mechanisms as those used in ameliorating salt stress (Etesami and Maheshwari, 2018). In particular, PGPR use microbial-induced systemic tolerance (MIST) mechanisms to alter phytohormones, stomatal conductivity, antioxidant activity, nutrient status and gene regulation of the plant (Kaushal, 2019). Several genera of PGPR, such as *Pseudomonas*, *Bacillus* and *Rhizobium*, have been reported to significantly improve crop tolerance to drought (Table 1) (Chandra et al., 2018; Jamil et al., 2018; Ullah et al., 2019; Yang et al., 2009).

Three ACC deaminase-producing *P. putida* strains (DPB13, 15 and 16) had a significantly positive effect on the root dry weight and elongation of drought-stressed finger millet (Chandra et al., 2018). Inoculation also increased foliar nutrient content and antioxidant content. Similarly, Armada et al. (2014) reported that *Bacillus thuringiensis* and *B. megaterium* were able to independently increase root dry weight in both French lavender and sage under osmotic stress. *B. thuringiensis* promoted lavender growth under drought stress by increasing K content, depressing stomatal conductance and controlling proline accumulation. PGPR-mediated drought tolerance was related to decreased glutathione reductase and ascorbate peroxidase activities, which indicated a reduction of ROS levels in the stressed plants.

Rhizobium leguminosarum (LR-30) was able to increase root dry weight and length in drought-stressed wheat (Hussain et al., 2014). The authors suggested that the beneficial effect may be due to the PGPR's ability to produce catalase, exopolysaccharides, siderophores, organic acid and auxin.

Jorge et al. (2019) reported the ability of *Methylobacterium oryzae* to stimulate the early growth of water-stressed lentil roots by 51%. The induction of drought tolerance by the PGPR was suggested to be at least partially due to their ability to produce and deliver cytokinins to the plants, as there were 4.5 times more cytokinins in inoculated compared to non-inoculated plants.

These studies on abiotic stress adaptation of plants mediated by PGPR demonstrate the diversity of bacteria that have promising plant-growth-promoting effects, but more field studies are needed to test their efficacy under these conditions. Field studies should include: (i) those environments that may be stressful for the PGPR, (ii) co-inoculation and (iii) efficacy with different crops and soil textures.

3 The effect of plant growth-promoting rhizobacteria on plant mineral nutrition

Nitrogen (N), phosphorus (P), potassium (K) and sulphur (S) are the four major macronutrients required for plant metabolism and growth (Gyanendranath, 2017; Jacoby et al., 2017). N, P and S are typically bound to organic molecules in the soil and are, thus, minimally bioavailable for plant uptake (Jacoby et al., 2017). To enhance plant growth, mineral fertilizers are often used to supplement soils; however, fertilization has been linked to soil degradation, water eutrophication and greenhouse gas emissions (Jacoby et al., 2017). Soil microbes, both bacteria and fungi, that are able to mineralize organic nutrient forms and release inorganic forms into the soil for plant uptake can be used as biofertilizers (Bargaz et al., 2018; Jacoby et al., 2017; Raklami et al., 2019; Van der Heijden et al., 2008). Biofertilization is an alternative to mineral fertilization, allowing the supplementation of crops with PGPR in place of, or in combination with, reduced levels of organic (such as compost, mulch or manure) or mineral fertilizers to increase nutrient availability, uptake and use efficiency. In a comprehensive meta-analysis to quantify the benefits of biofertilization in field studies, Schutz et al. (2018) found that biofertilizers were most effective in dry climates, and those with both N-fixing and P-solubilizing traits had the highest potential to improve crop yields. The yield response of crops due to biofertilizer application was small at low soil P levels, and the efficacy increased with increasing soil P levels with the optimum for P solubilizers being midway between those for arbuscular mycorrhizal fungi (AMF) and N fixers.

To avoid overlap with the literature of the well-documented plant-growth-promoting effects of AMF and nodulating/nitrogen-fixing bacteria, our focus will be on other important microbial players in the rhizosphere microbiome (Table 2).

3.1 Nutrient uptake

While the application of PGPR inoculants has been shown to increase nutrient uptake in crops (Adesemoye and Kloepper, 2009; Pii et al., 2015), it is hypothesized that this is primarily an indirect effect of increased root surface area (Vacheron et al., 2013). However, PGPR can also affect the root's nutrient uptake and transport systems, but few studies have been published in this regard (Bertrand et al., 2000; Vacheron et al., 2013). Calvo et al. (2019) reported recently that PGPR mixtures, primarily comprised of the genus *Bacillus*, not only enhanced the overall growth and nutrient uptake of *A. thaliana* grown in a soil-like system but also increased the transcript levels of nitrate and ammonium uptake and transport genes in roots. These data suggest a potential alternative explanation for the observed plant growth promotion and increased N content in PGPR-inoculated plants in addition to that of enhanced root surface area, but the mechanism by which these bacteria induce N uptake and transport genes is not yet understood.

There is considerable interest in using PGPR as a means to enhance the efficiency of nutrient uptake and reduce the application of fertilizers (Adesemoye and Kloepper, 2009; Jacoby et al., 2017). Many studies have shown that the maximum growth promotion of crops by PGPR occurs in the absence of N fertilization (Veresoglou and Menexes, 2010). Nguyen et al. (2019) carried out a study of the biostimulant effects of three *Bacillus velezensis* strains on spring wheat in a greenhouse using non-sterile field soil under varying mineral N fertilizer rates. Growth promotion was highest at a moderate N rate (50% of the recommended N dose), and there was no significant effect of the inoculants in the absence of N fertilizer. Total nutrient contents and nutrient uptake efficiency were increased in PGPR treatments due to increased plant biomass (Nguyen et al., 2019).

In another study the PGPR strain, *Enterobacter* sp. Fs-11, was tested for its ability to enhance sunflower yield and oil content in the field in conjunction with 50% reduced conventional fertilizer application (Shahid et al., 2019). Inoculation increased sunflower growth, yield, oil content and nutrient uptake. There was

Table 2 Summary of selected recent studies of the effect of PGPR inoculation on nutrient uptake and nutrient use efficiency in various crops

Nutrient (uptake or efficiency)	Inoculation method, experiment type	Crop	PGPR	Effect	Ref.
Uptake	Seed, field	Sunflower	*Enterobacter* sp. Fs-11 in combination with 50% reduced conventional fertilizer	Increased plant height and head diameter, increased P uptake of achenes and leaves	Shahid et al. (2019)
Uptake	Seed, field	Faba bean and wheat	Consortium: Rhizobacteria: BS17 (*Actinobacter* sp.), PGP27 (*Rahnella aquatilis*) Rhizobia: RhOF4 (*Ensifer meliloti*), RhOF155 (*E. meliloti*) AMF: *Glomus* sp., *Sclerocystis* sp., *Acaulospora* sp.	Consortia increased shoot and root dry weight, number of leaves and flowers, shoot's N, P, K and Na content, sugar and protein content, soil organic matter and soil total carbon	Raklami et al. (2019)
Uptake	Seed, growth chamber	*Arabadopsis thaliana*	Three consortia mixtures containing one or more of: *Bacillus pumilus*, *B. subtilis*, *B. safensis*, *B. amyloliquefaciens*, *Lysinibacillus xylanilyticus*, *Paenibacillus peoriae*	All mixtures increased fresh shoot and root weight, chlorophyll content, nutrient uptake and transcript level of nitrate and ammonia uptake genes	Calvo et al. (2019)
Uptake	Grain, greenhouse	Spring wheat	*Bacillus amyloliquefaciens*, *B. megaterium*, *Azospirillum brasilense* in combination with reduced and full fertilizer	Increased biomass, nutrient uptake and nutrient	Nguyen et al. (2019)
Efficiency	Soil, growth chamber	Kikuyu grass	PGPR bacterium SOS3 (*Paraburkholderia*) in combination with organic fertilizer	Reduced mineral N leachate, reduced soil mineral N, maintain shoot and root growth	Paungfoo-Lonhienne et al. (2019)

(*Continued*)

Table 2 (*Continued*)

Nutrient (uptake or efficiency)	Inoculation method, experiment type	Crop	PGPR	Effect	Ref.
Efficiency	Seed, field	*Amaranthus hypochondriacus*	Consortium of Bacillus sp. in conjunction with organic fertilizer	Increased nutrient uptake and increased nutrient use efficiency of N, P and K	Pandey et al. (2018)
Efficiency	Soil, greenhouse	Cucumber	*Bacillus cereus*, in combination with composted leguminous shrub (*Caragana microphylla* straw)	Enhanced nutrient uptake efficiency under reduced N input	Zhang et al. (2019)
Efficiency	Fertilizer, pot	Wheat	*Bacillus* sp. strain KAP6	Increased plant high, root length and biomass and increased N and P content and use efficiency	Ahmad et al. (2017)
Efficiency	Fertilizer, pot	Maize	*Pseudomonas putida*	Increased root and shoot dry weights and increased N, P and K uptake and use efficiencies	Noor et al. (2017)

no significant difference in these parameters between plants inoculated at the reduced fertilizer rate and non-inoculated plants with full dose fertilization, suggesting that *Enterobacter* sp. Fs-11 has a potential to be an alternative to conventional fertilization (Shahid et al., 2019).

Rhizobacteria, rhizobia and AMF are known to directly stimulate plant growth and nutrient uptake, but studies on consortia containing both PGPR and AMF, particularly in the field, are lacking. Raklami et al. (2019) assessed the impact of AMF and a bacterial consortium on faba bean and wheat in the field. The consortium, containing two PGPR isolates, two rhizobia isolates and three AMF isolates, significantly increased the dry weight of shoots and roots and the numbers of leaves and flowers of both crops. The consortium also increased the N, P, K, sodium, sugar and protein content of shoots and the total organic matter and total carbon of the soil. The authors suggested that the improvement in crop mineral nutrient content was due to the plant-growth-promoting capabilities of the microbes and the ability of AMF hyphae to explore more soil; higher mineral nutrient content was correlated with high AMF abundance. The improvement in soil organic matter and carbon may be related to the ability of the consortium to metabolize compounds produced by plant roots (Raklami et al., 2019).

3.2 Nutrient use efficiency

The effect of three *Bacillus* sp. isolates on the nutrient use efficiency of *Amaranthus hypochondriacus* was evaluated in a field study conducted by Pandey et al. (2018). In addition to significant plant growth in the field when inoculated with farmyard manure and the PGPR consortium, there was a significant increase in soil nutrients (N, P, K) and their uptake efficiency.

Lassaletta et al. (2014) reported that there are generally higher levels of nutrient use efficiency in countries that rely on symbiotic N fixation than in those relying on synthetic fertilizers as the major source of N inputs. Using this knowledge, Zhang et al. (2019) studied the effect of *B. cereus* in conjunction with a composted nitrogen-fixing shrub and varying levels of chicken manure to improve the N status of cucumber and enhance its nutrient use efficiency. Shoot and root biomass was influenced by the presence of the composted shrub but not *B. cereus*. However, nutrient use efficiency was increased by *B. cereus* and was highest in the presence of the composted shrub and 50% of the traditional manure treatment, suggesting that N input could be reduced while enhancing crop yield and N use efficiency.

In a similar approach, a growth chamber study by Paungfoo-Lohienne et al. (2019) showed a promising result in which soil inoculated with PGPR bacterium SOS3 (*Paraburkholderia* sp.) and amended with 50% organic and 50% conventional N fertilizers produced similar yields of kikuyu grass to those obtained using 100% conventional fertilizer. The combination also reduced mineral N leachate by 95% and decreased soil mineral N by 65% compared to non-inoculated conventionally fertilized soil, suggesting that PGPR enhanced plant N acquisition and assimilation. The authors hypothesized that reduced N loss from leaching was likely due to the PGPR stimulated increase in nutrient uptake by the plant. Despite having lower mineral N, plants in the combination treatment had similar shoot and root growth to those in the conventional treatment, signifying improved nutrient use efficiency.

Ahmad et al. (2017) studied the effect of bacterial-impregnated fertilizers on wheat yield and nutrient use efficiency in a pot experiment. Instead of inoculating seed or soil, the researchers inoculated fertilizers with *Bacillus* sp. strain KAP6 and compost or humic acid (to support PGPR survival) and assessed both N and P use efficiency of wheat. Fertilizers impregnated with the PGPR and either compost or humic acid improved plant height, root length, root biomass and N and P uptake compared to the control. Nitrogen and phosphorus use efficiencies both were increased by 20% in the PGPR and compost treatment applied at the highest rate compared to the control.

In a similar pot study on maize, Noor et al. (2017) impregnated diammonium phosphate (DAP) fertilizer with *P. putida* and compost and molasses and evaluated phosphate use efficiency. Besides increasing root and shoot dry weights, the *P. putida*-impregnated fertilizer applied at label rate increased P, N and K uptakes by 33%, 45% and 43%, respectively. N and K use efficiencies increased with increasing rates of impregnated fertilizer; in contrast, P use efficiency increased with decreasing rates of impregnated fertilizer. P use efficiency was 62% higher than the control when impregnated fertilizer was applied at 50% of the recommended rate.

Sustainable crop production remains a major challenge facing the world today. These studies highlight the potential of PGPR to enhance crop production. Biofertilizers offer the ability to decrease the use of mineral fertilizer while enhancing plant nutrient uptake and use efficiency. Although biofertilizers are a new technology and are being used with an increasing frequency, more studies are needed to determine which crop type and PGPR partner best enhance plant nutrition (Jacoby et al., 2017).

4 Future trends in research

Climate change, environmental degradation, the expanding human population and the need to increase crop productivity more sustainably to feed that population are all factors driving the interest in harnessing plant-microbe interactions to enhance crop production. This, coupled with the progress in opening the 'black box' of the soil-microbe-plant root interaction using culture-independent, high-throughput genomic, transcriptomic and metabolomic analytical tools, has led to an exponential increase in research in this field and a strong interest by industry to apply this knowledge (Finkel et al., 2017; Garcia and Kao-Kniffen, 2018; Compant et al., 2019).

While these 'omics' tools are allowing us to better understand the complexity of plant–microbe interactions and, in particular, the structure and diversity of these communities, there are still many challenges to overcome in linking structure to function and in moving this fundamental research to the field (Sergaki et al., 2018). Application of the concept of the plant holobiont and its hologenome, defined as the host and all of its microbial symbionts and the collective genomes of this unit, respectively (Rosenberg and Zilber-Rosenberg, 2016), is helpful in developing more holistic approaches to studying the root microbiome. For example, Garcia and Kao-Kniffen (2018) propose applying the evolutionary biology theory of multi-level selection, defined as the natural selection acting on whole groups of organisms in addition to individuals, to the plant rhizosphere as a new framework for testing how group dynamics in the rhizosphere impact the plant phenotype. Network analysis of metagenomic data is being employed to predict species interactions (Faust and Raes, 2012) and time series analysis can provide insight into temporal patterns in microbial communities (Faust et al., 2015).

Another approach that reduces the complexity of the root ecosystem and facilitates the characterization of community interactions is the generation of synthetic microbial communities (Grosskopf and Soyer, 2014; Finkel et al., 2017). Candidate members are identified via primer-based or metagenomic sequencing; representatives of highly abundant keystone species or taxa are combined into synthetic communities for inoculation into gnotobiotic plant systems, and their dynamics and interactions followed (Hartman et al., 2017). However, microbial taxa that occur in low abundance, satellite taxa, may also be important drivers of key functions in ecosystems and should not be ignored (Jousset et al., 2017).

Toju et al. (2018) outlined a strategy for developing 'core microbiomes' to optimize microbial functions in agricultural systems using informatics and network theory-based approaches in combination with multi-omics and artificial intelligence-based diagnostic platforms for site-adapted agriculture. They defined core microbiomes as sets of microorganisms that form cores of interactions that can be used to optimize microbial functions at the individual plant and ecosystem levels. They are key in recruiting an assembly of native plant-associated microorganisms with desirable physiological functions rather than just promoting plant growth and would be the first inoculated onto seeds or seedlings in order to block potential pathogens or pests.

Application of these fundamental approaches to enhancing the management of the root microbiome in the field is also emerging. Academics and industry are engaged in the large-scale isolation of plant-associated microorganisms from many crops and environments, coupled with genome sequencing and high-throughput functional screening of these isolates (Finkel et al., 2017). These collections should yield more diverse PGPR and identification of novel plant-growth-promoting traits that may be suitable for commercial use. Researchers and the inoculant industry are moving away from single-species inoculants to the development of microbial consortia with diverse plant-growth-promoting properties or that are adapted for specific crops or environmental conditions (reviewed by Dessaux et al., 2016; Parnell et al., 2016; Finkel et al., 2017; Compant et al., 2019). A recent report by Bradacova et al. (2019) demonstrated the superior performance of a microbial consortium product relative to single-strain inoculants in the production of tomato grown in the Negev desert under environmental stress conditions, while no differences were observed between single- and multi-strain inoculants when plants were grown in a protected greenhouse.

Plant breeding has traditionally been conducted without consideration of the plant microbiome, but inclusion of the rhizosphere microbiome in selection could yield plant phenotypes with enhanced growth under abiotic or biotic stress conditions (Dessaux et al., 2016; Garcia and Kao-Kniffen, 2018; Compant et al., 2019). Rouphael et al. (2018) summarize the potential for using non-destructive, high-throughput phenotyping technologies to screen for novel biostimulants, including PGPR in controlled environment and field conditions, although the technology available currently in the field is restricted to above-ground morphological and physiological features. Automated and semi-automated phenotyping platforms are available for the quantitative and dynamic analysis of root growth and architecture but are destructive and labour intensive for conventionally grown plants in soil (Paez-Garcia et al., 2015).

As these new approaches and technologies develop novel solutions for managing the root microbiome to enhance root function, nutrient uptake and ultimately crop yield, research is required to move this technology to the field. Practical consideration must be given to producing, formulating, storing and applying novel strains and consortia on farm to maintain their viability and efficacy (Parnell et al., 2016; Timmusk et al., 2017; Backer et al., 2018; Compant et al., 2019). The move to microbial consortia may necessitate the development of novel fermentation and formulation strategies to ensure the survival of microorganisms with differing nutritional and storage requirements. Production and formulation must be efficient and economic to maintain a viable inoculant industry. Vassilev et al. (2015) and Lobo et al. (2019) review some of the current and potential future solutions to these issues. New application technologies and crop management approaches may be necessary to optimize microbial benefits under varying soil, biotic and abiotic conditions (Sergaki et al., 2018). The current registration process for PGPR products is complex, lengthy and variable in many developed countries (Timmusk et al., 2017; Backer et al., 2018; Woo and Pepe, 2018), and it is likely that the registration of microbial consortia may be more challenging than that for single-strain products. The safety of novel strains and consortia for humans and the environment should be considered early on in their development as potential PGPR products (Martinez-Hidalgo et al., 2019; Ferreira et al., 2019).

5 Conclusion

Although PGPR have been studied for many years as biofertilizers, it is only recently that their potential for more sustainably enhancing crop productivity, while reducing the use of mineral fertilizers, has become a priority in research and investment by industry. The diversity of rhizobacteria, our limited ability to culture them and the complexity of their interactions with plants, other rhizosphere microbiota, the soil and the abiotic environment limited our ability previously to understand and manage them for practical purposes. The availability of new culture-independent metagenomic and metabolomic tools and of approaches for assessing and unravelling complex and dynamic community interactions is rapidly expanding our knowledge of the plant microbiome and offers new possibilities to optimize its function for crop productivity. Many challenges remain in applying this knowledge to the field, both theoretical and practical. In this review we have summarized these challenges and demonstrated how the new 'omics' tools are providing greater insight into the role of PGPR in enhancing root function, nutrient uptake and tolerance to abiotic stress. These studies, coupled with community-focused research tools such as network modelling and the analysis of synthetic microbial communities, will allow us to better predict the complex and dynamic interactions occurring in the plant microbiome. Approaches such as these will enable rational design and testing of microbial consortia/crop species that consistently enhance crop nutrition and productivity under real-world conditions.

6 Where to look for further information

The following articles provide a good overview of the subject:

- Jacoby, R., Peukert, M., Succurro, A., Koprivova, A. and Kopriva, S. (2017), 'The role of soil microorganisms in plant mineral nutrition – current knowledge and future directions', *Front. Plant Sci.* 8, 1–19.
- Rosier, A., Medeiros, F. H. V. and Bais, H. P. (2018), 'Defining plant growth promoting rhizobacteria molecular and biochemical networks in beneficial plant-microbe interactions', *Plant Soil* 428, 35–55.
- Vacheron, J., Desbrosses, G., Bouffard, M.-L., Toraine, B., Moenne-Loccoz, Y., et al. (2013), 'Plant growth-promoting rhizobacteria and root system functioning', *Front. Plant Sci.* 4, 356.

Key conferences in this field:

- PGPR workshops: Are held every 3 years; the most recent was the 11th International Plant Growth-Promoting Rhizobacteria Workshop, June 17-21, 2018, Victoria, British Columbia, Canada. The next workshop will be held in Toulouse, France, in 2021.
- Rhizosphere conferences: Are held every 4 years; the most recent was Rhizosphere 5, July 7-11, 2019, Saskatoon Saskatchewan, Canada. The location of Rhizosphere 6 has not yet been announced.
- International Symposia on Microbial Ecology: Are held every 2 years; the most recent was the 17th International Symposium on Microbial Ecology, August 12-17, 2018, Leipzig, Germany. ISME 18 will be held in Cape Town, South Africa in August 2021.

Key societies:

- International Society for Microbial Ecology (https://www.isme-microbes.org/).
- International Society for Plant-Microbe Interactions (https://www.ismpmi.org/).

7 References

Abd_Allah, E. F., Alqarawi, A. A., Hashem, A., Radhakrishnan, R., Al-Huqail, A., et al. (2017), 'Endophytic bacterium *Bacillus subtilis* (BERA 71) improves salt tolerance in chickpea plants by regulating the plant defense mechanisms', *J. Plant Interact.* 13, 37-44.

Adesemoye, A. O. and Kloepper, J. W. (2009), 'Plant-microbes interactions in enhanced fertilizer-use efficiency', *Appl. Microbiol. Biotechnol.* 85, 1-12.

Ahmad, S., Imran, M., Hussain, S., Mahmood, S., Hussain, A., et al. (2017), 'Bacterial impregnation of mineral fertilizers improves yield and nutrient use efficiency of wheat', *J. Sci. Food Agr.* 97, 3685-90.

Alexandratos, N. and Bruinsma, J. (2012), 'World agriculture towards 2030/2050: the 2012 revision', ESA Working paper No. 12-03. FAO, Rome, Italy.

Armada, E., Roldán, A. and Azcon, R. (2014), 'Differential activity of autochthonous bacteria in controlling drought stress in native *Lavandula* and *Salvia* plants species under drought conditions in natural arid soil', *Plant Microbe Interact.* 67, 410-20.

Asari, S., Tarkowska, D., Rolcik, J., Novak, O., Velazquez Palermo, D., et al. (2017), 'Analysis of plant growth-promoting properties of *Bacillus amyloliquefaciens* UCMB5113 using *Arabidopsis thaliana* as host plant', *Planta* 245, 15-30.

Backer, R., Rokem, J. S., Ilangumaran, G., Lamont, J., Preslickova, D., et al. (2018), 'Plant growth-promoting rhizobacteria: context, mechanisms of action, and roadmap to commercialization of biostimulants for sustainable agriculture', *Front. Plant Sci.* 9, 1473.

Bargaz, A., Lyamlouli, K., Chtouki, M., Zeroual, Y. and Dhiba, D. (2018), 'Soil microbial resources for improving fertilizers efficiency in an integrated plant nutrient management system', *Front. Microbiol.* 9, 1–25.

Bertrand, H., Plassard,, C., Pinochet, X., Touraine, B., Normand, P., et al. (2000), 'Stimulation of the ionic transport system in *Brassica napus* by a plant growth-promoting rhizobacterium (*Achromobacter* sp.)', *Can. J. Microbiol.* 46, 229–36.

Bradacova, K., Florea, A. S., Bar-Tal, A., Minz, D., Yermiyahu, U., et al. (2019), 'Microbial consortia versus single-strain inoculants: an advantage in PGPM-assisted tomato production?', *Agron.* 9, 105.

Bruto, M., Prigent-Combaret, C., Muller, D. and Moenne-Loccoz, Y. (2014), 'Analysis of genes contributing to plant-beneficial functions in plant growth-promoting rhizobacteria and related Proteobacteria', *Sci. Rep.* 4, 6261.

Calvo, P., Zebelo, S., McNear, D., Kloepper, J. and Fadamiro, H. (2019), 'Plant growth-promoting rhizobacteria induce changes in *Arabidopsis thaliana* gene expression of nitrate and ammonium uptake genes', *J. Plant Interact.* 14(1), 224–31.

Carro, L. and Nouioui, I. (2017), 'Taxonomy and systematics of plant probiotic bacteria in the genomic era', *AIMS Microbiol.* 3, 383–412.

Cassan, F. and Diaz-Zorita, M. (2016), '*Azospirillum* sp. in current agriculture: from the laboratory to the field', *Soil Biol. Biochem.* 103, 117–30.

Chandra, D., Srivastava, R., Glick, B. R. and Sharma, A. K. (2018), 'Drought-tolerant *Pseudomonas* spp. improve the growth performance of Finger Millet (*Eleusine coracana* (L.) Gaertn.) under non-stressed and drought-stressed conditions', *Pedosphere* 28, 227–40.

Compant, S., Samad, A., Faist, H. and Sessitsch, A. (2019), 'A review on the plant microbiome: ecology, functions, and emerging trends in microbial application', *J. Adv. Res.* 19, 29–37.

Dessaux, Y., Grandclement, C. and Faure, D. (2016), 'Engineering the rhizosphere', *Trends Plant Sci.* 21(3), 266–78.

Diaz-Zorita, M., Fernandez-Canigia, M. V., Bravo, O. A., Berger, A. and Satorre, E. H. (2015), 'Field evaluation of extensive crops inoculated with *Azospirillum* sp.', in: Cassan, F. D., Okon, Y. and Creus, C. M. (Eds), *Handbook for Azospirillum, Technical Issues and Protocols*. Springer International Publishing, Switzerland, pp. 435–45.

Dodd, I. C. and Ruiz-Lozano, J. M. (2012), 'Microbial enhancement of crop resource use efficiency', *Curr. Opin. Biotechnol.* 23, 236–42.

Egamberdieva, D., Wirth, S., Jabborova, D., Räsänen, L. A. and Liao, H. (2017), 'Coordination between *Bradyrhizobium* and *Pseudomonas* alleviates salt stress in soybean through altering root system architecture', *J. Plant Interact.* 12, 100–7.

Eida, A. A., Alzubaidy, H. S., De Zélicourt, A., Synek, L., Alsharif, W., et al. (2018), 'Phylogenetically diverse endophytic bacteria from desert plants induce transcriptional changes of tissue-specific ion transporters and salinity stress in *Arabidopsis thaliana*', *Plant Sci.* 280, 228–40.

El-Esawi, M., Alaraidh, I., Alsahli, A., Alamri, S., Ali, H., et al. (2018a), '*Bacillus firmus* (SW5) augments salt tolerance in soybean (*Glycine max* L.) by modulating root system architecture, antioxidant defense systems and stress-responsive genes expression', *Plant Physiol. Biochem.* 132, 375–84.

El-Esawi, M., Alaraidh, I., Alsahli, A., Alzahrani, S., Ali, H., et al. (2018b), '*Serratia liquefaciens* KM4 improves salt stress tolerance in maize by regulating redox potential, ion homeostasis, leaf gas exchange and stress-related gene expression', *Int. J. Mol. Sci.* 19, 1–15.

Etesami, H. and Maheshwari, D. K. (2018), 'Use of plant growth promoting rhizobacteria (PGPRs) with multiple plant growth promoting traits in stress agriculture: action mechanisms and future prospects', *Excotoxicol. Environ.* 156, 225–46.

Faust, K. and Raes, J. (2012), 'Microbial interactions; from networks to models', *Nat. Rev. Microbiol.* 10, 538–50.

Faust, K., Lahti, L., Gonze, D., de Vos, W. M. and Raes, J. (2015), 'Metagenomics meets time series analysis: unraveling microbial community dynamics', *Curr. Opin. Microbiol.* 25, 56–66.

Ferreira, C. M. H., Soares, H. M. V. M. and Soares, E. V. (2019), 'Promising bacterial genera for agricultural practices: an insight on plant growth-promoting properties and microbial safety aspects', *Sci. Total Environ.* 682, 779–99.

Finkel, O. M., Castrillo, G., Harrera Paredes, S., Salas Gonzalez, I. and Dangl, J. L. (2017), 'Understanding and exploiting plant beneficial microbes', *Curr. Opin. Plant Biol.* 38, 155–63.

Gamez, R. M., Rodriquez, F., Vidal, N. M., Ramirez, S., Alvarez, R. V., et al. (2019), 'Banana (*Musa acuminate*) transcriptome profiling in response to rhizobacteria: *Bacillus amyloliquefaciens* Bs006 and *Pseudomonas fluorescens* Ps006', *BMC Genomics* 20, 378.

Garcia, J. and Kao-Kniffen, J. (2018), 'Microbial group dynamics in plant rhizospheres and their implications on nutrient cycling', *Front. Microbiol.* 9, 1516.

Glick, B. R. (2012), 'Plant growth-promoting bacteria: mechanisms and applications', *Scientifica* 2012, Article ID 963401.

Gomes, D. G., Radi, A. J. and Silva de Aquino, G. (2018), 'Growth-promoting bacteria change the development of aerial part and root system of canola', *Semina: Cienc. Agrar., Londrina*, 39, 2375–84.

Grosskopf, T. and Soyer, O. S. (2014), 'Synthetic microbial communities', *Curr. Opin. Microbiol.* 18,72–7.

Gyanendranath, M. (2017), 'Essential plant nutrients and recent concepts about their uptake', in: Naeem, N., Ansari, A. and Gill, S. (Eds), *Essential Plant Nutrients*. Springer International Publishing AG, Switzerland, pp. 3–36.

Hao, H.-T., Zhao, X., Shang, Q.-H., Wang, Y., Guo, Z.-H., et al. (2016), 'Comparative digital gene expression analysis of the *Arabidopsis* response to volatiles emitted by *Bacillus amyloliquefaciens*', *PLoS ONE* 11(8), e0158621.

Hartman, K., van der Heijden, M. G. A., Roussely-Provent, V., Walser, J. C. and Schlaeppi, K. (2017), 'Deciphering composition and function of the root microbiome of a legume plant', *Microbiome* 5, 2.

Hussain, M. B., Zahir, Z. A., Asghar, H. N. and Asgher, M. (2014), 'Can catalase and exopolysaccharides producing rhizobia ameliorate drought stress in wheat?', *Int. J. Agr. Biol.* 16, 3–13.

Ibort, P., Imai, H., Utemura, M. and Aroca, R. (2018), 'Proteomic analysis reveals that tomato interaction with plant growth promoting bacteria is highly determined by ethylene perception', *J. Plant Physiol.* 220, 43–59.

Ilangumaran, G. and Smith, D. L. (2017), 'Plant growth promoting rhizobacteria in amelioration of salinity stress: a systems biology perspective', *Front. Plant Sci.* 8, Article 1768.

Jacoby, R., Peukert, M., Succurro, A., Koprivova, A. and Kopriva, S. (2017), 'The role of soil microorganisms in plant mineral nutrition–current knowledge and future directions', *Front. Plant Sci.* 8, 1–19.

Jamil, M., Ahamd, M., Anwar, F., Zahir, Z. A., Kharal, M. A., et al. (2018), 'Inducing drought tolerance in wheat through combined use of L-tryptophan and *Pseudomonas fluorescens*', *Pak. J. Agr. Sci.* 55, 331–7.

Jorge, G., Kisiala, A., Morrison, El., Aoki, M., Nogueira, A., et al. (2019), 'Endosymbiotic *Methylobacterium oryzae* mitigates the impact of limited water availability in lentil (*Lens culinaris* Medik.) by increasing plant cytokinin levels', *Environ. Exp. Bot.* 162, 525–40

Jousset, A., Bienhold, C., Chatzinotis, A., Gallien, I., Gobet, A., et al. (2017), 'Where less may be more; how the rare biosphere pulls ecosystem strings', *ISME J.* 11, 853–62.

Kaushal, M. (2019), 'Microbes in cahoots with plants: MIST to hit the jackpot of agricultural productivity during drought', *Int. J. Mol. Sci.* 20, 1–22.

Klikno, J. and Kutschera, U. (2017), 'Regulation of root development in *Arabidopsis thaliana* by phytohormone-secreting epiphytic methylobacteria', *Protoplasma* 254, 1867–77.

Kwon, Y. S., Lee, D. Y., Rakwal, R., Baek, S.-B., Lee, J. H., et al. (2016), 'Proteomic analyses of the interaction between the plant-growth promoting rhizobacterium *Paenibacillus polymyxa* E681 and *Arabidopsis thaliana*', *Proteomics* 16, 122–35.

Lade, S. B., Roman, C., Cueto-Ginzo, A. I., Maneiro, L., Munoz, P., et al. (2018), 'Root development in agronomically-distinct six-row barley (*Hordeum vulgare*) cultivars inoculated with *Azospirillum brasilense* Sp7', *Plant Breed.* 137, 338–45.

Lally, R. D., Galbally, P., Moreira, A. S., Spink, J., Ryan, D., et al. (2017), 'Application of endophytic *pseudomonas fluorescens* and a bacterial consortium to *Brassica napus* can increase plant height and biomass under greenhouse and field conditions', *Front. Plant Sci.* 8, 2193.

Lassaletta, L., Billen, G., Grizzetti, B., Anglade, J. and Garnier, J. (2014), '50 year trends in nitrogen use efficiency of world cropping systems: the relationship between yield and nitrogen input to cropland', *Environ. Res. Lett.* 9, 1–9.

Lobo, C. B., Tomas, M. S. J., Viruel, E., Ferrero, M. A. and Lucca, M. E. (2019), 'Development of low-cost formulations of plant growth-promoting bacteria as inoculants in beneficial agricultural technologies', *Microbiol. Res.* 219, 12–25.

Majeed, A., Muhammad, Z. and Ahmad, H. (2018), 'Plant growth-promoting bacteria; role in soil improvement, abiotic and biotic stress management of crops', *Plant Cell Rep.* 37, 1599–609.

Martinez-Hidalgo, P., Maymon, M., Pule-Meulenberg, F. and Hirsch, A. M. (2019), 'Engineering root microbiomes for healthier crops and soils using beneficial, environmentally safe bacteria', *Can. J. Microbiol.* 65, 91–104.

Nelson, L. M. (2004), 'Plant growth-promoting rhizobacteria: prospects for new inoculants', *Crop Manag.* 3(1) doi:10.1094/CM-2004-0301-05-RV.

Nguyen, M. L., Glaes, J., Spaepen, S., Bodson, B., Du Jardin, P., et al. (2019), 'Biostimulant effects of *Bacillus* strains on wheat from *in vitro* towards field conditions are modulated by nitrogen supply', *J. Plant Nutr. Soil Sci.* 182, 325–34.

Noor, S., Yaseen, M., Naveed, M. and Ahmad, R. (2017), 'Effectiveness of diammonium phosphate impregnated with *Pseudomonas putida* for improving maize growth and phosphorus use efficiency', *J. Anim. Plant Sci.* 27, 1588–95.

Ortiz-Castro, R. and Lopez-Bucio, J. (2019), 'Phytostimulation and root architectural responses to quorum-sensing signals and related molecules from rhizobacteria', *Plant Sci.* 284, 135–42.

Paez-Garcia, A., Motes, C. M., Scheible, W.-R., Chen, R. and Blancaflor, E. B. (2015), 'Root traits and phenotyping strategies for plant improvement', *Plants* 4, 334–55.

Pandey, C., Kumar Negi, Y., Maheshwari, D. K., Rawat, D. and Prabha, D. (2018), 'Potential of native cold tolerant plant growth promoting bacilli to enhance nutrient use efficiency and yield of *Amaranthus hypochondriacus*', *Plant Soil* 428, 307–20.

Parida, A. K. and Das, A. B. (2005), 'Salt tolerance and salinity effects on plants: a review', *Ecotoxicol. Environ. Saf.* 60, 324–49.

Parnell, J. J., Berka, R., Young, H. A., Sturino, J. M., Kang, Y., et al., (2016), 'From the lab to the farm: an industrial perspective of plant beneficial microorganisms', *Front. Plant Sci.* 7, Article 1110.

Parray, J. A., Jan, S., Kamili, A. N., Qadri, R. A., Egamberdieva, D., et al. (2016), 'Current perspectives on plant growth-promoting rhizobacteria', *J. Plant Growth Regul.* 35, 877–902.

Paterson, J., Jahanshah, G., Li, Y., Wang, Q., Mehnaz, S. and Gross, H. (2017), 'The contribution of genome mining strategies to the understanding of active principles of PGPR strains', *FEMS Microbiol. Ecol.* 93, 1–31.

Paungfoo-Lonhienne, C., Redding, M., Pratt, C. and Wang, W. (2019), 'Plant growth promoting rhizobacteria increase the efficiency of fertilisers while reducing nitrogen loss', *J. Environ. Manage.* 233, 337–41.

Pii, Y., Mimmo, T., Tomasi, N., Terzano, R., Cesco, S., et al. (2015), 'Microbial interactions in the rhizosphere: beneficial influences of plant growth-promoting rhizobacteria on nutrient acquisition process: a review', *Biol. Fertil. Soil* 51, 403–15.

Raklami, A., Bechtaoui, N., Tahiri, A. I., Anli, M., Meddich, A., et al. (2019), 'Use of rhizobacteria and mycorrhizae consortium in the open field as a strategy for improving crop nutrition, productivity and soil fertility', *Front. Microbiol.* 10, 1–11.

Reed, M. L. E. and Glick, B. R. (2013), 'Applications of plant growth-promoting bacteria for plant and soil systems', in: Gupta, V. K., Schmoll, M., Maki, M., Tuchy, M. and Mazutti, M. A. (Eds), *Applications of Microbial Engineering*. CRC Press, Boca Raton, FL, pp. 181–228.

Rosenberg, E. and Zilber-Rosenberg, I. (2016), 'Microbes drive evolution of animals and plants: the hologenome concept', *mBio* 7, e01395-15.

Rosier, A., Medeiros, F. H. V. and Bais, H. P. (2018), 'Defining plant growth promoting rhizobacteria molecular and biochemical networks in beneficial plant-microbe interactions', *Plant Soil* 428, 35–55.

Rouphael, Y., Spichal, L., Panzarova, K., Casa, R. and Colla, G. (2018), 'High-throughput plant phenotyping for developing novel biostimulants: from lab to field or from field to lab?', *Front. Plant Sci.* 9, 1197.

Royal Society (2009), *Reaping the Benefits: Science and the Sustainable Intensification of Agriculture*, Marcel Dekker, Inc., London, UK, 72p.

Sapre, S., Gontia-Mishra, I. and Tiwari, S. (2018), '*Klebsiella* sp. confers enhanced tolerance to salinity and plant growth promotion in oat seedlings (*Avena sativa*)', *Microbiol. Res.* 206, 25–32.

Schutz, L., Gattinger, A., Meier, M., Muller, A., Boller, T., et al. (2018), 'Improving crop yield and nutrient use efficiency via biofertilization – a global meta-analysis', *Front. Plant Sci.* 8, 2204.

Scotti, R., D'Agostino, N. and Zaccardelli, M. (2019), 'Gene expression profiling of tomato roots interacting with *Pseudomonas fluorescens* unravels the molecular reprogramming that occurs during the early phases of colonization', *Symbiosis* 78, 177–92.

Sergaki, C., Lagunas, B., Lidbury, I., Gifford, M. L. and Schafer, P. (2018), 'Challenges and approaches in microbiome research: from fundamental to applied', *Front. Plant Sci.* 9, 1205.

Sessitsch, A., Pfaffenbichler, N. and Mitter, B. (2019), 'Microbiome applications from lab to field: facing complexity'. *Trends Plant Sci.* 24, 194–8.

Shahid, M., Hameed, S., Zafar, M., Tahir, M., Ijaz, M., et al. (2019), '*Enterobacter* sp. strain Fs-11 adapted to diverse ecological conditions and promoted sunflower achene yield, nutrient uptake, and oil contents', *Environ. Microbiol.* 50, 459–69.

Shahzad, S., Khan, M., Zahir, Z. A., Asghar, H. N. and Chaudhry, U. K. (2017), 'Comparative effectiveness of different carriers to improve the efficacy of bacterial consortium for enhancing wheat production under salt affected field conditions', *Pak. J. Bot.* 49, 1523–30.

Sharifi, R. and Ryu, C.-M. (2018), 'Revisiting bacterial volatile-mediated plant growth promotion: lessons from the past and objectives for the future', *Ann. Bot.* 122, 349–58.

Timmusk, S., Behers, L., Muthoni, J., Muraya, A. and Aronsson, A.-C. (2017), 'Perspectives and challenges of microbial application for crop improvement', *Front. Plant Sci.* 8, Article 49.

Toju, H., Paey, J. G., Yamamichi, M., Narisawa, K., Hiruma, K., et al. (2018), 'Core microbiomes for sustainable agroecosystems', *Nat. Plants* 4, 247–57.

Ullah, S., Ashraf, M., Asghar, H. N., Iqbal, Z. and Ali, R. (2019), 'A review: plant growth promoting rhizobacteria mediated amelioration of drought in crop plants', *Soil Environ.* 38, 1-20.

Vacheron, J., Desbrosses, G., Bouffard, M.-L., Toraine, B., Moenne-Loccoz, Y., et al. (2013), 'Plant growth-promoting rhizobacteria and root system functioning', *Front. Plant Sci.* 4,356.

Vacheron,, J., Desbrosses, G., Renoud, S., Padilla, R., Walker, V., et al. (2018), 'Differential contribution of plant-beneficial functions from *Pseudomonas kilonensis* F113 to root system architecture alterations in *Arabidopsis thaliana* and *Zea mays*', Mol. Plant Microbe Inter. 31, 212–23.

Van der Heijden, M. G. A., Bardgett, R. D. and Van Straalen, N. M. (2008), 'The unseen majority: soil microbes as drivers of plant diversity and productivity in terrestrial ecosystems', *Ecol. Lett.* 11, 296–310.

Vassilev, N., Vassileva, M., Lopez, A., Martos, V., Reyes, A., et al. (2015), 'Unexploited potential of some biotechnological techniques for biofertilizer production and formulation', *Appl. Microbiol. Biotechnol.* 99, 4983–96.

Verbon, E. H. and Liberman, L. M. (2016), 'Beneficial microbes affect endogenous mechanisms controlling root development', *Trends Plant Sci.* 21, 218–29.

Veresoglou, S. D. and Menexes, G. (2010), 'Impact of inoculation with *Azospirillum* spp. on growth properties and seed yield of wheat: a meta-analysis of studies in the ISI Web of Science from 1981 to 2008', *Plant Soil* 337, 469–80.

Wang, J., Zhang, Y., Jin, J., Li, Q., Li, Q., et al. (2018), 'An intact cytokinin-signaling pathway is required for *Bacillus* sp. LZR216-promoted plant growth and root system architecture alteration in *Arabidopsis thaliana* seedlings', *Plant Growth Reg.* 84, 507–18.

Woo, S. L. and Pepe, O. (2018), 'Microbial consortia: promising probiotics as plant biostimulants for sustainable agriculture', *Front. Plant Sci.* 9, 1801.

Xie, J., Shi, H., Du, Z., Wang, T., Liu, X. and Chen, S. (2016), 'Comparative genomic and functional analysis reveal conservation of plant growth promoting traits in *Paenibacillus polymxya* and its closely related species', *Sci. Rep.* 6, 21329.

Yang, J., Kloepper, J. W. and Ryu, C. M. (2009), 'Rhizosphere bacteria help plants tolerate abiotic stress', *Trends Plant Sci.* 14, 1–4.

Zhang, X., Liu, X., Lan, Z., Tian, Y. and Li, J. (2019), 'Conjunctive use of composted leguminous shrub *Caragana microphylla*-straw and *Bacillus cereus* reduces nitrogen input but enhances nitrogen use efficiency and cucumber yields in alkaline soils', *Appl. Soil Ecol.* 139, 69–78.

Zhao, C. Z., Huang, J., Gyaneshwar, P. and Zhan, D. (2018), '*Rhizobium* sp. IRBG74 alters *Arabidopsis* root development by affecting auxin signaling', *Front. Microbiol.* 8, 2556.

Chapter 16

The use of arbuscular mycorrhizal fungi to improve root function and nutrient-use efficiency

Tom Thirkell, Grace Hoysted, Ashleigh Elliott and Katie Field, University of Leeds, UK; and Tim Daniell, University of Sheffield, UK

1 Introduction

With the combined pressures of an increasing human population, environmental degradation, pollution, and depletion of finite essential resources, urgent changes in the agricultural sector are required to ensure sustainable future food security (Godfray et al., 2010, Fears et al., 2019). The world's population is projected to increase to 9 billion by 2050, while reserves of natural resources essential for food production (such as rock phosphate) simultaneously decline (Cordell et al., 2009).

Global mismanagement of agricultural soils means there is further significant risk of irreparable degradation, reducing the capacity of the earth's landmasses to support the scale of food production required for the twenty-second century (Oldeman, 1994, Gibbs and Salmon, 2015). These challenges are set against a backdrop of a changing global climate, where extreme weather events occur with greater intensity and frequency (Coumou and Rahmstorf, 2012, Lesk et al., 2016, Lipper et al., 2014), and concentrations of greenhouse gases in the atmosphere rise (Meinshausen et al., 2011). Although

http://dx.doi.org/10.19103/AS.2020.0075.23

agricultural innovation and intensification throughout the last century have increased staple crop production, food security remains one of humankind's most pressing problems (Foley et al., 2011).

Enhancing the efficiency with which crops acquire nutrients and water from the soil is a critical step on the pathway to food security. In conventional, high-input systems, advances in crop phenotyping and breeding have led to root systems which are optimised for nutrient capture and water-use efficiency (Wasson et al., 2012). However, these improvements in root function have largely been achieved while considering the soil as little more than an inert store of water and nutrients, and with little consideration of the interactions between plants, their soil environment and the organisms therein. Recent research efforts have further highlighted the importance of considering the soil microbiota to improve sustainability in agriculture (Rillig et al., 2019, Ryan et al., 2019).

Soil micro-organisms are essential for a wide range of key ecosystem processes including nutrient cycling, organic matter decomposition and soil aggregate stabilisation. Through their influence on these processes and their interactions with plant roots, soil microbes can significantly influence the productivity and sustainability of agroecosystems, and may show great potential for exploitation (Philippot et al., 2013, Bardgett and van der Putten, 2014, Bender et al., 2016). The potential benefits of agronomically useful plant-associated microbes could conceivably dwarf any benefits which may be achieved by improvements focussed solely on root traits, but this area remains chronically unstudied, especially in the field (Lekberg and Helgason, 2018). Integrating 'microbe management' into agricultural systems may offer great potential to improve the efficiency of crop plant production (Bakker et al., 2012, Schlaeppi and Bulgarelli, 2015, Mueller and Sachs, 2015, Nogales et al., 2015, Busby et al., 2017); this is an area of pressing need for further research.

Arbuscular mycorrhizal fungi (AMF) encompass soil fungi within the family Glomeromycotina and form what is perhaps the most important and widespread symbiosis on earth (Parniske, 2008). AMF are obligate biotrophs that engage in symbioses with the roots of a wide range of host plant species (Smith and Read, 2008), relying on the transfer of carbon-rich lipids and carbohydrates (Ryan and Graham, 2002, Luginbuehl et al., 2017) from their host plant to complete their life cycle. In return, AMF usually confer a number of benefits to the associated host plant, including enhanced access and assimilation of mineral nutrients, improved plant growth, alleviation of water stress (Ruiz-Lozano et al., 2016, Delavaux et al., 2017, Gehring et al., 2017), enhanced tolerance of salinity, pH and heavy metals, increased tolerance and/or resistance to pests and diseases (Jung et al., 2012, Berdeni et al., 2018) and bio-fortification of grain with micronutrients (Watts-Williams and Gilbert, 2020).

AMF account for somewhere between 5% and 50% of soil microbial biomass (Olsson et al., 1999), and they may be useful tools in the sustainable intensification of agriculture (Cameron, 2010, Thirkell et al., 2017, Sosa-Hernández et al., 2019). The exploitation of AMF as a 'biofertilizer' is not a new concept; AMF inoculum has been produced for use in agriculture, horticulture, landscape restoration and site remediation for almost three decades (Hamel, 1996, Igiehon and Babalola, 2017).

The potential benefits to crops of AMF colonisation very often fail to translate the realised benefits for host plants; many agriculturally important crops experience neutral or even negative growth responses to colonisation by AMF (Smith and Smith, 2011a). In part, this may be due to a relative incompatibility between plants and fungi or, in other cases, is the result of intense management practices. It would be a gross oversimplification to claim that AMF are generally beneficial or deleterious to crop plants, and the agricultural relevance of AMF remains the subject of intense debate in the literature (Thirkell et al., 2017, Ryan and Graham, 2018, Ryan et al., 2019, Rillig et al., 2019). Substantial gaps in understanding remain, and significant research efforts are directed at understanding the conditions under which agricultural crop production may be influenced by AMF.

Our limited understanding of the significance of microbes in agricultural landscapes is further highlighted by recent molecular and physiological characterisation of the fine root endophyte fungi belonging to the family Mucoromycotina (MFRE). MFRE appear to be widespread across the land plant phylogeny (Hoysted et al., 2019), engaging in nutritional mutualism with plant hosts (Orchard et al., 2017, Hoysted et al., 2019). Despite increased recognition of the host range, identity and function of MFRE-plant symbioses, the importance of MFRE in agroecosystems remains critically under-studied. In this chapter, we highlight advances in our understanding of how AMF and associated positive interactions with the soil matrix may improve root function in agricultural systems, with particular focus on crop nutrient acquisition. We also discuss the functional diversity apparent in plant response to AMF colonisation and how this imposes limitations on AMF benefits to agricultural crops. The extent to which modern farming practices often present unintended barriers to realising these potential benefits is also highlighted. We then propose what we believe to be future research priorities for AMF within a sustainable agriculture framework.

2 Mycorrhizal nutrient acquisition

2.1 Symbiotic fungi improve phosphorus assimilation in crops

On a global scale, phosphorus (P) is one of the nutrients most limiting to plant growth. The majority of soil P, however, exists in forms unavailable for plant uptake (Marschner, 2011). The most abundant plant-available P in soils, the orthophosphate ion (PO_4^{3-}), is present in relatively low concentrations due to rapid immobilisation by metal ions (Tinker and Nye, 2000). As such, soils are often depleted in plant-available P (Barber, 1995), even when total P contents appear much higher (Vance et al., 2003). Agricultural soils under intensive, high-input management are normally supplemented with phosphorus fertiliser annually. Due to adsorption, precipitation and conversion to organic forms, only 10–30% of the applied mineral phosphate fertiliser is recovered by crops (Holford, 1997, Zhu et al., 2003, Syers et al., 2008). Given the finite supply of raw materials required for P-fertiliser production and the high financial and environmental costs, it is becoming increasingly clear that maintaining current rates of synthetic fertiliser addition is not an efficient, cost-effective or sustainable way of increasing crop production (Tilman et al., 2002).

For the plant host, the most well-established benefit of associating with AMF is the increased access to and acquisition of P from inorganic and organic soil pools (Sato et al., 2015). Increased P acquisition by mycorrhizal hyphae is related to both increased physical exploration of the soil and modification of the root environment, such as increased root branching and root hair length (Bolan, 1991, Smith and Read, 2008, Tinker and Nye, 2000). Plant P acquisition from soils is most often limited because of low mobility. As root P uptake may exceed the rate at which P can move by diffusion into the rhizosphere, sharply defined depletion zones of P can form around roots (Fig. 1), in some cases only millimetres from the root surface (Marschner, 2011, Smith et al., 2011). By engaging in symbiosis with AMF, the plant root may effectively increase the volume of soil from which P may be obtained, particularly where the hyphae grow significant distances from the root surface, far beyond the P-depletion zone (Jansa et al., 2003).

The absorptive area over which P may be acquired by extraradical AMF mycelium may be 10–40 times more extensive than that of roots (Giovannetti et al., 2001). This substantial extraradical hyphal length, in addition to small hyphal radii, allows AMF to explore soil volumes beyond the reach of plant roots alone (Tinker et al., 1992). AMF have been advocated as a potential tool, not only to improve plant P assimilation, but also to reduce the build-up of plant-inaccessible P in soils (Syers et al., 2008). This approach may improve root function while also enhancing the health, quality and function of agricultural soils. Phosphorus uptake by plants occurs either directly from the soil by epidermal phosphate transporters and root hairs (hereafter the 'direct pathway', after (Smith et al., 2011) or via a fungal partner (hereafter the 'mycorrhizal pathway') (Karandashov and Bucher, 2005, Yang et al., 2012, Péret et al., 2014) (Fig. 1).

In both direct and mycorrhizal pathways, inorganic P is mostly acquired as negatively charged PO_4^{3-} ions, which present problems of charge balancing, since the inner surface of the cell membrane has a negative electric potential and the concentration inside the cell is c.1000 times higher than in the soil

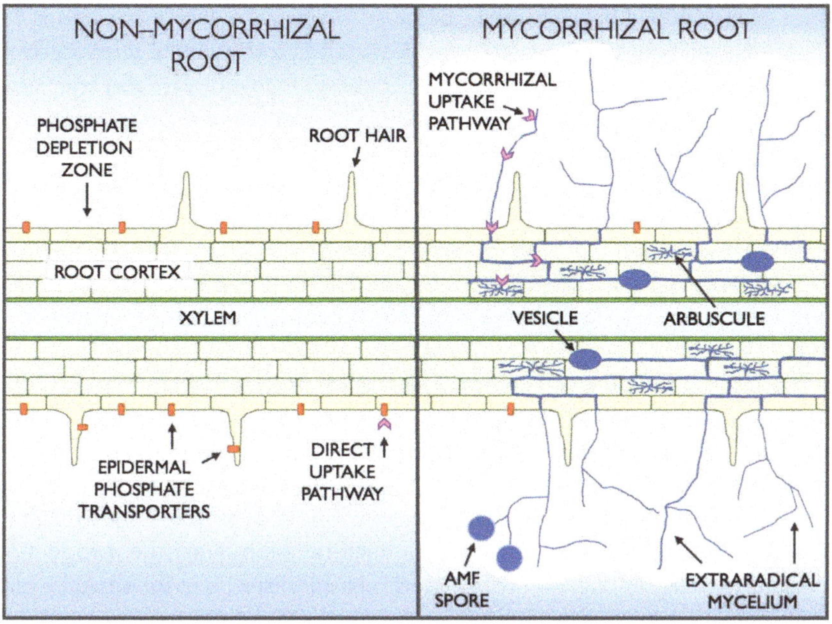

Figure 1 Plant root shown both in non-colonised and arbuscular mycorrhizal-colonised states, showing typical root and fungal structures. Around the non-colonised root, direct phosphate uptake by epidermal phosphate transporters (purple arrow) has resulted in a sharp P-depletion zone, from which P uptake can no longer take place. In the mycorrhizal root, P uptake also takes place by the mycorrhizal pathway, while plant root epidermal P transporters may be downregulated. By extending into the soil beyond the rhizosphere, the extraradical mycelium of the fungus increases the volume of soil from which nutrients can be acquired.

solution (Marschner, 2011). As such, inorganic phosphorus (Pi) transport across the plasma membrane is metabolically costly, requiring high-affinity Pi transporters (Smith et al., 2011). The mycorrhizal P uptake pathway is often reduced at high soil P concentrations, which corresponds with the general observation that AM colonisation is inhibited by a high Pi supply (Olsson et al., 1997, Breuillin et al., 2010, Kobae et al., 2016). However, regulation of symbiotic P transport is controlled at least partially by the host plant, as at high Pi, despite sufficient root colonisation, inhibition of the mycorrhizal P uptake pathway is accompanied by downregulation of the mycorrhiza-inducible Pi-transporter genes (Drissner et al., 2007, Nagy et al., 2009).

2.2 The role of root-associated soil fungi in crop N nutrition

Until relatively recently, AMF were assumed to play very little role in host plant N acquisition (Read, 1991), despite some earlier evidence to the contrary

(Ames et al., 1983). This view arose because while plant-AMF associations can occur in N-deficient soils, the AM symbiosis tends to dominate in ecosystems where decomposition and nitrification processes are favoured (Read and Perez-Moreno, 2003). As highly mobile nitrate (NO_3^-) ions are the principal form of plant-available N in such ecosystems, it was argued that sharp N-depletion zones would be unlikely to form around roots, and therefore AMF-facilitated plant N acquisition would be redundant (Read, 1991). This view has changed radically, and numerous experimental studies suggest that substantial N acquisition may be facilitated by arbuscule-forming soil symbiotic fungi (Hodge et al., 2001, Leigh et al., 2009, Hodge and Fitter, 2010, Thirkell et al., 2016, Hoysted et al., 2019).

As with P, the uptake efficiency of mineral N fertiliser by crops is generally very poor, with much lost to leaching, runoff and microbial immobilisation. Most N fertiliser is generated by the Haber-Bosch process, using H_2 (mostly from natural gas) and atmospheric N_2. While the supply of N_2 is effectively unlimited, natural gas is a finite resource which also experiences supply issues resulting from both economic and geopolitical factors. Moreover, the production, distribution and application of mineral N fertiliser is economically and energetically expensive and is often associated with substantial environmental impacts (Goucher et al., 2017). The ecological and agronomic significance of AMF contributions to plant N uptake strongly suggests that further research is warranted in this area (Bücking and Kafle, 2015, Hodge and Storer, 2015).

Recent investigations of the structural, functional and genomic diversity of mycorrhiza-forming fungi across a range of non-crop plants has revealed that plants, including crops, exploit a wider range of symbiotic options than previously thought (Field et al., 2015a, Rimington et al., 2016, Hoysted et al., 2018). This diversity in fungal endosymbioses includes the arbuscule-forming, fine root endophyte fungi (*Glomus tenue* or *Planticonsortium tenue* (Walker et al., 2018)), neglected since the early 1990s (McGonigle and Fitter, 1990) but which have recently been putatively reclassified into the Mucoromycotina (Orchard et al., 2017, Walker et al., 2018). Initially it was thought Mucoromycotina fine root endophyte (MFRE) symbioses were limited in distribution to the early-divergent, spore-forming lineages (i.e. liverworts, hornworts and ferns) (Bidartondo et al., 2011). However, the most recent research (Orchard et al., 2017, Field and Pressel, 2018, Hoysted et al., 2019) has revealed that colonisation by MFRE extends from liverworts to lycophytes and into the angiosperms (Fig. 2a–b), with emerging evidence of MFRE colonisation of grasses (Fig. 2b) and cereal crops.

Both AMF and MFRE have been shown to engage in carbon-for-phosphorus exchange with liverworts (Field et al., 2012, 2015b, 2016), but, as with AMF, the role of MFRE in plant nitrogen assimilation has remained equivocal. The latest research shows that MFRE are able to access nitrogen directly in inorganic forms and from organic matter in the soil (Field et al., 2019), and to transfer significant quantities to their host plant (Fig. 3). Together with new evidence of a role for MFRE in the N nutrition of vascular plants (Hoysted et al., 2019), it appears that functional diversity and complementarity of fungal symbionts may increase the source range and nutrient pool available to plants – both important considerations within agricultural systems.

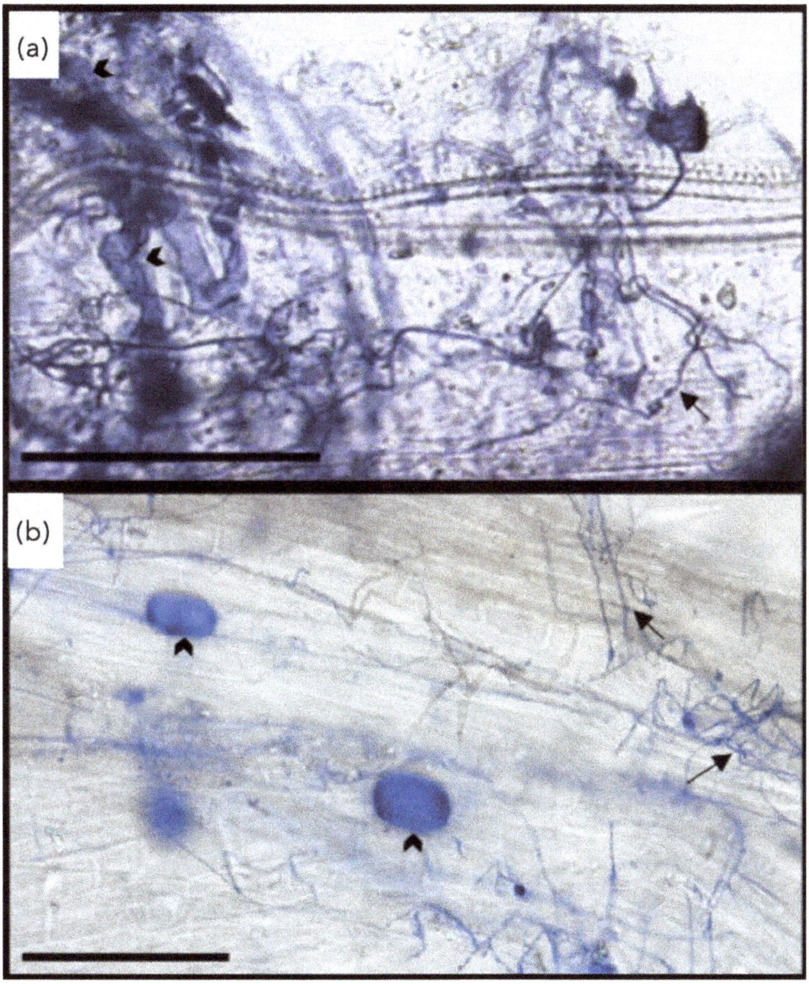

Figure 2 Light micrographs of Trypan blue-stained tissues. (a): *Trifolium pratense* (red clover) colonised by both Mucoromycotina fine root endophyte (MFRE, arrows) and Glomeromycotina (AMF, arrowheads) and (b): fine hyphae with small swellings/vesicles and large vesicles in a root of the grass *Holcus lanatus* colonised by both MFRE (arrows) and AMF (arrowheads) (Hoysted et al. 2019). Scale bars: (a) 100 um, (b) 50 um.

Substantial variation in the extent of AMF contribution to host plant N assimilation is reported in the literature (Bücking and Kafle, 2015, Hodge and Storer, 2015). We argue that this may be partly due to the use of non-pure inocula. It is possible (or indeed probable) that studies which have used unsterilised soils and/or fungal inocula generated by non-sterile propagation techniques (e.g. greenhouse pot cultures) may inadvertently contain both AMF and MFRE spores/hyphal fragments. How this contamination might affect observed mycorrhizal functioning is unclear, but may be partly responsible for conflicting reports observed among data sets. If MFRE contribute to crop N assimilation in plants forming dual associations with Glomeromycota and Mucoromycotina fungal symbionts, MFRE associations would represent a prime, untapped target trait for exploitation in crops to enhance access to mineral and organic soil N. This may be particularly significant for enhancing crop N uptake during early plant development stages, as MFRE are reported to frequently colonise plants inhabiting disturbed, nutrient-depleted habitats (Orchard et al., 2017). With higher disturbance tolerance, MFRE may not be subjected to the same selection pressures as AMF in agricultural systems, where reduced AMF diversity is common (see Section 4.2). Exploitation of MFRE in agricultural systems may allow greater crop yield outputs alongside reduced agrochemical inputs, and is a promising opportunity for future investigation.

2.3 The role of AMF in crop micronutrient uptake

Although N, P and K are the elements most commonly supplemented in agriculture, micronutrient limitation in crops remains a significant problem

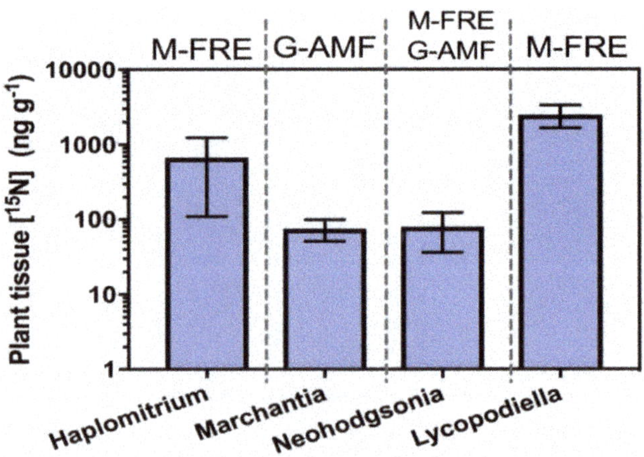

Figure 3 Concentration of fungal acquired ^{15}N (ng g^{-1}) in plant tissues of the liverworts *Haplomitrium, Marchantia* (Field et al., 2015b), *Neohodgsonia* (Field et al., 2016) and the lycopod *Lycopodiella* (Hoysted et al., 2019). Error bars = SEM, n = 6.

globally (Alloway, 2008). Decades of research show that AMF may aid plant uptake of relatively immobile metal micronutrients, such as copper (Cu), zinc (Zn), iron (Fe) and manganese (Mn) (Faber et al., 1990, Kothari et al., 1990, Kucey and Janzen, 1987, Li et al., 1991). Interestingly, AMF may act as a buffer for plant uptake of a number of mineral elements (e.g. Zn, Cu, Fe, Mn and Co), aiding acquisition where soils are impoverished while limiting plant accumulation where these micronutrients are present at potentially toxic concentrations (Ferrol et al., 2016, Watts-Williams et al., 2019b). At present, the mechanisms by which this buffering occurs remain unclear and warrant further attention. The role of MFRE in crop micronutrient uptake is unknown, but it is entirely possible that previous studies have inadvertently included MFRE in their assessments of AMF function.

AMF contribution to Cu uptake has been extensively researched and confirmed for a number of plant-fungus combinations (Manjunath and Habte, 1988, Lambert and Weidensaul, 1991). Indeed, AMF contributions to plant Cu uptake may be substantial. For example, Li et al. (1991) demonstrated that the AMF *Funneliformis mosseae* (Schüßler and Walker, 2010) supplied up to 62% of the Cu demand in white clover (*Trifolium repens* L.). Similarly, Lee and George (2005) showed the same fungal species facilitated 75% of Cu acquired by cucumber (*Cucumis sativus*). Conflicting reports demonstrate plant uptake of Mn and Fe may be enhanced (Clark and Zeto, 1996, Liu et al., 2000, Eivazi and Weir, 1989, Lehmann and Rillig, 2015) or reduced (Kothari et al., 1990, Liu et al., 2000, Corrêa et al., 2014) in mycorrhizal plants relative to non-mycorrhizal counterparts. Soil chemical properties such as pH and oxidation-reduction potential appear to moderate the extent to which AMF might enhance plant acquisition of these elements (Liu et al., 2000).

The importance of AMF in plant Zn nutrition has long been recognised (and extensively reviewed, see (Thompson, 1990, Cavagnaro, 2014, Cavagnaro, 2008)). Most experimental evidence of Zn uptake by AMF comes from studies using a ^{65}Zn tracer in compartments which exclude root growth but remain accessible to AMF hyphae (Cooper and Tinker, 1978, Bürkert and Robson, 1994, Jansa et al., 2003). As with Fe and Mn uptake mentioned above, mycorrhizal plant acquisition of Zn appears to be dependent on soil chemical properties; P concentrations have been shown repeatedly to influence AM Zn uptake (Chen et al., 2003, Ryan and Angus, 2003, Watts-Williams et al., 2013).

3 AMF effects on root architecture

Modifications to root system architecture (RSA) represent a significant potential for improving crop nutrient and water capture, but remain a relatively unexplored area of research (Rogers and Benfey, 2015). Ideotypes for root architecture have been proposed for optimisation of nutrient capture on different soils (White et al., 2013). As P is usually concentrated in topsoil layers, densely branched root systems with many lateral roots in upper soil horizons should maximise P uptake (Brown et al., 2013, Rose et al., 2012). In contrast, N (especially nitrate) is usually found in greater concentrations deeper in soils due to its greater mobility. Root ideotypes for N capture therefore have a 'bottom heavy' rooting architecture (Lynch, 2013, Lynch and Wojciechowski, 2015).

By influencing RSA, mycorrhization can exert indirect effects on crop nutrient acquisition. A significant body of research has been carried out investigating interactions between AMF and RSA in host plants (Jung and McCouch, 2013, Li et al., 2016, Rogers and Benfey, 2015, Yang et al., 2015). Typically, AMF colonisation increases lateral root formation (Gutjahr et al., 2009, Yao et al., 2009, Wu et al., 2011, Liu et al., 2014a), which may itself facilitate increased plant P uptake, independent of the mycorrhizal pathway. It has, however, proved difficult to isolate the direct effects of AMF on RSA from the secondary effects which AMF may induce, such as increased water availability and nutrient acquisition.

Numerous biotic and abiotic factors may modify plant RSA response to AMF; plant species, growth habit, soil nutrient status and fungal identity may simultaneously affect RSA (Jung and McCouch, 2013). AMF may indirectly affect RSA by alterations to plant root phosphate transporters (Paszkowski and Gutjahr, 2013). When these transporters are downregulated following AMF colonisation, the plants can no longer sense Pi concentrations in the soil, prompting a P starvation response including lateral root proliferation (Paszkowski and Gutjahr, 2013). AMF are also known to secrete diffusible signalling molecules into the soil, which alter root branching patterns in *Medicago truncatula* (Oláh et al., 2005) and in rice (Chiu et al., 2018). Whether this improves root function, or simply increases the opportunity for AMF colonisation of root cortex, remains unclear.

Studies using CT scanning of intact rhizosphere systems have shown the extent to which root proliferation may be inhibited by unfavourable soil conditions (Tracy et al., 2012a,b). Soil compaction remains a perennial problem in agricultural management; root elongation is limited in many fields in the UK, largely due to compaction (Valentine et al., 2012). Plants with restricted root elongation subsequently show reduced nutrient acquisition (Tracy et al., 2012a,b). AMF extra-radical hyphal networks can improve soil structure (Mardhiah et al., 2016) and root proliferation through compacted soil (Miransari et al., 2008, Miransari et al., 2007), in addition to making direct contributions to plant nutrient acquisition. By physically binding soil particles together, AMF can influence the distribution of pores of different sizes, increasing overall porosity (Martin et al., 2012). The full potential for application of AMF to reduce compaction and improve the structure of less-degraded agricultural soils and the contribution of fungal diversity remains unknown.

4 Barriers to AMF utilisation in agriculture

An incomplete understanding in many, if not all, areas of mycorrhizal biology mean that the prospect of utilising symbiotic fungi in agriculture remains a contentious issue. Substantial functional diversity exhibited across the symbiosis (in terms of plant benefit) clearly demonstrates that the AM relationship is not simply mutualistic (Feddermann et al., 2010, Smith and Smith, 2011b). From the many and varied published datasets available, it is clear that association with mycorrhiza-forming fungi is not always beneficial to the host plant. In extreme cases, the fungi may be entirely parasitic - satisfying all of their carbon and mineral nutritional needs by colonising plant roots, while providing no benefit to their host in return (Smith and Read, 2008). Even where they make contributions to plant nutrient acquisition, such fungi may elicit negative growth responses in their hosts (Hetrick et al., 1996, Smith et al., 2003). In extreme cases, breeding for reduced mycorrhizal receptivity has even been suggested for some cereal crops which are negatively influenced by AMF colonisation (Ryan et al., 2005).

4.1 Crop growth responses to symbiotic fungal colonisation

Negative mycorrhizal growth responses among crop plants appears to be fairly common, including in the most important staples (Smith and Smith, 2011a); there is mounting evidence to suggest that negative growth responses to mycorrhization are linked to reduced activity of root epidermal phosphorus transporters following AMF colonisation. It was previously understood that high affinity P transporters were constantly expressed at the epidermis of the root apex and root hairs, providing a stable contribution to P uptake, and that limiting soil P availability promoted additional P contribution from the mycorrhizal pathway. This has since been disproven, the direct pathway and the mycorrhizal pathway are not additive as had previously been assumed (Li et al., 2006, Smith et al., 2003, 2004). In many cases, the mycorrhizal pathway replaces the direct pathway as the main route for P acquisition, and root epidermal phosphate transporters may be downregulated almost completely (Smith et al., 2004). In this scenario, the mycorrhizal pathway alone is responsible for plant P uptake, while no direct uptake occurs (Li et al., 2008, Smith et al., 2011). Where mycorrhizal pathway contributions to plant P do not sufficiently substitute for the reduced or inactive direct pathway, total P acquisition by the root is thereby reduced (Smith et al., 2011).

Mycorrhizal downregulation of the direct pathway of plant N uptake may similarly decrease crop nutrient acquisition. In winter wheat, local and systemic downregulation of plant root N transporters has been demonstrated following AMF colonisation (Duan et al., 2015). Alarmingly, it appears that NO_3^- and Pi transporter downregulation can take place independent of nutrient transfer from fungus to plant (Tian et al., 2017), suggesting regulation by some other, non-nutritional factor. Clearly, further research is needed to determine the causes and extent of this phenomenon. Should differences in mycorrhizal responsiveness among cultivars of these crops be due to disparities in transporter downregulation, selection for mycorrhizal-associated traits in plant breeding may allow improvements in mycorrhizal responsiveness.

Excessive carbon movement from the host plant into mycorrhizal fungi was previously considered the most likely explanation for mycorrhizal growth suppression of hosts (Johnson et al., 1997). Recent evidence, again, suggests this may not be the case. For example, AMF can increase the C sink strength of host plant roots, preventing the accumulation of photosynthate in leaves, and lifting an inhibitory pressure on the rate of photosynthesis (Gavito et al., 2019). If this phenomenon is widespread, plant C limitation by excessive allocation to AMF may not be a significant problem, and AMF carbon demand may not be the cause of plant growth depression. Further work is now required to determine how far, and under what circumstances plant-fungal carbon transfer influences plant growth responses to mycorrhization.

Considering the prevalence of mycorrhizal inocula in soils both geographically (Davison et al., 2015) and through the soil profile (Sosa-Hernández et al., 2018), it would be very unusual for a susceptible plant to remain uncolonised by AMF in the field (Smith and Smith, 2011b). As such, colonisation by mycorrhizal fungi may reduce plant P uptake by the direct pathway in a significant proportion of agricultural crop plants. The effect of this is likely to be beneficial for the host plant where the mycorrhizal pathway compensates (or overcompensates) for reduced activity of the direct pathway, but will be detrimental where it does not. The mechanisms responsible for AM-induced plant transporter downregulation are not clearly understood, and the phenomenon raises intriguing questions about which partner (fungus or plant) is 'in control' of the symbiosis. Certainly, this must remain a research priority. Improved understanding here may eventually lead to the development of crops where mycorrhizal and direct routes for P (and perhaps N and other nutrients) are additive (Smith et al., 2011).

Exploring the genetic diversity of crop plants may provide valuable insights into mycorrhizal receptivity and responsiveness. Although a number of studies and reviews in the literature suggest that crop plants usually experience neutral or negative growth responses to mycorrhization (Ryan et al., 2005, Rillig et al., 2019), substantial variation between plant-fungal partner combinations is clearly evident. Indeed, positive responses have been shown in number of crop species (Zhang et al., 2019, Watts-Williams et al., 2019a). How far this diversity in mycorrhizal responsiveness is linked to microbial diversity in soils is unclear. This distinction is more than of academic interest; as we detail below, modern agricultural practices can be deleterious not only to the abundance but also to the diversity of AMF communities in agricultural soils (Helgason et al., 1998).

4.2 Detrimental effects of agricultural management on AMF

4.2.1 Tillage

It has been known for some years that tillage is highly disruptive to AMF mycelia in soil, and that fungal taxa may be extirpated following this disturbance (Bowles et al., 2017, García de León et al., 2018, Banerjee et al., 2019). Even very short-term disturbances may influence AMF community composition (Peyret-Guzzon et al., 2016). As the growing season in annual cropping systems is shorter than in perennial grassland, it is likely that ruderal species are selected for in these systems (Sosa-Hernández et al., 2019). It might be assumed that these disturbance-tolerant AMF may have rapid life cycles and minimize nutrient exchange with host plants. However, Duan et al. (2011) found that disturbance did not affect AMF abundance or contribution to P uptake by wheat, indicating AMF may maintain their function even in disturbed systems. There is evidence that MFRE proliferate in disturbed systems (Orchard et al., 2017), and so may be abundant in agricultural landscapes, potentially even replacing some AMF functions. Wider adoption of reduced tillage is likely to improve AMF proliferation and stabilization in soils, although the subsequent effects on root function will be difficult to predict. As with other factors, there is likely to be a great deal of context-dependence, influenced by elements such as plant and fungal genotype, fertiliser regime and others (Field and Pressel, 2018, Thirkell et al., 2017).

4.2.2 Fertiliser application

Macronutrient fertiliser application is widely believed to reduce the benefit of engaging in mycorrhizal symbiosis for plants (Smith and Read, 2008). Plentiful nutrient supply, it is argued, should allow sufficient direct pathway nutrient acquisition to satisfy plant demand (Olsson et al., 1997, Breuillin et al., 2010, Kobae et al., 2016). More recently, it has become clear that this view is overly simplistic, and that more nuanced soil nutrient stoichiometry may dictate the degree of mycorrhizal benefit received by plant hosts (Johnson et al., 2015). Specifically, high-N and limited-P in soils may in fact provide the conditions in which maximum benefit is afforded to plants by AMF (Johnson, 2010, Johnson et al., 2015, Williams et al., 2017, Thirkell et al., 2019). Experimental validation of this hypothesis remains limited, and is not universally supported by emerging experimental data sets (Jiang et al., 2018). Further experimental work is required to determine whether fertilisation regimes can be employed under which AMF benefit and yield are maintained.

Recent advances in molecular identification of AMF and the affordability of profiling techniques have improved our understanding of the effects of soil nutrient addition on AMF communities. In many cases, it is apparent that N and P fertilisation alters AMF community structure (Hassan et al., 2013, Liu et al., 2014b). Nutrient addition may even be more influential on AMF communities than underlying soil type (Bainard et al., 2015), although consistent trends are not apparent; fertilisation has variously been shown to increase (Phillips et al., 2019) or decrease (Antoninka et al., 2011, Chen et al., 2014, Williams et al., 2017, Jiang et al., 2018) AMF abundance and community richness. Indeed, in some cases, fertilisation seemingly has little effect on AMF community composition (Tian et al., 2013, Williams et al., 2013, Jansa et al., 2014). Our ignorance of how fertilisation affects AMF functioning in agricultural systems is perhaps even more profound and requires urgent experimental attention.

A greater understanding of how AMF function responds to the disturbance characteristics of agricultural systems will improve chances of ensuring benefit of AMF to crops. It is now understood that AMF readily colonise roots at least 1 m below the surface, in distinct sub-soil communities (Sosa-Hernández et al., 2018). As tillage rarely reaches deeper than 20 cm into the bulk soil, it is not clear how disturbance of the upper soil horizons might affect these deep mycorrhizal fungal communities. Cereals roots will commonly grow more than 1 m into the soil to access water and labile soil nutrients such as NO_3^- (Fan et al., 2016); early evidence suggests these subsoil roots are colonised by AMF, but their role in nutrient uptake remains to be investigated.

5 Adapt, replace, restore (or ignore?)

Even where strong evidence exists to promote the use of AMF in agriculture, challenges are still apparent when considering strategies for implementation. Here, we suggest three approaches that could be employed to exploit symbiotic fungi (particularly AMF) in agriculture, which we have termed as *adapt, replace* and *restore*.

5.1 Adapt – *exploiting genetic diversity in AMF-plant interactions*

Plant and fungal genetic diversity have considerable influence in moderating the degree to which plants benefit from mycorrhizal symbiosis. Understanding and being able to control these mechanisms through traditional crop breeding or genetic engineering may allow us to adapt current crops to derive greater benefit from AM symbioses. This may be possible even with the limited range of fungal types commonly present in agricultural systems, to elicit improved plant growth responses across a number of different host plant species or cultivars.

Contrasting growth responses to AMF colonisation are apparent among cultivars of many important crop species, including maize, rice, wheat, barley and sorghum (Ellouze et al., 2015, Sawers et al., 2017, Watts-Williams et al., 2019a, Elliott et al., 2020). Multiple factors are likely to be responsible for this inter- and intra-specific diversity in mycorrhizal functioning. For example, genotypic variation among maize lines results in contrasting transcript numbers of the AM-specific PHT1 phosphate transporters in colonised maize plants, which directly affect phosphorus uptake efficiency (Sawers et al., 2017).

Recent physiological and molecular studies have demonstrated genome regions of host plants which are associated with the degree of AMF colonisation in roots (Lehnert et al., 2017), and even the degree of plant growth response to colonisation (Lehnert et al., 2018, Huang et al., 2020). While further work is required to explore the extent to which these genetic components drive functional variation in the symbiosis, such advances represent among the most exciting avenues of research in the field at present. Breeding for this improved mycorrhizal responsiveness using variation in existing elite germplasm may allow improved crop nutrient uptake and yield, and reduce the need for synthetic fertiliser addition (Lefebvre, 2019).

5.2 Replace - *field inoculation with AMF*

As discussed, it is well established that different AMF species may elicit strongly contrasting growth responses from the same host plant species (Klironomos, 2003, Smith et al., 2004). Intra-specific diversity has even been demonstrated among isolates of single fungal species, which may differentially affect host plant nutrient uptake and growth (Munkvold et al., 2004, Ceballos et al., 2019). The tailoring of a combination of crop and AMF may, it is suggested, be an approach by which benefit might be optimised. Needless to say, this will necessitate the development of affordable inoculum of sufficient quantity and quality to allow practical application to agricultural systems.

Commercially produced AMF inoculum has been available for almost three decades, and is mainly applied in specialist cases for land restoration, or high-value crops such as ornamental flowers and fruit trees (Vosátka et al., 2012). Due to the pressing need to make agriculture more sustainable, an increasing number of companies have recently emerged producing and selling AMF inoculants (Vosátka et al., 2012, Faye et al., 2013).

Examples of successful application of these inoculants on large scales, particularly in the most commonly grown crops such as cereals, are few and far between. There are promising accounts where AMF inoculation of agricultural soil has been successful in increasing yields and profitable for farmers (Hijri, 2016). In this study, *Rhizophagus irregularis* was applied at planting as a liquid suspension of spores, requiring no specialist farming equipment, and increased potato yields by an average of 9.5% across 231 field trails (Hijri, 2016). However, in 14.6% of cases inoculation still proved disadvantageous and resulted in yield reductions, suggesting inocula require further optimisation for a general solution to result.

Such variability in inoculum success is a widespread issue, and in this case is particularly problematic as no links could be made to potato variety, soil type or region to suggest where inoculum application should be avoided (Hijri, 2016). Therefore, assessments may need to be made on a case-by-case basis. AMF inoculants have also been successfully applied to cassava-increasing yields irrespective of P fertilisation levels (Ceballos et al., 2013). A recent meta-analysis shows that field inoculation with AMF leads to an average increase in crop yields of 16%, with positive effects on yield across seven cereal crop cultivars of importance for human nutrition (Zhang et al., 2019). Indeed, positive effects on crop yields after field inoculation with AMF are supported by most published meta-analyses (Lekberg and Koide, 2005, Pellegrino et al., 2015). Variability and unpredictability of effectiveness, and negative responses to inoculation still drive debate in the literature about exactly how beneficial AMF inocula will prove to be, however (Ryan and Graham, 2018, Rillig et al., 2019, Ryan et al., 2019, Hart et al., 2018).

5.2.1 Inoculum production

If AMF inoculants are to achieve widespread application in agriculture, they must be effective, reliable, commercially viable and easy to apply. Currently, the cost of inoculants is a significant limiting factor, explaining why most current usage is limited to high-value crops (Vosátka et al., 2012). The high cost is partly due to the lengthy and expensive process of registering new microbial products (Malusá et al., 2012). The time and space needed to produce large quantities of inocula are also problematic; AMF are biotrophic and therefore can only be propagated at a large-scale in association with a suitable host plant.

Currently, the majority of inoculants are produced in greenhouses by growing mycorrhizal plants in inert substrates, which can produce hundreds of thousands of spores per litre of substrate and then be diluted and sold (Malusá et al., 2012). Studies testing inoculants produced via unsterile greenhouse production have often reported contamination with unwanted microbes including other fungal species (Vosátka et al., 2012, Faye et al., 2013). Axenic mycorrhizal cultures (grown with transformed roots) may be advantageous over those produced in a greenhouse, as sterile, mass production of propagules can be achieved in a shorter time and in less space space (Adholeya et al., 2005), albeit at a higher cost.

The process of extracting AMF spores from the soil, sterilising and successfully associating these with a transformed host root to produce an axenic AMF culture has only been achieved in a limited number of species. The most extensively cultivated species is *Rhizophagus irregularis*, largely because of the ease with which it proliferates - thousands of spores can be produced after as little as three months. A recent breakthrough has been made in the production of spores through pure AMF culture, by adding fatty acids (myristate) to a culture medium; in this system AMF (*R. irregularis*) growth and spore production were achieved in the absence of a host plant (Sugiura et al., 2019). Obvious challenges remain - propagating those species or isolates of AMF which are most conducive to large-scale production will be of limited benefit if these are not the most appropriate for use in individual situations.

Finding suitable methods for application of inoculants remains a key challenge in integrating AMF into sustainable management systems, especially if the range of AMF species in inocula is to be increased or improved. Inoculants must come in a form which is easy to apply and requires minimal modifications to current agricultural practices. Broadcasting or in-furrow application when sowing seeds is one such method, where inoculum and seeds can be spread quickly over a large area requiring no specialised farming equipment (Adholeya et al., 2005).

Where large quantities of inoculum are required to ensure seeds are in close vicinity of the AMF propagules, cost is likely to become a limiting factor. However, if initial applications are used to restore AMF diversity and propagule abundance to agricultural soil and then combined with non-harmful management strategies, repeat application may not be required. Alternatively, seed coating with AMF inoculum can be highly successful at increasing crop colonisation by AMF with substantially reduced inoculum quantities, as propagules are ideally positioned to colonise the germinating seed (Oliveira et al., 2016).

5.2.2 Reliability of inoculant outcomes

Despite progress in the production and application of AMF inoculants, variability in AMF establishment in plant roots and inconsistent ability to produce the desired yield enhancements could damage the reputation of AMF inoculants within the agricultural sector and prove a significant barrier to their widespread adoption. Inoculants often fail to establish (Vosátka et al., 2012, Berruti et al., 2016), and growth increases observed may be attributed to nutrients within the inoculant carrier substrate (Vosátka et al., 2012). Inoculum propagules can lose viability during storage or shipping, (Corkidi et al., 2004, Wiseman et al., 2009), suggesting the choice of inoculum carrier could be an important consideration to enhance shelf-life (Malusá et al., 2012).

Alternatively, AMF species within the inoculum may fail to establish if they are not compatible with the target environment (Verbruggen et al., 2012); AMF show varied tolerance to disturbance, soil nutrient availability (Schnoor et al., 2011, Zhao et al., 2017), soil pH, soil type (Lekberg et al., 2007, Oehl et al., 2010) and levels of toxic compounds such as heavy metals (Hassan et al., 2011). Designing inoculants suitable for the target environment and small-scale tests of inoculants that establish successfully in the soil will be essential. Requirements to produce inocula specific to crop identity or soil conditions are likely to prove an additional barrier to successful application.

Unsuccessful establishment could be combatted by using native AMF species which are already adapted to local soil conditions or generalist, near ubiquitous species such as *Rhizophagus irregularis* (Rosendahl, 2008, Rodriguez and Sanders, 2015). *R. irregularis* is considered a model species for AMF research and has been successfully used to enhance crop growth in several crops and environments (Ceballos et al., 2013, Hijri, 2016, Köhl et al., 2016). *R. irregularis* is also the first species to be successfully produced in axenic culture on a mass scale and have its genome published, making it the easiest to trace in the soil after inoculation (Rodriguez and Sanders, 2015). Partial genome sequencing of many *R. irregularis* isolates may allow identification of polymorphisms which could be used as markers to distinguish native and introduced isolates (Rodriguez and Sanders, 2015). This is especially relevant for *R. irregularis*, which is a common species in many soils including arable systems. Previous research has shown substantial levels of genetic variation between isolates (Stockinger et al., 2009), suggesting future research to identify isolate specific markers may be promising.

However, genetic variability within a single AMF species can be high; sequencing *R. irregularis* strains cultured in different laboratories demonstrated a substantial number of variations between isolates (Stockinger et al., 2009), and many isolate-specific genes (Chen et al., 2018). Genetic differences between isolates can translate into significantly different effects on host plant biomass or nutrient uptake (Angelard et al., 2010, Mensah et al., 2015, Koch et al., 2017). However, the underlying mechanism behind variation in symbiotic outcomes due to plant-fungal specificity is still largely unknown, and could not be related to any differences in fungal morphology or growth patterns which are highly conserved within species (Koch et al., 2017).

There is some concern that AMF introduced into the environment via inoculants could become invasive and suppress native AMF communities (Schwartz et al., 2006, Antunes et al., 2009, Hart et al., 2017). Commercially available inoculants usually contain a single generalist species (commonly *R. irregularis*) which is selected to be a competitive coloniser with a high sporulation rate to maximise the capacity to produce the inoculant and chance of successful establishment in the host. However, traits such as high competitive ability, high dispersal rates and a broad host range are often associated with other invasive species (Schwartz et al., 2006).

Outcompeting native AMF species and reducing their abundance in soil would be problematic in scenarios where the native species is of greater benefit to the plant host than the introduced species (Klironomos et al., 2005, Pellegrino et al., 2011, Séry et al., 2016). So far, it is not clear that *R. irregularis* inoculation alters the diversity of the native AMF community, although reductions in diversity were noted in some cases (Mummey et al., 2009, Koch et al., 2011, Jin et al., 2013). Further studies show AMF inoculum can cause shifts away from the dominant native species to *R. irregularis* (Elliott et al., 2020). The persistence and spread of AMF species introduced by inoculation must be the focus of long-term study; this represents a clear research gap.

In some cases, native AMF species appear to be more beneficial for their host plants than species introduced by inoculation (Estrada et al., 2013, Middleton et al., 2015, Séry et al., 2016). A range of plant growth responses (from parasitic to mutualistic) have been observed when inoculating plants with both native and foreign fungal genotypes (Klironomos, 2003). In this case, the relative frequency of positive plant growth response decreased with foreign AMF genotypes (Klironomos, 2003), suggesting native AMF may be most beneficial to host plants. Consequently, if commercial strains of AMF outcompete native AMF species and reduce plant root colonisation, they could unintentionally reduce AMF-acquired benefits to the plant.

Even where AMF inoculants are beneficial in terms of crop growth and nutrient uptake in the target environment and crop plants, there are concerns about the ability of AMF spores to spread into the wider natural environment (Hart et al., 2017, Hart et al., 2018). Studies under controlled conditions and at field scale would help to determine the extent to which AMF inocula might spread from intended areas of application and the impact this may have on local ecosystems.

5.3 Restore (or ignore) – *why make AMF the target of our efforts?*

In a recent review, Ryan and Graham (2018) argue that there is insufficient experimental evidence to suggest that farmers should prioritise agricultural management to accommodate mycorrhizal fungi. While there are many potential benefits, there is no plant physiological criterion which is unequivocally or universally improved by association with AMF. Moreover, plants which receive a benefit from AMF association in one factor, such as phosphorus uptake, may suffer negative responses in another, for example, attractiveness to herbivores (Hartley and Gange, 2009). In this case, increased herbivory may outweigh the benefits afforded by improved nutrition, and prove deleterious to crop performance. Such trade-offs necessarily arise in complex systems; determining whether increased AMF colonisation would provide a net benefit or net drawback remains difficult.

Critically, the problems facing agricultural sustainability are not only numerous, but also inter-connected and inter-dependent. Soil compaction, water retention, organic matter content, nutrient addition, tillage and others may all impact root function. It is clear that no single factor (AMF or otherwise) in management, breeding or policy can overcome the apparent challenges alone. It may in fact be prudent, as suggested by Ryan et al. (2019), to strive for enhanced sustainability in agricultural management more holistically, rather than focussing on a sole factor such as AMF.

Sustainable management practices, which improve the chances of AMF colonising (e.g. min-/no-tillage, minimised fertiliser addition, minimised soil compaction, diverse planting schemes and cover cropping), are also factors which can enhance ecosystem function. Vegetation diversity, in spatial and temporal scales (by the integration of cover cropping and intercropping) may enhance soil biodiversity, multi-functionality and ecosystem function (Brooker et al., 2015, Finney and Kaye, 2017, Ranaldo et al., 2020). Moreover, improved soil microbial diversity can lead to improved aboveground productivity (van der Heijden et al., 1998). To understand these complex interactions, large, multi-factorial studies assessing the effects of management practices on soil microbial diversity and ecosystem function are required.

6 Conclusion

Substantial progress has been made in the last decade in the fields of root and mycorrhizal research. We believe the discipline is at an exciting juncture where improved understanding of soil-microbe-plant processes may soon allow significant improvements in crop performance under more sustainable production. Causes of functional diversity in mycorrhizal associations are understood better than ever, and future research can only improve upon this. Symbiotic soil fungi offer many potential advantages to improving key root traits associated with nutrient and water assimilation, as well as improving soil structure and function. Our ability to accurately predict the situations in which these fungi are significantly beneficial remains limited, however. Owing to the ubiquity of AMF in agricultural soils, a combined approach might be most advantageous – enhancing sustainable management, exploiting well-understood benefits (e.g. proven cases where inoculation improves plant performance) and breeding for improved mycorrhizal growth responses. In order to realise these benefits however, more research is essential in many areas, in particular to assess compatibility, function, and the ways in which mycorrhizas may be effectively exploited in agricultural systems.

7 Future trends in research

We have identified several research questions (numbered 1–5, below) which we believe represent important gaps in understanding of the arbuscular mycorrhizal symbiosis, particularly in relation to agricultural crop root function. Within this discipline, we suggest these topics represent the most fertile avenues for research in the near future, and in answering these questions, rapid improvements could be made. Fundamentally, future research in these areas must aim to enhance the benefit of AMF without jeopardising agricultural sustainability.

1 What is the distribution of AMF (and other symbiotic soil fungi) within agricultural systems at local and wider spatial scales?
 Only relatively recently have molecular techniques advanced sufficiently to enable researchers to identify AMF species and isolates with confidence. There is substantial scope to improve our understanding of the complexity of the fungal communities present in agricultural fields. It also remains to be understood how the distribution of these fungi is determined by location, soil type, crop identity, nutrient availability, management practice (e.g. tillage, rotations, cover cropping) and other factors. Understanding the mechanisms driving soil fungal community differences will likely prove to be important; fungal species/isolate identity is a major determinant of the extent to which AMF benefit plant nutrition and growth. Investigations into the soil fungal community should not focus exclusively on Glomeromycotina (AMF), but must also include those clades which have recently been shown to have active roles in improving plant nutrition, for example, Mucoromycotina.
2 Which are the most influential symbiotic fungi in the soil community?
 Identifying the symbiotic fungi which are present in agricultural soils at a range of spatial scales can be advantageous only if we can also begin to understand the roles these fungi play in crop nutrient acquisition and growth. Combinations of field trials and controlled-environment studies will be required to identify those species/isolates of AMF (and other soil fungi) which are of notable benefit to their plant hosts. The identification and physiological categorisation of a fungal isolate which is beneficial in a number of crops and soil types, for example, may facilitate the development of high-quality inoculum.
3 How, why and to what extent do AMF suppress direct root uptake of mineral nutrients?
 Arbuscular mycorrhizal fungal colonisation commonly suppresses plant P uptake via root epidermal transporters, but the mechanisms by which this takes place are not well understood. Understanding how this

phenomenon is regulated will represent a major step in improving crop nutrient uptake and fertiliser efficiency. Routine breeding techniques and/or genetic modification may then allow the development of crops where the mycorrhizal and direct pathways of nutrient uptake are additive, rather than substitutive. Whether mycorrhizal downregulation of nitrate and ammonium transporters is common across crop species remains unknown, as does the effect of AMF colonisation on direct plant uptake of other mineral nutrients. Undoubtedly there are significant advances to make here, which may substantially improve crop root function.

4 Can we develop 'mycorrhiza-ready' crop varieties?
 The ubiquity of AMF in agricultural soils and the susceptibility of most plant species to colonisation mean that mycorrhization is the default state of most crops. The extent to which AMF are beneficial for crop hosts often varies considerably among cultivars, even when inoculated with a single isolate of fungi. We urge the research community to focus efforts on identifying and understanding the mechanistic basis of this functional diversity among crop genotypes. Identifying genetic determinants of mycorrhizal responsiveness may simultaneously allow negative growth responses to colonisation to be bred out of elite crop lines and the inclusion of beneficial traits. Further work is urgently needed to explain why the growth response of an individual cultivar varies markedly following inoculation with different AMF species/isolates.

5 How does agricultural management affect the benefit (or otherwise) of AMF?
 Many of the standard practices employed in high-yielding agricultural systems are known to be deleterious to AMF community diversity, abundance and the degree to which plants benefit from the symbiosis. While generalisations can be drawn from the literature (e.g. reducing tillage and avoiding excessive P fertiliser application are beneficial for mycorrhizas), our understanding is not sufficient to provide clear advice to farmers and growers as to best practice to ensure mycorrhizal benefit to crops. Field-scale experimental trials will be essential to identify thresholds for agricultural practices, beyond which management intensity negates any benefit of AMF colonisation and thus diminishes returns.

8 Where to look for more information

The standard reference text on mycorrhizas is S. E. Smith and D. J. Read, *Mycorrhizal Symbiosis* (see Smith and Read, 2008 in the References for full details), which provides an insightful, thorough and accessible overview of the subject. A review and opinion paper by F. A. Smith and S. E. Smith, *What is the significance of the arbuscular mycorrhizal colonisation of many economically important crop plants?* (full details can be found in the References section under Smith and Smith, 2011) succinctly introduces a number of the most important questions to be resolved in this field. A number of recent review papers (listed 1-5, below) offer the reader further academic perspectives on the current subject of mycorrhizas in agriculture.

1 Ryan et al., 2019. "Research must use a systems agronomy approach if management of the arbuscular mycorrhizal symbiosis is to contribute to sustainable intensification" *New Phytologist* 222 (3) 1176–1178.
2 Rillig et al., 2019. "Why farmers should manage the arbuscular mycorrhizal symbiosis". *New Phytologist* 222(3) 1171–1175.
3 Thirkell et al., 2017. "Are mycorrhizal fungi our sustainable saviours? Considerations for achieving food security". *Journal of Ecology* 105(4) 921–929.
4 Bender et al., 2016. "An underground revolution: biodiversity and soil ecological engineering for agricultural sustainability". *Trends in Ecology and Evolution* 31 (6) 440–452.
5 van der Heijden et al., 2015. "Mycorrhizal ecology and evolution: the past, the present and the future". *New Phytologist* 205 (4) 1406–1423.

Through this chapter, we have included a number of papers from the academic journals *New Phytologist, Journal of Ecology* and *Plant and Soil* which we recommend as sources of research, review and opinion articles. These journals provide the reader with a regular source of publications on the subject of mycorrhizas, ranging in focus from genetics and molecular biology to field-scale agronomic trials.

9 References

Adholeya, A., Tiwari, P. and Singh, R. 2005. Large-scale inoculum production of arbuscular mycorrhizal fungi on root organs and inoculation strategies. In: Stéphane Declerck, J. André Fortin and Désiré-Georges Strullu (Eds), *In Vitro Culture of Mycorrhizas*. Springer, Heidelberg, Germany.

Alloway, B. J. 2008. Micronutrients and crop production: an introduction. In: Brian J. Alloway (Ed.), *Micronutrient Deficiencies in Global Crop Production*. Springer, Dordrecht, Netherlands.

Ames, R. N., Reid, C. P. P., Porter, L. K. and Cambardella, C. 1983. Hyphal uptake and transport of nitrogen from two ^{15}N-labelled sources by *Glomus mosseae*, a vesicular-arbuscular mycorrhizal fungus. *New Phytologist* 95(3), 381–396.

Angelard, C., Colard, A., Niculita-Hirzel, H., Croll, D. and Sanders, I. R. 2010. Segregation in a mycorrhizal fungus alters rice growth and symbiosis-specific gene transcription. *Current Biology* 20(13), 1216–1221.

Antoninka, A., Reich, P. B. and Johnson, N. C. 2011. Seven years of carbon dioxide enrichment, nitrogen fertilization and plant diversity influence arbuscular mycorrhizal fungi in a grassland ecosystem. *New Phytologist* 192(1), 200–214.

Antunes, P. M., Koch, A. M., Dunfield, K. E., Hart, M. M., Downing, A., Rillig, M. C. and Klironomos, J. N. 2009. Influence of commercial inoculation with *Glomus intraradices* on the structure and functioning of an AM fungal community from an agricultural site. *Plant and Soil* 317(1-2), 257-266.

Bainard, L. D., Dai, M., Gomez, E. F., Torres-Arias, Y., Bainard, J. D., Sheng, M., Eilers, W. and Hamel, C. 2015. Arbuscular mycorrhizal fungal communities are influenced by agricultural land use and not soil type among the Chernozem great groups of the Canadian Prairies. *Plant and Soil* 387(1-2), 351-362.

Bakker, M. G., Manter, D. K., Sheflin, A. M., Weir, T. L. and Vivanco, J. M. 2012. Harnessing the rhizosphere microbiome through plant breeding and agricultural management. *Plant and Soil* 360(1-2), 1-13.

Banerjee, S., Walder, F., Büchi, L., Meyer, M., Held, A. Y., Gattinger, A., Keller, T., Charles, R. and van der Heijden, M. G. A. 2019. Agricultural intensification reduces microbial network complexity and the abundance of keystone taxa in roots. *The ISME Journal* 13(7), 1722-1736.

Barber, S. A. 1995. *Soil Nutrient Bioavailability: A Mechanistic Approach*. John Wiley & Sons, New York, USA.

Bardgett, R. D. and van der Putten, W. H. 2014. Belowground biodiversity and ecosystem functioning. *Nature* 515(7528), 505-511.

Bender, S. F., Wagg, C. and van der Heijden, M. G. A. 2016. An underground revolution: biodiversity and soil ecological engineering for agricultural sustainability. *Trends in Ecology and Evolution* 31(6), 440-452.

Berdeni, D., Cotton, T. E. A., Daniell, T. J., Bidartondo, M. I., Cameron, D. D. and Evans, K. L. 2018. The effects of arbuscular mycorrhizal fungal colonisation on nutrient status, growth, productivity, and canker resistance of apple (*Malus pumila*). *Frontiers in Microbiology* 9, 1461.

Berruti, A., Lumini, E., Balestrini, R. and Bianciotto, V. 2016. Arbuscular mycorrhizal fungi as natural biofertilizers: let's benefit from past successes. *Frontiers in Microbiology* 6, 1559.

Bidartondo, M. I., Read, D. J., Trappe, J. M., Merckx, V., Ligrone, R. and Duckett, J. G. 2011. The dawn of symbiosis between plants and fungi. *Biology Letters* 7(4), 574-577.

Bolan, N. S. 1991. A critical review on the role of mycorrhizal fungi in the uptake of phosphorus by plants. *Plant and Soil* 134(2), 189-207.

Bowles, T. M., Jackson, L. E., Loeher, M. and Cavagnaro, T. R. 2017. Ecological intensification and arbuscular mycorrhizas: a meta-analysis of tillage and cover crop effects. *Journal of Applied Ecology* 54(6), 1785-1793.

Breuillin, F., Schramm, J., Hajirezaei, M., Ahkami, A., Favre, P., Druege, U., Hause, B., Bucher, M., Kretzschmar, T., Bossolini, E., Kuhlemeier, C., Martinoia, E., Franken, P., Scholz, U. and Reinhardt, D. 2010. Phosphate systemically inhibits development of arbuscular mycorrhiza in *Petunia hybrida* and represses genes involved in mycorrhizal functioning. *The Plant Journal* 64(6), 1002-1017.

Brooker, R. W., Bennett, A. E., Cong, W. F., Daniell, T. J., George, T. S., Hallett, P. D., Hawes, C., Iannetta, P. P. M., Jones, H. G., Karley, A. J., Li, L., McKenzie, B. M., Pakeman, R. J., Paterson, E., Schob, C., Shen, J. B., Squire, G., Watson, C. A., Zhang, C. C., Zhang, F. S.,

Zhang, J. L. and White, P. J. 2015. Improving intercropping: a synthesis of research in agronomy, plant physiology and ecology. *New Phytologist* 206(1), 107–117.

Brown, L. K., George, T. S., Dupuy, L. X. and White, P. J. 2013. A conceptual model of root hair ideotypes for future agricultural environments: what combination of traits should be targeted to cope with limited P availability? *Annals of Botany* 112(2), 317–330.

Bücking, H. and Kafle, A. 2015. Role of arbuscular mycorrhizal fungi in the nitrogen uptake of plants: current knowledge and research gaps. *Agronomy* 5(4), 587–612.

Bürkert, B. and Robson, A. 1994. ^{65}Zn uptake in subterranean clover (*Trifolium subterraneum* L.) by three vesicular-arbuscular mycorrhizal fungi in a root-free sandy soil. *Soil Biology and Biochemistry* 26(9), 1117–1124.

Busby, P. E., Soman, C., Wagner, M. R., Friesen, M. L., Kremer, J., Bennett, A., Morsy, M., Eisen, J. A., Leach, J. E. and Dangl, J. L. 2017. Research priorities for harnessing plant microbiomes in sustainable agriculture. *PLoS Biology* 15(3), e2001793.

Cameron, D. D. 2010. Arbuscular mycorrhizal fungi as (agro)ecosystem engineers. *Plant and Soil* 333(1–2), 1–5.

Cavagnaro, T. R. 2008. The role of arbuscular mycorrhizas in improving plant zinc nutrition under low soil zinc concentrations: a review. *Plant and Soil* 304(1–2), 315–325.

Cavagnaro, T. R. 2014. Impacts of compost application on the formation and functioning of arbuscular mycorrhizas. *Soil Biology and Biochemistry* 78, 38–44.

Ceballos, I., Mateus, I. D., Peña, R., Peña-Quemba, D. C., Robbins, C., Ordoñez, Y. M., Rosikiewicz, P., Rojas, E. C., Thuita, M. and Mlay, D. P. 2019. Using variation in arbuscular mycorrhizal fungi to drive the productivity of the food security crop cassava. *bioRxiv*, 830547.

Ceballos, I., Ruiz, M., Fernandez, C., Pena, R., Rodriguez, A. and Sanders, I. R. 2013. The in vitro mass-produced model mycorrhizal fungus, *Rhizophagus irregularis*, significantly increases yields of the globally important food security crop cassava. *PLoS ONE* 8(8), e70633.

Chen, B. D., Li, X. L., Tao, H. Q., Christie, P. and Wong, M. H. 2003. The role of arbuscular mycorrhiza in zinc uptake by red clover growing in a calcareous soil spiked with various quantities of zinc. *Chemosphere* 50(6), 839–846.

Chen, E. C. H., Morin, E., Beaudet, D., Noel, J., Yildirir, G., Ndikumana, S., Charron, P., St-Onge, C., Giorgi, J., Krüger, M., Marton, T., Ropars, J., Grigoriev, I. V., Hainaut, M., Henrissat, B., Roux, C., Martin, F. and Corradi, N. 2018. High intraspecific genome diversity in the model arbuscular mycorrhizal symbiont *Rhizophagus irregularis*. *New Phytologist* 220(4), 1161–1171.

Chen, Y. L., Zhang, X., Ye, J. S., Han, H. Y., Wan, S. Q. and Chen, B. D. 2014. Six-year fertilization modifies the biodiversity of arbuscular mycorrhizal fungi in a temperate steppe in Inner Mongolia. *Soil Biology and Biochemistry* 69, 371–381.

Chiu, C. H., Choi, J. and Paszkowski, U. 2018. Independent signalling cues underpin arbuscular mycorrhizal symbiosis and large lateral root induction in rice. *New Phytologist* 217(2), 552–557.

Clark, R. B. and Zeto, S. K. 1996. Iron acquisition by mycorrhizal maize grown on alkaline soil. *Journal of Plant Nutrition* 19(2), 247–264.

Cooper, K. M. and Tinker, P. B. 1978. Translocation and transfer of nutrients in vesicular-arbuscular mycorrhizas: II. Uptake and translocation of phosphorus, zinc and sulphur. *New Phytologist* 81(1), 43–52.

Cordell, D., Drangert, J.-O. and White, S. 2009. The story of phosphorus: global food security and food for thought. *Global Environmental Change* 19(2), 292-305.

Corkidi, L., Allen, E. B., Merhaut, D., Allen, M. F., Downer, J., Bohn, J. and Evans, M. 2004. Assessing the infectivity of commercial mycorrhizal inoculants in plant nursery conditions. *Journal of Environmental Horticulture* 22(3), 149-154.

Corrêa, A., Cruz, C., Pérez-Tienda, J. and Ferrol, N. 2014. Shedding light onto nutrient responses of arbuscular mycorrhizal plants: nutrient interactions may lead to unpredicted outcomes of the symbiosis. *Plant Science* 221-222, 29-41.

Coumou, D. and Rahmstorf, S. 2012. A decade of weather extremes. *Nature Climate Change* 2(7), 491-496.

Davison, J., Moora, M., Öpik, M., Adholeya, A., Ainsaar, L., Bâ, A., Burla, S., Diedhiou, A. G., Hiiesalu, I., Jairus, T., Johnson, N. C., Kane, A., Koorem, K., Kochar, M., Ndiaye, C., Pärtel, M., Reier, Ü, Saks, Ü, Singh, R., Vasar, M. and Zobel, M. 2015. Global assessment of arbuscular mycorrhizal fungus diversity reveals very low endemism. *Science* 349(6251), 970-973.

Delavaux, C. S., Smith-Ramesh, L. M. and Kuebbing, S. E. 2017. Beyond nutrients: a meta-analysis of the diverse effects of arbuscular mycorrhizal fungi on plants and soils. *Ecology* 98(8), 2111-2119.

Drissner, D., Kunze, G., Callewaert, N., Gehrig, P., Tamasloukht, M. B., Boller, T., Felix, G., Amrhein, N. and Bucher, M. 2007. Lyso-phosphatidylcholine is a signal in the arbuscular mycorrhizal symbiosis. *Science* 318(5848), 265-268.

Duan, J., Tian, H., Drijber, R. A. and Gao, Y. 2015. Systemic and local regulation of phosphate and nitrogen transporter genes by arbuscular mycorrhizal fungi in roots of winter wheat (*Triticum aestivum* L.). *Plant Physiology and Biochemistry* 96, 199-208.

Duan, T., Facelli, E., Smith, S. E., Smith, F. A. and Nan, Z. 2011. Differential effects of soil disturbance and plant residue retention on function of arbuscular mycorrhizal (AM) symbiosis are not reflected in colonization of roots or hyphal development in soil. *Soil Biology and Biochemistry* 43(3), 571-578.

Eivazi, F. and Weir, C. C. 1989. Phosphorus and mycorrhizal interaction on uptake of P and trace elements by maize. *Fertilizer Research* 21(1), 19-22.

Elliott, A. J., Daniell, T. J., Cameron, D. D. and Field, K. J. 2020. A commercial arbuscular mycorrhizal inoculum increases root colonisation across wheat cultivars but does not increase assimilation of mycorrhiza-acquired nutrients. *Plants, People, Planet* 00, 1-12.

Ellouze, W., Hamel, C., Depauw, R. M., Knox, R. E., Cuthbert, R. D. and Singh, A. K. 2015. Potential to breed for mycorrhizal association in durum wheat. *Canadian Journal of Microbiology* 62(3), 263-271.

Estrada, B., Aroca, R., Barea, J. M. and Ruiz-Lozano, J. M. 2013. Native arbuscular mycorrhizal fungi isolated from a saline habitat improved maize antioxidant systems and plant tolerance to salinity. *Plant Science* 201-202, 42-51.

Faber, B. A., Zasoski, R. J., Burau, R. G. and Uriu, K. 1990. Zinc uptake by corn as affected by vesicular-arbuscular mycorrhizae. *Plant and Soil* 129(2), 121-130.

Fan, J., McConkey, B., Wang, H. and Janzen, H. 2016. Root distribution by depth for temperate agricultural crops. *Field Crops Research* 189, 68-74.

Faye, A., Dalpé, Y., Ndung'u-Magiroi, K., Jefwa, J., Ndoye, I., Diouf, M. and Lesueur, D. 2013. Evaluation of commercial arbuscular mycorrhizal inoculants. *Canadian Journal of Plant Science* 93(6), 1201-1208.

Fears, R., Ter Meulen, V. and Von Braun, J. 2019. Global food and nutrition security needs more and new science. *Science Advances* 5(12), eaba2946.

Feddermann, N., Finlay, R., Boller, T. and Elfstrand, M. 2010. Functional diversity in arbuscular mycorrhiza - the role of gene expression, phosphorous nutrition and symbiotic efficiency. *Fungal Ecology* 3(1), 1-8.

Ferrol, N., Tamayo, E. and Vargas, P. 2016. The heavy metal paradox in arbuscular mycorrhizas: from mechanisms to biotechnological applications. *Journal of Experimental Botany* 67(22), 6253-6265, erw403.

Field, K. J., Bidartondo, M. I., Rimington, W. R., Hoysted, G. A., Beerling, D. J., Cameron, D. D., Duckett, J. G., Leake, J. R. and Pressel, S. 2019. Functional complementarity of ancient plant-fungal mutualisms: contrasting nitrogen, phosphorus and carbon exchanges between Mucoromycotina and Glomeromycotina fungal symbionts of liverworts. *New Phytologist* 223(2), 908-921.

Field, K. J., Cameron, D. D., Leake, J. R., Tille, S., Bidartondo, M. I. and Beerling, D. J. 2012. Contrasting arbuscular mycorrhizal responses of vascular and non-vascular plants to a simulated Palaeozoic CO_2 decline. *Nature Communications* 3, 835.

Field, K. J. and Pressel, S. 2018. Unity in diversity: structural and functional insights into the ancient partnerships between plants and fungi. *New Phytologist* 220(4), 996-1011.

Field, K. J., Pressel, S., Duckett, J. G., Rimington, W. R. and Bidartondo, M. I. 2015a. Symbiotic options for the conquest of land. *Trends in Ecology and Evolution* 30(8), 477-486.

Field, K. J., Rimington, W. R., Bidartondo, M. I., Allinson, K. E., Beerling, D. J., Cameron, D. D., Duckett, J. G., Leake, J. R. and Pressel, S. 2015b. First evidence of mutualism between ancient plant lineages (Haplomitriopsida liverworts) and Mucoromycotina fungi and its response to simulated Palaeozoic changes in atmospheric CO_2. *New Phytologist* 205(2), 743-756.

Field, K. J., Rimington, W. R., Bidartondo, M. I., Allinson, K. E., Beerling, D. J., Cameron, D. D., Duckett, J. G., Leake, J. R. and Pressel, S. 2016. Functional analysis of liverworts in dual symbiosis with Glomeromycota and Mucoromycotina fungi under a simulated Palaeozoic CO_2 decline. *The ISME Journal* 10(6), 1514-1526.

Finney, D. M. and Kaye, J. P. 2017. Functional diversity in cover crop polycultures increases multifunctionality of an agricultural system. *Journal of Applied Ecology* 54(2), 509-517.

Foley, J. A., Ramankutty, N., Brauman, K. A., Cassidy, E. S., Gerber, J. S., Johnston, M., Mueller, N. D., O'Connell, C., Ray, D. K., West, P. C., Balzer, C., Bennett, E. M., Carpenter, S. R., Hill, J., Monfreda, C., Polasky, S., Rockström, J., Sheehan, J., Siebert, S., Tilman, D. and Zaks, D. P. 2011. Solutions for a cultivated planet. *Nature* 478(7369), 337-342.

García de León, D., Davison, J., Moora, M., Öpik, M., Feng, H., Hiiesalu, I., Jairus, T., Koorem, K., Liu, Y., Phosri, C., Sepp, S. K., Vasar, M. and Zobel, M. 2018. Anthropogenic disturbance equalizes diversity levels in arbuscular mycorrhizal fungal communities. *Global Change Biology* 24(6), 2649-2659.

Gavito, M. E., Jakobsen, I., Mikkelsen, T. N. and Mora, F. 2019. Direct evidence for modulation of photosynthesis by an arbuscular mycorrhiza-induced carbon sink strength. *New Phytologist* 223(2), 896-907.

Gehring, C. A., Swaty, R. and Deckert, R. 2017. Mycorrhizas, drought, and host-plant mortality. In: Nancy Collins Johnson, Catherine Gehring and Jan Jansa (Eds), *Mycorrhizal Mediation of Soil*. Elsevier, Amsterdam, Netherlands.

Gibbs, H. K. and Salmon, J. M. 2015. Mapping the world's degraded lands. *Applied Geography* 57, 12-21.

Giovannetti, M., Fortuna, P., Citernesi, A. S., Morini, S. and Nuti, M. P. 2001. The occurrence of anastomosis formation and nuclear exchange in intact arbuscular mycorrhizal networks. *New Phytologist* 151(3), 717-724.

Godfray, H. C. J., Beddington, J. R., Crute, I. R., Haddad, L., Lawrence, D., Muir, J. F., Pretty, J., Robinson, S., Thomas, S. M. and Toulmin, C. 2010. Food security: the challenge of feeding 9 billion people. *Science* 327(5967), 812-818.

Goucher, L., Bruce, R., Cameron, D. D., Lenny Koh, S. C. and Horton, P. 2017. The environmental impact of fertilizer embodied in a wheat-to-bread supply chain. *Nature Plants* 3, 17012.

Gutjahr, C., Casieri, L. and Paszkowski, U. 2009. *Glomus intraradices* induces changes in root system architecture of rice independently of common symbiosis signaling. *New Phytologist* 182(4), 829-837.

Hamel, C. 1996. Prospects and problems pertaining to the management of arbuscular mycorrhizae in agriculture. *Agriculture, Ecosystems and Environment* 60(2-3), 197-210.

Hart, M. M., Antunes, P. M. and Abbott, L. K. 2017. Unknown risks to soil biodiversity from commercial fungal inoculants. *Nature Ecology and Evolution* 1(4), 0115.

Hart, M. M., Antunes, P. M., Chaudhary, V. B. and Abbott, L. K. 2018. Fungal inoculants in the field: is the reward greater than the risk? *Functional Ecology* 32(1), 126-135.

Hartley, S. E. and Gange, A. C. 2009. Impacts of plant symbiotic fungi on insect herbivores: mutualism in a multitrophic context. *Annual Review of Entomology* 54, 323-342.

Hassan, S. E. D., Boon, E., St-Arnaud, M. and Hijri, M. 2011. Molecular biodiversity of arbuscular mycorrhizal fungi in trace metal-polluted soils. *Molecular Ecology* 20(16), 3469-3483.

Hassan, S. E. D., Liu, A., Bittman, S., Forge, T. A., Hunt, D. E., Hijri, M. and St-Arnaud, M. 2013. Impact of 12-year field treatments with organic and inorganic fertilizers on crop productivity and mycorrhizal community structure. *Biology and Fertility of Soils* 49(8), 1109-1121.

Helgason, T., Daniell, T. J., Husband, R., Fitter, A. H. and Young, J. P. 1998. Ploughing up the wood-wide web? *Nature* 394(6692), 431.

Hetrick, B. A. D., Wilson, G. W. T. and Todd, T. C. 1996. Mycorrhizal response in wheat cultivars: relationship to phosphorus. *Canadian Journal of Botany* 74(1), 19-25.

Hijri, M. 2016. Analysis of a large dataset of mycorrhiza inoculation field trials on potato shows highly significant increases in yield. *Mycorrhiza* 26(3), 209-214.

Hodge, A., Campbell, C. D. and Fitter, A. H. 2001. An arbuscular mycorrhizal fungus accelerates decomposition and acquires nitrogen directly from organic material. *Nature* 413(6853), 297-299.

Hodge, A. and Fitter, A. H. 2010. Substantial nitrogen acquisition by arbuscular mycorrhizal fungi from organic material has implications for N cycling. *Proceedings of the National Academy of Sciences of the United States of America* 107(31), 13754-13759.

Hodge, A. and Storer, K. 2015. Arbuscular mycorrhiza and nitrogen: implications for individual plants through to ecosystems. *Plant and Soil* 386(1-2), 1-19.

Holford, I. C. R. 1997. Soil phosphorus: its measurement, and its uptake by plants. *Soil Research* 35(2), 227-240.

Hoysted, G. A., Jacob, A. S., Kowal, J., Gieseman, P., Bidartondo, M. I., Duckett, J. G., Gebauer, G., Rimington, W. R., Schornack, S., Pressel, S. and Field, K. J. 2019. Mucoromycotina fine root endophyte fungi form nutritional mutualisms with vascular plants. *bioRxiv*, 531103.

Hoysted, G. A., Kowal, J., Jacob, A., Rimington, W. R., Duckett, J. G., Pressel, S., Orchard, S., Ryan, M. H., Field, K. J. and Bidartondo, M. I. 2018. A mycorrhizal revolution. *Current Opinion in Plant Biology* 44, 1-6.

Huang, R., Li, Z., Mao, C., Zhang, H., Sun, Z., Li, H., Huang, C., Feng, Y., Shen, X., Bucher, M., Zhang, Z., Lin, Y., Cao, Y. and Duanmu, D. 2020. Natural variation at Os CERK 1 regulates arbuscular mycorrhizal symbiosis in rice. *New Phytologist* 225(4), 1762-1776.

Igiehon, N. O. and Babalola, O. O. 2017. Biofertilizers and sustainable agriculture: exploring arbuscular mycorrhizal fungi. *Applied Microbiology and Biotechnology* 101(12), 4871-4881.

Jansa, J., Erb, A., Oberholzer, H. R., Smilauer, P. and Egli, S. 2014. Soil and geography are more important determinants of indigenous arbuscular mycorrhizal communities than management practices in Swiss agricultural soils. *Molecular Ecology* 23(8), 2118-2135.

Jansa, J., Mozafar, A. and Frossard, E. 2003. Long-distance transport of P and Zn through the hyphae of an arbuscular mycorrhizal fungus in symbiosis with maize. *Agronomie* 23(5-6), 481-488.

Jiang, S., Liu, Y., Luo, J., Qin, M., Johnson, N. C., Öpik, M., Vasar, M., Chai, Y., Zhou, X., Mao, L., Du, G., An, L. and Feng, H. 2018. Dynamics of arbuscular mycorrhizal fungal community structure and functioning along a nitrogen enrichment gradient in an alpine meadow ecosystem. *New Phytologist* 220(4), 1222-1235.

Jin, H., Germida, J. J. and Walley, F. L. 2013. Impact of arbuscular mycorrhizal fungal inoculants on subsequent arbuscular mycorrhizal fungi colonization in pot-cultured field pea (*Pisum sativum* L.). *Mycorrhiza* 23(1), 45-59.

Johnson, N. C. 2010. Resource stoichiometry elucidates the structure and function of arbuscular mycorrhizas across scales. *New Phytologist* 185(3), 631-647.

Johnson, N. C., Graham, J. H. and Smith, F. A. 1997. Functioning of mycorrhizal associations along the mutualism-parasitism continuum. *New Phytologist* 135(4), 575-585.

Johnson, N. C., Wilson, G. W. T., Wilson, J. A., Miller, R. M. and Bowker, M. A. 2015. Mycorrhizal phenotypes and the Law of the Minimum. *New Phytologist* 205(4), 1473-1484.

Jung, J. K. H. M. and McCouch, S. R. M. 2013. Getting to the roots of it: genetic and hormonal control of root architecture. *Frontiers in Plant Science* 4, 186.

Jung, S. C., Martinez-Medina, A., Lopez-Raez, J. A. and Pozo, M. J. 2012. Mycorrhiza-induced resistance and priming of plant defenses. *Journal of Chemical Ecology* 38(6), 651-664.

Karandashov, V. and Bucher, M. 2005. Symbiotic phosphate transport in arbuscular mycorrhizas. *Trends in Plant Science* 10(1), 22-29.

Klironomos, J. N. 2003. Variation in plant response to native and exotic arbuscular mycorrhizal fungi. *Ecology* 84(9), 2292-2301.

Klironomos, J. N., Allen, M. F., Rillig, M. C., Piotrowski, J., Makvandi-Nejad, S., Wolfe, B. E. and Powell, J. R. 2005. Abrupt rise in atmospheric CO_2 overestimates community response in a model plant-soil system. *Nature* 433(7026), 621-624.

Kobae, Y., Ohmori, Y., Saito, C., Yano, K., Ohtomo, R. and Fujiwara, T. 2016. Phosphate treatment strongly inhibits new arbuscule development but not the maintenance of arbuscule in mycorrhizal rice roots. *Plant Physiology* 171(1), 566–579.

Koch, A. M., Antunes, P. M., Barto, E. K., Cipollini, D., Mummey, D. L. and Klironomos, J. N. 2011. The effects of arbuscular mycorrhizal (AM) fungal and garlic mustard introductions on native AM fungal diversity. *Biological Invasions* 13(7), 1627–1639.

Koch, A. M., Antunes, P. M., Maherali, H., Hart, M. M. and Klironomos, J. N. 2017. Evolutionary asymmetry in the arbuscular mycorrhizal symbiosis: conservatism in fungal morphology does not predict host plant growth. *New Phytologist* 214(3), 1330–1337.

Köhl, L., Lukasiewicz, C. E. and van der Heijden, M. G. 2016. Establishment and effectiveness of inoculated arbuscular mycorrhizal fungi in agricultural soils. *Plant, Cell and Environment* 39(1), 136–146.

Kothari, S. K., Marschner, H. and Römheld, V. 1990. Direct and indirect effects of VA mycorrhizal fungi and rhizosphere microorganisms on acquisition of mineral nutrients by maize (*Zea mays* L.) in a calcareous soil. *New Phytologist* 116(4), 637–645.

Kucey, R. M. N. and Janzen, H. H. 1987. Effects of VAM and reduced nutrient availability on growth and phosphorus and micronutrient uptake of wheat and field beans under greenhouse conditions. *Plant and Soil* 104(1), 71–78.

Lambert, D. H. and Weidensaul, T. C. 1991. Element uptake by mycorrhizal soybean from sewage-sludge-treated soil. *Soil Science Society of America Journal* 55(2), 393–398.

Lee, Y.-J. and George, E. 2005. Contribution of mycorrhizal hyphae to the uptake of metal cations by cucumber plants at two levels of phosphorus supply. *Plant and Soil* 278(1–2), 361–370.

Lefebvre, B. 2019. An opportunity to breed rice for improved benefits from the arbuscular mycorrhizal symbiosis? *New Phytologist* 225(4), 1404–1406.

Lehmann, A. and Rillig, M. C. 2015. Arbuscular mycorrhizal contribution to copper, manganese and iron nutrient concentrations in crops – a meta-analysis. *Soil Biology and Biochemistry* 81, 147–158.

Lehnert, H., Serfling, A., Enders, M., Friedt, W. and Ordon, F. 2017. Genetics of mycorrhizal symbiosis in winter wheat (*Triticum aestivum*). *New Phytologist* 215(2), 779–791.

Lehnert, H., Serfling, A., Friedt, W. and Ordon, F. 2018. Genome-wide association studies reveal genomic regions associated with the response of wheat (*Triticum aestivum* L.) to mycorrhizae under drought stress conditions. *Frontiers in Plant Science* 9, 1728.

Leigh, J., Hodge, A. and Fitter, A. H. 2009. Arbuscular mycorrhizal fungi can transfer substantial amounts of nitrogen to their host plant from organic material. *New Phytologist* 181(1), 199–207.

Lekberg, Y. and Helgason, T. 2018. *In situ* mycorrhizal function – knowledge gaps and future directions. *New Phytologist* 220(4), 957–962.

Lekberg, Y. and Koide, R. T. 2005. Is plant performance limited by abundance of arbuscular mycorrhizal fungi? A meta-analysis of studies published between 1988 and 2003. *New Phytologist* 168(1), 189–204.

Lekberg, Y., Koide, R. T., Rohr, J. R., Aldrich-Wolfe, L. and Morton, J. B. 2007. Role of niche restrictions and dispersal in the composition of arbuscular mycorrhizal fungal communities. *Journal of Ecology* 95(1), 95–105.

Lesk, C., Rowhani, P. and Ramankutty, N. 2016. Influence of extreme weather disasters on global crop production. *Nature* 529(7584), 84–87.

Li, H., Smith, S. E., Holloway, R. E., Zhu, Y. and Smith, F. A. 2006. Arbuscular mycorrhizal fungi contribute to phosphorus uptake by wheat grown in a phosphorus-fixing soil even in the absence of positive growth responses. *New Phytologist* 172(3), 536-543.

Li, H. Y., Smith, S. E., Ophel-Keller, K., Holloway, R. E. and Smith, F. A. 2008. Naturally occurring arbuscular mycorrhizal fungi can replace direct P uptake by wheat when roots cannot access added P fertiliser. *Functional Plant Biology* 35(2), 124-130.

Li, X.-L., Marschner, H. and George, E. 1991. Acquisition of phosphorus and copper by VA-mycorrhizal hyphae and root-to-shoot transport in white clover. *Plant and Soil* 136(1), 49-57.

Li, X. X., Zeng, R. S. and Liao, H. 2016. Improving crop nutrient efficiency through root architecture modifications. *Journal of Integrative Plant Biology* 58(3), 193-202.

Lipper, L., Thornton, P., Campbell, B. M., Baedeker, T., Braimoh, A., Bwalya, M., Caron, P., Cattaneo, A., Garrity, D., Henry, K., Hottle, R., Jackson, L., Jarvis, A., Kossam, F., Mann, W., McCarthy, N., Meybeck, A., Neufeldt, H., Remington, T., Sen, P. T., Sessa, R., Shula, R., Tibu, A. and Torquebiau, E. F. 2014. Climate-smart agriculture for food security. *Nature Climate Change* 4(12), 1068-1072.

Liu, A., Hamel, C., Hamilton, R. I., Ma, B. L. and Smith, D. L. 2000. Acquisition of Cu, Zn, Mn and Fe by mycorrhizal maize (*Zea mays* L.) grown in soil at different P and micronutrient levels. *Mycorrhiza* 9(6), 331-336.

Liu, C.-Y., Huang, Y.-M., Zou, Y.-N. and Wu, Q.-S. 2014a. Regulation of root length and lateral root number in trifoliate orange applied by peroxide hydrogen and arbuscular mycorrhizal fungi. *Notulae Botanicae Horti Agrobotanici Cluj-Napoca* 42(1), 94-98.

Liu, W., Jiang, S., Zhang, Y., Yue, S., Christie, P., Murray, P. J., Li, X. and Zhang, J. 2014b. Spatiotemporal changes in arbuscular mycorrhizal fungal communities under different nitrogen inputs over a 5-year period in intensive agricultural ecosystems on the North China Plain. *FEMS Microbiology Ecology* 90(2), 436-453.

Luginbuehl, L. H., Menard, G. N., Kurup, S., Van Erp, H., Radhakrishnan, G. V., Breakspear, A., Oldroyd, G. E. D. and Eastmond, P. J. 2017. Fatty acids in arbuscular mycorrhizal fungi are synthesized by the host plant. *Science* 356(6343), 1175-1178.

Lynch, J. P. 2013. Steep, cheap and deep: an ideotype to optimize water and N acquisition by maize root systems. *Annals of Botany* 112(2), 347-357.

Lynch, J. P. and Wojciechowski, T. 2015. Opportunities and challenges in the subsoil: pathways to deeper rooted crops. *Journal of Experimental Botany* 66(8), 2199-2210.

Malusá, E., Sas-Paszt, L. and Ciesielska, J. 2012. Technologies for beneficial microorganisms inocula used as biofertilizers. *The Scientific World Journal* 2012, 491206.

Manjunath, A. and Habte, M. 1988. Development of vesicular-arbuscular mycorrhizal infection and the uptake of immobile nutrients in *Leucaena leucocephala*. *Plant and Soil* 106(1), 97-103.

Mardhiah, U., Caruso, T., Gurnell, A. and Rillig, M. C. 2016. Arbuscular mycorrhizal fungal hyphae reduce soil erosion by surface water flow in a greenhouse experiment. *Applied Soil Ecology* 99, 137-140.

Marschner, H. 2011. *Mineral Nutrition of Higher Plants*. Academic Press, London.

Martin, S. L., Mooney, S. J., Dickinson, M. J. and West, H. M. 2012. The effects of simultaneous root colonisation by three *Glomus* species on soil pore characteristics. *Soil Biology and Biochemistry* 49, 167-173.

McGonigle, T. P. and Fitter, A. H. 1990. Ecological specificity of vesicular-arbuscular mycorrhizal associations. *Mycological Research* 94(1), 120-122.

Meinshausen, M., Smith, S. J., Calvin, K., Daniel, J. S., Kainuma, M. L. T., Lamarque, J. F., Matsumoto, K., Montzka, S. A., Raper, S. C. B., Riahi, K., Thomson, A., Velders, G. J. M. and Van Vuuren, D. P. P. 2011. The RCP greenhouse gas concentrations and their extensions from 1765 to 2300. *Climatic Change* 109(1-2), 213-241.

Mensah, J. A., Koch, A. M., Antunes, P. M., Kiers, E. T., Hart, M. and Bucking, H. 2015. High functional diversity within species of arbuscular mycorrhizal fungi is associated with differences in phosphate and nitrogen uptake and fungal phosphate metabolism. *Mycorrhiza* 25(7), 533-546.

Middleton, E. L., Richardson, S., Koziol, L., Palmer, C. E., Yermakov, Z., Henning, J. A., Schultz, P. A. and Bever, J. D. 2015. Locally adapted arbuscular mycorrhizal fungi improve vigor and resistance to herbivory of native prairie plant species. *Ecosphere* 6(12), 1-16.

Miransari, M., Bahrami, H. A., Rejali, F. and Malakouti, M. J. 2008. Using arbuscular mycorrhiza to alleviate the stress of soil compaction on wheat (*Triticum aestivum* L.) growth. *Soil Biology and Biochemistry* 40(5), 1197-1206.

Miransari, M., Bahrami, H. A., Rejali, F., Malakouti, M. J. and Torabi, H. 2007. Using arbuscular mycorrhiza to reduce the stressful effects of soil compaction on corn (*Zea mays* L.) growth. *Soil Biology and Biochemistry* 39(8), 2014-2026.

Mueller, U. G. and Sachs, J. L. 2015. Engineering microbiomes to improve plant and animal health. *Trends in Microbiology* 23(10), 606-617.

Mummey, D. L., Antunes, P. M. and Rillig, M. C. 2009. Arbuscular mycorrhizal fungi pre-inoculant identity determines community composition in roots. *Soil Biology and Biochemistry* 41(6), 1173-1179.

Munkvold, L., Kjoller, R., Vestberg, M., Rosendahl, S. and Jakobsen, I. 2004. High functional diversity within species of arbuscular mycorrhizal fungi. *New Phytologist* 164(2), 357-364.

Nagy, R., Drissner, D., Amrhein, N., Jakobsen, I. and Bucher, M. 2009. Mycorrhizal phosphate uptake pathway in tomato is phosphorus-repressible and transcriptionally regulated. *New Phytologist* 181(4), 950-959.

Nogales, A., Nobre, T., Valadas, V., Ragonezi, C., Döring, M., Polidoros, A. and Arnholdt-Schmitt, B. 2015. Can functional hologenomics aid tackling current challenges in plant breeding? *Briefings in Functional Genomics* 15(4), 288-297.

Oehl, F., Laczko, E., Bogenrieder, A., Stahr, K., Bosch, R., van der Heijden, M. and Sieverding, E. 2010. Soil type and land use intensity determine the composition of arbuscular mycorrhizal fungal communities. *Soil Biology and Biochemistry* 42(5), 724-738.

Oláh, B., Brière, C., Bécard, G., Dénarié, J. and Gough, C. 2005. Nod factors and a diffusible factor from arbuscular mycorrhizal fungi stimulate lateral root formation in *Medicago truncatula* via the DMI1/DMI2 signalling pathway. *The Plant Journal* 44(2), 195-207.

Oldeman, L. R. 1994. The global extent of soil degradation. In: D. J. Greenland and I. Szabolcs (Eds), *Soil Resilience and Sustainable Land Use*. CAB International, Wallingford.

Oliveira, R. S., Rocha, I., Ma, Y., Vosátka, M. and Freitas, H. 2016. Seed coating with arbuscular mycorrhizal fungi as an ecotechnological approach for sustainable agricultural production of common wheat (*Triticum aestivum* L.). *Journal of Toxicology and Environmental Health, Part A* 79(7), 329-337.

Olsson, P. A., Bååth, E. and Jakobsen, I. 1997. Phosphorus effects on the mycelium and storage structures of an arbuscular mycorrhizal fungus as studied in the soil and roots by analysis of fatty acid signatures. *Applied and Environmental Microbiology* 63(9), 3531-3538.

Olsson, P. A., Thingstrup, I., Jakobsen, I. and Bååth, E. 1999. Estimation of the biomass of arbuscular mycorrhizal fungi in a linseed field. *Soil Biology and Biochemistry* 31(13), 1879-1887.

Orchard, S., Hilton, S., Bending, G. D., Dickie, I. A., Standish, R. J., Gleeson, D. B., Jeffery, R. P., Powell, J. R., Walker, C., Bass, D., Monk, J., Simonin, A. and Ryan, M. H. 2017. Fine endophytes (*Glomus tenue*) are related to Mucoromycotina, not Glomeromycota. *New Phytologist* 213(2), 481-486.

Parniske, M. 2008. Arbuscular mycorrhiza: the mother of plant root endosymbioses. *Nature Reviews. Microbiology* 6(10), 763-775.

Paszkowski, U. and Gutjahr, C. 2013. Multiple control levels of root system remodeling in arbuscular mycorrhizal symbiosis. *Frontiers in Plant Science*, 4, 204.

Pellegrino, E., Bedini, S., Avio, L., Bonari, E. and Giovannetti, M. 2011. Field inoculation effectiveness of native and exotic arbuscular mycorrhizal fungi in a Mediterranean agricultural soil. *Soil Biology and Biochemistry* 43(2), 367-376.

Pellegrino, E., Öpik, M., Bonari, E. and Ercoli, L. 2015. Responses of wheat to arbuscular mycorrhizal fungi: a meta-analysis of field studies from 1975 to 2013. *Soil Biology and Biochemistry* 84, 210-217.

Péret, B., Desnos, T., Jost, R., Kanno, S., Berkowitz, O. and Nussaume, L. 2014. Root architecture responses: in search of phosphate. *Plant Physiology* 166(4), 1713-1723.

Peyret-Guzzon, M., Stockinger, H., Bouffaud, M. L., Farcy, P., Wipf, D. and Redecker, D. 2016. Arbuscular mycorrhizal fungal communities and *Rhizophagus irregularis* populations shift in response to short-term ploughing and fertilisation in a buffer strip. *Mycorrhiza* 26(1), 33-46.

Philippot, L., Raaijmakers, J. M., Lemanceau, P. and van der Putten, W. H. 2013. Going back to the roots: the microbial ecology of the rhizosphere. *Nature Reviews. Microbiology* 11(11), 789-799.

Phillips, M. L., Weber, S. E., Andrews, L. V., Aronson, E. L., Allen, M. F. and Allen, E. B. 2019. Fungal community assembly in soils and roots under plant invasion and nitrogen deposition. *Fungal Ecology* 40, 107-117.

Ranaldo, M., Carlesi, S., Costanzo, A. and Bàrberi, P. 2020. Functional diversity of cover crop mixtures enhances biomass yield and weed suppression in a Mediterranean agroecosystem. *Weed Research* 60(1), 96-108.

Read, D. J. 1991. Mycorrhizas in ecosystems. *Experientia* 47(4), 376-391.

Read, D. J. and Perez-Moreno, J. 2003. Mycorrhizas and nutrient cycling in ecosystems - a journey towards relevance? *New Phytologist* 157(3), 475-492.

Rillig, M. C., Aguilar-Trigueros, C. A., Camenzind, T., Cavagnaro, T. R., Degrune, F., Hohmann, P., Lammel, D. R., Mansour, I., Roy, J., van der Heijden, M. G. A. and Yang, G. W. 2019. Why farmers should manage the arbuscular mycorrhizal symbiosis. *New Phytologist* 222(3), 1171-1175.

Rimington, W. R., Pressel, S., Field, K. J., Strullu-Derrien, C., Duckett, J. G. and Bidartondo, M. I. 2016. Reappraising the origin of mycorrhizas. In: Francis Martin (Ed.), *Molecular Mycorrhizal Symbiosis*. John Wiley & Sons, Hoboken, New Jersey, USA, pp. 31-32.

Rodriguez, A. and Sanders, I. R. 2015. The role of community and population ecology in applying mycorrhizal fungi for improved food security. *The ISME Journal* 9(5), 1053–1061.

Rogers, E. D. and Benfey, P. N. 2015. Regulation of plant root system architecture: implications for crop advancement. *Current Opinion in Biotechnology* 32, 93–98.

Rose, T. J., Impa, S. M., Rose, M. T., Pariasca-Tanaka, J., Mori, A., Heuer, S., Johnson-Beebout, S. E. and Wissuwa, M. 2012. Enhancing phosphorus and zinc acquisition efficiency in rice: a critical review of root traits and their potential utility in rice breeding. *Annals of Botany* 112(2), 331–345.

Rosendahl, S. 2008. Communities, populations and individuals of arbuscular mycorrhizal fungi. *New Phytologist* 178(2), 253–266.

Ruiz-Lozano, J. M., Aroca, R., Zamarreño, Á. M., Molina, S., Andreo-Jiménez, B., Porcel, R., García-Mina, J. M., Ruyter-Spira, C. and López-Ráez, J. A. 2016. Arbuscular mycorrhizal symbiosis induces strigolactone biosynthesis under drought and improves drought tolerance in lettuce and tomato. *Plant, Cell and Environment* 39(2), 441–452.

Ryan, M. H. and Angus, J. F. 2003. Arbuscular mycorrhizae in wheat and field pea crops on a low P soil: increased Zn-uptake but no increase in P-uptake or yield. *Plant and Soil* 250(2), 225–239.

Ryan, M. H. and Graham, J. H. 2002. Is there a role for arbuscular mycorrhizal fungi in production agriculture? *Plant and Soil* 244, 263–271.

Ryan, M. H. and Graham, J. H. 2018. Little evidence that farmers should consider abundance or diversity of arbuscular mycorrhizal fungi when managing crops. *New Phytologist* 220(4), 1092–1107.

Ryan, M. H., Graham, J. H., Morton, J. B. and Kirkegaard, J. A. 2019. Research must use a systems agronomy approach if management of the arbuscular mycorrhizal symbiosis is to contribute to sustainable intensification. *New Phytologist* 222(3), 1176–1178.

Ryan, M. H., Van Herwaarden, A. F., Angus, J. F. and Kirkegaard, J. A. 2005. Reduced growth of autumn-sown wheat in a low-P soil is associated with high colonisation by arbuscular mycorrhizal fungi. *Plant and Soil* 270(1), 275–286.

Sato, T., Ezawa, T., Cheng, W. G. and Tawaraya, K. 2015. Release of acid phosphatase from extraradical hyphae of arbuscular mycorrhizal fungus *Rhizophagus clarus*. *Soil Science and Plant Nutrition* 61(2), 269–274.

Sawers, R. J., Svane, S. F., Quan, C., Grønlund, M., Wozniak, B., Gebreselassie, M. N., González-Muñoz, E., Chávez Montes, R. A., Baxter, I., Goudet, J., Jakobsen, I. and Paszkowski, U. 2017. Phosphorus acquisition efficiency in arbuscular mycorrhizal maize is correlated with the abundance of root-external hyphae and the accumulation of transcripts encoding PHT1 phosphate transporters. *New Phytologist* 214(2), 632–643.

Schlaeppi, K. and Bulgarelli, D. 2015. The plant microbiome at work. *Molecular Plant-Microbe Interactions* 28(3), 212–217.

Schnoor, T. K., Lekberg, Y., Rosendahl, S. and Olsson, P. A. 2011. Mechanical soil disturbance as a determinant of arbuscular mycorrhizal fungal communities in semi-natural grassland. *Mycorrhiza* 21(3), 211–220.

Schüßler, A. and Walker, C. 2010. *The Glomeromycota: A Species List With New Families and New Genera*. The Royal Botanic Garden Kew, Botanische Staatssammlung Munich, and Oregon State University 19.

Schwartz, M. W., Hoeksema, J. D., Gehring, C. A., Johnson, N. C., Klironomos, J. N., Abbott, L. K. and Pringle, A. 2006. The promise and the potential consequences of the global transport of mycorrhizal fungal inoculum. *Ecology Letters* 9(5), 501–515.

Séry, D. J., Kouadjo, Z. G., Voko, B. R. and Zézé, A. 2016. Selecting native arbuscular mycorrhizal fungi to promote cassava growth and increase yield under field conditions. *Frontiers in Microbiology* 7, 2063.

Smith, F. A. and Smith, S. E. 2011. What is the significance of the arbuscular mycorrhizal colonisation of many economically important crop plants? *Plant and Soil* 348(1–2), 63–79.

Smith, S. E., Jakobsen, I., Grønlund, M. and Smith, F. A. 2011. Roles of arbuscular mycorrhizas in plant phosphorus nutrition: interactions between pathways of phosphorus uptake in arbuscular mycorrhizal roots have important implications for understanding and manipulating plant phosphorus acquisition. *Plant Physiology* 156(3), 1050–1057.

Smith, S. E. and Read, D. J. 2008. *Mycorrhizal Symbiosis*. Academic Press, London.

Smith, S. E. and Smith, F. A. 2011b. Roles of arbuscular mycorrhizas in plant nutrition and growth: new paradigms from cellular to ecosystem scales. *Annual Review of Plant Biology* 62, 227–250.

Smith, S. E., Smith, F. A. and Jakobsen, I. 2003. Mycorrhizal fungi can dominate phosphate supply to plants irrespective of growth responses. *Plant Physiology* 133(1), 16–20.

Smith, S. E., Smith, F. A. and Jakobsen, I. 2004. Functional diversity in arbuscular mycorrhizal (AM) symbioses: the contribution of the mycorrhizal P uptake pathway is not correlated with mycorrhizal responses in growth or total P uptake. *New Phytologist* 162(2), 511–524.

Sosa-Hernández, M. A., Leifheit, E. F., Ingraffia, R. and Rillig, M. C. 2019. Subsoil arbuscular mycorrhizal fungi for sustainability and climate-smart agriculture: a solution right under our feet? *Frontiers in Microbiology* 10, 744.

Sosa-Hernández, M. A., Roy, J., Hempel, S., Kautz, T., Kopke, U., Uksa, M., Schloter, M., Caruso, T. and Rillig, M. C. 2018. Subsoil arbuscular mycorrhizal fungal communities in arable soil differ from those in topsoil. *Soil Biology and Biochemistry* 117, 83–86.

Stockinger, H., Walker, C. and Schüßler, A. 2009. '*Glomus intraradices* DAOM197198', a model fungus in arbuscular mycorrhiza research, is not *Glomus intraradices*. *New Phytologist* 183(4), 1176–1187.

Sugiura, Y., Akiyama, R., Tanaka, S., Yano, K., Kameoka, H., Kawaguchi, M., Akiyama, K. and Saito, K. 2019. Myristate as a carbon and energy source for the asymbiotic growth of the arbuscular mycorrhizal fungus *Rhizophagus irregularis*. *bioRxiv*, 731489.

Syers, J., Johnston, A. and Curtin, D. 2008. Efficiency of soil and fertilizer phosphorus use. FAO Fertilizer and Plant Nutrition Bulletin 18.

Thirkell, T. J., Cameron, D. D. and Hodge, A. 2016. Resolving the 'nitrogen paradox' of arbuscular mycorrhizas: fertilization with organic matter brings considerable benefits for plant nutrition and growth. *Plant, Cell and Environment* 39(8), 1683–1690.

Thirkell, T. J., Cameron, D. D. and Hodge, A. 2019. Contrasting nitrogen fertilisation rates alter mycorrhizal contribution to barley nutrition in a field trial. *Frontiers in Plant Science* 10, 1312.

Thirkell, T. J., Charters, M. D., Elliott, A. J., Sait, S. M. and Field, K. J. 2017. Are mycorrhizal fungi our sustainable saviours? Considerations for achieving food security. *Journal of Ecology* 105(4), 921–929.

Thompson, J. P. 1990. Soil sterilization methods to show VA-mycorrhizae aid P and Zn nutrition of wheat in vertisols. *Soil Biology and Biochemistry* 22(2), 229-240.

Tian, H., Drijber, R. A., Zhang, J. L. and Li, X. L. 2013. Impact of long-term nitrogen fertilization and rotation with soybean on the diversity and phosphorus metabolism of indigenous arbuscular mycorrhizal fungi within the roots of maize (*Zea mays* L.). *Agriculture, Ecosystems and Environment* 164, 53-61.

Tian, H., Yuan, X., Duan, J., Li, W., Zhai, B. and Gao, Y. 2017. Influence of nutrient signals and carbon allocation on the expression of phosphate and nitrogen transporter genes in winter wheat (*Triticum aestivum* L.) roots colonized by arbuscular mycorrhizal fungi. *PLoS ONE* 12(2), e0172154.

Tilman, D., Cassman, K. G., Matson, P. A., Naylor, R. and Polasky, S. 2002. Agricultural sustainability and intensive production practices. *Nature* 418(6898), 671-677.

Tinker, P., Jones, M. and Durall, D. 1992. A functional comparison of ecto- and endomycorrhizas. In: D. J. Read, D. Lewis, A. Fitter and I. J. Alexander (Eds), *Mycorrhizas in Ecosystems*. CAB International, Wallingford, UK, pp. 303-310.

Tinker, P. B. and Nye, P. H. 2000. *Solute Movement in the Rhizosphere*. Oxford University Press, Oxford, UK.

Tracy, S. R., Black, C. R., Roberts, J. A., McNeill, A., Davidson, R., Tester, M., Samec, M., Korošak, D., Sturrock, C. and Mooney, S. J. 2012a. Quantifying the effect of soil compaction on three varieties of wheat (*Triticum aestivum* L.) using X-ray micro computed tomography (CT). *Plant and Soil* 353(1-2), 195-208.

Tracy, S. R., Black, C. R., Roberts, J. A., Sturrock, C., Mairhofer, S., Craigon, J. and Mooney, S. J. 2012b. Quantifying the impact of soil compaction on root system architecture in tomato (*Solanum lycopersicum*) by X-ray micro-computed tomography. *Annals of Botany* 110(2), 511-519.

Valentine, T. A., Hallett, P. D., Binnie, K., Young, M. W., Squire, G. R., Hawes, C. and Bengough, A. G. 2012. Soil strength and macropore volume limit root elongation rates in many UK agricultural soils. *Annals of Botany* 110(2), 259-270.

van der Heijden, M. G. A., Klironomos, J. N., Ursic, M., Moutoglis, P., Streitwolf-Engel, R., Boller, T., Wiemken, A. and Sanders, I. R. 1998. Mycorrhizal fungal diversity determines plant biodiversity, ecosystem variability and productivity. *Nature* 396(6706), 69-72.

Vance, C. P., Uhde-Stone, C. and Allan, D. L. 2003. Phosphorus acquisition and use: critical adaptations by plants for securing a nonrenewable resource. *New Phytologist* 157(3), 423-447.

Verbruggen, E., van der Heijden, M. G. A., Weedon, J. T., Kowalchuk, G. A. and Roling, W. F. M. 2012. Community assembly, species richness and nestedness of arbuscular mycorrhizal fungi in agricultural soils. *Molecular Ecology* 21(10), 2341-2353.

Vosátka, M., Látr, A., Gianinazzi, S. and Albrechtová, J. 2012. Development of arbuscular mycorrhizal biotechnology and industry: current achievements and bottlenecks. *Symbiosis* 58(1-3), 29-37.

Walker, C., Gollotte, A. and Redecker, D. 2018. A new genus, *Planticonsortium* (Mucoromycotina), and new combination (*P. tenue*), for the fine root endophyte, *Glomus tenue* (basionym *Rhizophagus tenuis*). *Mycorrhiza* 28(3), 213-219.

Wasson, A. P., Richards, R. A., Chatrath, R., Misra, S. C., Prasad, S. V., Rebetzke, G. J., Kirkegaard, J. A., Christopher, J. and Watt, M. 2012. Traits and selection strategies to improve root systems and water uptake in water-limited wheat crops. *Journal of Experimental Botany* 63(9), 3485-3498.

Watts-Williams, S. J., Emmett, B. D., Levesque-Tremblay, V., MacLean, A. M., Sun, X. P., Satterlee, J. W., Fei, Z. J. and Harrison, M. J. 2019a. Diverse *Sorghum bicolor* accessions show marked variation in growth and transcriptional responses to arbuscular mycorrhizal fungi. *Plant, Cell and Environment* 42(5), 1758–1774.

Watts-Williams, S. J. and Gilbert, S. E. 2020. Arbuscular mycorrhizal fungi affect the concentration and distribution of nutrients in the grain differently in barley compared with wheat. *Plants, People, Planet* 00, 1–11.

Watts-Williams, S. J., Jewell, N., Brien, C., Berger, B., Garnett, T. and Cavagnaro, T. R. 2019b. Using high-throughput phenotyping to explore growth responses to mycorrhizal fungi and zinc in three plant species. *Plant Phenomics* 2019, 5893953.

Watts-Williams, S. J., Patti, A. F. and Cavagnaro, T. R. 2013. Arbuscular mycorrhizas are beneficial under both deficient and toxic soil zinc conditions. *Plant and Soil* 371(1–2), 299–312.

White, P. J., George, T. S., Gregory, P. J., Bengough, A. G., Hallett, P. D. and McKenzie, B. M. 2013. Matching roots to their environment. *Annals of Botany* 112(2), 207–222.

Williams, A., Börjesson, G. and Hedlund, K. 2013. The effects of 55 years of different inorganic fertiliser regimes on soil properties and microbial community composition. *Soil Biology and Biochemistry* 67, 41–46.

Williams, A., Manoharan, L., Rosenstock, N. P., Olsson, P. A. and Hedlund, K. 2017. Long-term agricultural fertilization alters arbuscular mycorrhizal fungal community composition and barley (*Hordeum vulgare*) mycorrhizal carbon and phosphorus exchange. *New Phytologist* 213(2), 874–885.

Wiseman, P., Colvin, K. H. and Wells, C. E. 2009. Performance of mycorrhizal products marketed for woody landscape plants. *Journal of Environmental Horticulture* 27(1), 41–50.

Wu, Q.-S., Li, G.-H. and Zou, Y.-N. 2011. Improvement of root system architecture in peach (*Prunus persica*) seedlings by arbuscular mycorrhizal fungi, related to allocation of glucose/sucrose to root. *Notulae Botanicae Horti Agrobotanici Cluj-Napoca* 39(2), 232–236.

Yang, H., Zhang, Q., Dai, Y., Liu, Q., Tang, J., Bian, X. and Chen, X. 2015. Effects of arbuscular mycorrhizal fungi on plant growth depend on root system: a meta-analysis. *Plant and Soil* 389(1–2), 361–374.

Yang, S. Y., Grønlund, M., Jakobsen, I., Grotemeyer, M. S., Rentsch, D., Miyao, A., Hirochika, H., Kumar, C. S., Sundaresan, V., Salamin, N., Catausan, S., Mattes, N., Heuer, S. and Paszkowski, U. 2012. Nonredundant regulation of rice arbuscular mycorrhizal symbiosis by two members of the phosphate transporter 1 gene family. *The Plant Cell* 24(10), 4236–4251.

Yao, Q., Wang, L. R., Zhu, H. H. and Chen, J. Z. 2009. Effect of arbuscular mycorrhizal fungal inoculation on root system architecture of trifoliate orange (*Poncirus trifoliata* L. Raf.) seedlings. *Scientia Horticulturae* 121(4), 458–461.

Zhang, S., Lehmann, A., Zheng, W., You, Z. and Rillig, M. C. 2019. Arbuscular mycorrhizal fungi increase grain yields: a meta-analysis. *New Phytologist* 222(1), 543–555.

Zhao, H., Li, X., Zhang, Z., Zhao, Y., Yang, J. and Zhu, Y. 2017. Species diversity and drivers of arbuscular mycorrhizal fungal communities in a semi-arid mountain in China. *PeerJ* 5, e4155.

Zhu, Y. G., Smith, F. A. and Smith, S. E. 2003. Phosphorus efficiencies and responses of barley (*Hordeum vulgare* L.) to arbuscular mycorrhizal fungi grown in highly calcareous soil. *Mycorrhiza* 13(2), 93–100.

Chapter 17

Using systems agronomy to exploit deep roots in crops

John Kirkegaard and Julianne Lilley, CSIRO Agriculture and Food, Australia

1 Introduction

The impetus for this book, and the wonderful science outlined in the preceding chapters highlight the global interest in exploiting improved root systems for the sustainable intensification of agriculture. Improvements in root measurement techniques, understanding of root and rhizosphere interactions, and the genetic control of root growth and development signal new opportunities to exploit improved root systems in agriculture. However, despite the enormous promise, demonstrated examples of novel varieties or management systems that have specifically emerged from improvements in crop root systems, particularly rooting depth or architecture, are relatively rare. In wheat, as in many other crops, there are certainly clear examples of specific genetic resistances to biotic and abiotic stresses such as salt (James et al.

http://dx.doi.org/10.19103/AS.2020.0075.24

2012) and acid soil tolerance (Pooniya et al. 2020) as well as resistance to soil-borne pathogens such as cereal cyst nematodes (Ogbonnaya et al. 2001) that have all been transformational. There are also some notable examples in crop improvement approaches such as the work to improve beans in Latin America (Lynch 2019), rice in Asia (Wissuwa et al. 2016) and maize in different areas (Reyes et al. 2015), and these examples, if anything, reinforce the challenges and long-time horizons needed to translate the promise of good fundamental science to demonstrable outcomes to improve root systems in farmers' fields. Roots are certainly difficult to study, but another reason for this nexus can be that those engaged in fundamental root science can often be isolated either institutionally or culturally from those with an understanding of the context of farming systems in which improved root function must operate (Thorup-Kristensen and Kirkegaard 2016). Better connection between fundamental root science, breeders and systems agronomists is likely to improve and hasten the translation of promising new root science insights for sustainable intensification of agriculture.

In this chapter we highlight an example of such an approach using a specific case study that evaluated and then designed strategies to capture benefits from deeper rooting in the semi-arid southern Australian wheat belt. The research work described has spanned more than 15 years, but was initially stimulated by an even earlier observation that wheat crops were maturing early from water stress despite apparently available water remaining in the root zone (Passioura 1983). Although the research involved took many turns, the theme of better capturing deep subsoil water with deeper and more effective root systems remained as the constant theme – and achieving that within the context of farming systems in southern Australian wheat systems was the challenge. The national scale of the potential benefits revealed recently (Hunt et al. 2019) makes it worthwhile to record and consider the importance and effectiveness of the systems agronomy approach taken in the research.

2 Wheat-based farming systems in south-eastern Australia – a brief overview

The recent evolution of southern Australian dryland farming systems has been comprehensively reviewed by Kirkegaard et al. (2011) and Kirkegaard and van Rees (2019), but specific background for the region of focus in this chapter is necessary for context. The focus for the research outlined is in the dryland mixed-farming area of southern New South Wales (NSW), in the south-eastern Wheatbelt region of Australia (Fig. 1). The area comprises around 2 million ha of cropland that has been cleared and farmed for more than 100 years. It is characterised by low hills in the east grading to riverine plains in the west. The soils in the east consist of acidic loamy topsoils grading into clay

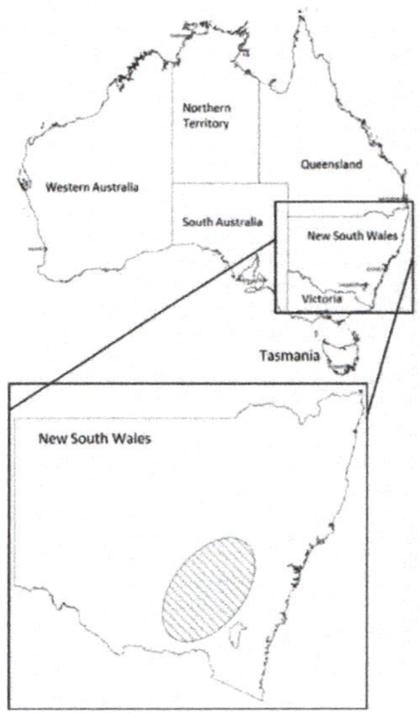

Characteristics of the Red Kandosol (Isbell 2002) at the experimental site

Horizon	Depth (m)	pH ($CaCl_2$)	EC (1:5) dS/m	Texture[A]	Bulk density (g/cm^3)
A11	0-0.10	5.12	0.47	SL	1.45
A12	0.10-0.15	4.28	0.14	SL	1.50
B21	0.15-0.40	5.61	0.04	LC	1.65
B22	0.40-0.70	5.91	0.03	LMC	1.63
B31	0.70-1.10	5.55	0.03	LMC	1.63
B32	1.10-1.30	6.47	0.09	LMC	1.60
D1	1.30-1.70	7.28	0.15	LMC	1.66
D2	1.70-2.00	7.66	0.11	LMC	1.65

[A] SL – silty clay; LC light clay; LMC light medium clay

Figure 1 The area of the southern New South Wales (NSW) wheatbelt comprises around 2 million ha of red Kandosol and Chromosol soils which have a dense, structured and gradational soil profile, acidic surface layers, but no chemical constraints in the subsoil, and typical spring wheat rooting depths of up to 1.8 m.

at depth (Kandosols), while soils on the plains are generally Chromosols and Sodosols (Isbell 2002) with loamy topsoil and a distinct texture contrast with lower horizons. Here we focus on the red Kandosols and Chromosols, which are typical of a significant proportion of the farmed area in the region, and which generally have no serious chemical constraints to crop root growth to depths of 1.5–2 m, but have high bulk density and soil strength below the cultivated layer (Fig. 1). The soils are predominately acidic in the surface and are either naturally deficient in, or depleted of major crop nutrients (N and P). The average annual rainfall is evenly distributed throughout the year and varies from an average of 650 mm in the east (elevation 500–600 m asl) to less than 400 mm on the plains (100 m asl), but is extremely variable by world standards (monthly rainfall can vary from <10 mm to > 150 mm throughout the year). The eastern part has undulating topography with both cropped areas and areas of stony and acid soils unsuitable for cropping that have served to maintain mixed farming systems where livestock graze non-arable areas, while on the plains to the west, more intensively cropped systems and larger farms lend themselves to economies of scale.

The farming systems comprise annual winter crops (3–5 years) phased with annual or perennial pastures (2–5 years) grazed by livestock. Although rainfall is evenly distributed (~40 mm per month), high temperatures and vapour pressure deficit generally preclude summer crops, and fields are generally retained fallow during the summer (December–April). Traditionally, spring-type crops are sown in May with varieties selected to flower after the last damaging frosts in spring (September), but before the onset of summer heat (November). In recent years, autumn rainfall has declined and spring temperatures have increased, both of which can reduce crop yield potential (Hochman et al. 2017). Wheat is the dominant crop grown in the area (~80%) and until the 1980s was rotated with annual grass-subterranean clover pastures. Broadleaf break crops remained a relatively low proportion of the cropping systems until the 1990s when adapted canola varieties (*Brassica napus*) became a significant component of the cropping system (Kirkegaard et al. 2016), usually following an application of lime. A range of other grain legumes are grown on much smaller areas.

Adoption of no-till and stubble retention increased during the 1990s as herbicides, rotation crops and improved machinery became available and most crops are now grown using no-till, stubble-retained systems with only strategic cultivation to periodically incorporate lime and control problem weeds. A general trend towards higher cropping intensity and shorter pasture phases observed nationally was also evident in southern NSW, in keeping with the trends in net farm worth (Kirkegaard et al. 2011). The prolonged dry period from 2002 to 2008 (now known as the Millennium drought) heralded a significant shift in the farming system with a refocus on water-use efficiency and an intensification of cereal production.

A strong focus in Australian broad-acre systems has been to capture, store and use rainfall as efficiently as possible to support crop growth, while managing inputs (including N) to achieve the water-limited productivity potential and quality requirements (Kirkegaard et al. 2014; Sadras et al. 2019). Benchmarking production against the potential to stimulate improvement has been a powerful tool at both the crop (French and Schultz 1984) and the farm level (Cary and Wilkinson 1994). Inadequate nitrogen supply remains one of the major current causes of yield gaps in southern Australian systems (Hochman and Horan 2018) as N application can be risky in seasons with limited spring rainfall due to 'haying-off' (van Herwaarden et al. 1998). On average, N application rates are low (approx. 40 kg ha^{-1} N), but reliance on fertiliser N has increased as pasture area has declined (Angus 2001).

As Australia emerged from the Millennium drought (1998–2008) there was a sharpened focus on strategies to utilise stored soil water more effectively. Increasing the capture of deep soil water using management and/or new genotypes with greater root vigour had been an active area of research for decades in Australia (O'Brien 1979; Richards and Passioura 1989; Tennant and Hall 2001) and elsewhere (Hurd 1968, 1974). However, the usefulness of increased rooting depth or other root traits is largely determined by the patterns of water availability as determined by seasonal rainfall distribution, soil type and crop management in specific environments. It was in this systems context, with a 'Genotype × Environment × Management' approach, that the research on ways to better capture subsoil water with improved root systems commenced.

3 Unused subsoil water

Passioura (1983) in his review of roots and drought resistance stimulated the impetus for this work with the observation that despite being severely water stressed, many droughted crops leave substantial amounts of apparently available water in the subsoil at maturity. He went on to consider the reasons for this conundrum by dissecting and analysing the issue with the help of his identity:

$$\text{yield} = \text{water used} \times \text{water use efficiency} \times \text{harvest index}$$

He used the identity to consider the ways in which root growth and shoot growth might be manipulated to improve the outcome in water-limited crops. He raised many questions pertinent to the work described in this chapter. Do the roots have insufficient time to extract the water? Are there too few, or too sparse roots clumped in macropores to extract the water effectively? Would investing in more roots come at the metabolic cost to shoot growth and possibly yield? Would extracting deep soil water more effectively come at a cost during grain-filling, or even impact on subsequent crops if the soil profile fails to recharge? A large part of the focus in his paper was on the potential to breed plants with root or shoot characteristics better able to capture water to improve water-limited yield, and he rightly pointed out that 'it is important to consider the long-term worth of putting a lot of effort into producing plants that can extract the last few centimetres of available water from the bottom of their root zones'.

We have explored many of these questions during the last 15 years at sites on acidic loam soils in southern NSW, Australia (Fig. 1) using a series of experiments and associated simulation modelling that has featured purpose-built rainout shelters, deep soil trenches for direct root observations, and by manipulating a range of soil and crop factors in the field. Kirkegaard and Lilley (2007) reviewed data on wheat rooting depth at 36 sites in southern NSW and found maximum rooting depths of spring wheat crops sown in mid-May on typical red acidic loam Kandosols was 1.8 m, with root penetration rates of 1.2 cm per day from sowing to anthesis. The major factors limiting rooting depth on specific soils were the depth of wetting and the sowing time. In order to determine the potential value of stored subsoil water on these soils, Kirkegaard et al. (2007) established an experiment with wheat using a rainout shelter and trickle irrigation with treatments wet to either 1.35 m or to 1.85 m where the latter had an extra 32 mm of plant available water (PAW) below 1.35 m (Fig. 2).

Rainout shelters excluded all rain from anthesis to maturity so that growth and yield relied on stored water alone. The treatment with access to the deep water utilised an extra 11mm of the 32mm available and yielded 0.6 t/ha more, representing a marginal water-use efficiency of the subsoil water of 60kg/ha/mm, around 3 times the value predicted for total seasonal water use (French and Schultz 1984). The experiment confirmed the failure of the crop to extract

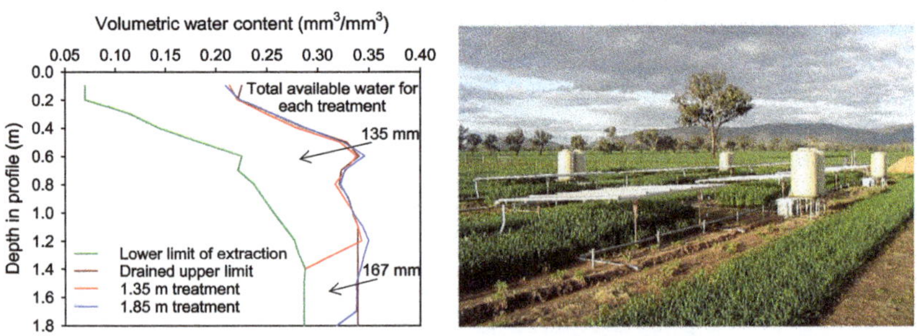

Figure 2 Wheat established under automatic rainout shelters with soil profiles wet to either 1.35 or 1.85 m providing an extra 32 mm of plant available water (PAW) in the subsoil during a terminal stress. The crop used 11 mm of the 32 mm available and yielded an extra 0.6 t/ha (from Kirkegaard et al. 2007).

all of the available subsoil water despite significant water stress, as 21 mm of apparently available water remained within the rooting depth. The seasonal conditions created, comprising complete absence of rainfall after anthesis and a full profile of water at sowing are in reality a rare occurrence, and the average value of the subsoil water across a range of seasons was of interest. Lilley and Kirkegaard (2007) used a simulation model to explore the impact of different seasonal rainfall distribution on the value of stored subsoil water. The simulation study suggested an average value of water stored below 1.3 m of 35 kg/ha/mm (range 0–100 kg/ha/mm), a value recently confirmed in a series of field experiments on the same soils in the region (Flohr et al. 2020). The study also suggested that deep stored water is often more valuable (in terms of t/ha of grain produced) in wetter seasons when it helps to achieve the higher yield potential that is set, rather than in dry seasons, when yield potential is low and the yield benefits are smaller. Other simulation studies on different soils and sites have also confirmed the significant value of stored subsoil water for grain yield (e.g. Manschadi et al. 2006).

Confirming the value of subsoil water on the red loam soils in southern NSW stimulated significant subsequent work to explore both the genetic and management options to capture more water from the subsoil.

4 Improving root systems to capture subsoil water

If breeders are to screen for more effective root systems, or agronomists to design management strategies to achieve more effective deeper rooting, a critical root length density (RLD) for water uptake must be known - partly to ensure that improvement in roots at depth will be effective, but also to avoid investment in too many subsoil roots, which may come at a carbon cost in developing reproductive sinks (i.e. grain). Passioura (1983) used a minimum RLD of 0.1 cm/cm^3 in his consideration of factors influencing water uptake, although the many unknowns in field soils made such an assumption difficult. In order to measure the potential capacity of deep roots to extract soil water, several experimental conditions must be met. First the roots must have an adequately wet soil profile to the full rootable depth so that no dry layers impede rooting descent or proliferation within subsoil layers; secondly the crop must be healthy (including root health) and managed with good agronomy to maximise the transpiration demand of the environment; and lastly there must be a sufficient period of water deficit generated to dry the surface layers so that the deepest roots are called upon to satisfy transpiration demand. Crops under terminal stress are known to preferentially dry the upper layers first (Dardanelli et al. 2004). We used a series of field experiments conducted with the rainout shelters shown in Fig. 2 that satisfied these criteria combined with frequent monitoring of soil water use and RLD measurements (as described in Kirkegaard et al. 2007) to identify critical RLD targets for effective subsoil water uptake in wheat. The subsoil RLD below 1 m depth varied from 1.0 ± 0.1 cm/cm^3 at 1 m depth to 0.01 ± 0.01 cm/cm^3 in the deepest layers (1.8 m). In all experiments, water uptake from the deepest layers equivalent to 10% PAW (1–2 mm from the deepest 0.1 m layer) was associated with measured RLD of between 0.02 and 0.05 cm/cm^3, while RLD of 0.05–0.1 cm/cm^3 were associated with water uptake of 2.5–5 mm, representing up to 50% of the PAW in the layer (Fig. 3). From this data we can conclude that wheat growing under terminal stress conditions on these soils in south-eastern Australia can extract significant water from the deepest subsoil layers when the measured RLD was as low as 0.05 ± 0.01 cm/cm^3. These low levels of RLD measured by root washing and image analysis are often undetectable with other methods such as core break (Wasson et al. 2014) and suggest that any detectable roots measured by currently available means are likely to be valuable for water uptake in seasons where it is required.

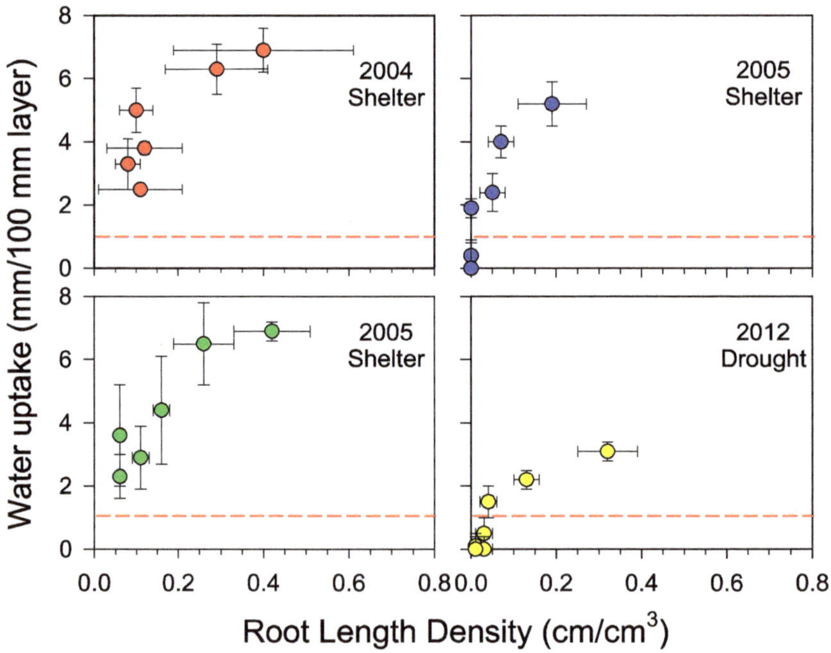

Figure 3 The relationship between root length density (RLD) and water uptake (mm/0.1 m layer) at the deepest layers of root occupancy in four experiments on wheat conducted with full starting profiles and under complete terminal stress under rainout shelters or natural drought at the Bethungra site in southern NSW. Horizontal (RLD) and vertical (water uptake) error bars represent SEM for individual data points. The red horizontal line depicts 1 mm (or 10%) of available water in a layer.

5 Increasing the depth of roots

The evidence provided above suggests that very low lengths of root in the subsoil can be valuable for crop growth. The question arises as to whether the slowing and cessation of downward growth post anthesis is related to soil-based constraints, or to plant-driven processes related to the greater sink strength of the developing reproductive organs. For wheat, most research suggests downward root growth ceases at or around anthesis despite no apparent soil constraints to growth (Kirkegaard and Lilley 2007; Thorup-Kristensen et al. 2009). Understanding the factors that may limit deeper root growth is important to target strategies for improvement. Based on the review of rooting depth on these soils (Kirkegaard and Lilley 2007) and the experiments mentioned above, the maximum rooting depth for spring wheat sown in May was around 1.8m - yet in many cases there appeared to be no specific soil impediments to deeper rooting below that depth. The subsoil was certainly dense at the site (Fig. 1) and the soil strength was extremely high. Indeed, the study of White and Kirkegaard (2010) showed that more than 90% of the roots below 30 cm were found in cracks or pores, which others have suggested will be the norm in subsoils due to the high strength generated by the soil overburden (Goa et al. 2016). However, White and Kirkegaard (2010) found that only 5% of the existing subsoil biopores were occupied by roots, suggesting there was significant scope for increased root growth in the remaining subsoil biopores, and for increased root growth at depth. This appeared to suggest that it was likely to be plant-based factors rather than soil-based factors that were limiting deeper root growth at the site.

To answer this question, we manipulated a range of agronomic and crop factors to ascertain the potential to increase the depth and proliferation of wheat roots in the dense structured clay subsoils. Evidence that plant-based, rather than soil-based constraints were limiting the depth and proliferation of wheat roots was provided in four ways as described below.

5.1 Wheat with a longer vegetative stage has deeper roots and increased subsoil water uptake

In their review of wheat rooting depth in southern NSW, Kirkegaard and Lilley (2007) found that earlier sowing with slower maturing wheat that extended the vegetative stage increased rooting depth although this evidence came from a survey of different experiments rather than a side-by-side comparison. In order to investigate this further we compared the rooting depth and subsoil water use of a winter wheat (cv Maroombi) sown early (4 April) with the spring wheat (cv Janz) sown in May in 2006. The soil was pre-wet to 2.5 m using trickle irrigation and both crops were managed to avoid nutritional or other

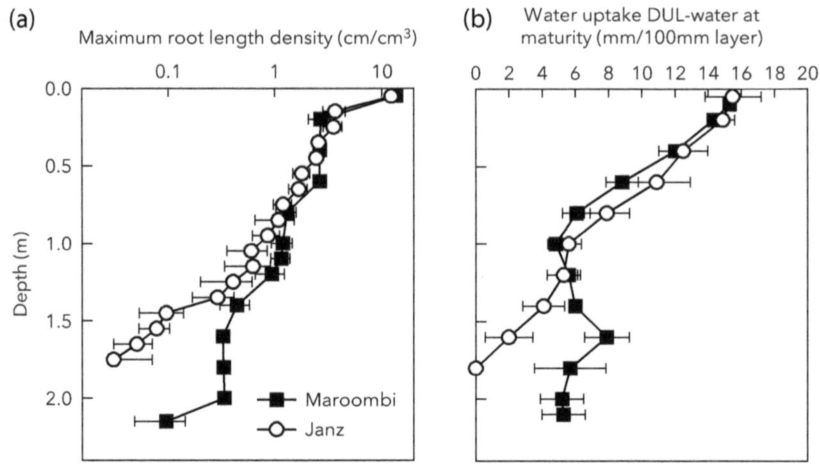

Figure 4 Rooting depth (a) and water uptake (b) for a May-sown spring wheat (cv Janz) and an early-April sown winter wheat (cv Maroombi) at Bethungra in 2006.

agronomic constraints. Drought conditions between flag leaf emergence and ear emergence ensured deep roots would be active in water uptake. Root length and soil water was monitored to a depth of 2.2 m using soil coring and neutron moisture meters. By August 10 (128 days after sowing) the winter wheat had reached a depth of 1.5 m (1.29 cm/day) and 2 weeks after anthesis roots had reached 2.2 m (1.3 cm/day) with no further change in root depth at maturity. The RLD in the winter wheat was higher than the spring wheat from a depth of 0.5 m and the maximum depth of rooting was 2.2 m compared to 1.8 m for the spring wheat. By harvest, both crops had extracted all of the PAW to around 1.2 m depth, but the winter wheat extracted significantly more water to the full depth of measurement (Fig. 4b). These data confirm that there were no specific soil constraints limiting deeper rooting in wheat, but that rooting depth was limited by the length of the vegetative phase of the faster maturing spring wheat. This difference in root depth between winter and spring wheat has been observed in other countries on different soil types (Thorup-Kristensen et al. 2009).

5.2 Higher seeding density can increase RLD at depth

Given that the seminal roots are generally the deepest roots for wheat grown in semiarid dryland environments (Watt et al. 2008), we hypothesised that increasing seed density should also increase the number of seminal roots, and consequently, the rooting density and potential water uptake from depth. Given the large number of unoccupied subsoil biopores, and presuming no other soil constraints, higher plant density should lead to greater RLD at depth. To test

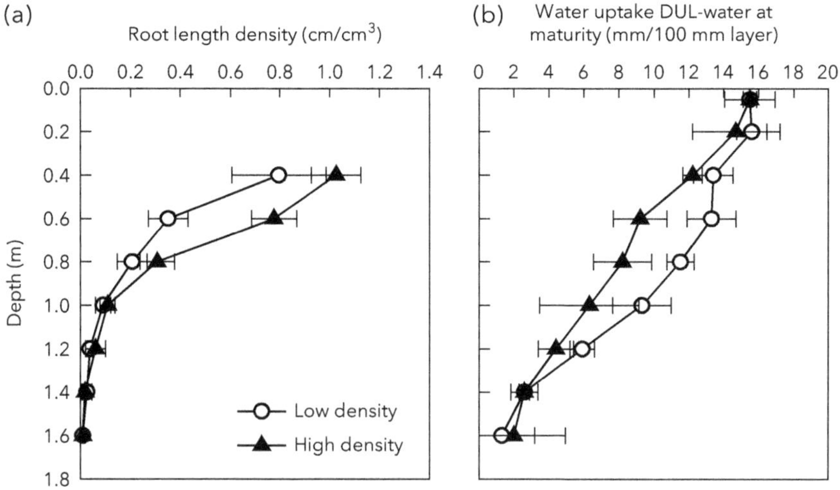

Figure 5 Increase in the root length density (a) and water uptake (b) in wheat under low (90 plants/m^2) and high (270 plants/m^2) planting density (note medium density (150 plants/m^2) was intermediate).

that hypothesis, we sowed the spring wheat (cv Janz) on 9 June at seeding rates to achieve target plant populations of 90, 180 and 270 plants/m^2. We measured RLD and soil water content using soil cores removed to 1.6 m at anthesis (16 October) and maturity (23 November) in order to assess the impact of plant density on subsoil root growth and water use during the post-anthesis period. Pre-wetting the soil to 1.6 m followed by a very dry season ensured significant demand for stored subsoil water. At anthesis, the roots had reached a depth of 1.6 m in all treatments (root penetration rate of 1.24 cm/d) and RLD increased with increased sowing density in the 0.6–1.2 m layers in the profile (Fig. 5). The mean RLDs in the 0.6–1.2 m layer were 0.34, 0.25 and 0.17 cm/cm^3 in the high, medium and low-density treatments, respectively. At anthesis there was no difference in PAW in the profile among the treatments (94 mm PAW remaining), and by maturity all treatments had dried the soil to wilting point to 0.6 m, but the high seed-density treatment had dried the soil profile to a much greater extent in the 0.7–1.2 m layer. The differences in the water uptake were most evident in the high and low-density treatments in the 0.6–1.0 m layer (Fig. 5).

5.3 Impact of ear removal on root growth in the post-anthesis period

The developing ear in a wheat plant becomes the major sink for assimilate in the post-anthesis period and it is estimated that carbon flow to the root system declines from 50% during the vegetative period to less than 5% in the post-anthesis period (Keith et al. 1986; Gregory and Atwell 1991). Removing the wheat ear at anthesis causes assimilate to be redirected to other organs including existing stems, and new shoot and root growth (King et al. 1967; Austin and Edrich 1975; Sharma-Natu and Ghildiyal 1994). The capacity of the wheat plant to increase root growth following ear removal at anthesis would demonstrate that for intact plants, internal regulation of assimilate distribution rather than soil or root factors per se, limit root growth during the post-anthesis period. We expanded the investigations from 2005 reported in Richards et al. (2007) in which RLD at maturity increased 10-fold from 0.03 to 0.3 cm/cm^3 in the 1.2–1.5 m layer in a wheat cultivar (H45) when the ear was removed at anthesis. In a subsequent experiment conducted in 2006, RLD increased at all depths when the ears were removed with increases of 0.3–1.0 cm/cm^3 in the 1.0–1.4 m layer in cultivar H45 during that season. The work provided further evidence that internal plant regulation, rather than soil conditions per se, limits root growth in these subsoils.

5.4 Deep ripping to reduce soil density and strength

An experiment investigating deep ripping to around 30 cm had no effect on the density of roots in the subsoil at the site, nor on growth or yield (data not shown). This was consistent with the review of deep ripping responses on these highly structured red loam soils in southern NSW over 30 years (Kirkegaard et al. 2008). It seems that although there may be some delay in root penetration in the dense surface soils, more obvious as root deformations in tap-rooted crops such as canola (Lisson et al. 2007), the roots utilise the considerable biopores and other structural features of the soil to penetrate compact layers, and then ramify through the subsoil using existing cracks and biopores (White and Kirkegaard 2010). On these well-structured acidic loam soils, free of significant sodicity, there have been few reports of economic responses to deep-ripping.

Taken together, these experiments support the hypothesis that plant-based, rather than soil-based limits to root growth in the subsoil are operating, and that improvements in rooting depth and density should focus on manipulating the plant, rather than the soil in the case of these highly structured acidic loam soils in southern NSW.

6 Capturing more subsoil water – manipulating the system rather than the soil

6.1 Earlier sowing systems

The potential to better capture subsoil water through the deeper rooting of earlier-sown crops (Fig. 4) seems to suggest a simple solution to the challenge of making better use of subsoil water. But it is important here to distinguish between earlier sowing systems, and an earlier-sown crop. Kirkegaard and Hunt (2010) used a simulation study to provide a framework that emphasised the need to modify several practices within the farming system simultaneously to best capture the potential from earlier-sown crops. The framework demonstrated that improving several aspects of the system as a whole (e.g. long-term soil management, crop rotation, summer fallow management, sowing time and varietal choice), rather than any individual component, were all required to maximise the potential benefits from the deeper rooting and increase subsoil water use by earlier-sown crops (Table 1). For example, simply sowing earlier, with no other changes (i.e. line 4, Table 1) improved yield from a baseline of 1.6 t/ha to 2.10 t/ha, which while significant, fell well short of the 4.01 t/ha average yield achieved when combined with no-till, fallow weed control and a pea break crop.

We briefly discuss how each of these system components can contribute to better water capture by deeper roots and its efficient use in early sowing systems.

6.2 Long-term soil management

Capturing and storing water in the subsoil relies on the maintenance of a stable surface structure to maximise infiltration of rainfall during each rainfall event. The widespread adoption of no-till and stubble-retained farming systems across southern Australia since the 1990s has now reached saturation with a majority of farms practising no-till-stubble retention (Llewellyn and Ouzman 2019). Pasture phases are also an important feature of the farming system that assist in maintaining high soil organic matter levels. Regular liming is required to avoid acidification and it appears that less liming during recent droughts combined with lack of lime incorporation may be increasing sub-surface acidity which is an issue of concern that can limit the rooting depth of susceptible

Table 1 The predicted impact of single and combined agronomic interventions on the predicted average farm wheat yield compared to a baseline scenario[a] of 1.6 t/ha in the Victorian Mallee (adapted from Kirkegaard and Hunt 2010)

	Mean yield (t/ha)	
System change	Single effect	Additive effect
1 No-till	1.84	1.84
2 Fallow weed control	2.37	2.80
3 Pea break crop	1.76	3.45
4 Sow earlier (from 25 April)	2.10	4.01
5 Long coleoptile wheat - sown on 25 April	1.45	4.54

[a] The baseline scenario consisted of burn/cultivate, grazed weedy fallow, continuous wheat, spring wheat sown after 25 May.

crops such as grain legumes and barley. Managing soil chemical and structural fertility with sound practices in the long-term ensures that capture, storage and utilisation of water and its effective storage at depth can be maximised when rainfall permits.

In addition to water infiltration and storage, maintaining good surface structure and avoiding soil compaction will also maximise the capacity of roots to penetrate the soil. Although the natural structural features of the red Kandosol and Chromosol soils in southern NSW are an advantage, excessive heavy tractor wheel traffic during wet conditions, excessive cultivation and lack of soil organic matter can still lead to compacted or hard-setting layers that will impede root growth. The move to wider, lower pressure tyres or tracked vehicles and to controlled traffic farming has reduced the level of compaction risk.

6.3 Crop sequence

The sequence of crops or pastures grown before a wheat crop can influence root growth in the subsoil both directly through impacts on root health, but also indirectly by influencing the residual water and nutrients left in the soil profile which can affect recharging of the soil profile. Angus et al. (2015) have reviewed the impact of different preceding crops on wheat yield, and found wheat after crops other than wheat yielded around 0.6 t/ha more grain than wheat after wheat. Thorup-Kristensen and Kirkegaard (2016) considered in detail how pre-crops and their management could influence the utilisation of deep resources by contrasting wheat farming systems in Denmark and Australia. Despite the seemingly obvious connection between poor root health as a result of increased root and shoot diseases in monocultures compared to improved crop rotations, there are relatively few published studies directly relating soil-borne diseases to root performance and water uptake. A relevant example from the red loam soils in southern NSW was provided by Kirkegaard et al. (1994) in which increased infection by the root rotting fungal disease Take-all (*Gaeumannomyces graminis*) reduced the uptake of water (by 30 mm) and nitrogen (20 kg N/ha) from the 1.2 to 1.8 m subsoil layer in wheat grown after wheat compared to wheat grown after canola. Stubble-borne foliar diseases can also cause reductions in root efficiency as they can reduce root growth as well as the overall plant growth and demand for subsoil resources (Ford et al. 2006).

Indirect effects of crop sequence on root growth and resource capture from the subsoil can also occur as a result of different residual water and nutrient profiles left by various crop and pasture species preceding wheat. Perennial crops or pastures such as lucerne (*Medicago sativa*) can have extreme effects on residual water so that the termination time of the lucerne prior to the wheat can be critical to avoid crop failure due to very low levels of subsoil water (Angus et al. 2000). In general, among annual crops, grain legumes will deplete the soil less than oilseeds (e.g. brassicas) or cereals, but crop duration can alter this general observation. For example, a late-maturing lupin crop can dry the soil more than an earlier maturing mustard crop (Kirkegaard and Ryan 2014), and the seasonal distribution of rainfall also has a significant influence on the residual soil water and N profiles following different crops.

6.4 Summer fallow management

In the mixed farming systems of southern Australia, there is a period of summer fallow following the harvest of crops in early summer (November) until the sowing of subsequent crops in the following autumn (May). Traditionally on mixed farms in the region, pasture availability is low over summer, so to provide a feed source to maintain animals, sheep would typically graze on the stubble post-harvest to capitalise on spilt grain and the relatively new cereal stubble. Weeds that emerged on summer rain during that period were generally considered to provide valuable green forage. As rainfall in the region was equi-seasonal or winter dominant, farmers relied on rainfall through autumn and winter to establish and grow the crops and summer rainfall was considered less important. However as crop yields increased with the adoption of improved tillage and crop sequence (leaving drier profiles at harvest), and the rainfall in autumn declined during the Millennium drought, the value of summer fallow rainfall was re-evaluated. A simulation study by Hunt and Kirkegaard (2011) suggested that with better capture of summer rainfall through strict summer weed control and maintenance of adequate crop stubble, the soil could store an extra 50 mm of water on average worth 1.5 t/ha to following wheat crops. The combination of regular summer rain, loam soils with good water-holding capacity and increasingly dry springs made improved fallow management highly valuable in the area. Subsequently, 9 experiments at 5 sites over 4 years in southern NSW demonstrated that strict weed control led to an average increase in soil storage of 50 mm water and 50 kg/ha N, and an average yield benefit of 1.0 t/ha (Kirkegaard et al. 2014).

Weed control was the major pathway to improved water storage in summer, but there was also interest in stubble management including grazing as growers were concerned about the potential damage done by sheep tramping to surface soil structure and reduced water infiltration. Studies by Bell et al. (2011) and Hunt et al. (2016) demonstrated that there was little impact of grazing stubble by sheep on soil water infiltration or crop yield, provided sheep did not reduce the stubble cover below the threshold of 2–3 t/ha (70% cover) below which the heavy rain can damage surface structure, reduce infiltration and increase run-off. The overall recommendation to maximise water and N during the summer fallow was to have strict fallow weed control (spray 12 days after each germinating rainfall event), and graze stubbles carefully to ensure that at least 2–3 t/ha of stubble cover was retained on the soil. Whole-farm simulation modelling suggested this spray-graze summer fallow strategy was the most economically optimum strategy which maximised water and N during the fallow while providing valuable summer feed for livestock without negative trade-offs (Moore and Hunt 2012).

6.5 Varietal selection and agronomy for earlier sowing

The combined effect of the good pre-crop management strategies listed above was to provide greater opportunities to establish crops earlier. The surface soil tended to have sufficient water for germination more frequently earlier in the season (March or April) compared to situations where summer weeds or low ground cover reduced water storage. However, as a result of the fact that the autumn 'breaking rains' were tending to occur later (Cai et al. 2012; Pook et al. 2009) the majority of Australian wheat varieties being released were faster-developing spring wheats, that could be sown later but still flower at the optimal time to avoid frost and heat. Kirkegaard and Hunt (2010) had predicted a significant yield increase associated with earlier-sown wheat combined with improved pre-crop agronomy, and this prediction was confirmed in experiments in southern NSW (Kirkegaard et al. 2014). Subsequently, Hunt (2017) described the opportunity to develop 'fast winter' wheat with a flexible sowing window, but stable flowering time by virtue of a unique combination of vernalisation and day length-sensitivity genes that had not previously been deployed in Australian wheat cultivars.

A series of papers by Flohr et al. (2017, 2018a,b) tested the flowering stability of these fast winter wheat in southern NSW. The development pattern demonstrated the potential to extend sowing periods while achieving 10–20% higher yields and flowering time stability. The use of phenology isolines, as well as experiments with multiple sowing dates to ensure comparisons of varieties were made at the optimal flowering times, confirmed the yield advantage of earlier-sown wheat overall, although the responses were variable. Recently Flohr et al. (2020) have demonstrated that the variable yield benefit of the early-sown wheat compared to traditional later-sown spring wheat was related to the presence of, and reliance on stored subsoil water. The yield advantage was only observed when deep subsoil water (>1 m depth) was present, and when dry spring conditions meant that the access to deep stored water provided a yield advantage of the earlier-sown wheat. The average marginal water-use efficiency of the water accessed from below 1 m depth was 35 kg grain/ha/mm, similar to that predicted in the earlier study of Lilley and Kirkegaard (2007).

Flohr et al. (2018a) have also considered further potential to capitalise on the G × M opportunities of early sowing systems by considering the interaction between management and a novel long-coleoptile trait in wheat (Rebetzke et al. 2007). The long coleoptile trait provides more successful establishment from earlier sowing of wheat by allowing crops to be sown into moisture that is stored deeper in the soil (~15 cm), beyond the usual maximum depth of seeding for current wheat varieties. Moisture-seeking tynes on seeding equipment can place the seed into moisture allowing earlier sowing opportunities. This novel trait was combined with long-fallowing to investigate the potential beneficial interactions between increased deep-water storage and earlier sowing opportunities with fast winter wheat. At all sites, the highest yielding strategies involved winter wheat with long coleoptiles established on stored subsoil moisture from the previous rotation, and achieved a predicted mean wheat yield increase of 1200 kg/ha or 42% relative to the baseline strategy of later sowing spring wheat. The results show promise for winter wheat with long coleoptiles to widen the sowing window, and remove the reliance on autumn rainfall for early establishment and thus stabilise flowering and maximise yield.

6.6 Grazing wheat - further income from early-sown crops

On mixed farms in south-eastern Australia, early-sown crops also provide opportunities to graze the vegetative biomass prior to stem elongation, providing a second source of income to increase profitability, flexibility and reduce risk in variable climates. A review of dual-purpose wheat indicated that significant grazing can be achieved with no loss in grain yield at yield levels up to 5 t/ha provided grazing ceases prior to stem elongation and sufficient residual biomass is retained (Bell et al. 2014; Bell et al. 2015; Sprague et al. 2018). The grazing value increased with earlier sowing of the crop with an estimated grazing benefit of 200 sheep grazing days/ha for every week earlier the crop could be sown. Given wheat crops are generally grazed during the vegetative stage when roots are descending rapidly into the soil, there was a concern that grazing may reduce rooting depth and limit access to deep soil water in grazed crops. Kirkegaard et al. (2015) showed no effect of grazing on the depth or density of roots at depth, but some reduction in the density of roots in the surface layers, presumably related to reduction in nodal roots as a result of tiller removal by grazing. The early-sown, grazed crops retain the advantage of deeper roots and access to deeper water to increase the potential grain yield while providing an added source of income from the grazed forage, and are a highly profitable component of mixed farming businesses.

7 Genetic improvements to rooting depth and density

Since Passioura's 1983 paper discussed the potential to modify crop roots systems for more effective water capture in semi-arid environments, there has been ongoing interest in the potential to genetically improve the rooting characteristics of wheat to better capture soil resources. Genetic differences in root morphology certainly exist (Gregory 2006) and both field (Christopher et al. 2008; McDonald et al. 2012) and root chamber (Manschadi et al. 2006) experiments have demonstrated that wheat with narrow root angle can develop deeper, denser and more evenly distributed roots to extract more soil water and extend the duration of green leaf area and yield. Genotypic variation in the vigour of spring root systems has also been demonstrated (Richards et al. 2007; Palta and Watt 2009) and considerable effort has been invested in screening large numbers of wheat lines in search of deeper and more extensive root systems (Wasson et al. 2012, 2014). Rich et al. (2016) reported that some Indian-bred lines that had been developed in environments where wheat relied on deep stored water for growth tended to have more extensive root length at depth, but the study emphasised the difficulties in phenotyping for roots in the field where soil type and seasonal conditions interacted strongly with root phenotypes. More recently Li et al. (2019) reported that airborne canopy temperature sensors which were capable of measuring many 1000s of wheat plots in the field showed promising correlations between deep roots, cooler canopy temperatures and high grain yield under terminal drought in Australia. These rapid phenotyping tools may provide a promising new approach to identify useful rooting traits, but the challenge remains to assess the value proposition within the farming system - rather than simply in an individual paddock in a dry year (Lilley and Kirkegaard 2016).

The benefit of improved root traits will be influenced by the pattern of water availability in the target environment as affected by seasonal rainfall distribution, soil type and crop management (Chenu et al. 2011, 2013). As a consequence, crop simulation studies have been used to predict the likely benefits of genetically improved root systems that grow deeper and proliferate more effectively in the soil which in some cases, such as for wheat in the US Great Plains, have shown more promise than various modified shoot traits (Sciarresi et al. 2019). Lilley and Kirkegaard (2011) conducted a simulation analysis investigating the potential benefits of deeper and denser root systems by modifying the rate of root descent and root proliferation. The study showed that increased capture of deep water can occur through selection of cultivars with improved root systems, but the impact on grain yield varied with site and season and interacted strongly with crop management, antecedent soil water content, seasonal rainfall distribution and soil type. On Kandosol soils in southern NSW, benefits of 0.3–0.4 t/ha were predicted from the faster and more efficient root systems and yield reductions were rare, although there was a considerable range in yield effects (−0.1–0.9 t/ha). The yield impacts of different management often exceeded or over-rode those of modified root systems (e.g. sowing date 0.1–0.9 t/ha; preceding crop or pasture 0.5–1.8 t/ha). For example, following lucerne there was little benefit to modified wheat roots systems (<0.1 t/ha) as the soil profile rarely rewet below 1 m. The benefits of deeper roots and the management or genetic strategies to achieve them must therefore be considered within a farming systems context (Lilley and Kirkegaard 2016; Thorup-Kristensen and Kirkegaard 2016), where carry over legacy effects from one season to the next will be important to understand in semi-arid dryland environments.

8 Accounting for legacy effects

The agronomic considerations regarding the benefits of deeper roots and utilisation of deep soil water to this point in the discussion have largely focussed on the agronomy of individual crops. In reality, a crop that accesses more water from the subsoil to gain a yield advantage will leave a legacy of a drier and potentially N-depleted subsoil, which may have an impact on subsequent crops if the distribution of subsequent rainfall means the soil profile fails to refill. Kirkegaard and Ryan (2014) showed that in a series of dry years on Kandosol soils in southern NSW, legacies of drier or N-depleted soil could persist up to 4 years through the crop sequence to influence the yield of subsequent crops. Although such legacies may be rare, they are more likely to persist in the subsoil as it may take some time for deeper soil profiles on soils with high plant-available water capacity to fully recharge. This issue of legacy effects has been investigated in recent studies using cropping systems simulations run continuously without resetting initial conditions, so that carry over effects of one crop to the next in a sequence of crops are captured. Lilley and Kirkegaard (2016) evaluated the importance of these legacy effects on wheat yield in early sowing systems at 8 sites throughout the semi-arid areas of Australia. In comparing the predicted benefits of deeper roots and earlier sowing systems, the study confirmed that previous investigations using simulation with annual resetting have significantly overestimated the benefit of deep roots by ignoring legacy effects. The range in predicted wheat yield benefits from deeper root systems dropped from 0.1–0.6 t/ha to 0–0.2 t/ha due to a legacy of drier soil at subsequent crop sowing. The study also revealed that when accounting for legacy effects, there was no benefit from root systems that were enhanced genetically to be deeper with higher root density on shallow soils (<1 m), and that the benefit of earlier sowing systems was also substantially reduced and unrelated to deep water access.

The study included two sites on deep red Kandosol soils in southern NSW that differed in annual rainfall (Cootamundra 619 mm and Ardlethan 471 mm).

Table 2 The predicted effect of a modified deeper root system (M), earlier sowing (E) and the combination of both (ME) on the change in average rooting depth, water uptake, pre-sowing soil water and yield compared to standard May-sown spring wheat at two sites on red Kandosol soils in southern NSW Australia (annual rainfall: Ardlethan 471 mm, Cootamundra 619 mm). Adapted from Lilley and Kirkegaard (2016)

	Root depth increase (m)			Increase in water uptake (mm)			Change in pre-sow water (mm)			Yield change (t/ha)		
Site	M	E	M+E	M	E	M+E	M	E	M+E	M	E	M+E
Ardlethan	0.13	0.15	0.13	10	26	33	−26	−11	−32	0.15	0.54	0.68
Cootamundra	0.26	0.27	0.32	16	31	45	−17	−7	−23	0.25	0.75	0.97

The soil depth was 2.2 m, and plant available water capacity (PAWC) was 228 mm. Table 2 shows a summary of the mean predicted changes in soil water at sowing and mean yield increase arising from the adoption of wheat varieties with genetically enhanced rooting depth, earlier sowing and the combination of both innovations. At both sites, modified roots and earlier sowing increased mean root depth, and water uptake, and reduced the PAW at sowing of subsequent crops, although in all cases there was a predicted yield benefit. At the drier Ardlethan site, the increases in rooting depth and water uptake were less than at Cootamundra, as was the predicted yield benefit, presumably related to the larger legacy effect and reduced starting PAW at sowing. At both sites, earlier sowing generated a larger yield improvement than modified root systems, while the benefits of the combined innovations were largely additive. Overall this study suggests that these loam soils in the equi-seasonal rainfall zone of southern NSW are a good target for both genetic and management innovations targeting increased capture of subsoil water, and that the increased yields observed in experiments are likely to persist across seasons, presumably by better capturing water that is either leached or evaporated in systems with crops sown in May.

9 Scaling up to whole farm benefits

On large farms in Australia, the benefits of earlier sowing systems at the paddock scale are accompanied by important flow-on benefits at the whole farm level. The large farm size in Australia (average 2000 ha) means that an entire farm-sowing programme may take up to 20 days. The capacity to start sowing some crops early with slow maturing varieties of suitable phenology, combined with improved pre-crop management means that all paddocks on the farm will potentially move into an earlier sowing window, and derive yield benefits accordingly. In a recent study, the benefits of earlier sowing were simulated at the farm level across all wheat-growing areas in Australia to determine the areas where yield benefits were likely, and the predicted benefit at a national scale

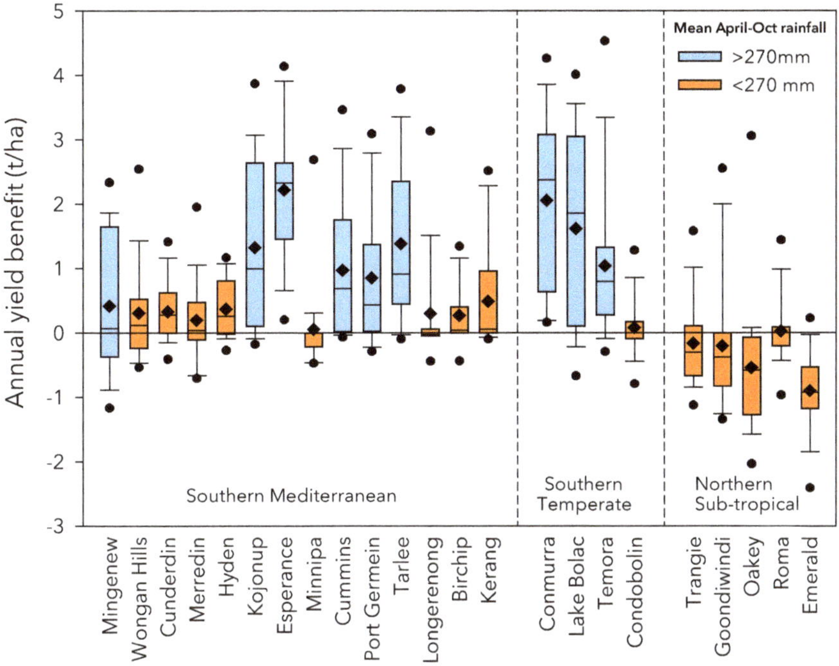

Figure 6 The predicted farm-level wheat yield benefit for a typical 2000 ha Australian farm arising from the integration of early-sown wheat cultivars across a 20-day sowing programme. The benefits are variable but positive in the southern Mediterranean and temperate areas, and negative in northern subtropical areas. In higher rainfall areas greater yield benefits arise due to more earlier sowing opportunities and higher yield potential. Adapted from Hunt et al. (2019).

(Hunt et al. 2019). The study predicted an average wheat yield benefit across Australia of 0.54 t/ha, equivalent to an extra $7 million tonnes of wheat per annum. The benefits were variable across seasons and mostly in the southern temperate and Mediterranean regions, while mean yield losses were predicted in northern, summer-dominant rainfall regions (Fig. 6). Yield benefits were generally increased in areas of high rainfall due to increased chances of early sowing and the higher yield potential, but even in the drier areas whole-farm yield increases were generally > 0.5 t/ha, and predicted yield losses were rare. Importantly the modelling assumes no disease (good rotations), full stubble retention (no-till farming systems) and good agronomy (adequate nutrition, weed and disease control) as essential parts of the early sowing system.

Much of this remarkable benefit from the earlier sowing system has been shown to derive from better access to subsoil water in earlier-sown crops (Flohr et al. 2020) as well as the farm-scale benefits of the entire sowing programme moving into a more favourable sowing window (Fletcher et al. 2019).

10 The future – using more of the soil and more of the season

A focus on the efficient capture, storage and use of rainfall in south-eastern Australia has been sharpened by the recent Millennium drought of 1998–2008 (Pook et al. 2009; Cai et al. 2012) and the prospects of drier and warmer conditions in the future as a result of anthropogenic climate change. Hochman et al. (2017) have shown that climate change has reduced the yield potential by 27% since 1990, while the national average wheat yield has plateaued. This suggests that Australian farmers have adopted significant innovations to offset the recent changes in climate during that period. Indeed Flohr et al. (2018b) report that average sowing dates for wheat have moved back by 1–1.5 days per year and Fletcher et al. (2019) report that up to 50% of Western Australian farmers sow a portion of their crops dry based on calendar date. Thus adaptation using earlier sowing systems is already a reality in Australia. Kirkegaard and Hunt (2010) used simulations restricted to the years of the Millennium drought to demonstrate that the package of agronomic innovations that includes earlier sowing systems would also improve productivity in the face of climate change, and Flohr et al. (2018a) predict that further genetic adaptations including longer coleoptiles may facilitate further improvements.

In most of the continuous simulation studies reported in this chapter the simulation rules were fixed, and a simulation of continuous wheat was assumed – while in reality, farmers can manage the crop sequence and sowing time dynamically, electing to sow crops that have a smaller water requirement following crops and seasons which leave a legacy of dry soil. For example, many legume crops or faster-maturing crops and drought-tolerant crops such as barley (*Hordeum vulgare*) can be chosen for later sowing to follow crops that leave a legacy of dry soil, providing a greater chance for soil profiles to re-fill and to achieve economic yields from less water. Chemical fallows where no crop is grown, but the soil is protected by retained crops residues, are also being reconsidered in southern Australia where very dry soil profiles at the start of the cropping season may dictate a season without a crop. Incorporating fallows can avoid crop losses, save input costs, provide greater timeliness on the remaining areas of large farms, and increase the storage of water and N for the subsequent crop, thus reducing economic risk (Cann et al. 2020; Flohr et al. 2018a). Information from soil water sensors and/or simple calculators of soil water availability are now available to growers via apps (SoilWater App) to assist in decisions regarding sowing according to stored water. Improvements in seasonal weather forecasts may also provide even further opportunities to refine management systems to avoid economic losses from crop and input management choices when soil water availability is low.

Basing sowing decisions and crop choice on stored soil water has always been a feature of the opportunity cropping systems in the north-eastern grain-growing areas in Australia, where sequences of summer crops, winter crops or fallow on deep clay soils with high water-holding capacity (>300 mm) are largely determined by the soil water availability on specific dates. As the seasons become drier and stored soil water makes up a greater proportion of the total water available to crops in southern Australia, similar approaches are beginning to evolve.

11 Conclusions

This chapter has described a body of research that was initially stimulated by the possibility that water remaining in the subsoil of annual wheat crops in southern Australia represented an under-utilised resource that could be captured by deeper root systems to improve productivity. Though initially stimulated by an article that focussed largely on breeding strategies (Passioura 1983), the research uncovered important interactions with management, and, in particular, earlier sowing systems, that to date have proven more effective than breeding to capitalise on opportunities to store, capture and utilise deep stored water more effectively. We have used a combination of field experiments and observations as well as systems simulation to predict and to confirm the benefits of deeper roots, and have answered, at least for this soil and cropping system, many of the original questions posed. Early-sown crops do have sufficient time to proliferate roots in the subsoil at densities that can capture significant amounts of stored water despite root clumping in structured soils, and the water is converted to grain at very high efficiency in seasons when it is called upon. Legacy effects are important, and the benefits of deeper roots are far greater on deeper soils and in wetter conditions where higher yield potentials are realised, and earlier sowing opportunities are more frequent. The predicted benefits of combinations of earlier sowing and genetically faster root growth were generally additive with genetics playing a smaller role. However, it may be that genetically faster root growth may be of greater benefit in later sown crops, where the time for roots to reach a greater soil depth is limited (Rebetzke et al. 2019). Above all, the research has demonstrated the benefits of linking root specialists with agronomists, breeders and farming systems simulators in the quest for impact from root science in farmer's fields.

12 Where to look for further information

The International Society for Root research holds a conference every 4 years and has a website with newsletters, proceedings and recent publications: https://www.rootresearch.org/.

Recent reviews on the topic of deep roots include:

- Lynch, J. P. and Wojciechowski, T. 2015. Opportunities and challenges in the subsoil: Pathways to deeper rooted crops. Journal of Experimental Botany, 66, 2199–2210. doi:10.1093/jxb/eru508.
- Thorup-Kristensen, K., Halberg, N., Nicolaisen, M., Olesen, J. E., Crews, T. E., Hinsinger, P., Kirkegaard, J., Pierret, A. and Dresbøll, D. B. 2020. Digging deeper for agricultural resources, the value of deep rooting. Trends in Plant Science 25, 406-417. doi: 10.1016/j.tplants.2019.12.007.

13 References

Angus, J. F., Gault, R. R., Good, A. J., Hart, A. B., Jones, T. D. and Peoples, M. B. 2000. Lucerne removal before a cropping phase. *Australian Journal of Agricultural Research* 51(7), 877–90. doi:10.1071/AR99183.

Angus, J. F. 2001. Nitrogen supply and demand in Australian agriculture. *Australian Journal of Experimental Agriculture* 41(3), 277–88. doi:10.1071/EA00141.

Angus, J. F., Kirkegaard, J. A., Hunt, J. R., Ryan, M. H., Ohlander, L. and Peoples, M. B. 2015. Break crops and rotations for wheat. *Crop and Pasture Science* 66(6), 523-52. doi:10.1071/CP14252.

Austin, R. B. and Edrich, J. 1975. Effects of ear removal on photosynthesis, carbohydrate accumulation and on the distribution of assimilated ^{14}C in wheat. *Annals of Botany* 39(2), 141–52. doi:10.1093/oxfordjournals.aob.a084922.

Bell, L. W., Kirkegaard, J. A., Swan, A., Hunt, J. R., Huth, N. I. and Fettell, N. A. 2011. Impacts of soil damage by grazing livestock on crop productivity. *Soil and Tillage Research* 113(1), 19–29. doi:10.1016/j.still.2011.02.003.

Bell, L. W., Moore, A. D. and Kirkegaard, J. A. 2014. Evolution in crop-livestock integration systems that improve farm productivity and environmental performance in Australia. *European Journal of Agronomy* 57, 10–20. doi:10.1016/j.eja.2013.04.007.

Bell, L. J., Lilley, J. M., Hunt, J. R. and Kirkegaard, J. A. 2015. Optimising grain yield and grazing potential of crops across Australia's high rainfall zone: a simulation analysis. 1. Wheat. *Crop and Pasture Science* 66, 332–48. doi:10.1071/CP14230.

Cai, W., Cowan, T. and Thatcher, M. 2012. Rainfall reductions over Southern Hemisphere semi-arid regions: the role of subtropical dry zone expansion. *Scientific Reports* 2, 702. doi:10.1038/srep00702.

Cann, D. J., Hunt, J. R. and Malcolm, B. 2020. Long fallows can maintain whole-farm profit and reduce the risk in semi-arid south-eastern Australia. *Agricultural Systems,* 178:102721. doi: 10.1016/j.agsy.2019.102721.

Cary, J. and Wilkinson, R. 1994. *An Evaluation of Farm Management 500: A Program Promoting Better Business Management*. The University of Melbourne, School of Agriculture and Forestry, Melbourne.

Chenu, K., Cooper, M., Hammer, G. L., Mathews, K. L., Dreccer, M. F. and Chapman, S. C. 2011. Environment characterization as an aid to wheat improvement: interpreting genotype-environment interactions by modelling water-deficit patterns in north-eastern Australia. *Journal of Experimental Botany* 62(6), 1743–55. doi:10.1093/jxb/erq459.

Chenu, K., Deihimfard, R. and Chapman, S. C. 2013. Large-scale characterization of drought pattern: a continent-wide modelling approach applied to the Australian wheatbelt--spatial and temporal trends. *New Phytologist* 198(3), 801–20. doi:10.1111/nph.12192.

Christopher, J. T., Manschadi, A. M., Hammer, G. L. and Borrell, A. K. 2008. Developmental and physiological traits associated with high yield and stay-green phenotype in wheat. *Australian Journal of Agricultural Research* 59(4), 354–64. doi:10.1071/AR07193.

Dardanelli, J. L., Ritchie, J. T., Calmon, M., Andriani, J. M. and Collino, D. J. 2004. An empirical model for root water uptake. *Field Crops Research* 87(1), 59–71. doi:10.1016/j.fcr.2003.09.008.

Fletcher, A., Flohr, B. and Harris, F. 2019. Evolution of early sowing systems in southern Australia. In: Pratley, J. and Kirkegaard, J. (Eds), *Australian Agriculture in 2020: From Conservation to Automation*. Agronomy Australia and Charles Sturt University, Wagga Wagga, pp. 47–64.

Flohr, B. M., Hunt, J. R., Kirkegaard, J. A. and Evans, J. R. 2017. Water and temperature stress define the optimal flowering period for wheat in south-eastern Australia. *Field Crops Research* 209, 108–19. doi:10.1016/j.fcr.2017.04.012.

Flohr, B. M., Hunt, J. R., Kirkegaard, J. A., Evans, J. R. and Lilley, J. M. 2018a. Genotype × management strategies to stabilise the flowering time of wheat in the south-eastern Australian wheatbelt. *Crop and Pasture Science* 69(6), 547–60. doi:10.1071/CP18014.

Flohr, B. M., Hunt, J. R., Kirkegaard, J. A., Evans, J. R., Trevaskis, B., Zwart, A., Swan, A., Fletcher, A. L. and Rheinheimer, B. 2018b. Fast winter wheat phenology can stabilise flowering date and maximise grain yield in semi-arid Mediterranean and temperate environments. *Field Crops Research* 223, 12–25. doi:10.1016/j.fcr.2018.03.021.

Flohr, B. M., Hunt, J. R., Kirkegaard, J. A., Trevaskis, B., Rheinheimer, B., Swan, A., Goward, L., Evans, J.R. and Bullock, M. 2020. Deep soil water-use determines the yield benefit of long-cycle wheat. *Frontiers of Plant Science* 11, 548–60. doi: 10.3389/fpls.2020.00548.

Ford, K. E., Gregory, P. J., Gooding, M. J. and Pepler, S. 2006. Genotype and fungicide effects on on late-season root growth of winter wheat. *Plant and Soil* 284(1–2), 33–44. doi:10.1007/s11104-006-0028-x.

French, R. J. and Schultz, J. E. 1984. Water use efficiency of wheat in a Mediterranean-type climate. I. The relation between yield, water use and climate. *Australian Journal of Agricultural Research* 35(6), 743–64. doi:10.1071/AR9840743.

Gao, W., Hodgkinson, L., Jin, K., Watts, C. W., Ashton, R. W., Shen, J., Ren, T., Dodd, I. C., Binley, A., Phillips, A. L., Hedden, P., Hawkesford, M. J. and Whalley, W. R. 2016. Deep roots and soil structure. *Plant, Cell and Environment* 39(8), 1662–8. doi:10.1111/pce.12684.

Gregory, P. J. and Atwell, B. J. 1991. The fate of carbon in pulse-labelled crops of barley and wheat. *Plant and Soil* 136(2), 205–13. doi:10.1007/BF02150051.

Gregory, P. J. 2006. *Plant Roots. Growth, Activity and Interaction with Soils*. Blackwell Publishing Ltd, Oxford.

Hochman, Z., Gobbett, D. L. and Horan, H. 2017. Climate trends account for stalled wheat yields in Australia since 1990. *Global Change Biology* 23(5), 2071–81. doi:10.1111/gcb.13604.

Hochman, Z. and Horan, H. 2018. Causes of wheat yield gaps and opportunities to advance the water-limited yield frontier in Australia. *Field Crops Research* 228, 20–30. doi:10.1016/j.fcr.2018.08.023.

Hunt, J. R. and Kirkegaard, J. A. 2011. Re-evaluating the contribution of summer fallow rain to wheat yield in southern Australia. *Crop and Pasture Science* 62(11), 915–29. doi:10.1071/CP11268.

Hunt, J. R., Swan, A. D., Fettell, N. A., Breust, P. D., Menz, I. D., Peoples, M. B. and Kirkegaard, J. A. 2016. Sheep grazing on crop residues do not reduce crop yields in no-till, controlled traffic farming systems in an equi-seasonal rainfall environment. *Field Crops Research* 196, 22–32. doi:10.1016/j.fcr.2016.05.012.

Hunt, J. R. 2017. Winter wheat cultivars in Australian farming systems: a review. *Crop and Pasture Science* 68(6), 501–15. doi:10.1071/CP17173.

Hunt, J. R., Lilley, J. M., Trevaskis, B., Flohr, B. M., Peake, A., Fletcher, A., Zwart, A. B., Gobbett, D. and Kirkegaard, J. A. 2019. Early sowing systems can boost Australian wheat yields despite recent climate change. *Nature Climate Change* 9(3), 244–7. doi:10.1038/s41558-019-0417-9.

Hurd, E. A. 1968. Growth of roots of 7 varieties of spring wheat at high and low moisture levels. *Agronomy Journal* 60(2), 201–5. doi:10.2134/agronj1968.00021962006000020018x.

Hurd, E. A. 1974. Phenotype and drought tolerance in wheat. *Agricultural Meteorology* 14(1–2), 39–55. doi:10.1016/0002-1571(74)90009-0.

Isbell, R. F. 2002. *The Australian Soil Classification* (revised 1st edn.). CSIRO Publishing, Melbourne.

James, R. A., Blake, C., Zwart, A. B., Hare, R. A., Rathjen, A. J. and Munns, R. 2012. Impact of ancestral wheat sodium exclusion genes *Nax1* and *Nax2* on grain yield of durum wheat on saline soils. *Functional Plant Biology: FPB* 39(7), 609–18. doi:10.1071/FP12121.

Keith, H., Oades, J. M. and Martin, J. K. 1986. Input of carbon to soil from wheat plants. *Soil Biology and Biochemistry* 18(4), 445–9. doi:10.1016/0038-0717(86)90051-9.

King, R. W., Wardlaw, I. F. and Evans, L. T. 1967. Effect of assimilate utilization on photosynthetic rate in wheat. *Planta* 77(3), 261–76. doi:10.1007/BF00385296.

Kirkegaard, J. A., Gardner, P. A., Angus, J. F. and Koetz, E. 1994. Effect of *Brassica* break crops on the growth and yield of wheat. *Australian Journal of Agricultural Research* 45(3), 529–45. doi:10.1071/AR9940529.

Kirkegaard, J. A. and Lilley, J. M. 2007. Root penetration rate – a benchmark to identify soil and plant limitations to rooting depth in wheat. *Australian Journal of Experimental Agriculture* 47(5), 590–602. doi:10.1071/EA06071.

Kirkegaard, J. A., Lilley, J. M., Howe, G. N. and Graham, J. M. 2007. Impact of subsoil water use on wheat yield. *Australian Journal of Agricultural Research* 58(4), 303–15. doi:10.1071/AR06285.

Kirkegaard, J. A., Angus, J., Swan, T., et al. 2008. Ripping yarns: 25 years of variable responses to ripping clay soils in south-eastern Australia. *Proceedings of the 14th Australian Agronomy Conference*, Adelaide, South Australia http://agronomyaustraliaproceedings.org/images/sampledata/2008/concurrent/managing-subsoils/5934_kirkegaardja.pdf.

Kirkegaard, J. A. and Hunt, J. R. 2010. Increasing productivity by matching management and genotype in water-limited environments. *Journal of Experimental Botany* 61(15), 4129–43. doi:10.1093/jxb/erq245.

Kirkegaard, J. A., Peoples, M. B., Angus, J. A. and Unkovich, M. 2011. Diversity and evolution in rain-fed farming systems in southern Australia. In: Tow, P., Cooper, I., Partridge, I. and Birch, C. (Eds), *Rainfed Farming Systems*. Springer, Dordrecht, Netherlands, pp. 715–54.

Kirkegaard, J. A., Hunt, J. R., McBeath, T. M., Lilley, J. M., Moore, A., Verburg, K., Robertson, M., Oliver, Y., Ward, P. R., Milroy, S. and Whitbread, A. M. 2014. Improving water productivity in the Australian Grains industry-a nationally coordinated approach. *Crop and Pasture Science* 65(7), 583-601. doi:10.1071/CP14019.

Kirkegaard, J. A. and Ryan, M. H. 2014. Magnitude and mechanisms of persistent crop sequence effects on wheat. *Field Crops Research* 164, 154–65. doi:10.1016/j.fcr.2014.05.005.

Kirkegaard, J. A., Lilley, J. M., Hunt, J. R., Sprague, S. J., Ytting, N. K., Rasmussen, I. S. and Graham, J. M. 2015. Effect of defoliation by grazing or shoot removal on the root growth of field-grown wheat (*Triticum aestivum*). *Crop and Pasture Science* 66(4), 249–59. doi:10.1071/CP14241.

Kirkegaard, J. A., Lilley, J. M. and Morrison, M. J. 2016. Drivers of trends in canola productivity in Australia. *Crop and Pasture Science* 67, i–ix. doi:10.1071/CPv67n4_FO.

Kirkegaard, J. A. and van Rees, H. 2019. Evolution of conservation agriculture in winter rainfall areas. In: Pratley, J. and Kirkegaard, J. (Eds), *Australian Agriculture in 2020: From Conservation to Automation*. Agronomy Australia and Charles Sturt University, Wagga Wagga, pp. 47–64. Available at: http://agronomyaustraliaproceedings.org/images/sampledata/specialpublications/Australian%20Agriculture%20in%202020.pdf.

Li, X., Ingvordsen, C. H., Weiss, M., Rebetzke, G. J., Condon, A. G., James, R. A. and Richards, R. A. 2019. Deeper roots associated with cooler canopies, higher normalized difference vegetation index, and greater yield in three wheat populations grown on stored soil water. *Journal of Experimental Botany* 70(18), 4963–74. doi:10.1093/jxb/erz232.

Lilley, J. M. and Kirkegaard, J. A. 2007. Seasonal variation in the value of subsoil water to wheat: simulation studies in southern New South Wales. *Australian Journal of Agricultural Research* 58(12), 1115–28. doi:10.1071/AR07046.

Lilley, J. M. and Kirkegaard, J. A. 2011. Benefits of increased soil exploration by wheat roots. *Field Crops Research* 122(2), 118–30. doi:10.1016/j.fcr.2011.03.010.

Lilley, J. M. and Kirkegaard, J. A. 2016. Farming systems context drives the value of deep wheat roots in semi-arid environments. *Journal of Experimental Botany* 67(12), 3665–81. doi:10.1093/jxb/erw093.

Lisson, S. N., Kirkegaard, J. A., Robertson, M. J. and Zwart, A. 2007. Canola yield decline in southern NSW: a diagnosis of causal factors. *Australian Journal of Experimental Agriculture* 47(12), 1435–45. doi:10.1071/EA07041.

Llewellyn, R. and Ouzman, J. 2019. Conservation agriculture in Australia: 30 years on. In: Pratley, J. and Kirkegaard, J. (Eds), *Australian Agriculture in 2020: From Conservation to Automation*. Agronomy Australia and Charles Sturt University, Wagga Wagga, pp. 21-32. Available at: http://agronomyaustraliaproceedings.org/images/sampledata/specialpublications/Australian%20Agriculture%20in%202020.pdf.

Lynch, J. P. 2019. Root phenotypes for improved nutrient capture: an underexploited opportunity for global agriculture. *New Phytologist* 223(2), 548–64. doi:10.1111/nph.15738.

Manschadi, A. M., Christopher, J., deVoil, P. and Hammer, G. L. 2006. The role of root architectural traits in adaptation of wheat to water-limited environments. *Functional Plant Biology* 33(9), 823–37. doi:10.1071/FP06055.

McDonald, G. K.,Taylor, J. D., Verbyla, A. and Kuchel, H. 2012. Assessing the importance of subsoil constraints to yield of wheat and its implications for yield improvement. *Crop and Pasture Science* 63(12), 1043–65. doi:10.1071/CP12244.

Moore, A. D. and Hunt, J. R. 2012. Does control of summer fallow weeds improve whole-farm productivity in both Mediterranean and temperate environments? *Proceedings of 16th Australian Agronomy Conference* Armidale, New South Wale. Available at: http://agronomyaustraliaproceedings.org/images/sampledata/2012/8145_5_moore.pdf.

O'Brien, L. 1979. Genetic variability of root growth in wheat (*Triticum aestivum* L.). *Australian Journal of Agricultural Research* 30(4), 587–95. doi:10.1071/AR9790587.

Ogbonnaya, F. C., Subrahmanyam, N. C., Moullet, O., de Majnik, Jd, Eagles, H. A., Brown, J. S., Eastwood, R. F., Kollmorgen, J., Appels, R. and Lagudah, E. S. 2001. Diagnostic DNA markers for cereal cyst nematode resistance in bread wheat. *Australian Journal of Agricultural Research* 52(12), 1367–74. doi:10.1071/AR01031.

Palta, J. and Watt, M. 2009. Vigorous crop root systems: form and function for improving the capture of water and nutrients. In: Sadras, V. O. and Calderini, D. F. (Eds), *Crop Physiology. Applications for Genetic Improvement and Agronomy*. Elsevier/Academic Press, Burlington, MA, pp. 309–25.

Passioura, J. B. 1983. Roots and drought resistance. *Agricultural Water Management* 7(1-3), 265–80. doi:10.1016/0378-3774(83)90089-6.

Pook, M., Lisson, S., Risbey, J., Ummenhofer, C. C., McIntosh, P. and Rebbeck, M. 2009. The autumn break for cropping in southeast Australia: trends, synoptic influences and impacts on wheat yield. *International Journal of Climatology* 29(13), 2012–26. doi:10.1002/joc.1833.

Pooniya, V., Palta, J. A., Chen, Y., Delhaize, E. and Siddique, K. H. M. 2020. Impact of the TaMATE1B gene on above and below-ground growth of durum wheat grown on an acid and Al^{3+}-toxic soil. *Plant and Soil* 447(1–2), 73–84. doi:10.1007/s11104-019-04231-6.

Rebetzke, G., Ingvordsen, C., Bovill, W., Trethowan, R. and Fletcher, A. 2019. Breeding evolution for conservation agriculture. In: Pratley, J. and Kirkegaard, J. (Eds), *Australian Agriculture in 2020: From Conservation to Automation*. Agronomy Australia and Charles Sturt University, Wagga Wagga, pp. 273–87. Available at: http://agronomyaustraliaproceedings.org/images/sampledata/specialpublications/Australian%20Agriculture%20in%202020.pdf.

Rebetzke, G. J., Richards, R. A., Fettell, N. A., Long, M., Condon, A. G., Forrester, R. I. and Botwright, T. L. 2007. Genotypic increases in coleoptile length improves stand establishment, vigour and grain yield of deep-sown wheat. *Field Crops Research* 100(1), 10–23. doi:10.1016/j.fcr.2006.05.001.

Reyes, A., Messina, C. D., Hammer, G. L., Liu, L., van Oosterom, E., Lafitte, R. and Cooper, M. 2015. Soil water capture trends over 50 years of single-cross maize (*Zea mays* L.) breeding in the US corn-belt. *Journal of Experimental Botany* 66(22), 7339–46. doi:10.1093/jxb/erv430.

Rich, S. M., Wasson, A. P., Richards, R. A., Katore, T., Prashar, R., Chowdhary, R., Saxena, D. C., Mamrutha, H. M., Zwart, A., Misra, S. C., Sai Prasad, S. V., Chatrath, R., Christopher, J. and Watt, M. 2016. Wheats developed for high yield on stored soil moisture have deep vigorous root systems. *Functional Plant Biology: FPB* 43(2), 173–88. doi:10.1071/FP15182.

Richards, R. A. and Passioura, J. B. 1989. A breeding program to reduce the diameter of the major xylem vessel in the seminal roots of wheat and its effect on grain-yield in rain-fed environments. *Australian Journal of Agricultural Research* 40(5), 943–50. doi:10.1071/AR9890943.

Richards, R. A., Watt, M. and Rebetzke, G. J. 2007. Physiological traits and cereal germplasm for sustainable agricultural systems. *Euphytica* 154(3), 409–25. doi:10.1007/s10681-006-9286-1.

Sadras, V. O., Kirkegaard, J. A. and Hunt, J. R. 2019. Water use in rainfed systems: physiology of grain yield and its agronomic implications. In: Pratley, J. and Kirkegaard, J. (Eds) *Australian Agriculture in 2020: From Conservation to Automation*. Agronomy Australia and Charles Sturt University, Wagga Wagga, pp. 205–20. Available at: http://agronomyaustraliaproceedings.org/images/sampledata/specialpublications/Australian%20Agriculture%20in%202020.pdf.

Sciarresi, C., Patrignani, A., Soltani, A., Sinclair, T. and Lollato, R. P. 2019. Plant traits to increase winter wheat yield in semiarid and subhumid environments. *Agronomy Journal* 111, 1–13 doi:10.2134/agronj2018.12.0766.

Sharma-Natu, P. and Ghildiyal, M. C. 1994. Photosynthesis and dry matter production in *T. monococcum* and *T. aestivum* wheat in response to ear removal. *Journal of Agronomy and Crop Science* 173(3–4), 218–24. doi:10.1111/j.1439-037X.1994.tb00557.x.

Sprague, S. J., Kirkegaard, J. A., Bell, L. W., Seymour, M., Graham, J. and Ryan, M. 2018. Dual-purpose cereals offer increased productivity across diverse regions of Australia's high rainfall zone. *Field Crops Research* 227, 119–31. doi:10.1016/j.fcr.2018.08.008.

Tennant, D. and Hall, D 2001. Improving water use of annual crops and pastures - limitations and opportunities in Western Australia. *Australian Journal of Agricultural Research* 52(2), 171–82. doi:10.1071/AR00005.

Thorup-Kristensen, K., Cortasa, M. S. and Loges, R. 2009. Winter wheat roots grow twice as deep as spring wheat roots, is this important for N uptake and N leaching losses? *Plant and Soil* 322(1–2), 101–14. doi:10.1007/s11104-009-9898-z.

Thorup-Kristensen, K. and Kirkegaard, J. A. 2016. Root system-based limits to agricultural productivity and efficiency – the farming systems context. *Annals of Botany* 118(4), 573–92. doi:10.1093/aob/mcw122.

van Herwaarden, A. F., Angus, J. F., Richards, R. A. and Farquar, G. D. 1998. Haying off, the negative grain yield response of dryland wheat to nitrogen fertilizer. II carbohydrate and protein dynamics. *Australian Journal of Agricultural Research* 49(7), 1083–93. doi:10.1071/A97040.

Wasson, A. P., Richards, R. A., Chatrath, R., Misra, S. C., Sai Prasad, S. V., Rebetzke, G. J., Kirkegaard, J. A., Christopher, J. and Watt, M. 2012. Traits and selection strategies to improve root systems and water uptake in water-limited wheat crops. *Journal of Experimental Botany* 63(9), 3485–98. doi:10.1093/jxb/ers111.

Wasson, A. P., Rebetzke, G. J., Kirkegaard, J. A., Christopher, J., Richards, R. A. and Watt, M. 2014. Soil coring at multiple field environments can directly quantify variation in deep root traits to select wheat genotypes for breeding. *Journal of Experimental Botany* 65(21), 6231–49. doi:10.1093/jxb/eru250.

Watt, M., Magee, L. J. and McCully, M. E. 2008. Types, structure and potential for axial water flow in the deepest roots of field-grown cereals. *New Phytologist* 178(1), 135–46. doi:10.1111/j.1469-8137.2007.02358.x.

White, R. G. and Kirkegaard, J. A. 2010. The distribution and abundance of wheat roots in a dense, structured subsoil - implications for water uptake. *Plant, Cell and Environment* 33(2), 133–48. doi:10.1111/j.1365-3040.2009.02059.x.

Wissuwa, M., Kretzschmar, T. and Rose, T. J. 2016. From promise to application: root traits for enhanced nutrient capture in rice breeding. *Journal of Experimental Botany* 67(12), 3605–15. doi:10.1093/jxb/erw061.

Chapter 18

Rootstocks to improve root function and resource-use efficiency

Francisco Pérez-Alfocea, CEBAS-CSIC, Spain; Stephen Yeboah, CSIR-Crops Research Institute, Ghana; and Ian C. Dodd, Lancaster Environment Centre, UK

1 Introduction

Lack of water and nutrients limits crop productivity worldwide. Farmers usually seek to overcome these limitations by applying irrigation and fertilisers. However, even when these resources are physically available, limited infrastructure or economic access may constrain farmer inputs. Moreover, society increasingly expects agriculture to be more sustainable, by decreasing irrigation and fertiliser inputs. Thus genotypes with efficient root systems should be selected to best acquire these resources. Many horticultural crops use grafted plants with genetically distinct shoot (scion) and root (rootstock) systems, focusing attention on rootstock traits to improve crop resource-use efficiency. Resource-use efficiency may be defined as crop yield or plant biomass per unit of resource input or resource uptake. Although rootstocks are commonly used in many tree crops (e.g. apple, citrus, stone fruit) and some vegetables, this chapter will predominantly review the latter, except in some cases where specific mechanisms have been elucidated in tree crops.

http://dx.doi.org/10.19103/AS.2020.0075.25

High-value vegetable crops (e.g. tomatoes, cucurbits) are commonly grown in protected cultivation in many countries. To some extent, this minimises water requirements as the humid greenhouse environment decreases evapotranspiration losses. Nevertheless, these crops usually rely on substantial fertiliser supplies to maintain optimal fruit production and quality. While these high-value vegetable crops can be grown on their own roots or self-grafted, much commercial production uses rootstocks to enhance pest and disease resistance and/or extend the growing season, offering much higher yields (Fig. 1) especially when rootzone conditions are suboptimal. These rootstocks also provide opportunities to enhance water and nutrient capture by altering root system architecture and/or uptake and transport mechanisms that affect plant growth and crop yield (Ghanem et al., 2011a; Gregory et al., 2013).

While it is well known that rootstock vigour can enhance the yield of crops grown under optimal conditions, relatively little is known about the underlying physiological mechanisms that could help in selecting more efficient rootstocks under suboptimal conditions (Albacete et al., 2015). Recent advances in root phenotyping (see relevant chapters in this book) and renewed interest in vegetable grafting (Colla et al., 2017) provide a timely opportunity to assess

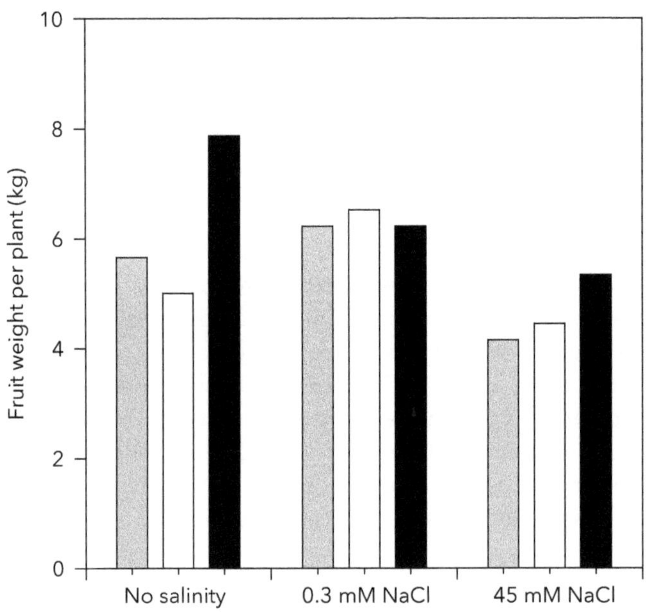

Figure 1 Individual plant fruit yield of own-rooted tomato plants (grey bars), self-grafted tomato plants (hollow bars), and those grafted onto *Beaufort* rootstock (filled bars) and grown in nutrient solution in glasshouses with no salinity, low (0.3 mM NaCl) salinity, and high (45 mM NaCl) salinity. Data re-plotted from Djidonou et al., 2019a and Savvas et al., 2011.

the physiological mechanisms by which rootstocks can improve resource-use efficiency, with attention given to the regulation of plant nutrient and water status. Since plant hormones regulate root architecture and shoot development, how scion and rootstock affect each other's hormonal status is also assessed. Lastly, we consider future perspectives on whether grafting (currently predominantly used within the technology-rich protected cultivation sector in some European, Asian and American countries) offers economic opportunities for smallholder farmers in developing countries of Africa, South America and Asia, by considering a case study in Ghana.

2 Nutrient relations

Although high-value vegetable crops are often grown in inert substrates and abundantly fertilised to minimise limitations in nutrient uptake, there is increasing interest in improving the efficiency of nutrient uptake. Developing a complex root architecture (the growth and branching of different root types) helps the root system to capture nutrients and ultimately transport these to the shoot. This function is particularly important in the infertile soils that limit yields of much of global agriculture. Rootstock mechanisms can provide the scion with opportunities to overcome intrinsic nutrient deficiencies in the substrate via:

- Increasing the rooting volume to enhance nutrient capture from the soil (e.g. root proliferation in nutrient-rich patches);
- increasing nutrient uptake by manipulating the rhizosphere environment (e.g. excretion of organic acids to mobilise phosphorus) and/or providing more efficient transport mechanisms;
- recruiting beneficial soil microbes that can enhance nutrient uptake via these mechanisms; and
- improving nutrient-use efficiency (growth or yield as a function of nutrient availability or uptake) with roots expending less carbon on respiration and/or nutrient re-allocation between different plant organs.

While it is not always clear which rootstock mechanism(s) improve scion nutrient status, the rootstock can buffer scion productivity against nutrient limitations. Rootstock-mediated changes in nitrogen (N) and phosphorus (P) uptake and use efficiency may allow growers to limit (expensive) fertiliser applications and/or maintain yield stability in low-fertility soils (Martínez-Andújar et al., 2020a).

2.1 Increasing the rooting volume

While many root phenotyping techniques have been developed to determine root architecture, relatively few studies have utilised these to determine whether variation in rootstock architecture affects crop nutrient acquisition. Optimal root systems for nutrient capture will depend on the spatial distribution of nutrients in the soil. Capture of relatively immobile nutrients such as phosphate ions, which are often accumulated in the upper soil layers, can be improved by selecting root systems with increased production of axial roots, shallower axial root growth angles, greater lateral root density and greater root hair length and density (Lynch, 2019). In contrast, relatively mobile nutrients such as ammonium (NH_4^+) and nitrate (NO_3^-) ions can be leached into lower soil layers, thus selecting for enhanced root vigour, steeper axial root growth angles and lateral root proliferation at depth should improve N uptake. The relative importance of these morphological changes likely depends on whether the root system is constrained (as typically occurs in soil-less production - usually with abundant nutrient supply) or unconstrained (as in open-field cultivation where nutrient supplies may be limiting). Many of these phenotyping protocols employ young recently germinated seedlings, and few studies have applied them to grafted plants which typically need to be older to allow root growth rates to stabilise following the formation of the graft union.

At anthesis, own-rooted tomatoes (*Solanum lycopersicum* cv. *Florida 47*) had up to twice the root biomass of grafted plants when grown under optimal nutrient supply (Djidonou et al., 2015), likely due to delayed root growth of the grafted plants following graft union establishment and/or differences in assimilate partitioning. By final fruit harvest almost 100 days later, root biomass did not statistically differ between own-rooted tomatoes and the commercial cultivar grafted on a range of commercial rootstocks (*Beaufort*, *Maxifort*, *Multifort* and *RST-04-105),* even though biomass of the *Beaufort* rootstock was 18% greater than the own-rooted plants (Djidonou et al., 2019a). Thus grafting-mediation limitation of root growth may be overcome with time, and it ultimately contributes to higher yields of grafted plants.

Comparing root architecture within 18 different own-rooted tomato rootstocks revealed significant variation in total root length (1.8-fold), root diameter (1.3-fold) and specific root length (1.5-fold) when grown under nutritionally optimal conditions. Although total root length and root diameter of the widely used tomato rootstock *Beaufort* was 45% and 15% lower respectively than the rootstock *RST-106*, its specific root length was 35% higher (Suchoff et al., 2017), which could enhance water and nutrient capture and may explain its wide commercial adoption. In this study, *Beaufort* and *Multifort* had the same root traits (total root length, root diameter, specific root length), consistent with similar yield performance and N requirements of grafted plants grown on these rootstocks in sandy soils at varying (56–336 kg N ha^{-1}) N rates (Djidonou et al., 2015). Greater N accumulation of *Florida 47* plants grafted on the *Beaufort* rootstock (compared to self-grafted controls) was associated with greater root length and surface area (Fig. 2) density (Djidonou et al., 2019b).

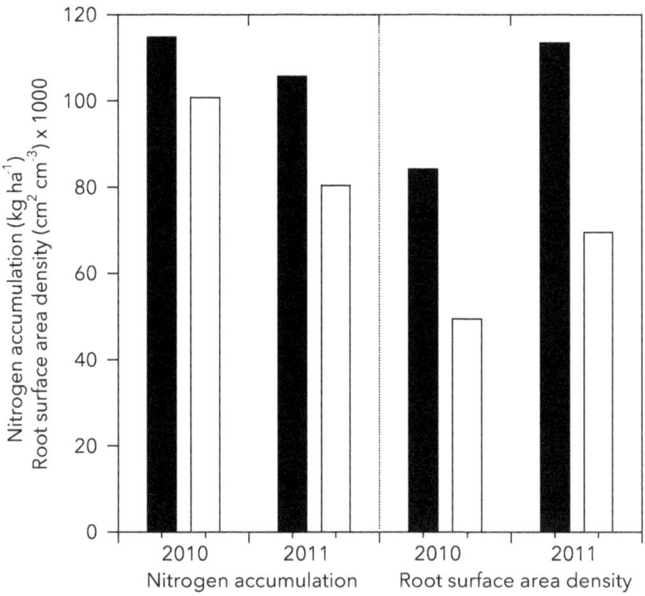

Figure 2 N accumulation and root surface area density of self-grafted *Florida 47* tomato (hollow bars) and grafted onto *Beaufort* rootstock (filled bars) and grown under a range of N application rates between 2010 and 2011. Data re-plotted from Djidonou et al., 2019b.

Similarly, when the commercial cultivar *Tribute* was grafted onto the *Maxifort* and *RST-106* rootstocks and grown under varying (0–224 kg N ha^{-1}) N rates that caused a 35% variation in yield, the yield was 6–15% greater when *Maxifort* was the rootstock (Suchoff et al., 2019). However, since there was no significant rootstock × N rate interaction, improved performance may be a constitutive rootstock trait, and not a specific adaptation to low N availability. This offers some hope that screening own-rooted plants for altered root traits can identify useful variation that can be exploited using selected genotypes as rootstocks.

Furthermore, in grafted plants (pepper (*Capsicum annuum*) grafted on 11 different wild pepper (*C. chinense*) rootstocks), scion biomass was significantly correlated with root length, surface area and volume (Nor et al., 2016). However, one rootstock increased scion biomass by 65% compared to self-grafted controls without significantly altering these root traits, suggesting that roots can regulate shoot growth by other, non-nutritional mechanisms.

Visible changes in root morphology may not be the only way of maximizing root absorptive area. When an F1 tomato hybrid was grafted onto two commercial rootstocks or self-grafted, *Beaufort* (*Solanum lycopersicum* × *S. habrochaites*) rootstocks had more root hairs than the self-grafted controls, along with increased root density (Oztekin et al., 2009). Although root hairs are especially important in enhancing P uptake, these rootstock differences were not associated with any differences in shoot biomass, probably because plants were grown under optimal nutrient supply in hydroponics that minimised diffusional limitations to nutrient uptake. Intrinsic rootstock differences in root hair length/density may be important in enhancing nutrient capture of soil-grown crops.

Changes in root system architecture to increase the volume of soil exploration are often associated with increased biomass partitioning to the root system as an adaptive mechanism to a specific nutrient deficiency. However, changes in root biomass did not explain the increased leaf K concentration and shoot growth of a tomato cultivar grafted onto contrasting rootstocks derived from a cross between *Solanum lycopersicum* and *S. pimpinellifolium* and cultivated under low-K (1 mM) conditions in a non-constrained semi-hydroponic-fed rootzone (Martínez-Andújar et al., 2016). Similarly, in a low-P (0.1 mM) root environment, rootstock-mediated changes in leaf P concentration and plant growth were not explained by altered root biomass (Martinez-Andújar et al., 2017). Moreover, root biomass was better correlated with xylem sap P concentration under optimal than suboptimal P supply, suggesting that other mechanisms mediate shoot growth and rootstock adaptations to nutrient limitation.

2.2 Increasing nutrient uptake efficiency

Nutrient uptake efficiency can be defined as the ratio between nutrient uptake and its availability in the rootzone (Hawkesford, 2012). Many studies revealed that some graft combinations were significantly more efficient in absorbing and transporting nutrients (e.g. N, P, K, Mg, Ca, Fe or other micronutrients) to the shoot than non-grafted plants in various species (reviewed in Savvas et al., 2010). Although nutrient uptake and assimilation seem important in determining rootstock-mediated vigour, rarely have increased nutrient concentrations been correlated with increased fruit yields due to the limited number of agronomic studies and the differential effects of rootstocks on vegetative and reproductive growth. Indeed, rootstock-mediated capacity to improve plant growth under a specific nutrient deficiency seems related to the ability to uptake and transport more macronutrients and micronutrients (e.g. Ca, K, S, Mn, B, Zn) to the shoot irrespective of the main growth-limiting nutrient (Martínez-Andújar et al., 2016, 2017).

Rootstocks can not only increase the uptake of beneficial nutrients, but also limit uptake of toxic elements (reviewed in Savvas et al., 2010). When used as a rootstock, the vigorous tomato hybrid *Solanum lycopersicum* × *S. habrochaites* decreased boron (B) uptake and foliar accumulation when plants were grown at toxic concentrations (10–15 mg L^{-1}), related to the decreased expression of B transporters in the root (Di Gioia et al., 2017). Although B is an essential nutrient for plants at low rootzone concentrations, it can be toxic in saline soils; thus, rootstock-mediated B exclusion may be an indirect but viable salt tolerance strategy. Interestingly, some B-excluding rootstocks have higher salinity tolerance, as in the common cucurbit F1 rootstock *Cucurbita maxima* × *C. moschata* (Colla et al., 2006). Nevertheless, maintaining productivity under salinity seems to depend not only on ionic regulation but also other (phytohormonal) root-to-shoot communication regulating shoot growth (Pérez-Alfocea et al., 2010; Sanoubar et al., 2013; Albacete et al., 2015).

Decreased absorption and transport of the toxic ion Na^+ to the leaves contributes to K/Na homeostasis and is a key factor for salt tolerance in many species (Munns and Tester, 2008; Pérez-Alfocea et al., 2010). Indeed, grafting the tomato cultivar *Jaguar* onto Na^+ excluding rootstocks increased yield by 80% under low and moderate salinities (25–75 mM NaCl) compared to the non-grafted plants, with yield under salinity negatively correlated with leaf concentration of toxic Na^+ and Cl^- ions (Estañ et al., 2005). However, fruit yield under salinity was correlated with improved transport of K^+ to the shoot in tomatoes grafted onto recombinant inbred lines (RILs) from a cross between the cultivated tomato *S. lycopersicum* and its salt-tolerant wild relative *S. pimpinellifolium* (Albacete et al., 2009). Thus rootstock-mediated enhancement of salt tolerance may result from both exclusion of toxic ions and increased uptake of essential nutrients.

Transgenic rootstocks provide further opportunities to manipulate nutrient uptake. Ectopic overexpression of the high-affinity potassium (K) transporter *CmHKT1;1* from *Cucurbita moschata* in cucumber reduces Na^+ transport to the shoot (Sun et al., 2018). Using this line as a rootstock could potentially limit foliar Na^+ accumulation, thereby increasing salt tolerance. Similarly, a recessive variant of the high-affinity K transporter *VisHKT1;1* was identified as the best candidate gene for a QTL controlling leaf Na^+ exclusion in grapevine hybrid rootstocks. Since this recessive allele with high capacity to reduce Na^+ transport was present in the wild relatives *Vitis champinii* and *V. rupestris*, rootstock breeding seems a promising strategy to increase salt tolerance of *V. vinifera* (Henderson et al., 2018). Overexpressing a modified Arabidopsis Ca^{2+}/H^+ exchanger CAX2B (providing enhanced Ca^{2+} substrate specificity and decreasing Mn^{2+} transport capability) in the rootstock bottle gourd (*Lagenaria siceraria*) increased growth and fruit quality (soluble solids) of melons (*Cucumis melo*) and watermelons (*Citrullus lanatus*) by increasing nutrient and/or water translocation to the scion (Han et al., 2009). Whether transgenic rootstocks will be commercially acceptable to consumers is unclear, even if the harvested portion is not transgenic.

Beyond specific transgenic approaches, there is some interest in selecting and breeding rootstocks for nutrient acquisition capacity in different species (Asins et al., 2017; Fazio et al., 2018). In a commercial tomato variety grafted onto

130 RILs from a cross between *Solanum lycopersicum* var. cerasiforme and *S. pimpinellifolium*, and cultivated under low N (1 mM NO_3^-), 62 rootstock-specific QTLs were detected (Asins et al., 2017). These included QTLs for vegetative growth, leaf N concentration and the concentrations of Mn and representatives of four phytohormone groups (cytokinins, gibberellins, salicylic acid and jasmonic acid) in the xylem sap. Furthermore, some nitrate, chloride and amino acid transporters; aquaporins; several purine permeases and phytohormone-related genes (cytokinins, gibberellins and salicylic acid) were identified as candidate genes for some of the detected QTLs. The heritabilities were high enough (0.33–0.39) to support selection of specific rootstock genotypes to improve N uptake under limited availability (Asins et al., 2017). While these are long-term (rootstock) breeding objectives, introgressing rootstock-mediated QTLs into elite varieties may prolong their commercial life and enhance their resource-use efficiency.

2.3 Recruiting beneficial soil microbes

Plant roots can substantially alter the rhizosphere microbiome, influencing the populations of beneficial microorganisms such as endophytic and arbuscular mycorrhizal fungi (AMF), and plant growth-promoting rhizobacteria (PGPR). This microbiome can enhance the root system's capacity to acquire water and nutrients. Different rootstocks can recruit different microbial populations, thereby affecting resource capture and hence crop yield and stability. Inoculating different own-rooted tomato genotypes (RILs from the crosss between *S. lycopersicum* var. cerasiforme and *S. pimpinellifolium)* with either an AMF *(Rhizophagus irregularis MUCL41833)* or a PGPR *(Variovorax paradoxus 5C-2)*, caused significant variation (2–5 fold range in % of root length colonised by the AMF and >100-fold range in colony forming units per gram of root for the PGPR) in microbial colonisation of the root system (Calvo-Polanco et al., 2016). Nevertheless, this variation was not clearly correlated with plant growth or leaf gas exchange of either well-watered plants or those exposed to drying soil. Similarly, limited variation in mycorrhizal colonisation (60–73% of the root length colonised by the AMF) was not correlated with significant rootstock effects on root dry weight or fruit yield of tomato plants grown under saline (6 dS m^{-1}) conditions (Otzekin et al., 2013). While rootstocks vary in their capacity to associate with beneficial microorganisms, identifying agronomic, physiological and genetic determinants of these rootstock-mediated traits is challenging (Plouznikoff et al., 2019). Thus most research on the interaction of rootstocks with beneficial soil organisms remains empirical, even if the commercial industry is increasingly employing both grafted plants and microbial inoculants.

2.4 Increasing the efficiency of nutrient utilization

Ultimately, nutrient-use efficiency (NUE) must be considered in relation to the production of the harvested organ (leaf or fruit) compared to the nutrient supplied to, or captured by, the plant. Increasing foliar nutritional status is of little value if yield is not increased, since it decreases NUE and depletes soil fertility, thereby increasing fertiliser requirements. Although it has been suggested that grafted plants should receive less fertilisers (one-half to two-thirds as much as currently recommended for non-grafted plants - Lee, 2003), optimal fertilization management of grafted plants may depend on specific rootstock × scion × environment interactions.

Rootstock-mediated redistribution of biomass and nutrients between vegetative and reproductive tissues also influences NUE, and maintaining a good balance between reproductive (generative) and vegetative vigour is essential for high yields. For example, pepper yields (both total and marketable) on the commercial rootstock *Creonte* were higher than the ungrafted commercial scion when grown under protected cultivation in a Mediterranean climate (López-Marin et al., 2013, 2017). This rootstock resulted in fruits with higher electrical conductivity despite increased fruit (and leaf) fresh weight, implying better partitioning of nutrients to the fruit, independent of root biomass. Irrespective of root vigour, differences in nutrient uptake efficiency and reallocation between different tissues also seem important in mediating shoot response.

Grafting cucurbit species seems to benefit their N nutrition, as foliar N concentration was positively correlated with fruit yield in melon and watermelon varieties grafted onto different *C. maxima* × *C. moschata* rootstocks (reviewed in Colla et al., 2010a). These rootstocks mediated NO_3^- uptake, translocation and assimilation (higher nitrate reductase activity and free amino acids, soluble proteins and total-N concentrations than non-grafted plants). Field experiments with melon plants grafted onto a *Cucurbita* hybrid rootstock confirmed that N uptake, NUE and marketable yield increased by 10–15% compared to the ungrafted plants (Colla et al., 2010a). Furthermore, grafting watermelon onto several experimental rootstocks increased NUE (biomass/N uptake) by 70% under optimum N supply (9 mM) in hydroponics, while one rootstock increased NUE by a similar extent under low N (0.2 mM) supply (Nawaz et al., 2018). Highly vigorous bottle gourd rootstocks increased NUE of a watermelon cultivar scion grown under low N (0.3 mM) using the nutrient film technique (Ulas et al., 2019). Although rootstocks can enhance shoot nutrient concentrations, further work is required to optimise yield and NUE by breeding and selecting rootstocks that can reduce fertiliser use without decreasing yields.

Increased P uptake occurred in eggplant (*Solanum melongena*) grafted on *Beaufort* (Leonardi and Giuffrida, 2006), cucumber on *Shintoza* (Rouphael et al.,

2008) and watermelon onto different *C. maxima* × *C. moschata* hybrids (Colla et al., 2010b). Increased P use efficiency was correlated with improved yield (11%) in open-field grafted tomatoes (Khah et al., 2006), but the underlying mechanisms remain largely unknown. In contrast, grafting can alter phosphorus utilisation, with a pumpkin (*Cucurbita pepo*) rootstock decreasing P uptake and translocation in the leaves of melon (Qi et al., 2006). Understanding P re-allocation within the plant, and its potential rootstock regulation, seems important to enhance phosphorus-use efficiency.

Rootstock-mediated variation in potassium (K) use efficiency has also been reported. Fruit yield of watermelon grafted onto different rootstocks from *Citrullus lanatus* spp., *Lagenaria siceraria* and *Cucurbita maxima* × *C. moschata* was less sensitive to low (0.1 mM) K nutrition than in self-grafted plants (Zhong et al., 2018). Changes in growth were positively correlated with increased shoot K status, the cytokinin zeatin riboside and chlorophyll concentrations. In tomato, the commercial rootstocks *Maxifort* and *Brigeor* can improve fruit yield by 15% under low K^+ supply (4 mM vs. 8 mM), although the response to the nutrient deficit varies greatly with the rootstock × scion combination (Schwarz et al., 2013).

3 Water relations

Rootstocks offer the potential to increase agronomic water-use efficiency (WUE - yield or biomass per unit of water lost) and drought tolerance of horticultural crops. Scion transpiration rate and its acclimation to water deficit are controlled genetically by the rootstock through different genetic architectures (Marguerit et al., 2012). Different mechanisms of rootstock-mediated drought resistance have been proposed, whereby dwarfing rootstocks restrict stomatal opening thereby limiting water loss, while invigorating rootstocks with a more extensive root system maintain water capture for an extended period (Tworkoski et al., 2016). These mechanisms are not mutually exclusive as the hormone transport between rootstock and scion (considered later) co-ordinates the balance between water capture and loss. For example, having more roots in drying soil enhances foliar ABA concentrations (Martin-Vertedor and Dodd, 2011) thereby promoting stomatal closure, while shoot export of ABA to the roots regulates root ABA concentrations and root architecture (McAdam et al., 2016). Rootstock mechanisms can provide the scion with opportunities to overcome limited soil water availability by:

- Increasing the rooting volume to enhance water capture from the soil.
- Regulating water loss from the leaves by hydraulic and non-hydraulic signals.
- Affecting canopy vigour and thus whole plant water use.

3.1 Root vigour and morphology

Rootstock morphology can affect scion responses to different irrigation treatments. Increased total root length and abundant fine roots of the *Beaufort* tomato rootstock increased agronomic WUE in high tunnel cultivation compared to other rootstocks and self-grafted controls with lower fine root abundance (Suchoff et al., 2018a,b). Plants grafted on *Beaufort* maintained better water relations, including higher relative water content (Fig. 3), leaf area, stomatal conductance and net CO_2 assimilation rate, related to the maintenance of root length even under reduced irrigation. Commercial cucurbit rootstocks from different genetic origins (pumpkin (*Cucurbita maxima*), squash (*C. pepo*), bottle gourd (*Lagenaria siceraria*) and an interspecific hybrid squash (*C. maxima* × *C. moschata*)) affected rootstock morphology (root length, diameter and area) when used in watermelon production (Bertucci et al., 2018). While it is commonly assumed that rootstocks with more vigorous root systems can maintain shoot water status if irrigation is restricted, further experiments are needed to understand the physiological mechanisms regulating root development.

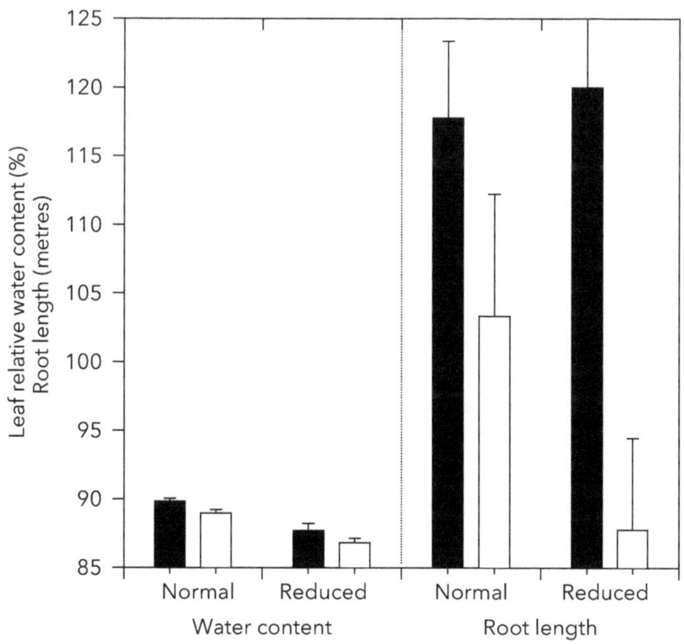

Figure 3 Leaf relative water content and root length of self-grafted *Cherokee Purple* tomato (hollow bars) and grafted onto *Beaufort* rootstock (filled bars) and grown under normal and reduced rate irrigation. Data re-plotted from Suchoff et al., 2018b.

3.2 Leaf gas exchange

Rootstocks can profoundly affect leaf gas exchange. Intrinsic water-use efficiency (WUEi) can be defined as the instantaneous ratio between leaf photosynthesis (*A*) and stomatal conductance (*gs*) and provides a physiological indicator that may be correlated with agronomic WUE. Grafting a drought-tolerant tomato landrace with low stomatal aperture and high WUEi (Galmés et al., 2011, 2013) onto vigorous commercial rootstocks increased scion *gs* up to 3-fold, but only increased *A* by 20%, probably because of other non-stomatal limitations to photosynthesis (Fullana-Pericàs et al., 2018, 2019). Similarly, grafting the tomato cultivar *Ramellet* onto commercial rootstocks (*Beaufort* and *Maxifort*) increased *gs* at the expense of WUEi (Fullana-Pericas et al., 2018). In contrast, when the same cultivar was grafted on the wild relative *Solanum pimpinellifolium*, *gs* decreased while WUEi increased (Fullana-Pericas et al., 2018). Thus rootstock selection can substantially change WUEi of the scion, even if further work is required to understand the physiological mechanisms by which this occurs.

Rootstock effects on stomatal conductance can vary with the duration of soil drying. When two different apple (*Malus* × *domestica*) cultivars were grafted on either the dwarfing M.9 or the semi-dwarfing MM.111 rootstocks, rootstock did not alter *gs* of well-watered plants grown in containers (Tworkoski et al., 2016). Withholding water caused much faster soil moisture depletion when both cultivars were grafted on the MM.111 rootstock, implying greater whole plant transpiration rates (caused by a larger canopy) as *gs* did not differ. Rootstock had no impact on midday leaf water potential (Ψ_{leaf}) or on the relationship between *gs* and leaf ABA concentration, even in plants where the stomata were almost fully closed. Pronounced rootstock effects on *gs* following re-watering (2.5-fold greater *gs* when plants were grafted on the MM.111 rootstock) could not be accounted for by differences in either Ψ_{leaf} or leaf ABA concentration, even though phaseic acid (an ABA metabolite) concentrations were significantly lower in plants grafted on the MM.111 rootstock. Since leaf ABA concentrations may be affected by both root and shoot water status, xylem hormone concentrations should also be measured to account for rootstock effects. In comparing another two apple rootstocks, M.9 exported higher xylem ABA concentrations from the root system than M.106 at the low sap flow rates (Else et al., 2018) that would occur with substantial soil drying, potentially explaining the lower *gs* of scions grafted on M.9 than MM.111 (Tworkoski et al., 2016). Further measurements of *gs* and xylem ABA concentrations seem necessary to support this interpretation, but rootstock-mediated variation in Ψ_{leaf} does not seem to explain the observed stomatal behaviour of apple.

In another comparison of rootstock physiology, M.9 showed a faster decrease in root cytokinin concentrations and root isopentenyl adenosine transferase (IPT) gene expression (responsible for *de novo* cytokinin synthesis) when exposed to an osmotic stress (100 mM NaCl for up to 16 hours) than the *Malus robusta* rootstock (Feng et al., 2019). Since this difference was ascribed to the deletion of a proline-responsive element upstream of a specific IPT gene, it would be interesting to measure root proline accumulation over the same time-course. Nevertheless, both decreased the export of cytokinins and enhanced ABA export from M.9 rootstocks would diminish *gs* of scions grafted to the rootstock, since cytokinins antagonise ABA-induced stomatal closure (Hansen and Dörffling, 2003).

3.3 Canopy vigour

Rootstock-mediated changes in canopy vigour can alleviate the effects of drought by reducing water use without affecting the yield. Low-vigour rootstocks derived from a *Solanum lycopersicum* × *S. pimpinellifolium* RIL population decreased leaf biomass of an elite tomato scion, thereby decreasing water use and increasing agronomic WUE by 40% compared to self-grafted plants (Cantero-Navarro et al., 2016). Alternatively, some vigorous rootstocks limit scion water use. Grafting the tomato cultivar *Durinta* onto the vigorous *Beaufort* rootstock decreased water use by 25% (compared to the non-grafted scion grown under optimal irrigation) when plants were grown at 75% of crop evapotranspiration (Al-Harbi et al., 2018). Reconciling these contrasting mechanisms of increasing agronomic WUE requires an understanding of rootstock regulation of stomatal conductance and leaf expansion. Furthermore, rootstock-mediated changes in canopy vigour can also alleviate the effects of salinity on fruit yield. Rootstocks that maintained vegetative vigour enhanced fruit yields compared to self-grafted controls when grown under moderate (75 mM NaCl) salinity (Albacete et al., 2009; Estañ et al., 2009), even though there were no differences in the absence of salinity. Thus rootstock modulation of leaf growth could increase yield (by affecting the partitioning of resources between vegetative and reproductive growth) and/or reduce irrigation requirements depending on the environmental conditions.

4 Phytohormones

Suboptimal rootzone conditions such as nutrient starvation or water deficits trigger changes in root phytohormone status to mitigate and/or adapt to the prevailing environment. These hormones play a major role not only in the local responses, but also in root-shoot coordination. Therefore rootstock hormone status can affect shoot processes in the grafted plant. Multi-analyte hormone measurements of root and xylem sap composition when a single scion is grafted on multiple rootstocks provide abundant correlations between plant physiological processes and xylem hormone profile (e.g. Albacete et al., 2008; Martinez-Andujar et al., 2016, 2017); yet, grafting experiments with hormone-deficient mutants or overexpressing mutants/transgenics provide much more specific evidence that individual hormones regulate specific processes.

4.1 ABA regulation of canopy growth and water use

When sweet orange (*Citrus sinensis* cv. Valencia) was grafted onto diploid or autotetraploid rootstocks of Rangpur lime (*C. limonia*), the latter rootstock decreased the stomatal conductance (*gs*) of well-watered plants by a third (Allario et al., 2013). Stomatal size or density was not affected by the rootstock, whereas increased leaf ABA concentrations occurred in scions grafted on the autotetraploid rootstocks, despite no change in foliar gene expression of the rate-limiting enzyme (NCED) for ABA biosynthesis. In contrast, substantially increased root ABA concentrations (5-fold) in the autotetraploid rootstock was associated with enhanced NCED gene expression especially in well-watered plants, indicating that root ABA biosynthesis and export to the shoot was important in mediating *gs* (Allario et al., 2013). While these observations suggest that root ABA production modulates root-to-shoot ABA signalling and shoot physiology, reciprocal grafting experiments with ABA-deficient mutants (Holbrook et al., 2002; Chen et al., 2003; Christmann et al., 2007; Dodd et al., 2009; Li et al., 2018), ABA-overproducing transgenics (Thompson et al., 2007; Martinez-Andujar et al., 2020b) and wild-type (WT) plants provide a different interpretation.

Scion genotype always had a stronger effect than rootstock genotype in such reciprocal grafts, with an ABA-deficient rootstock having limited effects on shoot water relations and *gs* of WT scions (Holbrook et al., 2002; Christmann et al., 2007; Dodd et al., 2009; see Li et al., 2018 for a meta-analysis). However, WT rootstocks partially restored growth and shoot ABA concentration of ABA-deficient scions (Chen et al., 2003; Dodd et al., 2009), indicating that root-synthesised ABA can influence shoot physiology. Although ABA-overproducing tomato rootstocks had no effect on *gs* of WT scions in well-watered pot-grown plants (Thompson et al., 2007), these rootstocks promoted vegetative vigour and fruit yield of salinized plants grown under saline (35 mM NaCl) conditions by altering root sensitivity to ABA under stress (Albacete et al., 2015). While such grafting experiments seem to offer unequivocal evidence of the efficacy of root-to-shoot ABA signalling, hormone crosstalk requires a more considered explanation, since ABA-deficient tomatoes have high-shoot ethylene evolution compared to WT plants (Sharp et al., 2000). Thus WT rootstocks also diminished xylem ACC concentrations of ABA-deficient *flacca* scions, normalising the shoot ethylene evolution (Dodd et al., 2009). Multi-analyte hormone quantification of xylem sap and shoot tissues seems necessary to unequivocally establish that altering biosynthesis of one hormone in the roots has direct (caused by that hormone) or indirect (caused by other hormonal changes) effects.

4.2 Cytokinin (CK) regulation of canopy senescence

Pioneering studies in cotton (*Gossypium hirsutum*) and tomato demonstrated that grafting can delay natural (Dong et al., 2008) and stress-induced (Albacete et al., 2009; Ghanem et al., 2011b; Wang et al., 2012) leaf senescence, thereby improving productivity, with CKs playing an important role. Grafting early-senescent cotton genotypes onto late-senescent genotypes delayed leaf senescence compared with the self-grafted early-senescent lines (Dong et al., 2008). Higher photosynthetic rates (1.5- fold to 1.8-fold), chlorophyll, and soluble protein contents were correlated with enhanced accumulation of the CKs Z+ZR (zeatin + zeatin riboside), and iP+iPA (isopentenyladenine + isopentenyladenosine) rather than DHZ+DHZR (dihydrozeatin + dihydrozeatin riboside). Similar results were reported with reciprocal grafts of two cotton genotypes with contrasting sensitivity to leaf senescence induction by K^+ deficit (0.01–0.03 mM; Wang et al., 2012). In tomato, rootstock-mediated changes in xylem and leaf xylem concentrations of the free-base cytokinin *trans*-zeatin (*t*-Z) were positively related with leaf growth and crop productivity under salinity (75 mM NaCl). In this case, the hormonal variation was obtained by either grafting a commercial tomato variety onto different rootstocks derived from a *Solanum lycopersicum* × *S. cheesmaniae* cross (Albacete et al., 2009, 2010) or grafting WT plants onto a CK-overproducing rootstock constitutively (35S cauliflower mosaic virus promoter) expressing the IPT gene responsible for *de novo* cytokinin synthesis (Ghanem et al., 2011a). The enhanced productivity under salinity was explained by CK-mediated improvement of source-sink relations, increasing photosynthesis through delaying stress-induced leaf senescence, reducing salt-induced flower abortion and increasing the individual fruit growth due to promoted sink metabolism (Ghanem et al., 2011a; Albacete et al., 2009, 2014). Increased root export of CKs (*t*-Z and iP type) was correlated with higher K^+ (20%) concentration in both leaf tissues (Ghanem et al., 2011a) and leaf xylem sap (Albacete et al., 2009), suggesting that CKs may regulate potassium uptake and/or transport.

4.3 ACC regulation of nutrient stress responses

The ethylene precursor 1-aminocyclopropane-1-carboxylic acid (ACC) seems to regulate rootstock-mediated responses to abiotic stresses in tomato (Albacete et al., 2008, 2009, 2010, 2014; Cantero-Navarro et al., 2016). Xylem sap ACC concentrations have been correlated with salinity-induced senescence and shoot growth inhibition (Albacete et al., 2008, 2009, 2010, 2014) and nutritional (K, P) deficiencies (Martinez-Andújar et al., 2016, 2017). Moreover, high constitutive ACC levels have been related to reduced vigour and leaf area in tomato genotypes grown as whole plants and when used as rootstocks of a commercial scion variety (Martínez-Andújar et al., 2016; Cantero-Navarro et al., 2016), with concurrent reductions in transpiration and thus enhanced agronomic WUE (Cantero-Navarro et al., 2016). Further experiments with tomato rootstocks antisensed for ACC synthase and ACC oxidase genes would provide further physiological assurance of the importance of ACC transport in regulating these processes.

Ethylene, very likely in combination with other hormones, seems to regulate plant response to low K by affecting reactive oxygen species (ROS) production, root morphology and gene expression of high affinity transporters (Armengaud et al., 2004; Jung et al., 2009). Under low K nutrition (1 mM), low root xylem ACC levels are proposed to promote the uptake and transport of K *versus* Na, and also Ca, S and micronutrients (B, Zn, Mn, Fe) in the vigorous grafted plants under low K. Additionally, converting *trans*-zeatin into *trans*-zeatin riboside and cytokinin interactions with ACC would lower ACC levels, thereby converting an elite tomato F1 cultivar into a (micro) nutrient-efficient phenotype, improving growth under reduced K fertilization (Martínez-Andújar et al., 2016).

P deficiency can also change hormone production, sensitivity and transport to regulate the expression of P-responsive genes and root architecture, thereby increasing P uptake and use efficiency (Lynch, 2011; Chiou and Lin, 2011), with root-shoot communication being essential for plant adaptation to P deprivation (Kapulnik and Koltai, 2016). P availability affects ethylene evolution and mediates systemic P signalling thus increasing root/shoot ratio during P deficiency (Nagarajan and Smith, 2012). Rootstocks can influence the adaptation to low P by inducing local and systemic responses. Rootstock-mediated growth promotion under low P (0.1 mM) in tomato was associated with the capacity to uptake and transport Ca, S, Mn and to the scion, with decreased P transport only explaining growth inhibition (Martínez-Andújar et al., 2017). Moreover, the cytokinin *t*-Z and the ethylene precursor ACC also seem to play an antagonistic positive (*t*-Z) and negative (ACC) role in rootstock-mediated regulation of shoot growth in response to low P nutrition, contributing to scion P use efficiency. Thus ACC and CKs play an important role in root-to-shoot signalling under P starvation (Martín et al., 2000; Zhang et al., 2014) and may comprise part of a systemic feedback signalling system adapting shoot growth to P availability. Therefore, low P uptake could modify the negative (ACC) and positive (*t*-Z) signals to adapt scion vegetative growth to rootstock P export.

5 Case study: Tomato rootstocks as a viable strategy to overcome abiotic stresses in Ghana

Tomato cultivation in Ghana remains as a subsistence farming activity and contributes significantly to the livelihoods of rural-based smallholder farmers. Tomato is critical in generating income for these farmers and provides a source of micronutrients important to prevent malnutrition (GhanaVeg reports,

2016). The demand for tomato in Ghana outstrips supply due to low yields principally caused by multiple abiotic stresses, and therefore, local production is augmented by imports from neighbouring countries. Tomato production is predominantly rainfed and irrigation infrastructure is limited, functioning only during the dry season. Production is constrained by limited water availability, drought, salinity and low soil fertility. These constraints are linked to limited varietal adaptation and low technology application in the tomato value chain. Drought and heat stress decreases tomato production and nutritional quality, especially for smallholders that produce with minimal inputs (IFPRI, 2013; Nyaku and Amissah, 2019). These factors limit the overall yearly production and make the country dependent on regional imports of up to 80 000 tons/year (FAOSTAT, 2014). Crop improvement options are desperately needed to transform the tomato industry for better quality traits such as flavour, taste, texture, firmness and longer shelf life.

Local breeding efforts for high-quality traits with high yield and extended shelf life have yielded limited results. The Crops Research Institute (CSIR), a national agriculture institute mandated to develop technologies to improve tomato production, is yet to release a variety after many years of research. Genetically modified tomato plants are not being considered due to socio-economic issues and lack of consumer acceptability. Grafting tomatoes can increase fruit yields and the quality and extend shelf life (Nkansah et al., 2013) and enhance nutrient uptake while improving water-use efficiency (Ofori, 2016; Nyaku and Amissah, 2019). Grafting two local tomato cultivars (*Fadeɛbege* and *Nkansah HT*) onto the *Black Beauty* eggplant rootstock in a greenhouse increased fruit yield by approximately 40% (due to increased fruit number) compared to non-grafted plants, as well as yield quality and shelf life (Ofori, 2016). Since eggplant is among the most abundant vegetables in Ghana, this makes grafting a viable option for producing the crop amidst increasing abiotic stresses.

Tomato production is mainly rainfed, but the crop can be irrigated in the dry season (Asante 2013). Salinity in either the soil or water can significantly affect production, decreasing yields especially during dry season production. About half the irrigated lands in Ghana are affected by salinity (Asare-Bediako et al., 2007). Grafting tomato onto resistant rootstocks such as *Radja* and *Pera* increased tomato production in moderately saline (7 dS m^{-1}) soils (Ofori, 2016). This option should be pursued in anticipation of better landscape management (retention and regeneration of native vegetation, adopting cropping practices that minimise deep drainage and maintenance of good pasture cover) to reduce excessive irrigation and leaching of soils in the absence of appropriate drainage systems that can reduce salinity in irrigated fields.

Climate change in Ghana has increased the variability of temperature and rainfall patterns. Mean annual temperature has increased by 1.0°C since 1960, while total annual rainfall in Ghana has decreased by an average of 2.3 mm per month (2.4%) per decade (Antwi-Agyei et al., 2013). The increased mid-season drought conditions and the associated heat stress, especially during the dry season should promote the grafting of tomatoes onto drought tolerant/resistant rootstocks in Ghana. Since eggplant roots are more effective in absorbing water than tomato roots (Abdelmageed and Gruda, 2009), grafting tomatoes onto eggplant rootstocks has the potential to improve yield and fruit quality of tomatoes (Kanyua, 2018). Grafting tomato onto selected heat-tolerant rootstocks also offers a certain degree of resistance against thermal stress (Guodaar, 2015; Kanyua, 2018).

In spite of considerable promise, vegetable grafting is still relatively unknown in Ghana. Its application suffers from the perception that grafting is a difficult practice and cannot be applied to vegetables, but only to tree crops. The operation of 'grafting schools' by the CSIR may be necessary to ensure that knowledge of grafting mechanics and benefits are disseminated to farmers.

6 Conclusion

Deciphering the physiological mechanisms by which rootstocks influence scion physiology and yield is difficult, because nutrient, water and phytohormone status are inextricably linked by multiple feedback mechanisms. Identifying rootstock-specific QTLs using recombinant inbred lines as rootstocks offers possibilities to uncover specific genetic mechanisms by which rootstocks alter shoot physiology. While rootstock × scion × environment interactions quickly escalate the scale of experiments required to determine the mechanisms by which rootstocks regulate scion physiology, the agronomic importance of grafted plants and increased commercial pressures for improved resource-use efficiency requires continued research efforts in determining rootstock mechanisms. Since yield gaps are greater in many developed countries, there is considerable need for better technology transfer to establish viable grafting-based industries.

7 Future trends

While current commercial production in intensive horticulture generally uses a relatively restricted range of rootstocks in many species, increased societal pressure to decrease the environmental footprint of both protected cultivation (Fernandez et al., 2018) and open-field production will stimulate the search for more nutrient and water-use efficient rootstocks. Current genetic diversity in many grafted vegetable species has scarcely been exploited when developing new rootstocks, in spite of large multinational projects (see Section 8). Identification of rootstock-mediated QTLs (Asins et al., 2010) in vegetable species has thus far attracted limited interest from commercial companies interested in rootstock breeding, at least with respect to improving nutrient and water-use efficiency. Complex rootstock-scion interactions may constrain commercial interest in this area.

In contrast, with extensive knowledge exchange activities, vegetable grafting is expected to gain increasing prominence in developing countries, where production is currently constrained by inadequate nutrient and water resources. Nevertheless, the costs of supplying grafted plants to interested farmers may restrict the development of this industry, with peer-to-peer learning of grafting techniques likely to play a major role in disseminating knowledge. While farmers are currently more likely to use grafted plants to limit the yield losses caused by pests (e.g. nematodes) and diseases (e.g. vascular wilts), sensitising them to this technology may result in its later adoption to overcome nutrient and water limitations in productivity.

8 Where to look for further information

Colla, G., Pérez-Alfocea, F. and Schwarz, D. 2017. *Vegetable Grafting: Principles and Practice.* CABI, Wallingford, UK.

Kumar, P., Rouphael, Y., Cardarelli, M. and Colla, G. 2017. Vegetable grafting as a tool to improve drought resistance and water use efficiency. *Frontiers of Plant Science* 8, 1130.

Martínez-Andújar, C., Albacete, A. and Pérez-Alfocea, F. 2020a. Rootstocks for increasing yield stability and sustainability in vegetable crops. *Acta Horticulturae* 273, 449–70. ISHS 2020. DOI 10.17660/ActaHortic.2020.1273.58 XXX IHC – 4.

International Society of Horticultural Science Workgroup on Vegetable Grafting: https://www.ishs.org/vegetable-grafting-0.

Scientia Horticulturae Special Issue Volume 149 (2013):

https://www.sciencedirect.com/journal/scientia-horticulturae/vol/149.

Scientia Horticulturae Special Issue Volume 127 (2010):

https://www.sciencedirect.com/journal/scientia-horticulturae/vol/127.

http://www.rootopower.eu/.

http://www.tomres.eu/.

USDA-SCRI Vegetable Grafting Project: http://www.vegetablegrafting.org/about/about-us/.

https://horticulture.ucdavis.edu/vegetable-grafting.

https://u.osu.edu/vegprolab/research-areas/grafting-2/.

World Vegetable Center (https://avrdc.org/tag/grafting/.)

9 Acknowledgements

FPA and ICD acknowledge funding from the EU FP7 program (ROOTOPOWER Project, Contract No. 289365) and from the Spanish MINECO-FEDER (project

RTI2018-099113-B-I00), while SY and ICD thank the Global Challenges Research Fund (Project RECIRCULATE, ES/P010857/1) project for supporting tomato research in Ghana.

10 References

Abdelmageed, A. H. A. and Gruda, N. 2009. Influence of grafting on growth, development and some physiological parameters of tomatoes under controlled heat stress conditions. *European Journal of Horticulture Science* 74(1), 16–20.

Albacete, A., Ghanem, M. E., Martínez-Andújar, C., Acosta, M., Sánchez-Bravo, J., Martínez, V., Lutts, S., Dodd, I. C. and Pérez-Alfocea, F. 2008. Hormonal changes in relation to biomass partitioning and shoot growth impairment in salinized tomato (*Solanum lycopersicum* L.) plants. *Journal of Experimental Botany* 59(15), 4119–31. doi:10.1093/jxb/ern251.

Albacete, A., Martínez-Andújar, C., Ghanem, M. E., Acosta, M., Sánchez-Bravo, J., Asins, M. J., Cuartero, J., Lutts, S., Dodd, I. C. and Pérez-Alfocea, F. 2009. Rootstock-mediated changes in xylem ionic and hormonal status are correlated with delayed leaf senescence, and increased leaf area and crop productivity in salinized tomato. *Plant, Cell and Environment* 32(7), 928–38. doi:10.1111/j.1365-3040.2009.01973.x.

Albacete, A., Ghanem, M. E., Dodd, I. C. and Pérez-Alfocea, F. 2010. Principal component analysis of hormone profiling data suggests an important role for cytokinins in regulating leaf growth and senescence of salinized tomato. *Plant Signaling and Behavior* 5(1), 45–8. doi:10.4161/psb.5.1.10120.

Albacete, A., Martínez-Andújar, C. and Pérez-Alfocea, F. 2014. Hormonal and metabolic regulation of source–sink relations under salinity and drought: from plant survival to crop yield stability. *Biotechnology Advances* 32(1), 12–30. doi:10.1016/j.biotechadv.2013.10.005.

Albacete, A., Martínez-Andújar, C., Martínez-Pérez, A., Thompson, A. J., Dodd, I. C. and Pérez-Alfocea, F. 2015. Unravelling rootstock × scion interactions to improve food security. *Journal of Experimental Botany* 66(8), 2211–26. doi:10.1093/jxb/erv027.

Al-Harbi, A. R., Al-Omran, A. M., Alqardaeai, T. A., Abdel-Rassak, H. S., Alharbi, K. R., Obadi, A. and Saad, M. A. 2018. Grafting affects tomato growth, productivity, and water use efficiency under different water regimes. *Journal of Agricultural Science and Technology* 20, 1227–41.

Allario, T., Brumos, J., Colmenero-Flores, J. M., Iglesias, D. J., Pina, J. A., Navarro, L., Talon, M., Ollitrault, P. and Morillon, R. 2013. Tetraploid Rangpur lime rootstock increases drought tolerance via enhanced constitutive root abscisic acid production. *Plant, Cell and Environment* 36(4), 856–68. doi:10.1111/pce.12021.

Antwi-Agyei, P., Dougill, A. J., Fraser, E. D. G. and Stringer, L. C. 2013. Characterising the nature of household vulnerability to climate variability: empirical evidence from two regions of Ghana. *Environment, Development and Sustainability* 15(4), 903–26. doi:10.1007/s10668-012-9418-9.

Armengaud, P., Breitling, R. and Amtmann, A. 2004. The potassium-dependent transcriptome of Arabidopsis reveals a prominent role of jasmonic acid in nutrient signaling. *Plant Physiology* 136(1), 2556–76. doi:10.1104/pp.104.046482.

Asante, A. V. 2013. *Smallholder Irrigation Technology in Ghana: Adoption and Profitability Analysis*. A thesis submitted to the Department of Agricultural Economics, Agribusiness and Extension. Kwame Nkrumah University of Science and Technology.

Asare-Bediako, E., Showemimo, F. A., Buah, J. N., Ushawu, Y. 2007. Tomato production constraints at Bontanga irrigation project in the northern region of Ghana. *Journal of Applied Sciences* 7(3), 459–61. doi:10.3923/jas.2007.459.461.

Asins, M. J., Bolarin, M. C., Pérez-Alfocea, F., Estan, M. T., Martínez-Andújar, C., Albacete, A., Villalta, I., Bernet, G. P., Dodd, I. C. and Carbonell, E. A. 2010. Genetic analysis of physiological components of salt tolerance conferred by Solanum rootstocks. What is the rootstock doing for the scion? *Theoretical and Applied Genetics* 121(1), 105–15. doi:10.1007/s00122-010-1294-9.

Asins, M. J., Albacete, A., Martinez-Andujar, C., Pérez-Alfocea, F., Dodd, I. C., Carbonell, E. A. and Dieleman, J. A. 2017. Genetic analysis of rootstock-mediated nitrogen (N) uptake and root-to-shoot signalling at contrasting N availabilities in tomato. *Plant Science* 263, 94–106. doi:10.1016/j.plantsci.2017.06.012.

Bertucci, M. B., Duchoff, D. H., Jennings, K. M., Monks, D. W., Gunter, C. C., Schultheis, J. R. and Louws, F. J. 2018. Comparison of root system morphology of cucurbit rootstocks for use in watermelon grafting. *HortTechnology* 28(5), 629–36. doi:10.21273/HORTTECH04098-18.

Calvo-Polanco, M., Sánchez-Romera, B., Aroca, R., Asins, M. J., Declerck, S., Dodd, I. C., Martínez-Andújar, C., Albacete, A. and Ruiz-Lozano, J. M. 2016. Exploring the use of recombinant inbred lines in combination with beneficial microbial inoculants (AM fungus and PGPR) to improve drought stress tolerance in tomato. *Environmental and Experimental Botany* 131, 47–57. doi:10.1016/j.envexpbot.2016.06.015.

Cantero-Navarro, E., Romero-Aranda, R., Fernández-Muñoz, R., Martínez-Andújar, C., Pérez-Alfocea, F. and Albacete, A. 2016. Improving agronomic water use efficiency in tomato by rootstock-mediated hormonal regulation of leaf biomass. *Plant Science* 251, 90–100. doi:10.1016/j.plantsci.2016.03.001.

Chen, G. X., Fu, S., Lips, H. and Sagi, M. 2003. Control of plant growth resides in the shoot, and not in the root, in reciprocal grafts of flacca and wild-type tomato (*Lycopersicon esculentum*), in the presence and absence of salinity stress. *Plant and Soil* 256(1), 205–15. doi:10.1023/A:1026279719242.

Chiou, T. J. and Lin, S. I. 2011. Signalling network in sensing phosphate availability in plants. *Annual Review of Plant Biology* 62(1), 185–206. doi:10.1146/annurev-arplant-042110-103849.

Christmann, A., Weiler, E. W., Steudle, E. and Grill, E. 2007. A hydraulic signal in root-to-shoot signalling of water shortage. *The Plant Journal* 52(1), 167–74. doi:10.1111/j.1365-313X.2007.03234.x.

Colla, G., Rouphael, Y., Cardarelli, M., Massa, D., Salerno, A. and Rea, E. 2006. Yield, fruit quality and mineral composition of grafted melon plants grown under saline conditions. *Journal of Horticultural Science and Biotechnology* 81(1), 146–52. doi:10.1080/14620316.2006.11512041.

Colla, G., Rouphael, Y., Cardarelli, M., Salerno, A. and Rea, E. 2010a. The effectiveness of grafting to improve alkalinity tolerance in watermelon. *Environmental and Experimental Botany* 68(3), 283–91. doi:10.1016/j.envexpbot.2009.12.005.

Colla, G., Rouphael, Y., Leonardi, C. and Bie, Z. 2010b. Role of grafting in vegetable crops grown under saline conditions. *Scientia Horticulturae* 127(2), 147–55. doi:10.1016/j.scienta.2010.08.004.

Colla, G., Pérez-Alfocea, F. and Schwarz, D. 2017. *Vegetable Grafting: Principles and Practices*. CABI Publishing, Wallingford.

Di Gioia, F., Aprile, A., Sabella, E., Santamaria, P., Pardossi, A., Miceli, A., De Bellis, L. and Nutricati, E. 2017. Grafting response to excess boron and expression analysis of genes coding boron transporters in tomato. *Plant Biology* 19(5), 728–35. doi:10.1111/plb.12589.

Djidonou, D., Lopiano, K., Zhao, X., Simonne, E. H., Erickson, J. E. and Koch, K. E. 2015. Estimating nitrogen nutritional crop requirements of grafted tomatoes under field conditions. *Scientia Horticulturae* 182, 18 –26. doi:10.1016/j.scienta.2014.10.034.

Djidonou, D., Zhao, X., Brecht, J. K. and Cordasco, K. M. 2019a. Influence of interspecific hybrid rootstocks on tomato growth, nutrient accumulation, yield and fruit composition under greenhouse conditions. *HortTechnology* 27(6), 868–77. doi:10.21273/HORTTECH03810-17.

Djidonou, D., Zhao, X., Koch, K. E. and Zotarelli, L. 2019b. Nitrogen accumulation and root distribution of grafted tomato plants as affected by nitrogen fertilization. *Hortscience* 54(11), 1907–14. doi:10.21273/HORTSCI14066-19.

Dodd, I. C., Theobald, J. C., Richer, S. K. and Davies, W. J. 2009. Partial phenotypic reversion of ABA-deficient flacca tomato (*Solanum lycopersicum*) scions by a wild-type rootstock: normalizing shoot ethylene relations promotes leaf area but does not diminish whole plant transpiration rate. *Journal of Experimental Botany* 60(14), 4029–39. doi:10.1093/jxb/erp236.

Dong, H., Niu, Y., Li, W. and Zhang, D. 2008. Effects of cotton rootstock on endogenous cytokinins and abscisic acid in xylem sap and leaves in relation to leaf senescence. *Journal of Experimental Botany* 59(6), 1295–304. doi:10.1093/jxb/ern035.

Else, M. A., Taylor, J. M., Young, S. and Atkinson, C. J. 2018. The effect of the graft union on hormonal and ionic signalling between rootstocks and scions of grafted apple (*Malus pumila* L. Mill.). *Environmental and Experimental Botany* 156, 325–36. doi:10.1016/j.envexpbot.2018.07.013.

Estañ, M. T., Martínez-Rodríguez, M. M., Pérez-Alfocea, F., Flowers, T. J. and Bolarín, M. C. 2005. Grafting raises the salt tolerance of tomato through limiting the transport of sodium and chloride to the shoot. *Journal of Experimental Botany* 56(412), 703–12. doi:10.1093/jxb/eri027.

Estañ, M. T., Villalta, I., Bolarin, M. C., Carbonell, E. A. and Asins, M. J. 2009. Identification of fruit yield loci controlling the salt tolerance conferred by Solanum rootstocks. *Theoretical and Applied Genetics* 118(2), 305–12. doi:10.1007/s00122-008-0900-6.

FAOSTAT. 2014.*Food and Agriculture Organization of the United Nations*. Available at: http://faostat.fao.org/site/567/default.aspx#ancor. Accessed 18 October 2019.

Fazio, G., Cheng, L., Lordan, J., Francescatto, P., Grusak, M. A. and Robinson, T. L. 2018. Breeding apple rootstocks for modulation of mineral nutrients in scions. *Acta Horticulturae* 1217(1217), 29–38. doi:10.17660/ActaHortic.2018.1217.3.

Feng, Y., Liu, J., Zhai, L., Gan, Z., Zhang, G., Yang, S., Wang, Y., Wu, T., Zhang, X., Xu, X. and Hann, Z. 2019. Natural variation in cytokinin maintenance improves salt tolerance in apple rootstocks. *Plant, Cell and Environment* 42(2), 424–36. doi:10.1111/pce.13403.

Fernandez, J. A., Orsini, F., Baeza, E., Oztekin, G. B., Munosz, P., Contreras, J. and Monterro, J. I. 2018. Current trends in protected cultivation in Mediterranean climates. *European Journal of Horticultural Science* 83(5), 294–305. doi:10.17660/eJHS.2018/83.5.3.

Fullana-Pericàs, M., Ponce, J., Conesa, M. À, Juan, A., Ribas-Carbó, M. and Galmés, J. 2018. Changes in yield, growth and photosynthesis in a drought-adapted Mediterranean tomato landrace (*Solanum lycopersicum* 'Ramellet') when grafted onto commercial rootstocks and *Solanum pimpinellifolium*. *Scientia Horticulturae* 233, 70–7. doi:10.1016/j.scienta.2018.01.045.

Fullana-Pericàs, M., Conesa, M. À, Pérez-Alfocea, F. and Galmés, J. 2019. The influence of grafting on crops' photosynthetic performance. *Plant Science* in press. doi:10.1016/j.plantsci.2019.110250.

Galmés, J., Conesa, M. À, Ochogavía, J. M., Perdomo, J. A., Francis, D. M., Ribas-Carbó, M., Savé, R., Flexas, J., Medrano, H. and Cifre, J. 2011. Physiological and morphological adaptations in relation to water use efficiency in Mediterranean accessions of *Solanum lycopersicum*. *Plant, Cell and Environment* 34(2), 245–60. doi:10.1111/j.1365-3040.2010.02239.x.

Galmés, J., Ochogavía, J. M., Gago, J., Roldán, E. J., Cifre, J. and Conesa, M. À. 2013. Leaf responses to drought stress in Mediterranean accessions of *Solanum lycopersicum*: anatomical adaptations in relation to gas exchange parameters. *Plant, Cell and Environment* 36(5), 920–35. doi:10.1111/pce.12022.

GhanaVeg Sector Reports. 2016. Update on food safety and plant health. sanitary and phytosanitary status of the vegetable sector. In: Gonzale, Y. S., et al. (Eds), *Ghana Vegetable Business Opportunities in Ghana: 2016*. Embassy of the Kingdom of the Netherlands, Accra, Ghana, pp. 38–40.

Ghanem, M. E., Hichri, I., Smigocki, A. C., Albacete, A., Fauconnier, M. L., Diatloff, E., Martinez-Andujar, C., Lutts, S., Dodd, I. C. and Pérez-Alfocea, F. 2011a. Root-targeted biotechnology to mediate hormonal signalling and improve crop stress tolerance. *Plant Cell Reports* 30(5), 807–23. doi:10.1007/s00299-011-1005-2.

Ghanem, M. E., Albacete, A., Smigocki, A. C., Frebort, I., Pospisilova, H., Martinez-Andujar, C., Acosta, M., Sanchez-Bravo, J., Lutts, S., Dodd, I. C. and Perez-Alfocea, F. 2011b. Root-synthesized cytokinins improve shoot growth and fruit yield in salinized tomato (*Solanum lycopersicum* L.) plants. *Journal of Experimental Botany* 62(1), 125–40. doi:10.1093/jxb/erq266.

Gregory, P. J., Atkinson, C. J., Bengough, A. G., Else, M. A., Fernández-Fernández, F., Harrison, R. J. and Schmidt, S. 2013. Contributions of roots and rootstocks to sustainable, intensified crop production. *Journal of Experimental Botany* 64(5), 1209–22. doi:10.1093/jxb/ers385.

Guodaar, L. 2015. *Effects of Climate Variability on Tomato Crop Production in the Offinso North District of Ashanti Region. Master of Philosophy*. Kwame Nkrumah University of Science and Technology, Kumasi, Ghana.

Han, J.-S., Park, S., Shigaki, T., Hirschi, K. D. and Kim, C. K. 2009. Improved watermelon quality using bottle gourd rootstock expressing a Ca2+/H+ antiporter. *Molecular Breeding* 24(3), 201–11. doi:10.1007/s11032-009-9284-9.

Hansen, H. and Dörffling, K. 2003. Root-derived trans-zeatin riboside and abscisic acid in drought-stressed and rewatered sunflower plants: interaction in the control of leaf diffusive resistance? *Functional Plant Biology* 30(4), 365–75. doi:10.1071/FP02223.

Hawkesford, M. J. 2012. *Improving Nutrient Use Efficiency in Crops*. Encyclopedia of Life Sciences. Available at: https://onlinelibrary.wiley.com/doi/pdf/10.1002/9780470015902.a0023734.

Henderson, S. W., Dunlevy, J. D., Wu, Y., Blackmore, D. H., Walker, R. R., Edwards, E. J., Gilliham, M. and Walker, A. R. 2018. Functional differences in transport properties of

natural HKT1;1 variants influence shoot Na+ exclusion in grapevine rootstocks. *New Phytologist* 217(3), 1113–27. doi:10.1111/nph.14888.
Holbrook, N. M., Shashidhar, V. R., James, R. A. and Munns, R. 2002. Stomatal control in tomato with ABA-deficient roots: response of grafted plants to soil drying. *Journal of Experimental Botany* 53(373), 1503–14. doi:10.1093/jxb/53.373.1503.
International Food Policy Research Institute (IFPRI). 2013. *Global Food Policy Report*. p. 141. ISBN 978-0-89629-562-9. doi:10.2499/9780896295629.
Jung, J. Y., Shin, R. and Schachtman, D. P. 2009. Ethylene mediates response and tolerance to potassium deprivation in Arabidopsis. *Plant Cell* 21(2), 607–21. doi:10.1105/tpc.108.063099.
Kanyua, S. I. 2018. *Evaluation of Grafting Technology for Management of Bacterial Wilt (Ralstonia solanacearum) of Tomato (Solanum lycopersicum L.)*. Department of Agricultural Science and Technology. Kenyatta University, Nairobi, Kenya.
Kapulnik, Y. and Koltai, H. 2016. Fine-tuning by strigolactones of root response to low phosphate. *Journal of Integrative Plant Biology* 58(3), 203–12. doi:10.1111/jipb.12454.
Khah, E. M., Kakava, E., Mavromatis, A., Chachalis, D. and Goulas, C. 2006. Effect of grafting on growth and yield of tomato (*Lycopersicon esculentum* Mill.) in greenhouse and open-field. *Journal of Applied Horticulture* 8, 3–7.
Lee, J.-M. 2003. Advances in vegetable grafting. *Chronica Horticulturae* 43, 13–9.
Leonardi, C. and Giuffrida, F. 2006. Variation of plant growth and macronutrient uptake in grafted tomatoes and eggplants on three different rootstocks. *European Journal of Horticultural Science* 71, 97–101.
Li, W., de Ollas, C. and Dodd, I. C. 2018. Long-distance ABA transport can mediate distal tissue responses by affecting local ABA concentrations. *Journal of Integrative Plant Biology* 60(1), 16–33. doi:10.1111/jipb.12605.
López-Marín, J., González, A., Pérez-Alfocea, F., Egea-Gilabert, C. and Fernández, J. A. 2013. Grafting is an efficient alternative to shading screens to alleviate thermal stress in greenhouse-grown sweet pepper. *Scientia Horticulturae* 149, 39–46. doi:10.1016/j.scienta.2012.02.034.
López-Marín, J., Gálvez, A., Del Amor, F. M., Albacete, A., Fernández, J. A., Egea-Gilabert, C. and Pérez-Alfocea, F. 2017. Selecting vegetative/generative/dwarfing rootstocks for improving fruit yield and quality in water stressed sweet peppers. *Scientia Horticulturae* 214, 9–17. doi: 10.1016/j.scienta.2016.11.012.
Lynch, J. P. 2011. Root phenes for enhanced soil exploration and phosphorus acquisition: tools for future crops. *Plant Physiology* 156(3), 1041–9. doi:10.1104/pp.111.175414.
Lynch, J. P. 2019. Root phenotypes for improved nutrient capture: an underexploited opportunity for global agriculture. *New Phytologist* 223(2), 548–64. doi:10.1111/nph.15738.
Marguerit, E., Brendel, O., Lebon, E., Van Leeuwen, C. and Ollat, N. 2012. Rootstock control of scion transpiration and its acclimation to water deficit are controlled by different genes. *The New Phytologist* 194(2), 416–29. doi:10.1111/j.1469-8137.2012.04059.x.
Martín, A. C., Del Pozo, J. C., Iglesias, J., Rubio, V., Solano, R., De La Peña, A., Leyva, A. and Paz-Ares, J. 2000. Influence of cytokinins on the expression of phosphate starvation responsive genes in Arabidopsis. *The Plant Journal* 24(5), 559–67. doi:10.1046/j.1365-313x.2000.00893.x.
Martin-Vertedor, A. I. and Dodd, I. C. 2011. Root-to-shoot signalling when soil moisture is heterogeneous: increasing the proportion of root biomass in drying soil inhibits

leaf growth and increases leaf ABA concentration. *Plant, Cell and Environment* 34(7), 1164–75. doi:10.1111/j.1365-3040.2011.02315.x.

Martínez-Andújar, C., Albacete, A., Martínez-Pérez, A., Pérez-Pérez, J. M., Asins, M. J. and Pérez-Alfocea, F. 2016. Root-to-shoot hormonal communication in contrasting rootstocks suggests an important role for the ethylene precursor aminocyclopropane-1-carboxylic acid in mediating plant growth under low-potassium nutrition in tomato. *Frontiers in Plant Science* 7, 1782. doi:10.3389/fpls.2016.01782.

Martínez-Andújar, C., Ruiz-Lozano, J. M., Dodd, I. C., Albacete, A. and Pérez-Alfocea, F. 2017. Hormonal and nutritional features in contrasting rootstock-mediated tomato growth under low-phosphorus nutrition. *Frontiers in Plant Science* 8, 533. doi:10.3389/fpls.2017.00533.

Martínez-Andújar, C., Albacete, A. and Pérez-Alfocea, F. 2020a. Rootstocks for increasing yield stability and sustainability in vegetable crops. *Acta Horticulturae*, 449–70. doi:10.17660/ActaHortic.2020.1273.58.

Martínez-Andújar, C., Martínez-Pérez, A., Ferrández-Ayela, A., Albacete, A., Martínez-Melgarejo, P. A., Dodd, I. C., Thompson, A. J., Pérez, J. M. and Pérez-Alfocea, F. 2020b. Impact of overexpression of 9-cis-epoxycarotenoid dioxygenase on growth and gene expression under salinity stress. *Plant Science* in press. https://doi.org/10.1016/j.plantsci.2019.110268.

McAdam, S. A., Brodribb, T. J. and Ross, J. J. 2016. Shoot-derived abscisic acid promotes root growth. *Plant, Cell and Environment* 39(3), 652–9. doi:10.1111/pce.12669.

Munns, R. and Tester, M. 2008. Mechanisms of salinity tolerance. *Annual Reviews of Plant Biology* 59, 651–81. doi:10.1146/annurev.arplant.59032607.092911.

Nagarajan, V. K. and Smith, A. P. 2012. Ethylene's role in phosphate starvation signaling: more than just a root growth regulator. *Plant and Cell Physiology* 53(2), 277–86. doi:10.1093/pcp/pcr186.

Nawaz, M. A., Han, X., Chen, C., Zheng, Z., Shireen, F., Bie, Z. and Huang, Y. 2018. Nitrogen use efficiency of watermelon grafted onto 10 wild watermelon rootstocks under low nitrogen conditions. *Agronomy* 8(11), 259. doi:10.3390/agronomy8110259.

Nkansah, G. O., Ahwireng, A. K., Amoatey, C. and Ayarna, A. 2013. Grafting onto African eggplant enhances growth, yield and fruit quality of tomatoes in tropical forest ecozones. *Journal of Applied Horticulture* 15, 16–20.

Nor, N. H. M., Fujinima, R., Menzies, N. W., Harper, S. and McGrath, D. 2016. Selection for root morphological traits improves the growth of grafted bell pepper. *Acta Horticulturae* 1127, 211–6.

Nyaku, S. T. and Amissah, N. 2019. *Grafting: an Effective Strategy for Nematode Management in Tomato Genotypes.* Recent Advances in Tomato Breeding and Production. Provisional Chapter, pages 1–15.

Ofori, E. 2016. Influence of grafting on the growth, yield, quality and shelf life of tomatoes (*Solanum lycopersicum* L.) grafted onto three Solanum species. Master of Philosophy, University of Ghana. Available at: http://ugspace.ug.edu.gh.

Oztekin, G. B., Giuffrida, F., Tuzel, Y. and Leonardi, C. 2009. Is the vigour of grafted tomato plants related to root characteristics ? *Journal of Food, Agriculture and Environment* 7, 364–8.

Oztekin, G. B., Tuzel, Y. and Tuzel, I. H. 2013. Does mycorrhiza improve salinity tolerance in grafted plants? *Scientia Horticulturae* 149, 55–60. doi:10.1016/j.scienta.2012.02.033.

Pérez-Alfocea, F., Albacete, A., Ghanem, M. E. and Dodd, I. C. 2010. Hormonal regulation of source-sink relations to maintain crop productivity under salinity: a case study of root-to-shoot signalling in tomato. *Functional Plant Biology* 37(7), 592–603. doi:10.1071/FP10012.

Plouznikoff, K., Asins, M. J., de Bulois, H. D., Carbonell, E. A. and Declerck, S. 2019. Genetic analysis of tomato root colonisation by arbuscular mycorrhizal fungi. *Annals of Botany* 124(6), 933–46. doi:10.1093/aob/mcy240.

Qi, H. Y., Liu, Y. F., Li, D. and Li, T. L. 2006. Effects of grafting on nutrient absorption, hormone content in xylem exudation and yield of melon (*Cucumis melo* L.). *Plant Physiology Communications* 42, 199–202.

Rouphael, Y., Cardarelli, M., Colla, G. and Rea, E. 2008. Yield, mineral composition, water relations, and water use efficiency of grafted mini-watermelon plants under deficit irrigation. *Hortscience* 43(3), 730–6. doi:10.21273/HORTSCI.43.3.730.

Sanoubar, R., Orsini, F. and Gianquinto, G. 2013. Ionic partitioning and stomatal regulation: dissecting functional elements of the genotypic basis of salt stress adaptation in grafted melon. *Plant Signaling and Behavior* 8(11), e27334, PMID: 24309549. doi:10.4161/psb.27334.

Savvas, D., Colla, G., Rouphael, Y. and Schwarz, D. 2010. Amelioration of heavy metal and nutrient stress in fruit vegetables by grafting. *Scientia Horticulturae* 127(2), 156–61. doi:10.1016/j.scienta.2010.09.011.

Savvas, D., Savva, A., Ntatsi, G., Ropokis, A., Karapanos, I., Krumbein, A. and Olympios, C. 2011. Effects of three commercial rootstocks on mineral nutrition, fruit yield, and quality of salinized tomato. *Journal of Plant Nutrition and Soil Science* 174(1), 154–62. doi:10.1002/jpln.201000099.

Schwarz, D., Öztekin, G. B., Tüzel, Y., Brückner, B. and Krumbein, A. 2013. Rootstocks can enhance tomato growth and quality characteristics at low potassium supply. *Scientia Horticulturae* 149, 70–9. doi:10.1016/j.scienta.2012.06.013.

Sharp, R. E., LeNoble, M. E., Else, M. A., Thorne, E. T. and Gherardi, F. 2000. Endogenous ABA maintains shoot growth in tomato independently of effects on plant water balance: evidence for an interaction with ethylene. *Journal of Experimental Botany* 51(350), 1575–84. doi:10.1093/jexbot/51.350.1575.

Suchoff, D. H., Gunter, C. C. and Louws, F. J. 2017. Comparative analysis of root system morphology of tomato rootstocks. *HortTechnology* 27(3), 319–24. doi:10.21273/HORTTECH03654-17.

Suchoff, D. H., Schultheis, J. R., Kleinhenz, M. D., Louws, F. J. and Gunter, C. C. 2018a. Rootstock improves high-tunnel tomato water-use efficiency. *HortTechnology* 28(3), 344–53. doi:10.21273/HORTTECH04015-18.

Suchoff, D. H., Gunter, C. C., Schultheis, J. R., Kleinhenz, M. D. and Louws, F. J. 2018b. Rootstock effect on grafted tomato transplant shoot and root responses to drying soils. *HortScience* 53(11), 1586–92. doi:10.21273/HORTSCI13215-18.

Suchoff, D. H., Gunter, C. C., Schultheis, J. R., Hassell, R. L. and Louws, F. J. 2019. The effect of grafting on nitrogen use in determinate field grown tomatoes. *The Journal of Horticultural Science and Biotechnology* 94(1), 102–9. doi:10.1080/14620316.2018.1450645.

Sun, J., Cao, H., Cheng, J., He, X., Sohail, H., Niu, M., Huang, Y. and Bie, Z. 2018. Pumpkin CmHKT1;1 controls shoot Na+ accumulation via limiting Na+ transport from rootstock to scion in grafted cucumber. *International Journal of Molecular Sciences* 19(9), 2648. doi:10.3390/ijms19092648.

Thompson, A. J., Andrews, J., Mulholland, B. J., McKee, J. M. T., Hilton, H. W., Horridge, J. S., Farquhar, G. D., Smeeton, R. C., Smillie, I. R. A., Black, C. R. and Taylor, I. B. 2007.

Overproduction of abscisic acid in tomato Increases transpiration efficiency and root hydraulic conductivity and influences leaf expansion. *Plant Physiology* 143(4), 1905–17. doi:10.1104/pp.106.093559.

Tworkoski, T., Fazio, G. and Glenn, D. M. 2016. Apple rootstock resistance to drought. *Scientia Horticulturae* 204, 70–8. doi:10.1016/j.scienta.2016.01.047.

Ulas, A., Doganci, E., Ulas, F. and Yetisir, H. 2019. Root-growth characteristics contributing to genotypic variation in nitrogen efficiency of bottle gourd and rootstock potential for watermelon. *Plants* 8(3), 77. doi:10.3390/plants8030077.

Wang, Y., Li, B., Du, M., Eneji, A. E., Wang, B., Duan, L., Li, Z. and Tian, X. 2012. Mechanism of phytohormone involvement in feedback regulation of cotton leaf senescence induced by potassium deficiency. *Journal of Experimental Botany* 63(16), 5887–901. doi:10.1093/jxb/ers238.

Zhang, Z., Liao, H. and Lucas, W. J. 2014. Molecular mechanisms underlying phosphate sensing, signaling, and adaptation in plants. *Journal of Integrative Plant Biology* 56(3), 192–220. doi:10.1111/jipb.12163.

Zhong, Y., Chen, C., Nawaz, M. A., Jiao, Y., Zheng, Z., Shi, X., Xie, W., Yu, Y., Guo, J., Zhu, S., Xie, M., Kong, Q., Cheng, F., Bie, Z. and Huang, Y. 2018. Using rootstock to increase watermelon fruit yield and quality at low potassium supply: a comprehensive analysis from agronomic, physiological and transcriptional perspective. *Scientia Horticulturae* 241, 144–51. doi:10.1016/j.scienta.2018.06.091.

Chapter 19

Delivering improved phosphorus acquisition by root systems in pasture and arable crops

Richard J. Simpson and Rebecca E. Haling, CSIRO Agriculture and Food, Australia; and Phillip Graham, Graham Advisory, Australia

1 Introduction

Intensification of food production since the 1940s has been underpinned by fertiliser use and is critical to our ability to feed the human population of the world (Stewart et al., 2005; Erisman et al., 2008). Phosphorus (P) is one of the handful of nutrients used in fertilisers that are now essential for global food security (Stewart et al., 2005; Sutton et al., 2013; Sattari et al., 2014). Most mineral P fertilisers are currently manufactured from phosphate rock, a non-renewable resource. Unfortunately, the global use of P is marked by inefficiencies in handling and use along the entire supply-demand chain (Weaver and Wong, 2011; Scholz and Wellmer, 2015; Scholz and Wellmer, 2018; Lun et al., 2018). In agriculture, only about half of the total P inputs to soil are recovered by crops. Roughly half of the unrecovered P is accumulated in fertilised soils, and

http://dx.doi.org/10.19103/AS.2020.0075.26

the other half becomes dispersed in the natural environment as a result of erosion, runoff or leaching (e.g. Vaccari et al., 2019; Lun et al., 2018; Bouwman et al., 2013; MacDonald et al., 2011; Cordell et al., 2009). Phosphorus lost from agriculture to the aquatic P cycle is contributing to eutrophication of waterways and ecosystem damage on an unprecedented scale throughout the world (Jarvie et al., 2013a,b; Campbell et al., 2017). Accelerated loss of P to the aquatic cycle also effectively ends our capacity to recycle this P for use in food production because the timeframe for P in the aquatic cycle is millennia (Filippelli, 2008; Föllmi, 1996).

Global estimates of P flows, such as these, clearly have their uses but they mask the diversity of farming systems across the world and, consequently, the opportunities for more sustainable use of the world's finite P reserves. There could not be a wider divergence in the P status of the soils that are used globally for agriculture. At one extreme are soils that are naturally P fertile or have been fertilised for so long (e.g. Vitousek et al., 2009; Sattari et al., 2012; Barrow and Debnath, 2014) that maximum yields can be achieved without continuous application of P fertiliser. Applications of P fertiliser to soils such as these, often exacerbate P losses to aquatic systems and contribute significantly to ecosystem damage (Jarvie et al., 2013b; Rowe et al., 2016). In contrast, vast areas of agricultural soil are P deficient for optimum crop growth, and production per hectare is either low, relative to the potential of the agricultural production system, or relies on the use of P fertiliser to achieve high yields (e.g. Stewart et al., 2005; Vitousek et al., 2009; Lynch, 2011).

In natural ecosystems, plants adapt to low-P soils using a variety of P acquisition mechanisms (e.g. acclimation of root architecture and morphology, mycorrhizal symbioses, specialised root structures, organic anion exudation, rhizosphere acidification), that either enhance the diffusion of P to the root or enable greater P interception and/or extraction (Lambers et al., 2008; Lambers et al., 2010; Richardson et al., 2011). Plants that evolve in these systems will often combine acclimation to low-P soil, with some form of P-use strategy that also confers a degree of low-P tolerance (e.g. slow growth, high internal P utilisation efficiency; Lambers et al., 2010; Waddell et al., 2015). However, sustainable, high crop growth rates are an imperative in an agricultural context because the social, environmental, and monetary costs of using land and water resources inefficiently for food and fibre production are too high (Tilman et al., 2011; Fischer and Connor, 2018).

It is in this context that agronomic practices can be used to enhance P acquisition by roots, and P-efficient plants can be used to produce more food and fibre from P-deficient soils, or to lower the fertiliser cost of correcting P deficiency for high production. Phosphorus-efficient plants can also help to mitigate the risk of P losses from agriculture to the wider environment.

2 Options for improving phosphorus acquisition by roots in pastures and arable crops

Phosphate in soil solution can be acquired by plants as a result of contact-exchange between the root and the soil (root interception), via mass flow in the movement of soil water towards the root, and by diffusion in the water-filled path between soil particles and the root. The relative contribution to P uptake by each mode of acquisition will vary depending on the concentration of available P in the soil. However, it is usual to expect that only 1% or less of available P will be supplied by root interception and 2–3% via mass flow due to the very low concentrations of phosphate in soil solution. The majority of P uptake occurs as a result of phosphate diffusion to the root (Barber, 1995; Tinker and Nye, 2000).

Diffusion of phosphate in soil may be approximated using Fick's first law of diffusion in which it is assumed that the rate of diffusion is stable over the time interval of measurement:

$$\mathbf{J} = -\mathbf{D} * \mathbf{A} * d\mathbf{C} / d\mathbf{x} \qquad (1)$$

where: ***J*** is the diffusive flux (μmol/s), ***A*** is the area of the absorbing root-soil interface (cm^2), ***D*** is the isothermic diffusion coefficient (cm^2/s), ***C*** is the phosphate concentration ($\mu mol/cm^3$) and $d\mathbf{C}/d\mathbf{x}$ is the concentration gradient created between the root-soil interface and the bulk soil when the concentration of phosphate at the root-soil interface is depleted as P is taken up by the root.

The rate of phosphate diffusion in soil is very low compared to other nutrients (e.g. nitrate or potassium) and is rate limiting for P acquisition from soils up to the point where soil P fertility is sufficient for maximum growth rates (Barber, 1995; Tinker and Nye, 2000). Consequently, successful agronomic and plant-based strategies to increase P acquisition by plant roots have been guided predominantly by the key principles embodied in the diffusion equation, rather than the possibility of modifying other steps in P uptake and metabolism by plants (e.g. rates at which P is transported across cell membranes during P uptake from soil solution: Mitsukawa et al., 1997; Rae et al., 2004; Richardson, 2009; Heuer et al., 2017).

The key principles are: (1) the rate of P diffusion to the root is directly proportional to the P concentration gradient between the bulk soil and the absorptive surface of the root system, and is (2) inversely proportional to the distance over which P diffusion occurs; and (3) the rate of P uptake by the plant is directly determined by the size of the absorptive surface area of the root system or, in the case of plants colonised by arbuscular mycorrhizal fungi (AMF), the effective surface area of the root-hyphae system.

2.1 Managing phosphate diffusion

The phosphate concentration in soil solution (***C***, Eqn 1) is most effectively manipulated by the application of P fertilisers (particularly soluble P fertilisers) to directly increase the solution P concentration, but it may also be improved by modification of a plant's rhizosphere (see comments below). Fertiliser efficiency is often greatly improved when P is banded close to where roots are expected to grow. In cropping systems this is typically achieved by placing P fertiliser below, or to the side of seeds at planting. However, in a non-arable system such as a permanent pasture, a relatively concentrated band of P may also be created by broadcasting P fertiliser onto the surface of the soil. The phosphorus chemistry of banded P is complex (Kar et al., 2012) and the effectiveness of bands is influenced by the P buffering capacity of the soil to which they have been applied (Dibb et al., 1990). In some instances, the way the fertiliser band is created must also take account of soil chemistry to ensure that the potential benefits are realised (e.g. compare the effectiveness of granular-P and fluid-P fertiliser bands in calcareous soils, Lombi et al., 2004).

Nevertheless, the basic principles are that: (i) P banding allows ***C*** to be elevated in the vicinity of the root at a lower rate of P application than would otherwise be necessary; and (ii) the effectiveness of the P band is enhanced because plants usually increase ***A*** (the absorptive surface area of their root system, Eqn 1) by proliferating root length in the P band (Duncan and Ohlrogge, 1958; Yao and Barber, 1986). Savings in P fertiliser can be very large (e.g. 30–60% less P fertiliser, Jarvis and Bolland, 1991; up to 75% less P fertiliser, Sanchez et al., 1991) compared with broadcast applications of P fertiliser, but the relative benefit depends on the initial P fertility of the soil. The greatest benefit from banding P is achieved in P-deficient soil and the least benefit in soils that are already close to the soil P fertility level needed for maximum crop yield (Sanchez et al., 1991). The volume of a fertiliser band may also limit its effectiveness for P acquisition if it is too small, and therefore limits the root length proliferation that can be achieved within the band (i.e. increase in ***A***). If it is necessary to apply higher rates of P, a larger band volume will be required to maximise P acquisition (e.g. Borkert and Barber, 1985; Duncan and Ohlrogge, 1958).

The diffusion coefficient and the distance over which P must diffuse (***D*** and *d**x***, Eqn 1) are primarily characteristics of the soil, but *d**x*** is strongly modified in drying soil because the tortuosity of the diffusion path is increased or the path is broken. Phosphorus concentrated near the soil surface is most vulnerable to soil drying and this accounts for numerous reports that deeper placement of P bands can substantially delay the impact of moisture stress on crop or pasture nutrition and growth (Scott, 1973; Cornish and Myers, 1977; Pinkerton and Simpson, 1986; Valizadeh et al., 2003)

2.2 Increasing the absorptive surface of a plant root system

Most plant-based examples of improved P-acquisition efficiency involve modification of ***A*** (the area of the P absorbing root-soil interface, Eqn 1). Increasing ***A***, particularly in P-enriched zones of a P-deficient soil profile: (i) directly improves the acquisition of P (Lynch, 2019) and (ii) enables the plant to achieve its maximum yield at a lower value of ***C*** (Haling et al., 2016a; Haling et al., 2016b; Sandral et al., 2018). The former is important when the principal objective is to achieve higher yields in P-impoverished soils and low-input agriculture (e.g. Burridge et al., 2019); the latter is the key to reducing fertiliser input costs associated with P accumulation in fertilised soil or P loss by leaching and runoff (Simpson et al., 2014).

A variety of root system traits contribute to a large ***A*** for P acquisition (e.g. Table 1). Plants allocate mass to nutrient foraging roots, root mass is converted to length via the specific root length, and root length is translated into ***A*** via the root radius and root hair length. Phosphorus uptake by genotypes within a plant species is often correlated with total root length alone (i.e. root length is a reasonable proxy for ***A***), but when comparing among species it becomes essential to also account for root radius and root hair length (e.g. McLachlan et al., 2019). It is, consequently, feasible to use the specific surface area of the root hair cylinder (specific SARHC = π * [root radius + root hair length]^2 * root length / root mass) as a first approximation for predicting which plant species are likely to have a lower 'critical'[1] P requirement and an improved capacity to acquire P from low-P soil (Yang et al., 2017; Sandral et al., 2018). However, specific SARHC alone may not predict P-efficient genotypes within a species because the acclimation of root traits to low-P soil (root plasticity) becomes increasingly important for improved P-acquisition efficiency within a species (e.g. Haling et al., 2016b; Zhu et al., 2010). The most successful strategy, to date, for identifying and developing crop genotypes and alternative plant species for improved P-acquisition efficiency in agriculture has been to utilise variation in nutrient foraging root traits (e.g. Table 1; Lynch, 2007; Lynch, 2011; Lynch, 2019; Burridge et al., 2019; Wissuwa et al., 2016; Gamuyao et al., 2012; Mori et al., 2016; Hufnagel et al., 2014; Manske et al., 2000; Zhao et al., 2004; Sandral et al., 2018; Sandral et al., 2019).

When a plant is colonised by AMF, ***A*** is also modified by the 'effective' surface for P acquisition of the mycorrhizal hyphae. However, the net effectiveness of the root-AMF association for P uptake may not be as large as anticipated from the potential reach of the hyphal network. In addition, selection for nutrient-foraging root traits appears to deliver improved

1 Here we use the term 'critical' P requirement to denote the P supply rate or soil test P concentration that supports near maximum crop yield in a P deficient soil. Typically, this is defined as the soil test P concentration corresponding with 90% or 95% of maximum yield (e.g. Gourley at al., 2019).

Table 1 Examples of roots traits that increase the root-soil interface and/or the efficiency of this interface for improved P acquisition from low-P soil.

Root trait	Impact/ benefits	Example(s)	Reference
Preferential allocation of mass to foraging roots (i.e. root proliferation)	Increased size of root system; topsoil foraging	Universal response of many species to nutrient deficiency including many temperate pasture legume species (e.g. *T. subterraneum*)	Brouwer (1962); Freschet et al. (2015); Haling et al. (2016a); Haling et al. (2018); Sandral et al. (2018)
Increased root number, branching, root length (e.g. basal, crown, lateral roots; high topsoil root length density)	Topsoil foraging	*Phaseolus vulgaris* L.; *Zea mays* L.; *Oryza sativa* L.; *Sorghum bicolor* (L.) Moench; *Triticum aestivum* L.	Miguel et al. (2013); Sun et al. (2018); Jia et al. (2018); Zhu and Lynch (2004); Gamuyao et al. (2012); Hufnagel et al. (2014); Manske et al. (2000)
Shallow root angles	Topsoil foraging	*Phaseolus vulgaris*; *Zea mays*; *Oryza sativa*; *Glycine max* (L.) Merr.	Miguel et al. (2015); Ho et al. (2005); Lynch (2011); Mori et al. (2016); Zhao et al. (2004)
High specific root length	Effective allocation of carbon to increased nutrient foraging;	*Ornithopus* spp.; *Trifolium subterraneum* L.; *Citrus sinensis* (L.) Osbeck	Haling et al. (2016a, 2018); Eissenstat (1991)
Long, dense root hairs	Effective allocation of carbon to increase root-soil interface	*Phaseolus vulgaris*; *Triticum aestivum* L.; *Hordeum vulgare* L.; *Oryza sativa*.; *Glycine max*; *Zea mays*	Miguel et al. (2015); Gahoonia et al. (1997); Nestler and Wissuwa (2016); Vandamme et al. (2013); Zhu et al. (2010)
Root cortical aerenchyma	Lower root metabolic costs	*Zea mays* (modelling)	Postma and Lynch (2011)
Root cortical senescence	Lower root metabolic costs	*Hordeum vulgare* (modelling); *Phaseolus vulgaris*	Schneider et al. (2017a); Strock et al. (2018)
Tolerance of biotic or abiotic constraints to root and/ or root hair growth	Increased size of root system/root-soil interface	*Hordeum vulgare* (Al-tolerance)	Delhaize et al. (2009)

P-acquisition efficiency irrespective of a plant's interactions with AMF (Strock et al., 2018; Lynch, 2019; McLachlan et al., 2020). Possible reasons for this are discussed below.

There is considerable diversity among plant species in the expression of P-efficient root traits, in root acclimation responses to localised P supply, and in the relative importance of the morphological and physiological responses of roots to P deficiency (e.g. Lyu et al., 2016). Consequently, a successful root trait strategy for improving P acquisition in one crop species may not always be ideal for improving P efficiency in another. For example, shallow axial root angles that increase the density of root length in topsoil layers have proven to be a useful way to improve P-acquisition efficiency in *Phaseolus vulgaris* L., *Glycine max* (L.) Merr. and *Zea mays* L. (Miguel et al., 2015; Zhao et al., 2004; Lynch, 2011) but were found to be unrelated to P-acquisition efficiency in *Sorghum bicolor* (L.) Moench (Marcus, 2013). Likewise, root length proliferation in a nutrient-rich patch of soil is a very common nutrient acquisition response (Robinson and Van Vuuren, 1998) yet some important species have relatively weak root proliferation responses to localised P supply (e.g. *Vicia faba* L., *Glycine max,* Li et al., 2014a; Lyu et al., 2016). Root hairs are often assumed to assist P acquisition by becoming longer in low-P soil (e.g. *Solanum lycopersicum* L., *Brassica napus* L., *Spinacia oleracea* L., *Zea mays* and *Arabidopsis thaliana* (L.) Heynh., Föhse and Jungk, 1983; Bates and Lynch, 1996; Zhu et al., 2010), but this is not the case in all species (e.g. *Trifolium* spp., Haling et al., 2016b; McLachlan et al., 2019; McLachlan et al., 2020). In contrast, flexible acclimation of root hairs to low-P soil exists among lines of *Zea mays*; root hairs may be relatively short with the ability to lengthen in low P conditions, short and unresponsive to P, or long and unresponsive to P (Zhu et al., 2010).

Combining beneficial P efficiency traits can have a synergistic impact on P acquisition. Miguel et al. (2015) demonstrated, using recombinant inbred lines of *P. vulgaris*, that varieties with long root hairs achieved 1.9–2.5-fold larger shoot yields than short root hair varieties and that shallow rooted varieties grew 1.6–2.1-fold larger than deep rooting varieties as a result of improved topsoil foraging and P acquisition in P deficient soil. However, varieties that combined long root hairs with shallow topsoil roots achieved the largest yields. The P acquisition benefit was larger than would be anticipated from the additive effect of these traits alone. However, it is unwise to assume that this will always be the case. Experiments that have compared the P acquisition of 'short' and 'long' root hair mutants (e.g. the *rht2* mutant of *Zea mays*, Wen and Schnable, 1994; or *RSL2* and *RSL3* mutants of *Brachypodium distachyon* (L.) P. Beauv., Zhang et al., 2018) have not always found longer, more dense root hairs to be beneficial (e.g. Zhang et al., 2018; Klamer et al., 2019). While adverse pleotropic effects of ectopic transgenes cannot be ruled out in these experiments, it is also observed that plants can be very plastic in their root system acclimation to P stress and may react to changes in one root trait by adjusting acclimation to P stress in another. This has been found to be the case in a short root hair variant of *Zea mays* which produces more fine roots than the long root hair variant and partially compensates for its shorter root hair length (Weber et al., 2018; Klamer et al., 2019). Consequently, traits that should confer improved P acquisition may not always appear to be associated with a net P-efficiency benefit (e.g. Nestler and Wissuwa, 2016).

2.3 Mycorrhizal fungi

With a few notable exceptions, AMF colonise the roots of a majority of agricultural plant species (Smith and Read, 2008) where they form potentially symbiotic associations that increase the surface area for P uptake (i.e. the mycorrhizosphere) through the development of an extensive root-hyphal network. In P-deficient soils, the potential increase in soil volume explored can be very large (e.g. ~15-fold, Mai et al., 2019) but the increase in yield due to improved P acquisition is not as large as might be anticipated (Richardson et al., 2011; McLachlan et al., 2020). Indeed, there is an on-going debate in agriculture concerning the benefit or otherwise of AMF in high production systems (Ryan and Graham, 2018). The discussion is multifacetted, but key issues that influence discussion of the degree to which AMF improve P-acquisition by plants include: (i) AMF can improve P acquisition by plants growing in P-deficient soil, but the P-acquisition benefit diminishes when P fertiliser is applied and crop yields may even be marginally depressed at optimum P supply (Ryan and Graham, 2018); this creates a mismatch in the optimum P supply for the AMF association versus that for high crop production (Mai et al., 2018), (ii) some crop species are clearly more AMF-dependent than others (Schweiger et al., 1995; Smith and Read, 2008; Ryan and Graham, 2018), (iii) yield and P uptake responses to AMF colonisation are often relatively large in sterilised, P-deficient soil, but the P acquisition benefits can range from negligible to significant when experiments are conducted using non-sterile field soils (Svenningsen et al., 2018; Cruz-Paredes et al., 2019). This is thought to be a consequence of competition with, or suppression of, AMF by other soil microbiota (Svenningsen et al., 2018; Cruz-Paredes et al., 2019). In addition to these issues, *P. vulgaris* variants that developed longer topsoil roots as a consequence of increased root cortical senescence have been shown to be more P-efficient despite significantly impaired AMF colonisation (Strock et al., 2018). The role of AMF in P acquisition is discussed further in the case study in Section 2.4.

2.4 Alleviation of constraints to root growth

Finally, it is important to remember that other agronomic strategies can also have a significant role in improving the absorptive capacity for P acquisition by crop roots. Alleviation of constraints to root growth such as hard, dry, or toxic (e.g. saline, acid) soil conditions will concurrently improve nutrient acquisition and, in some circumstances, may prove to be a more cost-effective and durable way to increase P-acquisition efficiency.

A clear example of this occurs in very acidic soils where aluminium (Al) concentrations are elevated in soil solution and become toxic for root growth. This in turn, restricts nutrient foraging and P acquisition (amongst other problems for a crop). Delhaize et al. (2009) demonstrated how the incorporation of a gene for tolerance of toxic aluminium restored root growth by *Hordeum vulgare* L. in a very acid soil and provided a partial restoration of P-acquisition efficiency. Only partial restoration of P acquisition was achieved because the gene protected the growth of axial and lateral root length, but did not protect the growth of root hairs and rhizosheath development which were also sensitive to Al-toxicity (Haling et al., 2010a,b). The application of lime to raise soil pH and eliminate Al-toxicity, consequently, provided a more complete solution for both root growth and P efficiency.

When applying these principles in the field, it is also essential to develop a wholistic understanding of the constraint to production. For example, the depth to which soil acidity extends in acid soil profiles varies substantially. It is often difficult to physically and economically address deep soil acidity by lime incorporation (Scott et al., 1997; Scott and Coombes, 2006). In these circumstances, the most effective and durable solution for improving nutrient acquisition is to use Al-tolerant varieties in combination with lime application (Scott et al., 1997; Scott et al., 2001).

3 Rhizosphere modifications that improve phosphorus acquisition

The phosphate concentration of soil solution (***C***, Eqn 1) may also be increased by plants that have a capacity to modify their rhizosphere in ways that promote desorption, solubilisation, or mineralisation of P from sparingly available soil P pools. The main ways this may be achieved include:

i Acidification of the rhizosphere by exudation of protons.
 Acidification of rhizosphere soil dissolves acid-soluble forms of P that are, otherwise, not readily available for plant uptake (Hedley et al., 1982; Li et al., 2003).

ii Exudation of organic acid anions from roots.

Organic acid anions increase the plant-availability of P by competing with inorganic P (Pi) and organic P (Po) for adsorption sites on soil particles, and mobilise sparingly soluble and sorbed P via ligand-promoted dissolution of P that is complexed with aluminium and iron oxides or calcium (Ryan et al., 2001; Wang and Lambers, 2020). When present in high enough concentrations, organic anions may modify the soil P chemistry significantly, reducing the binding strength of adsorbed phosphate, decreasing the chemical affinity of Al^{3+} for phosphate; (Wang and Lambers, 2020) and increasing the diffusion coefficient (i.e. ***D***, Eqn 1) of phosphate in the soil solution by orders of magnitude (Gerke, 1994). These impacts are modified by soil/rhizosphere pH and soil type (Oburger et al., 2011; Wang and Lambers, 2020).

iii Exudation of phosphatases.

Soils have considerable phosphatase activity, but activity is enhanced substantially in the rhizosphere where it is associated with depletion of soil-organic P (Tarafdar and Jungk, 1987; Gahoonia and Nielsen, 1992; Chen et al., 2002; George et al., 2002). Although these assessments do not distinguish between plant- and microbially derived phosphatase activities, a wide range of plants are known to exude phosphatase from their roots. When grown in sterile culture, many plants can hydrolyse a variety of organic P sources (e.g. glucose-1-P) for subsequent uptake and use. However, some forms of organic P common in soils, such as inositol phosphates, are relatively poorly utilised by plant roots (e.g. Hayes et al., 2000; Richardson et al., 2001). It was shown that this particular deficiency in root capability could be remedied by introducing a culture of soil microorganisms to the plant growth medium (Richardson et al., 2001); this highlights the importance of co-operativity between plant roots and soil microorganisms.

iv Stimulation of microbial activity (other than that of the AMF) in the rhizosphere.

Microorganisms are crucial to the cycling of Pi and Po in the detritusphere and the rhizosphere with significant proportions of total culturable bacteria and fungi reported to have Pi-solubilising/mobilising activity in laboratory media (e.g. Kucey et al., 1989; Bowen and Rovira, 1999; Schneider et al., 2017b). Microbial biomass and activity in the rhizosphere is many-fold greater than in bulk soil as a consequence of the release of mucilages, sugars, organic acid anions, and more complex carbon derived from root turnover and sloughed cells (Jones et al., 2009), and the amount of P held in the microbial biomass is considerably larger than that in the soil solution P pool (Seeling and Zasoski, 1993). Phosphorus is cycled rapidly and continuously through the microbial biomass (Oehl et al., 2001) and although it is often assumed that rhizosphere microbial activity provides a net P mineralisation (or solubilisation) benefit to plants, this is not guaranteed given the low P environment of the rhizosphere and will depend on the C:P ratio of the organic residues supporting microbial activity, the ability of the microbiome to solubilise P from the mineral phase, and the dynamics of microbial turnover (Richardson and Simpson, 2011; Meyer et al., 2019). Indeed, the stable microbial turnover of P from the soil solution Pi pool observed by Oehl et al. (2001) under 'steady-state' microbial conditions suggests that a net microbial release of P to the rhizosphere may only occur when the microbial biomass is in decline. As with other rhizosphere modifications, the high abundance of microorganisms and elevated rhizosphere enzyme activities are typically localised to within only a few millimetres of the root surface (Tarafdar and Jungk, 1987; Kandeler et al., 2001; Chen et al., 2002; Marschner et al., 2012).

3.1 Increasing phosphorus acquisition by rhizosphere management

With the exception of enhanced rhizosphere microbial activity, the mechanisms used by plants to modify P acquisition in the rhizosphere are typically enhanced as part of a plant species' general response to growth in P-deficient soil (Vance et al., 2003; Richardson et al., 2011; Wang and Lambers, 2020). Rhizosphere modifications typically influence the soil within only a few millimetres of the surface of the root/root hair cylinder (Marschner et al., 2012). Consequently, for many plant species, rhizosphere modifications are complimentary to, and may not replace the importance of a large root system for effective soil exploration (e.g. Mat Hassan et al., 2013; Mori et al., 2016). The limitations of the localised impact of rhizosphere modification are also recognised in intercropping research, because 'facilitated' P acquisition (where an intercropped species increases the size of the available P pool and facilitates improved P access for its neighbour) requires 'intimate intermingling' of roots by the two intercropped species (e.g. *Vicia faba* - *Zea mays* intercropped on alkaline soils, Li et al., 2003; Hinsinger et al., 2011). However, this is sometimes forgotten in P-acquisition research where it is often assumed that beneficial changes in rhizosphere biochemistry will translate to significantly improved P acquisition in the field. Confidence in the value of organic acid anion secretion from roots is provided by species like *Lupinus albus* L. which develops clusters of specialised rootlets with long root hairs that exude very high concentrations of citrate (Dinkelaker et al., 1989), sufficient to influence soil ~6 mm or more from the cluster root (Dessureault-Rompre et al., 2007). The formation of each cluster of rootlets is transient and other exudates (phosphatases, protons, secondary metabolites and microbial inhibitors) are released along with citrate from cluster roots. The rootlets are concentrated in small volumes of soil; a strategy that is obviously not conducive to 'nutrient foraging' and has, instead, been described as a specialised 'P mining' strategy (Lambers et al., 2006; Lambers et al., 2010). Phosphate is very effectively extracted from sparingly available and residual soil P pools by *L. albus* (e.g. Mat Hassan et al., 2013). When transferring P-acquisition aspirations for organic anion exudation to other crop species that do not form cluster roots, it is important to question whether the magnitude of organic anion release will be sufficient to mobilise sparingly available P (e.g. Ryan et al., 2014).

Although mobilisation of sparingly available P in the rhizosphere, as a result of rhizosphere phosphatase or organic anion release, has been demonstrated experimentally (e.g. enhanced rhizosphere phosphatase activity (Ye et al., 2018), organic acid anion release (Mat Hassan et al., 2013)), these studies are in contrast with other glasshouse and field experiments where differences in rhizosphere phosphatase activity, or organic anion release have not underpinned improvement, or explained differences in P acquisition by key crop species (e.g. George et al., 2008; Ryan et al., 2014).

There is a need for more wholistic studies that attempt to partition P acquisition quantitatively among all components of a plant's P acquisition strategy (e.g. root architecture, root morphology, rhizosphere modification) to guide plant improvement efforts along the most rewarding path as this is likely to be species-specific. Something akin to this has been attempted in experiments examining the root morphological and physiological responses of several crop species to low soil P (Mori et al., 2016; Lyu et al., 2016; Wang et al., 2016; Wen et al., 2019). Lyu et al. (2016) differentiated crop species on the basis of their primary reliance on root morphological plasticity (e.g. *Zea mays, Triticum aestivum* L.), as opposed to physiological plasticity (rhizosphere modification) (e.g. *Lupinus albus, Cicer arietinum* L.). Species differences were further defined by their reliance on rhizosphere phosphatase activity, organic anion release, and mycorrhizal colonisation (Wang et al., 2016; Wen et al., 2019). These approaches are a welcome improvement to assumptions that rhizosphere modifications are equally beneficial strategies for P acquisition, but they still do not quantify the proportions of P uptake that are due to the contrasting P acquisition strategies represented within a species (e.g. something similar to that attempted by Mori et al., 2016); this will be challenging to achieve.

4 Understanding the agronomic context in which improved phosphorus acquisition by roots can deliver benefits

The agricultural and environmental benefits of improving P acquisition by plant roots depend on the context in which improved P acquisition is achieved. It is important to evaluate the key elements of the P balance of a farming system (Fig. 1a) in relation to the P retention characteristics of the soil to understand how benefits in P efficiency may be achieved (Simpson et al., 2014). Perfect efficiency in the use of P fertiliser is achieved when P inputs to a farming system as fertiliser equal the removal of P in farm products. The metric we use to define this is referred to as the Phosphorus Balance Efficiency (PBE) where:

$$\text{PBE}(\%) = 100 * \left(\frac{\text{P output in agricultural products}}{\text{P inputs in fertiliser and imported feed}} \right) \quad (2)$$

In a fertilised farm system, when P inputs are at, or approach equality with P outputs, this equation returns a value that approaches, or equals 100%. However, an anomaly of the calculations is that it is also possible to achieve, or exceed 100% in very low production systems that are not adequately fertilised because the output of P in products is depleting soil P reserves. This is clearly not a sustainable option. Here we confine discussion of the PBE metric to when it is applied to fertilised agriculture.

Johnston et al. (2014) describe P fertiliser application to crops grown in the UK and the USA, on soils with optimum P fertility, which achieve PBEs of 94–100%. By definition, these farming systems must have no P loss or P accumulation because P inputs equal P output in products (Fig. 1a). The 'P-sorption'[2] capacity of these soils must be low, with sufficient P retention within the soil profile to prevent leaching of P from the root zone. These systems would also have low soil loss and minimal P in runoff. In a system operating with near-perfect PBE, using plants with an improved capacity to acquire P from soil can have no beneficial impact on P use. Theoretically, P use in these farming systems can only be reduced by improving the internal P-utilisation efficiency of the crop in ways that reduce the amount of P exported from fields (e.g. Veneklaas et al., 2012).

4.1 Agricultural fields with low to very low phosphorus-sorbing soils

Very often low P-sorption capacity coincides with low P-buffering (e.g. sandy soils which have low phosphate-adsorption capacity and a poor ability to buffer changes in the P concentration of soil solution; Ozanne, 1980). This can result in large losses of P applied as fertiliser due to leaching and/ or runoff. Accumulation of P in these soils after fertiliser addition can be a relatively small component of the farming system's P budget (e.g. Russell, 1960; Ozanne et al.,

2 Here we use the term 'P sorption' to imply the net process of phosphate movement from soil solution to the solid phase of the soil and ultimately into forms of phosphate that are sparingly available for plant uptake as defined by Barrow (1999).

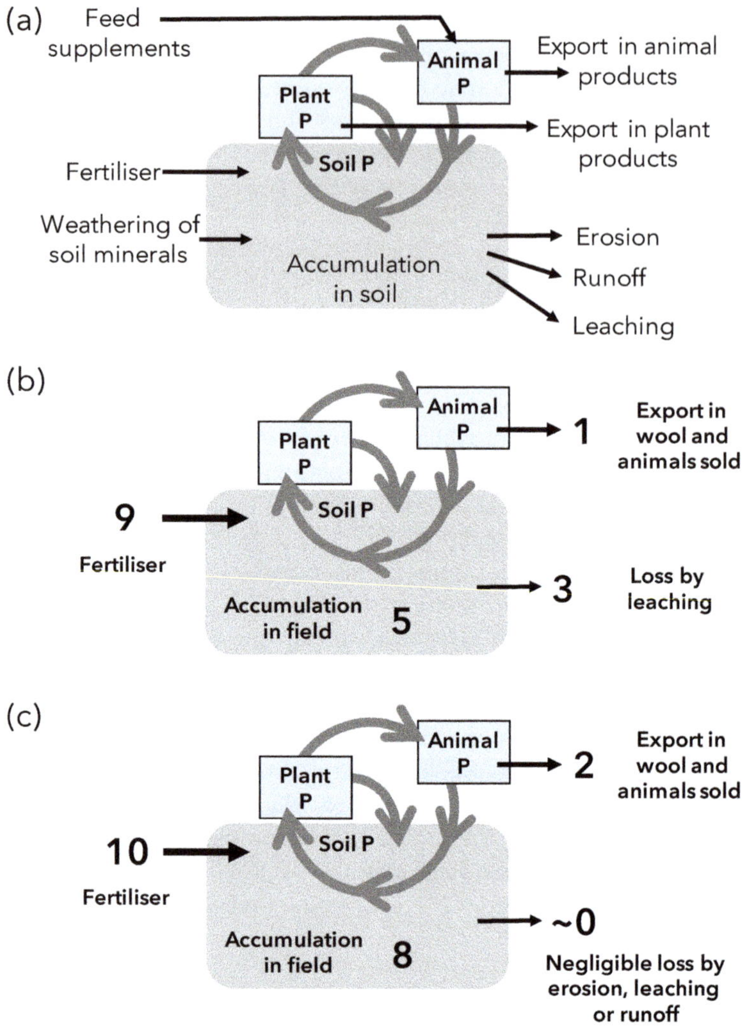

Figure 1 (a) The main inputs, outputs, and sinks for phosphorus (P) in a P-fertilised farming system. (b) Major net flows and accumulations of P in a permanent annual grass-annual legume pasture system growing on a low P-buffering, sandy soil near Willalooka, South Australia, that was grazed by sheep for wool and mutton production (data derived from Lewis et al., 1987). (c) Major net flows and accumulations of P in a permanent perennial grass-annual legume pasture growing on a soil with moderate P-buffering capacity (Phosphorus Buffering Index = 50; Burkitt et al., 2002) near Canberra, Australia. The pasture was grazed continuously by sheep for wool and sheep meat production and was being maintained at a soil test P concentration (0–10 cm depth) that was near the optimum for pasture growth (P1SR18 treatment from Simpson et al., 2015).

1961). For example, Fig. 1b shows the P budget of a fertilised, sheep grazing system in the Willalooka region of South Australia and demonstrates how P loss can contribute significantly to P inefficiency in a farm system based on sandy soils. The PBE of this sheep grazing system was only 13%; a third of the P applied as fertiliser was leached below the root zone each year and, of the P retained in the soil, about 55% was associated with soil organic matter (Lewis et al., 1987).

P losses of this nature can be substantially reduced by using P sources that are less soluble and release P slowly. However, there is often a trade-off between production and P loss because the ability of plants to access P from a less soluble P fertiliser is compromised (Ozanne et al., 1961). Using plant genotypes with lower external P requirements for maximum yield can also help mitigate the risk of P loss because the magnitude of loss via runoff and leaching is correlated with the soil test P (STP) concentrations of topsoil (Sharpley, 1995; McDowell et al., 2001; Melland et al., 2008; Bai et al., 2013; Withers et al., 2019). However, a benefit is only achieved when the use of these plants is combined with management of P inputs to reduce the soil's available-P concentration to the lower 'critical' STP requirement of the P-efficient plant. A strong nutrient foraging response, deeper rooted plants, and perennial species with deep, established root systems will also assist in capturing P that has been leached to depth (i.e. similar features to those of plants used to capture water and N that is leached to depth in a soil profile, Ridley et al., 1997; Dunbabin et al., 2003).

4.2 Agricultural fields with moderate to high phosphorus-sorbing soils

In contrast to soils with a low sorption capacity, soils with a moderate to high P-sorption capacity continue to slowly react with phosphate in ways that render it sparingly available for plant uptake (Barrow, 1974; Sample et al., 1980; Bolland, 1986a; Barrow, 1999; McLaughlin et al., 2011). In these soils, the rate of P diffusion to roots (and hence P acquisition) is improved by applying fertiliser to build the plant-available P concentration of the soil to the critical requirement of the crop or pasture (i.e. the STP concentration at which near-maximum yield is achieved). However, a component of the applied P inevitably accumulates in the soil as sparingly-available Pi. P application also results in the accumulation of Po (McLaughlin et al., 1990; Simpson et al., 2014; McLaren et al., 2020), only a proportion of which is immediately bio-available (Jarosch et al., 2015; Jarosch et al., 2019). Maintenance of stable soil-P fertility in these soils requires P inputs that cover P removal in agricultural products plus enough P to counteract the accumulations of P in the soil. Inevitably, budgets of P inputs and P outputs to fertilised agriculture on P-sorbing soils indicate that there is a P surplus. Indeed, management for optimum crop production (i.e. when soil fertility is maintained at the critical STP requirement of the crop) requires that there be a P surplus.

Figure 1c illustrates the key P flows in a southern Australian grassland system experiment managed with near-optimum soil P fertility, under continuous grazing. P accumulation in the field (mostly in sparingly-available pools of soil Pi and Po; McLaren et al., 2015) accounted for ~80% of the P fertiliser that it was necessary to apply annually to maintain near-maximum pasture yield (Simpson et al., 2015). Consequently, the PBE of the system was very low (20%).

Fertilising to cover P accumulations in high P-sorbing soils should not be confused with the problem of 'over-fertilising', when land is fertilised in excess of the P requirement for maximum crop yield (i.e. above the critical P requirement). Over-fertilising also results in a P budget surplus and low P-use efficiency (e.g. Simpson et al., 2015), but excessive accumulation of P in the soil is unwarranted and incurs an unnecessarily high risk of P loss to the environment (Gourley and Weaver, 2012; Jarvie et al., 2013a). Unfortunately, global budgets of P use in agriculture do not distinguish between P surpluses caused by 'best-practice' fertiliser use on P-sorbing soils, and excessive P-fertiliser use (e.g. MacDonald et al., 2011; Lun et al., 2018). Indeed, the tendency is to always assume that a P surplus represents poor fertiliser practice.

Excessive P application to soil must be recognised for what it is and should be addressed by reducing P inputs until critical soil P fertility levels for production and/or the environment are re-established (Gourley and Weaver, 2012).

Plants with root traits that enhance nutrient 'foraging' or 'mining' are associated with lower critical P requirements. In P-deficient soils they can deliver improved crop yields (Lynch, 2007; Lynch, 2019; Richardson et al., 2011) and in fertilised soils that accumulate P, the P inputs required for high production can be reduced (Simpson et al., 2011a, 2014).

5 Critical phosphorus benchmarks for soil phosphorus management

The most productive use of P fertiliser occurs when a P-deficient soil is fertilised to achieve the critical P requirement of the soil-plant system. Critical P specifications in '*amounts of P applied*' are not suitable benchmarks for crop management because they are not transferable between soils that differ in their P fertility status or P buffering capacities. However, critical STP benchmarks are potentially transferable and are widely promoted (e.g. Gourley et al., 2019; Bell et al., 2013; Bai et al., 2013; Johnston et al., 2013; Poulton et al., 2013; Dodd and Mallarino, 2005). They are essential for effective management of crop nutrition, soil fertility, and the environmental risks associated with fertiliser use

(Rowe et al., 2016) provided they have been determined reliably and are used correctly.

Conceptually, a critical STP benchmark represents the available-P concentration of the soil that can support sufficient diffusion of P to the crop's root system for maximum growth. It is, by necessity, a rudimentary approximation of the complex interplay between a growing root system and diffusion of P to it. In practice the critical STP requirement of a crop is, therefore, influenced by many variables including plant species, stage of crop growth, soil type (unless P buffering capacity of the soil is accounted for e.g. Ozanne and Shaw, 1967; Gourley et al., 2019), soil sampling depth and depth distribution of P (Helyar and Price, 1999; Probert and Jones, 1977). Difficulty accounting for these factors during the empirical definition of a critical STP benchmark, or when making fertiliser decisions, can impact adversely on the perceived reliability of STP benchmarks. This may lead to doubt about their usefulness (e.g. Rowe et al., 2016), use that is restricted to the region in which a benchmark was determined, or fertiliser practice that reverts to unguided rules-of-thumb (e.g. as described in Reuter et al., 1995) with haphazard soil P fertility outcomes (e.g. see Fig. 2). In some instances, reliable critical nutrient benchmarks have been determined but they are ignored (e.g. Gourley et al., 2015). These scenarios can result in over-fertilisation (Sattari et al., 2012; Rowe

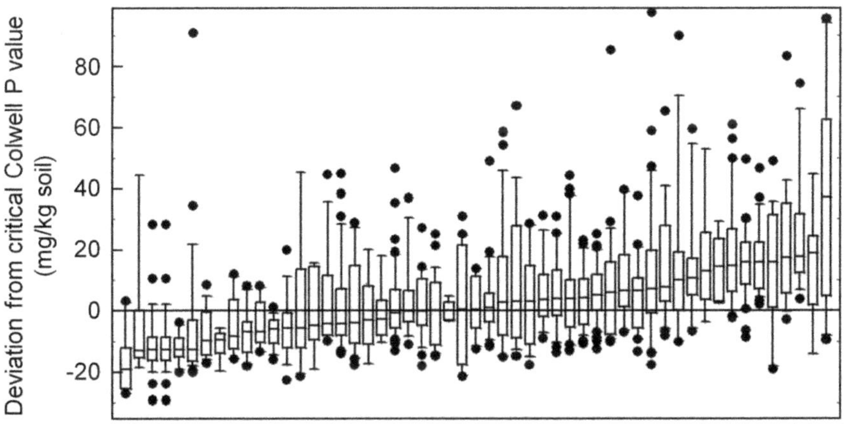

Figure 2 Box plots of the deviation in extractable P (Colwell, 1963) from the critical Colwell P concentration of soil in fields from 53 farms on the southern or central tablelands of New South Wales, Australia that were sampled (0–10 cm depth) in Spring 2016. Positive values indicate soil P fertility in the supra-optimal range; negative values indicate fields that were P-deficient. The graph shows the ranges in soil P fertility that existed on many farms prior to commencement of a regular soil fertility monitoring program. The box contains results from the middle 50% of fields, 80% of values occur inside the 'whiskers', and closed circles indicate fields with very high or very low STP concentrations. Bars show the median STP concentration. Redrawn from Simpson et al. (2017).

et al., 2016; Withers et al., 2019) which cannot be mitigated by improvements in plant P-acquisition efficiency.

6 Case study: improving the phosphorus efficiency of sheep and beef grassland farming in southern Australia

In this section we present a case study to demonstrate how diffusion of phosphate to roots in a grassland system can be managed to deliver effective use of P fertiliser in moderate to high P-sorbing soils. This was achieved by application of P banding technology, combined with annual monitoring of STP concentrations in the P 'band' to guide the pragmatic achievement of critical STP benchmarks for high grassland production. Improving knowledge of the P cycle in fertilised pastures and the factors that influence P sorption and accumulation in fertilised soil have shown how the introduction of novel plants with improved P-acquisition efficiency may enable a further substantive improvement in P-use efficiency in this grassland system.

6.1 Farming system context

Grassland farming is conducted on vast areas of P-deficient soil in the temperate zone of southern Australia (approximately delineated as south of latitude 32 S on the east coast to latitude 29 S on the west coast of Australia; Moore, 1970a). The climate is Mediterranean (winter-dominant rainfall with hot dry summers) in Western Australia and South Australia, with rainfall becoming more evenly distributed through the year towards the east coast of Australia. The soils are also deficient in N, and often other nutrients (S, K, and micronutrients depending on soil type; Williams and Andrew, 1970). Nitrogen (N) fertilisers are generally not used except in irrigated and high-rainfall, high-return dairy pastures, or occasionally as a strategic application to boost dryland pasture growth rates in winter. Instead, N deficiency is corrected by growing pasture legumes that are sown with grass in mixed pastures, or are oversown or encouraged to invade in natural grasslands. A variety of legumes of Mediterranean origin are used; their areas of use being determined primarily by annual rainfall (i.e. length of growing season) and by soil type (Donald, 1970). The most important pasture legumes are annual *Medicago* spp. (annual medics; low rainfall, neutral to alkaline soils), *Trifolium subterraneum* L. (subterranean clover; 300 mm – >900 mm annual rainfall, acid to neutral soils), *T. repens* L. (white clover; >900 mm rainfall, acid to neutral soils). *Medicago sativa* L. (lucerne or alfalfa) is used to a lesser extent as both a pasture and forage species across a wide climatic area provided soils are not subject to waterlogging or extreme acidity (Nichols et al., 2012).

Of these key legumes, *T. subterraneum* is the most widely used, with cultivars selected to fit differing growing-season lengths across the steep rainfall gradient of the temperate region in which it is grown (Donald, 1970; Nichols et al., 2013). This clover alone underpins the productivity of about 29.3 million hectares of dryland pasture (Hill and Donald, 1998), an area the size of Italy. The area of pasture production sown to *T. subterraneum* is the primary focus of this case study. Water (rainfall) is the key limiting resource in the grassland system. It determines total potential production and financial return per hectare; indeed, farms are often compared on the basis of production per 100 mm rainfall received (e.g. McEachern and Brown, 2010). High variability in rainfall contributes significantly to business risk and limits management options. In these environments, pastures are re-sown to improved varieties relatively infrequently. The high risk of a re-sown pasture failing to establish due to drought or because of low pasture renovation skill, combined with long periods (5–9 years) to repay the capital cost of pasture improvement are major disincentives (Lewis et al., 2012; Jackson and Malcolm, 2018). As a result, pastures are essentially permanent and are anticipated to remain in production for up to 20 years (Malcolm et al., 2014). New pasture varieties are selected for production combined with persistence reflecting this reality (e.g. Culvenor and Simpson, 2016).

Low soil fertility limits the efficiency of rainfall use (Mills et al., 2006) and application of P fertiliser is the easiest and most economical means to achieve high land-use efficiency and realise the productive potential of most farms (Carter and Day, 1970; Curll, 1977; Cayley et al., 1999; Osman et al., 1991; Smith et al., 2012; Lean et al., 1997). However, the PBE of sheep and beef production systems is very low and farmers need to apply 5- to 9-fold more P as fertiliser than they remove in animal products (median PBE = 9–19%; Weaver and Wong, 2011). This is due mainly to the moderate to high P-sorbing soils of the region which accumulate large proportions of the P that is applied as fertiliser (e.g. McLaughlin et al., 2011; Weaver and Wong, 2011; Simpson et al., 2015). Consequently, P fertiliser is a significant production cost for dryland grazing farms. It typically accounts for 20–25% of annual variable costs and is often the largest cost after labour and debt-servicing (McEachern and Brown, 2010).

6.2 Fertiliser practices to improve phosphorus acquisition on grassland farms

The most effective practices that can be employed to improve P uptake from these intrinsically P-deficient soils relate to the management of P diffusion to roots and include: use of soluble P fertilisers; localised banding of P at high concentrations within the root zone; and development and maintenance of an optimum soil available-P concentration. Here we outline these practices and explain the practical issues that define their use on farms.

6.2.1 Soluble phosphorus fertilisers

To achieve a rapid plant growth response when fertilising P-deficient soil, it is necessary to increase the phosphate concentration of the soil solution. This is most effectively done by applying soluble forms of P fertiliser to the soil (most often as single superphosphate in these systems). Experiments demonstrate repeatably that soluble P fertilisers deliver the most rapid increases in pasture production, and the largest yield gains per unit P applied (e.g. Sale et al., 1997; Lewis et al., 1997; Leech et al., 2019). Even when comparing among different P fertiliser products (including manures), their relative performance can usually be predicted by knowing their soluble P content (Leech et al., 2019). However, in the right circumstances there can be a role for less soluble P fertilisers (e.g. 'highly reactive' rock phosphate on very acid soils (Sale et al., 1997) and, particularly, on soils prone to P leaching (Lewis et al., 1997)).

6.2.2 Phosphorus banding

Subsurface banding of P that increases the concentration of phosphate close to a developing root system is a proven way of improving P fertiliser efficiency in crop systems (Jarvis and Bolland, 1991; Sanchez et al., 1991; see Section 2.1 'Managing phosphate diffusion'). It has also been shown, experimentally, to offer some advantages in pasture systems when surface soils become dry (Scott, 1973; Cornish and Myers, 1977; Pinkerton and Simpson, 1986), but it is neither practical nor economic to create subsurface P bands in permanent pastures. Nevertheless, P is usually relatively concentrated in the topsoil layer under pasture because the soil is subject to minimal disturbance and P in excreta and leaf-litter is deposited on the soil surface. This 'natural' surface P band is further enhanced by broadcasting soluble P fertiliser onto the soil surface in granular form.

When applied in a granule, phosphate moves rapidly from the granule into the soil, laterally and vertically, forming a P-enriched hemisphere below the granule (Fig. 3, Benbi and Gilkes, 1987; Hedley and McLaughlin, 2005). The concentration of phosphate in soil solution is highest near the granule and can exceed the P adsorption capacity of the soil (precipitates may form). P buffering is reduced in the high P zone and the high solution phosphate concentrations encourage diffusion of P into the P banded zone. In contrast, when a low solubility P source is applied to soil, the P concentration of the soil solution remains relatively low; this does not assist P diffusion to roots to the same extent (Lewis et al., 1997; Hedley and McLaughlin, 2005). The extent and speed of P movement into the soil is influenced by the P buffering

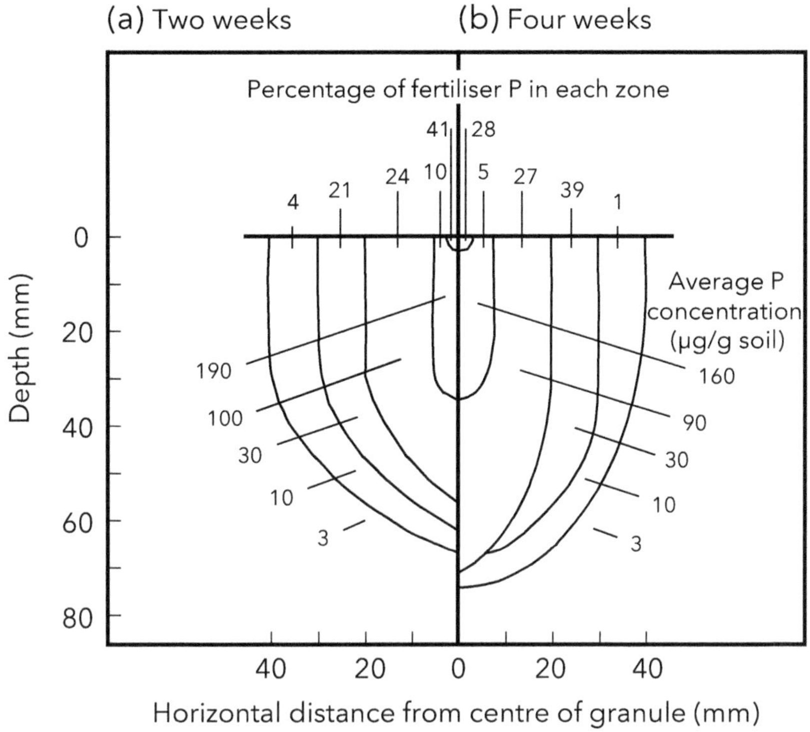

Figure 3 Distribution of phosphorus (P) in the soil (a podzolised sand) adjacent to a 135 mg granule of triple superphosphate (19.9% total P; 19.1% citrate-soluble; 16.0% water-soluble) at two and four weeks after its application to the soil surface. Adapted by permission from Springer Nature Customer Service Centre GmbH: Springer, Fertilizer Research, The movement into soil of P from superphosphate grains and its availability to plants, Benbi and Gilkes (1987).

capacity of the soil (Benbi and Gilkes, 1987). The dimensions of the band are initially shaped by the diffusion of P from P-fertiliser granules placed on the soil surface. However, empirical evidence indicates that ~70–80% of the increase in plant-available P after fertiliser application occurs in the topmost 10 cm of most soil profiles; the remaining impact is usually measured in the 10–20 cm layer of the topsoil even after a century of fertiliser use (Fig. 4; McCaskill and Cayley, 2000; Simpson et al., 2015; Schefe et al., 2015). Pasture species preferentially proliferate root length in the P-enriched surface soil band (Denton et al., 2006; Haling et al., 2016a,b). Heavy grazing, which can reduce total root production and root length at depth, does not diminish nutrient foraging in the surface soil layer (e.g. Greenwood and Hutchinson, 1998).

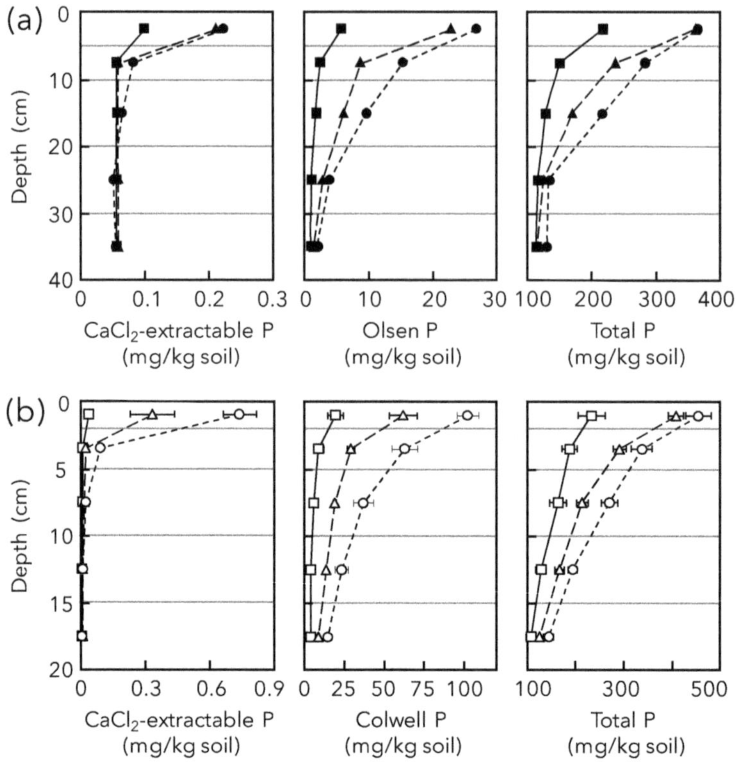

Figure 4 (a) Extractable P (0.01 M $CaCl_2$), Olsen extractable P (Olsen et al., 1954), and total P (nitric-perchloric acid digest) profiles of topsoil in the Permanent Top-Dressed pasture experiment at the Rutherglen Research Station, Rutherglen, Australia established as an unreplicated demonstration of the value of superphosphate applications in 1914 (Schefe et al., 2015). The data show soil P profiles in the unfertilised field (closed squares), a field to which 125 kg single superphosphate (~8.8% P) was applied per hectare every second year (closed triangles) and a field to which 250 kg single superphosphate was applied per hectare every second year (closed circles). The soil was sampled at 0–5, 5–10, 10–20, 20–30, and 30–40 cm depth intervals in November 2013 after 100 years of fertiliser use. The Phosphorus Buffering Index (PBI; Burkitt et al., 2008) of the top 10 cm of the P-fertilised soil profiles was ~115 (redrawn from Schefe et al., 2015). (b) Extractable P (0.005 M $CaCl_2$), Colwell extractable P (Colwell, 1963), and total P (sulfuric acid digest) profiles of topsoil from three soil P management treatments in a long-term grazed pasture experiment commenced near Canberra, Australia, in 1994 (Simpson et al., 2015). The soil P treatments were: unfertilised except for P inputs in supplementary feed for sheep (total P input over 20 year experiment = 5 kg P/ha; open squares), near-optimal soil P management (276 kg P/ha/20 years; open triangles) and supra-optimal P management for pasture growth (390 kg P/ha/20 years; open circles). The soil was sampled at 0–2, 2–5, 5–10, 10–15, and 15–20 cm depths in October 2014. Bars represent ± standard error; n= 3 replicate fields. The PBI of the top 10 cm of the soil profile was 50 (R. J. Simpson et al., unpublished data).

6.2.3 Development and maintenance of optimum phosphorus fertility

The plant-available P concentrations of surface P bands are managed by annual STP monitoring that is guided by empirically-derived STP benchmarks for high pasture production. Many of the practical problems that can prevent widespread application of soil test benchmarks to guide fertiliser management are overcome or avoided in the following ways:

i The most widely used pasture legumes in southern Australia have similar, relatively high, critical STP requirements compared to the grasses with which they are grown (Ozanne et al., 1969; Helyar and Anderson, 1970; Hill et al., 2005; Sandral et al., 2019). Atmospheric N fixed by the legume is the dominant source of N for growth by the pasture system. The critical P requirement of a mixed pasture is determined, consequently, by its legume component and this determines the benchmark STP concentration for optimum soil P fertility and high pasture production (Gourley et al., 2019).

ii Shallow testing of topsoils (0–10 cm) directly assesses ~70–80% of the plant available P in soil profiles (i.e. the surface P band; Fig. 4)[3]. Long-term experiments have demonstrated that in soils that are moderately P buffered, there is little leaching of P down the soil profile even after a century of regular P fertiliser applications, or after high rates of P applied over shorter timeframes (McCaskill and Cayley, 2000; Schefe et al., 2015; Simpson et al., 2015).

iii The STP benchmarks for pasture production are defined as the topsoil (0–10 cm depth) concentration of extractable P that corresponds with 95% of maximum growth *in spring* (Gourley et al., 2019). At this time of the year, temperature and soil moisture conditions are conducive to rapid growth rates (e.g. Cullen et al., 2008) and the demand for P by the crop is at its greatest. Farmers are encouraged to test pasture soils in spring and respond to the tests with a P fertiliser application in autumn after the summer drought and close to the opening of the next season's rainfall. Testing in spring removes the need to predict how P will be released during the growing season and assesses soil P fertility at the time of maximum crop demand. For a pasture soil that is being maintained close to its critical STP concentration, the fertiliser application creates a transient spike in P availability (see Fig. 5a; Simpson

3 The assumption that available P is found mainly in the uppermost layer of the soil breaks down in light-textured soils with very low P-buffering capacity (e.g. PBI <35) because soil P leaches from the uppermost soil layer (e.g. Lewis et al., 1981; Ritchie and Weaver 1993). Under these circumstances critical topsoil nutrient benchmarks may not be reliable indicators of pasture yield potential; testing soil samples taken to an appropriate depth must be considered (e.g. Probert and Jones, 1977).

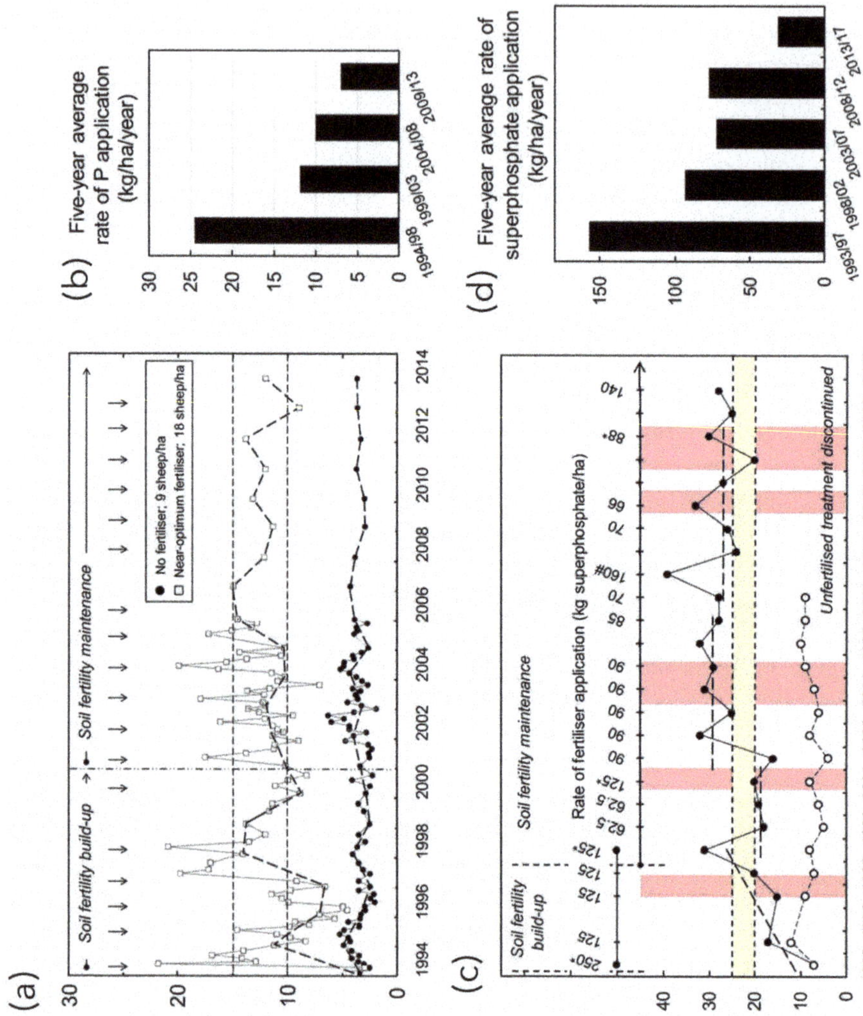
(a)
Olsen P (mg/kg soil)
Soil fertility build-up
Soil fertility maintenance
No fertiliser; 9 sheep/ha
Near-optimum fertiliser; 18 sheep/ha
(b)
Five-year average rate of P application (kg/ha/year)
(c)
Colwell P (mg/kg soil)
Soil fertility build-up
Soil fertility maintenance
Rate of fertiliser application (kg superphosphate/ha)
Unfertilised treatment discontinued
(d)
Five-year average rate of superphosphate application (kg/ha/year)

Figure 5 (a) The extractable P (Olsen et al., 1954) concentration of topsoil (0–10 cm) in selected grazing system treatments at Hall, ACT, Australia (adapted from Simpson et al., 2015). Three phases of the experiment are shown: 1994–2000, a soil fertility building phase, and 2001–2006, a soil fertility maintenance phase during which pasture was grazed continuously and STP was monitored at approximately 6-week intervals; and 2007–2014, a phase in which soil P fertility continued to be maintained, but with changed grazing management and only annual monitoring of the STP concentration. The fields that received no P fertiliser (closed circles) were stocked with 9 sheep/ha. Phosphorus was applied to the fertilised fields (open squares) with the intention of entering the spring period of pasture growth with a soil test P (STP) level within an Olsen P target band of 10–15 mg P/kg soil. These fields were stocked with 18 sheep/ha. Dashed horizontal lines delineate the target range for STP management. Arrows indicate when P fertiliser was applied. Each symbol represents the mean STP concentration of three replicate fields. Soil test P monitoring points are joined by a solid grey line to illustrate seasonal variability in soil test results. The dashed black line joins soil tests that were taken in Jan/Feb. These STP values were used to estimate the amount of P to apply each year. The dashed black line illustrates the variation in STP results that would be typical of annual monitoring at the same time each year. Phosphorus fertiliser was typically applied in autumn close to the break of each season. Error bars are not shown to improve the clarity of the figure. However, average coefficients of variation for the extractable P concentrations were 9.2%. (b) Five-year average rates of P application for the duration of the experiment. (c) Fertiliser application history and results of annual STP monitoring (Colwell, 1963) in a Grazing Systems Demonstration at, Bookham, NSW, Australia. Shaded vertical panels indicate years in which spring droughts occurred. The fertilised field carried 12–15 sheep/ha once the target STP concentration was achieved (zone delineated by dashed horizontal lines). Soil test results from an adjacent unfertilised paddock grazed continuously by 6 sheep/ha are also shown. The dashed lines show 'trends' in the data used at intervals to gauge progress in STP management. Sources: Graham (2006); Graham (2017); Simpson et al. (2009); R. P. Graham, unpublished data. (d) Five-year average rates of superphosphate (9% P) application for the duration of the Grazing Systems Demonstration.

et al., 2015) that exceeds the P requirement for early season growth and, provided the application rate for STP maintenance has been calculated correctly, the critical P level will be achieved at the peak of spring growth when demand for P is also peaking.

iv Because it is common for soils to have more than one potentially limiting nutrient (e.g. Trotter et al., 2014), successful application of the critical nutrient benchmarks also requires that there are no other limiting factors, or that they will be corrected concurrently. Critical soil test benchmarks for the most common secondary macronutrients limitations (e.g. S and K; Gourley et al., 2019) and soil toxicities (e.g. low pH/high Al; Slattery et al., 1999) have been determined and can be checked routinely.

6.3 Some pitfalls that have impeded progress

6.3.1 Confidence in the critical soil test phosphorus benchmarks for pasture production

It has long been recognised that optimum soil fertility management requires soil P fertility to be built by fertiliser application until the critical P benchmark for the soil-plant system is achieved, and for soil P management to transition to a maintenance phase during which the critical level of soil fertility is maintained (e.g. Fig. 5; McLachlan, 1965; Rudd, 1972; Holford and Crocker, 1988; Reuter et al., 1995). However, STP targets for fertiliser use were not generally promoted despite soil testing becoming generally accessible during the 1970s. This was partly because critical Colwell P concentrations (the most popular extractable P test; Colwell, 1963) were known to differ with soil type (e.g. Rudd, 1972). These issues were not resolved until 2007 (Gourley et al., 2007; Moody, 2007), when a major effort was made to reconcile historic data from ~650 fertiliser experiments conducted between 1955 and 2006 (Gourley et al., 2019). Earlier research had established the P requirement for maximum yield was related to the P buffering capacity of the soil (Ozanne and Shaw, 1967; Helyar and Spencer, 1977). It was demonstrated, using the historic data, that the critical Colwell P concentrations of pasture grown in different soil types could be predicted from the P buffering capacity of the soil. Concurrent testing for extractable P and Phosphorus Buffering Index (PBI) (a one-step test; Burkitt et al., 2002; Burkitt et al., 2008) was adopted almost immediately by the industry (Gourley et al., 2019).

6.3.2 Understanding and managing variability in soil test phosphorus concentrations

Targeted application of fertiliser P using a critical STP benchmark is illustrated in Fig. 5. There is always considerable seasonal variation in STP concentrations associated with time of fertiliser application (Fig. 5a), soil moisture/temperature conditions which stimulate bursts of P release from the microbial biomass, and P mineralisation/immobilisation as soils wet and dry (Perrott et al., 1990; Perrott et al., 1992; Blackwell et al., 2009). In addition, it is inevitable that sampling error, lack of replication in farm testing regimes, and error associated with the test itself will contribute to variability in soil test results. Before it was recognised that high variability in STP concentrations was eroding confidence in the use of soil tests, it was not uncommon for soil tests to be conducted at any time of the year, on an irregular basis and by irregularly using different P soil test methods. Farmers are now encouraged to test soils at the same time each year to minimise seasonal variability in test results, to set a pragmatic target range for soil P management that is guided by the critical P requirement of their soil-pasture system but is also cognisant of the intrinsic seasonal variability of soil test results, and to commence an annual soil-fertility monitoring program that will reveal temporal trends in soil P fertility (Simpson et al., 2009). As much as possible, fertiliser decisions are based on trends in the data (Fig. 5c), rather than one-off soil test results which may provide an unrealistic fertility assessment due to sampling and/or seasonal errors.

This approach is proven to maintain P supply to pasture roots for sustained high pasture and animal production and has been delivering a slow, but steady reduction in the fertiliser cost of production over time (Fig. 5b and c; Simpson et al., 2015). The reduction in costs can be attributed to at least two factors: (i) once the critical P fertility level is achieved, a lower fertiliser rate is required to maintain the target soil fertility than that required to build soil P fertility and this will deliver an obvious cost saving, and (ii) application of P to soil with moderate to high P buffering and sorption capacities has a 'P-sparing effect' (Barrow, 2015; Barrow et al., 2018). It is argued that the P-sparing arises because the negative charge on reactive surfaces in the soil is increased when sorbed P diffusively penetrates the soil particle; this reduces the P buffering capacity of the soil and the rate of further sorption reactions. Put simply, each application of P is expected to slowly improve the effectiveness of subsequent P applications (Bolland and Baker, 1998; Barrow et al., 1998, 2018).

In the examples provided (Fig. 5b and d), the initial step-change in P application rates was associated with the shift from building, to maintenance of soil P fertility. Thereafter, it is anticipated that P-sparing has contributed to the declining rate of maintenance P. However, it is difficult to determine the extent to which this was so, because Australian farmers also deal with a highly variable and drought-prone farming environment. Dry seasons are common and result in less P utilisation (less pasture growth), lower microbial activity and less P sorption; this means it is not necessary to apply as much fertiliser to maintain optimum fertility immediately after dry seasons. This can be seen in Fig. 5c where STP monitoring improved confidence in P fertiliser decisions and, increasingly, the decision was taken to skip fertiliser applications after dry seasons based on STP monitoring that indicated soil P-fertility had been conserved.

6.4 Progress towards the next step-change in the efficiency of phosphorus use for grassland production

Even with the use of soluble fertilisers, effective surface banding of fertiliser and good management of optimal P fertility, the PBE of grassland farms in southern Australia is low. Cropping systems in which soluble P fertiliser is banded with the seed can achieve median P balance efficiencies of ~48%. However, median P output in animal products from sheep and beef farms on similar soils is only 11–19% of P inputs (Weaver and Wong, 2011). Radiotracer experiments have demonstrated that the proportion of P captured by *T. subterraneum* pasture from a current superphosphate application is high (~42-50%; McLaren et al., 2017a) and is equivalent to, or better than that of crops (e.g. McLaughlin et al., 1988; McBeath et al., 2012). The P balance efficiency of grazed pasture is low because the pasture is not harvested like a crop. P removal in animal products is relatively small and most of the P acquired by the pasture is recycled (via the deposition of excreta, and death and decay of pasture) to the soil where it is re-exposed to P sorption/accumulation reactions (Donald and Williams, 1954; McLaughlin et al., 1990; Simpson et al., 2015; McLaren et al., 2017b). Direct phosphorus loss from these fields by leaching or runoff is relatively low (<1 kg/ha/year; Ridley et al., 2003) and does not account for the low PBE of the grassland system.

6.4.1 Factors that influence phosphorus accumulation in grazed fields

The accumulation of phosphate in acid soil can be mimicked in isothermic laboratory incubations and is described by the following empirical relationship:

$$Ps = a * C^{b1} * t^{b2} \tag{3}$$

where: ***Ps*** is the rate of phosphate sorption, ***a*** approximates the amount of sorbing material in a soil, ***C*** is the concentration of phosphate in soil solution, ***t*** is the time over which the reaction continues, and ***b1*** and ***b2*** are coefficients that describe the shape of the sorption relationship.

The ***b1*** and ***b2*** coefficients of Equation 3 vary widely between soils. However, they are reasonably well correlated when compared across a wide range of soils (Barrow, 1980a,b). Consequently, P sorption in the field is

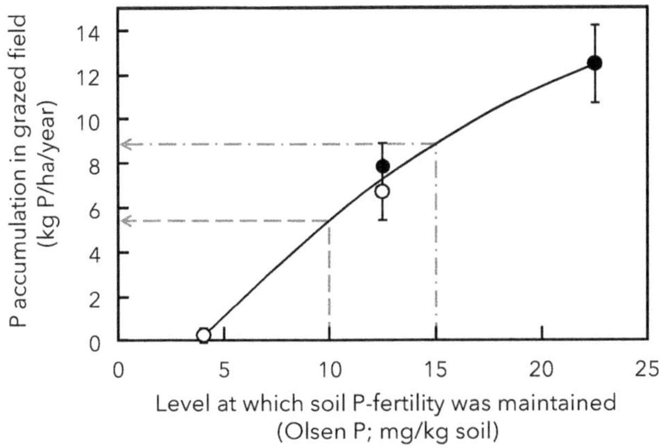

Figure 6 Average annual rates of phosphorus (P) accumulation in fields with a pasture based on *Trifolium subterraneum* (subterranean clover), *Phalaris aquatica* (phalaris) and annual grasses (e.g. *Bromus* spp., *Vulpia* spp.). The soil was acid (pH_{Ca} 4.6) and had a moderate P-buffering capacity (P buffering index = 50, Burkitt et al. 2002). The pasture was grazed continuously by 9 sheep/ha (○) or 18 sheep/ha (●) and maintained at three levels of extractable P in the topsoil (0–10 cm depth) over 6 years (data derived from Simpson et al., 2015). Soil fertility levels are the midpoint soil test value (Olsen et al., 1954) of the target range for soil fertility management. The expected critical Olsen P concentration for near maximum pasture growth in this system was 15 mg P/kg. Bars represent ± standard error. The dash-dot line indicates the expected annual accumulation of P if the soil is maintained at the soil test P (STP) concentration recommended for *T. subterraneum*-based pasture; the dashed line indicates the expected accumulation of P if soil P fertility is maintained near the critical STP concentration of *Ornithopus* spp. (see Fig. 7 and Sandral et al., 2019). Redrawn from Simpson et al. (2014).

expected to be proportional to the plant-available phosphate concentration at which a soil is maintained.

However, P accumulation in a fertilised grazing system soil is more complex than this because it also involves accumulation of P in soil organic matter and to a lesser extent in very small areas of fields where excreta accumulates due to animal congregation behaviours. The proportions of inorganic P (Pi) and organic P (Po) accumulating in the soil vary depending on the P status of the soil. Typically, more Po is accumulated than Pi when a P-deficient pasture is fertilised and pasture growth is being increased; similar proportions of Po and Pi are accumulated when P levels are near optimal for pasture production, and Pi accumulation exceeds Po accumulation under conditions of supra-optimal P fertility (Simpson et al., 1974; McLaren et al., 2015; McLaren et al., 2020).

Results from a long-term, P fertility management experiment grazed by sheep have been used to assess whether management of plant-available

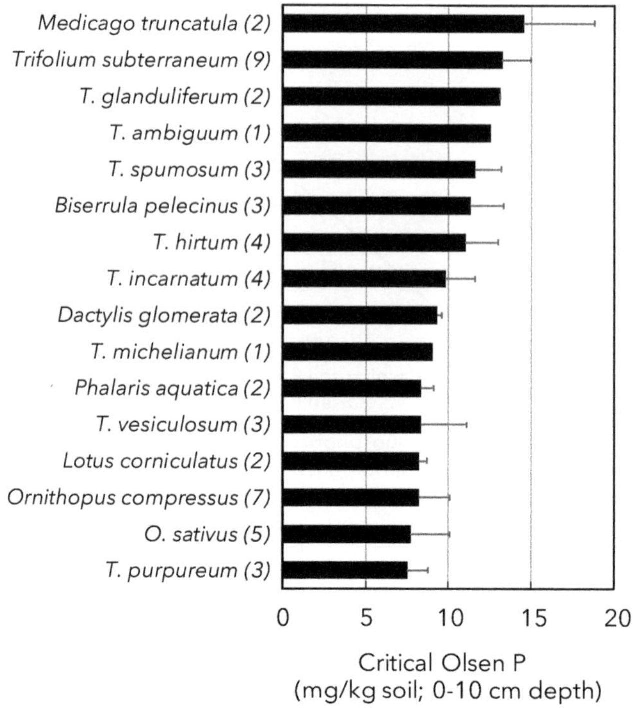

Figure 7 Average 'critical' extractable phosphorus (P) values (Olsen et al., 1954) for 14 pasture legumes and 2 grasses from a range of independent 'cultivar-site-year' assessments over three years and four field sites (Phosphorus Buffering Index range: 40-80; Burkitt et al., 2002). Critical soil test P values support 95% of maximum yield in spring. *Medicago sativa* was also included in these experiments but its critical P requirement could not be determined reliably because it often exceeded soil test P levels in the highest P treatments. Numbers in parentheses indicate the number of 'site-year' assessments made for each species. Bars representing 1x standard deviation are shown as a measure of the repeatability of the critical P determinations. Data were derived from Sandral et al. (2019).

phosphate was a dominant factor in the net accumulation of P in fertilised fields under grazing (Fig. 6; Simpson et al., 2014; Simpson et al., 2015). A positive (curvilinear) relationship between the STP target for soil P management and the annual rates of P accumulation in grazed fields was observed. Evidence from other long-term grazing experiments, while limited, supports this observation (Simpson et al., 2014). As described above, the experiment confirmed that maintaining soil P fertility close to the critical P requirement of the *T. subterraneum*-based pasture delivered the most effective use of P fertiliser for livestock production. Importantly, it also demonstrated that significantly higher rates of P fertiliser were needed to maintain supra-optimal STP concentrations because the rate of P accumulation in the soil was increased (Simpson et al.,

2015; McLaren et al., 2015, 2017b). This further underlined the folly of fertilising soils to levels well in excess of the critical P requirement of a farming system.

6.4.2 A clue to the development of a more phosphorus-efficient grazing system

The counterpoint to the previous observation was that using a productive, alternative pasture legume with a lower critical P requirement should enable the grazing system to be maintained at a lower STP concentration without impacting production adversely. This, in turn, would reduce P accumulation in grazed fields and less P fertiliser would be needed to maintain productivity (Fig. 6; Simpson et al., 2014).

The projected savings in the amount of P fertiliser necessary to maintain a lower critical STP concentration appeared to be substantial. For example, if it were possible to shift the target for soil P management from an Olsen P concentration (Olsen et al., 1954) of 15 mg P/kg soil (the optimum for *T. subterraneum* pastures: Gourley et al., 2019; Sandral et al., 2019) to 10 mg P/kg soil (i.e. close to the optimum subsequently found for serradellas (*Ornithopus* spp. – a P-efficient, alternative legume; Fig. 7, Sandral et al., 2019), the maintenance P cost of optimum pasture production would be reduced by about 30% (Simpson et al., 2014). To put this in context, the projected proportional reduction in the use of fertiliser P by this strategy is as large as the projected proportional benefit of recovering P from waste streams in Europe using novel and emerging technologies (17–31%; Tonini et al., 2019). Obviously, the quantum of the benefits in global terms and the degree-of-difficulty in implementing either strategy are probably not comparable.

6.4.3 The search for phosphorus-efficient pasture legumes

The task of finding alternative, P-efficient legume(s) to underpin improved P use efficiency was made potentially feasible by the fact that the P requirement of a single species of pasture legume (*T. subterraneum*) effectively determined the P requirement of this grassland farming system. However, the unparalleled success and wide adaptation of *T. subterraneum* also underlined the difficulty of the task. *Trifolium subterraneum* is grown in mixtures with various annual and perennial grasses and is well adapted to Mediterranean and temperate climates with highly variable rainfall and regular droughts. It is valued for its high productivity and persistence (Morley 1961; Rossiter, 1966; Nichols et al., 2012), feeding value for livestock (Doyle et al., 1993), N-fixation (Peoples et al., 2012), grazing tolerance (Nichols et al., 2012), and has now undergone about a century of selection and cultivar improvement (Nichols et al., 2013).

Much of the Australian flora have evolved in P-impoverished soils (Handreck, 1997) and this has led to interest in the suitability of native perennial legumes as an alternative to *T. subterraneum*. However, while there is a wide range in the critical P requirements among both native legumes and grasses, with some capable of reasonable growth in very P deficient media (Pang et al., 2010; Waddell et al., 2015), P-efficiency among the legumes was not accompanied by high potential yields in moderately fertile soils (e.g. Pang et al., 2010; Robinson et al., 2007). It is surmised that there is a potential role for these legumes, especially as ephemeral or opportunistic crops and pastures in arid and semi-arid environments (e.g. Nicol et al., 2013; Bell et al., 2012), but it is unlikely that they provide a viable P-efficient alternative to *T. subterraneum*.

Significant effort has also been expended to identify alternative pasture legumes from the Mediterranean basin as a way of diversifying the pasture base for animal production systems in southern Australia (Nichols et al., 2007; Nichols et al., 2012). The emphasis on legumes naturalised from the Mediterranean region reflects their successful acclimation to Australian soils and climate, their responsiveness to fertiliser applications, high herbage yields and seed production, and their persistence in managed grasslands (Donald, 1970). Nichols et al. (2013) detail 33 annual legume species registered for use in southern Australia in addition to *T. subterraneum*. Although 23 of these species were suited to acid soils and are, therefore, potential alternatives to *T. subterraneum*, 15 of the 23 species were also regarded as only suitable for localised use, rarely used, or of limited commercial value.

With this background in mind, Sandral et al. (2018) commenced a glasshouse experiment to determine whether there was any evidence for differing critical P requirements among 10 alternative annual legumes and *T. subterraneum*. Precautions were taken to mimic field conditions (e.g. the legumes were grown as micro-swards with P stratified in the topsoil to reflect surface P banding) to obtain realistic assessments of critical P requirements relative to *T. subterraneum*. However, final assessment of legumes truly capable of delivering lower critical P requirements was based on their subsequent growth responses to P fertiliser at four field sites under contrasting seasonal conditions (Fig. 7; Sandral et al., 2019). Nine of the ten species tested in the glasshouse were found to have a significantly lower critical P requirement than *T. subterraneum* (Sandral et al., 2018), but this was reduced to five legume species (out of 14 alternative legumes tested) that had consistently lower critical P requirements than *T. subterraneum* in field environments (Sandral et al., 2019).

Three of the legumes with lower critical P requirements were forage legumes species (crimson clover, *T. incarnatum* L.; purple clover, *T. purpureum* Loisel; arrowleaf clover, *T. vesiculosum* Savi). These species are suited to short-term pastures but have a limited role in the target farming zone where the high

Table 2 Critical P requirements (concentration of P applied to a topsoil layer (mg/kg soil) that corresponded with 90% of maximum yield), and comparative root morphology of nutrient foraging (topsoil) roots (mean ± s.e.) among three alternative pasture legumes measured at the soil P level (30 mg P applied/kg topsoil layer) where root proliferation by *T. subterraneum* was at its peak in a controlled-environment study of P acquisition (Haling et al., 2016a,b).

Parameter	subterranean clover cv. Leura (*Trifolium subterraneum*)	French serradella cv. Margurita (*Ornithopus sativus*)	yellow serradella cv. Santorini (*O. compressus*)
Critical P requirement (mg P/kg topsoil)	89 ± 1	38 ± 2	25 ± 2
Root mass (g/microsward)	0.61 ± 0.01	0.33 ± 0.01	0.27 ± 0.01
Root mass fraction (g/g)	0.24 ± 0.01	0.11 ± 0.004	0.08 ± 0.003
Specific root length (m/g root)	338 ± 23	459 ± 37	542 ± 26
Root length density (cm/cm^3)	53 ± 4	35 ± 3	35 ± 1
Root hair length (mm)	0.23 ± 0.004	0.63 ± 0.03	0.58 ± 0.02
Surface area of root hair cylinder (cm^2/microsward)	3359 ± 156	4686 ± 509	4245 ± 188

cost of re-sowing pastures carries a high financial risk relative to permanent pastures (Malcolm et al., 2014). However, soil P management recommendations for these P-efficient forage legumes can now be modified to guide fertiliser use in forage legume crops, during a pasture renovation sequence, or as a legume rotation phase in a cropping sequence. Only two pasture legume species with lower critical P requirements (*O. compressus* L., yellow serradella; *O. sativus* Brot., French (or pink) serradella) were considered potentially suitable for use in permanent pastures.

The field experiments demonstrated that all of the P-efficient alternatives could be grown with an Olsen STP concentration (0–10 cm) of 10 mg P/kg to achieve near-maximum spring yields in soils where it is recommended that *T. subterraneum* be grown with an Olsen STP concentration of 15 mg/kg (Sandral et al., 2019; Gourley et al., 2019). To date, our search within the *T. subterraneum* genome for P-efficient clover lines has not identified genotypes that can achieve the step-change in P-acquisition efficiency that is possible with the *Ornithopus* spp. (McLachlan et al., 2019).

6.4.4 How *Ornithopus* species achieve a lower critical phosphorus requirement

The critical P requirement among a large number of pasture legumes is determined by their ability to develop a large root-soil interface for P acquisition (Haling et al., 2016b; Sandral et al., 2018). In practice, a low critical P requirement has been associated with: (i) high capacity for proliferation of root mass in soil patches where P is relatively concentrated (e.g. in the topsoil P band created by broadcasting fertiliser), (ii) long specific root length, which enables root mass to be converted effectively into root length, and (iii) long root hairs, which convert root length into a large root cylinder volume. Comparisons of the root morphology and low-soil-P acclimation responses of *Ornithopus* spp. and *T. subterraneum* (Table 2; Haling et al., 2016b; Yang et al., 2017; Sandral et al., 2018) indicate that although the clover is very effective at proliferating root length in low-P soil, its relatively short specific root length means that this is only achieved by allocating a large amount of root dry mass to nutrient foraging. The proportion of plant mass allocated to nutrient foraging by *T. subterraneum* can be twice that allocated by an *Ornithopus* spp. when growing with a similar low P supply (e.g. Table 2). In addition, the clover has relatively short root hairs (0.23 mm) which prevent it from developing a comparable large root cylinder volume to that of the serradella with long root hairs (~0.6 mm), even when the clover has proliferated longer total length of nutrient foraging roots (e.g. Table 2; Haling et al., 2016a,b). Although other factors, such as exudation of organic anions, may play some role in P acquisition, comparative assessment of anion release from roots has not indicated a large role in P acquisition by these species and does not explain their contrasting critical P requirements (Kidd et al., 2016).

6.4.5 The role of arbuscular mycorrhizal fungi

Like many pasture legume species, the roots of *T. subterraneum* and *Ornithopus* spp. are usually found to be highly colonised by AMF (Smith et al., 2015; Sandral et al., 2019). Growth and P acquisition by *T. subterraneum* in P-deficient soil is usually improved by its association with AMF, but the benefit diminishes when soil P fertility and the P status of the host plant are improved (Abbott and Robson, 1977; Schweiger et al., 1995; Hill et al., 2010; McLachlan et al., 2020). In some cases, AMF colonisation of *T. subterraneum* has been associated with a lower critical P requirement relative to non-mycorrhizal plants (e.g. Schweiger et al., 1995), while in other cases the impact of AMF on critical P requirement has been marginal or non-existent, even when growth in P-deficient soil has clearly been improved (e.g. Abbot and Robson, 1977; Bolan et al., 1987; McLachlan et al., 2020).

Mycorrhizae colonise the roots of *Ornithopus* spp. at frequencies similar to those of *T. subterraneum* (e.g. Sandral et al., 2019). However, *Ornithopus* spp. appear to gain only a comparatively small P acquisition benefit from their AMF association (Schweiger et al., 1995; McLachlan et al., 2020). It has been hypothesised that the contrast in the AMF benefit obtained by these two legumes is due to the difference in their root hair lengths (i.e. the species with longer root hairs benefits less; Schweiger et al., 1995). However, it is difficult to reconcile this hypothesis with the observation that P acquisition by AMF occurs at much greater distances from the root surface than the length of root hairs. For example, most P acquisition by AMF occurs within ~1 cm of the root surface, with some P uptake also feasible at many centimetres distance (Jakobsen et al., 1992).

Likewise, the large difference in root hair length between *T. subterraneum* and *Ornithopus* spp. appears to largely explain their contrasting abilities to acquire P from low-P soil (Haling et al., 2016b). However, the ability of AMF to acquire P from soil at distances from the root that are a magnitude greater than the length of serradella root hairs (Jakobsen et al., 1992) should also negate the difference in P acquisition efficiency by these legumes. This does not occur. Subterranean clover exhibits significantly poorer P-acquisition efficiency and a significantly higher critical P requirement than serradella when colonised by AMF in the field (Sandral et al., 2019), and in glasshouse experiments in pasteurised media with and without AMF colonisation (Schweiger et al., 1995; McLachlan et al., 2020). The answer to this apparent anomaly lies, in part, in the fact that the clover reacts to AMF colonisation by allocating less root mass to nutrient foraging. This counteracts some of the anticipated P acquisition benefit of the AMF, and the differences in critical P requirements between *T. subterraneum* and serradella are preserved (McLachlan et al., 2020). It is notable that lower critical-P requirements detected for other alternative legumes in glasshouse experiments with low AMF colonisation rates, were not always detected consistently in the field (Fig. 7; Sandral et al., 2018; Sandral et al., 2019). This may be a consequence of AMF colonisation in the field, but is also likely to be a consequence of the many other challenges to root growth (soil hardness, soil acidity, dry topsoil, root diseases, etc.) in field environments (e.g. Watt et al., 2006; Simpson et al., 2011b). These challenges can reduce soil exploration and nutrient foraging by roots and may not be resisted or tolerated equally by different plant genotypes (e.g. Delhaize et al., 2009).

6.5 Is a region-wide shift to phosphorus-efficient pastures possible?

6.5.1 Requirements for soil phosphorus fertility management

As outlined already, soil test data are intrinsically variable due mainly to seasonal fluctuations in extractable P concentrations (Fig. 5). Farmers set a target range for their management of soil P that is guided by the critical P requirement of the grazing system. This typically has a peak-to-peak amplitude of about five Olsen P units (e.g. Fig. 5a; Simpson et al., 2009). The difference between the critical P requirements of *T. subterraneum* and the alternative P-efficient legumes identified by Sandral et al. (2019) was of similar size (Fig. 7) and, consequently, also suited the practical realities of managing soil fertility. The maximum benefit of using a P-efficient legume in a fertilised production system will only be realised if the target for soil P management is reduced to achieve the lower critical P requirement of the novel legume. It is questionable whether critical STP differences, less than five Olsen P units (even if statistically proven in field experiments), could be managed effectively enough on farms to deliver real savings in P-fertiliser costs or reduced environmental risks.

6.5.2 Edaphic requirements

Yellow serradella (*O. compressus*) was the first species of serradella to be commercially released in Australia (late 1950s). The early cultivars were intended for use on acid, infertile sandy soils where their deep roots, high tolerance of acid soils and hardseeds (transiently impermeable seed coats that protect the seeds against premature germination after summer storms) allowed them to be productive and persistent where *T. subterraneum* was failing (Bolland and Gladstones, 1987; Loi et al., 2005; Freebairn, 1996). Cultivars of French serradella (*O. sativus*) have been released for use since the mid 1990s also with the intent that they be used mainly on light textured soils. Serradella use was initially constrained by the relatively high cost of seed. However, the development of improved yellow and French serradella cultivars has underpinned their expanded use in rotation with crops since about 2009 (Nichols et al., 2012). A major breakthrough being the development of the first hardseeded cultivars of French serradella (Nutt, 2004a,b). This is an essential characteristic needed for persistence of the species in pastures that regularly experience drought (Dear et al., 2002). The expanding use of *Ornithopus* spp. in cropping systems demonstrated that the *Ornithopus* spp. were not 'niche' species only suited to sandy soils as thought originally. It is now recognised that they can be grown successfully on a much wider range of soil types where they can be highly productive, nodulate prolifically, and provide a forage with high feeding value that is potentially 'bloat-free' (Freebairn, 1996; Hackney et al., 2013).

6.5.3 Yield and persistence

In the experiments of Sandral et al. (2019) all of the P-efficient alternative legumes, except cultivars of *O. compressus*, produced peak spring yields that were equivalent to, or exceeded subterranean clover. However, *O. compressus* has been found to yield as well as *T. subterraneum* in other soil type/climatic districts (e.g. Bolland and Paynter, 1992). This highlights the continuing need to consider the fit of the P-efficient species to specific environments. In early experiments, the lower P-fertiliser requirement of *O. compressus* relative to *T. subterraneum* was also noted (e.g. Bolland, 1986b; Paynter, 1990; Bolland and Paynter, 1992). However, this did not lead to changed recommendations for soil P management because critical STP concentrations were not determined until the experiments of Sandral et al. (2019).

6.5.4 Further steps to achieving region-wide adoption of phosphorus-efficient serradella pastures

Phosphorus efficiency will only be captured by managing soil P fertility at the lower critical STP concentration suitable for *Ornithopus* spp. Research and extension to inform farm advisors and farmers who already grow *Ornithopus* spp. pastures that they can now adopt revised, species-specific STP benchmarks (Sandral et al., 2019) is progressing. However, expansion of *Ornithopus* spp. use into much larger areas of the permanent pasture zone where they have not been widely used, will be necessary to achieve a region-wide change in P use efficiency. This will depend on having cultivars that can deliver high yields and persistence across the diverse climatic zones of southern Australia. While there is evidence that some serradella cultivars can yield at least as well as *T. subterraneum* (Sandral et al., 2019), concerns that current cultivars may lack persistence in the new farming zones to which they are being targeted (e.g. Hayes et al., 2015) have not yet been addressed. Persistence of annual legumes is largely a function of their seed production, seed protection mechanisms, and seedling emergence and survival characteristics. Successful cultivars display stable flowering dates (irrespective of wide and unpredictable variations in the time of germination each year) that avoid frost damage at flowering and optimise seed production (Donald, 1970). Stable flowering dates among cultivars of *T. subterraneum* are regulated by their responses to long days (photoperiodism) and cold winter temperatures (vernalisation), and the optimum time for flowering is also well-defined (Aitken and Drake, 1941; Donald, 1970). Early research into these aspects of persistence among serradella cultivars indicates that some current cultivars may not display equivalent flowering date stability (Boschma et al., 2019). Seeds must also be protected against false breaks to the growing season by being initially impermeable to moisture (hardseeded), yet responsive to reliable rainfall when it occurs. Work on these cultivar traits has been in process for some time (e.g. Loi et al., 2005), but adaptations that are a better fit to new districts may also need to be pursued.

6.6 Summary: how research has contributed to the development of a more phosphorus-efficient and sustainable production system

In this case study, there were a number of essential steps necessary for the conception, development, and realisation of more P-efficient grassland farming.

6.6.1 Identifying opportunities to improve the efficiency of phosphorus fertiliser use

It is hard to know where to commence the discourse on how P efficiency has been improved in the grasslands of southern Australia because there has been a long history of incremental improvement in pasture production punctuated by step changes in progress as 'break-through' technologies emerged, all of which have ensured improving P use efficiency. Significant steps in this revolution have included the realisation that *T. subterraneum* was highly adapted to the southern Australian environment and could be widely adapted for use by developing cultivars to fit the rainfall gradient (i.e. length of growing season) that exists across the farming zone. Subsequently, there has been an effort to identify alternative legumes to diversify pasture composition and extend the N-fixing benefit of legumes to wider environments. Combining the use of P fertiliser and pasture legumes dramatically boosted P and N fertility across the region and delivered a step-change in production. Further innovations included recognition and alleviation of micronutrient deficiencies in soils, development of productive and persistent companion grasses, and the implementation of various fertiliser technologies to underpin more effective use of P on soils that are almost universally P-deficient and accumulate P when fertilised. Readers are referred to other sources for a more complete discussion of the revolution that has occurred over the last century (e.g. Rossiter, 1966; Moore, 1970b; Morley, 1961; Costin and Williams, 1983; Smith, 2000; Nichols et al., 2007, 2012; McLaughlin et al., 2011).

While the initial emphasis of research was to correct soil-nutrient deficiencies to enable improved pasture production and livestock carrying capacity, the emphasis in the modern era has shifted to the optimal use of P, guided by the critical P requirement of the soil-pasture-livestock system (e.g. Figs 2 and 5; Simpson et al., 2009; Gourley et al., 2019). Removing soil fertility limitations ensures the optimal use of Australia's most-limiting resources: water and land (e.g. Mills et al., 2006) and underpins the profitability and sustainability of farm businesses (Lean et al., 1997). Over-use of P fertiliser still occurs in some sectors, but it is actively discouraged (e.g. Gourley and Weaver, 2012).

Optimal management of P inputs to pastures growing on P-sorbing soils inevitably required P inputs that are greater than the amount of P that is exported in agricultural products. In many ways this is a 'technical' inefficiency that has seemed to be unavoidable because of the moderate to high P-sorbing nature of the most Australian soils. However, inefficiency creates opportunity. Steady improvement in the understanding of how and why P accumulates in fertilised soils, and the form and lability of accumulated P (e.g. Barrow, 1974, 1999; Simpson et al., 2015; Jarosch et al., 2015; McLaren et al., 2017b; Barrow et al., 2018) has opened a new opportunity to manage P accumulations in soil by using plants with roots that acquire P more effectively and support a lower critical P requirement for pasture production (e.g. Sandral et al., 2019).

6.6.2 Development of knowledge and tools to enable application of the science

Pastures that have a lower critical P requirement only deliver a reduced P-fertiliser requirement when the soils in which they are growing are fertilised to the lower critical STP concentration of the novel pasture. The saving in P-fertiliser is realised indirectly because at the lower STP concentration, P accumulation reactions in soil are slower and the risk of P losses via runoff and leaching is reduced. Consequently, it was crucial that farmers were equipped with the knowledge and tools to be able to fertilise soils to meet critical STP targets (e.g. Fig. 5). This is now promoted as best practice in the target area of this case study (e.g. Simpson et al., 2009). However, it is sobering to remember the timeframe that was required for research, development and adoption of this essential step. Soil tests were developed in the 1950–1960s (Olsen et al., 1954; Colwell, 1963), development and adoption of soil testing was an incremental process (e.g. Reuter et al., 1995), and their application on farms Australia-wide was finally enabled in 2007 with the release of reliable, widely applicable STP benchmarks (Gourley et al., 2019).

6.6.3 Identifying legumes that can underpin a step-change in P efficiency of the farming system

A search for alternative, high-yielding pasture legumes with substantially lower critical STP requirements than that of *T. subterraneum* was commenced soon after their potential P-efficiency advantages were realised (Simpson et al., 2014). Success in this task was only possible, however, because alternative pasture legumes had already been trialled over many years in southern Australia with the objective of diversifying the feed base for animal production (Nichols et al., 2007; Nichols et al., 2012). Despite a large number of potential candidates, the number of legumes that proved to be consistently P-efficient under field conditions was limited (Sandral et al., 2019). Only two P-efficient species potentially suited to use in permanent pastures were identified (*O. compressus* and *O. sativus*). Some of the barriers to their adoption were readily crossed because they were already in limited use with acknowledged agronomic and animal feed value (Nichols et al., 2007; Nichols et al., 2012). Widespread adoption will be essential for a real step-change in P efficiency of the grassland system. This now depends on the identification or development of cultivars that are adapted to the newer soil and climatic environments in which they will need to be grown and assurance that they can meet modern, socially fit-for-purpose criteria (e.g. nutritious, food-chain safe, low weed risk, etc.; Revell and Revell, 2007).

7 Conclusion and future trends in research

The pursuit of high PBE by improving nutrient foraging has natural limitations. For example, examination of a wide range of *T. subterraneum* genotypes and among phylogenetically allied *Trifolium* spp. (McLachlan et al., 2019; McLachlan et al., 2020; and authors' unpublished data) has failed, so far, to find variation in P-efficient root morphology traits (especially root hair length) that are equivalent to those of the serradella species. There are natural limits to the range in phenes within a species that reflect the environments in which they have evolved. In addition to this, there are practical limits imposed by the agricultural system being improved. In the mixed (legume-grass-forb) grasslands of southern Australia, the critical P requirement for production is effectively determined by the pasture component with the highest P requirement (currently the legume component; Hill et al., 2010). If widespread replacement of the dominant legume (*T. subterraneum*) with *Onithopus* spp. can be achieved (as discussed above), the critical P requirement of the grassland system will be reduced and will be roughly similar to that of key companion grasses (e.g. *Phalaris aquatica* L., *Dactylis glomerata* L.; Sandral et al., 2019). Although there are several other grassland component species with even lower critical P requirements (Hill et al., 2010), pursuing further improvement in legume P-acquisition efficiency beyond that delivered by *Ornithopus* spp. would be naive and futile because such a change would also demand new varieties of *P. aquatica* and *D. glomerata* with equally improved P efficiency. This is impractical for the foreseeable future.

Mixed grasslands are natural intercrops which already thrive through complementary access to N-fixation delivered via their legume component. Consequently, an alternative approach to further reduce the critical P requirement of the 'pasture system' as a whole, may be to introduce species that utilise a 'P-mining' acquisition strategy with the aim of capturing complementary and/or facilitated P acquisition for grassland production (e.g. analogous to the *Vicia faba-Zea mays* intercrop system; Li et al., 2003). Exploratory experiments using *Lupinus albus* as a test species suggest that mixing plants with 'P-mining' and 'P-foraging' strategies may lower the critical P requirement of a pasture, but a field experiment, whilst promising, was not conclusive (e.g. Becquer et al., 2017; Sandral et al., 2019). There is also evidence from numerous experiments of shared P access among interplanted crop species (e.g. Li et al., 2014b; Gardner and Boundy, 1983; Horst et al., 2001; Cu et al., 2005), but further work in pastures is limited by a lack of conventional pasture species with unequivocal P-mining attributes, and generally because of inconclusive or negative attempts to improve P acquisition of crop species *in soil* by increasing organic acid anion or phosphatase secretion from roots (e.g. for further discussion, see Richardson et al., 2011; Ryan et al., 2014).

Achievement of unequivocal 'P-mining' attributes in conventional crop and pasture varieties and even the integration of P-mining benefits into current agricultural systems (using species that include *L. albus*, etc.; e.g. Mat Hassan et al., 2013; Doolette et al., 2019) remain as major research challenges for improved P-acquisition efficiency in agriculture.

8 Where to look for further information

For classic texts on P supply and P acquisition by plants:

- Barber, S. A. (1995), *Soil nutrient bioavailability; a mechanistic approach*. John Wiley and Sons, New York, USA.
- Tinker, P. B. and Nye, P. H. (2000), *Solute movement in the rhizosphere*. Oxford University Press Inc., New York, USA.

For a broader understanding of the global P cycle and the impacts of anthropogenic activity:

- Filippelli, G. M. (2008), The global phosphorus cycle: past, present, and future. *Elements* 4, 89-95.
- Föllmi, K. B. (1996), The phosphorus cycle, phosphogenesis and marine phosphate-rich deposits. *Earth-Science Reviews* 40, 55-124.

For practical guides on the management of P fertiliser and the use of critical soil test values:

- Gourley, C. J. P., Melland, A. R., Waller, R. A., Awty, I. M., Smith, A. P., Peverill, K. I. and Hannah, M. C. (2007), *Making better fertiliser decisions for grazed pastures in Australia*. Victorian Government Department of Primary Industries, Melbourne. Currently available on the CSIRO Australian Soil Resource Information System website at: https://www.asris.csiro.au/downloads/BFD/Making%20Better%20Fertiliser%20Decisions%20for%20Grazed%20Pastures%20in%20Australia.pdf.

  Results supporting this resource are found in: Gourley, C. J. P., Weaver, D. M., Simpson, R. J., Aarons, S. R., Hannah, M. M. and Peverill, K. I. (2019), The development and application of functions describing pasture yield responses to phosphorus, potassium, and sulfur in Australia using meta-data analysis and derived soil-test calibration relationships. *Crop Pasture Sci.* 70, 1065-1079.
- *Making Better Fertiliser Decisions for Cropping Systems in Australia*. An online resource and tool including specific case studies for phosphorus. Currently available: www.bfdc.com.au. Associated material is available on

the Grains Research and Development Corporation website at: www.grdc.com.au (e.g. November 2013 Crop Nutrition Fact Sheet "Northern, Southern and Western Regions: Better Fertiliser Decisions for Crop Nutrition"; https://grdc.com.au/__data/assets/pdf_file/0025/194443/grdc_fs_better-fertiliser-decisions-for-crop-nutrition_high-res-pdf.pdf.pdf).

- Simpson, R., Graham, P., Davies, L., Zurcher, E. (2009). *Five easy steps to ensure you are making money from superphosphate*. Decision support tool. CSIRO & Industry and Investment NSW: Sydney. Currently available on the Meat and Livestock Australia website at: https://www.mla.com.au/globalassets/mla-corporate/generic/extension-training-and-tools/5-easy-steps-guide.pdf.

For a useful general discussion of the principles of P use efficiency in agriculture and the principle of the Four Rs of nutrient stewardship:

- Roberts, T. L., Johnston, A. E. (2015), Phosphorus use efficiency and management in agriculture. *Resour. Conserv. Recycl.* 105, 275-281.
  (*Note: the definition of critical P requirement used by Roberts and Johnston (2015) is different to that used in the current chapter and can only be achieved in soils with low or negligible P-sorption capacities*).
- *The Four Rs of nutrient stewardship* (developed collaboratively by The Fertilizer Institute, International Plant Nutrition Institute, Fertilizer Canada, International Fertilizer Association). Information currently available at: http://www.ipni.net/4R.

For a comprehensive overview of the pasture systems of Australia, including those described in the case study of this chapter:

- Moore, R. M. (1970), Australian Grasslands. Australian National University Press: Canberra, ACT.

For information regarding improvements to the internal P use efficiency of plants, a topic not covered by the current chapter:

- Veneklaas, E. J., Lambers, H., Bragg, J., Finnegan, P. M., Lovelock, C. E., Plaxton, W. C., Price, C. A., Scheible, W. R., Shane, M. W., White, P. J. and Raven, J. A. (2012), Opportunities for improving phosphorus-use efficiency in crop plants. *New Phytol.* 195, 306-320.
- Rose, T. J., Liu, L. and Wissuwa, M. (2013), Improving phosphorus efficiency in cereal crops: Is breeding for reduced grain phosphorus concentration part of the solution? *Front. Plant. Sci.* 4, 444.
- Rose, T. J. and Wissuwa, M. (2012), Rethinking internal phosphorus utilization efficiency: a new approach is needed to improve PUE in grain crops. *Adv. Agron.* 116, 185-217.

9 Acknowledgements

RJS and REH are members of a research team examining the development of 'Phosphorus Efficient Pastures' with funding from the Australian Government Department of Agriculture, Water and the Environment as part of its Rural R&D for Profit program, Meat and Livestock Australia, Dairy Australia, Australian Wool Innovations Ltd, and the participating research organisations and farmer groups.

10 References

Abbott, L. K. and Robson, A. D. (1977). Growth stimulation of subterranean clover with vesicular arbuscular mycorrhizas. *Aust. J. Agric. Res.* 28(4), 639–649.

Aitken, Y. and Drake, F. R. (1941). Studies on the varieties of subterranean clover. *Proc. R. Soc. Victoria* 53, 342–393.

Bai, Z., Li, H., Yang, X., Zhou, B., Shi, X., Wang, B., Li, D., Shen, J., Chen, Q., Qin, W., Oenema, O. and Zhang, F. (2013). The critical soil P levels for crop yield, soil fertility and environmental safety in different soil types. *Plant Soil* 372(1–2), 27–37.

Barber, S. A. (1995). *Soil Nutrient Bioavailability: A Mechanistic Approach*. New York: John Wiley and Sons.

Barrow, N. J. (1974). The slow reactions between soil and anions. I. Effects of soil, temperature and water content of a soil on the decrease in effectiveness of phosphate for plant growth. *Soil Sci.* 118(6), 380–386.

Barrow, N. J. (1980a). Differences among some North American soils in the rate of reaction with phosphate. *J. Environ. Qual.* 9(4), 644–648.

Barrow, N. J. (1980b). Differences amongst a wide-ranging collection of soils in the rate of reaction with phosphate. *Aust. J. Soil. Res.* 8, 215–224.

Barrow, N. J. (1999). The four laws of soil chemistry: the Leeper lecture 1998. *Soil Res.* 37(5), 787–829.

Barrow, N. J. (2015). Soil phosphate chemistry and the P-sparing effect of previous phosphate applications. *Plant Soil* 397(1–2), 401–409.

Barrow, N. J., Bolland, M. D. A. and Allen, D. G. (1998). Effect of previous additions of superphosphate on sorption of phosphate. *Soil Res.* 36(3), 359–372.

Barrow, N. J. and Debnath, A. (2014). Effect of phosphate status on the sorption and desorption properties of some soils of northern India. *Plant Soil* 378(1–2), 383–395.

Barrow, N. J., Debnath, P. B. A. and Debnath, A. (2018). Three residual benefits of applying phosphate fertilizer. *Soil Sci. Soc. Am. J.* 82(5), 1168–1176.

Bates, T. R. and Lynch, J. P. (1996). Stimulation of root hair elongation in *Arabidopsis thaliana* by low phosphorus availability. *Plant Cell Environ.* 19(5), 529–538.

Becquer, A., Haling, R. E., Stefanski, A., Richardson, A. E. and Simpson, R. J. (2017). Complementary phosphorus acquisition strategies of interplanted subterranean clover and white lupin increase sward yield in a low phosphorus soil. *Proceedings of the 18th Aust. Soc. Agron. Conference*, Ballarat, Australia. Available at: http://agronomyaustraliaproceedings.org/images/sampledata/2017/199_ASA2017_Becquer_Adeline_Final.pdf.

Bell, L. W., Ryan, M. H., Bennett, R. G., Collins, M. T. and Clarke, H. J. (2012). Growth, yield and seed composition of native Australian legumes with potential as grain crops. *J. Sci. Food Agric.* 92(7), 1354–1361.

Bell, M. J., Moody, P. W., Anderson, G. C. and Strong, W. (2013). Soil phosphorus - crop response calibration relationships and criteria for oilseeds, grain legumes and summer cereal crops grown in Australia. *Crop Pasture Sci.* 64(5), 499–513.

Benbi, D. K. and Gilkes, R. J. (1987). The movement into soil of P from superphosphate grains and its available to plants. *Fert. Res.* 12(1), 21–36.

Blackwell, M. S. A., Brookes, P. C., de la Fuente-Martinez, N., Murray, P. J., Snars, K. E., Williams, J. K. and Haygarth, P. M. (2009). Effects of soil drying and rate of re-wetting on concentrations and forms of phosphorus in leachate. *Biol. Fertil. Soils* 45(6), 635–643.

Bolan, N. S., Robson, A. D. and Barrow, N. J. (1987). Effects of vesicular-arbuscular mycorrhiza on the availability of iron phosphates to plants. *Plant Soil* 99(2–3), 401–410.

Bolland, M. D. A. (1986a). Residual value of phosphorus from superphosphate for wheat grown on soils of contrasting texture near Esperance, Western Australia. *Aust. J. Exp. Agric.* 26, 209–215.

Bolland, M. D. A. (1986b). Efficiency with which yellow serradella and subterranean clover use superphosphate on a deep sandy soil near Esperance, Western Australia. *Aust. J. Exp. Agric.* 26, 675–679.

Bolland, M. D. A. and Baker, M. J. (1998). Phosphate applied to soil increases the effectiveness of subsequent applications of phosphate for growing wheat shoots. *Aust. J. Exp. Agric.* 38(8), 865–869.

Bolland, M. D. A. and Gladstones, J. S. (1987). Serradella (*Ornithopus* spp.) as a pasture legume in Australia. *J. Aust. Inst. Agric. Sci.* 53, 5–10.

Bolland, M. D. A. and Paynter, B. H. (1992). Comparative responses of annual pasture legume species to superphosphate applications in medium and high rainfall areas of Western Australia. *Fert. Res.* 31(1), 21–33.

Borkert, C. M. and Barber, S. A. (1985). Predicting the most efficient phosphorus placement for soybeans. *Soil Sci. Soc. Am. J.* 49(4), 901–904.

Boschma, S., Kidd, D., Newell, M., Stefanski, A., Haling, R., Hayes, R., Ryan, M. and Simpson, R. (2019). Flowering time responses of serradella cultivars. *Proceedings of the 19th Aust. Soc. Agron. Conference*, Wagga Wagga, Australia. Available at: http://agronomyaustraliaproceedings.org/images/sampledata/2019/2019ASA_Simpson_Richard_169.pdf.

Bouwman, L., Goldewijk, K. K., Van Der Hoek, K. W., Beusen, A. H. W., Van Vuuren, D. P., Willems, J., Rufinoe, M. C. and Stehfesta, E. (2013). Exploring global changes in nitrogen and phosphorus cycles in agriculture induced by livestock production over the 1900–2050 period. *Proc. Natl Acad. Sci. U.S.A.* 110(52), 20882–20887.

Bowen, G. D. and Rovira, A. D. (1999). The rhizosphere and its management to improve plant growth. *Adv. Agron.* 66, 1–102.

Brouwer, R. (1962). Nutritive influences on the distribution of dry matter in the plant. *Neth. J. Agric. Sci.* 10(5), 399–408.

Burkitt, L. L., Moody, P. W., Gourley, C. J. P. and Hannah, M. C. (2002). A simple phosphorus buffering index for Australian soils. *Aust. J. Soil Res.* 40, 1–18.

Burkitt, L. L., Sale, P. W. G. and Gourley, C. J. P. (2008). Soil phosphorus buffering measures should not be adjusted for current phosphorus fertility. *Aust. J. Soil Res.* 46(8), 676–685.

Burridge, J. D., Findeis, J. L., Jochua, C. N., Miguel, M. A., Mubichi-Kut, F. M., Quinhentos, M. L., Xerinda, S. A. and Lynch, J. P. (2019). A case study on the efficacy of root phenotypic selection for edaphic stress tolerance in low-input agriculture: common bean breeding in Mozambique. *Field Crops Res.* 244, 107512. doi:10.1016/j.fcr.2019.107612.

Campbell, B. M., Beare, D. J., Bennett, E. M., Hall-Spencer, J. M., Ingram, J. S. I., Jaramillo, F., Ortiz, R., Ramankutty, N., Sayer, J. A. and Shindell, D. (2017). Agriculture production as a major driver of the Earth system exceeding planetary boundaries. *Ecol. Soc.* 22(4), 8. doi:10.5751/ES-09595-220408.

Carter, E. D. and Day, H. R. (1970). Interrelationships of stocking rate and superphosphate rate on pasture as determinants of animal production. I. Continuously grazed old pasture land. *Aust. J. Agric. Res.* 21(3), 473–491.

Cayley, J. W. D., Kearney, G. A., Saul, G. R. and Lescun, C. L. (1999). The long-term influence of superphosphate and stocking rate on the production of spring-lambing Merino sheep in the high rainfall zone of southern Australia. *Aust. J. Agric. Res.* 50(7), 1179–1190.

Chen, C. R., Condron, L. M., Davis, M. R. and Sherlock, R. R. (2002). Phosphorus dynamics in the rhizosphere of perennial ryegrass (*Lolium perenne* L.) and radiata pine (*Pinus radiata* D.Don). *Soil Biol. Biochem.* 34(4), 487–499.

Colwell, J. D. (1963). The estimation of the phosphorus fertilizer requirements of wheat in southern New South Wales by soil analysis. *Aust. J. Exp. Agric.* 3(10), 190–197.

Cordell, D., Drangert, J.-O. and White, S. (2009). The story of phosphorus: global food security and food for thought. *Glob. Environ. Chang.* 19(2), 292–305.

Cornish, P. S. and Myers, L. F. (1977). Low pasture productivity of a sedimentary soil in relation to phosphate and water supply. *Aust. J. Exp. Agric.* 17(88), 776–783.

Costin, A. B. and Williams, C. H. (1983). *Phosphorus in Australia*. Canberra, Australia: Australian National University, Centre for Resource and Environmental Studies.

Cruz-Paredes, C., Svenningsen, N. B., Nybroe, O., Kjoller, R., Froslev, T. G. and Jakobsen, I. (2019). Suppression of arbuscular mycorrhizal fungal activity in a diverse collection of non-cultivated soils. *FEMS Microbiol. Ecol.* 95(3), 1–10.

Cu, S. T. T., Hutson, J. and Schuller, K. A. (2005). Mixed culture of wheat (*Triticum aestivum* L.) with white lupin (*Lupinus albus* L.) improves the growth and phosphorus nutrition of the wheat. *Plant Soil* 272(1–2), 143–151.

Cullen, B. R., Eckard, R. J., Callow, M. N., Johnson, I. R., Chapman, D. F., Rawnsley, R. P., Garcia, S. C., White, T. and Snow, V. O. (2008). Simulating pasture growth rates in Australian and New Zealand grazing systems. *Aust. J. Agric. Res.* 59(8), 761–768.

Culvenor, R. A. and Simpson, R. J. (2016). Interaction of plant genotype and management in the persistence of a perennial grass exposed to grazing and soil fertility stresses. *Grass Forage Sci.* 71(4), 540–558.

Curll, M. L. (1977). Superphosphate on perennial pastures. I. Effect of a pasture response on sheep production. *Aust. J. Agric. Res.* 28(6), 991–1005.

Dear, B. S., Wilson, B. C. D., Rodham, C. A., McCaskie, P. and Sandral, G. A. (2002). Productivity and persistence of *Trifolium hirtum*, *T. michelianum*, *T. glanduliferum* and *Ornithopus sativus* sown as monoculture or in mixtures with *T. subterraneaum* in the south-eastern Australian wheat belt. *Aust. J. Exp. Agric.* 42(5), 549–556.

Delhaize, E., Taylor, P., Hocking, P. J., Simpson, R. J., Ryan, P. R. and Richardson, A. E. (2009). Transgenic barley (*Hordeum vulgare* L.) that express the wheat aluminium resistance gene (*TaALMT1*) show enhanced phosphorus nutrition and grain production when grown on an acid soil. *Plant Biotechnol. J.* 7(5), 391–400.

Denton, M. D., Sasse, C., Tibbett, M. and Ryan, M. H. (2006). Root distributions of Australian herbaceous perennial legumes in response to phosphorus placement. *Funct. Plant Biol.* 33(12), 1091–1102.

Dessureault-Rompre, J., Nowack, B., Schulin, R. and Luster, J. (2007). Spatial and temporal variation in organic acid anion exudation and nutrient anion uptake in the rhizosphere of *Lupinus albus* L. *Plant Soil* 301(1–2), 123–134.

Dibb, D. W., Fixen, P. E. and Murphy, L. S. (1990). Balanced fertilization with particular reference to phosphates: interaction of phosphorus with other inputs and management practices. *Fert. Res.* 26(1–3), 29–52.

Dinkelaker, B., Romheld, B. V. and Marschner, H. (1989). Citric acid excretion and precipitation of Ca citrate in the rhizosphere of white lupin (*Lupinus albus* L.). *Plant Cell. Environ.* 12, 285–292.

Dodd, J. R. and Mallarino, A. P. (2005). Soil-test phosphorus and crop grain yield responses to long-term phosphorus fertilisation for corn-soybean rotations. *Soil Sci. Soc. Am. J.* 69(4), 1118–1128.

Donald, C. M. (1970). Temperate pasture species. In: Moore, R. M. (Ed.), *Australian Grasslands*. Canberra, Australia: Australian National University Press, pp. 303–320.

Donald, C. M. and Williams, C. H. (1954). Fertility and productivity of a podzolic soil as influenced by subterranean clover (*Trifolium subterraneum* L.) and superphosphate. *Aust. J. Agric. Res.* 5(4), 664–687.

Doolette, A., Armstrong, R., Tang, C., Guppy, C., Mason, S. and McNeill, A. (2019). Phosphorus uptake benefit for wheat following legume break crops in semi-arid Australian farming systems. *Nutr. Cycl. Agroecosyst.* 113(3), 247–266.

Doyle, P. T., Grimm, M. and Thompson, A. N. (1993). Grazing for pasture and sheep management in the annual pasture zone. In: Kemp, D. R. and Michalk, D. L. (Eds) *Pasture Management Technology for the 21st Century*. Melbourne: CSIRO Australia, pp. 71–90.

Dunbabin, V., Diggle, A. and Rengel, Z. (2003). Is there an optimal root architecture for nitrate capture in leaching environments? *Plant Cell Environ.* 26(6), 835–844.

Duncan, W. G. and Ohlrogge, A. J. (1958). Principles of nutrient uptake from fertilizer bands. II. Root development in the band. *Agron. J.* 50, 605–608.

Eissenstat, D. M. (1991). On the relationship between specific root length and rate of root proliferation: a field study using citrus rootstocks. *New Phytol.* 118, 63–68.

Erisman, J. W., Sutton, M. A., Galloway, J. N., Klimont, Z. and Winiwarter, W. (2008). How a century of ammonia synthesis changed the world. *Nat. Geosci.* 1(10), 636–639.

Filippelli, G. M. (2008). The global phosphorus cycle: past, present, and future. *Elements* 4(2) , 89–95.

Fischer, R. A. and Connor, D. J. (2018). Issues for cropping and agricultural science in the next 20 years. *Field Crops Res.* 222, 121–142.

Föhse, D. and Jungk, A. (1983). Influence of phosphate and nitrate supply on root hair formation of rape, spinach and tomato plants. *Plant Soil* 74(3), 359–368.

Föllmi, K. B. (1996). The phosphorus cycle, phosphogenesis and marine phosphate-rich deposits. *Earth Sci. Rev.* 40(1–2), 55–124.

Freebairn, R. D. (1996). *The History of Serradella (Ornithopus* spp.*) in NSW – A Miracle Pasture for Light Soils. Agdex 137*. Sydney: NSW Agriculture and Fisheries.

Freschet, G. T., Swart, E. M. and Cornelissen, J. H. C. (2015). Integrated plant phenotypic responses to contrasting above- and below-ground resources: key roles of specific leaf area and root mass fraction. *New Phytol.* 206(4), 1247–1260.

Gahoonia, T. S. and Nielsen, N. E. (1992). The effect of root induced pH changes on the depletion of inorganic and organic phosphorus in the rhizosphere. *Plant Soil* 143(2), 185–191.

Gahoonia, T. S., Care, D. and Nielsen, N. E. (1997). Root hairs and phosphorus acquisition of wheat and barley cultivars. *Plant Soil* 191(2), 181–188.

Gamuyao, R., Chin, J. H., Pariasca-Tanaka, J., Pesaresi, P., Dalid, C., Slamet-Loedin, I., Tecson-Mendoza, E. M., Wissuwa, M. and Heuer, S. (2012). The protein kinase OsPSTOL1 from traditional rice confers tolerance of phosphorus deficiency. *Nature* 488, 535–539.

Gardner, W. K. and Boundy, K. A. (1983). The acquisition of phosphorus by *Lupinus albus* L. IV. The effect of interplanting wheat and white lupin on the growth and mineral composition of the two species. *Plant Soil* 70(3), 391–402.

George, T. S., Gregory, P. J., Robinson, J. S., Buresh, R. J. and Jama, B. (2002). Utilisation of soil organic P by agroforestry and crop species in the field, western Kenya. *Plant Soil* 246(1), 53–63.

George, T. S., Hocking, P. J., Gregory, P. J. and Richardson, A. E. (2008). Variation of root-associated phosphatase in wheat cultivars explains their ability to utilise organic P substrates in-vitro, but does not effectively predict P-nutrition in a range soils. *Environ. Exp. Bot.* 64, 239–249.

Gerke, J. (1994). Kinetics of soil phosphate desorption as affected by citric acid. *Z. Pflanzenernaehr. Bodenk.* 157(1), 17–22.

Gourley, C. J. P., Aarons, S. R., Hannah, M. C., Awty, I. M., Dougherty, W. J. and Burkitt, L. L. (2015). Soil phosphorus, potassium and sulphur excesses, regularities and heterogeneity in grazing-based dairy farms. *Agric. Ecosyst. Environ.* 201, 70–82.

Gourley, C. J. P., Melland, A. R., Waller, R. A., Awty, I. M., Smith, A. P., Peverill, K. I. and Hannah, M. C. (2007). Making better fertiliser decisions for grazed pastures in Australia. Melbourne: Victorian Government Department of Primary Industries. Available at: http://www.asris.csiro.au/downloads/BFD/Making%20Better%20Fertiliser%20Decisions%20for%20Grazed%20Pastures%20in%20Australia.pdf.

Gourley, C. J. P. and Weaver, D. M. (2012). Nutrient surpluses in Australian grazing systems: management practices, policy approaches, and difficult choices to improve water quality. *Crop Pasture Sci.* 63(9), 805–818.

Gourley, C. J. P., Weaver, D. M., Simpson, R. J., Aarons, S. R., Hannah, M. M. and Peverill, K. I. (2019). The development and application of functions describing pasture yield responses to phosphorus, potassium and sulfur in Australia using meta-data analysis and derived soil-test calibration relationships. *Crop Pasture Sci.* 70(12), 1065–1079.

Graham, R. P. (2006). Bookham grazing demonstration results. *NSW DPI Sheep Conference*. Orange: New South Wales Department of Primary Industries, pp. 211–216.

Graham, P. (2017). Bookham grazing demonstration results. New South Wales: Department of Primary Industries. Available at: http://www.fscl.org.au/wp-content/uploads/2017/09/Bookham-Grazing-demo-results-P-Graham.pdf.

Greenwood, K. L. and Hutchinson, K. J. (1998). Root characteristics of temperate pasture in New South Wales after grazing at three stocking rates for 30 years. *Grass Forage Sci.* 53(2), 120–128.

Hackney, B., Rodham, C. and Piltz, J. (2013). *Using French Serradella to Increase Crop and Livestock Production*. Sydney: Meat and Livestock Australia Ltd.

Haling, R. E., Brown, L. K., Stefanski, A., Kidd, D. R., Ryan, M. H., Sandral, G. A., George, T. S., Lambers, H. and Simpson, R. J. (2018). Differences in nutrient foraging among *Trifolium subterraneum* cultivars deliver improved P-acquisition efficiency. *Plant Soil* 424(1-2), 539-554.

Haling, R. E., Richardson, A. E., Culvenor, R. A., Lambers, H. and Simpson, R. J. (2010a). Root morphology, root-hair development and rhizosheath formation on perennial grass seedlings is influenced by soil acidity. *Plant Soil* 335(1-2), 457-468.

Haling, R. E., Simpson, R. J., Delhaize, E., Hocking, P. J. and Richardson, A. E. (2010b). Effect of lime on root growth, morphology and the rhizosheath of cereal seedlings growing in an acid soil. *Plant Soil* 327(1-2), 199-212.

Haling, R. E., Yang, Z. J., Shadwell, N., Culvenor, R. A., Stefanski, A., Ryan, M. H., Sandral, G. A., Kidd, D. R., Lambers, H. and Simpson, R. J. (2016a). Growth and root dry matter allocation by pasture legumes and a grass with contrasting external critical phosphorus requirements. *Plant Soil* 407(1-2), 67-79.

Haling, R. E., Yang, Z. J., Shadwell, N., Culvenor, R. A., Stefanski, A., Ryan, M. H., Sandral, G. A., Kidd, D. R., Lambers, H. and Simpson, R. J. (2016b). Root morphological traits that determine phosphorus-acquisition efficiency and critical external phosphorus requirement in pasture species. *Funct. Plant Biol.* 43(9), 815-826.

Handreck, K. A. (1997). Phosphorus requirements of Australian native plants. *Aust. J. Soil Res.* 35(2), 241-289.

Hayes, J. E., Simpson, R. J. and Richardson, A. E. (2000). The growth and utilisation of plants in sterile media when supplied with inositol hexaphosphate, glucose 1-phosphate or inorganic phosphate. *Plant Soil* 220(1/2), 165-174.

Hayes, R. C., Sandral, G. A., Simpson, R., Price, A., Stefanski, A. and Newell, M. T. (2015). A preliminary evaluation of alternative annual legume species under grazing on the Southern Tablelands of NSW. *Proceedings of the 17th Aust. Soc. Agron. Conference, Hobart*, Australia. Available at: http://agronomyaustraliaproceedings.org/images/sampledata/2015_Conference/pdf/agronomy2015final00332.pdf.

Hedley, M. and McLaughlin, M. (2005). Reactions of phosphate fertilizers and by-products in soils. In: Sims, J. T. and Sharpley, A. N. (Eds), *Phosphorus: Agriculture and the Environment, Agronomy Monograph No. 46*. Madison, WI: American Society of Agronomy, Crop Science Society of America, Soil Science Society of America, pp. 181-252.

Hedley, M. J., White, R. E. and Nye, P. H. (1982). Plant induced changes in the rhizosphere of rape (*Brassica napus* var. Emerald) seedlings. III Changes in L value, soil phosphate fractions and phosphatase activity. *New Phytol.* 91, 45-56.

Helyar, K. R. and Anderson, A. J. (1970). Responses of five pasture species to phosphorus, lime, and nitrogen on an infertile acid soil with a high phosphate sorption capacity. *Aust. J. Agric. Res.* 21(5), 677-692.

Helyar, K. and Spencer, K. (1977). Sodium bicarbonate soil test values and the phosphate buffering capacity of soils. *Aust. J. Soil Res.* 15(3), 263-273.

Helyar, K. R. and Price, G. H. (1999). Making recommendations based on soil tests. In: Peverill, K. I., Sparrow, L. A. and Reuter, D. J. (Eds), *Soil Analysis: An Interpretation Manual*. Collingwood, Australia: CSIRO Publishing.

Heuer, S., Gaxiola, R., Schilling, R., Herrera-Estrella, L., Lopez-Arredondo, D., Wissuwa, M., Delhaize, E. and Rouached, H. (2017). Improving phosphorus use efficiency: a complex trait with emerging opportunities. *Plant J.* 90(5), 868-885.

Hill, J. O., Simpson, R. J., Ryan, M. H. and Chapman, D. F. (2010). Root hair morphology and mycorrhizal colonisation of pasture species in response to phosphorus and nitrogen nutrition. *Crop Pasture Sci.* 61(2), 122-131.

Hill, J. O., Simpson, R. J., Wood, J. T., Moore, A. D. and Chapman, D. F. (2005). The phosphorus and nitrogen requirements of temperate pasture species and their influence on grassland botanical composition. *Aust. J. Agric. Res.* 56(10), 1027-1039.

Hill, M. J. and Donald, G. E. (1998). *Australian Temperate Pastures Database*. Perth, Western Australia: National Pasture Improvement Coordinating Committee/CSIRO Division of Animal Production (CD-ROM).

Hinsinger, P., Betencourt, E., Bernard, L., Brauman, A., Plassard, C., Shen, J., Tang, X. and Zhang, F. (2011). P for two, sharing a scarce resource: soil phosphorus acquisition in the rhizosphere of intercropped species. *Plant Physiol.* 156(3), 1078-1086.

Ho, M. D., Rosas, J. C., Brown, K. M. and Lynch, J. P. (2005). Root architectural tradeoffs for water and phosphorus acquisition. *Funct. Plant Biol.* 32(8), 737-748.

Holford, I. and Crocker, G. (1988). Efficacy of various soil phosphate tests for predicting phosphate responsiveness and requirements of clover pastures on acidic tableland soils. *Aust. J. Soil Res.* 26(3), 479-488.

Horst, W. J., Kamh, M., Jibrin, J. M. and Chude, V. O. (2001). Agronomic measures for increasing P availability to crops. *Plant Soil* 237(2), 211-223.

Hufnagel, B., de Sousa, S. M., Assis, L., Guimaraes, C. T., Leiser, W., Azevedo, G. C., Negri, B., Larson, B. G., Shaff, J. E., Pastina, M. M., Barros, B. A., Weltzien, E., Rattunde, H. F., Viana, J. H., Clark, R. T., Falcão, A., Gazaffi, R., Garcia, A. A., Schaffert, R. E., Kochian, L. V. and Magalhaes, J. V. (2014). Duplicate and conquer: multiple homologs of PHOSPHORUS-STARVATION TOLERANCE1 enhance phosphorus acquisition and sorghum performance on low phosphorus soils. *Plant Physiol.* 166(2), 659-677.

Jackson, T. and Malcolm, B. (2018). Using returns, risks and learning to understand innovation adoption. *Aust. Agribusiness Rev.* 26, 32-47.

Jakobsen, I., Abbott, L. K. and Robson, A. D. (1992). External hyphae of vesicular-arbuscular mycorrhizal fungi associated with *Trifolium subterraneum* L. 2. Hyphal transport of ^{32}P over defined distances. *New Phytol.* 120(4), 509-516.

Jarosch, K. A., Doolette, A. L., Smernik, R. J., Tamburini, F., Frossard, E. and Bünemann, E. K. (2015). Characterisation of soil organic phosphorus in NaOH-EDTA extracts: a comparison of ^{31}P NMR spectroscopy and enzyme addition assays. *Soil Biol. Biochem.* 91, 298-309.

Jarosch, K. A., Kandeler, E., Frossard, E. and Bünemann, E. K. (2019). Is the enzymatic hydrolysis of soil organic phosphorus compounds limited by enzyme or substrate availability? *Soil Biol. Biochem.* 139, 107628. doi:10.1016/j.soilbio.2019.107628.

Jarvie, H. P., Sharpley, A. N., Withers, P. J., Scott, J. T., Haggard, B. E. and Neal, C. (2013a). Phosphorus mitigation to control river eutrophication: murky waters, inconvenient truths, and "postnormal" science. *J. Environ. Qual.* 42(2), 295-304.

Jarvie, H. P., Sharpley, A. N., Spears, B., Buda, A. R., May, L. and Kleinman, P. J. A. (2013b). Water quality remediation faces unprecedented challenges from legacy phosphorus. *Environ. Sci. Technol.* 47(16), 8997-8998.

Jarvis, R. J. and Bolland, M. D. A. (1991). Lupin grain yields and fertiliser effectiveness are increased by banding superphosphate below the seed. *Aust. J. Exp. Agric.* 31(3), 357-366.

Jia, X., Liu, P. and Lynch, J. P. (2018). Greater lateral root branching density in maize (*Zea mays* L.) improves phosphorus acquisition from low phosphorus soil. *J. Exp. Bot.* 69(20), 4961–4970.

Johnston, A. E., Poulton, P. R. and White, R. P. (2013). Plant-available soil phosphorus. Part II: the response of arable crops to Olsen P on a sandy clay loam and a silty clay loam. *Soil Use Manag.* 29(1), 12–21.

Johnston, A. E., Poulton, P. R., Fixen, P. E. and Curtin, D. (2014). Phosphorus: its efficient use in agriculture. *Adv. Agron.* 123, 177–228.

Jones, D. L., Nguyen, C. and Finlay, R. D. (2009). Carbon flow in the rhizosphere: carbon trading at the soil-root interface. *Plant Soil* 321(1–2), 5–33.

Kandeler, E., Marschner, P., Tscherko, D., Gahoonia, T. S. and Nielsen, N. E. (2001). Microbial community composition and functional diversity in the rhizosphere of maize. *Plant Soil* 238, 301–312.

Kar, G., Peak, D. and Schoenau, J. J. (2012). Spatial distribution and chemical speciation of soil phosphorus in a band application. *Soil Sci. Soc. Am. J.* 76(6), 2297–2306.

Kidd, D. R., Ryan, M. H., Haling, R. E., Lambers, H., Sandral, G. A., Yang, Z., Culvenor, R. A., Cawthray, G. R., Stefanski, A. and Simpson, R. J. (2016). Rhizosphere carboxylates and morphological root traits in pasture legumes and grasses. *Plant Soil* 402(1–2), 77–89.

Klamer, F., Vogel, F., Li, X., Bremer, H., Neumann, G., Neuhäuser, B., Hochholdinger, F. and Ludewig, U. (2019). Estimating the importance of maize root hairs in low phosphorus conditions and under drought. *Ann. Bot.* 124(6), 961–968.

Kucey, R. M. N., Janzen, H. H. and Leggett, M. E. (1989). Microbially mediated increases in plant-available phosphorus. *Adv. Agron.* 42, 199–228.

Lambers, H., Shane, M. W., Cramer, M. D., Pearse, S. J. and Veneklaas, E. J. (2006). Root structure and functioning for efficient acquisition of phosphorus: matching morphological and physiological traits. *Ann. Bot.* 98(4), 693–713.

Lambers, H., Raven, J. A., Shaver, G. R. and Smith, S. E. (2008). Plant nutrient-acquisition strategies change with soil age. *Trends Ecol. Evol. (Amst.)* 23(2), 95–103.

Lambers, H., Brundrett, M. C., Raven, J. A. and Hopper, S. D. (2010). Plant mineral nutrition in ancient landscapes: high plant species diversity on infertile soils is linked to functional diversity for nutritional strategies. *Plant Soil* 334(1–2), 11–31.

Lean, G. R., Vizard, A. L. and Webb Ware, J. K. (1997). Changes in productivity and profitability of wool-growing farms that follow recommendations from agricultural and veterinary studies. *Aust. Vet. J.* 75(10), 726–731.

Leech, F. J., Richardson, A. E., Kertesz, M. A., Orchard, B. A., Banerjee, S. and Graham, P. (2019). Comparative effect of alternative fertilisers on pasture production, soil properties and soil microbial community structure. *Crop Pasture Sci.* 70(12), 1110–1127.

Lewis, C., Malcolm, B., Farquharson, R., Leury, B., Behrendt, R. and Clark, S. (2012). Economic analysis of improved perennial pasture systems. *Aust. Farm Bus. Manag. J.* 9, 37–55.

Lewis, D. C., Clarke, A. L. and Hall, W. B. (1981). Factors affecting the retention of phosphorus applied as superphosphate to the sandy soils in south-eastern South Australia. *Aust. J. Soil Res.* 19(2), 167–174.

Lewis, D. C., Clarke, A. L. and Hall, W. B. (1987). Accumulation of plant nutrients and changes in soil properties of sandy soils under fertilized pasture in southeastern South Australia. 1. Phosphorus. *Aust. J. Soil Res.* 25(2), 193–202.

Lewis, D. C., Bolland, M. D. A., Gilkes, R. J. and Hamilton, L. J. (1997). Review of Australian phosphate rock research. *Aust. J. Exp. Agric.* 37(8), 845-859.

Li, H., Ma, Q., Li, H., Zhang, F., Rengel, Z. and Shen, J. (2014a). Root morphological responses to localized nutrient supply differ among crop species with contrasting root traits. *Plant Soil* 376(1-2), 151-163.

Li, L., Zhang, F., Li, X., Christie, P., Sun, J., Yang, S. and Tang, C. (2003). Interspecific facilitation of nutrient uptake by intercropped maize and faba bean. *Nutr. Cycl. Agroecosyst.* 65(1), 61-71.

Li, L., Tilman, D., Lambers, H. and Zhang, F. S. (2014b). Plant diversity and overyielding: insights from belowground facilitation of intercropping in agriculture. *New Phytol.* 203(1), 63-69.

Loi, A., Howieson, J. G., Nutt, B. J. and Carr, S. J. (2005). A second generation of annual pasture legumes and their potential for inclusion in Mediterranean type farming systems. *Aust. J. Exp. Agric.* 45(3), 289-299.

Lombi, E., McLaughlin, M. J., Johnston, C., Armstrong, R. D. and Holloway, R. E. (2004). Mobility and lability of phosphorus from granular and fluid monoammonium phosphate differs in a calcareous soil. *Soil Sci. Soc. Am. J.* 68(2), 682-689.

Lun, F., Liu, J., Ciais, P., Nesme, T., Chang, J., Wang, R., Goll, D., Sardans, J., Peñuelas, J. and Obersteiner, M. (2018). Global and regional phosphorus budgets in agricultural systems and their implications for phosphorus-use efficiency. *Earth Syst. Sci. Data* 10(1), 1-18.

Lynch, J. P. (2007). Roots of the second green revolution. *Aust. J. Bot.* 55(5), 493-512.

Lynch, J. P. (2011). Root phenes for enhanced soil exploration and phosphorus acquisition: tools for future crops. *Plant Physiol.* 156(3), 1041-1049.

Lynch, J. P. (2019). Root phenotypes for improved nutrient capture: an underexploited opportunity for global agriculture. *New Phytol.* 223(2), 548-564.

Lyu, Y., Tang, H., Li, H., Zhang, F., Rengel, Z., Whalley, W. R. and Shen, J. (2016). Major crop species show differential balance between root morphological and physiological responses to variable phosphorus supply. *Front. Plant Sci.* 7, 1939. doi: 10.3389/fpls.2016.01939.

MacDonald, G. K., Bennett, E. M., Potter, P. A. and Ramankutty, N. (2011). Agronomic phosphorus imbalances across the world's croplands. *Proc. Natl Acad. Sci. U.S.A.* 108(7), 3086-3091.

Mai, W., Xue, X., Feng, G., Yang, R. and Tian, C. (2018). Can optimization of phosphorus input lead to high productivity and high phosphorus use efficiency of cotton through maximization of root/mycorrhizal efficiency in phosphorus acquisition? *Field Crops Res.* 216, 100-108.

Mai, W., Xue, X., Feng, G., Yang, R. and Tian, C. (2019). Arbuscular mycorrhizal fungi - 15-fold enlargement of the soil volume of cotton roots for phosphorus uptake in intensive planting conditions. *Eur. J. Soil Biol.* 90, 31-35.

Malcolm, B., Smith, K. F. and Jacobs, J. L. (2014). Perennial pasture persistence: the economic perspective. *Crop Pasture Sci.* 65(8), 713-720.

Manske, G. G. B., Ortiz-Monasterio, J. I., Van Ginkel, M., González, R. M., Rajaram, S., Molina, E. and Vlek, P. L. G. (2000). Traits associated with improved P-uptake efficiency in CIMMYT's semidwarf spring bread wheat grown on an acid andisol in Mexico. *Plant Soil* 221(2), 189-204.

Marcus, O. O. (2013). Genetic assessment and mapping of QTLs for biomass, P uptake, crown root angle and mycorrhiza colonization for adaptation to low phosphorus conditions in west African *Sorghum bicolor* L. Master Thesis, University of Hohenheim, Germany.

Marschner, P., Marhan, S. and Kandeler, E. (2012). Microscale distribution and function of soil microorganisms in the interface between rhizosphere and detritusphere. *Soil Biol. Biochem.* 49, 174–183.

Mat Hassan, H., Hasbullah, H. and Marschner, P. (2013). Growth and rhizosphere P pools of legume-wheat rotations at low P supply. *Biol. Fertil. Soils* 49(1), 41–49.

McBeath, T. M., McLaughlin, M. J., Kirby, J. K. and Armstrong, R. D. (2012). The effect of soil water status on fertiliser, topsoil and subsoil phosphorus utilisation by wheat. *Plant Soil* 358(1–2), 337–348.

McCaskill, M. R. and Cayley, J. W. D. (2000). Soil audit of a long-term phosphate experiment in south-western Victoria: total phosphorus, sulfur, nitrogen, and major cations. *Aust. J. Agric. Res.* 51(6), 737–748.

McDowell, R. W., Sharpley, A. N., Condron, L. M., Haygarth, P. M. and Brookes, P. C. (2001). Processes controlling soil phosphorus release to runoff and implications for agricultural management. *Nutr. Cycl. Agroecosyst.* 59(3), 269–284.

McEachern, S. F. J. and Brown, D. (2010). *AgInsights 2009*. Wagga Wagga, Australia: Holmes Sackett Pty Ltd.

McLachlan, J. W., Haling, R. E., Simpson, R. J., Li, X., Flavel, R. J. and Guppy, C. N. (2019). Variation in root morphology and P-acquisition efficiency among *Trifolium subterraneum* genotypes. *Crop Pasture Sci.* 70(11), 1015–1032.

McLachlan, J. W., Becquer, A., Haling, R. E., Simpson, R. J., Flavel, R. J. and Guppy, C. N. (2020). Intrinsic root morphology determines the phosphorus acquisition efficiency of five annual pasture legumes irrespective of mycorrhizal colonisation. *Funct. Plant Biol.* https://doi.org/10.1071/FP20007.

McLachlan, K. D. (1965). The nature of available phosphorus in some acid pasture soils and a comparison of estimating procedures. *Aust. J. Exp. Agric.* 5(17), 125–132.

McLaren, T. I., Simpson, R. J., McLaughlin, M. J., Smernik, R. J., McBeath, T. M., Guppy, C. N. and Richardson, A. E. (2015). An assessment of various measures of soil phosphorus and the net accumulation of phosphorus in fertilized soils under pasture. *J. Plant Nutr. Soil Sci.* 178(4), 543–554.

McLaren, T. I., McBeath, T. M., Simpson, R. J., Richardson, A. E., Stefanski, A., Guppy, C. N., Smernik, R. J., Rivers, C., Johnston, C. and McLaughlin, M. J. (2017a). Direct recovery of ^{33}P-labelled fertiliser phosphorus in subterranean clover (*Trifolium subterraneum*) pastures under field conditions – the role of agronomic management. *Agric. Ecosyst. Environ.* 246, 144–156.

McLaren, T. I., Smernik, R. J., Simpson, R. J., McLaughlin, M. J., McBeath, T. M., Guppy, C. N. and Richardson, A. E. (2017b). The chemical nature of organic phosphorus that accumulates in fertilized soils of a temperate pasture as determined by solution ^{31}P NMR spectroscopy. *J. Plant Nutr. Soil Sci.* 180(1), 27–38.

McLaren, T. I., Smernik, R. J., McLaughlin, M. J., McBeath, T. M., McCaskill, M. R., Robertson, F. A. and Simpson, R. J. (2020). Soil phosphorus pools with addition of fertiliser phosphorus in a long-term grazing experiment. *Nutr. Cycl. Agroecosyst.* 116(2), 151–164.

McLaughlin, M. J., Alston, A. M. and Martin, J. K. (1988). Phosphorus cycling in wheat pasture rotations. I. The source of phosphorus taken up by wheat. *Aust. J. Soil Res.* 26(2), 323–331.

McLaughlin, M. J., Baker, T. G., James, T. R. and Rundle, J. A. (1990). Distribution and forms of phosphorus and aluminum in acidic topsoils under pastures in south-eastern Australia. *Aust. J. Soil Res.* 28(3), 371–385.

McLaughlin, M. J., McBeath, T. M., Smernik, R., Stacey, S. P., Ajiboye, B. and Guppy, C. (2011). The chemical nature of P-accumulation in agricultural soils - implications for fertiliser management and design: an Australian perspective. *Plant Soil* 349(1–2), 69–87.

Melland, A. R., McCaskill, M. R., White, R. E. and Chapman, D. F. (2008). Loss of phosphorus and nitrogen in runoff and subsurface drainage from high and low input pastures grazed by sheep in Australia. *Aust. J. Soil Res.* 46, 161–172.

Meyer, G., Maurhofer, M., Frossard, E., Gamper, H. A., Mäder, P., Mészáros, É., Schönholzer-Mauclaire, L., Symanczik, S. and Oberson, A. (2019). *Pseudomonas protegens* CHA0 does not increase phosphorus uptake from ^{33}P labeled synthetic hydroxyapatite by wheat grown on calcareous soil. *Soil Biol. Biochem.* 131, 217–228.

Miguel, M. A., Widrig, A., Vieira, R. F., Brown, K. M. and Lynch, J. P. (2013). Basal root whorl number: a modulator of phosphorus acquisition in common bean (*Phaseolus vulgaris*). *Ann. Bot.* 112(6), 973–982.

Miguel, M. A., Postma, J. A. and Lynch, J. P. (2015). Phene synergism between root hair length and basal root growth angle for phosphorus acquisition. *Plant Physiol.* 167(4), 1430–1439.

Mills, A., Moot, D. J. and McKenzie, B. A. (2006). Cocksfoot pasture production in relation to environmental variables. *Proc. N. Z. Grassland Assoc.* 68, 89–94.

Mitsukawa, N., Okumura, S., Shirano, Y., Sato, S., Kato, T., Harashima, S. and Shibata, D. (1997). Overexpression of an *Arabidopsis thaliana* high-affinity phosphate transporter gene in tobacco cultured cells enhances cell growth under phosphate-limited conditions. *Proc. Natl Acad. Sci. U.S.A.* 94(13), 7098–7102.

Moody, P. W. (2007). Interpretation of a single-point P buffering index for adjusting critical levels of the Colwell soil P test. *Aust. J. Soil Res.* 45(1), 55–62.

Moore, R. M. (1970a). *Australian Grasslands*. Canberra, Australia: Australian National University Press.

Moore, R. M. (1970b). Australian grasslands. In: Moore, R. M. (Ed.), *Australian Grasslands*. Canberra, Australia: Australian National University Press, pp. 85–100.

Mori, A., Fukuda, T., Vejchasarn, P., Nestler, J., Pariasca-Tanaka, J. and Wissuwa, M. (2016). The role of root size versus root efficiency in phosphorus acquisition in rice. *J. Exp. Bot.* 67(4), 1179–1189.

Morley, F. H. W. (1961). Subterranean clover. *Adv. Agron.* 13, 57–123.

Nestler, J. and Wissuwa, M. (2016). Superior root hair formation confers root efficiency in some, but not all, rice genotypes upon P deficiency. *Front. Plant Sci.* 7, 1935. doi:10.3389/fpls.2016.01935.

Nichols, P. G. H., Loi, A., Nutt, B. J., Evans, P. M., Craig, A. D., Pengelly, B. C., Dear, B. S., Lloyd, D. L., Revell, C. K., Nair, R. M., Ewing, M. A., Howieson, J. G., Auricht, G. A., Howie, J. H., Sandral, G. A., Carr, S. J., de Koning, C. T., Hackney, B. F., Crocker, G. J., Snowball, R., Hughes, S. J., Hall, E. J., Foster, K. J., Skinner, P. W., Barbetti, M. J. and You, M. P. (2007). New annual and short-lived perennial pasture legumes for Australian agriculture - 15 years of revolution. *Field Crops Res.* 104(1–3), 10–23.

Nichols, P. G. H., Revell, C. K., Humphries, A. W., Howie, J. H., Hall, E. J., Sandral, G. A., Ghamkhar, K. and Harris, C. A. (2012). Temperate pasture legumes in Australia – their history, current use, and future prospects. *Crop Pasture Sci.* 63(9), 691–725.

Nichols, P. G. H., Foster, K. J., Piano, E., Pecetti, L., Kaur, P., Ghamkhar, K. and Collins, W. J. (2013). Genetic improvement of subterranean clover (*Trifolium subterraneum* L.). 1. Germplasm, traits and future prospects. *Crop Pasture Sci.* 64(4), 312–346.

Nicol, D. L., Finlayson, J., Colmer, T. D. and Ryan, M. H. (2013). Opportunistic Mediterranean agriculture - Using ephemeral pasture legumes to utilize summer rainfall. *Agric. Syst.* 120, 76–84.

Nutt, B. J. (2004a). French serradella (*Ornithopus sativus*) 'Erica'. *Plant Var. J.* 17, 313–315.

Nutt, B. J. (2004b). French serradella (*Ornithopus sativus*) 'Margurita'. *Plant Var. J.* 17, 310–312.

Oburger, E., Jones, D. L. and Wenzel, W. W. (2011). Phosphorus saturation and pH differentially regulate the efficiency of organic acid anion-mediated P solubilization mechanisms is soil. *Plant Soil* 341(1–2), 363–382.

Oehl, F., Oberson, A., Probst, M., Fliessbach, A., Roth, H. and Frossard, E. (2001). Kinetics of microbial phosphorus uptake in cultivated soils. *Biol. Fertil. Soils* 34(1), 31–41.

Olsen, S. R., Cole, C. V., Watanabe, F. S. and Dean, L. A. (1954). *Estimation of Available Phosphorus in Soils by Extraction with Sodium Bicarbonate*. Washington, DC: United States Department of Agriculture, Circular No. 939.

Osman, A. E., Cocks, P. S., Russi, L. and Pagnotta, M. A. (1991). Response of Mediterranean grassland to phosphate and stocking rates: biomass production and botanical composition. *J. Agric. Sci.* 116(1), 37–46.

Ozanne, P. G. (1980). Phosphate nutrition of plants – A general treatise. In: Khasawneh, F. E., Sample, E. C. and Kamprath, E. J. (Eds), *The Role of Phosphorus in Agriculture*. Madison, WI: American Society of Agronomy/Crop Science Society of America/Soil Science Society of America, pp. 559–589.

Ozanne, P. G., Kirton, D. J. and Shaw, T. C. (1961). The loss of phosphorus from sandy soils. *Aust. J. Agric. Res.* 12(3), 409–423.

Ozanne, P. G. and Shaw, T. C. (1967). Phosphate sorption by soils as a measure of the phosphate requirement for pasture growth. *Aust. J. Agric. Res.* 18(4), 601–612.

Ozanne, P. G., Keay, J. and Biddiscombe, E. F. (1969). The Comparative applied phosphate requirements of eight annual pasture species. *Aust. J. Agric. Res.* 20(5), 809–818.

Pang, J., Tibbett, M., Denton, M. D., Lambers, H., Siddique, K. H. M., Bolland, M. D. A., Revell, C. K. and Ryan, M. H. (2010). Variation in seedling growth of 11 perennial legumes in response to phosphorus supply. *Plant Soil* 328(1–2), 133–143.

Paynter, B. H. (1990). Comparative phosphate requirements of yellow serradella (*Ornithopus compressus*), burr medic (*Medicago polymorpha* var *brevispina*) and subterranean clover (*Trifolium subterraneum*). *Aust. J. Exp. Agric.* 30(4), 507–514.

Peoples, M. B., Brockwell, J., Hunt, J. R., Swan, A. D., Watson, L., Hayes, R. C., Li, G. D., Hackney, B., Nuttall, J. G., Davies, S. L. and Fillery, I. R. P. (2012). Factors affecting the potential contributions of N_2 fixation by legumes in Australian pasture systems. *Crop Pasture Sci.* 63(9), 759–786.

Perrott, K. W., Sarathchandra, S. U. and Waller, J. E. (1990). Seasonal storage and release of phosphorus and potassium by organic matter and the microbial biomass in a high-producing pastoral soil. *Aust. J. Soil Res.* 28(4), 593–608.

Perrott, K. W., Sarathchandra, S. U. and Dow, B. W. (1992). Seasonal and fertilizer effects on the organic cycle and microbial biomass in a hill country soil under pasture. *Soil Biol. Biochem.* 30(3), 383–394.

Pinkerton, A. and Simpson, J. R. (1986). Interactions of surface drying and subsurface nutrients affecting plant growth on acidic soil profiles from an old pasture. *Aust. J. Exp. Agric.* 26(6), 681–689.
Postma, J. A. and Lynch, J. P. (2011). Root cortical aerenchyma enhances the growth of maize on soils with suboptimal availability of nitrogen, phosphorus, and potassium. *Plant Physiol.* 156(3), 1190–1201.
Poulton, P. R., Johnston, A. E. and White, R. P. (2013). Plant-available soil phosphorus. Part I: the response of winter wheat and spring barley to Olsen P on a silty clay loam. *Soil Use Manag.* 29(1), 4–11.
Probert, M. E. and Jones, R. K. (1977). The use of soil analysis for predicting the response to sulphur of pasture legumes in the Australian tropics. *Aust. J. Soil Res.* 15(2), 137–146.
Rae, A. L., Jarmey, J. M., Mudge, S. R. and Smith, F. W. (2004). Over-expression of a high-affinity phosphate transporter in transgenic barley plants does not enhance phosphate uptake rates. *Funct. Plant Biol.* 31(2), 141–148.
Reuter, D. J., Dyson, C. B., Elliott, D. E., Lewis, D. C. and Rudd, C. L. (1995). An appraisal of soil phosphorus testing data for crops and pastures in South Australia. *Aust. J. Exp. Agric.* 35(7), 979-995.
Revell, C. and Revell, D. (2007). Meeting 'duty of care' obligations when developing new pasture species. *Field Crops Res.* 104(1-3), 95–102.
Richardson, A. E. (2009). Regulating the phosphorus nutrition of plants: molecular biology meeting agronomic needs. *Plant Soil* 322(1-2), 17–24.
Richardson, A. E., Hadobas, P. A., Hayes, J. E., O'Hara, C. P. and Simpson, R. J. (2001). Utilization of phosphorus by pasture plants supplied with myo-inositol hexaphosphate is enhanced by the presence of soil microorganisms. *Plant Soil* 229(1), 47–56.
Richardson, A. E., Lynch, J. P., Ryan, P. R., Delhaize, E., Smith, F. A., Smith, S. E., Harvey, P. R., Ryan, M. H., Veneklaas, E. J., Lambers, H., Oberson, A., Culvenor, R. A. and Simpson, R. J. (2011). Plant and microbial strategies to improve the phosphorus efficiency of agriculture. *Plant Soil* 349(1-2), 121–156.
Richardson, A. E. and Simpson, R. J. (2011). Soil microorganisms mediating phosphorus availability update on microbial phosphorus. *Plant Physiol.* 156(3), 989–996.
Ridley, A. M., White, R. E., Simpson, R. J. and Callinan, L. (1997). Water use and drainage under phalaris, cocksfoot, and annual ryegrass pastures. *Aust. J. Agric. Res.* 48(7), 1011–1023.
Ridley, A. M., Christy, B. P., White, R. E., McLean, T. and Green, R. (2003). North-East Victoria SGS National Experiment site: water and nutrient losses from grazing systems on contrasting soil types and levels of inputs. *Aust. J. Exp. Agric.* 43(8), 799–815.
Ritchie, G. S. P. and Weaver, D. M. (1993). Phosphorus retention and release from sandy soils of the Peel-Harvey catchment. *Fert. Res.* 36(2), 115–122.
Roberts, T. L. and Johnston, A. E. (2015). Phosphorus use efficiency and management in agriculture. *Resour. Conserv. Recycl.* 105, 275–281.
Robinson, D. and Van Vuuren, M. M. I. (1998). Responses of wild plants to nutrient patches in relation to growth rate and life-form. In: Lambers, H., Poorter, H. and Van Vuuren, M. M. I. (Eds), *Inherent Variation in Plant Growth. Physiological Mechanisms and Ecological Consequences*. Leiden: Backhuys Publishers, pp. 237–257.
Robinson, K., Bell, L. W., Bennett, R. G., Henry, D. A., Tibbett, M. and Ryan, M. H. (2007). Perennial legumes native to Australia – a preliminary investigation of nutritive value and response to cutting. *Aust. J. Exp. Agric.* 47(2), 170–176.

Rose, T. J. and Wissuwa, M. (2012). Rethinking internal phosphorus utilization efficiency: a new approach is needed to improve PUE in grain crops. *Adv. Agron.* 116, 185–217.

Rose, T. J., Liu, L. and Wissuwa, M. (2013). Improving phosphorus efficiency in cereal crops: is breeding for reduced grain phosphorus concentration part of the solution? *Front. Plant. Sci.* 4, 444.

Rossiter, R. C. (1966). Ecology of the Mediterranean annual type pasture. *Adv. Agron.* 18, 1–56.

Rowe, H., Withers, P. J. A., Baas, P., Chan, N. I., Doody, D., Holiman, J., Jacobs, B., Li, H., MacDonald, G. K., McDowell, R., Sharpley, A. N., Shen, J., Taheri, W., Wallenstein, M. and Weintraub, M. N. (2016). Integrating legacy soil phosphorus into sustainable nutrient management strategies for future food, bioenergy and water security. *Nutr. Cycl. Agroecosyst.* 104(3), 393–412.

Rudd, C. L. (1972). Response of annual medic pasture to superphosphate applications and correlation with available soil phosphorus. *Aust. J. Exp. Agric. Anim. Husb.* 12, 43–48.

Russell, J. S. (1960). Soil fertility changes in the long term experimental plots at Kybybolite, South Australia. II. Changes I phosphorus. *Aust. J. Agric. Res.* 11(6), 926–947.

Ryan, M. H. and Graham, J. H. (2018). Little evidence that farmers should consider abundance or diversity of arbuscular mycorrhizal fungi when managing crops. *New Phytol.* 220(4), 1092–1107.

Ryan, P., Delhaize, E. and Jones, D. (2001). Function and mechanisms of organic anion exudation from roots. *Annu. Rev. Plant. Physiol. Plant. Mol. Biol.* 52(1), 527–560.

Ryan, P. R., James, R. A., Weligama, C., Delhaize, E., Rattey, A., Lewis, D. C., Bovill, W. D., McDonald, G., Rathjen, T. M., Wang, E., Fettell, N. A. and Richardson, A. E. (2014). Can citrate efflux from roots improve phosphorus uptake by plants? Testing the hypothesis with near-isogenic lines of wheat. *Physiol. Plant.* 151(3), 230–242.

Sale, P. W. G., Gilkes, R. J., Bolland, M. D. A., Simpson, P. G., Lewis, D. C., Ratkowsky, D. A., Gilbert, M. A., Garden, D. L., Cayley, J. W. D. and Johnson, D. (1997). The agronomic effectiveness of reactive phosphate rocks. 1. Effect of the pasture environment. *Aust. J. Exp. Agric.* 37(8), 921–936.

Sample, E. C., Soper, R. J. and Racz, G. J. (1980). Reactions of phosphate fertilizers in soils. In: Khasawneh, F. E., Sample, E. C. and Kamprath, E. J. (Eds), *The Role of Phosphorus in Agriculture*. Madison WI: American Society of Agronomy, Crop Science Society of America, Soil Science Society of America, pp. 263–310.

Sanchez, C. A. , Porter, P. S. and Ulloa, M. F. (1991). Relative efficiency of broadcast and banded phosphorus for sweet corn produced on histosols. *Soil Sci. Soc. Am. J.* 55(3), 871–875.

Sandral, G. A., Haling, R. E., Ryan, M. H., Price, A., Pitt, W. M., Hildebrand, S. M., Fuller, C. G., Kidd, D. R., Stefanksi, A., Lambers, H. and Simpson, R. J. (2018). Intrinsic capacity for nutrient foraging predicts critical external phosphorus requirement of 12 pasture legumes. *Crop Pasture Sci.* 69(2), 174–182.

Sandral, G. A., Price, A., Hildebrand, S. M., Fuller, C. G., Haling, R. E., Stefanski, A., Yang, Z., Culvenor, R. A., Ryan, M. H., Kidd, D. R., Diffey, S., Lambers, H. and Simpson, R. J. (2019). Field benchmarking of the critical external phosphorus requirements of pasture legumes for southern Australia. *Crop Pasture Sci.* 70(12), 1080–1096.

Sattari, S. Z., Bouwman, A. F., Giller, K. E. and Van Ittersum, M. K. (2012). Residual soil phosphorus as the missing link in the global phosphorus crisis puzzle. *Proc. Natl Acad. Sci. U.S.A.* 109(16), 6348–6353.

Sattari, S. Z., van Ittersum, M. K., Bouwman, A. F., Smit, A. L. and Janssen, B. H. (2014). Crop yield response to soil fertility and N, P, K inputs in different environments: testing and improving the QUEFTS model. *Field Crops Res.* 157, 35–46.

Schefe, C. R., Barlow, K. M., Robinson, N. J., Crawford, D. M., McLaren, T. I., Smernik, R. J., Croatto, G., Walsh, R. D. and Kitching, M. (2015). 100 years of superphosphate addition to pasture in an acid soil - current nutrient status and future management. *Soil Res.* 53(6), 662–676.

Schneider, H. M., Postma, J. A., Wojciechowski, T., Kuppe, C. and Lynch, J. P. (2017a). Root cortical senescence improves growth under suboptimal availability of N, P, and K. *Plant Physiol.* 174(4), 2333–2347.

Schneider, K. D., Voroney, R. P., Lynch, D. H., Oberson, A., Frossard, E. and Bünemann, E. K. (2017b). Microbially-mediated P fluxes in calcareous soils as a function of water-extractable phosphate. *Soil Biol. Biochem.* 106, 51–60.

Scholz, R. W. and Wellmer, F.-W. (2015). Losses and use efficiencies along the phosphorus cycle. Part 1: Delemmata and losses on the mines and other nodes of the supply chain. *Resour. Conserv. Recycl.* 105, 216–234.

Scholz, R. W. and Wellmer, F.-W. (2018). Although there is no physical short-term scarcity of phosphorus, its resource efficiency should be improved. *J. Indust. Ecol.* 23(2), 313–318.

Schweiger, P. F., Robson, A. D. and Barrow, N. J. (1995). Root hair length determines beneficial effect of a *Glomus* species on shoot growth of some pasture species. *New Phytol.* 131(2), 247–254.

Scott, B. J. (1973). The response of barrel medic pasture to topdressed and placed superphosphate in central western New South Wales. *Aust. J. Exp. Agric.* 13(65), 705–710.

Scott, B. J., Conyers, M. K., Poile, G. J. and Cullis, B. R. (1997). Subsurface acidity and liming affect yield of cereals. *Aust. J. Agric. Res.* 48(6), 843–854.

Scott, B. J., Fisher, J. A. and Cullis, B. R. (2001). Aluminium tolerance and lime increase wheat yield on the acidic soils of central and southern New South Wales. *Aust. J. Exp. Agric.* 41(4), 523–532.

Scott, B. J. and Coombes, N. E. (2006). Poor incorporation of lime limits grain yield response in wheat. *Aust. J. Exp. Agric.* 46(11), 1481–1487.

Seeling, B. and Zasoski, R. J. (1993). Microbial effects in maintaining organic and inorganic solution phosphorus concentrations in a grassland topsoil. *Plant Soil* 148(2), 277–284.

Sharpley, A. N. (1995). Dependence of runoff phosphorus on extractable soil phosphorus. *J. Environ. Qual.* 24(5), 920–926.

Simpson, J. R., Bromfield, S. M. and Jones, L. (1974). Effects of management on soil fertility under pasture. 3. Changes in total soil nitrogen, carbon, phosphorus and exchangeable cations. *Aust. J. Exp. Agric.* 14(69), 487–494.

Simpson, R., Graham, P., Davies, L. and Zurcher, E. (2009). Five easy steps to ensure you are making money from superphosphate. Canberra and Orange, Australia: CSIRO & Industry and Investment NSW. Available at: https://www.mla.com.au/globalassets/mla-corporate/generic/extension-training-and-tools/5-easy-steps-guide.pdf.

Simpson, R., Haling, R., Virgona, J. and Ferguson, N. (2017). Managing the phosphorus cycle in clover-based pasture for more effective use of P-fertiliser inputs. *Proceedings of the 58th Ann. Conf. Grassld Soc. Southern Aust.* Carrum Downs, Victoria, Australia: Grassland Society of Southern Australia Inc., pp. 17–21.

Simpson, R. J., Oberson, A., Culvenor, R. A., Ryan, M. H., Veneklaas, E. J., Lambers, H., Lynch, J. P., Ryan, P. R., Delhaize, E., Smith, F. A., Smith, S. E., Harvey, P. R. and Richardson, A. E. (2011a). Strategies and agronomic interventions to improve the phosphorus-use efficiency of farming systems. *Plant Soil* 349(1–2), 89–120.

Simpson, R. J., Richardson, A. E., Riley, I. T., McKay, A. C., McKay, S. F., Ballard, R. A., Ophel-Keller, K., Hartley, D., O'Rourke, T. A., Li, H., Sivasithamparam, K., Ryan, M. H. and Barbetti, M. J. (2011b). Damage to roots of *Trifolium subterraneum* L. (subterranean clover), failure of seedlings to establish and the presence of root pathogens during autumn–winter. *Grass Forage Sci.* 66(4), 585–605.

Simpson, R. J., Richardson, A. E., Nichols, S. N. and Crush, J. R. (2014). Pasture plants and soil fertility management to improve the efficiency of phosphorus fertiliser use in temperate grassland systems. *Crop Pasture Sci.* 65(6), 556–575.

Simpson, R. J., Stefanski, A., Marshall, D. J., Moore, A. D. and Richardson, A. E. (2015). Management of soil phosphorus fertility determines the phosphorus budget of a temperate grazing system and is the key to improving phosphorus-balance efficiency. *Agric. Ecosyst. Environ.* 212, 263–277.

Slattery, W. J., Conyers, M. K. and Aitken, R. L. (1999). Soil pH, aluminium, manganese and lime requirement. In: Peverill, K. I., Sparrow, L. A. and Reuter, D. J. (Eds), *Soil Analysis: An Interpretation Manual*. Collingwood, Australia: CSIRO Publishing, pp. 103–128.

Smith, D. F. (2000). *Natural Gain: In the Grazing Lands of Southern Australia*. Sydney, Australia: University of New South Wales Press.

Smith, L. C., Moss, R. A., Morton, J. D., Metherell, A. K. and Fraser, T. J. (2012). Pasture production from a long-term fertiliser trial under irrigation. *N. Z. J. Agric. Res.* 55(2), 105–117.

Smith, S. E. and Read, D. J. (2008). *Mycorrhizal Symbiosis*. London: Academic Press.

Smith, S. E., Anderson, I. C. and Smith, F. A. (2015). Mycorrhizal associations and phosphorus acquisition: from cells to ecosystems. *Ann. Plant Rev.* 48, 409–440.

Stewart, W. M., Dibb, D. W., Johnston, A. E. and Smyth, T. J. (2005). The contribution of commercial fertilizer nutrients to food production. *Agron. J.* 97(1), 1–6.

Strock, C. F., de la Riva, L. M. and Lynch, J. P. (2018). Reduction in root secondary growth as a strategy for phosphorus acquisition. *Plant Physiol.* 176(1), 691–703.

Sun, B., Gao, Y. and Lynch, J. P. (2018). Large crown root number improves topsoil foraging and phosphorus acquisition. *Plant Physiol.* 177(1), 90–104.

Sutton, M. A., Bleeker, A., Howard, C. M., Bekunda, M., Grizzetti, B., de Vries, W., van Grinsven, H. J. M., Abrol, Y. P., Adhya, T. K., Billen, G., Davidson, E. A., Datta, A., Diaz, R., Erisman, J. W., Liu, X. J., Oenema, O., Palm, C., Raghuram, N., Reis, S., Scholz, R. W., Sims, T., Westhoek, H. and Zhang, F. S., with contributions from Ayyappan, S., Bouwman, A. F., Bustamante, M., Fowler, D., Galloway, J. N., Gavito, M. E., Garnier, J., Greenwood, S., Hellums, D. T., Holland, M., Hoysall, C., Jaramillo, V. J., Klimont, Z., Ometto, J. P., Pathak, H., Plocq Fichelet, V., Powlson, D., Ramakrishna, K., Roy, A., Sanders, K., Sharma, C., Singh, B., Singh, U., Yan, X. Y. and Zhang, Y. (2013). Our Nutrient World: the challenge to produce more food and energy with less pollution. In: *Global Overview of Nutrient Management*. Centre for Ecology and Hydrology, Edinburgh on behalf of the Global Partnership on Nutrient Management and the International Nitrogen Initiative. Available at: https://www.unenvironment.org/resources/report/our-nutrient-world-challenge-produce-more-food-and-energy-less-pollution.

Svenningsen, N. B., Watts-Williams, S. J., Joner, E. J., Battini, F., Efthymiou, A., Cruz-Paredes, C., Nybroe, O. and Jakobsen, I. (2018). Suppression of the activity of arbuscular mycorrhizal fungi by the soil microbiota. *ISME J.* 12(5), 1296–1307.

Tarafdar, J. C. and Jungk, A. (1987). Phosphatase activity in the rhizosphere and its relation to the depletion of soil organic phosphorus. *Biol. Fertil. Soils* 3(4), 199–204.

Tinker, P. B. and Nye, P. H. (2000). *Solute Movement in the Rhizosphere*. New York: Oxford University Press Inc.

Tilman, D., Balzer, C., Hill, J. and Befort, B. L. (2011). Global food demand and the sustainable intensification of agriculture. *Proc. Natl Acad. Sci. U.S.A.* 108(50), 20260–20264.

Tonini, D., Saveyn, H. G. M. and Huygens, D. (2019). Environmental and health co-benefits for advanced phosphorus recovery. *Nat. Sustainability* 2(11), 1051–1061.

Trotter, M., Guppy, C., Haling, R., Trotter, T., Edwards, C. and Lamb, D. (2014). Spatial variability in pH and key soil nutrients: is this an opportunity to increase fertiliser and lime-use efficiency in grazing systems? *Crop Pasture Sci.* 65(8), 817–827.

Vaccari, D. A., Powers, S. M. and Liu, X. (2019). Demand-driven model for global phosphate rock suggests paths for phosphorus sustainability. *Environ. Sci. Technol.* 53(17), 10417–10425.

Valizadeh, G. R., Rengel, Z. and Rate, A. W. (2003). Response of wheat genotypes efficient in P utilization and genotypes responsive to P fertilization to different P banding depths and watering regimes. *Aust. J. Agric. Res.* 54(1), 59–65.

Vance, C. P., Uhde-Stone, C. and Allan, D. L. (2003). Phosphorus acquisition and use: critical adaptations by plants for securing a non-renewable resource. *New Phytol.* 157(3), 423–447.

Vandamme, E., Renkens, M., Pypers, P., Smolders, E., Vanlauwe, B. and Merckx, R. (2013). Root hairs explain P uptake efficiency of soybean genotypes grown in a P-deficient ferralsol. *Plant Soil* 369(1–2), 269–282.

Veneklaas, E. J., Lambers, H., Bragg, J., Finnegan, P. M., Lovelock, C. E., Plaxton, W. C., Price, C. A., Scheible, W. R., Shane, M. W., White, P. J. and Raven, J. A. (2012). Opportunities for improving phosphorus-use efficiency in crop plants. *New Phytol.* 195(2), 306–320.

Vitousek, P. M., Naylor, R., Crews, T., David, M. B., Drinkwater, L. E., Holland, E., Johnes, P. J., Katzenberger, J., Martinelli, L. A., Matson, P. A., Nziguheba, G., Ojima, D., Palm, C. A., Robertson, G. P. , Sanchez, P. A., Townsend, A. R. and Zhang, F. S. (2009). Nutrient imbalances in agricultural development. *Science* 324(5934), 1519 –1520.

Waddell, H. A., Simpson, R. J., Henderson, B., Ryan, M. H., Lambers, H., Garden, D. L. and Richardson, A. E. (2015). Differential growth response of *Rytidosperma* species (wallaby grass) to phosphorus application and implications for grassland management. *Grass Forage Sci.* 71, 245–258.

Wang, Y., Krogstad, T., Clarke, J. L., Hallama, M., Øgaard, A. F., Eich-Greatorex, S., Kandeler, E. and Clarke, N. (2016). Rhizosphere organic anions play a minor role in improving crop species' ability to take up residual phosphorus (P) in agricultural soils low in P availability. *Front. Plant Sci.* 7, 1664. doi:10.3389/fpls.2016.01664.

Wang, Y. and Lambers, H. (2020). Root-released organic anions in response to low phosphorus availability: recent progress, challenges and future perspectives. *Plant Soil* 447(1–2), 135–156.

Watt, M., Silk, W. K. and Passioura, J. B. (2006). Rates of root and organism growth, soil conditions, and temporal and spatial development of the rhizosphere. *Ann. Bot.* 97(5), 839–855.

Weaver, D. M. and Wong, M. T. F. (2011). Scope to improve phosphorus (P) management and balance efficiency of crop and pasture soils with contrasting P status and buffering indices. *Plant Soil* 349(1–2), 37–54.

Weber, N. F., Herrmann, I., Hochholdinger, F., Ludewig, U. and Neumann, G. (2018). PGPR-induced growth stimulation and nutrient acquisition in maize: do root hairs matter? *Sci. Agric. Bohem.* 49(3), 164–172.

Wen, T. J. and Schnable, P. S. (1994). Analyses of mutants of three genes that influence root hair development in *Zea mays* (Gramineae) suggest that root hairs are dispensable. *Am. J. Bot.* 81(7), 833–842.

Wen, Z., Li, H., Shen, Q., Tang, X., Xiong, C., Li, H., Pang, J., Ryan, M. H., Lambers, H. and Shen, J. (2019). Tradeoffs among root morphology, exudation and mycorrhizal symbioses for phosphorus-acquisition strategies of 16 crop species. *New Phytol.* 223(2), 882–895.

Williams, C. H. and Andrew, C. S. (1970). Mineral nutrition of pastures. In: Moore, R. M. (Ed.), *Australian Grasslands*. Canberra, Australia: Australian National University Press, pp. 321–338.

Wissuwa, M., Kretzschmar, T. and Rose, T. J. (2016). From promise to application: root traits for enhanced nutrient capture in rice breeding. *J. Exp. Bot.* 67(12), 3605–3615.

Withers, P. J. A., Vadas, P. A., Uusitalo, R., Forber, K. J., Hart, M., Foy, R. H., Delgado, A., Dougherty, W., Lilja, H., Burkitt, L. L., Rubæk, G. H., Pote, D., Barlow, K., Rothwell, S. and Owens, P. R. (2019). A global perspective on integrated strategies to manage soil phosphorus status for eutrophication control without limiting land productivity. *J. Environ. Qual.* 48(5), 1234–1246.

Yang, Z., Culvenor, R. A., Haling, R. E., Stefanski, A., Ryan, M. H., Sandral, G. A., Kidd, D. R., Lambers, H. and Simpson, R. J. (2017). Variation in root traits associated with nutrient foraging among temperate pasture legumes and grasses. *Grass Forage Sci.* 72(1), 93–103.

Yao, J. and Barber, S. A. (1986). Effect of one phosphorus rate placed in different soil volumes on P uptake and growth of wheat. *Commun. Soil Sci. Plant Anal.* 17(8), 819–827.

Ye, D., Zhang, X., Li, T., Xu, J. and Chen, G. (2018). Phosphorus-acquisition characteristics and rhizosphere properties of wild barley in relation to genotypic differences as dependent on soil phosphorus availability. *Plant Soil* 423(1–2), 503–516.

Zhang, C., Simpson, R. J., Kim, C. M., Warthmann, N., Delhaize, E., Dolan, L., Byrne, M. E., Wu, Y. and Ryan, P. R. (2018). Do longer root hairs improve phosphorus uptake? Testing the hypothesis with transgenic *Brachypodium distachyon* lines overexpressing endogenous RSL genes. *New Phytol.* 217(4), 1654–1666.

Zhao, J., Fu, J., Liao, H., He, Y., Nian, H., Hu, Y., Qiu, L., Dong, Y. and Yan, X. (2004). Characterization of root architecture in an applied core collection for phosphorus efficiency of soybean germplasm. *Chin. Sci. Bull.* 49(15), 1611–1620.

Zhu, J. and Lynch, J. P. (2004). The contribution of lateral rooting to phosphorus acquisition efficiency in maize (*Zea mays*) seedlings. *Funct. Plant Biol.* 31(10), 949-958.

Zhu, J., Zhang, C. and Lynch, J. P. (2010). The utility of phenotypic plasticity of root hair length for phosphorus acquisition. *Funct. Plant Biol.* 37(4), 313–322.

Index

Note: Page numbers in italics denote figures.

www.ingramcontent.com/pod-product-compliance
Ingram Content Group UK Ltd.
Pitfield, Milton Keynes, MK11 3LW, UK
UKHW021824100926
8556IPUK00003B/136

9 781786 763600